UNIT OPERATIONS AND PROCESSES IN ENVIRONMENTAL ENGINEERING

THE PWS-KENT SERIES IN CIVIL ENGINEERING

UNIT OPERATIONS AND PROCESSES IN ENVIRONMENTAL ENGINEERING

Tom D. Reynolds

Texas A&M University

PWS PUBLISHING COMPANY
Boston

PWS PUBLISHING COMPANY
20 Park Plaza, Boston, MA 02116-4324

International Thomson Publishing
The trademark ITP is used under license.

PWS Publishing Company is a division of Wadsworth, Inc.

Printed in the United States of America
10   9

Library of Congress Cataloging in Publication Data:

Reynolds, Tom D.
    Unit operations and processes in environmental engineering.

    Previously published as: Unit operations and processes of environmental engineering. 1977.
    Includes bibliographies and index.
    1. Environmental engineering.   I. Title.
TD145.R48   1982   628.1'62   81–12308
ISBN 0–8185–0493–5   AACR2

ISBN 0-8185-0493-5

Sponsoring Editor: Ray Kingman
Signing Representative: Ragu Raghaven
Production: Del Mar Associates, Del Mar, California
Manuscript Editor: Lois Oster
Interior Design: John Odam
Cover Design: John Odam
Typesetting: Graphic Typesetting Service
Production Services Coordinator: Stacey C. Sawyer
Coordinating Designer: Jamie Sue Brooks

Preface

This text is for an advanced undergraduate course or a graduate-level course in Environmental Engineering. It has been written primarily for the Civil Engineering curriculum; however, it may be used in other fields, such as Chemical Engineering, where water and wastewater treatment are taught. The book presents water and wastewater treatment under a single cover using the unit operation and process approach. The fundamentals of each operation and process are first presented in each chapter, then the application in water treatment and wastewater treatment fields is given. Some operations or processes are used in only one field, such as activated sludge, and are presented accordingly. The undergraduate civil engineer's background in chemistry, mathematics, and fluid mechanics, and a prior introductory course in environmental engineering, are adequate preparation for use of the text. In nearly all chapters, numerous example problems are given along with practice problem sets at the ends of the chapters, since an engineering discipline is most rapidly learned by problem solving. The book is oriented toward engineering design based on fundamentals.

In using the book, Chapter 1 should be covered first because it is introductory and has typical flowsheets of water, wastewater, and advanced wastewater treatment plants. After Chapter 1, the chapters are not presented in chronological order and each chapter is written to be complete within itself. This allows the instructor to select the sequence of chapters or parts of chapters in any manner desired. If the instructor wants to cover water treatment in the first part of a semester and wastewater treatment in the latter part, he or she may choose the sequence of the chapters or parts of chapters accordingly. The material coverage for a water treatment, wastewater treatment, and advanced wastewater treatment approach would include the following chapters or parts of chapters and major topics.

Water Treatment

Chapter	Topic
16	Screening
2	Coagulation and Flocculation
3	Sedimentation
4	Filtration
6	Adsorption
17	Disinfection
7	Ion Exchange
8	Membrane Processes
15	Solids Handling

Wastewater Treatment

Chapter	Topic
16	Screening and Shredding, Grit Removal
3	Sedimentation
9	Activated Sludge
10	Oxygen Transfer and Mixing
11	Trickling Filters and Rotary Biological Contactors
12	Stabilization Ponds and Aerated Lagoons
17	Disinfection
13	Anaerobic Digestion
14	Aerobic Digestion
15	Solids Handling

Advanced Wastewater Treatment

Chapter	Topic
2	Coagulation and Flocculation
3	Sedimentation
5	Ammonia Stripping
4	Filtration
6	Adsorption
8	Membrane Processes
7	Ion Exchange
5	Ammonia Removal by Ion Exchange, Breakpoint Chlorination, and Biological Means

If the instructor wants to cover the chapters based on the fundamental principle involved—that is, physical, chemical, or biological—the chapters may be chosen accordingly. The material coverage for this approach would include the following chapters or parts of chapters and major topics.

Physical Treatment

Chapter	Topic
16	Screening, Screening and Shredding, Grit Removal, and Flow Equalization
3	Sedimentation
4	Filtration
5	Ammonia Stripping
6	Adsorption
8	Dialysis and Reverse Osmosis
10	Oxygen Transfer and Mixing

Chemical Treatment

Chapter	Topic
16	Neutralization
2	Coagulation and Flocculation
7	Ion Exchange
8	Electrodialysis
17	Disinfection
5	Ammonia Removal by Ion Exchange and Breakpoint Chlorination

Biological Treatment

Chapter	Topic
9	Activated Sludge
11	Trickling Filters and Rotary Biological Contactors
12	Stabilization Ponds and Aerated Lagoons
13	Anaerobic Digestion
14	Aerobic Digestion
5	Ammonia Removal by Nitrification-Denitrification

Solids handling can best be covered as a separate subject because it involves both physical and chemical treatments.

Another approach to using the book is to cover the topics in the order of their occurrence, with the most widely used topics being covered first and the less common topics last. In using this approach, the instructor may choose the sequence of topics to his or her own liking.

The book covers more material than can be conveniently covered in a one-semester, three-credit-hour course; thus, the instructor may choose to limit the depth of coverage of some topics. However, the material that is not covered in class but is included in the book can serve as a valuable reference source for the student.

The glossary includes most terms that are covered in an introductory Environmental Engineering course. The Appendix contains tables that give common conversion factors, atomic numbers and weights, properties of water, dissolved oxygen concentrations at various temperatures, chloride concentrations and atmospheric pressures, a graph that gives the oxygen concentration in air at various temperatures and elevations, and three diagrams of reactors.

Although the book is intended as a textbook, it will be useful as a reference book for practicing engineers and for engineers who wish to do self-study. The example problems in the metric system (International System of Units, abbreviated SI) will help practicing engineers, as well as students, in using this dimensional system.

The author gratefully acknowledges the thorough reading of the manuscript and the meaningful suggestions by the reviewers, who were Dr. Dragoslav Misic, California Polytechnic State University; Dean Earnest F. Gloyna, University of Texas at Austin; Dr. K. Keshavan, Worcester Polytechnic Institute; Dean E. Joe Middlebrooks, Utah State University; and Dr. Christopher Uchrin, Rutgers State University.

Tom D. Reynolds

Contents

Chapter One

INTRODUCTION

The unit operations and processes used in environmental engineering may be classified as physical, chemical, or biological treatments according to their functional principle. In the strict sense, a unit operation is a physical treatment and a unit process is a chemical or biological treatment; however, the terms *unit operations* and *unit processes* are frequently used interchangeably. Some typical unit operations are sedimentation, flotation, and granular bed filtration. Some typical unit processes are coagulation, flocculation, carbon adsorption, ion exchange, chlorination, activated sludge, trickling filters, aerobic digestion, and anaerobic digestion. Many unit operations and processes, such as coagulation, flocculation, and sedimentation, are used in both water and wastewater treatment. Consequently, these operations or processes may be studied by investigating their fundamentals and then studying their application in water and wastewater treatment. Some unit operations or processes are limited to one field—that is, water or wastewater treatment. However, these may still be studied by investigating their fundamentals and then their application. This approach to studying the various unit operations and processes is presented in this text.

The degree to which a water must be treated depends on the raw water quality and the desired quality of the finished water. Likewise, the degree to which a wastewater must be treated depends on the raw wastewater quality and the required effluent quality. Since the degree of treatment determines the number and types of unit operations and processes to be used, there are numerous flowsheets employed in water treatment and, in particular, in wastewater treatment. In order to illustrate the integration of unit operations and unit processes in the overall plant design, the following discussion presents some of the most common flowsheets used in water and wastewater treatment and, also, a brief description of the unit operations and unit processes involved.

Water Treatment Plants

The most common treatment plants for surface waters are rapid sand filtration plants and lime-soda softening plants. Ground waters usually have a much better quality than surface waters; consequently, for ground waters the most common plants are gas stripping and chlorination plants and softening plants, either lime-soda or ion exchange type.

The flowsheet for a rapid sand filtration plant, shown in Figure 1.1, consists of coarse and fine screens, chemical coagulation, flocculation, sedimentation, granular media filtration, and chlorination. The coarse screens remove large debris, whereas the finer traveling screens remove smaller debris. Chemical coagulation and flocculation produce a precipitate or floc which enmeshes most of the colloidal solids. Most of the floc is removed in the settling basins. The granular media filters remove most of the fine nonsettling floc, and disinfection kills any pathogenic organisms present. After disinfection, the water is stored in the clear well and is pumped to the distribution system by the high service pumps. The clear well provides storage so that the plant may operate at a constant rate on the day of

Figure 1.1. Rapid Sand Filtration Plant

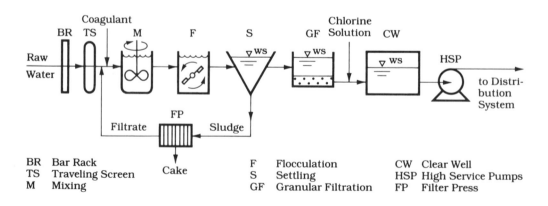

BR	Bar Rack			F	Flocculation	CW	Clear Well
TS	Traveling Screen	Cake		S	Settling	HSP	High Service Pumps
M	Mixing			GF	Granular Filtration	FP	Filter Press

maximum demand; that is, when the hourly demand is greater than the hourly water production, the water required is provided from storage in the clear well. The solids handling system, for disposal of the sludge from the clarifier, consists of a filter press to dewater the sludge. The cake is disposed of by sanitary landfill, although other solids handling systems may be used.

The flowsheet for a lime-soda softening plant, shown in Figure 1.2, consists of coarse and fine screens, chemical precipitation, flocculation, sedimentation, recarbonation, granular media filtration, and disinfection. The addition of slaked lime (calcium hydroxide) and soda ash (sodium carbonate) precipitates the calcium and magnesium ions as calcium carbonate and magnesium hydroxide. During flocculation, the precipitate or floc formed enmeshes most of the colloidal solids, and most of the floc is removed by the settling basins. Recarbonation by carbon dioxide lowers the pH and stabilizes the water so that further precipitation does not occur. The granular media filters remove most of the fine nonsettling floc, and disinfection kills any pathogenic organisms present. Lime-soda softening, in addition to removing hardness, produces coagulation and settling of colloidal solids, although sometimes a coagulant is added to improve performance. The solids handling system consists of centrifugation and disposal of the dewatered cake by sanitary landfill; however, other solids handling systems may be used. Sometimes the lime-soda softening process uses two flocculators and two settling basins. For this case,

Figure 1.2. Lime-Soda Softening Plant

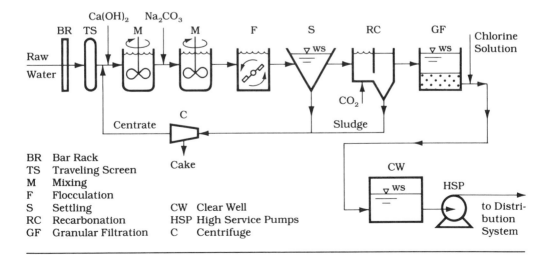

BR Bar Rack
TS Traveling Screen
M Mixing
F Flocculation
S Settling   CW Clear Well
RC Recarbonation HSP High Service Pumps
GF Granular Filtration C Centrifuge

the flowsheet will have the first flocculator, the first settling basin, the second flocculator, and then the second settling basin. This is done when the chemicals are added and flocculated in a stagewise manner. Also, dual units are used when stabilization is accomplished in two stages.

Frequently, ground waters are of such high quality that the only treatments required are gas stripping, to remove gases such as carbon dioxide when they are present in supersaturated amounts, and chlorination, to provide a residual in the distribution system. The supply and treatment plant for such a water would consist of the well, a gas stripping unit, disinfection facilities, a ground storage reservoir, and a high service pump station. If a ground water has sufficient hardness, softening by the lime-soda process or by the ion exchange process is required. Also, if a ground water has sufficient iron or manganese content, special treatment is required for their removal.

Industries usually require process water for manufacturing their products, boiler feed water for their boilers, and cooling water for their condensers. Usually, process water is provided by the previously described flowsheets; however, some industries may require specialized treatments such as demineralization for process water. Boiler feed water must be demineralized if high pressure boilers are used. Demineralization is usually accomplished by cation and anion exchangers operated on the hydrogen and hydroxyl cycles. Cooling water is usually drinking water quality except in those cases in which surface, well, or salt waters are used without treatment.

Wastewater Treatment Plants

The most common municipal wastewater treatment plants are primary and secondary treatment plants, tertiary treatment plants, and physical-chemical treatment plants.

Primary treatment consists of removing a substantial amount of the suspended solids from a wastewater. The collected solids must be treated, in most cases, followed by proper disposal. Secondary treatment consists of bio-oxidizing the remaining organic suspended solids and the organic dissolved solids. The flowsheet of a conventional activated sludge plant, shown in Figure 1.3, consists of screening, grit removal, primary clarification, activated sludge treatment, and chlorination. The coarse solids are removed by screening, and the sand and silt are removed by the grit removal system. Primary clarification removes as many suspended solids as possible, and the primary effluent is mixed with the return activated sludge. The mixed liquor then flows to the aeration tank. Bio-oxidation of most of the remaining organic matter occurs in the aera-

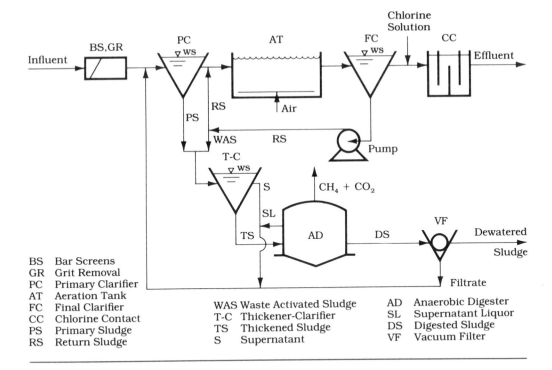

Chlorine
Solution

BS,GR   PC   AT   FC   CC

Influent   Effluent

WS   WS

PS   RS   Air
WAS   RS   Pump
T-C
WS   S $CH_4 + CO_2$

SL   VF

TS   AD   DS   Dewatered
                         Sludge

BS   Bar Screens
GR   Grit Removal
PC   Primary Clarifier   Filtrate
AT   Aeration Tank
FC   Final Clarifier   WAS Waste Activated Sludge   AD   Anaerobic Digester
CC   Chlorine Contact   T-C Thickener-Clarifier   SL   Supernatant Liquor
PS   Primary Sludge   TS   Thickened Sludge   DS   Digested Sludge
RS   Return Sludge   S   Supernatant   VF   Vacuum Filter

tion tank, and the final clarifier removes the biological solids, which are returned to mix with the incoming primary
effluent. The effluent from the final clarifier is disinfected
to kill pathogenic organisms and then discharged to the
receiving body of water. The primary clarifier sludge and
the waste activated sludge (in other words, the excess activated sludge produced by the microbial solids, which has
to be wasted from the system) are mixed together, then
thickened to increase the solids content. The thickened
sludge is sent to the anaerobic digester for bio-oxidation of
the organic solids. The digested sludge is dewatered by vacuum filtration and the dewatered sludge is disposed of in
a sanitary landfill. Minor flows, such as the thickener supernatant, the anaerobic digester supernatant, and the
vacuum filter filtrate are returned to the head of the plant.
Although the solids handling system shown consists of
thickening, anaerobic digestion, and vacuum filtration,
other solids handling systems, such as aerobic digestion
and centrifugation, are used. This flowsheet gives about 85
to 95 percent five-day biochemical oxygen demand (BOD_5)
and suspended solids removal.

Tertiary treatment of a secondary effluent consists
of providing further treatment to increase the quality of

Figure 1.3. Activated Sludge Plant for a Municipal Wastewater the effluent. The flowsheet for a tertiary treatment plant is shown in Figure 1.4. It consists of lime coagulation, flocculation, sedimentation, ammonia stripping, recarbonation, sedimentation, multimedia filtration, carbon adsorption, and breakpoint chlorination. The coagulant used is quicklime (calcium oxide), which is reacted with water to produce slaked lime (calcium hydroxide), which is added ahead of the mixing basin. Lime coagulation, flocculation, and sedimentation at a high pH removes most of the suspended solids and phosphorus. Ammonia stripping at a high pH removes most of the ammonia. Recarbonation is provided to lower the pH and stabilize the wastewater. The settling basin downline from the recarbonation basin removes the calcium carbonate precipitated by recarbonation. Multimedia filtration removes most of the nonsettling floc, and carbon adsorption removes most of the remaining dissolved organic compounds. In addition to disinfection, breakpoint chlorination chemically oxidizes the remaining ammonia to chloramines and the remaining organic matter to other end products. The solids handling system shown permits recovery of the quicklime coagulant, thus reducing the lime requirements and the amount of lime sludge to be disposed. The coagulant is recovered by lime recalcination

Figure 1.4. Tertiary Treatment of a Secondary Effluent by Physical-Chemical Methods

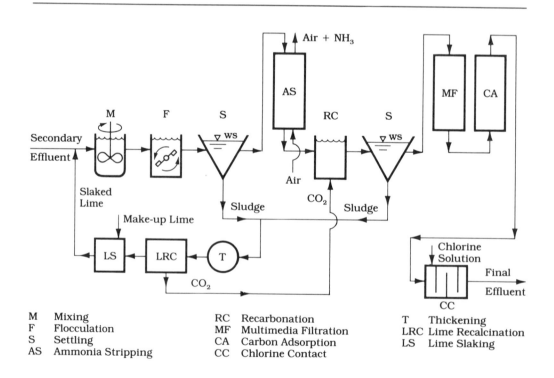

M	Mixing	RC	Recarbonation	T	Thickening
F	Flocculation	MF	Multimedia Filtration	LRC	Lime Recalcination
S	Settling	CA	Carbon Adsorption	LS	Lime Slaking
AS	Ammonia Stripping	CC	Chlorine Contact		

in which the calcium carbonate precipitate in the sludge is heated at a high temperature to produce the coagulant, calcium oxide. In addition to coagulant recovery, the organic solids in the sludge are incinerated. The flowsheet in Figure 1.4 will produce an effluent approaching drinking water quality when treating municipal secondary effluents.

In recent years, the success of physical-chemical treatment in tertiary treatment has led to the use of physical-chemical treatment of raw municipal wastewaters, in lieu of conventional biological treatments, since a higher quality effluent can be obtained. The flowsheet of a physical-chemical treatment plant for raw municipal wastewaters is shown in Figure 1.5. It consists of lime coagulation, flocculation, sedimentation, recarbonation, sedimentation, multimedia filtration, carbon adsorption, and breakpoint chlorination. Lime coagulation, flocculation, and sedimentation remove most of the suspended solids, phosphorus, and organic nitrogen. Recarbonation lowers the pH and stabilizes the wastewater. The sedimentation basin downline from the recarbonation unit removes most of the calcium carbonate precipitated by recarbonation. Multimedia filtration removes most of the fine, nonsettling floc, and carbon adsorption removes most of the remaining organic

Figure 1.5. Physical-Chemical Treatment of Raw Municipal Wastewater

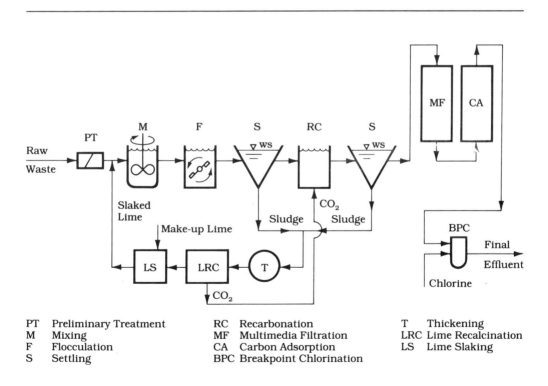

PT	Preliminary Treatment	RC	Recarbonation	T	Thickening
M	Mixing	MF	Multimedia Filtration	LRC	Lime Recalcination
F	Flocculation	CA	Carbon Adsorption	LS	Lime Slaking
S	Settling	BPC	Breakpoint Chlorination		

compounds. In addition to disinfection, breakpoint chlorination chemically oxidizes the remaining ammonia to chloramines and the remaining organic materials to other end products. The solids handling system shown permits recovery of the quicklime coagulant and reduces the amount of lime sludge to be disposed of in addition to incinerating the organic solids in the lime sludge. The flowsheet shown removes from 96 to 99 percent of the BOD_5 and suspended solids when treating municipal wastewaters.

Industrial wastewaters may be broadly classified as organic and inorganic wastewaters. Even for biodegradable organic industrial wastewaters, the quality parameters and the flow rates vary appreciably between the numerous industries in this category. Therefore, a wide variation exists in the various flowsheets employed. For organic industrial wastewaters with low to moderate suspended solids content, the flowsheet shown in Figure 1.6 is frequently used. It consists of a completely mixed activated sludge unit and

Figure 1.6. Completely Mixed Activated Sludge Plant for an Industrial Wastewater

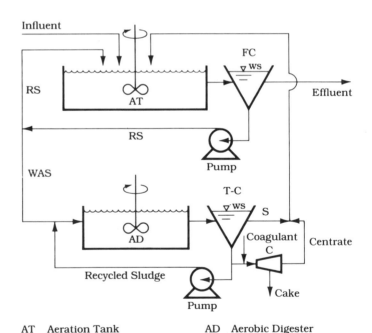

AT	Aeration Tank	AD	Aerobic Digester
FC	Final Clarifier	T-C	Thickener-Clarifier
RS	Return Sludge	S	Supernatant
WAS	Waste Activated Sludge	C	Centrifuge

ENVIRONMENTAL ENGINEERING

an aerobic digestion system for treating the waste activated sludge. A completely mixed activated sludge unit has a reactor basin which is usually square or circular in plan view. The influent wastewater, on entering, is spread throughout the reactor basin volume in a very short period of time. The digested sludge from the aerobic digestion system is chemically conditioned to release water and is then dewatered by centrifugation. The cake is usually disposed of by sanitary landfill, and the centrate is returned to the head of the plant. For organic industrial wastewaters with low to moderate suspended solids content in which the organic content varies appreciably during the day, a constant level equalization basin may be required as shown in Figure 1.7. This flowsheet consists of an equalization basin, a dispersed plug-flow activated sludge unit, and an incinerator for disposal of the waste activated sludge. A dispersed plug-flow activated sludge unit has a reactor basin that is usually rectangular in plan view and that has significant longitudinal dispersion of fluid elements throughout the length of the basin. The waste activated sludge is chemically conditioned, then dewatered by a filter press. The sludge cake is incinerated and the ash disposed of in a sanitary landfill. The mixing system for the completely mixed reactor and aerobic digester in Figure 1.6 and for the equalization basin in Figure 1.7 is shown schematically as a propeller. In actual basins, the mixing is usually provided by the aeration system. If appreciable land area is available, organic industrial wastewaters with low to moderate suspended solids content may be treated by the aer-

Figure 1.7. Dispersed Plug-Flow Activated Sludge Plant for an Industrial Wastewater

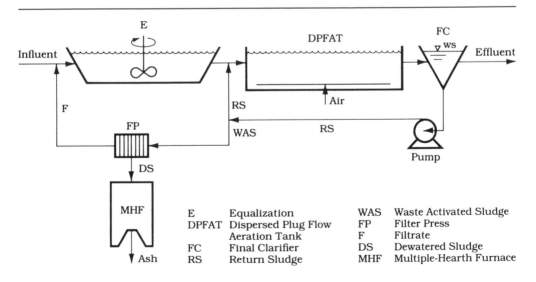

E	DPFAT	Equalization	WAS	Waste Activated Sludge
DPFAT		Dispersed Plug Flow Aeration Tank	FP	Filter Press
			F	Filtrate
FC		Final Clarifier	DS	Dewatered Sludge
RS		Return Sludge	MHF	Multiple-Hearth Furnace

**Figure 1.8. Aerated Lagoon
System for an Industrial
Wastewater**

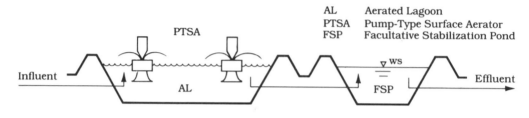

ated lagoon flowsheet shown in Figure 1.8. It consists of an aerated lagoon, which is essentially an activated sludge unit without recycle, and a facultative stabilization pond, which serves as a final clarifier. The biological solids that settle in the stabilization pond undergo anaerobic decomposition on the pond bottom, and the solids are usually removed once every several years. Frequently, a settling tank with sludge rakes for continuous sludge removal is used for final clarification; the removed sludge is treated to stabilize it.

For inorganic industrial wastewaters, the unit operations and processes used for treatment depend on the wastewater characteristics. For instance, industrial wastewaters from plating industries, which contain heavy metallic ions such as copper, zinc, and cadmium, may be treated by ion exchange. To locate flowsheets used for a particular industrial wastewater, the reader is referred to the various texts and literature on industrial waste treatment.

Design Flow Rates and Parameters

Water and wastewater treatment plants for municipalities are usually designed for a flow rate that will occur from 10 to 25 years in the future. This requires a population projection for the design period and also an estimate of the water demand or wastewater flow per capita for the future period. The most accurate way to determine the water demand per capita is by studying long-term water pumpage records (at least 12 months' duration). Pumping records will also allow determination of the minimum, average, and maximum day demands in addition to the maximum hourly demand. Usually, in the United States, the average annual water demand for municipalities is from about 100 to 200 gal/capita-day. Frequently, the existing demand is increased slightly to obtain the demand per capita in the future, since the demand per capita increases as the population increases. The percent increase is usually taken as 0.1

times the percent increase in population. Thus, if the existing demand is 145 gal/capita-day, and the population is expected to increase 80 percent in the future, the future demand would be (145)[100% + (0.1)80%] or 157 gal/capita-day. Long-term analyses on the water supply will give quality parameters, such as pH, turbidity, color, odor, taste, hardness, alkalinity, dissolved solids, and coliform count. Short-term analyses on parameters subject to variation, such as the turbidity of a surface water, should be used with caution.

The most accurate way to determine the wastewater flow per capita is by studying long-term flow records. Usually for municipalities with separate sewer systems in the United States, the wastewater flow is approximately equal to the water demand. The future flow rate per capita is increased over the present flow rate in a similar manner as is done for future water demands. Also, flow records will allow determination of the ratio of the peak hourly flow to the average hourly flow, the ratio of the minimum hourly flow to the average hourly flow, and the amount of storm water infiltration. Long-term analyses of the wastewater will give quality parameters, such as the five-day biochemical oxygen demand (BOD_5), suspended solids concentration, chemical oxygen demand (COD), organic nitrogen, ammonia nitrogen, nitrite nitrogen, nitrate nitrogen, grease content, and pH. For separate sewer systems in the United States, the average BOD_5 is about 200 mg/ℓ and the average suspended solids concentration is about 300 mg/ℓ. Arbitrarily assumed design parameters, such as the population equivalent of 0.17 lb BOD_5/capita-day, should be used with caution and should be substantiated by several months of analyses and flow gauging.

Water and wastewater treatment plants for industries are usually designed to accommodate water demands and wastewater flows for a future flow projection of 5 to 10 years, and the plants are designed to facilitate future expansions. This is done because industries usually prefer stagewise expansions; these require less initial expense than plants for longer design periods. Also, the uncertainty of the future market for their product necessitates prudent capital expenditures. Water demands for an industry are usually expressed in gallons per production unit. For example, for a pulp and paper mill it would be gallons used per ton of paper produced. By knowing the number of production units expected in the future, the future water demand can be determined.

In designing a wastewater treatment plant for an industry, it is desirable to have an industrial waste survey done for the industry. In a waste survey, a flowsheet for the industry is developed that shows the operations used in producing their product in sequential order. All sources of water use and wastewater generation are located and the flow rate of each is determined. Wastewater streams are sampled and analysed to give quality parameters such as BOD_5 and suspended solids. The final result of the survey is a flow and material balance diagram showing the flow rates and quality parameters for each waste stream and the outfall sewer. In studying the diagram, places are located where reuse and usable byproduct recovery is feasible, where a reduction in water usage is possible, where beneficial process modifications can be done, and where segregation of flows is desirable or necessary. An example of reuse is the reusing of washwaters instead of using them on a once-through basis. An example of usable byproduct recovery is the recovery of oil from waste streams within a refinery by the use of gravity-type oil-water separation tanks. The oil that floats to the top of the tanks is skimmed off and the recovered oil is sent back for processing. An example of a beneficial process modification is the substitution of low-BOD_5 detergents for high-BOD_5 soaps in the wash operations done in a woolen textile mill. Flow segregation should be done for streams having a low BOD_5 concentration, inert materials, or characteristics that make them incompatible with the main wastewater flow. Some segregated streams may be eliminated from the main wastewater flow and not require any treatment. Other segregated flows may require separate treatment from the main wastewater flow, or, in some cases, they may be pretreated and the pretreated stream sent back to the main wastewater flow. For instance, an acidic stream with a significant BOD_5 concentration may be neutralized, then remixed with the main wastewater flow. Once the flows are adjusted for reuse, usable byproduct recovery, flow reduction, beneficial process modifications, and flow segregation, the final flow in gallons per day for the outfall sewer and the final quality parameters, such as pounds of BOD_5 and pounds of suspended solids per day, can be determined. These can be correlated with production units to give gallons of wastewater, pounds of BOD_5, and pounds of suspended solids per production unit. Then, knowing the number of production units in the future, the design parameters and design flow for the industrial wastewater treatment plant can be determined. In most cases, an industrial waste survey significantly reduces the amount of wastewater flow to be treated, thus decreasing the costs of treatment.

Drinking water standards for water treatment plants

and effluent standards for wastewater treatment plants are established by governmental regulations. Primary drinking water standards for public water supplies are based on the National Interim Primary Drinking Water Regulations, which were issued on December 24, 1975. These are based on the National Safe Drinking Water Act, Public Law 93-523, which was signed into law on December 16, 1974. Public Law 93-523 gives the Environmental Protection Agency (EPA) the responsibility for establishing primary drinking water standards relating to health. Secondary regulations relating to taste, odor, and appearance of drinking water have been recommended by the EPA, and most states have adopted these or a modification of them.

Comprehensive federal water pollution control legislation was enacted on October 18, 1972 under Public Law 92-500. This law requires the EPA to establish effluent guidelines from which individual states issue discharge permits and impose specific effluent limitations. Under Public Law 92-500, municipalities are required to have secondary treatment.

Chapter Two

COAGULATION AND FLOCCULATION

Coagulation and flocculation consist of adding a floc-forming chemical reagent to a water or wastewater to enmesh or combine with non-settleable colloidal solids and slow-settling suspended solids to produce a rapid-settling floc. The floc is subsequently removed in most cases by sedimentation. *Coagulation* is the addition and rapid mixing of a coagulant, the resulting destabilization of the colloidal and fine suspended solids, and the initial aggregation of the destabilized particles. *Flocculation* is the slow stirring or gentle agitation to aggregate the destabilized particles and form a rapid-settling floc.

In water treatment the principal use of coagulation and flocculation is to agglomerate solids prior to sedimentation and rapid sand filtration. In municipal wastewater treatment, coagulation and flocculation are used to agglomerate solids in the physical-chemical treatment of raw wastewaters and primary or secondary effluents. In industrial waste treatment coagulation is employed to coalesce solids in wastewaters that have an appreciable suspended solids content. In water treatment, the principal coagulants used are aluminum and iron salts, although polyelectrolytes are employed to some extent. In wastewater treatment, aluminum and iron salts, lime, and polyelectrolytes are used. Chemical precipitation, which is closely related to chemical coagulation, consists of the precipitation of unwanted ions from a water or wastewater. In the coagulation of municipal wastewaters, not only does coagulation of solids occur, but also the chemical precipitation of much of the phosphate ion takes place.

15

A Three-Compartment Paddle-Wheel Flocculation Basin at a Water Treatment Plant. Note that the flow is parallel to the paddle-wheel shafts, as shown in Figure 2.17. The flow passes through the orifices in the walls. Tapered flocculation is provided by varying the size of the compartments and the number and length of the blades on the paddle wheels.

Theory of Coagulation

A portion of the dispersed solids in surface waters and wastewaters are nonsettleable suspended materials that have a particle size ranging from 0.1 millimicron (10^{-7} mm) to 100 microns (10^{-1} mm). Since colloids have a particle size ranging from one millimicron (10^{-6} mm) to one micron (10^{-3} mm), a significant fraction of the nonsettleable matter is colloidal particulates. The supracolloidal fraction has a particle size ranging from one micron (10^{-3} mm) to 100 microns (10^{-1} mm). Many of the particles in this fraction have certain colloidal characteristics such as negligible settling velocities.

Colloidal Characteristics
Colloidal dispersions are classified according to the dispersed phase and the dispersed medium. The principal systems involved in both water and wastewater treatment are solids dispersed in liquids (sols) and liquids dispersed in

liquids (emulsions). When suspended in water, organic matter, such as microbes, and inorganic matter, such as clays, are examples of a system consisting of solids dispersed in a liquid. An oil dispersed in water is an example of a liquid dispersed in a liquid. In water and wastewater treatment, solids dispersed in liquids (sols) are of particular interest. An important feature of a solid colloid dispersed in water is that the solid particles will not settle by the force of gravity. When a solid colloid stays in suspension and does not settle, the system is in a stable condition.

Colloids have an extremely large surface area per unit volume of the particles—that is, a large specific surface area. Because of the large surface area, colloids tend to adsorb substances, such as water molecules and ions, from the surrounding water. Also, colloids develop or have an electrostatic charge relative to the surrounding water.

Colloidal solids in water may be classified as *hydrophilic* or *hydrophobic* according to their affinity for water. Hydrophilic colloids have an affinity for water due to the existence of water-soluble groups on the colloidal surface. Some of the principal groups are the amino, carboxyl, sulfonic, and hydroxyl. Since these groups are water soluble, they promote hydration and cause a water layer or film to collect and surround the hydrophilic colloid. Frequently, this water layer or film is termed the *water of hydration* or *bound water.* Usually, organic colloids, such as proteins or their degradation products, are hydrophilic. Hydrophobic colloids have little, if any, affinity for water; as a result, they do not have any significant water film or water of hydration. Usually inorganic colloids, such as clays, are hydrophobic.

Colloidal particles have electrostatic forces that are important in maintaining a dispersion of the colloid. The surface of a colloidal particle tends to acquire an electrostatic charge due to the ionization of surface groups and the adsorption of ions from the surrounding solution. Also, colloidal minerals, such as clays, will have an electrostatic charge due to the ionic deficit within the mineral lattice. Hydrophilic colloids, such as proteinaceous materials and microbes, have charges due to the ionization of such groups as the amino $(-NH_2)$ and the carboxyl $(-COOH)$, which are located on the colloidal surface. When the pH is at the isoelectric point, the net or overall charge is zero since the amino group is ionized $(-NH_3^+)$ and also the carboxyl group is ionized $(-COO^-)$. At a pH below the isoelectric point, the carboxyl group is not ionized $(-COOH)$, and the colloid is positively charged due to the ionized amino group $(-NH_3^+)$. At a pH above the isoelectric point, the amino group loses a hydrogen, producing a neutral group ($-NH_2$), and the colloid is negatively charged due to the ionized carboxyl group ($-COO^-$). In general, most naturally occurring hydrophilic colloids, such as proteinaceous matter and microbes, have a negative charge if the pH is at or above the neutral range. Some colloidal materials, such as oil droplets and some other chemically inert substances, will preferentially adsorb negative ions, particularly the hydroxyl ion, from their surrounding solution and become negatively charged. Colloidal minerals, such as clays, have more nonmetallic atoms than metallic atoms within their crystalline structure, resulting in a net negative charge. Usually most naturally occurring hydrophobic colloids, such as clays, are inorganic materials and have a negative charge. The sign and magnitude of the charge of a colloid will depend on the type of colloidal matter and the characteristics of the surrounding solution.

In most colloidal systems, the colloids are maintained in suspension (in other words, stabilized) due to the elec-

A Turbine-Type Flocculation Basin at a Tertiary Wastewater Treatment Plant. Note baffle fence for outlet flow.

ENVIRONMENTAL ENGINEERING

trostatic forces of the colloids themselves. Since most naturally occurring colloids are negatively charged and like charges are repulsive, the colloids remain in suspension due to the action of the repulsive forces.

A negative colloidal particle will attract to its surface ions of the opposite charge—counterions—from the surrounding water, as depicted in Figure 2.1. The compact layer of counterions is frequently termed the *fixed layer;* outside the fixed layer is the *diffused layer.* Both layers will contain positive and negative charged ions; however, there will be a much larger number of positive ions than negative

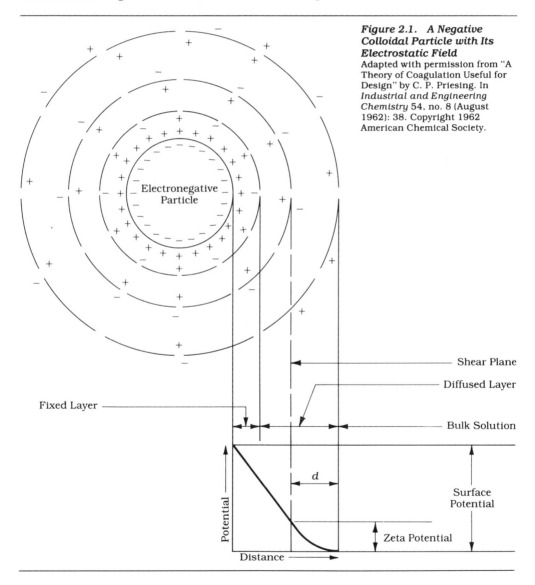

Figure 2.1. A Negative Colloidal Particle with Its Electrostatic Field
Adapted with permission from "A Theory of Coagulation Useful for Design" by C. P. Priesing. In *Industrial and Engineering Chemistry* 54, no. 8 (August 1962): 38. Copyright 1962 American Chemical Society.

ions. The two layers represent the region surrounding the particle where there is an electrostatic potential due to the particle, as illustrated in Figure 2.1. The concentration of the counterions is greatest at the particle surface; it decreases to that of the bulk solution at the outer boundary of the diffused layer. The shear plane or shear surface surrounding the particle encloses the volume of water (in other words, the bound water or water envelope) that moves with the particle. The *zeta potential* is the electrostatic potential at the shear surface as shown in Figure 2.1. This potential is usually related to the stability of a colloidal suspension.

A colloidal suspension is stable if the particles remain in suspension and do not coagulate. The colloidal stability depends on the relative magnitude of the forces of attraction and the forces of repulsion. The forces of attraction are due to van der Waals' forces, which are effective only in the immediate neighborhood of the colloidal particle. The forces of repulsion are due to the electrostatic forces of the colloidal dispersion. The magnitude of these forces is measured by the zeta potential, which is

$$\zeta = \frac{4\pi q d}{D} \qquad (2.1)$$

where

ζ = zeta potential;
q = charge per unit area;
d = thickness of the layer surrounding the shear surface through which the charge is effective, as shown in Figure 2.1;
D = dielectric constant of the liquid.

Thus, the zeta potential measures the charge of the colloidal particle, and it is dependent on the distance through which the charge is effective. It follows that the greater the zeta potential, the greater are the repulsion forces between the colloids and, therefore, the more stable is the colloidal suspension. Also, the presence of a bound water layer and its thickness affects colloidal stability, since it prevents the particles from coming into close contact.

Hydrophilic colloids have a shear surface at the outer boundary of the bound water layer. Hydrophobic colloids have a shear surface near the outer boundary of the fixed layer.

Coagulation of Colloids (Destabilization)
When a coagulant is added to a water or wastewater, destabilization of the colloids occurs and a coagulant floc is formed. Since the chemistry involved is very complex, the

following discussion is intended to be introductory and will illustrate the known interactions that occur. These interactions are: (1) the reduction of the zeta potential to a degree where the attractive van der Waals' forces and the agitation provided cause the particles to coalesce; (2) the aggregation of particles by interparticulate bridging between reactive groups on the colloids; and (3) the enmeshment of particles in the precipitate floc that is formed. Figure 2.2 shows the interparticulate forces acting on a colloidal particle. The repulsive forces are due to the electrostatic zeta potential, and the attractive forces are due to

Figure 2.2. Colloidal Interparticulate Forces versus Distance

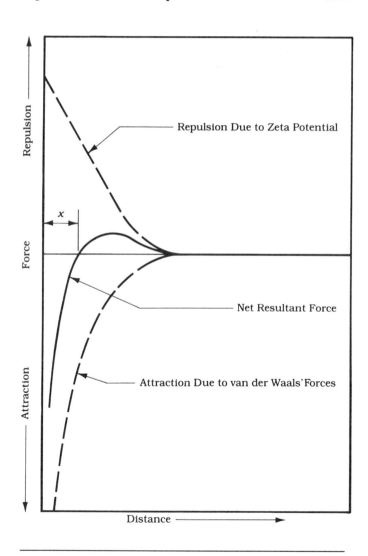

van der Waals' forces acting between the particles. The net resultant force is attractive out to the distance, x. Beyond this point the net resultant force is repulsive.

When a coagulant salt is added to a water, it dissociates, and the metallic ion undergoes hydrolysis and creates positively charged hydroxo-metallic ion complexes. Usually a coagulant is an aluminum salt, such as $Al_2(SO_4)_3$, or an iron salt, such as $Fe_2(SO_4)_3$. There are numerous species of hydroxo-metallic complexes formed because the complexes, which are hydrolysis products, tend to polymerize (Stumm, W., & Morgan, J. J., 1962; Stumm, W., & O'Melia, C. R., 1968). The generalized expression for these complexes is $Me_q(OH)_p^{z+}$. For an aluminum salt, some of the resulting polymers are $Al_6(OH)_{15}^{+3}$, $Al_7(OH)_{17}^{+4}$, $Al_8(OH)_{20}^{+4}$, and $Al_{13}(OH)_{34}^{+5}$. For an iron salt, some of the resulting polymers are $Fe_2(OH)_2^{+4}$ and $Fe_2(OH)_4^{+5}$. The hydroxo-metallic complexes are polyvalent, possess high positive charges, and are adsorbed to the surface of the negative colloids. This results in a reduction of the zeta potential to a level where the colloids are destabilized. The destabilized particles, along with their adsorbed hydroxo-metallic complexes, agg.egate by interparticulate attraction due to van der Waals' forces. These forces are aided by the gentle agitation of the water. In the aggregation process, the agitation is very important since it causes the destabilized particles to come in close vicinity or collide and then coalesce.

The aggregation of the destabilized particles also occurs by interparticulate bridging involving chemical interactions between reactive groups on the destabilized particles. The agitation of the water is also important in this type of aggregation, since it causes interparticulate contacts.

The dosages of coagulant salts used in coagulating waters and wastewaters are usually in an appreciable excess of the amount required to produce the necessary positive hydroxo-metallic complexes. The excess complexes continue to polymerize until they form an insoluble metallic hydroxide, $Al(OH)_3$ or $Fe(OH)_3$, and the solution will be supersaturated with the hydroxide. In the formation of the metallic hydroxide, there is enmeshing of the negative colloids with the precipitate as it forms. This enmeshment type of coagulation is sometimes referred to as *precipitate* or *sweep coagulation*.

Originally it was thought that the zeta potential reduction was caused by the adsorption of the metallic ions from the coagulant salt. However, it is now known that the principal action is the adsorption of the highly positively charged hydroxo-metallic complexes. The species of polyvalent metallic ion complexes are much more effective in coagulating a colloidal dispersion than are monovalent complexes; thus, polyvalent metallic salts are always used in coagulation.

For dilute colloidal suspensions, the rate of coagulation may be extremely slow because of the low particulate concentrations, which cause an inadequate number of interparticulate contacts. In many water and wastewater treatment plants, it has been found that recycling a small portion of the settled sludge, before or after rapid mixing, maintains the colloidal concentration at a level where rapid coagulation and flocculation occur. For dilute suspensions, relatively large coagulant dosages may cause restabilization of the colloids. When this occurs the negatively charged colloids become positively charged; this is believed to be due to existing positively charged reactive sites on the colloidal surfaces.

The coagulation of colloids by organic polymers occurs by a chemical interaction or bridging. The polymers have ionizable groups such as carboxyl, amino, and sulfonic, and these groups bind with reactive sites or groups on the surfaces of the colloids. In this manner several colloids may be bound to a single polymer molecule to form a bridging structure. Bridging between particles is optimum when the colloids are about one-half covered with adsorbed segments of the polymers.

Frequently in coagulation, the terms *electrokinetic*, *perikinetic*, and *orthokinetic coagulation* are encountered. Electrokinetic coagulation refers to coagulation that is a result of zeta potential reduction. Perikinetic coagulation refers to coagulation in which the interparticle contacts result from Brownian movement. In orthokinetic coagulation the interparticle contacts are caused by fluid motion due to agitation.

Proper coagulation and flocculation of a water is very important in the settling of the coagulated water, because the settling velocity is proportional to the square of the particle diameter. Thus, the production of large floc particles results in rapid settling.

Coagulants

The most widely used coagulants in water treatment are aluminum sulfate and iron salts. Aluminum sulfate (filter alum) is employed more frequently than iron salts because it is usually cheaper. Iron salts have an advantage over filter alum because they are effective over a wider pH range. In the lime-soda softening process, the lime serves as a coagulant since it produces a heavy floc or precipitate consisting of calcium carbonate and magnesium hydroxide. This precipitate has coagulating and flocculating properties. The most widely used coagulants in wastewater treatment are filter alum and lime. Sometimes coagulant aids, such as recycled sludge or polyelectrolytes, are required to produce a rapid-settling floc. The principal factors affecting the coagulation and flocculation of water or wastewater are turbidity, suspended solids, temperature, pH, cationic and anionic composition and concentration, duration and degree of agitation during coagulation and flocculation, dosage and nature of the coagulant, and, if required, the coagulant aid. The selection of a coagulant requires the use of laboratory or pilot plant coagulation studies, since a given water or wastewater may show optimum coagulation results for a particular coagulant. Usually laboratory studies using the jar test are adequate for selecting a coagulant for a water treatment plant, whereas laboratory and frequently pilot studies are required for wastewaters. As shown in the previous section, coagulation chemistry is complex, and theoretical chemical equations to determine the amount of the metallic hydroxides produced give only approximate results. In this section, the common coagulants, their chemical reactions with alkalinity, and their characteristics are discussed.

Aluminum Sulfate
Sufficient alkalinity must be present in the water to react with the aluminum sulfate to produce the hydroxide floc. Usually for the pH ranges involved, the alkalinity is in the form of the bicarbonate ion. The simplified chemical reaction to produce the floc is

$$Al_2(SO_4)_3 \cdot 14H_2O + 3Ca(HCO_3)_2 \rightarrow$$
$$2Al(OH)_3 + 3CaSO_4 + 14H_2O + 6CO_2 \qquad (2.2)$$

Certain waters may not have sufficient alkalinity to react with the alum, so alkalinity must be added. Usually alkalinity in the form of the hydroxide ion is added by the addition of calcium hydroxide (slaked or hydrated lime). The coagulation reaction with calcium hydroxide is

$$Al_2(SO_4)_3 \cdot 14H_2O + 3Ca(OH)_2 \rightarrow$$
$$2Al(OH)_3 + 3CaSO_4 + 14H_2O \qquad (2.3)$$

Alkalinity may also be added in the form of the carbonate ion by the addition of sodium carbonate (soda ash). Most waters have sufficient alkalinity, so that no chemical needs to be added other than aluminum sulfate. The optimum pH range for alum is from about 4.5 to 8.0, since aluminum

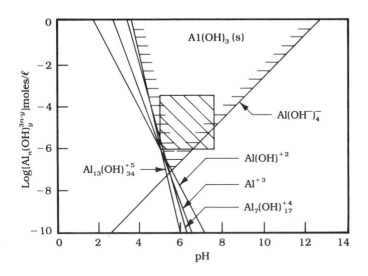

Figure 2.3. Solubility of Aluminum Hydroxide. (Shaded area is the usual operating region used in water treatment.) Adapted from *Journal of American Water Works Association*, Volume 60, Number 5 (May 1968), by permission. Copyright 1968, the American Water Works Association.

hydroxide is relatively insoluble within this range, as shown in Figure 2.3. Also illustrated in this figure is the usual range of aluminum hydroxide concentrations for dosages used in water treatment. These usually produce an oversaturated solution of aluminum hydroxide. Aluminum sulfate is available in the dry or liquid form; however, the dry form is most common. The dry chemical may be granular, powdered, or lump form, with the granular being the most widely used. The granules, which are 15 to 22 percent Al_2O_3, contain approximately 14 waters of crystallization, weigh from 60 to 63 lb/ft^3, and may be dry fed. The dry chemical may be shipped in bags, barrels, or bulk (carload). The liquid form is 50 percent alum and is shipped by tank car or tank truck.

Ferrous Sulfate

Ferrous sulfate requires alkalinity in the form of the hydroxide ion in order to produce a rapid reaction. Consequently, slaked or hydrated lime, $Ca(OH)_2$, is usually added to raise the pH to a level where the ferrous ions are precipitated as ferric hydroxide. This reaction is an oxidation-reduction reaction requiring some dissolved oxygen in the water. In the coagulation reaction, the oxygen is reduced and the ferrous ion is oxidized to the ferric state, where it precipitates as ferric hydroxide. The simplified chemical reaction is

$$2FeSO_4 \cdot 7H_2O + 2Ca(OH)_2 + \tfrac{1}{2}O_2 \rightarrow$$
$$\underline{2Fe(OH)_3} + 2CaSO_4 + 13H_2O \qquad (2.4)$$

For the above reaction to occur, the pH must be raised to about 9.5, and sometimes stabilization is required for the excess lime employed. Ferrous sulfate and lime coagulation is usually less expensive than alum. In general, the precipitate formed, ferric hydroxide, is a dense, quick-settling floc. Ferrous sulfate is available in dry or liquid form; however, the dry form is most common. The dry chemical may be granules or lumps, with the granules being the most widely used. The granules, which are 55 percent $FeSO_4$, contain seven waters of crystallization, weigh from 63 to 66 lb/ft^3, and are usually dry fed. The dry chemical may be shipped in bags, barrels, or bulk.

Chlorinated copperas treatment is another method to use ferrous sulfate. In this process ferrous sulfate is reacted with chlorine, and the ferrous ion is oxidized to the ferric ion as follows:

$$3FeSO_4 \cdot 7H_2O + 1.5Cl_2 \rightarrow$$
$$Fe_2(SO_4)_3 + FeCl_3 + 21H_2O \qquad (2.5)$$

This reaction occurs at a pH as low as about 4.0. The products, ferric sulfate and ferric chloride, are very effective coagulants, and discussions of these are presented in the following two sections.

Ferric Sulfate

The simplified reaction of ferric sulfate with natural bicarbonate alkalinity to form ferric hydroxide is

$$Fe_2(SO_4)_3 + 3Ca(HCO_3)_2 \rightarrow$$
$$\underline{2Fe(OH)_3} + 3CaSO_4 + 6CO_2 \qquad (2.6)$$

The reaction usually produces a dense, rapid-settling floc. If the natural alkalinity is insufficient for the reaction, slaked lime may be used instead. The optimum pH range for ferric sulfate is from about 4 to 12, since ferric hydroxide is relatively insoluble within this range, as shown in Figure 2.4. Also illustrated in Figure 2.4 is the usual range of ferric hydroxide concentrations for dosages used in water treatment; these usually produce an oversaturated solution of ferric hydroxide. Ferric sulfate is available in dry form as

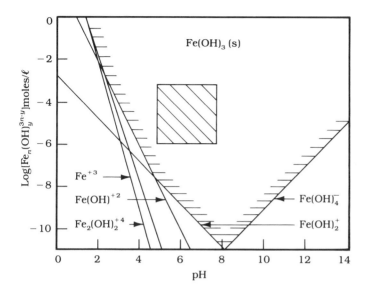

Figure 2.4. Solubility of Ferric Hydroxide. (Shaded area is the usual operating region used in water treatment.)
Adapted from *Journal of American Water Works Association*, Volume 60, Number 5 (May 1968), by permission. Copyright 1968, the American Water Works Association.

granules or as a powder, with the granules being the most common. The granules are 90 to 94 percent $Fe_2(SO_4)_3$, contain nine waters of crystallization, and weigh from 70 to 72 lb/ft^3. The chemical is usually dry fed. The dry chemical may be shipped in bags or barrels.

Ferric Chloride

The simplified reaction of ferric chloride with natural bicarbonate alkalinity to form ferric hydroxide is

$$2FeCl_3 + 3Ca(HCO_3)_2 \rightarrow$$
$$\underline{2Fe(OH)_3} + 3CaSO_4 + 6CO_2 \qquad (2.7)$$

If the natural alkalinity is insufficient for the reaction, slaked lime may be added to form the hydroxide, as given by the equation

$$2FeCl_3 + 3Ca(OH)_2 \rightarrow$$
$$\underline{2Fe(OH)_3} + 3CaCl_2 \qquad (2.8)$$

The optimum pH range for ferric chloride is the same as for ferric sulfate, which is from about 4 to 12. The floc formed is generally a dense, rapid-settling floc. Ferric chloride is available in dry or liquid form. The dry chemical may be in powder or lump form, with the lump form being the most common. The lumps, which are 59 to 61 percent $FeCl_3$, contain six waters of crystallization and weigh from 60 to 64 lb/ft^3. The lumps are very hydroscopic and are usually solution fed. Upon absorbing water they decompose to yield hydrochloric acid. The powdered or anhydrous form is 98 percent $FeCl_3$, contains no water of crystallization, and weighs from 85 to 90 lb/ft^3. The liquid form is 37 to 47 percent $FeCl_3$. The dry form is shipped in barrels, and the solution form in bulk.

Lime

High lime treatment is frequently used in wastewater treatment, and the lime may be in the form of slaked lime or hydrated lime. Slaked lime (milk of lime), $Ca(OH)_2$, is produced by reacting quicklime, CaO, with water in lime-slaking equipment. Quicklime is available in the dry form as granules or lumps, with the lump form being the most common. Quicklime lumps are usually 70 to 96 percent CaO and weigh from 55 to 70 lb/ft^3. Quicklime is shipped in bags, barrels, or bulk. If the amount of lime required is rather small, it may be desirable to use hydrated lime, $Ca(OH)_2$, since the feeders are less expensive than slaking equipment. Hydrated lime is from 82 to 99 percent $Ca(OH)_2$, and it is available as a powder in two different densities. Light hydrated lime weighs from 24 to 48 lb/ft^3, whereas dense hydrated lime weighs from 40 to 70 lb/ft^3. It is available in bags, barrels, or bulk.

Coagulant Aids

In both water and wastewater treatment, coagulant aids are sometimes used to produce a quick-forming, dense, rapid-settling floc and to insure optimum coagulation.

Alkalinity addition is required to aid coagulation if the natural alkalinity is insufficient to produce a good floc. Lime is usually used, and it may be fed as slaked lime (milk of lime) or hydrated lime. Soda ash (sodium carbonate) is used to a lesser extent than lime since it is more expensive than quicklime. Soda ash is available as a powder and may be dry fed. The powder is 99.4 percent Na_2CO_3 and is available in three specific weights—23, 35, and 65 lb/ft^3. It comes in bags, barrels, or bulk.

Polyelectrolytes are also used to obtain optimum coagulation, and these aids may be classified according to their ionic characteristics. There are anionic (negatively charged), cationic (positively charged), and polyampholites, which have both positively and negatively charged groups. Polyelectrolytes may be of natural origin, such as starch or polysaccharide gums, or they may be synthetic in origin. Most polyelectrolytes used in water and wastewater treatment are synthetic organic chemicals. When used as coagulant aids they assist in coagulation, principally by chemical bridging or interaction between reactive groups on the polyelectrolyte and the floc. In some cases, polyelectrolytes may be used as the sole coagulant, with no other chemicals required. Polyelectrolytes are frequently in powder form and may require specific procedures to prepare aqueous solutions for feeding. Usually the dosage is less than about 0.3 mg/ℓ. Activated silica is an inorganic, anionic polyelectrolyte made from sodium silicate. It has been used to some degree as a coagulant aid.

Turbidity addition, such as recycling some chemically precipitated sludge ahead of the mixing or flocculation basins, is occasionally required to furnish sufficient particulate concentrations for rapid coagulation. Adequate particulate concentrations provide sufficient interparticulate collisions to yield optimum coagulation. Clays are sometimes used for turbidity addition instead of recycled sludge.

Adjustment of pH is required if the pH of the coagulated water does not fall within the pH range for minimum solubility of the metallic hydroxide. Increasing the pH is usually done by addition of lime; pH reduction is usually accomplished by the addition of a mineral acid, such as sulfuric acid.

Jar Tests

The laboratory technique of the jar test is usually used to determine the proper coagulant and coagulant aid, if needed, and the chemical dosages required for the coagulation of a particular water. In this test, samples of the water are poured into a series of glass beakers, and various dosages of the coagulant and coagulant aid are added to the beakers. The contents are rapidly stirred to simulate rapid mixing, then gently stirred to simulate flocculation. After a given time, the stirring is ceased and the floc formed is allowed to settled. The most important aspects to note are the time for floc formation, the floc size, its settling characteristics, the percent turbidity and color removed, and the final pH of the coagulated and settled water. The chemical dosage determined from the procedure gives an estimate of the dosage required for the treatment plant. A detailed outline of the procedure for the jar test may be found in numerous publications (Black, A. P., Buswell, A. M., Eidsness, F. A., & Black, A. L., 1957; Black, A. P., & Harris, R. J., 1969; Camp, T. R., 1968 & 1952). Camp

(1968) developed a modification of the jar test that gives the G and GT parameters needed for the design of rapid-mixing and flocculation basins.

Chemical Feeders

Chemical feeders may be classified as solution- or dry-feed type according to the manner of measuring the chemical dosage. It is preferable to use dry feeders because less equipment and labor are required than for solution feeders. For solution feeders, a solution of a known concentration of the chemical is prepared in a storage tank. The feed consists of metering the solution by a metering pump as it is fed to the mixing basins. Frequently, two storage tanks are used so that one may be charged while the other is on-line. Solution feeding is not as desirable as dry feeding; however, some chemicals, such as ferric chloride, which is hydroscopic, must be fed in solution. Dry feeders measure the chemical dosage by metering the dry chemical as it is transferred from a storage bin into a dissolving chamber. A minor stream of water dissolves the chemical; then the solution is fed to the mixing basins. The measurement of the dry chemical feed rate may be done by gravimetric or volumetric methods; numerous types of equipment are available. Thus, the dry-feed system consists of a hopper, a proportioning mechanism (gravimetric or volumetric), the dissolving chamber, and the chemical feed lines to the mixing basins. If a large volume of chemical is stored, usually the bin is directly above the feeder, and the hopper is mounted on the bottom of the storage bin.

Quicklime slakers are usually dry feeders. They may be one of two types: the paste (pug-mill) type and the slurry or detention type. For the pug-mill type, the dry feeder feeds the quicklime to a pug mill, where the quicklime and water are agitated to form a very thick slurry or paste of 30 to 40 percent calcium hydroxide. The paste is then diluted with water to form a slurry of about 10 percent calcium hydroxide (slaked lime or milk of lime), and this slurry is fed to the rapid-mix basins. For the detention type, the dry feeder feeds the quicklime to a slaking chamber, where the quicklime and water are agitated to form a 16 to 20 percent slurry of calcium hydroxide. This slurry is diluted to about 10 percent calcium hydroxide, then fed to the rapid-mix basins. The efficiency of the slaker depends, in part, on the temperature of the quicklime and water mixture under agitation. Although heat is evolved during slaking, some slakers require the temperature in the slaking chamber to be up to 170°F, which may necessitate preheating of the water fed to the slaker, particularly during startup. Ap-

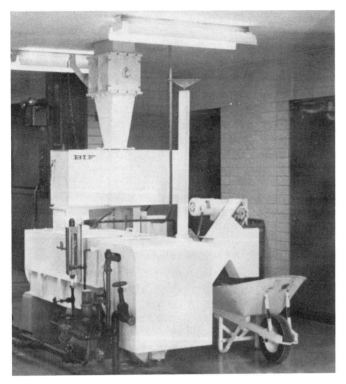

Lime Slaker at a Lime-Soda Softening Plant. The calcium oxide moves from the hopper onto an endless, horizontal moving belt mounted on scales in the box with the BIF emblem. The calcium oxide then falls from the belt into the dissolving box below. The inert materials are removed by moving buckets and fall into the wheelbarrow.

proximately 500 Btu of heat are evolved per pound of calcium oxide slaked, and the time required for slaking depends partially on the quality of the quicklime. The slaking time usually ranges from several minutes to as long as one hour.

Rapid Mixing and Flocculation

In the rapid-mix basins, intense mixing or agitation is required to disperse the chemicals uniformly throughout the basin and to allow adequate contact between the coagulant and the suspended particles. By the time the water leaves the rapid-mix basins, the coagulation process has progressed sufficiently to form microfloc.

In the flocculation basins, the fine microfloc begins to agglomerate into larger floc particles. This aggregation process (flocculation) is dependent on the duration and amount of gentle agitation applied to the water. By the time the water leaves the flocculation basins, the floc has agglomerated into large, dense, rapid-settling floc particles.

The types of devices usually used to furnish the agitation required in both rapid mixing and flocculation may be generally classified as: (1) mechanical agitators, such as paddles, (2) pneumatic agitators, and (3) baffle basins. The mechanical type is the most common.

A rational approach to evaluate mixing and to design basins employing mixing has been developed by T. R. Camp (1955). Camp realized that rapid mixing and flocculation are basically mixing operations and, consequently, are governed by the same principles and require similar design parameters. According to his research, the degree of mixing is based on the power imparted to the water, which he measured by the *velocity gradient*. The velocity gradient of two fluid particles which are 0.05 ft apart and have a velocity relative to each other of 2.0 fps (feet per second) is 2.0 divided by 0.05 or 40 fps/ft. The equation for the velocity gradient for mechanical or pneumatic agitation is

$$G = \sqrt{\frac{W}{\mu}} = \sqrt{\frac{P}{\mu V}} \qquad (2.9)$$

where

G = velocity gradient, fps/ft or sec^{-1};
W = power imparted to the water per unit volume of the basin, ft-lb/sec-ft^3;
P = power imparted to the water, ft-lb/sec;
V = basin volume, ft^3;
μ = absolute viscosity of the water, lb-force-sec/ft^2 (at 50°F, $\mu = 2.73 \times 10^{-5}$ lb-sec/ft^2).

The velocity gradient for baffle basins is given by

$$G = \sqrt{\frac{\gamma h_L}{\mu T}} \qquad (2.10)$$

where

γ = density of water, 62.4 lb/ft^3;
h_L = head loss due to friction, turbulence, and so on;
T = detention time.

The rate of particulate collisions is proportional to the velocity gradient, G; therefore the gradient must be sufficient to furnish the desired rate of particulate collisions. The velocity gradient is also related to the shear forces in the water; thus, large velocity gradients produce appreciable shear forces. If the velocity gradient is too great, excessive shear forces will result and prevent the desired floc formation. The total number of particle collisions is proportional to the product of the velocity gradient, G, and the detention time, T. Thus the value of GT is of importance in design. Figure 2.5 depicts the relationship between the velocity gradient, the water temperature, and the power imparted to the water per unit volume.

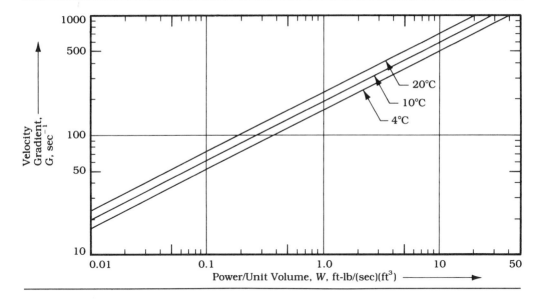

*Figure 2.5. Mixing Power
Requirements*
Adapted from *Water Treatment
Plant Design,* by permission.
Copyright 1969, the American
Water Works Association.

Camp (1968) developed a jar-test procedure using beakers with and without inside baffles. From mixing and flocculation tests using his technique, it is possible to determine the optimum G and T values for a particular coagulant and a given water or wastewater.

Rapid Mixing

Although power for rapid mixing may be imparted to the water by mechanical agitation, pneumatic agitation, and baffle basins, the power required for each method must be the same if the mixing is to be at the same intensity.

Mechanical agitation is the most common method for rapid mixing since it is reliable, very effective, and extremely flexible in operation. Usually rapid mixing employs vertical-shaft rotary mixing devices such as turbine impellers, paddle impellers, or, in some cases, propellers. All of the rotary mixing devices impart motion to the water in addition to turbulence. Types of rapid-mixing chambers or basins are shown in Figure 2.6 (a)–(e). The in-line mixer is the most compact method and is increasing in popularity. Since the optimum velocity gradient may vary with respect to time, it is desirable to have equipment with variable-speed drives. A speed variation of 1:4 is commonly used. Numerous variable drive devices are commercially available. If only one chemical is added, a mixing basin with only one compartment may be used. If, however, more than one chemical is required, sequential application and dispersion of each chemical is desirable, necessitating multiple compartments. Mechanical mixing basins are not

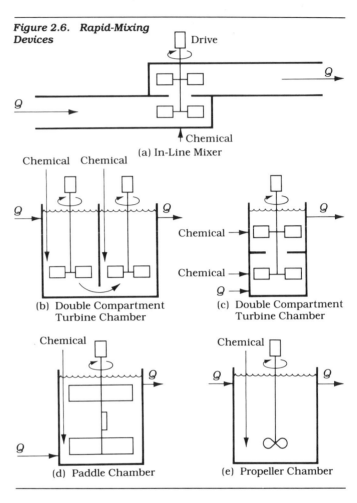

Figure 2.6. Rapid-Mixing Devices

Drive

Q

Q

Chemical

(a) In-Line Mixer

Chemical   Chemical

Q

Q

(b) Double Compartment
Turbine Chamber

Chemical →

Chemical →

Q →

Q

(c) Double Compartment
Turbine Chamber

Chemical

Q

Q

(d) Paddle Chamber

Chemical

Q

Q

(e) Propeller Chamber

affected to any extent by variation in the flow rate and have low head losses. Typical rapid-mixing basins have detention times and velocity gradients as shown in Table 2.1 (AWWA, 1969).

**Table 2.1 Detention Times and Velocity Gradients
of Rapid-Mixing Basins**

Detention Time (sec)	G (fps/ft or sec^{-1})
20	1000
30	900
40	790
50 or more	700

Detention times from 20 to 60 sec are generally used, although some mixing basins have had detention times as small as 10 sec or as long as 2 to 5 min. To obtain high

Figure 2.7. Types of Turbine Impellers

Adapted from *Unit Operations of Chemical Engineering* by W. L. McCabe and J. C. Smith. Copyright © 1976 by McGraw-Hill Book Co., Inc. Reprinted by permission.

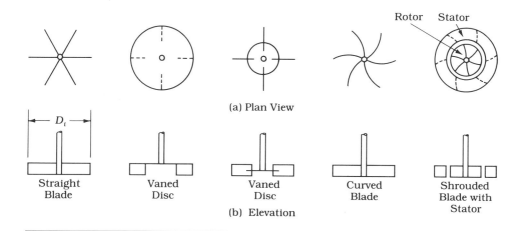

(a) Plan View

Straight Blade | Vaned Disc | Vaned Disc | Curved Blade | Shrouded Blade with Stator

(b) Elevation

velocity gradients, such as 700 to 1000 fps/ft, requires relatively high mixing power levels. Single compartment mixing basins are usually circular or square in plan view, and the fluid depth is 1.0 to 1.25 times the basin diameter or width. Tanks may be baffled or unbaffled; however, small baffles are desirable because they minimize vortexing and rotational flow.

Rotary mixing devices may be classified as turbines, paddle impellers, or propellers according to McCabe and Smith (1976). The types of *turbine impellers,* as shown in Figure 2.7, are the straight blade, vaned disc, curved blade, or shrouded curved blade with a stationary diffuser, with the vaned disc being the most widely used. The stationary vanes of the shrouded turbine prevent rotational flow. The impeller blades may be pitched or vertical, but the vertical are the most common. The diameter of the impeller is usually from 30 to 50 percent of the tank diameter or width, and the impeller is usually mounted one impeller diameter above the tank bottom. The speeds range from 10 to 150 rpm, and the flow is radially outward from the turbine. It divides at the tank wall, giving a flow pattern as shown in Figure 2.8 (a). Small baffles extending into the tank a distance of 0.10 times the tank width or diameter will minimize vortexing and rotational flow and, consequently, cause more power to be imparted to the liquid. This results in greater turbulence, which is desirable for agitation. Turbines are the most effective of all the mechanical agitation or mixing devices because they produce high shear, turbulence, and velocity gradients.

Adapted from "Mixing—Present Theory and Practice: Parts I and II" by J. H. Rushton and J. Y. Oldshue. In *Chemical Engineering Progress* 46, no. 4 (April 1953):161; and 49, no. 5 (May 1953):267. Reprinted by permission.

Figure 2.8. Flow Regime in

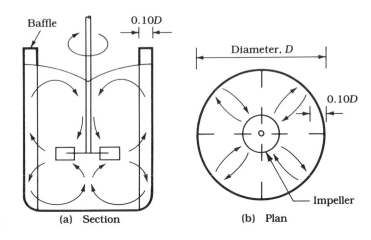

(a) Section   (b) Plan

Paddle impellers usually have two or four blades. The blades may be pitched or vertical, the vertical type the most common. The diameter of a paddle impeller is usually from 50 to 80 percent of the tank diameter or width, and the width of a paddle is usually $\frac{1}{6}$ to $\frac{1}{10}$ of the diameter. The paddles usually are mounted one-half of a paddle diameter above the tank bottom, as shown in Figure 2.9 (a). The flow regime for a two-blade paddle, which is similar to the turbine impeller, is depicted in Figure 2.10. The paddle speeds range from 20 to 150 rpm, and baffling is required to minimize vortexing and rotational flow except at very slow speeds. The paddle is not as efficient as the turbine type since it does not produce as much turbulence and shear forces.

The *propeller impeller*, shown in Figure 2.11, may have either two or three blades, and the blades are pitched to impart axial flow to the liquid. The rotation of a propeller traces out a helix in the liquid, and the pitch is defined as the distance the liquid moves axially during one revolution, divided by the propeller diameter. Usually the pitch is 1.0 or 2.0, and the maximum propeller diameter is about 18 in. The axial flow, as depicted in Figure 2.12 (a), strikes the bottom of the tank and divides and imparts a flow regime as shown. For deep tanks two propellers may be mounted on the same shaft and may produce liquid motion in the same direction or in opposite directions. The propeller speed is ordinarily from 400 to 1750 rpm, and baffling is required in large tanks to minimize vortexing and rotational flow. In small tanks the propeller may be mounted off-

Figure 2.9. Types of Paddle Impellers

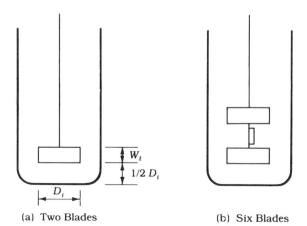

(a) Two Blades   (b) Six Blades

Figure 2.10. Flow Regime in a Paddle-Impeller Tank

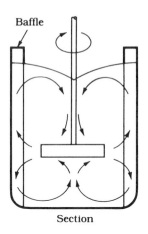

Section

Figure 2.11. Types of Propeller Impellers
Adapted from *Unit Operations of Chemical Engineering* by W. L. McCabe and J. C. Smith. Copyright © 1976 by McGraw-Hill Book Co., Inc. Reprinted by permission.

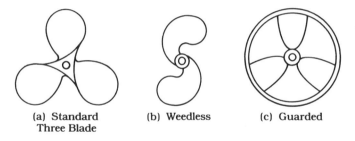

(a) Standard   (b) Weedless   (c) Guarded
Three Blade

Figure 2.12. Flow Regime in a Propeller-Impeller Tank
Adapted from "Mixing—Present Theory and Practice: Parts I and II," by J. H. Rushton and J. Y. Oldshue. In *Chemical Engineering Progress* 46, no. 4 (April 1953):161; and 49, no. 5 (May 1953):267.

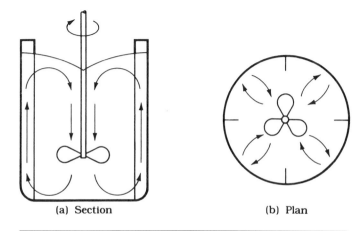

(a) Section   (b) Plan

center to avoid rotational flow. Propeller agitators are very effective in large tanks because of the high velocities imparted to the liquid.

The power imparted to the liquid by various impellers may be determined using relationships developed by Rushton (Rushton, J. H., 1952; Rushton, J. H., Bissell, E. S., Hesse, H. C., & Everett, H. J., 1947; Rushton, J. H., Costich, E. W., & Everett, H. J., 1950; Rushton, J. H., & Oldshue, J. Y., 1953; Rushton, J. H., & Mahoney, L. H., 1954) for impellers employed in chemical process industries. For *turbulent flow* ($N_{Re} > 10,000$), the power imparted by an impeller in a baffled tank is given by the following equation:

$$P = \frac{K_T n^3 D_t^5 \gamma}{g_c} \qquad (2.11)$$

where

P = power, ft-lb/sec;
K_T = impeller constant for turbulent flow;
n = rotational speed, rps;
D_t = impeller diameter, ft;
γ = density of the liquid, lb/ft^3;
g_c = acceleration due to gravity, 32.17 ft/sec^2.

If the flow is *laminar* ($N_{Re} < 10$ to 20), the power imparted by an impeller in either a baffled or unbaffled tank is

$$P = \frac{K_L n^2 D_t^3 \mu}{g_c} \qquad (2.12)$$

where

K_L = impeller constant for laminar flow;
μ = absolute viscosity of the liquid, lb-mass (weight)/ ft-sec.

The Reynolds number for impellers is given by

$$N_{Re} = \frac{D_t^2 n \gamma}{\mu} \qquad (2.13)$$

where

N_{Re} = Reynolds number, dimensionless.

For laminar flow, the power imparted in a tank is independent of the presence of baffles. In turbulent flow, however, the power imparted in an unbaffled tank may be as low as one-sixth the power imparted in the same tank with baffles. Values of the impeller constants, K_T and K_L, for various types of impellers, are given in Table 2.2 for circular tanks having four baffles. For turbulent flow it has been found that the power required for agitation in a baffled vertical square tank is the same as in a baffled vertical circular tank having a diameter equal to the width of the square tank. In an unbaffled square tank the power imparted is about 75 percent of that imparted in a baffled square or a baffled circular tank. Also, two straight blade turbines mounted one turbine diameter apart on the same shaft impart about 1.9 times as much power as one turbine alone. In nearly all cases of rapid mixing for coagulation, the flow regime is well within the turbulent range.

Table 2.2. Values of Constants K_L and K_T in Eqs. (2.12) and (2.13) for Baffled Tanks Having Four Baffles at Tank Wall, with Width Equal to 10 Percent of the Tank Diameter

Type of Impeller	K_L	K_T
Propeller, pitch of 1, 3 blades	41.0	0.32
Propeller, pitch of 2, 3 blades	43.5	1.00
Turbine, 4 flat blades, vaned disc	71.0	6.30
Turbine, 6 flat blades, vaned disc	71.0	6.30
Turbine, 6 curved blades	70.0	4.80
Fan turbine, 6 blades at 45°	70.0	1.65
Shrouded turbine, 6 curved blades	97.5	1.08
Shrouded turbine, with stator, no baffles	172.5	1.12
Flat paddles, 2 blades (single paddle), $D_t/W_t = 4$	43.0	2.25
Flat paddles, 2 blades, $D_t/W_t = 6$	36.5	1.60
Flat paddles, 2 blades, $D_t/W_t = 8$	33.0	1.15
Flat paddles, 4 blades, $D_t/W_t = 6$	49.0	2.75
Flat paddles, 6 blades, $D_t/W_t = 6$	71.0	3.82

From: (1) "Mixing of Liquids in Chemical Processing" by J. H. Rushton. In *Industrial and Engineering Chemistry* 44, no. 2 (December 1952): 2931, copyright 1952, American Chemical Society; and (2) "Mixing—Present Theory and Practice" by J. H. Rushton and J. Y. Oldshue. In *Chemical Engineering Progress* 46, no. 4 (April 1953):161. Reprinted by permission.

Pneumatic mixing basins employ tanks and aeration devices somewhat similar to those used in the activated sludge process, as depicted in Figure 2.13. The detention times and velocity gradients are of the same magnitude and range as those used for mechanical rapid mixing. Variation of the velocity gradient may be obtained by varying the air flow rate. Pneumatic mixing is not affected to any extent by variations in the influent flow rate, and the hydraulic head losses are relatively small. By selecting the design velocity gradient, G, it is possible to determine the power required

Figure 2.13.
Pneumatic Rapid Mixing

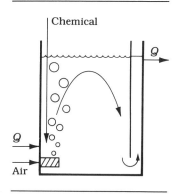

Figure 2.14. Baffle Basin Rapid Mixing

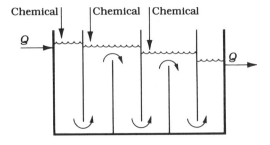

by Eq. (2.9) or Figure 2.5. The basin volume, V, may be determined from the flow rate and detention time, T. The air flow rate to impart the desired power to the water may then be determined by the equation

$$P = 81.5 \, G_a \log \left(\frac{h + 34}{34} \right) \qquad (2.14)$$

where

P = power, ft-lb/sec;
G_a = air flow rate at operating temperature and pressure, cfm;
h = depth to the diffusers, ft.

The baffle-type mixing basins, as depicted in Figure 2.14, depend on hydraulic turbulence to furnish the desired velocity gradient. The velocity gradient imparted to the water is given by Eq. (2.10), and the volume is determined from the flow rate and the detention time, T. The head loss usually varies from 1 to 3 feet, and these basins have very little short circuiting. Baffle basins are not suitable where there is a wide variation in the flow rate, and it is not possible to vary the velocity gradient to any extent. Because of these disadvantages, baffle basins are presently not widely used. In both mechanical and pneumatic mixing, varying the power imparted results in a variation of the velocity gradient.

Example Problem 2.1
Rapid Mixing

A square rapid-mixing basin, with a depth of water equal to 1.25 times the width, is to be designed for a flow of 2.0 million gallons per day (MGD). The velocity gradient is to be 790 fps/ft, the detention time is 40 sec, the operating temperature is 50°F (10°C), and the turbine shaft speed is 100 rpm. Determine:

1. The basin dimensions.
2. The horsepower required.
3. The impeller diameter if a vane-disc impeller with four flat blades is employed and the tank has four vertical baffles (one on each tank wall). The impeller diameter is to be 30 to 50 percent of the tank width.
4. The impeller diameter if no vertical baffles are used.
5. The air required if pneumatic mixing is employed and the diffusers are 0.5 ft above the tank bottom.

Solution

The volume is

$$V = \frac{2.0 \times 10^6 \, \text{gal}}{1440 \, \text{min}} \left| \frac{1 \, \text{min}}{60 \, \text{sec}} \right| \frac{1 \, \text{ft}^3}{7.48 \, \text{gal}} \left| \frac{40 \, \text{sec}}{} \right.$$

$$= 123.79 \, \text{ft}^3$$

The dimensions are given by

$$(W)(W)(1.25 \, W) = 123.79 \text{ ft}^3 \text{ or } W = 4.63 \text{ ft or}$$
$$4 \text{ ft-9 in.}$$

From this, $H = (1.25)(4.75 \text{ ft}) = 5.94$ ft or 6 ft-0 in.

Use W = 4 ft-9 in., H = 6 ft-0 in.

Volume = $(4.75)^2(6.0) = 135.38 \text{ ft}^3$

Equation (2.9) may be rearranged to give the power imparted as $W = G^2\mu$ or $W = (790/\text{sec})^2 (2.73 \times 10^{-5}$ lb-sec/ft^2) = 17.04 ft-lb/sec-ft^3. The value of the total power, P, is

$$P = \frac{17.04 \, \text{ft-lb}}{\text{sec-ft}^3} \left| \frac{135.38 \, \text{ft}^3}{} \right. = 2307 \text{ ft-lb/sec}$$

$$P = \frac{2307 \, \text{ft-lb}}{\text{sec}} \left| \frac{\text{sec}}{550 \, \text{ft-lb}} \right| \frac{\text{hp}}{} = 4.19 \, \text{hp}$$

The impeller speed, n, is 100 rpm or 100/60 or 1.667 rps. Assume turbulent flow; thus, from Table 2.2, K_T = 6.30. The viscosity at 50°F is 1.310 centipoise; thus, the viscosity in terms of lb-mass (weight) is

$$\mu = \frac{1.310 \, \text{centipoise}}{} \left| \frac{6.72 \times 10^{-4} \, \text{lb/ft-sec}}{1 \, \text{centipoise}} \right.$$

$$= 8.803 \times 10^{-4} \, \text{lb/ft-sec}$$

Rearranging the power equation to determine the impeller diameter gives

$$D_l = \left(\frac{Pg_c}{K_T n^3 \gamma}\right)^{1/5}$$

$$= \left[\frac{2307\,\text{ft-lb}}{\text{sec}} \middle| \frac{32.17\,\text{ft}}{\text{sec}^2} \middle| 6.30 \middle| (1.667\,\text{rps})^3\right.$$

$$\left. \times \frac{\text{ft}^3}{62.4\,\text{lb}}\right]^{1/5} = \underline{2.10\,\text{ft}}\; D_l/W = 2.10/4.75$$

$$= 0.442 \text{ or } 44.2 \text{ percent.}$$

Check on the Reynolds number, N_{Re}:

$$N_{Re} = \frac{nD_l^2 \gamma}{\mu}$$

$$= \frac{1.667}{\text{sec}} \middle| \frac{(2.10\,\text{ft})^2}{} \middle| \frac{\text{ft-sec}}{8.803 \times 10^{-4}\,\text{lb}} \middle| \frac{62.4\,\text{lb}}{\text{ft}^3}$$

$$= 521,000 >>> 10,000. \text{ Thus, the flow regime}$$
is turbulent.

If no vertical baffles are used, the power imparted is 75 percent of that for a baffled tank. Therefore, to impart the same power requires a larger impeller. The value $K_T = 0.75(6.30) = 4.725$. The impeller diameter is given by

$$D_l = \left(\frac{Pg_c}{K_T n^3 \gamma}\right)^{1/5}$$

$$= \left[\frac{2307\,\text{ft-lb}}{\text{sec}} \middle| \frac{32.17\,\text{ft}}{\text{sec}^2} \middle| 4.725 \middle| (1.667\,\text{rps})^3\right.$$

$$\left. \times \frac{\text{ft}^3}{62.4\,\text{lb}}\right]^{1/5}$$

$$D_l = \underline{2.22\,\text{ft}}$$

$$D_l/W = 2.22/4.75 = 0.468 = 46.8 \text{ percent}$$

If pneumatic mixing is used, the power equation for diffused air may be rearranged to yield

$$G_a = \frac{P/81.5}{\log\left(\dfrac{h+34}{34}\right)}$$

and $h = 6.00\,\text{ft} - 0.50\,\text{ft} = 5.50\,\text{ft}$.
Thus,

$$G_a = \frac{2307/81.5}{\log\left(\dfrac{5.50 + 34}{34}\right)}$$

$$G_a = \underline{435\,\text{cfm}}$$

The air flow required is at operating conditions.

Flocculation

The power required for the gentle agitation or stirring of the water during flocculation may be imparted by mechanical and pneumatic agitation, with mechanical agitation being the most common. Formerly, baffle basins were used for flocculation; however, since the available range of G and GT values is limited, these are not employed at present to any extent. Most mechanical agitators are paddle wheels, as shown in Figure 2.15, although turbines and propellers are also used.

The degree of completion of the flocculation process depends on the relative ease and rate by which the small microfloc aggregate into large floc particles, and the total number of particulate collisions during flocculation. Thus, the degree of completion is dependent on the floc characteristics, the velocity gradient, G, and the value of GT (Culp, R. L., & Culp, G. L., 1978). The magnitude of the dimensionless parameter, GT, is related to the total number of collisions during aggregation in the flocculation process. A high GT value indicates a large number of collisions during aggregation. A more accurate parameter is GCT, where C is the ratio of the floc volume to the total water volume being flocculated. If the velocity gradient in flocculation is too great, the shear forces will prevent the formation of a large floc. If the velocity gradient is insufficient, adequate interparticulate collisions will not occur and a proper floc will not be formed. If the water is difficult to coagulate, the floc will be fragile and a final velocity gradient less than 5 fps/ft may be required. If, however, the water coagulates readily, a high strength floc usually results and the final velocity gradient may be as large as 100 fps/ft (AWWA, 1969).

Flocculation basins are frequently designed to provide for tapered flocculation in which the flow is subjected to decreasing G values as it passes through the flocculation basin. This produces a rapid buildup of small dense floc, which subsequently aggregates at lower G values into larger, dense, rapid-settling floc particles. Tapered flocculation is usually accomplished by providing a high G value during

Figure 2.15. Horizontal-Shaft Paddle Wheel for Flocculation

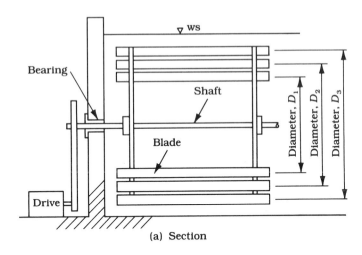

(a) Section

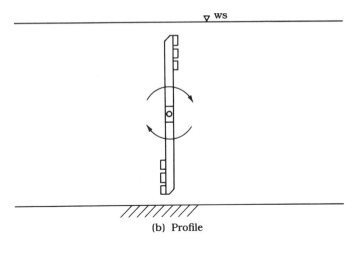

(b) Profile

the first third of the flocculation period, a lower G value during the next third, and a much lower G value during the last third. For example, a typical series of G values could be 80, 40, and 20 fps/ft. Although many basins are designed which do not have tapered flocculation, optimum flocculation usually necessitates its use.

Typical arrangements for flocculators employing paddle wheels on horizontal shafts are shown in Figure 2.16 and 2.17. In the cross-flow pattern, Figure 2.16 (a), the paddle-wheel shafts are mounted at right angles to the overall water flow. In the axial-flow pattern, Figure 2.17 (a), the paddle-wheel shafts are parallel to the flow. At least three consecutive compartments, as shown in Figures 2.16 and 2.17, are required to minimize short circuiting. The partitions between compartments are usually wood baffle fences made of horizontal wood slats with spacings between the slats for passage of the flow, or concrete walls with orifices. The wood baffle fences are more flexible since different slat spacings may be used. Multiple compartments, in addition to minimizing short circuiting, facilitate tapered flocculation design. For the cross-flow pattern, tapered flocculation may be provided by varying the paddle size, the number of paddles, and the diameter of the paddle wheels on the various horizontal shafts. Also, it can be attained by varying the rotational speed of the various horizontal shafts. For the axial-flow pattern, tapered flocculation may be obtained by varying the paddle size and the number of paddles on each paddle wheel having a common horizontal shaft. All mechanical flocculation devices should be equipped

Figure 2.16. Horizontal-Shaft Paddle-Wheel Flocculator (Cross-Flow Pattern)

Adapted from *Water Treatment Plant Design*, by permission. Copyright 1969, the American Water Works Association.

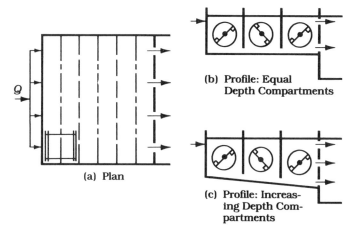

(a) Plan

(b) Profile: Equal Depth Compartments

(c) Profile: Increasing Depth Compartments

Figure 2.17. Horizontal-Shaft Paddle-Wheel Flocculator (Axial-Flow Pattern)

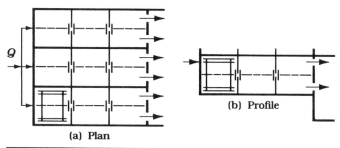

(a) Plan

(b) Profile

with variable-speed drives having a range up to 1:4 to meet variations in the quality of the feed water. If the compartments are separated by concrete walls with orifices, the orifices should have circular deflector plates immediately upstream and downstream to minimize short circuiting. The velocity gradient through each orifice should not exceed the gradient in the compartment immediately upstream. The velocity gradient may be estimated using the head loss, h, from the orifice equation, $Q = 0.60 A \times \sqrt{2gh}$, where Q is in cfs, A is the area in ft^2, and h is in ft. The velocity and the time of passage through the orifice, T, may be computed, and the velocity gradient may be determined from Eq. (2.10), which is $G = (\gamma h_L / \mu T)^{1/2}$, where $h_L = h$ for an orifice.

Vertical-shaft devices, such as the paddle wheels shown in Figures 2.18 (a) and (b), are sometimes used. A typical flocculator layout employing these units is shown in Figure 2.18 (c). It should be noted from the layout that the compartments are arranged in series to minimize short circuiting and also to facilitate tapered flocculation design.

The power imparted to the water by paddle wheels may be determined using Newton's law for the drag force exerted by a submerged object moving in a liquid. The drag force for a paddle-wheel blade is

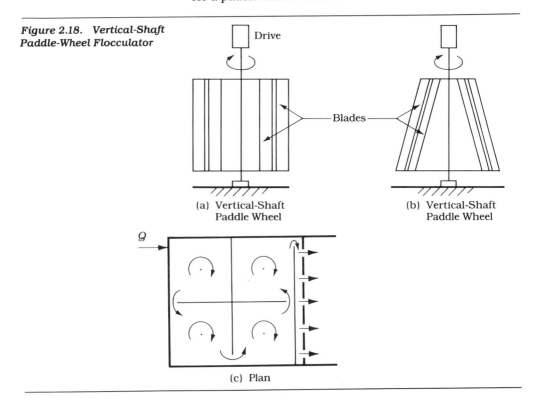

Figure 2.18. Vertical-Shaft Paddle-Wheel Flocculator

Drive

Blades

(a) Vertical-Shaft
Paddle Wheel

(b) Vertical-Shaft
Paddle Wheel

(c) Plan

$$F_D = C_D A \rho \frac{v^2}{2} \tag{2.15}$$

where

F_D = drag force of the paddle, lb;
C_D = coefficient of drag;
A = paddle-blade area at right angle to the direction of movement, ft^2;
ρ = mass of the water, $\rho = \gamma/g_c$;
v = velocity of the paddle blade relative to the water, fps.

Since the power is equal to the force times the velocity, the power is given by

$$P = F_D v = C_D A \rho \frac{v^3}{2} = C_D A \left(\frac{62.4}{64.4}\right) v^3$$

$$= 0.97 \, C_D A v^3 \tag{2.16}$$

where

P = power, ft-lb/sec.

The drag coefficient depends basically on the geometry of the paddle. Values for various paddle dimensions are given in Table 2.3 (Rouse, H., 1950).

Table 2.3. Values for Various Paddle Dimensions

Length-Width Ratio	C_D
5	1.20
20	1.50
∞	1.90

Practice has shown that the peripheral velocity of the paddle blades should range from 0.3 to 3.0 fps, and the velocity of a paddle blade relative to the water is approximately three-fourths the peripheral blade velocity. Also, the total paddle-blade area on a horizontal shaft should not exceed 15 to 20 percent of the total basin cross-sectional area, or excessive rotational flow will result.

In the design of a horizontal-shaft paddle-wheel flocculator, usually a design velocity gradient, G, and detention time, T, are determined from laboratory or pilot plant studies, and the GT value is computed to insure that it is within an acceptable GT range. Then the flow rate and detention time are used to compute the basin volume, V. Basin dimensions are calculated by knowing the volume of the basin and the number of horizontal shafts selected. If tapered flocculation is used, the G values for each compartment must be determined from laboratory or pilot plant studies.

COAGULATION AND FLOCCULATION   **47**

Then, a paddle-wheel design is assumed and the peripheral speeds of the paddles to impart the desired power are computed. If the resulting peripheral speeds are excessive, larger paddles or more paddles are assumed, and the peripheral speeds are redetermined to see if they are satisfactory.

Other mechanical agitators such as the walking-beam flocculator, turbines, and propellers have been used occasionally; however, the paddle wheels are by far the most widely used.

Pneumatic flocculation has been used, at times, in lieu of mechanical agitation. Knowing the design G value, the required power to be imparted per unit basin volume, P/V, may be determined from Eq. (2.9). Using the design GT value, the detention time, T, is computed and, using the flow rate, the basin volume is determined. The air flow, G_a, needed to impart the desired power may then be determined using Eq. (2.14).

The velocity of flocculated water, in orifices or ports in baffle fences and in conduits (when employed) that lead from the flocculator, should not be greater than 0.5 to 1.0 fps to avoid shearing apart the floc.

Example Problem 2.2
Flocculation

A cross-flow, horizontal-shaft, paddle-wheel flocculation basin is to be designed for a flow of 6.5 MGD, a mean velocity gradient of 30 $\sec^{-1}$ (at 50°F), and a detention time of 50 min. The GT value should be from 50,000 to 100,000. Tapered flocculation is to be provided, and three compartments of equal depth in series are to be used, as shown in Figure 2.16 (b). The G values determined from laboratory tests for the three compartments are: $G_1 = 50 \sec^{-1}$, $G_2 = 25 \sec^{-1}$, and $G_3 = 15 \sec^{-1}$. These give an average G value of 30 $\sec^{-1}$. The compartments are to be separated by slotted, redwood baffle fences, and the floor of the basin is level. The basin should be 50 ft in width to adjoin the settling basin. The speed of the blades relative to the water is three-quarters of the peripheral blade speed. Determine:

1. The GT value.
2. The basin dimensions.
3. The paddle-wheel design.
4. The power to be imparted to the water in each compartment.
5. The rotational speed of each horizontal shaft in rpm.
6. The rotational speed range if 1:4 variable-speed drives are employed.
7. The peripheral speed of the outside paddle blades in fps.

Solution

The *GT* value is

$$GT = \frac{30}{\text{sec}} \left| \frac{50 \text{ min}}{} \right| \frac{60 \text{ sec}}{\text{min}}$$

$$= \underline{90{,}000}$$

Since the *GT* value is between 50,000 and 100,000, the detention time of 50 min is satisfactory.
Basin volume, *V*, is given by

$$V = \frac{6.5 \times 10^6 \text{ gal}}{24 \text{ hr}} \left| \frac{\text{hr}}{60 \text{ min}} \right| \frac{50 \text{ min}}{} \left| \frac{\text{ft}^3}{7.48 \text{ gal}} \right.$$

$$= 30{,}173 \text{ ft}^3$$

Profile area = 30,173 ft³/50 ft = 603.46 ft²
Assume compartments are square in profile, and *x* is the compartment width and depth. Thus,

$$(3x)(x) = 603.46 \text{ ft}^2 \quad \text{or} \quad x = 14.18 \quad \text{or} \quad 14 \text{ ft-3 in.}$$
$$3x = 3(14 \text{ ft-3 in.}) = 42 \text{ ft-9 in.}$$

Use width = depth = 14 ft-3 in., length = 42 ft-9 in.

$$\text{Volume} = (14.25)(42.75)(50.0) = 30{,}459 \text{ ft}^3$$

Assume a paddle-wheel design as shown in Figure 2.15 (a), with $D_1 = 5.5$ ft, $D_2 = 8.5$ ft, and $D_3 = 11.5$ ft. Use four paddle wheels per shaft, and assume the blades are 6 in. × 10 ft. The space between blades is 12 in.

$$\text{Blade area per shaft} = (0.5 \text{ ft})(10 \text{ ft})(6)(4)$$
$$= 120 \text{ ft}^2$$

$$\text{Percent of cross-sectional area} = \frac{120}{50} \left| \frac{100}{14.25} \right.$$

$$= 16.84 \text{ percent}$$

Since this is between 15 to 20 percent, make the trial design using the assumed paddle-wheel design. The power, *P*, is given by

$$G = \sqrt{\frac{P}{\mu V}} \quad \text{or} \quad P = \mu G^2 V$$

Power for first compartment, *P*, is given by

$$P = \frac{2.72 \times 10^{-5} \text{ lb-sec}}{\text{ft}^2} \left| \frac{(50)^2}{\text{sec}^2} \right| \frac{30{,}459 \text{ ft}^3}{3}$$

$$= 690.40 \text{ ft-lb/sec} = 1.26 \text{ hp}$$

Power per wheel = (690.40 ft-lb/sec) 1/4 = 172.60 ft-lb/sec, and

$$P = 0.97 \, C_D A v^3$$

The length/width = 10/0.5 = 20, thus $C_D = 1.50$. The blade velocity relative to water is

$$v = (\text{rps}) \left(\frac{\pi D}{\text{rev}} \right) 0.75$$

Thus,

$$v_1 = (\text{rps})(\pi)(11.5 \text{ ft})0.75 = 27.08 \, (\text{rps})$$

In a like manner, $v_2 = 20.03$ (rps), $v_3 = 12.96$ (rps). The power per wheel is

$$P = 0.97 \, C_D A_1 v_1^{\,3} + 0.97 \, C_D A_2 v_2^{\,3} + 0.97 \, C_D A_3 v_3^{\,3}$$

Since $A_1 = A_2 = A_3$,

$$P = (0.97 \, C_D)(A_1) \left[(27.08)^3 (\text{rps})^3 \right.$$
$$\left. + (20.03)^3 (\text{rps})^3 + (12.96)^3 (\text{rps})^3 \right]$$

or

$$172.60 = (0.97)(1.5)(0.5)(10)(2)(19{,}858 + 8{,}036$$
$$+ 2{,}177)(\text{rps})^3$$

From this, rps = 0.073 and rpm = (0.073)60 = 4.38. Peripheral speed = (πD)rps; thus, for the outside blade,

$$v_1' = (\pi)(11.5 \text{ ft})(0.073) = 2.64 \text{ fps}$$

Since $v_1' < 3.0$ fps, the design of the paddle wheels is satisfactory. The maximum rotational speed is 4.38 rpm; therefore, the minimum rotational speed = (1/4) × 4.38 = 1.10 rpm. Thus,

Rotational speed for 1:4 drive = 1.10 to 4.38 rpm

In a like manner, the power, rotational speed, peripheral blade speed, and rotational speed range are computed for the second and third compartments. Following is a summary of the values:

First compartment:

P = 690.40 ft-lb/sec = 1.26 hp
rpm = 4.38; range = 1.10 to 4.38 rpm
fps = 2.64

Second compartment:

P = 172.60 ft-lb/sec = 0.31 hp
rpm = 2.76; range = 0.69 to 2.76 rpm
fps = 1.66

Third compartment:

P = 62.14 ft-lb/sec = 0.11 hp
rpm = 1.98; range = 0.50 to 1.98 rpm
fps = 1.19

A cross-flow, horizontal shaft, paddle-wheel flocculation basin is to be designed for a flow of 25,000 m^3/day, a mean velocity gradient of 30 s^{-1} (at 10°C), and a detention time of 50 min. The *GT* value should be from 50,000 to 100,000. Tapered flocculation is to be provided and three compartments of equal depth in series are to be used, as shown in Figure 2.16 (b). The *G* values determined from laboratory tests for the three compartments are: G_1 = 50 s^{-1}, G_2 = 25 s^{-1}, and G_3 = 15 s^{-1}. These give an average *G* value of 30 s^{-1}. The compartments are to be separated by slotted, redwood baffle fences, and the floor of the basin is level. The basin should be 15.0 m in width to adjoin the settling basin. The speed of the blades relative to the water is three-quarters of the peripheral blade speed. Determine:

1. The *GT* value.
2. The basin dimensions.
3. The paddle-wheel design.
4. The power to be imparted to the water in each compartment.
5. The rotational speed of each horizontal shaft in rpm.
6. The rotational speed range if 1:4 variable-speed drives are employed.
7. The peripheral speed of the outside paddle blades in m/s.

Solution

The *GT* value is

$$GT = \frac{30}{s} \left| \frac{50 \text{ min}}{} \right| \frac{60 \text{ s}}{\text{min}}$$

$$= \underline{90,000}$$

Since the *GT* value is between 50,000 and 100,000, the detention time of 50 min is satisfactory. Basin volume, *V*, is given by

$$V = \frac{25,000 \text{ m}^3}{24 \text{ h}} \left| \frac{\text{h}}{60 \text{ min}} \right| \frac{50 \text{ min}}{}$$

$$= 868.06 \text{ m}^3$$

Profile area $= 868.06 \text{ m}^3/15.0 \text{ m} = 57.87 \text{ m}^2$
Assume compartments are square in profile, and *x* is the compartment width and depth. Thus,

$$(3x)(x) = 57.87 \text{ m}^2 \quad x^2 = 19.3 \text{ m}^2 \quad x = 4.39 \text{ m}$$

$$3x = 3(4.39 \text{ m}) = 13.17 \text{ m}$$

<u>Use width = depth = 4.39 m, length = 13.17 m</u>

Volume $= (4.39 \text{ m})(13.17 \text{ m})(15.0 \text{ m}) = 867.2 \text{ m}^3$

Assume a paddle-wheel design as shown in Figure 2.15 (a) with $D_1 = 1.70$ m, $D_2 = 2.60$ m, and $D_3 = 3.50$ m. Use four paddle wheels per shaft and assume the blades are 15 cm wide and 3.00 m long.

$$\text{Blade area per shaft} = (0.15 \text{ m})(3.00 \text{ m})(6)(4)$$
$$= 10.8 \text{ m}^2$$

$$\text{Percent of cross-sectional area} = \frac{10.8}{15} \left| \frac{100}{4.39} \right.$$

$$= 16.4 \text{ percent}$$

Since this is between 15 to 20 percent, make the trial design using the assumed paddle-wheel design. The power, *P*, is given by

$$G = \sqrt{\frac{P}{\mu V}} \quad \text{or} \quad P = \mu G^2 V$$

Absolute viscosity, μ, at 10°C = 1.307 centipoise (cp). Power for first compartment, *P*, is given by

$$P = \frac{1.307\,\text{cp}}{} \left|\frac{}{1\,\text{cp}}\right| \frac{10^{-2}\text{g}}{\text{cm-s}} \left|\frac{100\,\text{cm}}{\text{m}}\right| \frac{\text{kg}}{1000\,\text{g}}$$

$$\times \frac{\text{N-s}^2}{\text{kg-m}} \left|\frac{(50)^2}{\text{s}^2}\right| \frac{867.2\,\text{m}^3}{3}$$

$$= \underline{944.5\ \text{N-m/s} = 944.5\ \text{J/s} = 944.5\ \text{W}}$$

Power per wheel = (944.5 N-m/s)1/4 = 236.1 N-m/s

$$P = C_D A \rho \frac{v^3}{2}$$

The length/width = 3.0/0.15 = 20, thus C_D = 1.50. The blade velocity relative to water is

$$v = (\text{rps})\left(\frac{\pi D}{\text{rev}}\right) 0.75$$

Thus,

$$v_1 = (\text{rps})(\pi)(3.50\,\text{m})(0.75) = 8.247\,(\text{rps})$$

In a like manner, v_2 = 6.126 (rps), v_3 = 4.006 (rps)
The power per wheel is

$$P = C_D A_1 \rho \frac{v_1^3}{2} + C_D A_2 \rho \frac{v_2^3}{2} + C_D A_3 \rho \frac{v_3^3}{2}$$

Since $A_1 = A_2 = A_3$, and all C_D values are equal,

$$P = C_D A_1 \frac{\rho}{2} (v_1^3 + v_2^3 + v_3^3)$$

$$236.1\,\text{N-m/s} = (1.5)(0.15\,\text{m})(3.0\,\text{m})(2)(999.7\,\text{kg/m}^3)$$
$$\times (1/2)[(8.247)^3(\text{rps})^3 + (6.126)^3$$
$$\times (\text{rps})^3$$
$$+ (4.006)^3(\text{rps})^3]\text{m}^3/\text{s}^3$$
$$236.1\,\text{N-m/s} = (674.8\,\text{kg-m}^2/\text{s}^3)(1\,\text{N-s/kg-m})$$
$$\times (560.9 + 229.9 + 64.3)(\text{rps})^3$$

From which, rps = 0.074 and thus $\underline{\text{rpm} = (0.074)60}$
$\underline{= 4.44}$
Peripheral speed = (πD)rps; thus, for the outside blade,

$$v_1' = (\pi)(3.5\,\text{m})(0.074) = \underline{0.81\ \text{m/s}}$$

Since v'_1 < 3.0/3.281 m/s, the design of the paddle wheels is satisfactory. The maximum rotational speed is 4.44 rpm; therefore, the minimum rotational speed

$= (1/4)4.44 = 1.11$ rpm. Thus,

Rotational speed for 1:4 drive $= 1.11$ to 4.44 rpm

In a like manner, the power, rotational speed, peripheral blade speed, and rotational speed range are computed for the second and third compartments. A summary of the values are as follows:

First compartment:

$P = 944.5$ N-m/s $= 944.5$ J/s $= 944.5$ W
rpm $= 4.44$; range $= 1.11$ to 4.44 rpm
velocity $= 0.81$ m/s

Second compartment:

$P = 236.0$ N-m/s $= 236.0$ J/s $= 236.0$ W
rpm $= 2.82$; range $= 0.71$ to 2.82 rpm
m/s $= 0.52$

Third compartment:

$P = 85.0$ N-m/s $= 85.0$ J/s $= 85.0$ W
rpm $= 2.00$; range $= 0.50$ to 2.00 rpm
m/s $= 0.50$

Solids-Contact Units

Solids-contact units, shown in Figure 2.19 (a) and (b), are frequently called upflow clarifiers. These units combine mixing, flocculation, and sedimentation into a single structural unit. These are designed to maintain a large volume of flocculated solids within the system, which enhances the flocculation of incoming solids since there are more interparticulate collisions. The volume of the solids in the contact zone may vary from 5 to 50 percent of the zone volume depending on the particular use. The volume of the solids is taken as the percent volume after 30 minutes settling in a test cylinder. Solids-contact units are of two basic designs: (1) the slurry-recirculation type and (2) the sludge-blanket filtration type.

In the slurry-recirculation type, as shown in Figure 2.19 (a), the large floc mass is maintained by recycling an appreciable portion of the floc through a center compartment by means of a pitched blade impeller. The mixing and flocculation are accomplished in the compartment surrounding the recycle center well. The flocculated water passes under the skirt upward into the clarification zone; however, it passes only through the top of the sludge blanket. After passing through the clarification section, the

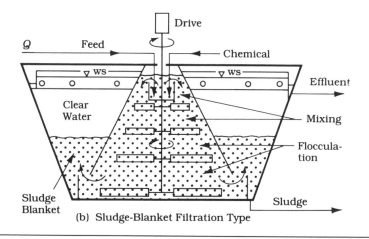

Figure 2.19. Solids-Contact Units

(a) Slurry-Recirculation Type

(b) Sludge-Blanket Filtration Type

water leaves by an effluent launder.

In the sludge-blanket filtration type, as shown in Figure 2.19 (b), mixing and flocculation are achieved in a center compartment. The necessary agitation and gentle stirring are provided by pitched blade impellers. The flocculated water, on leaving the flocculation section, passes upward through the sludge blanket to obtain floc removal by contact with flocculated solids in the blanket. The water continues to flow upward through the clarification section into the effluent launder.

The main advantage of the contact-solids units over conventional mixing, flocculation, and clarification units is their reduced size. Consequently, the units are more compact and occupy less land space. The units are best suited to treat a feed water that has a relatively constant quality with respect to time. Because of the short mixing, flocculation, and clarification detention times, it is difficult to treat a feed water that has a wide variation in quality. They are frequently used to coagulate lake waters and soften well waters. The basic designs shown in Figure 2.19 (a) and (b) are available from numerous manufacturers.

Lime-Soda Softening

Hardness may be defined as the ability of a water to consume excessive amounts of soap prior to forming a lather and to produce scale in hot water heaters, boilers, or other units in which the temperature of the water is significantly increased. It is due to the presence of polyvalent metallic ions, principally calcium and magnesium. Calcium and magnesium ions will react with soap to form insoluble organic salts that are present as scum on the water surface. Once all the calcium and magnesium ions have been precipitated, a lather can be formed. Calcium and magnesium may be removed by the lime-soda softening process in which the unwanted ions are precipitated by adding slaked lime and soda ash. These reagents produce a voluminous precipitate of calcium carbonate and magnesium hydroxide, which acts by sweep coagulation, thus enmeshing the suspended particles as the precipitate is formed. Occasionally a coagulant, such as ferrous sulfate, is added along with the slaked lime and soda ash to aid in the coagulation and flocculation process. The total hardness (TH) is the sum

Solids-Contact Unit at a Water Treatment Plant. Unit is empty so that painting and maintenance can be done.

of the calcium and magnesium ion concentrations. It is usually expressed as meq/ℓ or mg/ℓ of equivalent $CaCO_3$. Alkalinity is due to the bicarbonate, carbonate, and hydroxyl ion concentrations. For natural waters with a pH less than 9, the alkalinity (Alk) is the bicarbonate and carbonate alkalinities, since the hydroxyl ion concentration is negligible below this pH. Alkalinity is also expressed as meq/ℓ or mg/ℓ equivalent $CaCO_3$. The carbonate hardness (CH) is that part of the total hardness that is chemically equivalent to the bicarbonate and carbonate alkalinities. The noncarbonate hardness (NCH) is equal to the total hardness minus the carbonate hardness. The amount of lime and soda ash required in softening depends on the concentration of the total hardness (TH), the carbonate hardness (CH), the noncarbonate hardness (NCH), the magnesium ion, and the carbon dioxide.

If the total hardness (TH) is greater than the alkalinity (Alk) when measured as meq/ℓ or mg/ℓ $CaCO_3$, then

Carbonate Hardness (CH) = Alkalinity (Alk)

and

Noncarbonate Hardness (NCH)
= Total Hardness (TH) − Carbonate Hardness (CH)

Solids-Contact Unit in Operation. Some floc is visible in the water.

If the total hardness (TH) is less than the alkalinity (Alk), then

$$\text{Carbonate Hardness (CH)} = \text{Total Hardness (TH)}$$

and

$$\text{Noncarbonate Hardness (NCH)} = 0$$

For the pH ranges that occur in natural waters, the alkalinity is usually in the form of the bicarbonate ion, HCO_3^{-1}. In fact, at pH 7.5 or below, the bicarbonate ion is essentially all the alkalinity.

As slaked lime is added to a water it will react with any carbon dioxide present as follows:

$$Ca(OH)_2 + CO_2 \rightarrow \underline{CaCO_3} + H_2O \qquad (2.17)$$

The lime reacts with carbonate hardness as shown by

$$Ca(OH)_2 + Ca(HCO_3)_2 \rightarrow \underline{2\ CaCO_3} + 2\ H_2O \qquad (2.18)$$

and

$$Ca(OH)_2 + Mg(HCO_3)_2 \rightarrow MgCO_3 \qquad (2.19)$$
$$+ \underline{CaCO_3} + 2\ H_2O$$

The product magnesium carbonate in Eq. (2.19) is soluble, but more lime will remove it by

$$Ca(OH)_2 + MgCO_3 \rightarrow \underline{CaCO_3} + \underline{Mg(OH)_2} \qquad (2.20)$$

Also, magnesium noncarbonate hardness, shown as magnesium sulfate, is removed as additional lime is added by the reaction

$$Ca(OH)_2 + MgSO_4 \rightarrow CaSO_4 + \underline{Mg(OH)_2} \qquad (2.21)$$

Although in Eq. (2.21) the magnesium is precipitated, an equivalent amount of calcium that has been added remains in solution. The water will now contain only the original calcium noncarbonate hardness and the calcium noncarbonate hardness produced from Eq. (2.21), which equals the magnesium noncarbonate hardness. The removal of this calcium may be accomplished by soda ash by the reaction

$$Na_2CO_3 + CaSO_4 \rightarrow Na_2SO_4 + \underline{CaCO_3} \qquad (2.22)$$

To precipitate calcium carbonate requires a pH of about 9.5; and to precipitate magnesium hydroxide requires a pH of about 10.8, which necessitates adding an excess of about 1.25 meq/ℓ of lime to raise the pH.

From the previous discussion, the meq/ℓ of lime required is equal to the sum of the carbon dioxide, carbonate hardness, and magnesium ion concentrations when expressed in meq/ℓ, plus 1.25 meq/ℓ excess to raise the pH. The meq/ℓ of soda ash required is equal to the noncarbonate hardness expressed as meq/ℓ.

After softening the water, it will contain the excess lime and the magnesium hydroxide and calcium carbonate that did not precipitate or settle. The excess lime and magnesium hydroxide are stabilized by carbon dioxide, which lowers the pH to about 9.5, resulting in

$$CO_2 + Ca(OH)_2 \rightarrow \underline{CaCO_3} + H_2O \qquad (2.23)$$

and

$$CO_2 + Mg(OH)_2 \rightarrow MgCO_3 + H_2O \qquad (2.24)$$

Further stabilization to about pH 8.5 will stabilize the calcium carbonate by

$$CO_2 + CaCO_3 + H_2O \rightarrow Ca(HCO_3)_2 \qquad (2.25)$$

It is not possible to remove all of the hardness from a water due to the slight solubility of calcium carbonate and magnesium hydroxide and, also, the presence of some of the precipitate as a very fine nonsettling floc. Once stabilization has been accomplished, the fine precipitate dissolves back into solution. In actual practice the lime-soda process will soften a water to about 50 to 80 mg/ℓ residual hardness as calcium carbonate.

Example Problem 2.3
Lime-Soda Softening

A water that is to be softened by the lime-soda process has the following carbon dioxide, calcium, magnesium, and bicarbonate concentrations:

CO_2	8 mg/ℓ
Ca^{+2}	65 mg/ℓ
Mg^{+2}	32 mg/ℓ
HCO_3^{-1}	260 mg/ℓ

Determine the calcium oxide and soda ash required per million gallons if the purities are 85 and 95 percent, respectively. Use meq/ℓ as the basis of computations for both the alkalinity and hardness instead of mg/ℓ as $CaCO_3$.

Solution

The weights of 1 meq of each of the ions are: Ca^{+2} = 40/2 = 20 mg; Mg^{+2} = 24/2 = 12 mg; HCO_3^{-1} = 61/1 = 61 mg. One meq of carbon dioxide weighs 22 mg. Thus, the total hardness (TH) is 65/20 + 32/12 or 3.25 + 2.67 = 5.92 meq/ℓ. The alkalinity (Alk) is 260/61 = 4.26 meq/ℓ. Since the total hardness (TH) is greater than the alkalinity (Alk), the carbonate hardness (CH) is 4.26 meq/ℓ and the noncarbonate hardness (NCH) is 5.92 − 4.26 = 1.66 meq/ℓ. The magnesium ion content is 2.67 meq/ℓ and the carbon dioxide is 8/22 or 0.36 meq/ℓ. Thus, the lime required in meq/ℓ for the various reactants, and the excess to raise the pH is given by the following summation:

CO_2	0.36 meq/ℓ
CH	4.26 meq/ℓ
Mg^{+2}	2.67 meq/ℓ
Excess	1.25 meq/ℓ
Total =	8.54 meq/ℓ

The soda ash required for the NCH is 1.66 meq/ℓ. One meq of CaO is (40 + 16)1/2 = 28 mg and one meq of Na_2CO_3 is [2(23) + 12 + 3(16)]1/2 = 53 mg. The lime required is (8.54)28 = 239.0 mg/ℓ and the soda ash is (1.66)53 = 88.0 mg/ℓ. Thus, the lime per million gals is

$$CaO = \frac{239}{10^6} \left| \frac{10^6 \, gal}{} \right| \frac{8.34 \, lb}{gal} \left| \frac{}{0.85} \right. = \underline{2,345 \, lb}$$

The soda ash per million gal is

$$Na_2CO_3 = \frac{88}{10^6} \left| \frac{10^6 \, gal}{} \right| \frac{8.34 \, lb}{gal} \left| \frac{}{0.95} \right. = \underline{773 \, lb}$$

Coagulation and Flocculation in Water Treatment

All of the previously discussed coagulants have been used in water treatment, with aluminum sulfate and ferrous sulfate and lime being the most common. The coagulant dosage depends on the turbidity and the relative ease by which coagulation occurs, and the usual chemical dosages vary from 5 to 90 mg/ℓ. In addition to the previously discussed coagulant aids, activated silica and turbidity addition by means of clay have been used. Polyelectrolyte coagulant aids, if required, are usually added after rapid mixing.

In the softening of water by the lime-soda process, the precipitates formed, calcium carbonate and magnesium hydroxide, enmesh particulates and remove considerable turbidity. Lime and soda ash used in this manner act as coagulants and usually yield a large, dense, fast-settling floc. Frequently, optimum floc settling characteristics are produced by using a coagulant, such as ferrous sulfate, in conjunction with these chemicals.

In water coagulation and water softening, the rapid-mix basins usually have detention times from 20 to 60 sec and G values from 1,000 to 700 sec^{-1}, as described in the previous section on rapid mixing. The flocculators used in water coagulation and water softening are usually of the paddle-wheel type employing horizontal shafts and a cross-flow pattern. For water coagulation, the flocculation time is usually from 20 to 60 min, the G values are from 35 to 70 sec^{-1}, and GT values are from 48,000 to 210,000 (Camp, T. R., 1955). Fragile flocs require low G values and long detention times. For water softening the precipitates formed are usually dense and rapid settling. For flocculation in lime-soda softening, the G values are 1.15 to 1.75 and the GT values are 1.60 to 1.70 times those used in the flocculation of coagulated waters.

The solids-contact units have been used successfully for feed waters with a relatively constant quality with respect to time; thus, they have been successfully used in the coagulation and the softening of lake waters and the softening of well waters. For coagulation, the solids concentration in the contact zone is usually 5 percent by volume; the G values vary from 75 to 175 sec^{-1} and GT values from 125,000 to 150,000. For water softening, using a 10 percent solids concentration by volume results in G values from 130 to 200 sec^{-1} and GT values from 200,000 to 250,000. For water softening, using a much higher solids concentration, such as 20 to 40 percent by volume, results in G values from 250 to 400 sec^{-1} and GT values from 300,000 to 400,000 (AWWA, 1969).

Coagulation and Flocculation in Wastewater Treatment

All of the previously discussed coagulants have been used in coagulating municipal wastewaters, and some have been employed in coagulating industrial wastewaters with sizable suspended solids concentrations. Municipal wastewaters, because of their relatively high turbidities and suspended solids, usually coagulate more readily than surface waters, although the chemical dosages required are much higher. For untreated or raw municipal wastewaters, coagulant dosages in the magnitude of 300 mg/ℓ or more may be required.

Aluminum sulfate, in addition to coagulating colloidal and suspended solids, removes an appreciable amount of the phosphorus from the wastewater. The chemistry of phosphorus removal by alum is not completely understood; however, polyphosphates and organic phosphorus are probably removed by complex reactions, enmeshment in and sorption by the floc. For simplicity purposes, it is assumed that the phosphorus remaining after the initial coagulation is orthophosphate, and the removal is represented by the simplified reaction (EPA, 1973a)

$$Al_2(SO_4)_3 + 2PO_4^{-3} \rightarrow 2AlPO_4 + 3SO_4^{-2} \qquad (2.26)$$

The precipitate formed is considered to be enmeshed in the floc. The removal is dependent on the pH, and optimum removal occurs between pH 5.5 to 6.5.

Iron salts, in addition to coagulating colloidal matter and suspended solids, also remove a substantial amount of the phosphorus from municipal wastewaters. The polyphosphates and organic phosphates are removed in a manner similar to the removal by alum—that is, complex reactions and enmeshment in or sorption by the floc. The ferric ion will react with orthophosphate to produce a precipitate, ferric phosphate, $FePO_4$, which will be removed by the floc. The ferrous ion also will remove orthophosphate as a precipitate; however, the chemistry is much more involved than removal by the ferric ion. The removal of phosphorus by iron salts is dependent on the pH and optimum removal is between 4.5 to 8.

Lime is frequently used as a coagulant for municipal wastewaters or effluents and, in addition to removing colloidal and suspended solids, excellent phosphate removal results. The precipitation reactions for lime are identical to the softening reactions by lime, which are

$$Ca(OH)_2 + Ca(HCO_3)_2 \rightarrow 2CaCO_3 + 2H_2O \qquad (2.27)$$

$$2Ca(OH)_2 + Mg(HCO_3)_2 \rightarrow 2CaCO_3 + Mg(OH)_2 + 2H_2O \qquad (2.28)$$

To precipitate calcium carbonate requires a pH of 9.5 or more, and to precipitate magnesium hydroxide requires a pH greater than 10.8. A high pH is beneficial since the amount of phosphate ion removal increases as the pH increases. The simplified reaction for phosphorus precipitation by lime is (EPA, 1973a)

$$5Ca^{+2} + 40H^{-1} + 3HPO_4^{-2} \rightarrow$$
$$\underline{Ca_5(OH)(PO_4)_3} + 3H_2O \qquad (2.29)$$

The solubility of the precipitate formed, calcium hydroxyapatite, decreases as the pH increases; at pH 9.0 or higher the maximum removal occurs. Another benefit of the high pH involved in phosphate precipitation is that subsequent treatment by air stripping will remove ammonia, with maximum removal occurring at pH 10.8 or greater. In lime coagulation, the calcium carbonate precipitate formed is granular, and the removal of colloidal and suspended solids is due to enmeshment in the precipitate. Conversely, magnesium hydroxide is a gelatinous precipitate that probably removes colloidal and suspended solids by enmeshment in the precipitate and sorption by the precipitate. After lime coagulation, the wastewater must be stabilized to lower the pH and to precipitate the excess lime. If stabilization is not accomplished, excessive encrustations of calcium carbonate will form on the media used in subsequent treatment by multimedia filtration. Usually stabilization is achieved by using carbon dioxide, and the stabilized wastewater will have a pH of 7.0 to 8.5. If lime coagulation precedes biological treatment, the pH must be lowered to 9.5 or lower to prevent inhibition of the biological processes.

The coagulant aids most commonly used in wastewater coagulation are polyelectrolytes, turbidity addition by means of recycled chemically precipitated sludge, and lime addition. Polyelectrolyte coagulant aids, when required, are usually added after rapid mixing. Since the aluminum and iron salts used in coagulation are acidic, relatively large coagulant dosages may require lime addition to avoid an unwanted pH drop. The coagulant solution and the milk of lime slurry must be added to the mixing basins by separate feed lines.

In the alum and iron salt coagulation of municipal wastewaters and effluents, the rapid-mix basins usually have detention times from one to two minutes to adequately disperse the chemicals and initiate the coagulation process (EPA, 1973b). Although municipal wastewaters and effluents usually coagulate more readily than surface waters, longer rapid-mixing times are usually required due to the high suspended solids concentrations and large coagulant dosages. The velocity gradients for rapid mixing are usually about 300 sec^{-1} (EPA, 1975) and are generally lower than encountered in water treatment due to the nature of the organic solids. Overmixing may rupture the existing waste-water solids into smaller particles, which require larger coagulant dosages and longer flocculation detention times. Since flocculation of wastewaters and effluents usually occurs with ease, the detention times and GT values required for flocculation are generally less than those used in water treatment. For alum and iron salt coagulation, the flocculation time is usually from 15 to 30 min, typical G values are from 20 to 75 sec^{-1}, and GT values range from 10,000 to 100,000 (Metcalf & Eddy, Inc., 1979). For mechanical flocculation the detention time is usually about 15 min (EPA, 1975). For lime coagulation, the detention times for rapid mixing are usually from 1 to 2 min. (EPA, 1975). The precipitates from lime coagulation, $CaCO_3$ and $Mg(OH)_2$, benefit very little from long flocculation times and a detention time of 5 to 10 min may be adequate. The precipitates formed have high strengths, and the G values for flocculation are usually 100 sec^{-1} or more (EPA, 1975). For the case of relatively large plants, paddle-wheel flocculation with horizontal shafts is generally used, and the peripheral speed of the paddles is usually less than 2.0 fps. If paddle-wheel flocculation is used, it is usually preferable to use tapered flocculation. For smaller plants pneumatic mixing and flocculation have been successfully used, particularly for lime coagulation.

Solids-contact units of the slurry-recirculation type have been used successfully in some plants for coagulating municipal wastewaters and effluents. The slurry-recirculation type units are not as sensitive to varying flow rates and varying loading as the sludge-blanket filtration type. Since solids-contact units require more skill in operation than conventional mixing, flocculation, and settling, this should be considered in process selection.

In the coagulation of industrial wastewaters, a rapid-mix time of 0.5 to 6 min and a flocculation time of 20 to 30 min have been reported (Eckenfelder, W. W., 1966 & 1970). Coagulant aids, such as polyelectrolytes, are usually added after rapid mixing. It should be understood that, due to the varying nature of industrial wastes, laboratory and frequently pilot plant studies are required to determine the most effective coagulant, the coagulant aid (if necessary), the chemical dosages, the optimum rapid-mixing and flocculation times, and the G and GT values.

References

American Water Works Association (AWWA). 1969. *Water Treatment Plant Design*. New York: AWWA.

American Water Works Association (AWWA). 1971. *Water Quality and Treatment*. New York: McGraw-Hill.

Black, A. P.; Buswell, A. M.; Eidsness, F. A.; and Black, A. L. 1957. Review of the Jar Test. *Jour. AWWA* 39, no. 11: 1414.

Black, A. P., and Harris, R. J. 1969. New Dimensions for the Old Jar Test. *Water & Wastes Engineering*, Dec.:49.

Camp, T. R. 1955. Flocculation and Flocculation Basins. *Trans. ASCE* 120:1.

Camp, T. R. 1968. Floc Volume Concentration. *Jour. AWWA* 60, no. 6:656.

Camp, T. R. 1952. "Water Treatment" in *The Handbook of Applied Hydraulics*, ed. C. V. Davis. New York: McGraw-Hill.

Cohen, J. M. 1957. Improved Jar Test Procedure. *Jour. AWWA* 49, no. 11:1425.

Conway, R. A., and Ross, R. D. 1974. *Handbook of Industrial Waste Disposal*. New York: Van Nostrand Reinhold.

Culp, G. L., and Culp, R. L. 1974. *New Concepts in Water Purification*. New York: Van Nostrand Reinhold.

Culp, R. L., and Culp, G. L. 1978. *Handbook of Advanced Wastewater Treatment*, 2nd ed. New York: Van Nostrand Reinhold.

Eckenfelder, W. W. 1966. *Industrial Water Pollution Control*. New York: McGraw-Hill.

Eckenfelder, W. W., Jr. 1980. *Principles of Water Quality Management*. Boston: CBI Publishing.

Eckenfelder, W. W. 1970. *Water Quality Engineering for Practicing Engineers*. New York: Barnes and Noble.

Environmental Protection Agency. (EPA). 1973a. *Phosphorous Removal*. EPA Process Design Manual, Washington, D.C.

Environmental Protection Agency (EPA). 1973b. *Physical-Chemical Wastewater Treatment Plant Design*. EPA Technology Transfer Seminar Publication, Washington, D.C.

Environmental Protection Agency (EPA). 1975. *Suspended Solids Removal*. EPA Process Design Manual, Washington, D.C.

Environmental Protection Agency (EPA). 1974. *Upgrading Existing Wastewater Treatment Plants*. EPA Process Design Manual. Washington, D.C.

Great Lakes–Upper Mississippi Board of State Sanitary Engineers. 1978. *Recommended Standards for Sewage Works*. Ten state standards, Albany, N.Y.

Hudson, H. E., Jr., and Wolfner, J. P. 1967. Design of Mixing and Flocculation Basins. *Jour. AWWA* 59, no. 10:1257.

McCabe, W. L., and Smith, J. C. 1976. *Unit Operations of Chemical Engineering*, 3rd ed. New York: McGraw-Hill.

Metcalf and Eddy, Inc. 1979. *Wastewater Engineering, Treatment, Disposal and Reuse*. New York: McGraw-Hill.

O'Melia, C. R. 1970. "Coagulation in Water and Wastewater Treatment" in *Advances in Water Quality Improvement by Physical and Chemical Processes*, ed. E. F. Gloyna and W. W. Eckenfelder, Jr. Austin, Tex.: University of Texas Press.

Priesing, C. P. 1962. A Theory of Coagulation Useful for Design. *Ind. and Eng. Chem.* 54, no. 8:38; and 54, no. 9:54.

Rich, L. G. 1963. *Unit Processes of Sanitary Engineering*. New York: Wiley.

Rouse, H. 1950. "Fundamental Principles of Flow" in *Engineering Hydraulics*, ed. H. Rouse. New York: Wiley.

Rushton, J. H. 1952. Mixing of Liquids in Chemical Processing. *Ind. and Eng. Chem* 44, no. 12:2931.

Rushton, J. H.; Bissell, E. S.; Hesse, H. C.; and Everett, H. J. 1947. Designing and Utilization of Internal Fittings for Mixing Vessels. *Chem. Engr. Progr.* 43, no. 12:649.

Rushton, J. H.; Costich, E. W.; and Everett, H. J. 1950. Power
Characteristics of Mixing Impellers; Part I and II. *Chem. Eng.
Progr.*, 46, no. 8:395; and 46, no. 9:467.

Rushton, J. H., and Mahoney, L. H. 1954. *Mixing Power and
Pumpage Capacity.* Annual Meeting of AIME, 15 February
1954, New York.

Rushton, J. H., and Oldshue, J. Y. 1953. Mixing—Present Theory
and Practice; Parts I and II. *Chem. Eng. Progr.* 49, no. 4:161;
and 49, no. 5:267.

Sanks, R. L. 1978. *Water Treatment Plant Design.* Ann Arbor,
Mich.: Ann Arbor Science Publishers.

Sawyer, C. N., and McCarty, P. L. 1978. *Chemistry for Environmental Engineers,* 3rd ed. New York: McGraw-Hill.

South Lake Tahoe Public Utility District. 1971. *Advanced Wastewater Treatment as Practiced at South Tahoe.* Tech. Report
for the EPA, Project 17010 ELQ (QPRD 52-01-67), August
1971.

Steel, E. W., and McGhee, T. J. 1979. *Water Supply and Sewerage.* New York: McGraw-Hill.

Stumm, W., and Morgan, J. J. 1962. Chemical Aspects of Coagulation. *Jour. AWWA* 54, no. 8: 971.

Stumm, W., and O'Melia, C. R. 1968. Stoichiometry of Coagulation. *Jour. AWWA* 60, no. 5:514.

Sundstrom, D. W., and Klei, H. E. 1979. *Wastewater Treatment.*
Englewood Cliffs, N.J.: Prentice-Hall.

Water Pollution Control Federation (WPCF). 1977. WPCF Manual of
Practice no. 8, Washington, D.C.

Weber, M. J., Jr. 1972. *Physiochemical Processes for Water Quality Control.* New York: Wiley-Interscience.

Problems

1. A rapid-mixing basin is to be designed for a water coagulation plant, and the design flow for the basin is 4.0
MGD. The basin is to be square with a depth equal to
1.25 times the width. The velocity gradient is to be 900
$\sec^{-1}$ (at 50°F), and the detention time is 30 sec. Determine:

   a. The basin dimensions.

   b. The input horsepower required.

   c. The impeller speed if a vane-disc impeller with four
   flat blades is employed and the tank is not baffled.
   The impeller diameter is to be 50 percent of the basin width.

2. A flocculation basin is to be designed for a water coagulation plant, and the design flow for the basin is
12.0 MGD. The basin is to be a cross-flow, horizontal-
shaft, paddle-wheel type with a mean velocity gradient
of 25 $\sec^{-1}$ (at 50°F), a detention time of 45 min, and
a *GT* value from 50,000 to 100,000. Tapered flocculation is to be provided, and three compartments of equal
depth in series are to be used as shown in Figure

2.16 (b). The compartments are to be separated by slotted, redwood baffle fences, and the basin floor is to be level. The G values for the compartments are to be 45, 20, and 10 sec^{-1}. The length of the basin is to be one-half its width. The paddle wheels are to have wood blades with a 6-in. width and a length of 10 ft. The outside blades should clear the floor by 1.5 ft and be 1.5 ft below the water surface. There are to be six blades per paddle wheel, and the blades should have a clear spacing of 12 in. Adjacent paddle wheels should have a clear spacing of 24 to 36 in. between blades. The wall clearance is 12 to 18 in. Determine:

a. The basin dimensions.
b. The paddle-wheel design.
c. The power to be imparted to the water in each compartment and the total power required for the basin.
d. The rotational speed range if 1:4 variable-speed drives are employed.

3. A pneumatic flocculation basin is to be designed for a tertiary treatment plant having a flow of 19,000 m^3/d. The plant is to employ high pH lime coagulation, and pertinent data for the flocculation basin are: detention time = 5 min, G = 150 s^{-1} (at 10°C), length = 2 times width, depth = 3.0 m, diffuser depth = 2.75 m, and air flow = 6.80 m^3/h per diffuser. Determine:

a. The basin dimensions.
b. The total air flow in m^3/h.
c. The number of diffusers.

4. An impeller-powered flocculation basin is to be designed for a tertiary treatment plant having a flow of 25 MGD. The plant is to employ alum coagulation, and pertinent data for the flocculation basin are: detention time = 20 min, G = 35 sec^{-1} (at 50°F), GT = 10,000 to 100,000, width = 1.25 times depth, length = twice width, no baffling, number of impellers = 2, number of blades per impeller = 6 pitched at 45°, impeller diameter = 30 percent of basin width, K_L = 70.0 and K_T = 1.65. Determine:

a. Basin dimensions.
b. Impeller diameter.
c. Speed of impellers in rpm.

5. A water is to be softened by the lime-soda ash process. It has 86 mg/ℓ Ca^{+2}, 35 mg/ℓ Mg^{+2}, 299 mg/ℓ HCO$_3{}^{-1}$, and 6 mg/ℓ CO$_2$. Determine the pounds of quicklime and soda ash required per million gallons treated if the commercial grade chemicals have a purity of 85 and 95 percent, respectively.

6. A water is to be softened by the lime-soda ash process. It has a hardness of 225 mg/ℓ as $CaCO_3$, an alkalinity of 178 mg/ℓ as $CaCO_3$, a Mg^{+2} content of 39 mg/ℓ, and 4 mg/ℓ of CO_2. Determine the pounds of quicklime and soda ash required per million gallons treated if the commercial chemicals have a purity of 85 and 95 percent, respectively.

Chapter Three

SEDIMENTATION

Sedimentation is a solid-liquid separation utilizing gravitational settling to remove suspended solids. It is commonly used in water treatment, wastewater treatment, and advanced wastewater treatment. In water treatment its main applications are:

1. plain settling of surface waters prior to treatment by a rapid sand filtration plant;
2. settling of coagulated and flocculated waters prior to rapid sand filtration;
3. settling of coagulated and flocculated waters in a lime-soda type softening plant;
4. settling of treated waters in an iron or manganese removal plant.

In wastewater treatment its main uses are:

1. grit or sand and silt removal;
2. suspended solids removal in primary clarifiers;
3. biological floc removal in activated sludge final clarifiers;
4. humus removal in trickling filter final clarifiers.

In advanced wastewater treatment and tertiary treatment, its main purpose is the removal of chemically coagulated floc prior to filtration.

Sedimentation is one of the earliest unit operations used in water or wastewater treatment. The principles of sedimentation are the same for basins used in either water or wastewater treatment; the equipment and operational methods are also similar.

Sedimentation basins are usually constructed of reinforced concrete and may be circular, square, or rectangular in plan view. Circular tanks may be from 15 to 300 ft in diameter and are usually from 6 to 16 ft deep. The most common sizes are from 35 to 150 ft in diameter and depths are usually 10 to 14 ft. Standard-size tanks have diameters with 5-ft intervals in order to accommodate commercially built sludge rake mechanisms. Square tanks have widths from 35 to 200 ft and depths from 6 to 19 ft. Standard-size square tanks have widths with 5-ft intervals. The freeboard for circular or square tanks is from 1 to 2.5 ft. Tanks that are not standard size may be furnished with specially built sludge rake mechanisms. Also, collectors for tanks with depths greater than those stated can be obtained by special order. Rectangular tanks usually have three types of sludge rake mechanisms: (1) sprocket and chain driven rakes, (2) rakes supported from a traveling bridge, and (3) tandem scrapers built for square basins. Rectangular tanks with sprocket and chain drives have widths from 5 to 20 ft, lengths up to about 250 ft, and depths greater than 6 ft. Widths up to 80 to 100 ft are possible by using four or five multiple bays with individual cleaning mechanisms. Rectangular tanks with traveling bridges that support the sludge rakes have widths from about 10 to 120 ft and lengths from 40 to 300 ft. Traveling bridges have rapid sludge removal, and the rakes may be removed for inspection or repair without draining the basin. Rectangular tanks using two square tank sludge rake mechanisms in tandem give a settling tank with a 2:1 length to width ratio. Tanks as large as 200 ft by 400 ft have been built in this manner, and this type of tank construction is particularly well suited for large water treatment plants. Rectangular tanks can use common wall construction and also occupy less land space than circular clarifiers of equal volume.

Coe and Clevenger (1916) presented a classification for the types of settling which may occur; this was later refined by Camp (1946) and Fitch (1956). This classification divides settling into four general types or classes which are based on the concentration of the particles and the ability of the particles to interact. A discussion of the various types of settling is presented in the following text.

Type I Settling

Type I settling, or *free settling*, is the settling of discrete, nonflocculent particles in a dilute suspension. The particles settle as separate units, and there is no apparent flocculation or interaction between the particles. Examples of type I settling are the plain sedimentation of surface waters and the settling of sand particles in grit chambers.

In type I settling, a particle will accelerate until the drag force, F_D, equals the impelling force, F_I; then settling occurs at a constant velocity, V_s. The impelling force, F_I, is

$$F_I = (\rho_s - \rho)gV \qquad (3.1)$$

where

$F_I =$ impelling force;
$\rho_s =$ mass density of the particle;
$\rho \ =$ mass density of liquid;
$V =$ volume of particle;
$g \ =$ acceleration due to gravity.

A Rectangular Settling Basin at a Water Treatment Plant. Two sludge rake mechanisms designed for square basins are used in tandem to form a rectangular basin with a length to width ratio of 2:1. The tops of the two center columns with drives are in foreground and background. View is towards basin inlet.

Sludge Rake Mechanisms in a Rectangular Settling Basin at a Water Treatment Plant. These mechanisms are in a basin identical to that shown in the top illustration. The rakes are for square basins and have hinged rakes for scraping the basin corners. View is towards basin outlet.

The drag force is given by Newton's law:

$$F_D = C_D A_c \rho \left(\frac{V_s^2}{2} \right)$$ (3.2)

where

F_D = drag force;
C_D = coefficient of drag, which is a function of the Reynolds number, N_{Re};
A_c = area in cross section at right angle to the velocity;
ρ = mass density of liquid;
V_s = settling velocity.

Combining Eqs. (3.1) and (3.2) gives

$$(\rho_s - \rho)gV = C_D A_c \rho \left(\frac{V_s^2}{2} \right)$$

or

$$V_s = \sqrt{\frac{2g}{C_D} \left(\frac{\rho_s - \rho}{\rho} \right) \frac{V}{A_c}}$$ (3.3)

For spheres of diameter, d, the volume, V, is

$$V = \left(\frac{\pi}{6} \right) d^3$$

The cross-section area, A_c, is

$$A_c = \left(\frac{\pi}{4} \right) d^2$$

From the preceding two equations it follows that for spheres

$$\frac{V}{A_c} = \left(\frac{\pi}{6} \right) d^3 \left(\frac{4}{\pi} \right) \frac{1}{d^2} = \frac{2}{3} d$$

Substituting the above expression for V/A_c into Eq. (3.3) gives the following equations:

$$V_s = \sqrt{\frac{4g}{3C_D} \left(\frac{\rho_s - \rho}{\rho} \right) d}$$ (3.4a)

or

$$V_s = \sqrt{\frac{4g}{3C_D} (S_s - 1)d}$$ (3.4b)

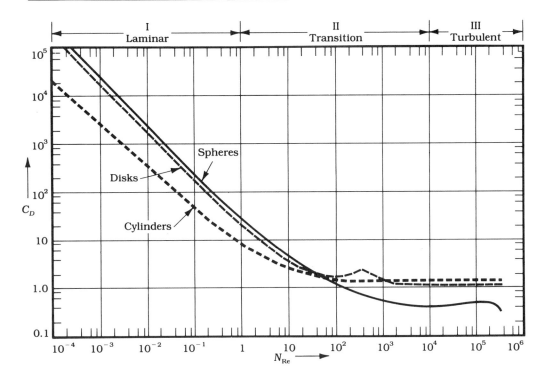

where S_s = specific gravity of the particles.

The numerical value of the drag coefficient depends on whether the flow regime around the particle is laminar or turbulent. Figure 3.1 shows the drag coefficient for various shapes as a function of the Reynolds number, N_{Re}. The Reynolds number for the spheres is defined as

Figure 3.1. Drag Coefficients versus N_{Re} Adapted from "Sedimentation and the Design of Settling Tanks" by T. R. Camp. In *Transactions of the American Society of Civil Engineers* 111 (1952):895. Reprinted by permission.

$$N_{Re} = \frac{V_s d}{\nu} = \frac{V_s d \rho}{\mu} \qquad (3.5)$$

where

ν = kinematic viscosity;
μ = dynamic viscosity;
ρ = mass density.

In Figure 3.1 there are three distinct regions: I, laminar flow; II, transition flow; and III, turbulent flow. For region I, laminar flow, N_{Re} is less than 1 and the viscous forces are more important than inertia forces. The relationship for the drag coefficient for spheres is

$$C_D = \frac{24}{N_{Re}} \qquad (3.6)$$

SEDIMENTATION

73

For region II, transition flow, N_{Re} equals 1 to 10^4 and viscous and inertia forces are of equal importance. The relationship for the drag coefficient for spheres is (Fair, G. M., Geyer, J. C., & Okun, D. A., 1968)

$$C_D = \frac{24}{N_{Re}} + \frac{3}{\sqrt{N_{Re}}} + 0.34 \qquad (3.7)$$

For region III, turbulent flow, N_{Re} is greater than 10^4 and inertia forces are the most important. The drag coefficient for spheres is

$$C_D = 0.4 \qquad (3.8)$$

For laminar flow (region I), Eq. (3.5) may be combined with Eq. (3.6) to eliminate N_{Re} giving

$$C_D = \frac{24\nu}{V_s d}$$

Substituting the above equation for C_D into Eq. (3.4b) gives the following expression:

$$V_s = \sqrt{\left(\frac{4g}{3}\right)\left(\frac{V_s d}{24\nu}\right)(S_s - 1)d} \qquad (3.9)$$

Squaring Eq. (3.9) and rearranging gives Stokes' law:

$$V_s = \frac{g}{18\nu}(S_s - 1)d^2 \qquad (3.10)$$

Or, since $\nu = \mu/\rho$, substitution into Eq. (3.10) yields

$$V_s = \frac{g}{18\mu}(\rho_s - \rho)d^2 \qquad (3.11)$$

which is another form of Stokes' law. Much of the settling of dilute suspensions in water and wastewater treatment follows Stokes' law.

For transition flow (region II), the determination of the settling velocity is a trial and error solution using Eq. (3.4a) or (3.4b) and Eq. (3.7).

For turbulent flow (region III), substituting Eq. (3.8) into Eq. (3.4b) gives

$$V_s = \sqrt{3.3g(S_s - 1)d} \qquad (3.12)$$

The settling velocity of sand in grit removal chambers can be determined using Eq. (3.12).

A Circular Primary Clarifier at an Activated Sludge Wastewater Treatment Plant.

A Circular Final Clarifier Being Repaired at an Activated Sludge Plant. Note surface skimmer, bottom sludge rakes, sludge hopper, and hydrostatic blowout plugs in bottom slab.

The solution for the settling velocity in regions I, II, and III can also be done using the graph in Figure 3.2 (Camp, T. R., 1952). This graph gives a direct solution for the settling velocity, V_s, if the diameter, specific gravity, and temperature are known.

The ideal basin theory by Camp (1946) assumes the following:

1. The settling is type I settling—in other words, discrete particles.
2. There is an even distribution of the flow entering the basin.
3. There is an even distribution of the flow leaving the basin.

Figure 3.2. Type I Settling of Spheres in Water at 10°C
Adapted from "Water Treatment" by T. R. Camp in *Handbook of Applied Hydraulics*, 2nd ed., ed. C. V. Davis. Copyright © 1952 by McGraw-Hill Book Co., Inc. Reprinted by permission.

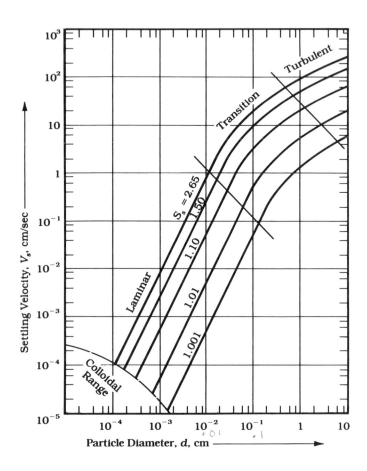

4. There are three zones in the basin: (1) the entrance zone, (2) the outlet zone, and (3) the sludge zone.
5. There is a uniform distribution of particles throughout the depth of the entrance zone.
6. Particles that enter the sludge zone remain there and particles that enter the outlet zone are removed.

Figure 3.3 shows an ideal rectangular settling basin of a length, L, a width, W, and a depth, H. V_0 is the settling velocity of the smallest particle size that is 100 percent removed. When a particle of this size enters the basin at the water surface, point 1, it has a trajectory as shown and

Figure 3.3. Ideal Rectangular Settling Basin

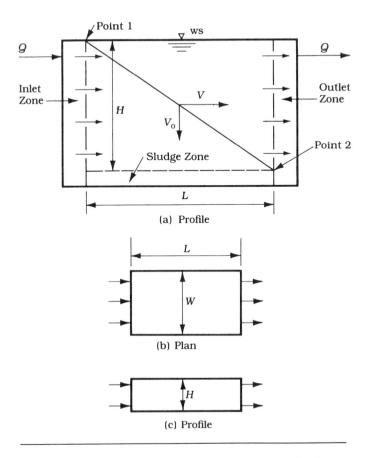

(a) Profile

(b) Plan

(c) Profile

intercepts the sludge zone at point 2, which is at the down-stream end. The detention time, t, is equal to the depth, H, divided by the settling velocity, V_0, or

$$t = \frac{H}{V_0} \tag{3.13}$$

The detention time, t, is also equal to the length, L, divided by the horizontal velocity, V, or

$$t = \frac{L}{V} \tag{3.14}$$

The horizontal velocity, V, is equal to the flow rate, Q, divided by the cross-sectional area, HW, or

$$V = \frac{Q}{HW} \tag{3.15}$$

Combining Eqs. (3.14) and (3.15) to eliminate V gives

$$t = \frac{LWH}{Q} \tag{3.16}$$

Since LWH equals the basin volume, Ψ, the detention time, t, is equal to the basin volume, Ψ, divided by the flow rate, Q, or

$$t = \frac{\Psi}{Q} \tag{3.17}$$

Equating Eqs. (3.16) and (3.13) gives

$$\frac{LWH}{Q} = \frac{H}{V_0} \tag{3.18}$$

Rearranging yields

$$V_0 = \frac{Q}{LW} \tag{3.19}$$

or

$$V_0 = \frac{Q}{A_p} = \text{overflow rate, gal/day-ft}^2 \tag{3.20}$$

where A_p is the plan area of the basin. Equation (3.20) shows that the overflow rate is equivalent to the settling velocity of the smallest particle size that is 100 percent removed.

The previous fundamentals also apply to an ideal circular settling basin, shown in Figure 3.4. The horizontal velocity, V, is given by

$$V = \frac{Q}{2\pi r H} \tag{3.21}$$

From inspection of Figure 3.4,

$$\frac{dh}{dr} = \frac{V_0}{V} \tag{3.22}$$

Substituting Eq. (3.21) into Eq. (3.22) gives

$$\frac{dh}{dr} = \frac{2\pi r H V_0}{Q} \tag{3.23}$$

Rearranging Eq. (3.23) and setting the integration limits yields

Figure 3.4. **Ideal Circular Settling Basin**

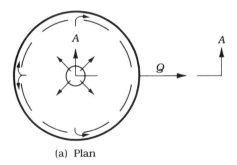

(a) Plan

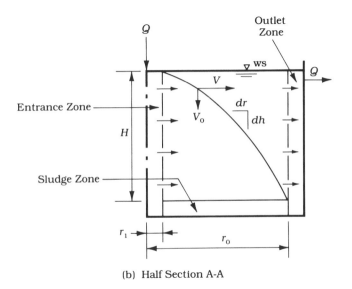

(b) Half Section A-A

$$\int_0^H dh = \frac{2\pi H V_0}{Q} \int_{r_1}^{r_0} r\,dr \qquad (3.24)$$

Integrating gives

$$H = \frac{2\pi H V_0}{Q} \left[\frac{r^2}{2} \right]_{r_1}^{r_0} \qquad (3.25)$$

or

$$H = \frac{\pi H V_0}{Q}(r_0^2 - r_1^2) = \frac{H A_p V_0}{Q} \qquad (3.26)$$

where A_p = plan area of the basin. Cancelling the H terms in Eq. (3.26) and rearranging yields

$$V_0 = \frac{Q}{A_p} = \text{overflow rate, gal/day-ft}^2 \qquad (3.27)$$

Equation (3.27) is identical to Eq. (3.20) for the rectangular basin.

The depth of the ideal rectangular or circular basin is given by

$$H = V_0 t \qquad (3.28)$$

where V_0 is the overflow rate expressed as a velocity. It can be shown that an overflow rate of 100 gal/day-ft^2 is equal to a settling velocity of 0.555 ft/hr. Also, a settling velocity of one cm/sec is equal to an overflow rate of 21,200 gal/day-ft^2. Using these conversion values, settling velocities at any overflow rate can be determined by proportionality.

Inspection of Figures 3.5 (Camp, T. R., 1946) and 3.6 shows that all particles with a settling velocity, V_1, greater than V_0 will be 100 percent removed since their trajectory intercepts the sludge zone. For particles with a settling velocity, V_2, less than V_0, the fraction removed, R_2, is equal to H_2/H or V_2/V_0. Thus,

$$R_2 = \frac{H_2}{H} = \frac{V_2}{V_0} \qquad (3.29)$$

A large variation in particle size will exist in a typical suspension of particles. Thus, one must evaluate the entire range of settling velocities in determining the overall removal for a given design settling velocity or overflow rate.

Figure 3.5. Profile through an Ideal Rectangular Basin

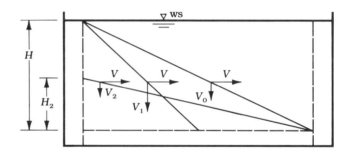

Figure 3.6.   Half Section through an Ideal Circular Basin

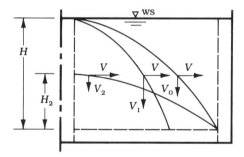

This requires experimental analyses usually employing the use of a settling column. In a batch settling column, samples are withdrawn at various times and various depths, and the solids concentrations are determined. An analysis of the data by an appropriate procedure will yield a settling velocity curve as shown in Figure 3.7. The fraction of the total particles removed for a design velocity, V_0, will be (Camp, T. R., 1946)

$$\text{Fraction removed} = (1 - F_0) + \frac{1}{V_0}\int_0^{F_0} VdF \qquad (3.30)$$

where

$$1 - F_0 = \text{fraction of particles with velocity } V \text{ greater than } V_0;$$

$$\frac{1}{V_0}\int_0^{F_0} VdF = \text{fraction of particles with velocity } V \text{ less than } V_0.$$

In summarizing the ideal settling basin theory, the removal of suspended solids is a function of the overflow rate

Figure 3.7.   Type I Settling Curve

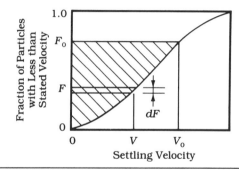

or design settling velocity, V_0, the detention time, t, and the depth, H. Although the ideal settling basin analysis is theoretical, it does give a rational method for the design of sedimentation tanks. An evaluation of vast amounts of operating and research data using the theoretical parameters gives a range of overflow rates and detention times that, in most cases, has been found to be satisfactory for municipal waters and wastewaters.

Type II Settling

Type II settling is the settling of flocculent particles in a dilute suspension. The particles flocculate during settling; thus they increase in size and settle at a faster velocity. Examples of type II settling are the primary settling of wastewaters and the settling of chemically coagulated waters and wastewaters.

Batch settling tests are usually required to evaluate the settling characteristics of a flocculent suspension. A schematic drawing of a batch settling column is shown in Figure 3.8. The column should be at least 5 to 8 inches in diameter to minimize side-wall effects, and the height should be at least equal to the depth of the proposed settling tank. Sampling ports are provided at equal intervals in height.

The suspension must be mixed thoroughly and poured rapidly into the column in order to insure that a uniform distribution of the particles occurs throughout the height of the column. To be representative the test must take place under quiescent conditions and the temperature should not vary more than 1°C throughout the column height in order to avoid convection currents. Samples are removed through the ports at periodic time intervals and the suspended solids concentrations are determined. The percent removal is calculated for each sample knowing the initial suspended solids concentration and the concentration of the sample. The percent removal is plotted on a graph as a number versus time and depth of collection for the sample. Interpolations are made between the plotted points, and curves of equal percent removal, R_A, R_B, and so on are drawn, as shown in Figure 3.9.

The overflow rates, V_0, are determined for the various settling times, t_a, t_b, and so on where the R curves intercept the x-axis. For example, for the curve R_C, the overflow rate is

Figure 3.8. Batch Settling Column Details for Type II Settling

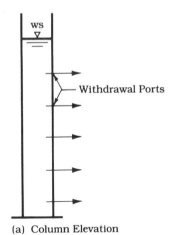

ws

— Withdrawal Ports

(a) Column Elevation

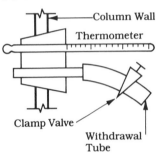

Column Wall

Thermometer

Clamp Valve

Withdrawal Tube

(b) Withdrawal Port Detail

$$V_0 = \frac{H}{t_c} \times \text{proper conversions} \qquad (3.31)$$

Figure 3.9. Settling Diagram for Type II Settling

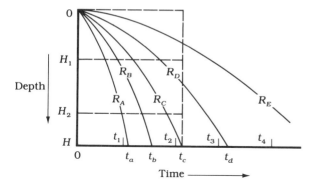

where H is the height of the column and t_c is the intercept of the R_C curve and the x-axis. The fractions of solids removed, R_T, for the times, t_a, t_b, and so on are then determined. For example, for time t_c the fraction removed, R_T, would be

$$R_T = R_C + \frac{H_2}{H}(R_D - R_C) + \frac{H_1}{H}(R_E - R_D) \qquad (3.32)$$

Figure 3.10.
Profile for Type II Settling

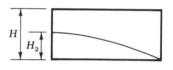

where H_2 represents the height that the particles of $(R_D - R_C)$ size would settle during t_c. These would intercept the sludge zone in a basin, as in Figure 3.10. By using the various times, t_a, t_b, and so on, the various overflow rates, V_0, and the various fractions removed, R_T, a graph of the overflow rates versus fractions removed can be constructed. Also, a graph of the fractions removed versus detention times can be made. In applying the curves to design a tank, scale-up factors of 0.65 for the overflow rate and 1.75 for detention time are used to compensate for the side-wall effects of the settling column (Eckenfelder, W. W., 1980).

Example Problem 3.1
Primary Clarifier

A primary clarifier is to be designed to treat an industrial wastewater having 320 mg/ℓ suspended solids and a flow of 2.0 MGD. A batch settling test was performed using an 8-in.-diameter column that was 10 ft long and had withdrawal ports every 2.0 ft. The reduced data giving the percent removals are as shown in Table 3.1.

Table 3.1. Percent Suspended Solids Removal at Given Depths

Time (min)	2 ft	4 ft	6 ft	8 ft	10 ft
0	0	0	0	0	0
10	28	18	18	12	a
20	48	39	25	27	a
30	68	50	34	31	a
45	70	56	53	41	a
60	85	66	59	53	a
90	88	82	73	62	a

a. Data showed an increase in solids concentration

Determine:

1. The design detention time and design overflow rate if 65 percent of the suspended solids are to be removed.
2. The design diameter and depth.

Solution

A plot of the percent removals at the various depths and times is shown in Figure 3.11. Interpolations have been

---

Figure 3.11.   Graph Showing Suspended Solids Removal (as a Percent) at Various Depths and Settling Times, for Example Problem 3.1.

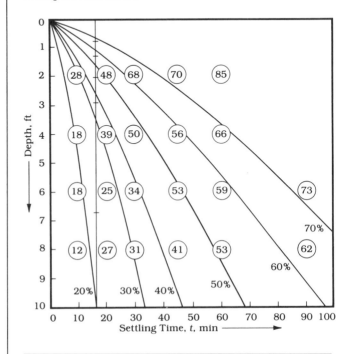

---

made to locate the 20, 30, 40, 50, 60, and 70 percent removal curves, and the curves have been drawn on the plot. The 20 percent curve intersects the x-axis at 16 minutes; thus, the overflow rate at that time is

$$V_0 = \frac{10\,ft}{16\,min} \left| \frac{1440\,min}{day} \right| \frac{7.48\,gal}{ft^3}$$
$$= 6730\,gal/day\text{-}ft^2$$

The detention time in hours is 16/60 or 0.27 hours. The point midway between the 20 and 30 percent curves at 16 min is located as shown and is at a depth of 6.7 ft. In a like manner, the points midway between the 30 and 40, 40 and 50, 50 and 60, and 60 and 70 percent curves are located and the respective depths are 2.9, 2.0, 1.3, and 0.8 ft. Using these values, the total fraction removed (R_T) at 16 min (0.27 hr) is

$$R_T = 20 + (6.7/10)(30-20) + (2.9/10)(40-30)$$
$$+ (2.0/10)(50-40) + (1.3/10)(60-50)$$
$$+ (0.8/10)(70-60)$$
$$= 33.7\,percent$$

Similarly, the overflow rates, detention times, and total fractions removed are computed for the 30, 40, 50, and 60 percent curves and a summary of the reduced data is shown in Table 3.2.

Table 3.2. Reduced Data for 30, 40, 50, and 60
Percent Curves

Time t (hr)	Overflow Rate V_0 (gal/day-ft^2)	Fraction Removed R_T (%)
0.27	6730	33.7
0.55	3260	48.7
0.77	2340	56.7
1.13	1590	63.8
1.60	1120	68.6

A plot of the fraction removed (R_T) versus detention time (t) is shown in Figure 3.12. Also, a plot of the fraction removed (R_T) versus overflow rate (V_0) is shown in Figure 3.13. For 65 percent removal the detention time is 1.22 hr; thus, the design detention time is $(1.22)1.75 = 2.14$ hr. For 65 percent removal the overflow rate is 1420 gal/day-ft^2; thus, the design overflow rate is $(1420)(0.65) = 923$ gal/day-ft^2. The required area is

$$A = \frac{2{,}000{,}000 \, \text{gal}}{\text{day}} \left| \frac{\text{day-ft}^2}{923 \, \text{gal}} \right. = 2167 \, \text{ft}^2$$

Thus, the diameter, D, is

$$D = \left[\frac{4}{\pi}(2167)\right]^{1/2} = 52.5 \, \text{ft or } \underline{55 \, \text{ft for standard size}}$$

The required depth, H, is

$$H = \frac{2{,}000{,}000 \, \text{gal}}{24 \, \text{hr}} \left| \, 2.14 \, \text{hr} \, \right| \left. \frac{\text{ft}^3}{7.48 \, \text{gal}} \right| \frac{4}{\pi} \left| \frac{1}{(55 \, \text{ft})^2} \right.$$

$$= 10.03 \, \text{ft} \quad \underline{\text{Use 10 ft-3 in.}}$$

Figure 3.12.   Suspended Solids Removal versus Detention Time, for Example Problem 3.1

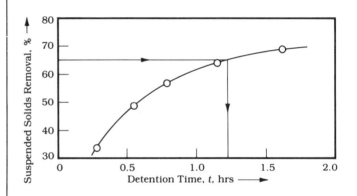

Figure 3.13.   Suspended Solids Removal versus Overflow Rate, for Example Problem 3.1

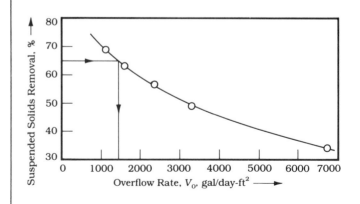

Type III settling, or *zone* or *hindered settling*, is the settling of an intermediate concentration of particles in which the particles are so close together that interparticle forces hinder the settling of neighboring particles. The particles remain in a fixed position relative to each other and all settle at a constant velocity. As a result, the mass of particles settle as a zone. At the top of the settling mass, there will be a distinct solid-liquid interface between the settling particle mass and the clarified liquid. An example of type III settling is the settling that occurs in the intermediate depths in a final clarifier for the activated sludge process. Type IV settling, or *compression settling*, is the settling of particles that are of such a high concentration that the particles touch each other and settling can occur only by compression of the compacting mass. An example of type IV settling is the compression settling that occurs in the lower depths of a final clarifier for the activated sludge process. Both discrete and flocculent particles may settle by zone or compression settling; however, flocculent particles are the most common type encountered.

The settling of a flocculent suspension of activated sludge placed in a graduate cylinder is as shown in Figure 3.14 (a). At first, time $t=0$ and the particles have zone settling (ZS). They have the same relative position with respect to each other. The concentration of particles is so great that they interfere with the velocity fields of each other and the rate of settling is a function of the solids concentration. At time $t=t_1$, the sludge mass has settled until a clear water zone exists above the sludge. Below the region of zone or hindered settling, the concentration of the particles has become so great that many of the particles have made physical contact with each other. This is transition settling (TS) from zone settling to compression settling (CS). Below the transition zone is the compression settling zone where all of the particles are in contact with each other and compression has begun. At time $t=t_2$, the zone settling region has disappeared and all particles are undergoing transition or compression settling. At time $t=t_3$, the transition zone has disappeared and all the particles are in a state of compression settling. At time $t=t_4$, the compression settling is almost complete. Figure 3.14 (b) shows the settling curve for the sludge water interface in the batch settling test.

Figure 3.15 (a) shows a cross section of a circular final clarifier for the activated sludge process illustrating the classes of settling which occur. The clear water zone is usually about 5 to 6 ft deep, and the total depth for zone or hindered, transition, and compression settling is usually about 5 to 7 ft.

Figure 3.14. Settling of a Concentrated Suspension

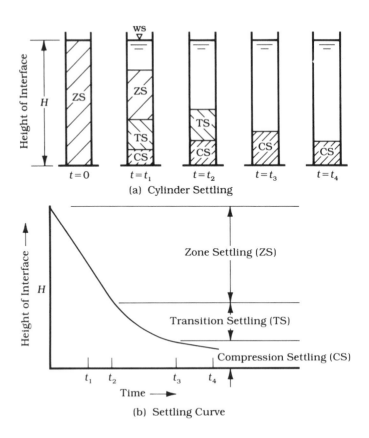

(a) Cylinder Settling

(b) Settling Curve

Figure 3.15. Settling in a Final Clarifier for the Activated Sludge Process

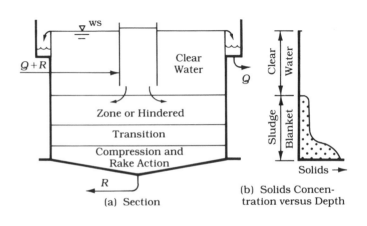

(a) Section

(b) Solids Concentration versus Depth

Batch settling tests as previously described can be used to obtain the parameters needed for the design of an activated sludge final clarifier by a method presented by Talmage and Fitch (1955); however, another method based on solids flux is available (Dick, R. I., 1970). In final clarifiers, both clarification of the liquid and thickening of the solids must be accomplished. Batch settling tests for design are usually done in a 1-liter graduate cylinder equipped with a slow stirring device rotating at four to six revolutions per hour to stimulate the raking action of the trusswork of the mechanical sludge rakes. The procedure for determining the area of a final clarifier after obtaining a settling curve, Figure 3.16, is as follows:

1. Determine the slope of the hindered settling region, V_0. This is the settling velocity required for clarification.
2. Extend tangents from the hindered settling region and the compression region and bisect the angle formed to locate point 1.
3. Draw a tangent to the curve at point 1.

Figure 3.16. *Analysis of a Batch Settling Curve, Concentrated Suspension*
Adapted and reprinted with permission from "Determining Thickener Unit Areas" by W. P. Talmage and E. B. Fitch. In *Industrial and Engineering Chemistry* 47, no. 1 (January 1955):38. Copyright © 1955 American Chemical Society.

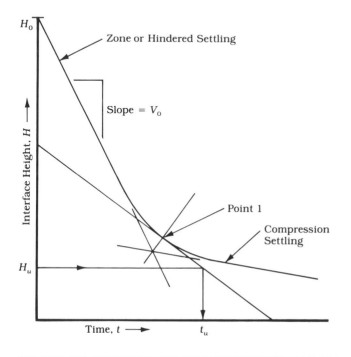

4. Knowing the initial sludge concentration, C_0, and the initial sludge height, H_0, select a design underflow concentration, C_u, and determine the interface height, H_u.

Since
$$C_u H_u = C_0 H_0 \qquad (3.33)$$

then

$$H_u = \frac{C_0 H_0}{C_u} \qquad (3.34)$$

5. Draw a horizontal line from H_u to intersect the tangent line and determine the time, t_u. This is the time required to reach the desired underflow concentration, C_u.
6. Determine the area required for thickening, A_t, from

$$A_t = 1.5\,(Q + R)\,\frac{t_u}{H_0} \qquad (3.35)$$

where
Q = flow to the aeration tank prior to junction with the recycled sludge line;
R = recycled sludge flow;
$Q + R$ = total flow to the final clarifier;
1.5 = scale-up factor (Eckenfelder, W. W., 1980).
7. Determine the area required for clarification, A_c, from

$$A_c = 2.0\,\frac{Q}{V_0} \qquad (3.36)$$

where
Q = flow to the aeration tank prior to junction with the recycled sludge line or the effluent flow from the final clarifier;
2.0 = scale-up factor (Eckenfelder, W. W., 1980).

One area, A_t or A_c, will be the largest and will be the controlling area for the final clarifier design. Settling tests should be made for a range in the mixed liquor suspended solids (MLSS) concentration to be expected in the design plant, and the test showing the most conservative design should be used.

In the design of final clarifiers, the flow rate, Q, is usually taken as the average daily flow. During the peak hours of the day, however, the flow rate may be considerably more and could be up to about five times the average flow for municipal wastewaters. This has caused many final clarifiers to have appreciable solids spill over the effluent weirs during the peak of the day. Therefore, the thickening and clarification area for the peak hour of the day should be checked. In many cases the peak flow condition controls the design area for the final clarifier for an activated sludge process.

Example Problem 3.2
Final Clarifier

A final clarifier is to be designed for an activated sludge plant treating an industrial wastewater having a design flow of 1.2 MGD. Batch settling studies have been performed in the laboratory using an acclimated culture of activated sludge and a graduate cylinder with a very slow rotating stirrer. The MLSS in the test was 2500 mg/ℓ. The interface height versus settling time is shown in Figure 3.17. The design MLSS is 2500 mg/ℓ and the design underflow concentration is 10,000 mg/ℓ. Determine:

Figure 3.17. *Graph for Example Problem 3.2*

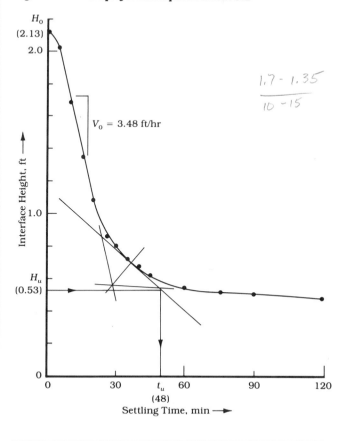

$$\frac{1.7 - 1.35}{10 - 15}$$

$V_0 = 3.48$ ft/hr

1. The area required for clarification.
2. The area required for thickening.
3. The design diameter.

Solution

A material balance for the recycle is

$$(1.2\,\text{MGD})(0) + (R)(10,000) = (1.2\,\text{MGD} + R)(2500)$$

Thus, $R = 0.40$ MGD. The interface height of the underflow is $H_u = C_0 H_0 / C_u = (2500)(2.13/10,000) = 0.53$ ft. The bisecting angle and tangent construction is shown on the graph, and the intersection within the H_u line gives a t_u value of 48 min. The settling velocity, V_0, is 3.48 ft/hr. The area required for thickening is

$$A_t = 1.5(Q + R)\frac{t_u}{H_0}$$

$$= \frac{1.5 \quad\left|\quad 1.60 \times 10^6\,\text{gal} \quad\right|\quad 48\,\text{min} \quad\right|\quad \text{ft}^3}{\quad\left|\quad 1440\,\text{min} \quad\right|\quad 2.13\,\text{ft} \quad\right|\quad 7.48\,\text{gal}}$$

$$= \underline{5021\,\text{ft}^2}$$

The area required for clarification is

$$A_c = 2.0\frac{Q}{V_0}$$

$$= \frac{2.0 \quad\left|\quad 1.2 \times 10^6\,\text{gal} \quad\right|\quad \text{hr} \quad\right|\quad \text{ft}^3}{\quad\left|\quad 24\,\text{hr} \quad\right|\quad 3.48\,\text{ft} \quad\right|\quad 7.48\,\text{gal}}$$

$$= \underline{3842\,\text{ft}^2}$$

The area for thickening controls, thus the diameter, is

$$D = \left[\frac{4}{\pi}(5021\,\text{ft}^2)\right]^{1/2}$$

$$= 80.0\,\text{ft} \qquad \underline{\text{Use 80.0 ft for standard size.}}$$

Another approach to the design of final clarifiers for the activated sludge processes and of sludge thickeners is based on the solids flux concept (Dick, R. I., 1970). The solids flux is the rate of solids thickening per unit area in plan view—in other words, the lb/hour-ft^2. As the solids settle in clarifiers and thickeners, they must be thickened from the initial concentration, C_0, to the underflow concentration, C_u, at the bottom of the tank. As the solids move downward, at some level in the tank a limiting solids flux, G_L, occurs. This flux must not be exceeded or solids will build up and spill over with the effluent from the tank. The movement of the solids downward occurs by hindered settling and also by the bulk flow downward due to the underflow. The data required for the flux design method are determined from batch settling tests. Numerous concentrations of the sludge are allowed to settle to obtain the hindered settling velocities. The hindered or zone settling velocities, V_0, are measured as illustrated in Example Problem 3.2 using a slowly stirred graduate cylinder. A plot is made of the hindered settling velocity, V_0, versus the solids concentration, C, as shown in Figure 3.18. At numerous concentrations, the solids flux is computed since it is obtained by multiplying the velocity by the solids concentration. The resulting curve of flux versus concentration is shown in Figure 3.19.

At any level in the settling tank, the movement of solids by settling is

$$G_s = C_t V_t \qquad\qquad (3.37)$$

where

G_s = solids flux by gravity;
C_t = solids concentration;
V_t = hindered settling velocity.

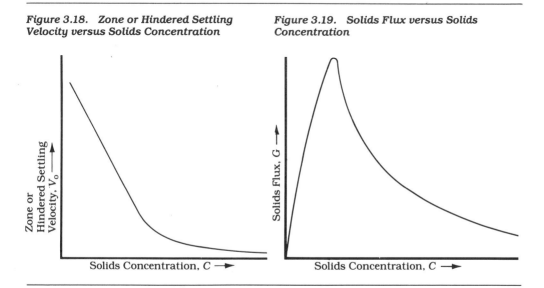

Figure 3.18. Zone or Hindered Settling Velocity versus Solids Concentration

Figure 3.19. Solids Flux versus Solids Concentration

The movement of the solids due to bulk flow is given by

$$G_b = C_t V_b \qquad (3.38)$$

where

G_b = bulk flux;
V_b = bulk velocity.

The total solids flux for gravity settling and bulk movement is therefore

$$G_t = G_s + G_b = C_t V_t + C_t V_b \qquad (3.39)$$

where

G_t = total flux.

The bulk velocity is given by

$$V_b = \frac{Q_u}{A} \qquad (3.40)$$

where

Q_u = flow rate of the underflow;
A = plan area of the tank.

The mass rate of solids settling—that is, the weight of the solids settling per unit time—is

$$M_t = Q_0 C_0 = Q_u C_u \qquad (3.41)$$

where

M_t = rate of solids settling;
Q_0 = influent flow rate to the tank;
C_0 = influent solids concentration.

The limiting cross-sectional area, A, required is given by

$$A = \frac{M_t}{G_L} = \frac{Q_0 C_0}{G_L} \qquad (3.42)$$

where

G_L = limiting flux.

Rearranging Eq. (3.41) gives $Q_u = M_t/C_u$, and combining this with Eq. (3.40) and Eq. (3.42) gives

$$V_b = \frac{Q_u}{A} = \frac{M_t}{C_u A} = \frac{G_L}{C_u} \qquad (3.43)$$

These relationships are shown in Figure 3.20. Selecting an underflow concentration, C_u, and drawing a tangent to the flux curve gives the y-axis intercept as G_L, the limiting flux value. The slope of the tangent is equal to V_b, the bulk velocity. The value of the gravity flux is G_s, whereas the value of the bulk flux is $G_L - G_s$. These concepts are illustrated in Example Problem 3.3 using batch settling data.

Figure 3.20. **Solids Flux versus Solids Concentration**

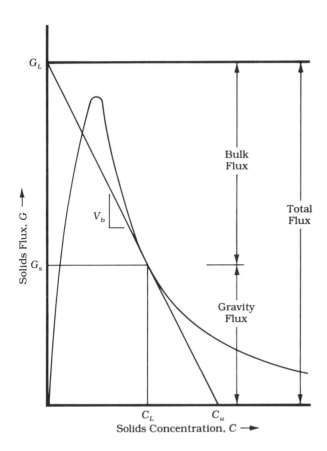

Batch settling tests have been performed using an acclimated activated sludge to give the data in Table 3.3.

Table 3.3. Concentrations, Settling Velocities, and Solids Flux for Various Tests

Test no.	C (mg/ℓ)	V (ft/hr)	$G = CV$ (lb/hr-ft^2)
1	12460	0.409	0.318
2	9930	0.817	0.506
3	7450	1.525	0.709
4	5220	3.281	1.068
5	3140	9.646	1.900
6	1580	13.710	1.351

The design mixed liquor flow to the final clarifier is 2530 gpm (gallons per minute), the MLSS is 2500 mg/ℓ, and the underflow concentration is 12,000 mg/ℓ. Determine the diameter of the final clarifier.

Solution

The settling curve showing the settling velocity versus solids concentration is shown in Figure 3.21. The flux curve showing the solids flux versus solids concentration is shown in Figure 3.22. A tangent to the curve drawn from $C_u = 12,000$ mg/ℓ gives a G_L value of 1.80 lb/hr-ft^2. Using a scale-up factor of 1.5 gives $G_L = 1.80/1.5$ or 1.20 lb/hr-ft^2. The rate at which the solids settle, M_t, is equal to $Q_0 C_0$, or $M_t = (2530$ gal/min$)(60$ min/hr$)(8.34$ lb/gal$)(2500/10^6)$ or 3165 lb/hr. From Eq. (3.42) the area required is M_t/G_L, or $A = (3165$ lb/hr$)/(1.20$ lb/hr-ft$^2)$ or 2638 ft^2. The required diameter is given by

Figure 3.21. Example Problem 3.3

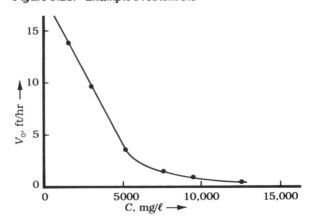

ENVIRONMENTAL ENGINEERING

$$D = \left[\frac{4}{\pi} (2638\,\text{ft}^2) \right]^{1/2}$$

$$= 58.0\,\text{ft} \quad \underline{\text{Use 60 ft for standard size.}}$$

Figure 3.22. Example Problem 3.3

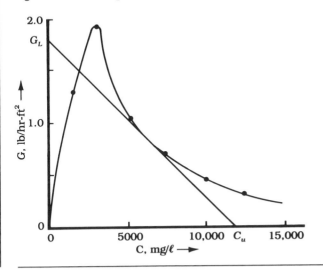

Example Problem 3.3 SI
Final Clarifier

Batch settling tests have been performed using an acclimated activated sludge to give the data in Table 3.4.

Table 3.4. Concentrations, Settling Velocities, and Solids Flux for Various Tests

Test no.	C (mg/ℓ)	V (m/h)	$G = CV$ (kg/h-m^2)
1	12460	0.125	1.558
2	9930	0.249	2.473
3	7450	0.465	3.464
4	5220	1.000	5.220
5	3140	2.941	9.235
6	1580	4.180	6.604

The design mixed liquor flow to the final clarifier is 160 ℓ/s, the MLSS is 2500 mg/ℓ, and the underflow concentration is 12,000 mg/ℓ. Determine the diameter of the final clarifier.

Solution

The settling curve showing the settling velocity versus

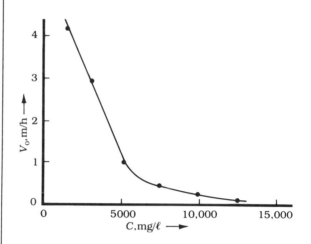

Figure 3.23. Example Problem 3.3 SI

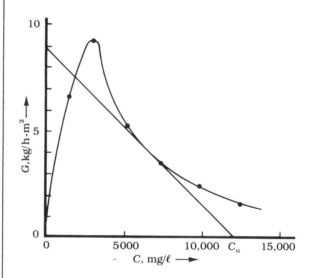

Figure 3.24. Example Problem 3.3 SI

solids concentration is shown in Figure 3.23. From the previous data, the flux curve showing the solids flux versus solids concentration is shown in Figure 3.24. A tangent to the curve drawn from $C_u = 12,000$ mg/ℓ gives a G_L value of 8.90 kg/h-m^2. Using a scale-up factor of 1.5 gives $G_L = 8.90/1.5$ or 5.93 kg/h-m^2. The rate at which the solids settle, M_t, is equal to Q_0C_0, or $M_t =$ (160 ℓ/s)(60 s/min)(60 min/h)(2.50 g/ℓ)(kg/1000 g) = 1440 kg/h. From Eq. (3.42) the area required is M_t/G_L, or $A = (1440 \text{ kg/h})(\text{h-m}^2/5.93 \text{ kg}) = 242.8 \text{ m}^2$. The required diameter is given by

$$D = \left[\frac{4}{\pi}(242.8 \text{ m}^2) \right]^{1/2}$$

$$= \underline{17.58 \text{ m}}$$

Actual settling basins are rectangular, square, or circular in plan area. A single rectangular basin will cost more than a circular basin of the same size; however, if numerous tanks are required the rectangular units can be constructed with common walls and be the most economical. Rectangular basins have a disadvantage if they have sprocket and chain drives for the sludge rakes because these will have more wear than the rotary type scraper mechanisms used for circular settling tanks.

Actual Sedimentation Basins

Figure 3.25 shows a schematic of the inlet to a rectangular tank if the unit is adjacent to a flocculation basin. This occurs frequently in the chemical coagulation of waters and wastewaters. The flocculation basin will be the same width as the settling tank but is usually not as deep. The two basins are separated by a wood baffle fence or a concrete wall with numerous ports. The inlet water will enter uniformly across the basin. This inlet arrangement closely approaches that of an ideal rectangular tank; the only difference is that the inlet zone does not extend down

Figure 3.25. Inlet and Outlet Details for a Rectangular Settling Tank with Orifice Flume Outlet Preceded by Flocculation

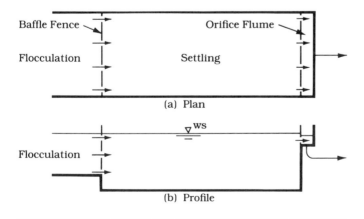

*Figure 3.26. Inlet and Outlet Details for a Rectangular
Settling Tank with Orifice Flume Inlet and Weir Channel
Outlet*

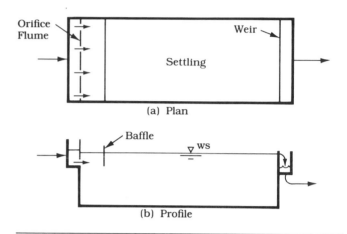

(a) Plan

(b) Profile

to the full depth of the settling tank but extends down to
the depth of the flocculator. If the rectangular basin does
not adjoin a flocculator, as shown in Figure 3.26, the inlet
water is distributed uniformly across the basin by a flume
with ports into the tank, and the inlet zone does not extend
down the full depth of the tank as depicted in an ideal tank.
However, a baffle in front of the flume will disperse the
water downward to give a deeper inlet zone. Figure 3.26
also shows one type of outlet for a rectangular basin. A weir
is used which spills into the effluent flume and extends
across the entire width of the basin. If, however, the water
is a chemically coagulated water, a weir should be avoided
because the turbulence will break up much of the fine floc
and result in poor filter performance. For chemically treated
waters it is best to have an orifice flume across the basin
width, as in Figure 3.25. An orifice flume does not have a
high degree of turbulence and will not break up fine floc. In
either of these outlet arrangements, the flow regime is conservative because the outlet zone does not extend down the
full depth as depicted in the ideal tank. Figure 3.27 shows
the details of a rectangular settling tank of the type used in
wastewater treatment. It has mechanical collection of the
sludge and also surface skimming. In water treatment surface skimmers are not required. Large rectangular tanks,
such as those used in water treatment plants having a capacity above about 2 MGD, frequently have length to width
ratios of 2:1. Each tank uses two rotary scrapers that are
designed for square basins.

Figure 3.27. Rectangular Settling Tank
Courtesy of Walker Process, Inc.

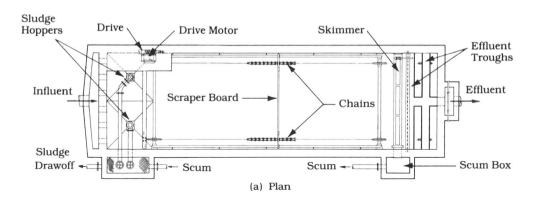

(a) Plan

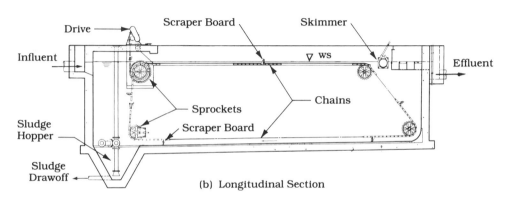

(b) Longitudinal Section

In circular tanks, the flow either enters the center of the tank (center feed) or the periphery of the tank (side feed). Figure 3.28 shows the inlets to center-feed tanks. If the tank is less than about 30 feet in diameter, the inlet pipe will enter through the wall and discharge into the baffle well, as shown in Figure 3.28 (b). Then, the flow enters in a downward direction. If the tank is greater than about 30 feet in diameter, the inlet pipe will run underneath the tank and discharge vertically in the center at the baffle well, as shown in Figure 3.28 (c). The outlets for both tanks consist of a weir channel around the periphery giving a uniform flow removal. The depth of the outlet zone is not as great as for an ideal basin; therefore, it is conservative. Figures 3.29 and 3.30 show the details of the center-feed circular clarifiers that are used in wastewater treatment. They have both mechanical sludge rakes and surface skim-

101

**Figure 3.28. Inlet and Outlet Details for Circular Tanks
(Center Feed)**

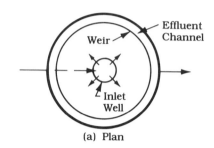

(a) Plan

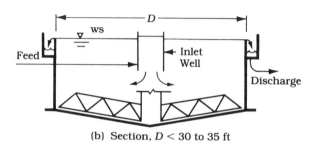

(b) Section, $D < 30$ to 35 ft

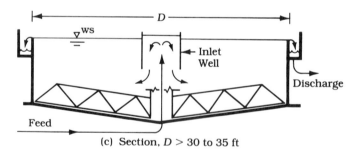

(c) Section, $D > 30$ to 35 ft

ming. The unit in Figure 3.29 has the influent pipe entering through the clarifier wall and extending to the baffle
well. Figure 3.30 shows a unit where the influent pipe runs
under the clarifier, rises at the center, and discharges into
the inlet well. Circular tanks used in water treatment are
similar to those used in wastewater treatment except that
surface skimmers are not required. The bottom of a circular
tank slopes to the center at a slope which is usually 1:12;
thus, it forms a flat inverted cone. In design, the volume of
the cone is not considered in the design volume, which is
taken as being the plan area times the depth of the water

Figure 3.29. Circular Settling tank (Center Feed by Pipe through Wall)
Courtesy of Infilco Degremont, Inc.

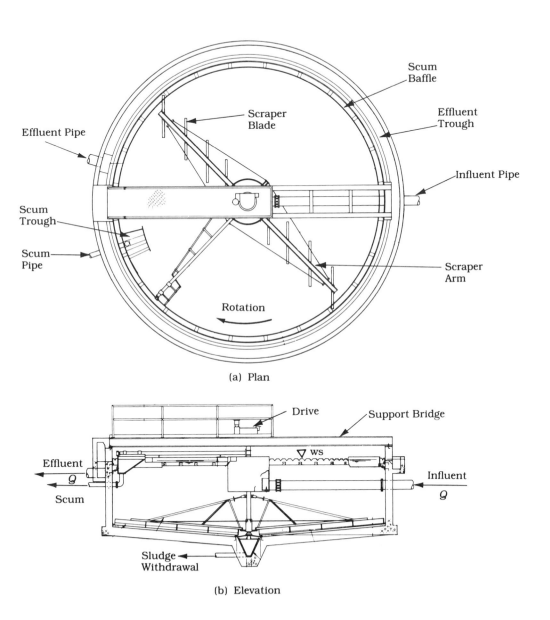

(a) Plan

(b) Elevation

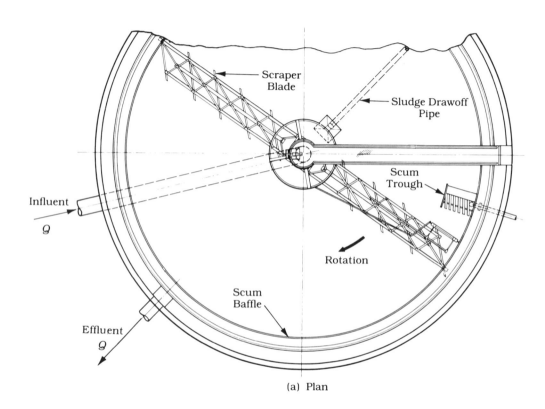

(a) Plan

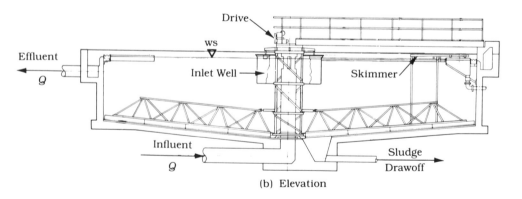

(b) Elevation

Figure 3.31. Inlet and Outlet Details for a Circular Tank (Peripheral Feed)

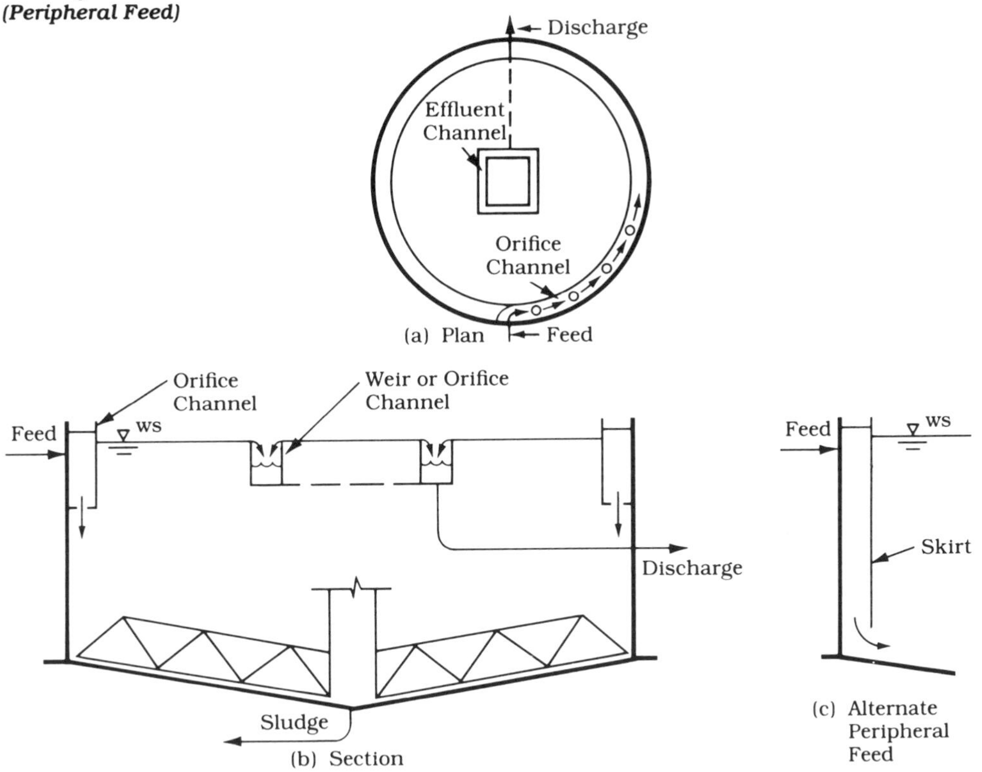

(a) Plan

(b) Section

(c) Alternate Peripheral Feed

at the side of the tank. The sludge is usually collected in a hopper near the center of the tank.

Figure 3.31 shows the inlet details for a periphery-feed tank. As the flow enters it is deflected so that it moves around the periphery in an orifice channel, as shown in Figure 3.31(a). From the channel the flow discharges through the orifices into the clarifier, as shown in Figure 3.31(b). Sometimes, instead of an orifice channel, there will simply be a skirt surrounding the inside of the tank and the liquid flows out under the skirt, as shown in Figure 3.31(c). Peripheral entry does not give as uniform a flow as the previously discussed tanks. The outlet consists of a weir channel in the center of the basin and, since the outlet zone does not extend the full depth of the tank, it is conservative.

Actual settling basins are affected by dead spaces in the basins, eddy currents, wind currents, and thermal currents. In the ideal settling basin all of the fluid elements pass through the basin at a time equal to the theoretical detention time, t, which is equal to V/Q. Actual basins, however, have most of the fluid elements passing through at a time shorter than the theoretical detention time, although some fluid elements take longer. Dead spaces and eddy currents have rotational flow and do very little sedimentation since the inflow to and the outflow from these spaces is very small. As a result, the net volume available for settling is reduced and the mean flow-through time for the fluid elements is decreased. Also, wind and thermal currents create flows that pass directly from the inlet to the outlet of the basin, which decreases the mean flow-through time. The magnitude of the effects of dead spaces, thermal currents, and so on, and the hydraulic characteristics of a basin, may be measured by using tracer studies. A slug of tracer is added to the influent and the tracer concentration is observed at the outlet, as shown in Figure 3.32. If there are dead spaces, the following relationship occurs (Camp, T. R., 1946 & 1952):

$$\frac{\text{Mean } t}{\text{Theoretical } t} < 1 \tag{3.44}$$

If there are no dead spaces, the relationship is

$$\frac{\text{Mean } t}{\text{Theoretical } t} = 1 \tag{3.45}$$

If short circuiting is occurring, the time relationship is

$$\frac{\text{Median } t}{\text{Mean } t} < 1 \tag{3.46}$$

Figure 3.32. Settling Basin Characteristics as Shown by Tracer Studies

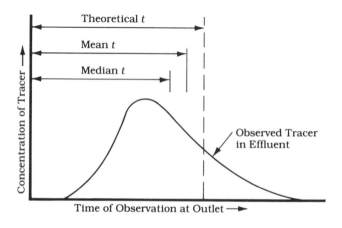

ENVIRONMENTAL ENGINEERING

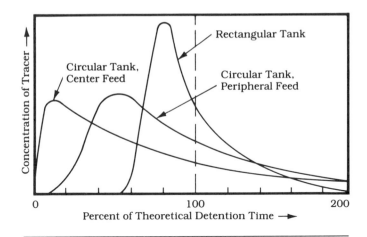

Figure 3.33. *Tracer Studies on Circular and Rectangular Settling Tanks*

If there is no short circuiting, the following results:

$$\text{Mean } t = \text{Median } t \qquad (3.47)$$

If a basin is unstable, the time-concentration plot cannot be reproduced in a series of tracer tests. Consequently, erratic basin performance can be expected.

Figure 3.33 shows the results of tracer studies on three types of settling basins. It can be seen that the rectangular basin approached the ideal more closely than the circular type. Of the circular type, the peripheral feed had better performance than the center-feed tank.

In water and wastewater engineering most of the suspensions are flocculant to a certain degree. Flocculant particles of the same initial size and density as discrete particles will intercept the sludge zone in a shorter time due to flocculation and more rapid settling. Therefore, if the ideal settling basin theory is applied to slightly flocculant particles, it will be conservative. Although there are differences between the ideal basin and actual basins, the ideal settling basin theory gives the most rational approach to design and reveals that the most important design parameters are: (1) the overflow rate or design settling velocity, (2) detention time, and (3) depth.

Sedimentation in Water Treatment Plants

In water treatment, sedimentation of both untreated waters (plain sedimentation) and chemically coagulated waters is practiced. If a water has a high turbidity due to silt, plain sedimentation may be used to reduce the turbidity. Plain sedimentation is frequently used for waters having consistent turbidities greater than 1000 mg/ℓ. Some rivers, such as the Mississippi, may have infrequent turbidities as high as 40,000 mg/ℓ. When plain sedimentation is used, the detention time may be as much as 30 days and, due to the extremely large volume, these basins are usually earthen and are constructed using dikes. In most cases, a water to be settled has been coagulated by the addition of chemicals such as those employed in rapid sand filtration plants and lime-soda softening plants.

The settling characteristics of the floc or precipitate depend upon the characteristics of the water, the coagulant used, and the degree of flocculation. The only method to accurately determine the settling velocities and the required overflow rates and detention times is to perform experimental settling tests. Generally, overflow rates of 500 to 800 gal/day-ft^2 and detention times of 2 to 8 hr are used for waters coagulated with alum or iron salts in rapid sand filtration plants, and weir loadings usually range from 12,000 to 22,000 gal/day-ft. The overflow rates, detention times, and weir loadings are based on the average daily flow. In lime-soda softening plants, the overflow rates are usually 500 to 1000 gal/day-ft^2, the detention times are 4 to 8 hr, and the weir loadings are usually from 22,000 to 26,000 gal/day-ft. Since turbulence due to weirs may shear apart fine floc, it is advisable to use orifice channels for the outlets or to use low weir loadings. Shearing of floc will result in poor filter performance and short filter runs.

Example Problem 3.4
Clarifier for Water
Treatment

A rectangular clarification basin is to be designed for a rapid sand filtration plant. The flow is 8 MGD, the overflow rate is 600 gal/day-ft^2, and the detention time is 6 hr. Two sludge scraper mechanisms for square tanks are to be used in tandem to give a rectangular tank with a length to width ratio of 2:1. Determine the dimensions of the basin.

Solution

The plan area required = $(8.0 \times 10^6$ gal/day)(day-ft^2/600 gal) or 13,333 ft^2. Since the length, L, is twice the width, W, $(2W)(W) = 13,333$ ft^2. From this, $W = 81.65$ ft; thus, the next standard size is 85 ft. Therefore, the plan dimensions of the basin are

Width = 85 ft, Length = 170 ft

ENVIRONMENTAL ENGINEERING

The actual overflow rate $= (8.0 \times 10^6$ gal/day$) \div (85$ ft$)(170$ ft$)$ or 553.6 gal/day-ft^2. Since the depth, H, is equal to the settling rate times the detention time, $H = (553.6$ gal/day-ft$^2)($ft$^3/7.48$ gal$)($day/24 hr$)(6$ hr$)$ or 18.5 ft. Thus,

Depth $= 18.50$ ft   Use 18 ft-6 in.

Sedimentation in Wastewater Treatment Plants

In conventional wastewater treatment plants primary sedimentation is used to remove as much settleable solids as possible from raw wastewaters. Secondary settling in activated sludge plants is employed to remove the MLSS and in trickling filter plants to remove any growths that may slough off the filters. As a result, good secondary settling produces a high-quality effluent low in suspended solids. In advanced or tertiary wastewater treatment plants sedimentation is used for coagulated wastewaters to remove flocculated suspended solids and/or chemical precipitates.

Primary Sedimentation

Recommended criteria for primary clarifiers treating municipal wastewaters are listed in Table 3.5.

Table 3.5. Overflow Rates and Depths for Primary Clarifiers

Type Treatment	Overflow Rate (gal/day-ft^2)		Depth (ft)
	Average	Peak	
Primary Settling Followed by Secondary Treatment	800–1200	2000–3000	10–12
Primary Settling with Waste Activated Sludge	600–800	1200–1500	12–15

Taken from EPA, *Suspended Solids Removal*, EPA Process Design Manual, January 1975.

The detention times based on the average daily flow are usually from about 45 min to 2 hr; however, the depths and overflow rates listed in Table 3.5 should control in design. Multiple tanks should be used when the flow exceeds 1.0 MGD. For plants having a capacity less than 1.0 MGD, peak weir loadings should not exceed 20,000 gal/day-ft. For plants having a capacity greater than 1.0 MGD, peak load-

Adapted from *Recommended Standards for Sewage Works* by the Upper Mississippi and Great Lakes Boards of Public Health Engineers, 1978.

Figure 3.34. Percent BOD₅ Removal versus Overflow Rate in Performance of Primary Clarifiers Treating Municipal Wastewaters.

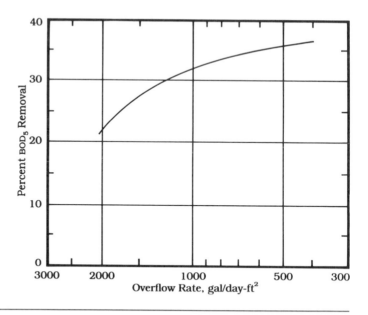

Adapted from *Water Supply and Sewage*, 4th ed., by E. W. Steel. Copyright © 1960 by McGraw-Hill Book Co., Inc. Reprinted by permission.

Figure 3.35. Percent BOD₅ and Suspended Solids Removal versus Detention Time in Performance of Primary Clarifiers Treating Municipal Wastewaters

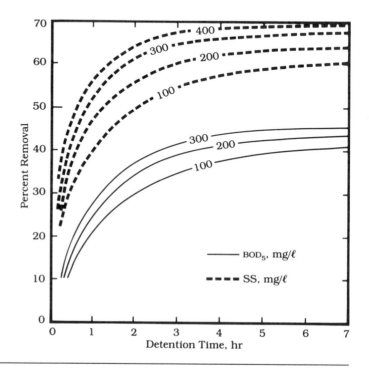

ings should not exceed 30,000 gal/day-ft. A surface skimmer and a baffle are necessary for primary clarifiers to remove scum from the water surface. Although individual tank performance varies, an estimate of the BOD_5 and suspended solids removal can be made using the detention time and overflow rate based on the average daily flow and Figures 3.34 and 3.35.

Example Problem 3.5
Primary Clarifier

A primary clarifier for a municipal wastewater treatment plant is to be designed for an average flow of 2.0 MGD. The state's regulatory agency criteria for primary clarifiers are: peak overflow rate = 2,200 gal/day-ft^2, average overflow rate = 900 gal/day-ft^2, and peak weir loading = 30,000 gal/day-ft. The ratio of the peak hourly flow to the average hourly flow = 2.75. Determine:

1. The clarifier diameter.
2. The weir loading.

Solution

The required area based on average flow = (2,000,000 gal/day) (day-ft^2/900 gal) = 2222.2 ft^2. The required area based on the peak flow = (2,000,000 gal/day)(2.75)(day-ft^2/2200 gal) = 2500 ft^2. The peak flow controls. Thus, 2500 ft^2 = $(\pi/4)D^2$ or D = 56.42 ft.

Use 60 ft for standard size.

The length of the peripheral weir = (π)(60 ft) or 188.50 ft. Thus, the peak weir loading = (2,000,000 gal/day)(2.75)/(188.50 ft) or

Peak weir loading = 29,177 gal/day-ft

Secondary Sedimentation

Recommended criteria for secondary clarifiers treating municipal wastewaters are shown in Table 3.6.

The detention time based on the average daily flow is usually from about 1.0 to 2.5 hr; however, the depths, overflow rates, and solids loadings listed in Table 3.6 should control in design. Multiple tanks should be used when the plant capacity exceeds 1.0 MGD, and peak weir loadings similar to those used for primary clarifiers should not be exceeded. Final clarifiers should be provided with skimmers and baffles to remove any floating materials.

Table 3.6 Overflow Rates, Solids Loadings, and Depths for Secondary Clarifiers

Type Treatment	Overflow Rate (gal/day-ft^2)		Solids Loading (lb/day-ft^2)		Depth (ft)
	Average	Peak	Average	Peak	
Activated Sludge (except Extended Aeration)	400–800	1000 –2000	20–30	50	12–15
Activated Sludge, Extended Aeration	200–400	800	20–30	50	12–15
Activated Sludge, Pure Oxygen	400–800	1000 –2000	25–35	50	12–15
Trickling Filters	400–600	1000 –2000	–	–	10–12

Taken from EPA, *Suspended Solids Removal*, EPA Process Design Manual, January 1975.

Final Clarifier for the Activated Sludge Process. Sludge removal is by suction withdrawal.
Courtesy of Dorr-Oliver, Inc.

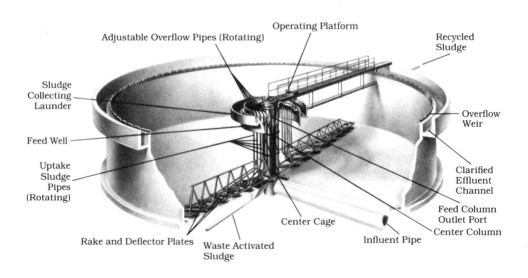

ENVIRONMENTAL ENGINEERING

Final clarifiers for the activated sludge process frequently use suction removal of the settled sludge. In these units, the scraper blades are mounted in pairs, with each pair forming a vee (V) in plan view. As the sludge rakes move, the sludge is collected in each vee and removed by a suction pipe mounted above each pair of blades. The sludge is discharged from the suction pipes into a collection well in the center of the unit. The sludge is removed from the collection well by gravity flow in the recycled sludge pipe. The water surface of the sludge in the collection well is slightly lower than the water surface in the clarifier, and this elevation difference is the head required to operate the suction pipes for sludge removal. Clarifiers having suction type sludge removal usually have a sludge blanket less than about 2 feet deep, and the residence time of the sludge in the clarifier is minimal.

Example Problem 3.6
Final Clarifier

A final clarifier is to be designed for an activated sludge treatment plant serving a municipality. The state's regulatory agency criteria for final clarifiers used for activated sludge are: peak overflow rate = 1400 gal/day-ft^2, average overflow rate = 600 gal/day-ft^2, peak solids loading = 50 lb/day-ft^2, and peak weir loading = 30,000 gal/day-ft. The flow to the aeration basin prior to junction with the recycle line = 3.0 MGD. The recycled sludge flow is 50 percent of the influent flow and is constant throughout the day. The MLSS is 3000 mg/ℓ, and the ratio of the peak hourly influent flow to the average hourly flow is 2.50. Determine:

1. The clarifier diameter.
2. The peak weir loading if peripheral effluent weirs are used.

Solution

The recycle is (0.50)(3.0 MGD) = 1.5 MGD. The average mixed liquor flow = 3.0 + 1.5 = 4.5 MGD. The peak mixed liquor flow = (2.50)(3.0) + 1.5 = 7.5 + 1.5 = 9.0 MGD. The area for clarification based on the average flow = (3,000,000 gal/day)(day-ft^2/600 gal) = 5000 ft^2. The area for clarification based on the peak flow = (7,500,000 gal/day)(day-ft^2/1400 gal) = 5357 ft^2. The peak solids flow = (9,000,000 gal/day)(8.34 lb/gal)(3000/10^6) = 225,180 lb/day. The area for solids loading = (225,180 lb/day)(day-ft^2/50 lb) = 4504 ft^2. Thus, the peak overflow rate controls. The diameter is given by $(\pi/4)(D^2) = 5357 \text{ ft}^2$ or

$$D = 82.59 \text{ ft} \qquad \underline{\text{Use 85.0 ft for standard size.}}$$

The peak weir loading = $(7{,}500{,}000 \text{ gal/day})/(\pi)(85.0 \text{ ft})$ or

$$\text{Peak weir loading} = \underline{28{,}086 \text{ gal/day-ft}}.$$

Chemical Treatment Sedimentation

The peak overflow rates used for coagulation in tertiary treatment or for coagulation of raw municipal wastewaters and secondary effluents depend mainly upon the type of coagulant employed. Recommended peak overflow rates are shown in Table 3.7.

Table 3.7. Peak Overflow Rates for Various Coagulants

Coagulant	Peak Overflow Rates (gal/day-ft^2)
Alum	500–600
Iron Salts	700–800
Lime	1400–1600

EPA, *Suspended Solids Removal*, EPA Process Design Manual, January 1975.

A detention time of at least two hours based on the average daily flow should be provided. Alum or iron salt coagulated wastewaters should not have a weir loading greater than 10,000 to 15,000 gal/day-ft based on the average daily flow. Lime coagulated wastewaters should not have an average weir loading greater than 20,000 to 30,000 gal/day-ft. The expected performance from chemical coagulation of wastewaters can best be determined from pilot plant studies. Chemical sludges produced from chemical coagulation may vary from 0.5 to more than 1.0 percent of the plant capacity and have solids concentrations from 1 to 15 percent, depending on the chemical used and basin efficiency.

Lime-feed or lime-sludge drawoff lines should be glass-lined or PVC pipe to facilitate cleaning of encrustations. Also, recycling of sludge from the bottom of the clarifiers to the rapid-mix tank should be provided to assist in coagulation. For lime-settling basins the mechanical sludge rakes should be the bottom scraper type and not the suction pickup type due to the dense sludge to be removed.

Inclined Settling Devices

Inclined settling devices include inclined-tube settlers and the Lamella separator, both of which have overflow rates much higher than conventional settling basins.

Inclined-Tube Settlers

Figure 3.36 (a) shows a module of inclined-tube settlers. Figure 3.37 shows the modules installed in a circular clarifier, whereas Figure 3.38 shows the modules installed in a rectangular clarifier. The water to be clarified passes upward through the tubes, and as settling occurs the solids are collected on the bottom of the tubes, as shown in Figure 3.36 (b). The tubes are inclined at an angle of 45 to 60 degrees, which is steep enough to cause the settled sludge to slide down the tubes. The sludge falls from the tubes to the bottom of the clarifier, where it is removed by the sludge rakes. The tube cross section may be of numerous geometric shapes; however, a square or rectangular cross section is the most common type.

Figure 3.36. Inclined-Tube Settler

Part (a) courtesy of Neptune Microfloc, Inc.

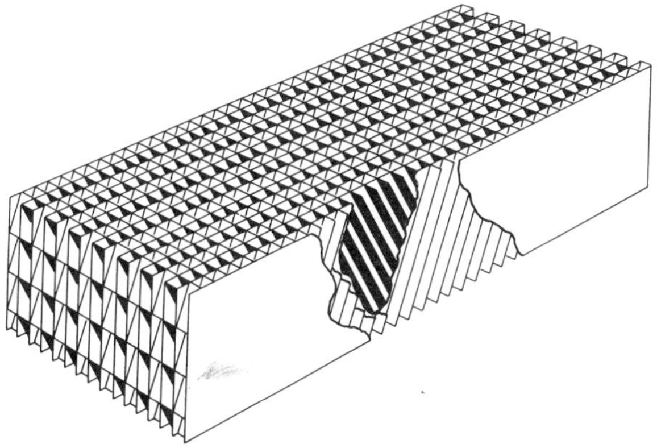

(a) Module of Inclined Tubes

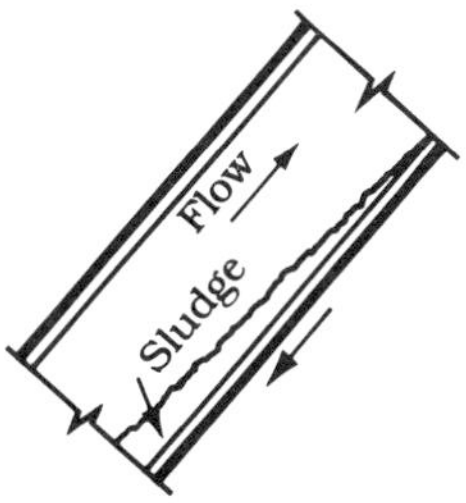

(b) Inclined-Tube Detail

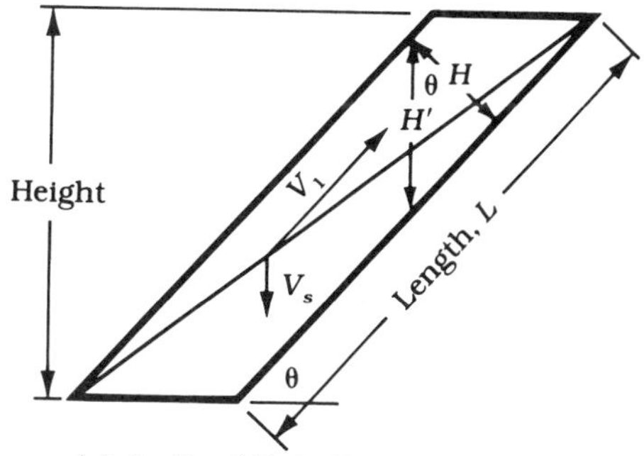

(c) Inclined-Tube Detail

Figure 3.37. Inclined-Tube Settlers in a Circular Clarifier

Figure 3.37. Inclined-Tube Settlers in a Circular Clarifier

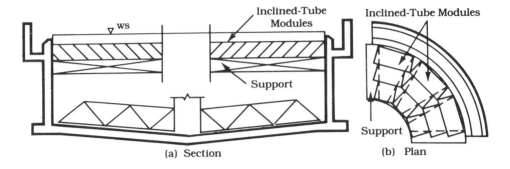

(a) Section   (b) Plan

Figure 3.38. Inclined-Tube Settlers in a Rectangular Clarifier

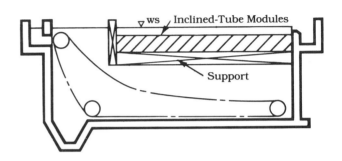

The advantage of a tube settler over a conventional tank can be illustrated using the ideal settling theory and a theoretical problem. If the flow to a conventional rectangular settling tank is 1.0 MGD, the detention is 2 hr, the overflow rate is 1,000 gal/day-ft^2, and the length to width ratio is 4:1, then the dimensions of the tank will be 63.25 feet long, 15.81 feet wide, and 11.1 feet deep. The overflow rate of 1,000 gal/day-ft^2 corresponds to a settling velocity of 5.55 ft/hr. According to the ideal settling basin theory, a settling particle must intersect the sludge zone before it reaches the outlet end of a settling chamber in order to be removed. Thus, if V_s is the settling velocity, H is the depth, V_1 is the horizontal velocity, and L is the chamber length, the critical trajectory of a settling particle is such that the following relationship holds true:

$$\frac{V_1}{L} = \frac{V_s}{H} \tag{3.48}$$

or

$$V_1 = V_s L/H \tag{3.49}$$

Applying Eq. (3.49) to the hypothetical problem gives

$$V_1 = \left(5.55\,\frac{\text{ft}}{\text{hr}}\right)\left(\frac{63.25\,\text{ft}}{11.10\,\text{ft}}\right) = 31.6\,\text{ft/hr}$$

Thus, the horizontal velocity is 31.6 ft/hr. For an inclined-tube settler, a particle must settle through the distance H', as shown in Figure 3.36 (c). Thus, for an inclined tube, Eq. (3.48) becomes

$$\frac{V_1}{L} = \frac{V_s}{H'} \qquad (3.50)$$

Assume, for the hypothetical problem, that modules of tube settlers are to be placed in the rectangular tank. The modules are 3.0 feet high and the tubes are 2 inches deep and inclined at a 45° angle. The value of H' is given by

$$H' = \frac{H}{\cos\theta} \qquad (3.51)$$

or

$$H' = \left(\frac{2\,\text{in.}}{\cos 45°}\right)\left(\frac{\text{ft}}{12\,\text{in.}}\right) = 0.236\,\text{ft}$$

The value of L is

$$L = \text{module height/sin}\,\theta \qquad (3.52)$$

or

$$L = 3.0\,\text{ft/sin}\,45° = 4.24\,\text{ft}$$

Now applying Eq. (3.50) to get the velocity through the tube gives the following:

$$V_1 = \left(\frac{L}{H'}\right)V_s = \left(\frac{4.24}{0.236}\right)\left(\frac{5.55\,\text{ft}}{\text{hr}}\right) = 99.7\,\text{ft/hr}$$

Thus, the velocity through the inclined-tube settler can be 99.7 ft/hr and still have the same degree of solids removal as the horizontal settling unit. This is 99.7/31.6, or 3.2 times as much flow as the conventional basin can accommodate. Thus, the advantages of the inclined-tube settlers are readily apparent.

Usually, the overflow rates used for inclined-tube settlers are from three to six times as great as those used for conventional settling tanks. Laminar flow is necessary for efficient settling since turbulent flow would scour the settled solids. Laminar conditions are made possible by the use of tubes with small hydraulic radii.

Figures 3.37 and 3.38 show modules of inclined-tube settlers installed in a new or existing circular or rectangular clarifier. In either type clarifier, a large portion of the plan area, usually 67 to 80 percent, is occupied by the tube modules. New settling tanks using inclined-tube settlers will have much less area requirements than conventional settling tanks; however, one of the most common uses of inclined tubes is to increase the capacity of existing clarifiers.

In water treatment where the water temperature is above 50°F, it has been reported that effluent turbidities from 1 to 7 JTU (Jackson Turbidity Unit) may be achieved, depending upon the overflow rate, if the raw water turbidity is less than 1000 JTU (Culp, G. L. & Culp, R. L., 1974). When the water temperature is below 40°F, expected turbidities are from 1 to 10 JTU if the raw water turbidity is less than 1000 JTU. Since the settling velocity is dependent upon the water temperature, better results are obtained with warm waters. The usual overflow rates based upon the area covered by the tube modules are from 3600 to 6000 gal/day-ft^2. The tube settlers are ideal for increasing the capacity of existing clarifiers.

In wastewater treatment, tube settlers have been successfully used in secondary settling for activated sludge and trickling filter plants and for settling of coagulated wastewaters. They are not well suited as primary clarifiers because biological growths develop within the tubes. In particular, they are useful in increasing the capacities of existing final clarifiers. When used for final clarifiers, the activated sludge mixed liquor or trickling filter effluent is discharged below the tube modules. Some sludge settles and the final effluent is clarified as it passes upward through the tube settlers. The sludge from the settlers falls from the modules to the bottom of the clarifier and is removed with the other sludge. Overflow rates as high as 4000 gal/day-ft^2 have been used, which is about five times the overflow rates normally used for conventional settling tanks. Some installations have had microbial slime build-ups inside the tube settlers. This can be minimized by installing an air grid underneath the modules and using air scouring to periodically clean off the growths.

Lamella Separator

The Lamella separator is similar to the inclined-tube settlers except that inclined plates are used to form the settling compartments and the sludge and water flow is cocurrent instead of countercurrent. The manufacturer recommends it only for use with coagulated waters and wastewaters.

To prevent short circuiting and basin instability, it is essential that the influent flow enter a sedimentation basin uniformly and also that the effluent flow leave uniformly. Figure 3.39 (a) shows the plan of a typical rectangular basin with an influent orifice flume and an effluent weir channel. Figure 3.39 (b) shows a cross-section through the influent flume, and Figure 3.39 (c) shows a cross section through the effluent channel. Figure 3.40 (a) shows a profile of the influent flume, whereas Figure 3.40 (b) shows a profile through the effluent channel. Although the profile shows the influent flume with circular orifices, square orifices are also used. The effluent weir may be a suppressed weir or a series of 90-degree V-notch weirs, as shown in Figure 3.40 (b) for a portion of the channel. If V-notch weirs are used, they are usually at about 8-in. centers. The elevation of the water surface in the effluent box, shown in Figure 3.40 (b), is set by the elevation of the water surface in the next downline treatment unit and the total head loss between the two water surfaces.

Inlet and Outlet Hydraulics

Figure 3.39. Inlet and Outlet Details for a Rectangular Tank

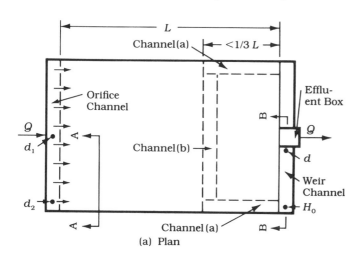

(a) Plan

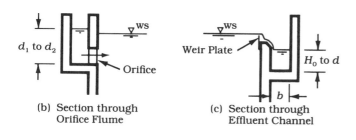

(b) Section through Orifice Flume

(c) Section through Effluent Channel

Figure 3.40. Sections through Orifice Flume and Effluent Channel

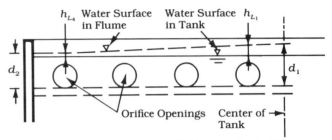

(a) Section A-A through Tank Showing Orifice Flume

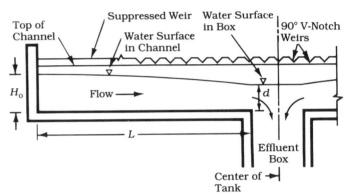

(b) Section B-B through Effluent Channel and Effluent Box

In the design of an orifice flume, as shown in Figures 3.39 (a) and (b) and 3.40 (a), good design practice requires that the discharge from the most distant orifice from the influent pipe be at least 90 percent of the discharge from the closest orifice. Usually, the friction and form losses in the flume are very small compared to the head losses through the orifices, which makes this criterion easy to satisfy. The discharge from an orifice is given by $Q = 0.6 \times A\sqrt{2gh}$, where Q = cfs, A = orifice area (ft^2), h = head loss (ft), and g = acceleration due to gravity. The velocity in the influent pipe and flume should be sufficient to maintain the suspended solids in suspension, yet not over 3 to 4 fps to avoid unnecessary head losses. If the influent is coagulated water, low velocities should be used to avoid shearing the floc.

In the design of a weir channel, as shown in Figures 3.39 (a) and (c) and 3.40 (b), the flow enters uniformly along the channel since the suppressed weir crest and the crests of the V-notch weirs are level. The discharge over a suppressed weir is given by $Q = 3.33LH^{3/2}$, where Q = cfs, L = weir length (ft), and H = head (ft). If 90-degree V-notch weirs are used, the discharge over each is given by $Q = 2.54\,LH^{5/2}$. Usually V-notch weirs are used for settling tanks used in wastewater treatment. The depth of flow in the effluent channel is given by the lateral spillway equation for a level channel (Camp, T. R., 1940):

$$H_0 = \sqrt{d^2 + \frac{2Q^2}{gb^2d} + \frac{fLQ^2}{12gb^2\bar{r}\bar{d}}} \qquad (3.53)$$

where

H_0 = upstream water depth, ft;
d = downstream water depth, ft;
Q = total discharge, cfs;
g = acceleration due to gravity, ft/sec^2;
b = channel width, ft;
f = Darcy friction factor, 0.03 to 0.12 for concrete;
L = channel length, ft;
$\bar{r}$ = mean hydraulic radius, ft;
$\bar{d}$ = mean depth, ft.

In using Eq. (3.53), a trial calculation ignoring friction may be made to determine the H_0 value without friction. The approximate H_0 value with friction may be estimated, since the friction head is from 6 to 16 percent of the water surface drawdown. Then the values of $\bar{r}$ and $\bar{d}$ are determined and a recalculation is made to determine H_0. Usually a freefall of 3 to 4 in. is allowed from the weir crest to the maximum water surface which occurs at H_0. The value of the downstream depth, d, is set by the elevation of the water surface in the effluent box. The minimum value of d is the critical depth, y_c, and this occurs when the water surface in the effluent box is low enough to cause $d = y_c$. The critical depth for a rectangular channel is given by $y_c = (q^2/g)^{1/3}$, where y_c = critical depth (ft), q = discharge per ft width of channel (that is, $q = Q/b$, cfs/ft), and g = acceleration due to gravity. If the water surface in the effluent box is low enough to cause a freefall (that is, the water surface is below the effluent channel), the value of d is y_c since critical depth occurs just before a freefall.* If the weir loading causes the required weir length to be greater than the tank width, the channel may be extended along the sides of the basin up to a length of one-third the basin length, as shown by

*Critical depth can be assumed at the freefall.

channel (a) in Figure 3.39 (a). If more channel is still required, a self-supporting channel may be constructed across the basin, as shown by channel (b) in Figure 3.39 (a). If a channel is constructed across the basin, it should have about 10- to 15-ft minimum clearance between it and the channel across the end of the tank. A head loss of $0.5\ V^2/2g$ occurs at 90-degree bends in a channel. If the effluent channel is an orifice channel, which is frequently the case in water treatment, it may still be designed using Eq. (3.53), and the discharge from the last orifice should be at least 90 percent of the discharge of the first orifice.

For a peripheral-feed circular clarifier with an influent orifice flume, the flume may be designed as described for a rectangular tank. The effluent channels may be designed using Eq. (3.53). For a center-feed circular clarifier, the entrance loss equals the velocity head. The peripheral effluent channel may be designed using Eq. (3.53) and assuming that the flow splits directly opposite the effluent box so that one-half the total flow enters each side of the effluent box.

Example Problem 3.7
Inlets and Outlets

A rectangular settling basin is 70 ft wide by 140 ft long and has a flow of 4.90 MGD (7.58 cfs). The inlet flume is an orifice channel with eight orifices that are 10 in. in diameter, each with an area of $0.545\ \text{ft}^2$. The difference in the elevations of the water surface at the influent pipe (that is, the center of the flume) and at the last orifice in the flume is 0.02 ft. This loss in head is due to friction and form losses in the flume. The effluent weir plate consists of 90-degree V-notch weirs spaced at 8-in. centers. The effluent channel is 160 ft in length and extends across the downstream end of the basin (70 ft) and 45 ft upstream along each side. The effluent channel is rectangular in cross section and is 1.75 ft in width. There is a 4-in. freefall between the V-notch weir crests and the maximum water depth. The friction factor is 0.08 and there is a freefall at the effluent box. Assume the friction head is approximately equal to 16 percent of the water surface drawdown. Determine:

1. The ratio of the flow from the last influent orifice to the flow from the influent orifice nearest to the influent pipe.
2. The head on the V-notch weirs.
3. The water depth in the effluent channel at the freefall into the effluent box and the depth at the upstream end of the channel.
4. The vertical distance from the crests of the V-notch weirs to the invert of the effluent channel.

Solution

The average flow through the orifices is 7.58 cfs/8 or 0.95 cfs per orifice. The head loss at the first orifice, h_{L_1}, is given by $0.95 = (0.6)(0.545)(2gh_{L_1})^{1/2}$, from which $h_{L_1} = 0.13$ ft. The head loss at the last orifice, h_{L_4}, is $h_{L_4} = h_{L_1} - 0.02$ or $h_{L_4} = 0.13 - 0.02 = 0.11$ ft. The flow from the last orifice is $Q_4 = (0.6)(0.545)[(2g)(0.11)]^{1/2}$, from which $Q_4 = 0.87$ cfs. The ratio of the flows through the orifices is $(Q_4/Q_1)(100$ percent$)$ or $(0.87/0.95)(100$ percent$)$ or 91.6 percent. The number of V-notch weirs is 160 ft/0.667 ft = 240. The flow for each is 7.58/240 or 0.0316 cfs. The head on the weirs is given by $0.0316 = 2.54\ H^{5/2}$ or $H = 0.17$ ft, which is 2.04 in. The discharge per ft width of channel is (7.58 cfs) (1/2)(1/1.75 ft) or 2.166 cfs/ft. Thus, the critical depth is $y_c = [(2.166)^2/32.2]^{1/3}$ or 0.53 ft. This is the water depth at the freefall into the effluent box. Using Eq. (3.53) and ignoring the friction term gives $H_0 = 0.91$ ft. The approximate value of H_0 with friction is given by $H_0 = 0.91 + 0.16\ (H_0 - 0.53)$ or $H_0 = 0.99$ ft. The mean depth is $\bar{d} = (0.99 + 0.53)1/2$ or 0.76 ft, and the computed mean hydraulic radius is 0.41 ft. Substituting for the terms in Eq. (3.53) gives

$$H_0 = \left[(0.53)^2 + \frac{(2)(3.79)^2}{(32.2)(1.75)^2(0.53)} + \frac{(0.08)(80)(3.79)^2}{(12)(32.2)(1.75)^2(0.41)(0.76)} \right]^{1/2}$$

From this $H_0 = 1.04$ ft. The flow at the 90-degree turn in the channel is (7.58 cfs)(1/2)(45)/(45 + 35) or 2.132 cfs. The approximate flow velocity at the turn is (2.132)/(1.75)(0.75) or 1.62 fps. Thus, the head loss is $(0.5)(1.62)^2/2g$ or 0.02 ft. The upstream channel depth is 1.04 + 0.02 or 1.06 ft. The difference in elevation from the V-notch weir crests to the invert of the effluent channel is 1.06 + 4/12 or 1.39 ft.

In the hydraulic design of a water or wastewater treatment plant, the total head losses between the various treatment units are determined for the maximum expected flow rate and are used in setting the water surface elevations in the various units. Take, for example, an activated sludge plant with a pump station between the preliminary treatment units and the primary clarifier and with a river as the receiving body of water. The total head losses are determined between the water surfaces of the primary clarifier and the aeration tank, the aeration tank and the final clarifier, the final clarifier and the chlorine contact tank, and the chlorine tank and the river. Then the maximum elevation of the river water surface is determined. To determine the minimum allowable water surface elevation of the chlorine contact tank, the head loss between the tank and the river is added to the elevation of the river water surface. To determine the minimum allowable water surface elevation of the final clarifier, the head loss between the clarifier and the contact tank is added to the elevation of the water surface in the chlorine contact tank. In a similar manner, the minimum allowable water surface elevations are determined for the aeration tank and the primary clarifier. The water surface elevations in the various units must be equal to or greater than the minimum allowable elevations in order to avoid backwater from the river. Usually the water surface elevations of the various units are above the minimum allowable elevations, but the drop in elevation between the units is only sufficient to accommodate the maximum expected flow.

References

American Water Works Association (AWWA). 1969. *Water Treatment Plant Design.* New York: AWWA.

American Water Works Association (AWWA). 1971. *Water Quality and Treatment.* New York: McGraw-Hill.

Camp, T. R. 1940. Lateral Spillway Channels. *Trans. ASCE* 105:606.

Camp, T. R. 1946. Sedimentation and the Design of Settling Tanks. *Trans. ASCE* 111:895.

Camp, T. R. 1952. Water Treatment. In *The Handbook of Applied Hydraulics.* 3rd ed., edited by C. V. Davis. New York: McGraw-Hill.

Coe, H. S., and Clevenger, G. H. 1916. Methods for Determining the Capacities of Slime-Settling Tanks. *Trans. Am. Inst. Mining Met. Engrs.* 55:356.

Conway, R. A., and Ross, R. D. 1980. *Handbook of Industrial Waste Disposal.* New York: Van Nostrand Reinhold.

Culp, G. L., and Conley, W. 1970. High-Rate Sedimentation with the Tube-Clarifier Concept. In *Advances in Water Quality Improvement by Physical and Chemical Processes,* edited by E. F. Gloyna and W. W. Eckenfelder. Austin, Tex.: University of Texas Press.

Culp, G. L., and Culp, R. L. 1974. *New Concepts in Water Purification.* New York: Van Nostrand Reinhold.

Culp, R. L., and Culp, G. L. 1978. *Handbook of Advanced Wastewater Treatment.* 2nd ed. New York: Van Nostrand Reinhold.

Dick, R. I. 1970. Role of Activated Sludge Final Settling Tanks. *Jour. SED* 96, SA2:423.

Eckenfelder, W. W. 1966. *Industrial Water Pollution Control.* New

York: McGraw-Hill.

Eckenfelder, W. W. 1980. *Principles of Water Quality Management.* Boston, Mass.: CBI Publishing.

Eckenfelder, W. W. 1970. *Water Quality Engineering for Practicing Engineers.* New York: Barnes and Noble.

Eckenfelder, W. W., and O'Connor, D. J. 1961. *Biological Waste Treatment.* London: Pergamon Press.

Environmental Protection Agency (EPA). 1973. *Phosphorus Removal.* EPA Process Design Manual, Washington, D.C.

Environmental Protection Agency (EPA). 1975. *Suspended Solids Removal.* EPA Process Design Manual, Washington, D.C.

Environmental Protection Agency (EPA). 1974. *Upgrading Existing Wastewater Treatment Plants.* EPA Process Design Manual, Washington, D.C.

Fair, G. M., Geyer, J. C., and Okun, D. A. 1968. *Waste and Wastewater Engineering. Vol. 2: Water Purification and Wastewater Treatment and Disposal.* New York: Wiley.

Fitch, E. B. 1956. Sedimentation Process Fundamentals. In *Biological Treatment of Sewage and Industrial Wastes.* Vol. 2, edited by J. McCabe and W. W. Eckenfelder, Jr. New York: Reinhold.

Great Lakes–Upper Mississippi Board of State Sanitary Engineers. 1978. Recommended Standards for Sewage Works. Ten state standards, Albany, N.Y.

Hazen, A. 1904. On Sedimentation. *Trans. ASCE* 53:45.

Ingersoll, A. C., McKee, J. E., and Brooks, N. H. 1956. Fundamental Concepts of Rectangular Settling Tanks. *Trans. ASCE* 121:1179.

Kynch, G. J. 1952. A Theory of Sedimentation. *Trans. Faraday Soc.* 48:161.

Lapple, C. E., and Shepherd, C. B. 1940. Calculation of Particle Trajectories. *Ind. and Eng. Chem.* 32, no. 5:605.

McCabe, W. L., and Smith, J. C. 1967. *Unit Operations of Chemical Engineering.* 2nd ed. New York: McGraw-Hill.

Metcalf and Eddy, Inc. 1979. *Wastewater Engineering, Treatment, Disposal and Reuse.* New York: McGraw-Hill.

O'Connor, D. J., and Eckenfelder, W. W. 1956. Evaluation of Laboratory Settling Data for Process Design. In *Biological Treatment of Sewage and Industrial Wastes.* Vol. 2, edited by J. McCabe and W. W. Eckenfelder, Jr. New York: Reinhold.

Perry, R. H., and Chilton, C. H. 1973. *Chemical Engineer's Handbook.* 5th ed. New York: McGraw-Hill.

Planz, P. 1969. Performance of (Activated Sludge) Secondary Sedimentation Basins. In *Proceedings of the Fourth International Conference.* Prague: International Association on Water Pollution Research.

Ramalho, R. S. 1977. *Introduction to Wastewater Treatment Processes.* New York: Academic Press.

Rich, L. G. 1971. *Unit Operations of Sanitary Engineering.* New York: Wiley.

Sanks, R. L. 1978. *Water Treatment Plant Design.* Ann Arbor, Mich.: Ann Arbor Science Publishers.

Schroeder, E. D. 1977. *Water and Wastewater Treatment.* New York: McGraw-Hill.

Steel, E. W., and McGhee, T. J. 1979. *Water Supply and Sewerage.* New York: McGraw-Hill.

Sundstrom, D. W., and Klei, H. E. 1979. *Wastewater Treatment.* Englewood Cliffs, N.J.: Prentice-Hall.

Talmage, W. P., and Fitch, E. B. 1955. Determining Thickener Unit Areas. *Ind. and Eng. Chem.* 47, no. 1:38.

Water Pollution Control Federation (WPCF). 1977. *Wastewater Treatment Plant Design.* WPCF Manual of Practice no. 8.

Weber, W. J. 1972. *Physicochemical Processes for Water Quality Control.* New York: Wiley-Interscience.

Problems

1. A circular primary clarifier is to be designed for a municipal wastewater having an average flow of 1.5 MGD; during the maximum flow during the day the flow rate is 2.6 times the average hourly flow. Pertinent data are: overflow rate based on average daily flow = 800 gal/day-ft^2, detention time based on average daily flow = 2.0 hr, overflow rate based on peak hourly flow = 75.0 gal/hr-ft^2, and detention time based on peak hourly flow = 0.5 hr. The minimum depth = 8.0 ft. Determine:

   a. The diameter and depth of the tank.

   b. The maximum depth of water over the effluent weirs if the weirs are 90-degree triangular type spaced at 8-in. centers. The flow over a 90-degree triangular weir is given by $Q = 2.48\,H^{5/2}$, where Q = discharge in cfs and H = head in ft.

   c. The weir depth if there is 1 in. of freeboard at maximum flow.

   d. The depth of water in the effluent channel at the effluent box if the effluent has a freefall into the box and the channel is 1 ft-3 in. wide. The depth will be the critical depth, y_c, for a rectangular channel. This is given by the equation $y_c = (q^2/g)^{1/3}$, where y_c = critical depth in ft, q = discharge per unit width of channel (cfs/ft), and g = acceleration due to gravity.

   e. The depth of water, H_0, in the effluent channel on the opposite side of the tank from the effluent box if the channel is level and the friction factor is 0.04. The relationship for the depth is $H_0 = [(y_c^2 + 2Q^2/gb^2 y_c) + (fLQ^2/12gb^2\,\bar{r}\,\bar{d})]^{1/2}$, where H_0 = depth in ft, y_c = critical depth at the freefall, Q = one-half the maximum flow to the tank, b = channel width in ft, f = friction factor, L = one-half the tank circumference, $\bar{r}$ = mean hydraulic radius, and $\bar{d}$ = mean depth.

   f. The depth of the effluent channel if the triangular weirs have a 4-in. freefall to the maximum water surface in the channel.

2. A batch settling test has been performed on an industrial wastewater having an initial suspended solids of 597 mg/ℓ to develop criteria for the design of a primary clarifier. The test column was 5 in. in diameter and 8

ft high, and sampling ports were located 2, 4, 6, and 8 ft from the water surface in the column. The suspended solids remaining after the various sampling times are given in Table 3.8.

Table 3.8

Depth (ft)	Time (min)				
	10	20	30	45	60
2	394	352	243	182	148
4	460	406	337	295	216
6	512	429	376	318	306
8	1018	1142	1208	1315	1405

The wastewater flow is 2.5 MGD. Determine:

a. The design overflow rate and detention time if 65 percent of the suspended solids are to be removed. Use a scale-up factor of 1.50 for the overflow rate and 1.75 for the detention time.

b. The diameter and depth if a circular clarifier is used.

3. A batch laboratory settling test has been performed on an acclimated activated sludge for an industrial wastewater treatment plant to obtain criteria to design the final clarifier. The mixed liquor sample had an MLSS concentration of 3000 mg/ℓ. A two-liter graduate cylinder having a graduated height of 65 centimeters was used as the settling column, and it was equipped with a rotary stirrer turning at four to six revolutions per hour to break up the sludge arching effect. The data giving the observed interface height versus time are shown in Table 3.9.

Table 3.9

Time (min)	Interface Height (cm)
0	65.0
2.5	56.2
5	50.1
7.5	40.0
10	33.2
12.5	25.4
15	20.8
20	17.2
25	15.6
30	15.0
35	14.0
40	13.7
45	13.0
50	13.0
55	12.0

The mixed liquor flow rate to the final clarifier is 2300 gpm, the design MLSS is 3000 mg/ℓ, and the return sludge is to have a suspended solids concentration of 12,000 mg/ℓ. Determine:
a. The area required for clarification.
b. The area required for thickening.
c. The diameter of the clarifier.

4. Assume that the primary clarifier in Problem 1 is for an activated sludge plant. Pertinent data are: water surface elevation in the effluent box is 3 in. below the invert of the effluent channel, the head loss between the water surfaces of the effluent box and the aeration tank = 0.65 ft, and the elevation of the water surface in the aeration tank is 320.87. Determine the elevation of the water surface in the effluent box and the water surface elevation in the primary clarifier at the maximum flow rate.

5. An activated sludge plant has a pump station between the preliminary treatment facilities and the primary clarifier. The flowsheet consists of a primary clarifier, aeration tank, final clarifier, and chlorine contact tank. The head losses in ft are as follows: head on primary weirs = 0.12, freefall to H_0 = 0.33, H_0 = 1.14, freefall to water surface in effluent box = 0.33, head loss between effluent box and aeration tank = 0.36, head loss between aeration tank and final clarifier = 0.39; head on final clarifier weirs = 0.12, freefall to H_0 = 0.33, H_0 = 1.14, freefall to water surface in effluent box = 0.33, head loss from final clarifier effluent box to chlorine contact tank = 0.32; head on weir in chlorine contact tank = 0.16, freefall to H_0 = 0.33, H_0 = 1.08, freefall to water surface in contact tank effluent box = 0.33, and head loss from contact tank effluent box to river water surface = 0.82. The maximum recorded river water surface elevation = 320.85. Determine the total head loss through the plant and the minimum allowable water surface elevations for the primary clarifier, the aeration tank, the final clarifier, and the chlorine contact tank.

6. Ferrous sulfate is used as a coagulant and lime as a coagulant aid at a water treatment plant. The specific gravity of the floc and adhered suspended material is 1.005 at a temperature of 10°C. Determine:
a. The overflow rate in gal/day-ft^2 if the flow is 7.0 MGD and the clarifier is 75 ft by 150 ft.
b. The settling rate in cm/sec for the overflow rate in (a).

c. The diameter in cm of the smallest floc particle that is of a size that is 100 percent removed if it is assumed that the particles are spherical in shape.

7. A batch settling test has been performed on an industrial wastewater having an initial suspended solids of 597 mg/ℓ to develop criteria for the design of a primary clarifier. The test column was 12.7 cm in diameter and 2.40 m high, and sampling ports were located 0.6, 1.2, 1.8, and 2.4 m from the water surface in the column. The suspended solids, mg/ℓ, remaining after the various sampling times are given in Table 3.10.

Table 3.10

Depth (m)	Time (min)				
	10	20	30	45	60
0.6	394	352	243	182	148
1.2	460	406	337	295	216
1.8	512	429	376	318	306
2.4	1018	1142	1208	1315	1405

The wastewater flow is 9500 m^3/d. Determine:
a. The design overflow rate and detention time if 65 percent of the suspended solids are to be removed. Use a scale-up factor of 1.50 for the overflow rate and 1.75 for the detention time.
b. The diameter and depth if a circular clarifier is used.

8. A primary clarifier is to be designed for a municipal wastewater treatment plant. The flow is 1.10 MGD and the peak flow during the day is 3.10 times the average hourly flow. Pertinent data are: peak overflow rate = 1800 gal/day-ft^2, detention time based on the peak flow = 30 min, maximum peak weir loading = 30,000 gal/day-ft, and minimum tank depth = 7 ft. Determine:
a. The clarifier diameter.
b. The clarifier depth.
c. The peak weir loading. Is it acceptable?

9. An existing rectangular clarifier at a water treatment plant is 18 ft wide and 72 ft long. It was designed for an overflow rate of 600 gal/day-ft^2 and is currently operating at the design capacity. Inclined-tube settlers are to be added and will occupy 80 percent of the plan area and will have a design overflow rate of 5000 gal/day-ft^2. Determine the present capacity in gal/day and the increased capacity in gal/day.

Chapter Four

FILTRATION

Filtration is a solid-liquid separation in which the liquid passes through a porous medium or other porous material to remove as much fine suspended solids as possible. It is used in water treatment to filter chemically coagulated and settled waters to produce a high-quality drinking water. In wastewater treatment it is used to filter (1) untreated secondary effluents, (2) chemically treated secondary effluents, and (3) chemically treated raw wastewaters. In all three of the uses in wastewater treatment, the objective is to produce a high-quality effluent.

Filters may be classified according to the types of media used as follows:

1. Single-medium filters: These have one type medium, usually sand or crushed anthracite coal.
2. Dual-media filters: These have two types of media, usually crushed anthracite and sand.
3. Multimedia filters: These have three types of media, usually crushed anthracite, sand, and garnet.

In water treatment all three types are used; however, the dual- and multimedia filters are becoming increasingly popular. In advanced and tertiary wastewater treatment, nearly all the filters are dual- or multimedia types.

The principles of filtration are the same for filters used in either water or wastewater treatment and, also, the filter structures, equipment, accessories, and method of operation are similar for both types of service.

Single-Medium Filters Rapid sand filters used in water treatment practice are usually of the gravity type and are commonly housed in open concrete basins. Figure 4.1 shows a perspective of three gravity filters and also a cutaway section showing the filter sand, the underlaying gravel, and the underdrain system. Figure 4.2 shows a schematic section that gives more details such as sand depth, gravel depth, the underdrain system, and so on. Figure 4.3 shows the filter layout, and Figure 4.4 shows the filter piping and underdrain system with all of the control valves and the rate of flow controllers. Although open gravity filters are the most common, pressure filters, shown in Figures 4.5 through 4.7, are also used. Frequently, crushed anthracite coal is employed instead of quartz sand. The sand bed is usually 24 to 30 inches in depth and the underlaying gravel is usually 15 to 24 inches thick.

Figure 4.1. Gravity Filters and Accessories
Courtesy of the National Lime Association.

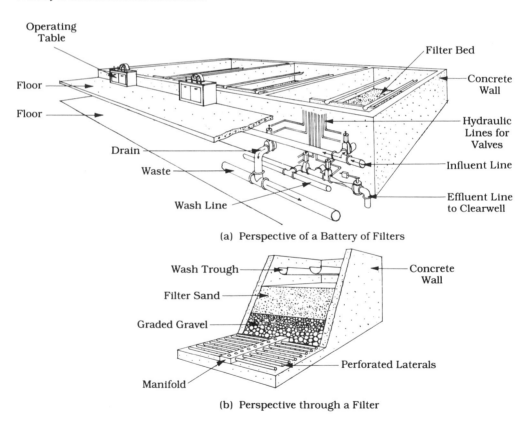

(a) Perspective of a Battery of Filters

(b) Perspective through a Filter

Figure 4.2. Section through a Rapid Sand Filter

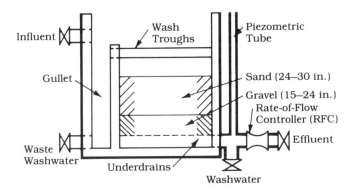

Figure 4.3. Wash System Layout

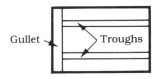

Figure 4.4. Filter Piping Layout

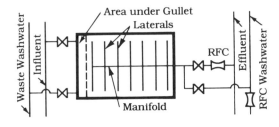

Figure 4.5. Profile through a Horizontal Pressure Filter

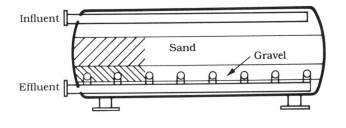

Figure 4.6. Cross Section
through a Horizontal
Pressure Filter

Figure 4.7. Cross Section through a Vertical Pressure Filter

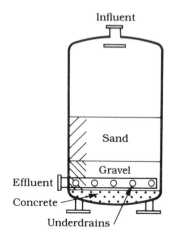

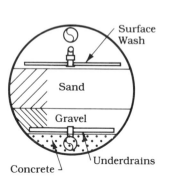

Figure 4.8. Schematic Section Showing a Filter during
Filtration

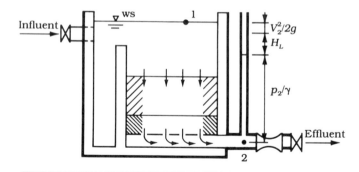

Figure 4.8 shows a schematic section through a filter during the filtration cycle. Approximately 3 to 4 feet of water is above the sand and the water passes downward through the media into the underdrain system. From there it flows through the rate of flow controller which controls the rate at which the water is filtered. During the filter run the valve positions, shown in Figure 4.4, are as follows: (1) Influent and effluent valves are open and (2) washwater and washwater waste valves are closed. The action of the sand in removing finely suspended floc smaller than the pore openings consists mainly of adhesion, flocculation, sedimentation, and straining. As the water moves downward through the pore spaces, some of the fine suspended floc collides with the sand surfaces and adheres to the sand particles. As the water passes through pore constrictions some of the fine floc is brought together, flocculation occurs, and the

Row of Rapid Sand Filters at a Water Treatment Plant

Pressure Filters at a Tertiary Wastewater Treatment Plant

Pipe Gallery at a Modern Water Treatment Plant

enlarged floc settles on the top of the sand particles immediately below the constrictions. Also, the buildup of floc that has been removed in the filter creates a straining action and some of the incoming floc is removed by straining. During a filter run the accumulated floc causes the pore spaces to become smaller, the velocities to increase, and some of the removed floc to be carried deeper within the filter bed. Straining may also occur at the surface of the filter if large particles of floc are strained and form a compressible cake that assists in filtering smaller particles. In summary, the removal of suspended solids is by surface removal at the top of the bed and depth removal within the filter bed itself (Baumann, E. R., & Oulman, C. S., 1970). For water treatment, depth removal is usually the most important in rapid sand filtration.

When a clean filter is put into operation, the floc accumulation is in the first few inches of the sand; however, as the time of operation increases, the floc accumulation extends deeper into the filter bed. The accumulated floc causes an increase in the hydraulic head loss. The magnitude of the head loss, H_L, is illustrated by writing Bernoulli's energy equation between point 1 on the water surface in Figure 4.8 and point 2 at the center of the effluent line. The resulting equation is

$$\frac{V_1^2}{2g} + \frac{p_1}{\gamma} + Z_1 = \frac{V_2^2}{2g} + \frac{p_2}{\gamma} + Z_2 + H_L \qquad (4.1)$$

where

V_1, V_2 = respective velocities;

p_1, p_2 = respective pressures;

Z_1, Z_2 = respective elevation heads;
γ = specific weight of water;
g = acceleration due to gravity;
H_L = head loss in feet.

If the relative pressure is used, p_1 = 0. Also, V_1 = 0 and the datum may be selected so that Z_2 = 0. Incorporating these values in Eq. (4.1) and rearranging gives

$$\frac{p_2}{\gamma} = Z_1 - \frac{V_2^2}{2g} - H_L \qquad (4.2)$$

Since it is common to have 4 ft of water over the sand, 2 ft-6 in. of sand, 1 ft-6 in. of gravel, and 1 ft depth for the underdrains, this gives a Z_1 value of $4.0 + 2.5 + 1.5 + 0.5$ or 8.5 ft. Also, pipes are usually designed for about 4 fps; therefore, Eq. (4.2) yields an expression for the gauge pressure head at point 2, which is

$$\frac{p_2}{\gamma} = 8.5\,\text{ft} - \frac{(4)^2}{2g} - H_L = 8.25\,\text{ft} - H_L \qquad (4.3)$$

When a clean filter is put on line the head loss, H_L, is about 0.5 to 1.5 ft depending upon the filtration rate but, as the filter run progresses, the head loss increases. From Eq. (4.3), it can be seen that when the head loss is 8.25 feet, the relative pressure at point 2 is zero. Any further increase in head loss creates a negative pressure which is undesirable. In practice, when the head loss reaches 6 to 8 ft, the filter is backwashed. The amount of washwater required is from 1 to 5 percent of the water filtered, with a typical value being 2 to 3 percent.

The shape of the curve showing the head loss as a function of filtrate volume for a particular filter is dependent upon the type of filter action, as depicted by Figures 4.9, 4.10, and 4.11 (Baumann, E. R., & Oulman, C. S., 1970). If the filter action is by surface removal of compressible solids, the head-loss curve will be exponential, as shown in Figure 4.9. This type curve is associated with fine grained media and low filtration rates. The filter action by micro-screens and diatomaceous earth filters is surface removal. If the filter action is depth removal of flocculated suspended solids, the head-loss curve will be rather flat, as shown in Figure 4.10. This type of curve is encountered with deep granular filters at relatively high filtration rates. If the filter action is by surface removal and depth removal of flocculated suspended solids, the head-loss curves will be as shown in Figure 4.11. At low filtration rates surface removal

Adapted from "Sand and Diatomite Filtration Practices" by E. R. Baumann and C. S. Oulman. In *Water Quality Improvement by Physical and Chemical Processes*, ed. E. F. Gloyna and W. W. Eckenfelder, Jr. Copyright © 1970 by the University of Texas Press. Reprinted by permission.

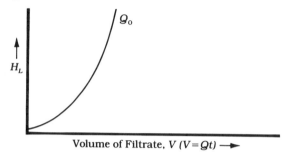

Figure 4.9. Head-Loss Curve for Surface Removal of Compressible Solids

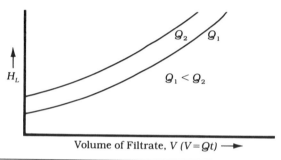

Figure 4.10. Head-Loss Curve for Depth Removal of Flocculent Solids

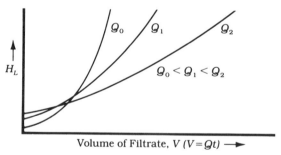

Figure 4.11. Head-Loss Curves for Combined Surface and Depth Removal of Flocculent Solids

is predominant and the curve is similar to that in Figure 4.9. At higher filtration rates the solids penetrate deep within the filter. The principal action is depth removal and the head-loss curves are similar to those in Figure 4.10. In rapid sand filtration at the usual filtration rates, depth removal is usually the main filter action and a flat head-loss curve results.

Surface removal will result when the feed water contains large floc and high turbidity. The top pore spaces will rapidly become clogged, resulting in relatively short filter runs. Depth removal will result when the feed water con-

Filter Console Table at a Water Treatment Plant. Gauges include rate of filtration, head-loss indicator, washwater flow rate, and gauges for the pneumatic control system for all valves. The levers are for controlling all valves.

Filter Floor at a Water Treatment Plant. Note consoles for all filters.

tains small floc and low turbidity resulting in deeper penetration within the filter bed and relatively long filter runs.

Filter sands are characterized by the *effective size* and the *uniformity coefficient*. The effective size is equal to the sieve size in millimeters that will pass 10 percent (by weight) of the sand. The uniformity coefficient is equal to the sieve size passing 60 percent of the sand divided by that size passing 10 percent. Most rapid sand filters have sands with an effective size of 0.35 to 0.50 mm; however, some have sand with an effective size as high as 0.70 mm. The uniformity coefficient, which is a measure of gradation, is generally not less than 1.3 or more than 1.7.

The gravel serves to support the sand bed and it is usually placed in several layers. The total depth may be from 6 to 24 inches; however, 18 inches is typical. The size of the top layer of gravel depends upon the sand size, whereas the size of the bottom gravel depends upon the type of underdrain system. Usually five layers are used and the gravel grades from less than $\frac{1}{16}$ inch at the top to 1 to 2 inches at the bottom.

The underdrain system serves to collect the filtered water from the bed during the filtration cycle. During the washing cycle, it serves to distribute the backwash water. The rate of flow of the backwash governs the hydraulic design of the filter since it is several times greater than the filtration rate. Underdrain systems are basically two types: (1) a manifold with perforated lateral pipes or (2) a false bottom. In the manifold and perforated pipe system, the perforations are directed downward so that the high velocity of the backwash water is dissipated by the filter bottom and the surrounding gravel. The false bottom consists of a perforated bottom with a waterway underneath that removes the filtered water and permits the backwash to enter the filter bed.

The standard rate of filtration has been considered as 2 gal/min-ft^2 of filter bed area since this is a common rate at which the first rapid sand filters were operated. Present coagulation and sedimentation practice allows the use of higher filtration rates. Frequently, plants are rated at 2 gal/min-ft^2, but provisions are made for operation at rates up to 5 gal/min-ft^2. Although most filters are operated at a constant filtration rate, a declining rate of filtration is sometimes employed. In this type of operation the rate of filtration is decreased as the filter run progresses and the degree of clogging increases. Frequently, this results in longer filtration runs and better effluent quality. It is limited to medium to large plants because the filters must be staggered in the degree of clogging to permit a constant rate in the total water production from the plant.

Gravity filters may employ a single filter, as shown in Figure 4.12, or a double filter within each concrete basin, as shown in Figure 4.13. The single filter arrangement is the most popular and the length to width ratio varies from 1:1.5 to 1:2. The double filter is built almost square with the length to width ratio of 1:1. The largest filters that have been built are about 2100 ft^2 in plan area.

If the filter operation is optimum, the maximum allowable head loss, H_a, occurs simultaneously with the maximum allowable effluent turbidity, C_a, as shown in Figure

Figure 4.12.   Layout and Underdrains for a Single Filter

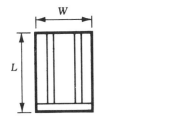 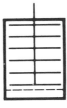

Figure 4.13.   Layout and Underdrains for a Double Filter

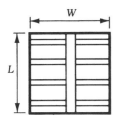 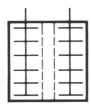

Figure 4.14.   Head Loss and Effluent Turbidity versus Filter Run Time for Optimum Performance

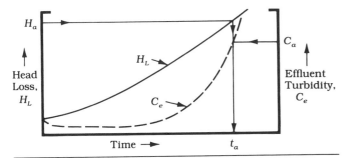

4.14. In many cases, this does not occur and the termination of the filter run is controlled by whichever occurs first—that is, H_a or C_a. The length of the filter run will depend upon the quality of the feed water, and filter runs may range from less than a day to several days. Backwashing removes the floc that has accumulated upon and within the filter bed. In modern filtration practice, a surface wash or air scour system is considered essential for high filter performance. To wash a filter, the influent valve, as shown in Figure 4.4, is closed and, when the water has filtered down below the wash troughs, the effluent valve is closed. The waste washwater valve is opened and the surface wash is started at a rate of about 0.5 gal/min-ft². After about 1 minute of surface washing, the backwash flow is initiated by gradually opening the washwater influent valve, and the bed is allowed to expand to the desired height, as shown in Figure 4.15 (a). The backwash flow should be from 15 to 20 gal/min-ft² and the bed expansion should be from 20 to 50 percent to suspend the bottom sand grains. The optimum backwash flow will depend upon the washwater temperature because a cold washwater will expand the bed more than a warm one. The backwashing is continued until the

---

Figure 4.15. Schematic Showing Filter during Backwashing

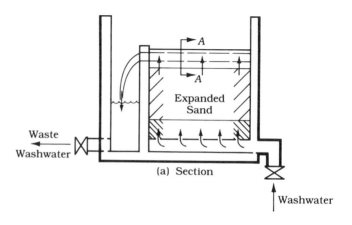

(a) Section

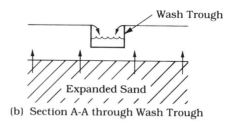

(b) Section A-A through Wash Trough

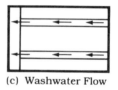

(c) Washwater Flow

Backwashing of Rapid Sand Filters at a Lime Softening Water Treatment Plant.

waste washwater appears relatively clear and the surface wash is terminated at 1 to 2 minutes prior to the end of the backwash. The surface wash system is misnamed because it not only washes the filter surface prior to backwashing but also scours the expanded bed during the backwash; thus, it really should be called an auxiliary scour system. Usually 3 to 10 minutes of backwash flow is required for complete washing and the total off-line time will be up to 20 minutes. After backwashing, the initial water filtered should be wasted until the effluent turbidity is acceptable. The washwater may be supplied by a pump which pumps directly from the clearwell or by an elevated storage tank at the plant site. Usually, the volume of washwater is about 1 to 5 percent of the water filtered, with 2 to 3 percent being typical.

Hydraulics of Filtration

The head loss through a clean bed of porous media having a relatively uniform diameter, as given by the Carman-Kozeny equation, may be developed starting with the Darcy-Weisbach equation, which is

$$h_L = f \frac{L V^2}{D_c 2g} \tag{4.4}$$

where

h_L = frictional head loss;
f = dimensionless friction factor;

L = conduit length;
D_c = conduit diameter;
V = mean conduit velocity;
g = acceleration due to gravity.

The flow channels in a porous bed are irregular; thus the diameter, D_c, may be replaced by the term $4r$, where r is the hydraulic radius for a conduit diameter. If D is the bed depth, substituting this and $D_c = 4r$ into Eq. (4.4) gives

$$h_L = f \frac{DV^2}{8rg}$$ (4.5)

If there are n particles in the bed and the particle volume is v_p, the total volume of particles is nv_p. If the porosity is ε, the total bed volume is $nv_p/(1 - \varepsilon)$. The total channel volume is the void space or $\varepsilon nv_p/(1 - \varepsilon)$. If the wetted surface is considered as the total surface of the particles, it is ns_p, where s_p is the surface area per particle. The hydraulic radius, r, is the total channel volume divided by the wetted surface, or

$$r = \left(\frac{\varepsilon}{1 - \varepsilon}\right) \frac{v_p}{s_p}$$ (4.6)

For spherical particles,

$$\frac{v_p}{s_p} = \frac{\pi d^3/6}{\pi d^2} = \frac{d}{6}$$ (4.7)

For irregularly shaped particles,

$$\frac{v_p}{s_p} = \phi \frac{d}{6}$$ (4.8)

where ϕ is the shape factor. Shape factors are 1 for spheres, 0.73 for crushed coal and angular sand, 0.82 for rounded sand, and 0.75 for average sand (Carman, P. C., 1937). The approach velocity, V_a, is equal to the flow, Q, divided by the filter surface, A. Thus, the velocity through the pore spaces is

$$V = \frac{V_a}{\varepsilon}$$ (4.9)

Substituting Eqs. (4.6), (4.8), and (4.9) into Eq. (4.5) gives

$$h_L = f' \frac{D}{\phi d} \frac{1 - \varepsilon}{\varepsilon^3} \frac{V_a^2}{g}$$ (4.10)

which is the Carman-Kozeny relationship where f' = dimensionless friction factor (Carman, P. C., 1937; Kozeny, G., 1927). The friction factor f' is given by (Ergun, S., 1952)

$$f' = 150 \left(\frac{1-\varepsilon}{N_{Re}}\right) + 1.75 \qquad (4.11)$$

The Reynolds number, N_{Re}, is defined as

$$N_{Re} = \frac{\phi d V_a}{\nu} = \frac{\phi \rho d V_a}{\mu} \qquad (4.12)$$

where ρ = mass density, μ = dynamic viscosity, and ν = kinematic viscosity. The head loss through a clean bed of porous media having a relatively uniform diameter is also given by the Rose equation (Rose, H. E., 1949), which is

$$h_L = \frac{1.067}{\phi} \frac{C_D}{g} D \frac{V_a^2}{\varepsilon^4} \frac{1}{d} \qquad (4.13)$$

where C_D is the coefficient of drag. The coefficient of drag for $N_{Re} < 1$ is given by

$$C_D = \frac{24}{N_{Re}} \qquad (4.14)$$

For $N_{Re} > 1$ but $< 10^4$, the drag coefficient is (Fair, G. M., Geyer, J. C., & Okun, D. A., 1968)

$$C_D = \frac{24}{N_{Re}} + \frac{3}{\sqrt{N_{Re}}} + 0.34 \qquad (4.15)$$

For beds with varying particle size, the Rose equation is

$$h_L = \frac{1.067}{\phi} \frac{C_D}{g} D \frac{V_a^2}{\varepsilon^4} \Sigma \frac{x}{d} \qquad (4.16)$$

where x is the weight fractions for particle sizes, d. For stratified beds with uniform porosity, the Rose equation is

$$h_L = \frac{1.067}{\phi} \frac{D}{g} \frac{V_a^2}{\varepsilon^4} \Sigma \frac{C_D x}{d} \qquad (4.17)$$

In both equations, the summation terms may be obtained from computations using sieve analyses.

The head loss through the underdrain system is usually negligible compared to the head loss through the bed. Although the Carman-Kozeny and Rose equations are limited to clean filter beds, they also illustrate the relationship between the head loss and the degree of clogging. As a filter bed clogs, the effective porosity, ε, decreases, which results in an increase in head loss, h_L.

Example Problem 4.1
Head Loss through a
Filter Bed

A rapid sand filter has a sand bed 24 in. in depth. Pertinent data are: specific gravity of the sand = 2.65, shape factor $(\phi) = 0.82$, porosity $(\varepsilon) = 0.45$, filtration rate = 2.5 gpm/ft^2, and operating temperature = 50°F (10°C). The sieve analysis of the sand is shown in columns (1) and (2) of Table 4.1. Determine the head loss for the clean filter bed using the Rose equation for a stratified bed.

Solution

The reduced data are given in Table 4.1.

Table 4.1. Stratified Data from Sieve Analysis

(1) Sieve Size	(2) Weight Retained (%)	(3) d (ft)	(4) N_{Re}	(5) C_D	(6) $C_D x/d$ (ft^{-1})
14–20	0.87	0.003283	1.065	22.54	59.7
20–28	8.63	0.002333	0.757	31.70	1172.6
28–32	26.30	0.001779	0.577	41.59	6148.5
32–35	30.10	0.001500	0.486	49.38	9908.2
35–42	20.64	0.001258	0.408	58.82	9650.6
42–48	7.09	0.001058	0.343	69.97	4688.9
48–60	3.19	0.000888	0.288	83.33	2995.2
60–65	2.16	0.000746	0.242	99.17	2872.2
65–100	1.02	0.000583	0.189	126.98	2220.5

$$\Sigma C_D x/d = 39,716.4 \text{ ft}^{-1}$$

The mean diameter, d, column (3), is the average of the sieve size openings. The Reynolds number, $N_{Re} = \phi d V_a/\nu$. The velocity, V_a, and the viscosity, ν, are

$$V_a = (2.5 \text{ gal/min})(\text{ft}^3/7.48 \text{ gal})(\text{min}/60 \text{ sec})$$
$$= 0.00557 \text{ ft/sec}$$

$$\nu = 1.3101 \text{ centistokes at } 50°F$$
$$= (1.3101)(1.075 \times 10^{-5})$$
$$= 1.4084 \times 10^{-5} \text{ ft}^2/\text{sec}$$

The N_{Re} for sieve size 14–20 is

$$N_{Re} = (0.82)(0.003283 \text{ ft})(0.00557 \text{ ft/sec})$$
$$\div (0.000014084 \text{ ft}^2/\text{sec})$$

$$= 1.065$$

The N_{Re} values for the other sieve sizes are computed the same way and the N_{Re} values are shown in column (4). The drag coefficient $C_D = 24/N_{Re}$ and the C_D values are shown in column (5). For the first sieve size, $N_{Re} = 1.065$, but since it is approximatley 1, the C_D value may still be computed using $C_D = 24/N_{Re}$. The $C_D x/d$ values are shown in column (6) and the $\Sigma C_D x/d = 39{,}716.4 \text{ ft}^{-1}$. The Rose equation is

$$h_L = \frac{1.067}{\phi} \frac{D}{g} \frac{V_a^2}{\varepsilon^4} \Sigma \frac{C_D x}{d}$$

$$h_L = \frac{1.067}{0.82} \left| \frac{2.0 \text{ ft}}{} \right| \frac{\sec^2}{32.17 \text{ ft}} \left| \frac{(0.00557 \text{ ft})^2}{\sec^2} \right|$$

$$\times \frac{1}{(0.45)^4} \left| \frac{39{,}716.4}{\text{ft}} \right.$$

$$= \underline{2.96 \text{ ft}}$$

Hydraulics of Expanded Beds

The hydraulics of expanded beds may be analyzed for both uniform and stratified beds. For a uniform bed of depth, D, backwashing will expand the bed to expanded depth D_e. During backwashing, the frictional resistance of the particles equals the head loss of the liquid expanding the bed (Fair, G. M., Geyer, J. C., & Okun, D. A., 1968; Fair, G. M., & Hatch, L. P., 1933). Thus,

$$h_L \rho g = (\rho_s - \rho)(1 - \varepsilon)(D_e)g \qquad (4.18)$$

where ρ_s = mass density of the particles and ε_e = porosity of the expanded bed. Canceling g and rearranging Eq. (4.18) gives,

$$h_L = \left(\frac{\rho_s - \rho}{\rho} \right) (1 - \varepsilon_e)(D_e) \qquad (4.19)$$

The value of ε_e can be determined from,

$$\varepsilon_e = \left(\frac{V_b}{V_s} \right)^{0.22} \qquad (4.20)$$

where V_b = upflow velocity of the backwash water and V_s = settling velocity of the particles. Consequently, a bed of uniform particles will expand when

$$V_b = V_s \varepsilon_e^{4.5} \tag{4.21}$$

The volume of the sand in an unexpanded bed will equal the volume of sand in an expanded bed or, stated mathematically,

$$(1 - \varepsilon)AD = (1 - \varepsilon_e) A D_e \tag{4.22}$$

where A = bed area. Rearranging gives

$$D_e = \left(\frac{1 - \varepsilon}{1 - \varepsilon_e}\right) D \tag{4.23}$$

Substituting Eq. (4.20) for ε_e in Eq. (4.23) gives,

$$D_e = \left[\frac{1 - \varepsilon}{1 - (V_b/V_s)^{0.22}}\right] D \tag{4.24}$$

For stratified beds, the smaller particles in the upper layers expand first. Once V_b is sufficient to fluidize the largest particles, the entire bed will be expanded. The expansion of the bed is represented by a modification of Eq. (4.23):

$$D_e = (1 - \varepsilon)D\Sigma\frac{x}{1 - \varepsilon_e} \tag{4.25}$$

where x is the weight fraction of particles with an expanded porosity, ε_e.

Example Problem 4.2
Filter Backwashing

A rapid sand filter having the same sand analysis as in Example Problem 4.1 is to be backwashed. Determine:

1. The backwash velocity required to expand the bed.
2. The backwash flow required to expand the bed.
3. The head loss at the beginning of the backwash.
4. The depth of the expanded sand bed.

Solution

The reduced data are given in Table 4.2.

Table 4.2. Reduced Data from Sieve Analysis

(1) Sieve Size	(2) Weight Retained (%)	(3) d (ft)	(4) V_s (ft/sec)	(5) ε_e	(6) $\dfrac{x}{1-\varepsilon_e}$
14–20	0.87	0.003283	0.498	0.454	0.016
20–28	8.63	0.002333	0.334	0.494	0.171
28–32	26.30	0.001779	0.245	0.529	0.558
32–35	30.10	0.001500	0.202	0.552	0.672
35–42	20.64	0.001258	0.164	0.578	0.489
42–48	7.09	0.001058	0.136	0.602	0.178
48–60	3.19	0.000888	0.111	0.630	0.086
60–65	2.16	0.000746	0.091	0.657	0.063
65–100	1.02	0.000583	0.068	0.701	0.034

$$\Sigma \frac{x}{1-\varepsilon_e} = 2.267$$

The backwash velocity to expand the bed requires the settling velocity of the largest particles. The settling velocity, V_s, is given by

$$V_s = \left[\frac{4g}{3C_D}(S_s - 1)d\right]^{1/2}$$

The drag coefficient, C_D, for the transition range that applies to this problem is given by

$$C_D = \frac{24}{N_{Re}} + \frac{3}{\sqrt{N_{Re}}} + 0.34$$

The N_{Re} value is

$$N_{Re} = \phi \frac{dV_s}{\nu}$$

For the first sieve size, $d = (0.003283\ \text{ft})(30.48\ \text{cm/ft})$ or 0.10 cm. From Figure 3.2, a particle 0.10 cm in diameter and having a specific gravity of 2.65 has a settling velocity, V_s, of about 14 cm/sec. The viscosity, ν, is 1.3101 centistokes $= 1.3101 \times 10^{-2}\ \text{cm}^2/\text{sec}$. The approximate N_{Re} is

$$N_{Re} = \frac{0.82 \mid 0.10 \mid 14 \mid 10^2}{\mid 1.3101 \mid} = 87.6$$

$$C_D = \frac{24}{87.6} + \frac{3}{\sqrt{87.6}} + 0.34 = 0.935$$

$$V_s = \left[\frac{4}{3} \left| \frac{32.17\,\text{ft}}{\text{sec}^2} \right| \frac{1}{0.935} \right| 2.65 - 1$$

$$\times \left. \frac{0.003283\,\text{ft}}{1} \right]^{1/2}$$

$$= 0.498\,\text{ft/sec}$$

$$V_b = V_s \varepsilon^{4.5}$$

$$= (0.498\,\text{ft/sec})(0.45)^{4.5}$$

$$= \underline{0.0137\,\text{ft/sec} \ = \ 0.822\,\text{ft/min}}$$

Rate $= (0.822\,\text{ft/min})(7.48\,\text{gal/ft}^3)$

$$= \underline{6.15\,\text{gpm/ft}^2}$$

To determine the head loss at the beginning of the back-wash, $1 - \varepsilon$ and D are substituted for $1 - \varepsilon_e$ and D_e, respectively, in Eq. (4.19) to give

$$h_L = \left(\frac{\rho_s - \rho}{\rho} \right)(1 - \varepsilon)(D)$$

$$= (S_s - 1)(1 - \varepsilon)(D)$$

$$= (2.65 - 1)(1 - 0.45)(2\,\text{ft})$$

$$= \underline{1.82\,\text{ft}}$$

Depth of expanded bed requires the expanded porosities, which are given by

$$\varepsilon_e = \left(\frac{V_b}{V_s} \right)^{0.22}$$

$$= \left(\frac{0.0137}{V_s} \right)^{0.22}$$

The settling velocity, V_s, for the first sieve size is 0.498 ft/sec from the previous calculations. In a like manner, the settling velocities for all sieve sizes were determined and are shown in column (4). The value of ε_e for the first sieve size is

$$\varepsilon_e = \left(\frac{0.0137}{0.498} \right)^{0.22}$$

$$= 0.454$$

In a like manner the ε_e values for all sieve sizes were determined and are shown in column (5). Dividing the weight fraction retained on the first sieve (0.0087) by (1 − 0.454) gives $x/(1 − \varepsilon_e)$ of 0.016. In a like manner, all values of $x/(1 − \varepsilon_e)$ were determined and are shown in column (6). The Σ value is 2.267. The expanded bed depth for the stratified bed is

$$D_e = (1 − \varepsilon)D\Sigma \frac{x}{1 − \varepsilon_e}$$

$$= (1 − 0.45)(2\,\text{ft})(2.267)$$

$$= \underline{2.49\,\text{ft}}$$

Note: The backwash rate in this problem is the minimum required to expand the bed. Actually, 15 to 20 gpm/ft^2 is used to expand a bed for proper agitation during cleansing.

Operational Problems

The major operating problems encountered in the use of rapid sand filtration are mud accumulations or mudballs, bed shrinkage, and air binding. Mudball formation is a condition which may occur when the filter feed contains a muddy floc and the filter is not adequately backwashed. The muddy floc will accumulate on the surface of the sand bed forming a muddy mat that will penetrate any cracks in the top of the sand. If a surface wash is not used, some of the mud may be pressed together to form small muddy balls during the backwash. With subsequent cycles of filtration and backwashing these balls enlarge and become caked with sand and may eventually settle to the gravel layer. They interfere with uniform filtration and cause inadequate backwashings. Mudball formation may be minimized by the use of a surface wash that breaks up any muddy mat formation.

Bed shrinkage may occur if the sand grains become covered with a soft slime coating. This causes the bed to compact as the filter run progresses and results in cracks in the bed surface and along the side walls of the filter. These cracks are undesirable because they may allow improperly filtered water to pass through the bed and fine muddy floc may accumulate in them to start mudball formation. Slime coatings on the filter sand may be minimized by the use of a surface wash system.

Air binding is caused by the release of air gases dissolved in the water, such as nitrogen and oxygen, thus creating air bubbles in the sand bed. Air binding usually results when a filter is operated under a negative head, and it may interfere with the rate of filtration. Also, at the beginning of the backwash, the violent agitation due to the rising air bubbles may cause a loss of sand. The principal method of control is the avoidance of negative head or pressures.

Multimedia Filters

These filters, which have more than one medium, may be open gravity filters, as shown in Figure 4.1, or pressure filters, as shown in Figures 4.5 and 4.7. In water treatment, they have become more popular in recent years. In advanced and tertiary waste treatment, they are the main type of filters that have been used successfully. Dual-media filter beds usually employ anthracite and sand; however, other materials have been used such as activated carbon and sand. Multimedia filter beds generally use anthracite, sand, and garnet. However, other materials have been used such as activated carbon, sand, and garnet. Also, dual- and multimedia filters using ion exchange resins as one of the media have been tried. In some of these filters, the media may have additional characteristics other than removing particulates. For example, activated carbon removes dissolved organic substances.

The main advantages of multimedia filters compared to single-medium filters are longer filtration runs, higher filtration rates, and the ability to filter a water with higher turbidity and suspended solids. The advantages of the multimedia filters are due to (1) the media particle size, (2) the different specific gravities of the media, and (3) the media gradation. These result in a filter with a larger percent of the pore volume being available for solids storage. In the single-medium filter, the pore volume available for solids storage is in the top portion of the bed, whereas, in the multimedia filter, the available pore volume is extended deep within the filter bed. Due to the deep penetration of accumulated floc, these filters are frequently referred to as "deep bed filters." The single-medium filters are rarely used in wastewater or advanced wastewater treatment due to short filter runs. Due to the large pore volume being available for floc storage, the multimedia filters can be used in advanced or tertiary wastewater treatment and still have a reasonable filter run.

Dual-Media Filters

The dual-media filter, consisting of a layer of coarse anthracite coal above a layer of fine sand, is one technique for increasing the pore volume of a filter. Figure 4.16 (b) shows the grain and pore size in a dual-media filter, and Figure 4.16 (a) shows these characteristics for a single-medium filter. It can be seen from the pore size profile that the available pore volume of the dual-media filter will be greater than the single-medium filter. The available pore volume, however, will not be as large as the total pore volume due to the fine to coarse gradation within each layer. Ideally, the available pore volume would be maximum at the top of the filter and gradually decrease to a minimum at the bottom of the filter.

Usually, a dual-media filter consists of an 18- to 24-in. layer of crushed anthracite coal overlaying a 6- to 12-in. layer of sand. Coal has a specific gravity of 1.2 to 1.6 and sand has a specific gravity of 2.65. During the first backwash, the sand layer remains below the coal due to its higher specific gravity and its grain size relative to the coal particles. After the first backwash, there will not be a distinct interface between the two layers but instead there will be a blended region of both coal particles and sand grains.

The size and characteristics of the anthracite and sand media and the thickness of the layers depend upon whether the filter is to be used for water or wastewater treatment. Filtration rates may vary from 2 to 10 gal/min-ft^2; however, a rate ranging from 3 to 6 gal/min-ft^2 is common.

Mixed-Media Filters

The ideal filter has a pore size and gradation as shown in Figure 4.16 (c). The pore size is greatest at the top of the bed and gradually decreases to a minimum at the bottom. The available pore volume, like the pore space, is maximum at the top of the bed and decreases to a minimum at the bottom. The media have a gradation which is from coarse at the top to fine at the bottom. The ideal filter may be approached by using a dual-media filter of crushed anthracite coal above sand and placing a third very dense medium below the sand. This allows the third medium to be very fine and still remain in the lower depths during backwashing. The resulting filter is referred to as a mixed-media filter since there is some intermixing between the layers during backwashing. Garnet, which has a specific gravity of about 4.2, has been found to be ideal as the third medium. Ilmenite, having a specific gravity of about 4.5, is also used but to a lesser extent. The anthracite, sand, and garnet or ilmenite are properly sized to allow some intermixing of the media during backwash. After backwashing there will be no distinct interface between the media layers. The filter bed will approach the ideal, as shown in Figure 4.16 (c), which has a gradual decrease in pore size with increasing depth.

Figure 4.16. Gradation and Pore Size in Various Filters
Courtesy of Neptune Microfloc, Inc.

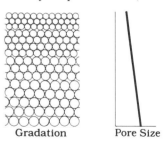

(a) Single-Medium Filter

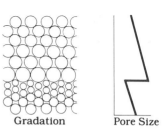

(b) Dual-Media Filter

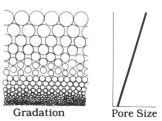

(c) Ideal Filter

Figure 4.17. Media Distribution in a Mixed-Media Filter
Adapted from *Mixed Media*, Bulletin no. KL 4206, by Neptune Microfloc, Inc. Copyright 1975 by Neptune Microfloc, Inc. Reprinted by permission.

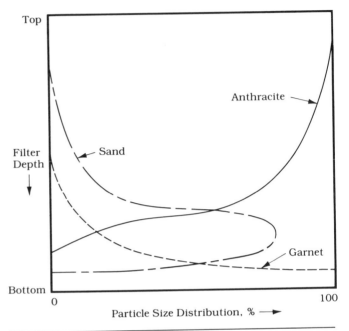

Since the pore size decreases from the top to the bottom of the filter, the filter will have a large available pore volume extending throughout the depth of the filter bed. Figure 4.17 shows the particle size distribution in a mixed-media filter. About 3 inches of coarse garnet or ilmenite are placed under the third layer to prevent fine particles from entering the underlaying gravel.

The size and characteristics of the media and the thickness of the layers placed in the filter depend upon the type of service for the filter—that is, whether it is for water or wastewater treatment. The filtration rates that have been used in water treatment or advanced waste treatment are from 2 to 12 gal/min-ft^2; however, a rate ranging from 3 to 6 gal/min-ft^2 is common.

Filter Layout, Appurtenances, and Details

Usually gravity filters are constructed of poured-in-place reinforced concrete. The minimum number of filters required is two; however, four are preferable. The filters are placed side by side in a row, and the pipe gallery, which contains all the necessary piping, valves, and so on, runs parallel to the filter row. In cold climates the filters are usually enclosed in a building; however, they may be in the open if a perforated pipe furnishing compressed air bubbles is placed around the perimeter of each filter immediately above the filter bed. The slight agitation by the air bubbles prevents freezing. The pipe gallery and operating floor are always enclosed to protect personnel and equipment. A dehumidified pipe gallery will reduce maintenance on controls, valves, and other equipment. Pressure filters are usually cylindrical in shape and are prefabricated of steel with a maximum diameter of 10 to 12 ft and a maximum length of 60 ft. They should be equipped with a sight glass for observation of the bed during backwashing and should have an access manhole for necessary maintenance. Hydraulic removal of the filter media should be provided.

The two basic types of control systems are the manual and fully automatic. The development of accurate and reliable equipment for measuring the effluent turbidity not only gives a continuous record of filter performance but greatly assists in filter operation. Fully automatic systems which have a programmer activated by the effluent turbidity or head loss are available. Once the effluent turbidity or head loss reaches a preset level, the programmer takes the filter off-line, backwashes it, and places it back on-line.

In most filters presently built, the control of the flow is by a rate of flow controller. The rate of flow controller maintains a uniform flow rate with a constant water depth over the filter bed. Uniform flow is maintained by varying the head loss between the filter bed surface and the downstream side of the rate of flow controller. The controller usually consists of a venturi with a variable opening diaphragm or butterfly valve on the downstream side. The valve is activated by the difference in pressure between the upstream side and the throat of the venturi. Another method for flow control without the use of a rate of flow controller consists of a weir in the inlet to the filter and a weir in the effluent channel discharging into the clearwell. The weir downstream of the filter keeps a minimum water depth over the filter bed. The weir in the inlet maintains a constant rate of flow to the filter, and its flow rate is independent of the depth of water over the filter bed since the weir crest is above the water surface. As the filter run progresses, the depth of water over the filter bed increases due to the increase in head loss. The major disadvantage of this type of flow control is that the filter walls must be from 5 to 6 feet deeper than required when a rate of flow controller is used (Baumann, E. R., & Oulman, C. S., 1970). If pressure filters are used, another way to control the filtration rate is by the use of pumps that pump at a relatively constant rate to the filters.

Another auxiliary scour system, termed *air scouring*, is available in lieu of the surface wash. To wash with this system, the water is filtered down to about 6 inches above the bed. Air is applied to the underdrain system at 2 to 5 cfm/ft^2 for 3 to 10 minutes, then the backwash is initiated at 2 to 5 gpm/ft^2. Once the water level is about 1 ft below the washwater troughs, the air is stopped and the backwash is operated at the normal rate for the usual period of time (Culp, R. L., & Culp, G. L., 1978).

Washwater troughs are usually spaced at a clear distance of 5 to 6 ft from each other. They serve to remove the backwash and, since the distance from the top of all troughs to the underdrains is a constant value, they aid in maintaining a uniform backwash. The trough bottoms should be at least 6 in. above the expanded bed during backwashing. Troughs are made of materials such as reinforced concrete, fiberglass, and enameled steel. Troughs are frequently of precast reinforced concrete and have a rectangular cross section, as shown in Figure 4.18 (a). Figure 4.18 (b) shows the washwater overflowing into a trough. The total discharge is twice that of a suppressed weir or

$$Q = 2(3.33)LH^{3/2} \qquad (4.26)$$

where

Q = total discharge, cfs;
L = trough length, ft;
H = head on weir, ft.

Figure 4.18 (c) shows the profile of a trough during backwashing. Shortly before the freefall into the gullet the water is at the critical depth, y_c.[*] For a rectangular channel the critical depth is

$$y_c = \sqrt[3]{\frac{q^2}{g}} \qquad (4.27)$$

where

y_c = critical depth, ft;
q = discharge per ft of width, $q = Q/b$, cfs/ft;
g = acceleration by gravity.

The upstream depth, H_0, for a level, rectangular, lateral spillway with a freefall and no friction is (Camp, T. R., 1952)

$$H_0 = \sqrt{y_c^2 + \frac{2Q^2}{gb^2 y_c}} \qquad (4.28)$$

[*]Critical depth can be assumed at the freefall.

Figure 4.18. Wash Trough Details

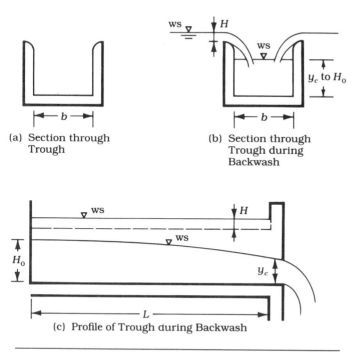

(a) Section through Trough

(b) Section through Trough during Backwash

(c) Profile of Trough during Backwash

where

H_0 = upstream depth, ft;
Q = total discharge, cfs;
b = trough width, ft.

The friction loss will increase the value of H_0 by 6 to 16 percent of the water surface drawdown. A distance of at least 3 in. should be provided from the maximum depth to the top of the trough.

Example Problem 4.3
Wash Trough Design

A rapid sand filter has washwater troughs that are 16 ft long and spaced 6.20 ft from center to center. The precast concrete troughs are rectangular in cross section with an inside width of 18 in. The backwash flow is 15 gpm/ft². It is desired to determine the water depth over the trough sides, H, the depth at the freefall, y_c, the upstream water depth, H_0, and the trough depth if the top of the trough is at least 3 in. above the maximum water surface in the trough.

Assume the friction loss is 16 percent of the surface drawdown.

Solution

The total flow Q, is (16 ft)(6.20 ft)(15 gpm/ft^2) or 1488 gpm, which is 3.32 cfs. Rearranging Eq. (4.26) gives

$$H = \left[\frac{Q}{2(3.33)L}\right]^{2/3} = [(3.32)/(2)(3.33)(16.0)]^{2/3}$$

$$= \underline{0.099\,\text{ft or }1.19\,\text{in.}}$$

The value of q is 3.32/1.5 or 2.21 cfs/ft. The depth of the water at the freefall is given by

$$y_c = \sqrt[3]{\frac{q^2}{g}} = \sqrt[3]{\frac{(2.21)^2}{32.2}}$$

$$= \underline{0.53\,\text{ft or }6.40\,\text{in.}}$$

The upstream water depth without friction is given by

$$H_0 = \sqrt{y_c^2 + \frac{2Q^2}{gb^2 y_c}} = \sqrt{(0.53)^2 + \frac{(2)(3.32)^2}{(32.2)(1.5)^2(0.53)}}$$

$$= 0.925\,\text{ft}$$

The value of H_0 with friction is given by $H_0 = 0.925 + 0.16\,(H_0 - 0.53)$ or $H_0 = 1.00$ ft. Thus, the upstream water depth including friction is

$$H_0 = \underline{1.00\,\text{ft or }12\,\text{in.}}$$

The trough depth should be

$$\text{Depth} = 12\,\text{in.} + 3\,\text{in.} = \underline{15\,\text{in.}}$$

Example Problem 4.3 SI
Wash Trough Design

A rapid sand filter has washwater troughs that are 5.0 m long and spaced at 1.9 m from center to center. The precast concrete troughs are rectangular in cross section with an inside width of 0.45 m. The backwash flow is 10 ℓ/s-m^2. It is desired to determine the water depth over the trough sides, H, the depth at the freefall, y_c, the upstream water depth, H_0, and the trough depth if the top of the trough is at least 7.5 cm above the maximum water surface in the trough.

Assume the friction loss is 16 percent of the surface drawdown.

Solution

The total flow, Q, is $(5.0 \text{ m})(1.9 \text{ m})(10 \text{ } \ell/\text{s-m}^2) = 95 \text{ } \ell/\text{s}$, or $Q = (95 \text{ } \ell/\text{s})(\text{m}^3/1000 \text{ } \ell) = 0.095 \text{ m}^3/\text{s}$. The discharge over a suppressed weir is given by $Q = 1.86LH^{3/2}$, where $Q = \text{m}^3/\text{s}$, $L = \text{m}$, and $H = \text{m}$. Rearranging this equation for two weirs gives

$$H = \left[\frac{Q}{2(1.86)L} \right]^{2/3} = [(0.095)/(2)(1.86)(5.0)]^{2/3}$$

$$= \underline{0.030 \text{ m or } 3.0 \text{ cm}}$$

The value of q is $(0.095 \text{ m}^3/\text{s})/0.45 \text{ m} = 0.211 \text{ m}^3/\text{s-m}$. The water depth at the freefall is

$$y_c = \sqrt[3]{\frac{q^2}{g}} = \sqrt[3]{\frac{(0.211)^2}{9.806}}$$

$$= \underline{0.166 \text{ m}}$$

The upstream water depth without friction is given by

$$H_0 = \sqrt{y_c^2 + \frac{2Q^2}{gb^2 y_c}}$$

$$= \sqrt{(0.166)^2 + \frac{(2)(0.095)^2}{(9.806)(0.45)^2(0.166)}}$$

$$= 0.287 \text{ m}$$

The value of H_0 with friction is given by $H_0 = 0.287 \text{ m} + 0.16 (H_0 - 0.166 \text{ m})$ or

$$H_0 = \underline{0.310 \text{ m}}$$

The trough depth should be

$$\text{Depth} = 0.310 + 0.075 = \underline{0.385 \text{ m}}$$

The slow sand filter, which was developed during the mid-1800s, was the first type filter used for water treatment. Plain sedimentation of the water prior to filtration was usually provided. These filters were single-medium filters having an effective sand size of about 0.2 to 0.4 mm and were operated at filtration rates of 0.05 to 0.15 gal/min-ft². **Filtration in Water Treatment**

The filters were cleaned manually, usually every four to six weeks, by scraping off the top layers of clogged sand and cleaning the sand with a scouring device. Because of the large land area requirements and the manual labor involved, the slow sand filter was replaced by the rapid sand filter.

The rapid sand filter is always preceded by chemical coagulation, flocculation, and sedimentation. The first filters, which operated at about 2 gal/min-ft^2, consisted of a quartz sand bed overlaying a gravel layer. The turbidity removals ranged from 90 to 98 percent if the feed water turbidity was between 5 to 10 JTU. Although the standard rate of filtration is generally considered as 2 gal/min-ft^2, most rapid sand filters are operated at 3 to 5 gal/min-ft^2 and have coarse sand beds. The primary filter action in rapid sand filtration is usually depth removal. As shown in Table 4.3, the sand beds are usually 24 to 30 in. thick and have an effective size of 0.35 to 0.70 mm and a uniformity coefficient less than 1.7.

Table 4.3. Single-Medium Filter Characteristics for Water Treatment

Characteristic	Value	
	Range	Typical
Sand Medium:		
Depth, in.	24–30	27
Effective size, mm	0.35–0.70	0.60
Uniformity coefficient	<1.7	<1.7
Filtration rate, gpm/ft^2	2–5	4
Anthracite Medium:		
Depth, in.	24–30	27
Effective size, mm	0.70–0.75	0.75
Uniformity coefficient	<1.75	<1.75
Filtration rate, gpm/ft^2	2–5	4

Calcium carbonate encrustations on the sand grains may occur when lime-soda softening is employed. These enlarge the sand grains, which is undesirable. It may be controlled by lowering the pH by carbonation prior to filtration to precipitate excess lime and stabilize the water. Also, stabilization using sodium hexametaphosphate or other such chemicals may be used. Many rapid sand filters employ crushed anthracite coal as a medium instead of quartz sand because anthracite is less susceptible to encrustations. Typical anthracite characteristics are shown in Table 4.3. To support the anthracite bed, gravel or graded anthracite of 12-in. thickness may be used.

Since the development of the dual-media and mixed-media filters, most new plants have these types of filters. The primary filter action is depth removal. The characteristics of the dual- and mixed-media filters used in water treatment are shown in Tables 4.4 and 4.5. The principal advantages of these filters over sand filters are higher filtration rates and longer filter runs because of the increased volume for floc storage within the filter. As a result, less backwash water is required per unit volume of filtrate produced.

Table 4.4. Dual-Media Filter Characteristics for Water Treatment

Characteristic	Value	
	Range	Typical
Anthracite:		
Depth, in.	18–24	24
Effective size, mm	0.9–1.1	1.0
Uniformity coefficient	1.6–1.8	1.7
Sand:		
Depth, in.	6–8	6
Effective size, mm	0.45–0.55	0.5
Uniformity coefficient	1.5–1.7	1.6
Filtration rate, gpm/ft^2	3–8	5

Table 4.5. Mixed-Media Filter Characteristics for Water Treatment

Characteristic	Value	
	Range	Typical
Anthracite:		
Depth, in.	16.5–21	18
Effective size, mm	0.95–1.0	1.00
Uniformity coefficient	1.55–1.75	<1.75
Sand:		
Depth, in.	6–9	9
Effective size, mm	0.45–0.55	0.50
Uniformity coefficient	1.5–1.65	1.60
Garnet:		
Depth, in.	3–4.5	3
Effective size, mm	0.20–0.35	0.20
Uniformity coefficient	1.6–2.0	<1.6
Filtration rate, gpm/ft^2	4–10	6

Filtration in Wastewater Treatment

Filtration in advanced waste treatment may be employed for the following purposes: (1) filtration of secondary effluents, (2) filtration of chemically treated secondary effluents, and (3) filtration of chemically treated primary or raw wastewaters. With the exception of some new filtration techniques and the intermittent sand filter, the filters used in advanced or tertiary wastewater treatment are usually dual-media or mixed-media filters. Characteristics of the dual-media filters are shown in Table 4.6; characteristics of the mixed-media filters are shown in Table 4.7. The principal difference between filters used in water treatment and filters used in wastewater treatment is the size of the media.

Table 4.6. Dual-Media Filter Characteristics for Advanced or Tertiary Wastewater Treatment

Characteristic	Value	
	Range	Typical
Anthracite:		
Depth, in.	12–24	18
Effective size, mm	0.8–2.0	1.2
Uniformity coefficient	1.3–1.8	1.6
Sand:		
Depth, in.	6–12	12
Effective size, mm	0.4–0.8	0.55
Uniformity coefficient	1.2–1.6	1.5
Filtration rate, gpm/ft^2	2–10	5

Table 4.7. Multimedia or Mixed-Media Filter Characteristics for Advanced or Tertiary Wastewater Treatment

Characteristic	Value	
	Range	Typical
Anthracite:		
Depth, in.	8–20	16
Effective size, mm	1.0–2.0	1.4
Uniformity coefficient	1.4–1.8	1.5
Sand:		
Depth, in.	8–16	10
Effective size, mm	0.4–0.8	0.5
Uniformity coefficient	1.3–1.8	1.6
Garnet:		
Depth, in.	2–6	4
Effective size, mm	0.2–0.6	0.3
Uniformity coefficient	1.5–1.8	1.6
Filtration rate, gpm/ft^2	2–10	5

Tables 4.6 and 4.7 adapted from *Wastewater Engineering, Treatment, Disposal and Reuse* by Metcalf and Eddy, Inc. Copyright © 1979 by McGraw-Hill Book Co., Inc. Reprinted by permission.

For wastewater treatment, the granules must be larger so that the filter will have the desired flow rate capacity and the required storage volume for the accumulated floc.

Of the numerous variables that affect filter performance in advanced wastewater treatment, two of the most important are the floc strength (the ability to withstand shear forces) and the concentration of the suspended solids. Biological flocs are usually more resistant to shear forces than chemical flocs, particularly flocs from alum and iron salt coagulation. In filtering untreated secondary effluents, the primary filter action is surface removal and, as a result, excessive head losses usually terminate the filter runs. The deterioration in the quality of the filtrate rarely determines the end of a filter run. Chemical flocs from alum and iron salt coagulation tend to penetrate deep into a filter bed; thus the main filtering action is depth removal. The termination of the filter run is usually due to filtrate quality deterioration and breakthrough usually occurs at relatively low head losses, such as 3 to 6 feet (Tchobanoglous, G., & Eliassen, R., 1970). Polymer filter aids may be added to the feed to a filter to strengthen alum or iron salt flocs prior to filtration, thus allowing higher filtration rates than usually employed and also longer filter runs. In coagulation with lime, the calcium carbonate precipitate is relatively strong and tends to be removed on the filter surface; thus surface removal is important in this case. The calcium carbonate may form a dense mat which must be broken up by the surface wash prior to backwashing.

In order to prevent microbial slime buildup on the filter media, an auxiliary scour system, either the surface wash or air scour type, is necessary when filtering wastewaters. Calcium carbonate encrustations on the filter media may occur when high pH lime coagulation is employed. Stabilization is necessary to prevent operational problems.

There is a general relationship between the filter influent suspended solids concentration, the filtration rate, and the filter run. For example, if the influent suspended solids are 20 mg/ℓ, the filtration rate is 4 gal/min-ft^2 and the filter run is 72 hr, then the filter run at 6 gal/min-ft^2 would be approximately (72 hr) $\times$ 4/6 = 48 hr. If the suspended solids were reduced to 10 mg/ℓ and the filtration rate were 4 gal/min-ft^2, the filter run would be approximately (72 hr) $\times$ 20/10 = 144 hr.

Filtration of Secondary Effluents
A review of the data from seven tertiary treatment plants (EPA, 1975; Zenz, D. R., Lue-Hing, C., & Obayashi, A., 1972) both pilot and full scale, shows an average suspended solids

removal of 66.2 percent, an average filter run of 15.6 hours, an average filtration rate of 3.7 gpm/ft^2, and an average suspended solids concentration of 18.3 mg/ℓ in the feed to the filters. Two of the plants had dual-media filters, whereas the remaining five had mixed-media filters. In general, the mixed-media filters gave better performance both in terms of suspended solids removal and duration of the filter runs.

The filtrability of the suspended solids in untreated secondary effluents increases with an increase in the mean cell residence time and the hydraulic detention time in activated sludge plants. Apparently the floc strength is increased with an increase in the mean cell residence time and the hydraulic detention time. The expected effluent quality indicated by Culp and Culp (1978) for untreated secondary effluents filtered by multimedia filtration is shown in Table 4.8.

Table 4.8. Expected Effluent Suspended Solids versus Type of Secondary Treatment

Effluent	Effluent Suspended Solids (mg/ℓ)
Extended Aeration	1–5
Conventional Activated Sludge	3–10
Contact Stabilization	6–15
Two-Stage	6–15
High-Rate Trickling Filter	10–20

Intermittent sand filters have been used to give a combined physical-biological treatment for effluents from oxidation ponds or lagoons (Marshall, G. R., & Middlebrooks, E. J., 1974; Reynolds, J. H., Harris, S. E., Hill, D. W., Filip, D. S., & Middlebrooks, E. J., 1976). The intermittent operation results in both aerobic digestion and dewatering of the filtered solids, thus reducing the required maintenance. Although the land requirements are quite large, the intermittent sand filters are a feasible method for treating effluents from small installations.

Filtration of Chemically Coagulated Effluents

A review of the data from four tertiary treatment plants (EPA, 1975; South Tahoe Public Utility District, 1971), both pilot and full scale, shows an average suspended solids removal of 74.2 percent, an average filter run of 33.7 hr, an average filtration rate of 3.0 gpm/ft^2, and an average suspended solids of 9.3 mg/ℓ in the feed to the filters. Lime clarification was practiced at three of the plants and alum clarification at one. Three of the plants had dual-media filters, whereas one had a mixed-media filter. The mixed-media filter gave better performance in terms of suspended solids removal, although the filter runs from both types of filters were about equal. The mixed-media filter, however, was filtering at an average rate of 3.4 gpm/ft^2, compared to 2.8 gpm/ft^2 for the dual-media filters.

Filtration of Chemically Treated Primary or Raw Wastewaters

A review of the data from four large pilot plants using physical-chemical treatments of primary or raw wastewater (Bishop, D. F., O'Farrell, T. P., & Stamberg, J. B., 1970 & 1972; EPA, 1975; Villers, R. V., Berg, E. L., Brunner, C. N., & Masse, A. N., 1971) showed an average suspended solids removal of 73.0 percent at an average filtration rate of 3.3 gpm/ft^2. All four plants employed lime clarification and dual-media filters, either 24 or 36 inches deep. The feed at two of the plants had an average suspended solids concentration of 122 mg/ℓ, while the feed at the other two plants had an average of 131 mg/ℓ. The plants with the lower feed suspended solids had an average filter run of 31 hours, while the two plants having the higher feed suspended solids had an average filter run of 24 hours.

Upflow Filtration

Another technique to obtain the ideal filter in which the direction of filtration is from coarse to fine media is the upflow filter, shown in Figure 4.19. This filter is a single-medium filter using sand and, once hydraulic gradation has occurred by backwashing, the water being filtered is

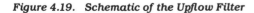

Figure 4.19.   Schematic of the Upflow Filter

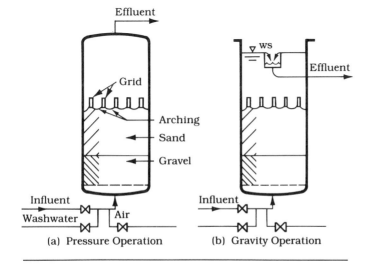

(a) Pressure Operation   (b) Gravity Operation

passed upward through the bed. Fluidization of the bed will occur if the head loss is sufficient to expand the medium and, as a result, floc breakthrough will occur. Fluidization may be avoided by making the bed extremely deep or by placing a restraining grid on the top of the bed, as shown in Figure 4.19. The bar spacing must be large enough to permit bed expansion during backwashing, and the arching action of the sand grains allows such spacing. In the cleaning cycle the bed is agitated with the air scour system to break the arching action and loosen deposits. Once this has occurred, the backwash water is started at a sufficient rate to further expand the bed, resulting in cleansing. Once the backwashing is complete, the washwater is discontinued and the sand is allowed to settle to its original position.

The use of the upflow filter for drinking waters is limited because of a possible floc breakthrough during filtration. However, it has been used for industrial water treatment and wastewater treatment. It may be used as a pressure or a gravity filter.

Typical design parameters for upflow sand filters used in wastewater treatment are as follows (EPA, 1975): (1) filtration rate from 2 to 3 gal/min-ft^2, (2) terminal head loss from 6 to 20 ft, (3) bed depth of 60 in. of 1- to 2-mm sand, 10 in. of 2- to 3-mm sand and 4 in. of 10- to 15-mm sand.

Upflow filters treating secondary effluents having an average suspended solids content of 17 mg/ℓ have shown an average solids removal of 64.6 percent (EPA, 1975; WPCF, 1977). The filtration rate was from 2 to 5 gal/min-ft^2, with an average of 4.4. Filter runs varied from 7 to 150 hours.

Miscellaneous Filters

Other filters such as the diatomaceous earth filter and microscreens are used for special purposes. The diatomaceous earth filter is a pressure filter employing filter elements coated with a precoat of diatomaceous earth. They are used for small-scale operations such as swimming pool filtration. The microscreen consists of a rotating cylindrical drum covered with a wire mesh filter cloth. The water being filtered passes from the inside to the outside of the drum, and the filtered deposits are washed away in a discharge trough by water jets. They have been used to a limited extent in tertiary treatment; however, difficulties have been encountered in processing a water with a fluctuating flow rate and a fluctuating suspended solids concentration. They have been successfully used to remove algal growths from drinking waters prior to rapid sand filtration.

American Water Works Association (AWWA). 1969. *Water Treatment Plant Design.* New York: AWWA.

American Water Works Association (AWWA). 1971. *Water Quality and Treatment.* New York: McGraw-Hill.

Baumann, E. R., and Oulman, C. S. 1970. Sand and Diatomite Filtration Practices. In *Water Quality Improvement by Physical and Chemical Processes,* edited by E. F. Gloyna and W. W. Eckenfelder, Jr. Austin, Tex.: University of Texas Press.

Bishop, D. F., O'Farrell, T. P., and Stamberg, J. B. 1972. Physical-Chemical Treatment of Municipal Wastewater. *Jour. WPCF* 44, no. 3: 361.

Bishop, D. F., O'Farrell, T. P., and Stamberg, J. B. 1970. Physical-Chemical Treatment of Municipal Wastewaters. Paper presented before the 43rd annual meeting, Water Pollution Control Federation, Boston, Mass.

Bishop, D. F., O'Farrell, T. P., and Stamberg, J. B. 1971. Advanced Waste Treatment Systems at the Environmental Protection Agency—District of Columbia Pilot Plant. Paper presented at the 68th national meeting of the American Institute of Chemical Engineers (AICh.E.), Houston, Tex.

Camp, T. R. 1940. Lateral Spillway Channels. *Trans. ASCE* 105: 606.

Camp, T. R. 1964. Theory of Water Filtration. *Jour. SED ASCE* 90, SA4: 1.

Camp, T. R. 1952. Water Treatment. In *The Handbook of Applied Hydraulics,* edited by C. V. Davis. New York: McGraw-Hill.

Carman, P. C. 1937. Fluid Flow through Granular Beds. *Trans. Inst. Chem. Engrs. (London)* 15: 150.

Cleasby, J. L. 1972. Filtration. In *Physicochemical Processes for Water Quality Control,* edited by W. J. Weber, Jr. New York: Wiley.

Conway, R. A., and Ross, R. D. 1980. *Handbook of Industrial Waste Disposal.* New York: Van Nostrand Reinhold.

Culp, G., and Conley, W. 1970. High Rate Filtration with the Mixed-Media Concept. In *Water Quality Improvement by Physical and Chemical Processes,* edited by E. F. Gloyna and W. W. Eckenfelder, Jr. Austin, Tex.: University of Texas Press.

Culp, R. L., and Culp, G. L. 1978. *Handbook of Advanced Wastewater Treatment.* 2nd ed. New York: Van Nostrand Reinhold.

Culp, G. L., and Culp, R. L. 1974. *New Concepts in Water Purification.* New York: Van Nostrand Reinhold.

Eckenfelder, W. W. 1980. *Principles of Water Quality Management.* Boston: CBI Publishing.

Environmental Protection Agency (EPA). 1973. *Phosphorus Removal.* EPA Process Design Manual, Washington, D.C.

Environmental Protection Agency (EPA). 1975. *Suspended Solids Removal.* EPA Process Design Manual, Washington, D.C.

Environmental Protection Agency (EPA). 1974. *Upgrading Existing Wastewater Treatment Plants.* EPA Process Design Manual, Washington, D.C.

Ergun, S. 1952. *Chem. Engr. Progress* 48: 89.

Fair, G. M., Geyer, J. C., and Okun, D. A. 1968. *Water and Wastewater Engineering.* Vol. 2. *Water Purification and Wastewater Treatment and Disposal.* New York: Wiley.

Fair, G. M., and Hatch, L. P. 1933. Fundamental Factors Governing the Streamline Flow of Water through Sand. *Jour. AWWA* 25:1551.

Great Lakes–Upper Mississippi Board of State Sanitary Engineers. 1978. Recommended Standards for Sewage Works. Ten state standards, Albany, N.Y.

Hoover, C. P. 1946. *Water Supply and Treatment*. National Lime Association, Washington, D.C.

Kozeny, G. 1927. *Sitzber. Akad. Wiss. Wein, Math.–Naturw. Kl., Abt. IIa* 136.

Marshall, G. R., and Middlebrooks, E. J. 1974. Intermittent Sand Filtration to Upgrade Existing Wastewater Treatment Facilities. Utah Water Research Laboratory, PRJEW 115-2, Utah State University, Logan, Utah.

McCabe, W. L., and Smith, J. C. 1967. *Unit Operations of Chemical Engineering*. 2nd ed. New York: McGraw-Hill.

Metcalf and Eddy, Inc. 1979. *Wastewater Engineering, Treatment, Disposal and Reuse*. New York: McGraw-Hill.

Neptune MicroFloc Inc. 1975. *Mixed Media*. Bulletin no. KL 4206, Corvallis, Oreg.

O'Melia, C. R., and Stumm, W. 1967. Theory of Water Filtration. *Jour. AWWA* 59, no. 11:1393.

Perry, R. H., and Chilton, C. H. 1973. *Chemical Engineer's Handbook*. 5th ed. New York: McGraw-Hill.

Reynolds, J. H., Harris, S. E., Hill, D. W., Filip, D. S., and Middlebrooks, E. J. 1976. Intermittent Sand Filtration for Upgrading Waste Stabilization Ponds. In *Ponds as a Wastewater Treatment Alternative*, edited by E. F. Gloyna, J. F. Malina, Jr., and E. M. Davis. Austin, Tex.: University of Texas Press.

Rich, L. G. 1971. *Unit Operations of Sanitary Engineering*. New York: Wiley.

Rose, H. E. 1945. *Proc. Inst. Mech. Engrs. (London)* 153:141, 154; also, 160:493 (1949).

Sanks, R. L. 1978. *Water Treatment Plant Design*. Ann Arbor, Mich.: Ann Arbor Science Publishers.

Schroeder, E. D. 1977. *Water and Wastewater Treatment*. New York: McGraw-Hill.

South Tahoe Public Utility District. 1971. *Advanced Wastewater Treatment as Practiced at South Tahoe*. Tech. Report for the EPA, Project 17010 ELQ (WPRD 52-01-67).

Steel, E. W., and McGhee, T. J. 1979. *Water Supply and Sewerage*. New York: McGraw-Hill.

Sundstrom, D. W., and Klei, H. E. 1979. *Wastewater Treatment*. Englewood Cliffs, N.J.: Prentice-Hall.

Tchobanoglous, G., and Eliassen, R. 1970. Filtration of Treated Sewage Effluent. *Jour. SED ASCE* 96, no. SA2:243.

Villers, R. V., Berg, E. L., Brunner, C. N., and Masse, A. N. 1971. Municipal Wastewater Treatment by Physical and Chemical Methods. *Water and Sewage Works*. Reference no. 1971, p. R-62.

Weber, W. J., Jr. 1972. *Physicochemical Processes for Water Quality Control*. New York: Wiley-Interscience.

Water Pollution Control Federation (WPCF). 1977. *Wastewater Treatment Plant Design*. WPCF Manual of Practice no. 8, Washington, D.C.

Zenz, D. R., Lue-Hing, C., and Obayashi, A. 1972. *Preliminary Report on Hanover Park Bay Project*. EPA Grant no. WPRD 92-01-68 (R-2).

1. A single-medium rapid sand filter has a maximum grain size of 0.22 cm, a porosity (ε) of 0.40, and the specific gravity of the sand is 2.65. The bed will be fluidized during backwashing when the superficial velocity of the backwash water, V_b, is equal to or greater than $\varepsilon^{4.5}V_s$, where V_s is the settling velocity of the largest size sand particle. Determine the minimum backwash rate in gpm/ft^2 to fluidize the bed. *Note:* To obtain sufficient bed expansion the actual backwash rate is usually several magnitudes times the minimum rate.

2. A single-medium rapid sand filter has a maximum grain size of 0.22 cm, a porosity (ε) of 0.40, and the specific gravity of the sand is 2.65. The bed will be fluidized during backwashing when the superficial velocity of the backwash water, V_b, is equal to or greater than $\varepsilon^{4.5}V_s$, where V_s is the settling velocity of the largest size sand particle. Determine the minimum backwash rate in ℓ/s-m^2 to fluidize the bed. *Note:* To obtain sufficient bed expansion the actual backwash rate is usually several magnitudes times the minimum rate.

3. A gravity-type mixed-media filter has washwater troughs that are level and are 18 ft long, 1.25 ft wide, and are spaced 6 ft apart center to center. The wash rate is 15 gpm/ft^2. Determine:
   a. The depth of water above the trough crest if the trough is level. The discharge over the weir on one side of the trough is given by $Q = 3.33\,LH^{3/2}$, where Q = cfs, L = trough length in ft, and H = depth of water over the crest in ft.
   b. The depth of the water in the trough where the water spills over into the gullet. For a freefall from a rectangular open channel, the water depth is equal to the critical depth, y_c. This is given by $y_c = (q^2/g)^{1/3}$, where q = the discharge per unit width (cfs/ft) and g = the acceleration due to gravity.
   c. The upstream water depth in the wash trough. The upstream depth, H_0, for a trough having a freefall discharge is given by $H_o = (y_c^2 + 2Q^2/gb^2y_c)^{1/2}$, where H_0 = upstream depth in ft, b = trough width in ft, and Q = total trough discharge in cfs. H_0 from the previous equation is for a frictionless channel. H_0 should be increased by 16 percent of the water surface drawdown to allow for friction.
   d. The trough depth if the trough crest is 3 in. above the maximum water depth in the trough, H_0.

4. A mixed-media filter at a water treatment plant is treating a clarified water having a turbidity of 5 JTU. The filtration rate has been 3 gpm/ft^2, which results in a 3-day filter run. The filter runs are terminated when the head loss reaches 7 feet. Determine:
   a. The expected filter run if the filtration rate is 5 gpm/ft^2.
   b. The expected filter run if the turbidity is 10 JTU and the filtration rate is 3 gpm/ft^2.
   c. The expected filter run if the turbidity is 8 JTU and the filtration rate is 4 gpm/ft^2.

5. A trimedia filter has an 18-in. crushed anthracite coal layer (specific gravity = 1.20), a 9-in. sand layer (specific gravity = 2.65), and a 3-in. garnet layer (specific gravity = 4.20). The average sizes of particles in the three layers are 1.5 mm, 0.80 mm, and 0.30 mm, respectively. The ϕ values are 0.9 and the porosities of the respective layers are 0.40, 0.45, and 0.50. The filtration rate is 5 gpm/ft^2 and the water temperature is 50°F (10°C). Determine the head loss for the clean filter.

6. A rapid sand filter has a sand bed 30 in. in depth. Pertinent data are: specific gravity of the sand = 2.65, shape factor (ϕ) = 0.75, porosity (ε) = 0.41, filtration rate = 2.25 gpm/ft^2, and operating temperature = 50°F (10°C). The sieve analysis of the sand is given in Table 4.9.

Table 4.9.

Sieve Size	Weight Retained (%)
14–20	0.44
20–28	14.33
28–32	42.78
32–35	27.07
35–42	9.76
42–48	4.22
48–60	0.54
60–100	0.42

Determine the head loss for the clean filter bed in a stratified condition using the Rose equation.

7. A rapid sand filter having the same sand analysis as in problem 6 is to be backwashed. Determine:
   a. The backwash velocity required to expand the bed.
   b. The backwash flow in gpm/ft^2 required to expand the bed.
   c. The head loss at the beginning of the backwash.
   d. The depth of the expanded bed.

Chapter Five

AMMONIA REMOVAL

U sually it is desirable to have a low ammonia concentration in a final effluent because it is toxic to fish at concentrations greater than about 3 mg/ℓ and also because it will be bio-oxidized by nitrifying microorganisms to form nitrates using molecular oxygen from the receiving water body. Nitrates are undesirable because they are nutrients that stimulate algal and aquatic growths. Ammonia removal may be accomplished by physical, chemical, or biological means.

Physical Operations

Induced-draft stripping towers and spray ponds are physical operations used to remove ammonia from wastewaters. In order for the ammonia to be stripped from a wastewater it must be in the dissolved gas form (NH_3), which requires a pH of about 10.8 or greater.

Induced-Draft Stripping Towers
Figure 5.1 shows a cross section through the stripping tower used at the Orange County Water District, Santa Ana, California, and Figure 5.2 shows a schematic drawing of this type of tower.

The type of tower shown in Figures 5.1 and 5.2 is a countercurrent system since the solvent gas (air) flows upward through the packing while the solvent liquid (water) flows downward. The tower consists of a fan, a packing to give intimate air and water contact, a tray or other distribution system to uniformly irrigate the top of the packing, a grid to support the packing and distribute the incoming air, a drift eliminator to collect liquid droplets in the exit

Figure 5.1. Ammonia Stripping Tower in Orange County, California

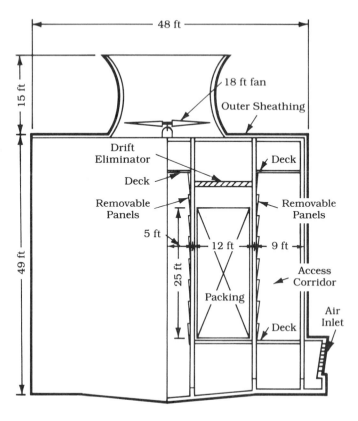

Figure 5.1. Ammonia Stripping Tower in Orange County, California

Figure 5.2. Schematic Drawing of an Ammonia Stripping Tower

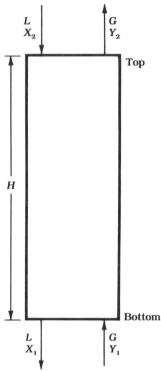

Figure 5.2. Schematic Drawing of an Ammonia Stripping Tower

H = Tower height
L = Wastewater mass flow rate
G = Air mass flow rate
X_1, X_2 = Ammonia concentrations in the wastewater (as mass ratios)
Y_1, Y_2 = Ammonia concentrations in the air (as mass ratios)

air, and the tower structure. In the induced-draft tower, the fan at the top of the tower draws the air from the tower; thus, the tower operates at a pressure slightly less than the outside pressure. Generally, a two-speed fan is used to give flexibility in operation. The pressure drop in the tower depends primarily upon the air and water flow rates, the viscosities, and the packing design. The open packing used in ammonia stripping towers usually has a pressure drop of 0.1 to 0.2 in. per foot depth.

The original packing in the ammonia stripping tower at South Tahoe was a wood-grid type using $\frac{3}{8}$ in. by $1\frac{1}{2}$-in. redwood slats having a clear spacing of about $1\frac{1}{2}$ inches. Over a period of time, calcium carbonate encrustations developed and, since there was no access to the packing for cleaning, it eventually clogged. To avoid the encrustation problem, a plastic type packing was designed and is in use in the ammonia stripping tower installation at the advanced wastewater treatment plant at the Orange County Water District, Santa Ana, California. The grid type of packing consists of $\frac{1}{2}$-in.-diameter PVC pipe placed horizontally at 3-in. center to center spacing and placed vertically at 2-in. center to center spacing. The packing is placed in alternate layers at right angles. Pilot studies at the Orange County Water District showed that encrustations could be washed off by high pressure hoses, and access corridors in the tower were incorporated to allow washing of the plastic packing while the packing is in place. Also, the packing was constructed in modular frames which could be removed from the tower for cleaning if necessary.

In order to derive equations for the design of ammonia stripping towers it is necessary to understand the equilibrium relationships for a gas-liquid system, the fundamentals of gas mass transfer, and the material balances for a stripping tower. The equilibrium relationship for ammonia is given by Henry's law. One form of this law is

$$p_A = mX \qquad (5.1)$$

where

p_A = partial pressure of the ammonia in the air mixture in contact with the solution at equilibrium;

m = constant;

X = concentration of the ammonia in the solution at equilibrium, expressed as a mole or mass ratio.

Henry's law is valid for a limited range in solute gas concentrations since it assumes a straight-line relationship between the partial pressure and the concentration. The equilibrium partial pressures for ammonia gas dissolved in water at various temperatures are given in Table 5.1.

Table 5.1. Ammonia Partial Pressure versus Temperature

Temperature (°C)	p_A (mm Hg)	X (gm NH$_3$/10^6 gm H$_2$O)
0	0.0112	50
10	0.0189	50
20	0.0300	50
25	0.0370	50
30	0.0479	50
40	0.0770	50
50	0.1110	50

Taken from J. Perry, *Chemical Engineers Handbook*, 4th ed. (New York: McGraw Hill, 1969).

The equilibrium ammonia concentration in an air mixture as expressed by mass or weight ratios is related to its partial pressure by

$$Y^* = \left(\frac{p_A}{P_t}\right)\left(\frac{M_A}{M_{Air}}\right) \qquad (5.2)$$

where

Y^* = mass or weight ratio of the ammonia;
P_t = total pressure of the atmosphere;
p_A = partial pressure of the ammonia;
M_A = molecular weight of the ammonia, 17 gm/gm mole;
M_{Air} = molecular weight of air, 29 gm/gm mole.

Knowing the ammonia solubility, its equilibrium partial pressure, and the atmospheric pressure, it is possible to determine the coordinates for an equilibrium curve which shows the equilibrium concentrations of ammonia in the air and in the water.

In ammonia stripping it is sometimes useful to know the percent ammonia in the solution that is in the form of ammonia gas. The ammonia gas is in equilibrium with the ammonium ion as given by

$$NH_3 + H_2O \rightleftharpoons NH_4^{+1} + OH^{-1} \qquad (5.3)$$

As the pH is increased, the equilibrium shifts more to the left. The percent ammonia in the gas form at 25°C is as follows (Metcalf and Eddy, 1979):

$$NH_3 \text{ (percent)} = \frac{100}{1 + 1.75 \times 10^{+9}[H^{+1}]} \qquad (5.4)$$

where H^{+1} = hydrogen ion concentration. At 25°C and pH 10.8, 97.3 percent of the ammonia will be in the form of ammonia gas molecules dissolved in water. Since the partial pressure of ammonia in air is essentially zero, ammonia stripping will occur at a neutral pH range; however, the operation would have an extremely poor efficiency since most of the ammonia is in the form of the ammonium ion. Raising the pH to about 10.8 causes the major portion of the ammonia to be in the form of ammonia gas molecules; thus, stripping readily occurs and the stripping towers will have high efficiencies.

An ammonia stripping tower is to be designed to operate at 750-mm atmospheric pressure and an average air and water temperature of 20°C (68°F). Determine the concentrations expressed as mass ratios for the equilibrium curve.

Solution

The equilibrium mass of ammonia in the air is given by Eq. (5.2), which is

$$Y^* = (p_A/P_t)(M_A/M_{Air}) = (p_A/750)(17/29)$$

From the previous data, $p_A = 0.030$ mm Hg for 50 parts ammonia per 10^6 parts of water; thus, $Y^* = (0.030/750)(17/29) = 2.34 \times 10^{-5}$ part of ammonia per part air. The coordinates for the equilibrium curve are $X_1 = 0$, $Y_1 = 0$ and $X_2 = 5 \times 10^{-5}$, $Y_2 = 2.34 \times 10^{-5}$.

To determine the theoretical air required for an ammonia stripping tower, consider the schematic drawing shown in Figure 5.2. At steady state a materials balance on the ammonia gives [Input] = [Output] since there is no chemical reaction. Thus materials balance is

$$LX_2 + GY_1 = LX_1 + GY_2 \tag{5.5}$$

which may be simplified to

$$L(X_2 - X_1) = G(Y_2 - Y_1) \tag{5.6}$$

If it is assumed that there is no ammonia in the air entering $(Y_1 = 0)$ and, for engineering purposes, the ammonia concentration in effluent is negligible $(X_1 \approx 0)$, then the water to air ratio is given by a simplification of Eq. (5.6):

$$\frac{L}{G} = \frac{Y_2}{X_2} \tag{5.7}$$

It can be seen from Eq. (5.7) that the theoretical water to air flow ratio is equal to the slope of the equilibrium curve for the temperature and elevation of the tower under consideration. The reciprocal of the slope,

$$\frac{G}{L} = \frac{1}{\text{slope}} = \frac{1}{(Y_2)/(X_2)} \tag{5.8}$$

gives the air to water flow ratio for the theoretical air flow required. Usually, the design air flow is 1.50 to 1.75 times the theoretical.

An induced-draft packed countercurrent ammonia stripping tower is to be designed for an advanced wastewater treatment plant using the equilibrium curve data from Example Problem 5.1. Pertinent data for the design are as follows: design flow = 2.0 MGD, hydraulic load = 2.00 gpm/ft^2, design air flow = 1.75 times the theoretical, average water and air temperature = 20°C (68°F), atmospheric pressure = 750-mm Hg, and the wastewater pH = 10.8. Determine:

1. Theoretical air required, lb/hr-ft^2.
2. Design air required, lb/hr-ft^2.
3. Design air required, cfm/ft^2 and ft^3/gal.

Solution

The slope of the equilibrium curve is Y_2/X_2 or (23.4 × 10^{-6}lb NH$_3$/lb air)/(50 × 10^{-6}lb NH$_3$/lb water) or 0.468 lb water/lb air. Thus the air to water ratio, G/L, is 1/0.468 = 2.14 lb air/lb water. The mass flow rate of the water, L, is (2.00 gal/min-ft^2)(8.34 lb/gal)(60 min/hr) = 1000.8 lb water/hr-ft^2. Therefore, the theoretical air flow required, G = (2.14 lb air/lb water)(1000.8 lb water/hr-ft^2), or

$$\underline{G(\text{theoretical}) = 2141.7\,\text{lb air/hr-ft}^2}$$

The design air flow is

$$G(\text{design}) = (2141.7)\,1.75$$

$$= \underline{3748.0\,\text{lb air/hr-ft}^2}$$

For 20°C (68°F or 528°R) and 750-mm Hg atmospheric pressure, the design air flow, Q_G, is (3748.0 lb air/hr-ft^2)(1 lb-mole/29 lb)(359 ft^3/1 lb-mole)(528°R/492°R)(760 mm/750 mm)(hr/60 min) or

$$Q_G\,(\text{design}) = \underline{840.9\,\text{cfm/ft}^2}$$
Air per gallon = 840.9/2.00 = $\underline{420.5\,\text{ft}^3/\text{gal}}$

An induced-draft packed countercurrent ammonia stripping tower is to be designed for an advanced wastewater treatment plant using the equilibrium curve data from Example Problem 5.2. Pertinent data for the design are as follows: design flow = 7600 m^3/d, hydraulic load = 4.90 m^3/h-m^2, design air flow = 1.75 times the theoretical, average water and air temperature = 20°C, atmospheric pressure = 750-mm Hg, and the wastewater pH = 10.8. Determine:

1. Theoretical air required, kg/h-m^2.
2. Design air required, kg/h-m^2.
3. Design air required, m^3/h-m^2 and m^3/ℓ.

Solution

The slope of the equilibrium curve is Y_2/X_2 or (23.4 × 10^{-6} kg NH_3/kg air)/(50 × 10^{-6} kg NH_3/kg water) or 0.468 kg water/kg air. Thus the air to water ratio, G/L, is 1/0.468 = 2.14 kg air/kg water. The mass flow rate of the water, L, is (4.90 m^3/h-m^2)(1000 ℓ/m^3)(kg/ℓ) = 4900 kg water/h-m^2. Therefore, the theoretical air flow required, G = (2.14 kg air/kg water)(4900 kg water/h-m^2) or

$\underline{G(\text{theoretical}) = 10,486 \text{ kg air/h-}m^2}$

The design air flow is

$G(\text{design}) = (10,486) \, 1.75$

$\underline{\phantom{G(\text{design})} = 18,350 \text{ kg air/h-}m^2}$

For 20°C (293°K) and 750-mm Hg atmospheric pressure, the design air flow, Q_G, is (18,350 kg air/h-m^2) × (1 g-mole/29 g)(1000 g/kg)(22.4 ℓ/g-mole) (m^3/1000 ℓ) (760 mm/750 mm)(293°K/273°K) or

$Q_G(\text{design}) = \underline{15,415 \, m^3/h\text{-}m^2}$
Air per liter = (15,415 m^3/h-m^2)/(4.90 m^3/h-m^2)
$\phantom{\text{Air per liter}} \times (1000 \, \ell/m^3)$
$\phantom{\text{Air per liter}} = \underline{3.15 \, m^3/\ell}$

The temperatures of the wastewater and the air (since it indirectly affects the wastewater temperature) are important in ammonia stripping tower operation because of two effects. First, if the wet bulb temperature drops to 32°F the tower will probably be affected by ice formation within the tower. The winter operational data from South Tahoe have shown the average daily exit water temperature from the tower (38.6°F) to be only 0.8°F below the average daily ambient air temperature (39.4°F). Consequently, the average daily exit water temperature can be expected to be near the average daily ambient air temperature. Second, because the solubility of ammonia increases with a decrease in temperature, more air is required for stripping as the temperature decreases. Figure 5.3 shows the ammonia equilibrium curves for various temperatures at a total atmospheric pressure of 760 mm. Figure 5.4 shows the theoretical air flow required versus temperature. At 5°C the air required is about 3.5 times that at 30°C. Thus, tower operations are limited to warm operating conditions. In cold climates, it may be feasible to use a stripping tower during warm months and, during cold months, use an alternate method for removal, such as spray ponds.

Figure 5.3. Equilibrium Curves for Ammonia in Air and Water at an Atmospheric Pressure of 760 mm

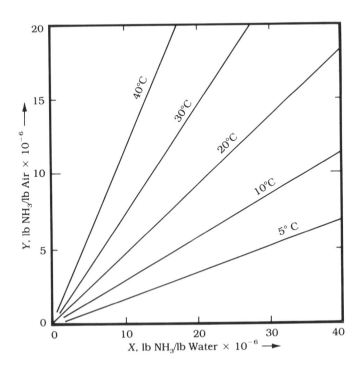

Design considerations for ammonia stripping towers are: unit hydraulic loading = 1 to 3 gpm/ft², with 1 to 2 gpm/ft² being typical; design air flow = 1.5 to 1.75 times the theoretical air flow; maximum air pressure drop for the entire tower < 2 to 3 in. of water; fan motor speed = 1 or 2 speed; fan tip speed = 9000 to 12,000 ft/min; packing depth = 20 to 25 ft; packing spacing = 2 to 4 in. horizontal and vertical, wood or plastic packing, and influent water pH = 10.8 to 11.5. The packing employed usually consists of an expanded grid made of pvc plastic pipe or redwood slats. Also, a polypropylene grid has been used.

Ponds

Gonzales and Culp (1973) have used high pH spray ponds for ammonia removal at South Tahoe. Some of the factors influencing the performance of a spray pond are: water temperature, air temperature, initial and final solute gas con-

Figure 5.4.   Theoretical Air Required for Ammonia Stripping at Various Temperatures at 760 mm, pH 11

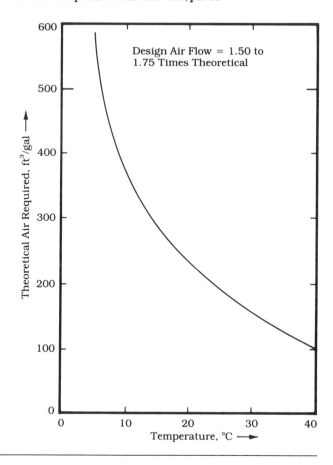

centration in the water, solute gas concentration in the air, air movement, pH, detention time, depth of pond, area of the pond, aeration time (spray height), droplet size, spray nozzle spacing, layout pattern of the spray nozzles, and the number of water turnovers that are sprayed. At South Tahoe, high pH spray ponds have been used as an alternate method for ammonia removal in cold weather. The sprays employed have large nozzles and produce coarse sprays with a height of about 15 ft; the pond depth is about 2.0 ft. It has been found that a 20-hour detention time and a water turnover of 5.2 cycles results in 52 percent removal of ammonia. Although the spray ponds do not give as good an ammonia removal as a stripping tower, pond operation at temperatures less than freezing is possible. Thus, high pH spray ponds are a feasible method for ammonia removal at low climatic temperatures.

Chemical Processes

Breakpoint chlorination and ion exchange (using clinoptilolite) are two unit processes that may be used for ammonia removal.

Breakpoint Chlorination

Chlorination by chlorine gas or hypochlorite salts will oxidize ammonia to form intermediate chloramines and, finally, to form nitrogen gas and hydrochloric acid. The reaction of chlorine gas with water to produce hypochlorous acid (HOCl) is

$$Cl_2 + H_2O \rightarrow HCl + HOCl \quad \underset{pH<7}{\overset{pH>8}{\rightleftharpoons}} \quad H^+ + OCl^- \qquad (5.9)$$

Hypochlorous acid will react with ammonia in a wastewater to produce monochloramine (NH_2Cl), dichloramine ($NHCl_2$), and trichloramine (NCl_3). These reactions are dependent upon the pH, temperature, reaction time, and initial chlorine to ammonia ratio. Monochloramine and dichloramine are formed in the pH range of 4.5 to 8.5. Above pH 8.5, the predominant form of the chloramines is monochloramine, whereas below pH 4.5, the predominant form is trichloramine. If the amount of chlorine added is greater than that required for the breakpoint and the pH is about 7 to 8, the intermediate formed monochloramine is oxidized to yield nitrogen gas which is liberated from the system. The series of steps required to oxidize the ammonia are as follows (Cassel, A. F., Pressley, T. A., Schuk, W. W., & Bishop, D. F., 1971):

$$Cl_2 + H_2O \rightarrow HOCl + HCl \qquad (5.10)$$

$$NH_4^+ + HOCl \rightarrow NH_2Cl + H_2O + H^+ \qquad (5.11)$$

$$2\,NH_2Cl + HOCl \rightarrow N_2 \uparrow + 3\,HCl + H_2O \qquad (5.12)$$

The overall reaction obtained from Eqs. (5.10), (5.11), and (5.12) is

$$3\,Cl_2 + 2NH_4^+ \rightarrow N_2 \uparrow + 6\,HCl + 2\,H^+ \qquad (5.13)$$

The stoichiometric amount of chlorine required is 3 moles of chlorine per 2 moles of ammonia or 7.5 lbs of chlorine per pound ammonia nitrogen. Studies have shown that for raw municipal wastewaters, secondary effluents, and lime clarified and filtered secondary effluents, the chlorine required per pound of ammonia nitrogen is 10:1, 9:1, and 8:1, respectively. The amount of chlorine required is independent of temperature for the range from 40° to 100°F and 95 to 99 percent of the ammonia will be oxidized to nitrogen gas. There may be some undesirable side reactions which produce dichloramine, trichloramine (which is very offensive in odor), and nitrate ions. If, however, the pH is maintained in the range from about 7 to 8, the side reactions are minimized. If the alkalinity is insufficient to maintain this pH range, a base such as NaOH must be added to neutralize the hydrochloric acid formed. The reactions given by Eqs. (5.10) through (5.13) occur rapidly and, under proper conditions, are completed within several minutes. Disadvantages of breakpoint chlorination are that the dissolved solids are increased, and the process is about twice as expensive as ammonia stripping. Also, due to adverse health implications of trihalomethanes that are formed during chlorination, the use of breakpoint chlorination for ammonia removal may decline in the future.

Ion Exchange

Cation exchange using a natural zeolite called clinoptilolite has been employed to remove ammonia (Mercer, B. W., Ames, L. L., Touhill, C. J., Van Slyke, W. J., & Dean, R. B., 1970). In this process the secondary effluent is treated by multimedia filtration, carbon adsorption, and then ion exchange using a fixed-bed column containing clinoptilolite. Upon exhaustion the clinoptilolite column is regenerated using a brine solution of about 2 percent NaCl, which is renovated by electrolysis. The spent regenerant contains mainly NH_4^{+1}, Ca^{+2}, Mg^{+2}, Na^{+1}, and Cl^{-1} ions. Prior to electrolysis, Na_2CO_3 and NaOH are added to precipitate as much $CaCO_3$ and $Mg(OH)_2$ as possible. The regenerant passes through electrolytic cells which produce Cl_2 at the anode and H_2 at the cathode. The chlorine reacts with the ammonia by breakpoint chlorination to produce N_2 gas, and N_2 and H_2 gases are removed as off-gases. After the electrolytic treatment the regenerant has been renovated so that it may be reused.

Biological Processes

The three-stage biological nitrification-denitrification process, as shown in Figure 5.5, removes ammonia in addition to organic, nitrite, and nitrate nitrogen. Also, phosphorus may be removed by adding a coagulant prior to the final clarifier for the first and third stages.

In the first stage, the principal biochemical action is the breakdown of carbonaceous and organic nitrogen matter to terminate the nitrogen in the form of ammonia. Usually a completely mixed reactor is used. If phosphorus removal is desired, a coagulant may be added prior to the final clarifier. The first stage will have a BOD_5 of about 50 mg/ℓ.

In the second stage, the main biochemical action is the conversion of ammonia nitrogen to the nitrate ion—that is, nitrification. A plug-flow reactor is usually used. Nitrification is accomplished by two aerobic autotrophic bacterial species, *Nitrosomonas* and *Nitrobacter*. *Nitrosomonas* converts the ammonium ion to the nitrite ion, whereas *Nitrobacter* converts the nitrite ion to the nitrate ion. The biochemical equation for nitrification proposed by McCarty (1970) is

---

Figure 5.5. Three-Stage Biological Nitrification-Denitrification Flowsheet

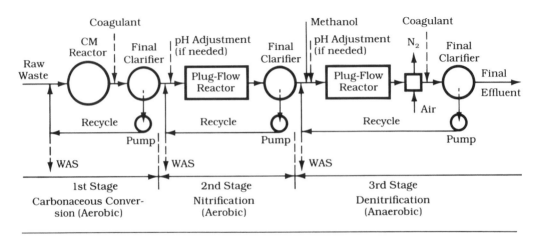

1st Stage Carbonaceous Conversion (Aerobic) | 2nd Stage Nitrification (Aerobic) | 3rd Stage Denitrification (Anaerobic)

$$NH_4^{+1} + 1.682\,O_2 + 0.182\,CO_2$$
$$+ 0.0455\,HCO_3^{-1} \rightarrow 0.0455\,C_5H_7NO_2$$
$$+ 0.955\,NO_3^{-1} + 0.909\,H_2O + 1.909\,H^{+1} \qquad (5.14)$$

From Eq. (5.14) it can be seen that the oxygen required for nitrification is appreciable and amounts to 2.99 mg oxygen per mg ammonium ion.

In the third stage, the main biochemical action is the anaerobic conversion of the nitrate ion to form nitrogen gas—that is, denitrification. A plug-flow reactor that has submerged mixing paddles is usually used. Since the carbon source is limited at the beginning of the third stage, a supplemental carbon source such as methanol is added. The biochemical equation proposed by McCarty for denitrification is (McCarty, P. L., Beck, L., & St. Amant, P., 1969)

$$NO_3^{-1} + 1.08\,CH_3OH + H^{+1} \rightarrow 0.065\,C_5H_7NO_2$$
$$+ 0.47N_2 + 0.76CO_2 + 2.44H_2O \qquad (5.15)$$

In the anaerobic mixing reactor, the nitrate ion is converted to nitrogen gas and no oxygen will be present in the reactor. Odors should not occur because the anaerobic bacteria prefer the nitrate ion (NO_3^{-1}) more than the sulfate ion (SO_4^{-2}) as a source of combined oxygen. After the mixture leaves the reactor it passes through a nitrogen stripping tank with a detention time of 5 to 10 minutes. Compressed air is released at the bottom to strip out the nitrogen gas. After stripping, the mixture passes to the final clarifier where the anaerobic sludge is separated and the effluent leaves. If a nitrogen stripping tank is not provided, floating sludge will occur in the final clarifier. If phosphorus removal is desired, a coagulant may be added prior to the final clarifier. In both the second and third stages, plug flow is approached by using long rectangular reactors with a large length to width ratio or rectangular reactors with baffles.

The optimum pH, the detention time based on the influent flow (that is, $\theta_t = V/Q$), and the mean cell residence time (θ_c) for the three stages are shown in Table 5.2.

The use of the three-stage nitrification-denitrification system employing chemical precipitation of phosphorus in the third-stage final clarifier has resulted in 86.3 percent COD, 95.5 percent suspended solids, 96.2 percent organic nitrogen, 97.3 percent ammonia nitrogen, and 88.1 percent phosphorus removals. The effluent contained 0.4 mg/ℓ organic, 0.3 mg/ℓ ammonia, 0.3 mg/ℓ nitrite, and 0.9

mg/ℓ nitrate nitrogen; the total phosphorus content was 1.5 mg/ℓ (EPA, 1971).

Table 5.2. Operational Parameters for the Three-Stage Biological Nitrification–Denitrification Process

	pH	θ_l (hrs)	θ_c (days)
1st Stage	6.5–8.0	<3	<5
2nd Stage	7.8–9.0	<3	10–25
3rd Stage	6.5–7.5	<2	<5

Taken from Environmental Protection Agency, Advanced Waste Treatment and Water Reuse Symposium, sessions 1–5, Dallas, Texas, 1971.

Design criteria for the three-stage activated sludge nitrification-denitrification process or any of the previously discussed process substitutions is presented in an EPA (1975) publication.

Miscellaneous Methods

Ammonia stripping, breakpoint chlorination, ion exchange, and the biological nitrification-denitrification process are the four main methods for ammonia removal. Oxidation ponds, however, may remove ammonia by anaerobic microbial denitrification in the lower pond depths, and some ammonia bio-oxidized to nitrate will be removed by assimilation in algal cells.

References

AWARE, Inc. 1974. *Process Design Techniques for Industrial Waste Treatment*, edited by C. E. Adams and W. W. Eckenfelder, Jr. Nashville, Tenn.: Enviro Press.

Barth, E. F., Brenner, R. C., and Lewis, R. F. 1968. Chemical-Biological Control of Nitrogen and Phosphorus in Wastewater Effluents. *Jour. WPCF* 40:2040.

Cassel, A. F., Pressley, T. A., Schuk, W. W., and Bishop, D. F. 1971. Physical-Chemical Nitrogen Removal from Municipal Wastewaters. Paper presented at the 68th national meeting of the American Institute of Chemical Engineers (AICh.E.), Houston, Tex.

Chilton, T. H., and Colburn, A. P. 1935. Distillation and Adsorption in Packed Columns. *Ind. and Eng. Chem.* 27, no. 3:255.

Colburn, A. P. 1939. The Simplified Calculation of Diffusional Processes, General Consideration of Two-Film Resistances. *Trans. AICh.E.* 35:211.

Conway, R. A., and Ross, R. D. 1980. *Handbook of Industrial Waste Disposal*. New York: Van Nostrand Reinhold.

Culp, G. L., and Hamann, C. L. 1974. Advanced Waste Treatment Process Selection; Parts 1, 2 and 3. *Public Works*. March, April, and May.

Culp, R. L., and Culp, G. L. 1978. *Handbook of Advanced Wastewater Treatment*. 2nd ed. New York: Van Nostrand Reinhold.

Eckenfelder, W. W., Jr. 1980. *Principles of Water Quality Management*. Boston: CBI Publishing.

Eliassen, R., and Tchobanoglous, G. 1969. Removal of Nitrogen and Phosphorus from Waste Water. *Environmental Science and Technology* 3, no. 6.

Environmental Protection Agency (EPA). 1971. Advanced Waste Treatment and Water Reuse Symposium, sessions 1–5, Dallas, Tex.

Environmental Protection Agency (EPA). 1974. Physical-Chemical Nitrogen Removal. EPA Technology Transfer Seminar Publication, Washington, D.C.

Environmental Protection Agency (EPA). 1975. *Nitrogen Control.* EPA Process Design Manual, Washington, D.C.

Gonzales, J. G., and Culp, R. L. 1973. New Developments in Ammonia Stripping. *Public Works.* May and June.

Lewis, W. K., and Whitman, W. G. 1924. Principles of Gas Absorption. *Ind. and Eng. Chem.* 16, no. 12:1215.

McCabe, W. L., and Smith, J. C. 1967. *Unit Operations of Chemical Engineering.* New York: McGraw-Hill.

McCarty, P. L. 1970. Biological Processes for Nitrogen Removal: Theory and Application. *Proceedings Twelfth Sanitary Engineering Conference.* Urbana, Ill.: University of Illinois.

McCarty, P. L., Beck, L., and St. Amant, P. 1969. Biological Denitrification of Wastewaters by Addition of Organic Materials. *Proceedings of the 24th Annual Purdue Industrial Waste Conference.* Part 2.

Mercer, B. W., Ames, L. L., Touhill, C. J., Van Slyke, W. J., and Dean, R. B. 1970. Ammonia Removal from Secondary Effluents by Selective Ion Exchange. *Jour. WPCF* 42, no. 2, part 2:R95.

Metcalf and Eddy, Inc. 1979. *Wastewater Engineering, Treatment, Disposal and Reuse.* New York: McGraw-Hill.

Mulbarger, M. C. 1971. Nitrification and Denitrification in Activated Sludge. *Jour. WPCF* 43, no. 10:2059.

O'Farrell, T. P., Frauson, F. P., Cassel, A. F., and Bishop, D. F. 1972. Nitrogen Removal by Ammonia Stripping. *Jour. WPCF* 44, no. 8:1527.

Perry, J. H. 1969. *Chemical Engineers Handbook.* 4th ed. New York: McGraw-Hill.

Pressley, T. A., Bishop, D. F., Pinto, A. P., and Cassel, A. F. 1973. *Ammonia-Nitrogen Removal by Breakpoint Chlorination.* Report prepared for the Environmental Protection Agency, contract no. 14-12-818.

Ramalho, R. S. 1977. *Introduction to Wastewater Treatment Processes.* New York: Academic Press.

Reynolds, T. D., and Westervelt, R. 1969. Ion Exchange as a Tertiary Treatment. Paper presented at the American Society of Civil Engineers annual meeting and national meeting on Water Resources Engineering, New Orleans, La.

Schroeder, E. D. 1977. *Water and Wastewater Treatment.* New York: McGraw-Hill.

South Tahoe Public Utility District. 1971. *Advanced Wastewater Treatment as Practiced at South Tahoe.* Tech. Report for the EPA, Project 17010 ELQ (WPRD 52-01-67).

Steel, E. W., and McGhee, T. J. 1979. *Water Supply and Sewerage.* New York: McGraw-Hill.

Sundstrom, D. W., and Klei, H. E. 1979. *Wastewater Treatment.* Englewood Cliffs, N.J.: Prentice-Hall.

Water Pollution Control Federation (WPCF). 1977. *Wastewater Treatment Plant Design.* WPCF Manual of Practice no. 8, Washington, D.C.

Weber, W. J., Jr. 1972. *Physicochemical Processes for Water Quality Control*. New York: Wiley-Interscience.

Wild, H. E., Sawyer, C. N., and McMahon, T. C. 1971. Factors Affecting Nitrification Kinetics. *Jour. WPCF* 43, no. 9:1845.

Problems

1. An induced-draft packed countercurrent ammonia stripping tower is to be designed for a tertiary treatment plant for a municipal wastewater. Pertinent data are: flow = 3.0 MGD, operating pressure = 710 mm, minimum operating water and air temperature = 15°C, design hydraulic loading = 2 gpm/ft^2, design air flow = 1.75 times the theoretical, and wastewater pH = 10.9. Determine:
   a. Theoretical air required, lb/hr-ft^2.
   b. Design air required, lb/hr-ft^2.
   c. Design air required, cfm/ft^2 and ft^3/gal.
   d. Plan dimensions if tower is square.

2. An induced-draft packed countercurrent ammonia stripping tower is to be designed for a tertiary treatment plant for a municipal wastewater. Pertinent data are: flow = 11,400 m^3/day, minimum operating water and air temperature = 15°C, design hydraulic loading = 4.85 m^3/h-m^2, design air flow = 1.75 times the theoretical, and wastewater pH = 10.9. Operating pressure = 710 mm. Determine:
   a. Theoretical air required, kg/h-m^2.
   b. Design air required, kg/h-m^2.
   c. Design air required, m^3/h-m^2 and m^3 of air/ℓ of wastewater.
   d. Plan dimensions if tower is square.

3. A tertiary treatment plant for a municipal wastewater treatment is to use breakpoint chlorination for ammonia removal. Pertinent data are: flow = 2.0 MGD and ammonia concentration = 28 mg/ℓ as NH$_3$- $\underline{N}$. Determine:
   a. The pounds of chlorine theoretically required per month.
   b. The pounds of chlorine actually required per month.

Chapter Six

ADSORPTION

Adsorption consists of using the capacity of an adsorbent to remove certain substances from a solution. Activated carbon is an adsorbent that is widely used in water treatment, advanced wastewater treatment, and the treatment of certain organic industrial wastewaters because it adsorbs a wide variety of organic compounds and also is economically feasible. In water treatment it is used to remove compounds which cause objectionable taste, odor, or color. In advanced wastewater treatment it is used to adsorb organic compounds, and in industrial wastewater treatment it is mainly used to adsorb toxic organic compounds. It is generally used in granular form either in batch, column (both fixed bed and countercurrent bed), or fluidized-bed operations, with fixed-bed columns being the most common. Occasionally, activated carbon is used in powdered form and is not recovered for regeneration; however, such application is usually limited to water treatment where the amounts of carbon used are not appreciable. Adsorbents other than activated carbon are used to a lesser extent in environmental engineering.

Adsorption

Adsorption is the collection of a substance onto the surface of the adsorbent solids, whereas *absorption* is the penetration of the collected substance into the solid. Since both of these frequently occur simultaneously, some choose to call the phenomena *sorption*. Although both adsorption and absorption occur in sorption by activated carbon and other solids, the unit operation is usually referred to as adsorption.

187

Adsorption may be classified as (1) *physical adsorption* and (2) *chemical adsorption.* Physical adsorption is primarily due to van der Waals' forces and is a reversible occurrence. When the molecular forces of attraction between the solute and the adsorbent are greater than the forces of attraction between the solute and the solvent, the solute will be adsorbed onto the adsorbent surface. An example of physical adsorption is the adsorption by activated carbon. Activated carbon has numerous capillaries within the carbon particles and the surface available for adsorption includes the surfaces of the pores in addition to the external surface of the particles. Actually, the pore surface area greatly exceeds the surface area of the particles and most of the adsorption occurs on the pore surfaces. For activated carbon the ratio of the total surface area to the mass is extremely large. In chemical adsorption, a chemical reaction occurs between the solid and the adsorbed solute and the reaction is usually irreversible. Chemical adsorption is rarely used in environmental engineering; however, physical adsorption is widely used.

Activated carbon is made from numerous materials such as wood, sawdust, fruit pits and coconut shells, coal, lignite, or petroleum base residues. The manufacture essentially consists of carbonization of the solids followed by activation using hot air or steam.

When activated carbon particles are placed in a solution containing an organic solute and the slurry is agitated or mixed to give adequate contact, the adsorption of the solute occurs. The solute concentration will decrease from an initial concentration, C_0, to an equilibrium value, C_e, if the contact time is sufficient during the slurry test. Usually, equilibrium occurs within about one to four hours. By employing a series of slurry tests it is usually possible to obtain a relationship between the equilibrium concentration (C_e) and the amount of organic substance adsorbed (x) per unit mass of activated carbon (m).

The Freundlich isotherm, which is an empirical formulation, frequently will represent the adsorption equilibrium over a limited range in solute concentration. One form of the equation is

$$\frac{x}{m} = X = K C_e^{1/n} \tag{6.1}$$

where

x = mass of solute adsorbed;

m = mass of adsorbent;

X = mass ratio of the solid phase—that is, the mass of adsorbed solute per mass of adsorbent;

C_e = equilibrium concentration of solute, mass/volume;

K,n = experimental constants.

Inspection of Eq. (6.1) shows that the adsorption isotherm should plot a straight line on log-log graph paper if the x-axis represents the solid-phase concentration, x/m, and the y-axis represents the liquid-phase concentration, C_e. The slope of the line will be $1/n$ and, once the value of $1/n$ is known, the value of K may be determined. One of the most important aspects of the Freundlich isotherm in relation to the feasibility of using carbon adsorption is the numerical value of n and the value of x/m when $C_e = C_0$. The n value is the same regardless of the units used for the equilibrium concentration. The constant, K, however, does fluctuate with different units employed for the equilibrium concentrations. The larger the n value and the x/m value (when $C_e = C_0$), the more economically feasible is the use of carbon adsorption. Usually, the x/m value is from 0.2 to 0.8 gm COD (chemical oxygen demand) per gm carbon.

Another isotherm, which frequently will represent adsorption equilibrium, is the Langmuir isotherm, which is

$$\frac{x}{m} = X = \frac{aKC_e}{1 + KC_e} \qquad (6.2)$$

where

a = mass of adsorbed solute required to completely saturate a unit mass of absorbent;

K = experimental constant.

The Langmuir isotherm was derived assuming that (1) there is a limited area available for adsorption, (2) the adsorbed solute material on the surface is only one molecule in thickness, and (3) the adsorption is reversible and an equilibrium condition is achieved. When the adsorbent is placed in a solution, the solute is adsorbed and, also, desorption occurs but the rate of adsorption is greater than the rate of desorption. Finally, an equilibrium condition is attained where the rate of adsorption is equal to the rate of desorption. The formulation, like the Freundlich isotherm, is valid only over a limited range in solute concentrations.

The rate of adsorption is limited by one of the various mass transport mechanisms involved. These are: (1) the movement of the solute from the bulk solution to the liquid film or boundary layer surrounding the adsorbent solid, (2) the diffusion of the solute through the liquid film, (3) the diffusion of the solute inward through the capillaries or pores within the adsorbent solid, and (4) the adsorption of the solute onto the capillary walls or surfaces. Step 2 is frequently termed *film diffusion* and step 3 is usually referred to as *pore diffusion*. Weber (1972) has found that in a stirred batch operation or in a continuous-flow operation operated at the velocities used in water or wastewater treatment, the rate of adsorption is usually limited by film diffusion (step 2) or, in some cases, by pore diffusion (step 3).

Example Problem 6.1
Batch-Type Adsorption
Tests

A phenolic wastewater, which has a phenol concentration of 0.400 gm/ℓ as TOC (total organic carbon), is to be treated by granular activated carbon. Batch tests (slurry-type) have been performed in the laboratory to obtain the relative adsorption. Determine the Freundlich isotherm constants.

Solution

The data from the batch tests are shown in columns (1), (2), and (3) of Table 6.1.

Table 6.1. Reduced Data from Batch or Slurry-Type Tests

(1) m (gm/ℓ)	(2) C_0 (gm/ℓ)	(3) C_e (gm/ℓ)	(4) $\Delta C, x$ (gm/ℓ)	(5) $X = x/m$ (gm/gm)
0.52	0.400	0.322	0.078	0.150
2.32	0.400	0.117	0.283	0.122
3.46	0.400	0.051	0.349	0.101
3.84	0.400	0.039	0.361	0.094
4.50	0.400	0.023	0.377	0.084
5.40	0.400	0.012	0.388	0.072
6.67	0.400	0.0061	0.3939	0.059
7.60	0.400	0.0042	0.3958	0.052
8.82	0.400	0.0011	0.3969	0.045

The reduced data are shown in columns (4) and (5). A plot of the data is shown in Figure 6.1 and the slope of the line is equal to 0.22. Therefore, $n = 1/\text{slope} = 1/0.22 = \underline{4.55}$. Taking the logarithm of both sides of Eq. (6.1) gives $\log X = \log K + 1/n \log C_e$. When $C_e = 1.0$, $X = K$, and thus from an extension of the graph, $\underline{K = 0.198}$. The isotherm is therefore

$$X = x/m = 0.198\,C_e^{1/4.55}$$

The maximum value of x/m is 0.166 gm TOC/gm carbon, which occurs when $C_e = C_0 = 0.400$ gm/ℓ TOC.

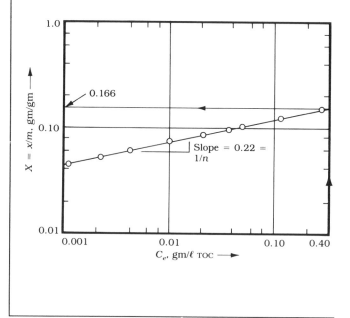

Figure 6.1. Graph for Determining Freundlich Isotherm Constants, Example Problem 6.1

Adsorption using a granular adsorbent may be accomplished by batch, column, or fluidized-bed operations. The usual contacting systems are fixed bed or countercurrent moving beds due to their lower labor costs and high utilization of the adsorption capacity of the adsorbent. The fixed beds may employ downflow or upflow of the liquid; however, downflow is more popular since the granular adsorbent bed may also serve as a filter for suspended solids in addition to the adsorption of organic substances. The countercurrent moving beds employ upflow of the liquid and downflow of the adsorbent solid since the adsorbent can be moved by the force of gravity. Both fixed beds and moving beds may use gravity or pressure liquid flow. The maximum available diameter of steel cylinders is about 12 feet, and pressurized flow is usually limited to these type vessels. If the contactor is to be larger than about 12 feet in diameter, it is usually built on the site and is of concrete and generally employs gravity flow.

Figure 6.2 shows a typical fixed-bed carbon column facility used in wastewater treatment employing a single column with downflow of the liquid. The column is similar to a pressure filter and has an inlet distributor, an underdrain

Column Contacting Techniques and Equipment

Figure 6.2. Fixed-Bed
Adsorption System Using a
Single Carbon Column

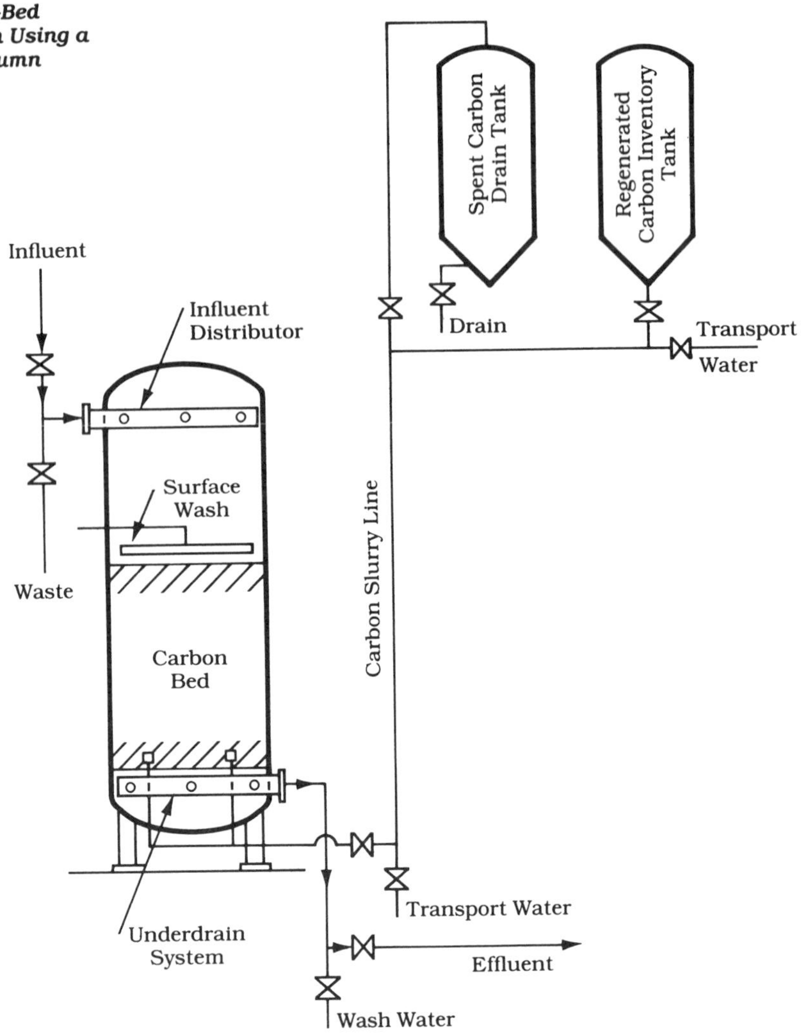

system, and a surface wash. During the adsorption cycle
the influent flow enters through the inlet distributor at the
top of the column, and the wastewater flows downward
through the bed and leaves through the underdrain system. The unit hydraulic flow rate employed is usually from
2 to 5 gpm/ft^2. When the head loss becomes excessive due
to the accumulated suspended solids, the column is taken
off-line and backwashed. Backwashing consists of operating the surface scour for a short duration, then using the
backwash water to expand the bed by 10 to 50 percent. This
will require a backwash of 10 to 20 gpm/ft^2 depending upon
the carbon particle size. To remove carbon for regeneration,

ENVIRONMENTAL ENGINEERING

the column is taken off-line and the bed is expanded by 20 to 35 percent using the backwash at a rate of 8 to 18 gpm/ft^2, depending upon the carbon particle size. The carbon-water slurry is removed from the bottom of the bed by opening the bottom valve and the transport water valve thus sending the carbon-water slurry to the spent carbon drain tanks. After draining, the spent carbon is regenerated using a furnace, which is usually a multiple hearth type, and the regenerated carbon is stored in the regenerated carbon inventory tank. In Figure 6.2, the regenerated carbon inventory tank is shown smaller than the carbon in the column for illustrative purposes; however, it actually holds enough carbon to fill an empty column. An empty column is refilled with carbon by using a carbon-water slurry transported from the regenerated carbon inventory tank. The slurry enters the bottom of the column and, once filled to the proper level, the column is placed back on-line. If the wastewater flow is continuous and no storage for the flow is provided, two columns are required since a column must be taken off-line to regenerate its carbon. In nearly all cases, two columns are provided and they are usually used in series.

Figure 6.3 shows a typical countercurrent moving-bed carbon column facility used in wastewater treatment employing upflow of the water. Two or more columns are usually provided and are always operated in series. The influent wastewater flow enters the bottom of the first column by means of a manifold system employing screens which uniformly distribute the flow across the bottom; the liquid flows upward through the column. The unit hydraulic flow rate employed is usually from 2 to 10 gpm/ft^2. The effluent is collected by a screen and manifold system at the top of the column and flows to the bottom manifold of the second column. The carbon flow is not continuous but instead is pulsewise. Spent carbon is removed and fresh carbon added when 5 to 10 percent of the total carbon in the column is spent. The wastewater flow does not have to be discontinued when carbon is withdrawn and more carbon added. To remove spent carbon, the inlet carbon valve at the top of the column is opened. If the effluent flow rises in the carbon filling chamber, some of the influent wastewater flow should be bypassed to the second column to lower the hydraulic gradient below the filling chamber. The jet water is started at the bottom of the column and the outlet carbon valve and transport water valve are opened. Due to the jet action and the weight of the carbon, the carbon-water slurry will flow from the column to the spent carbon drain tank, where the water is removed as drainage. Once

Figure 6.3. Moving-Bed Adsorption System Using Two Carbon Columns

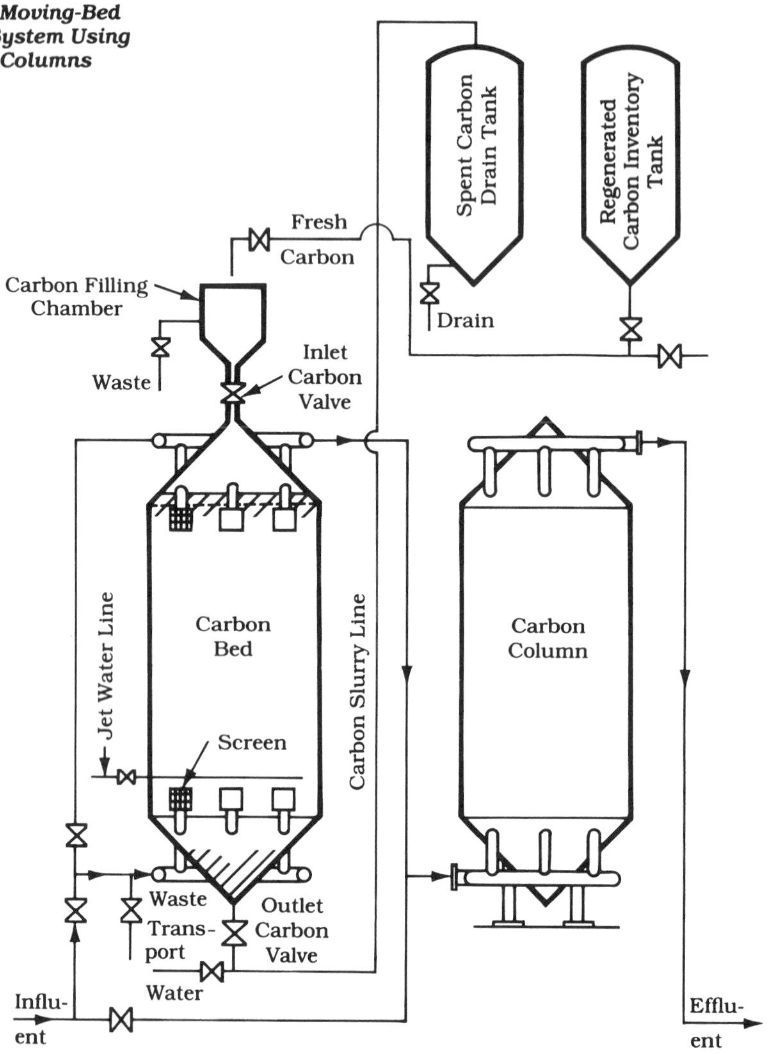

sufficient carbon has been withdrawn, the outlet carbon valve, the transport water valve, and the jet water valve are closed. To add fresh carbon, the fresh carbon-water slurry is allowed to flow into the carbon filling chamber. The fresh carbon flows downward into the column and the transport water from the filling chamber flows to waste. After fresh carbon is added, the inlet carbon valve is closed and, if any wastewater has been bypassed to the second column, the bypass flow is discontinued. To wash the column to remove accumulated suspended solids which have been filtered consists of reversing the flow through the column. The influent is temporarily sent to the second column although

ENVIRONMENTAL ENGINEERING

Figure 6.4. Typical Carbon Regenerating Facility

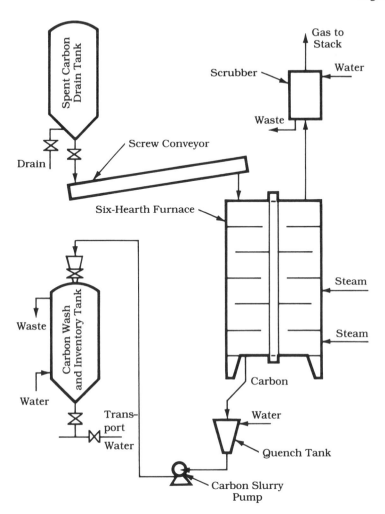

some of the influent is temporarily sent downward through the first column to backwash it. The backwash is for about 10 minutes at a rate of 10 to 12 gpm/ft^2, and the waste washwater leaves through the bottom manifold. After backwashing, the column is placed back on-line and is again operated in series with the second column. Although the type of column shown in Figure 6.3 is usually operated as a countercurrent moving bed, it may also be employed as a fixed bed.

Figure 6.4 shows typical equipment required to regenerate spent granular activated carbon. It consists of the spent carbon drain tank, a multiple-hearth furnace (usually six hearths), a quench tank, and the carbon wash and inventory tank. The furnace is heated to 1500 to 1700°F, then the spent carbon is fed by a screw conveyer to the multiple-hearth furnace. As the carbon passes through the first two hearths, it is dried and the adsorbed organic matter is baked. By the time the carbon has reached the third hearth the adsorbed organic materials have been ignited, thus evolving as gases and leaving a carbon residue in the pores of the activated carbon. The activating step consists of oxidizing the remaining carbon residue by the combined effects of heat, steam produced by evaporating water, oxygen and carbon dioxide present, and additional steam added. The regenerated carbon is discharged from the furnace bottom into the quench tank, where cold water is introduced to cool the carbon. The activated carbon slurry is then pumped to the carbon wash and inventory tank. The upward flowing wash water in the tank removes the fines which have been produced by the breakdown of larger carbon particles during regeneration. The regenerated carbon is stored in the carbon inventory tank and is sent to the carbon columns as a slurry.

Fixed-Bed Adsorption Columns

A fixed-bed adsorption column is shown in Figure 6.5 and a typical breakthrough curve for this type of column is shown in Figure 6.6. The breakthrough curve for a column shows the solute concentration in the effluent on the y-axis versus the effluent throughput volume on the x-axis. The length of the column in which adsorption occurs is termed the *sorption tone*, Z_s. It is in this zone that the solute is transferred from the liquid to the solid phase—that is, mass transfer of the solute occurs. Frequently, the sorption zone is termed the *mass transfer zone*. Above this zone the solute in the liquid phase is in equilibrium with that sorbed on the solid phase, and the solute concentration is C_0. Above the sorption zone the equilibrium solid-phase concentration is X_0 (q_0). The value of X_0 or q_0 is equal to the x/m value from slurry test data when $C_e = C_0$ (see Example Problem 6.1). The mass transfer fronts for the liquid and solid phases are shown at two different times, t_1 and t_2, in Figure 6.5. Both fronts have the same geometry at t_2 as they had at t_1; however, they have moved downward. The liquid flow in Figure 6.5 is shown moving downward through the column; however, if the flow is upward through the column the discussed fundamentals remain the same. As the sorption zone passes down the column, the concentration of the solute in the effluent is theoretically zero. Actually, a slight amount of leakage or bypassing of the organic solute

Figure 6.5. A Fixed-Bed Column and Its Liquid- and Solid-Phase Concentrations

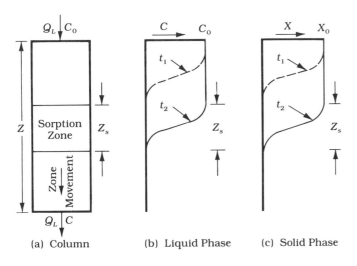

(a) Column   (b) Liquid Phase   (c) Solid Phase

Figure 6.6. Typical Breakthrough Curve

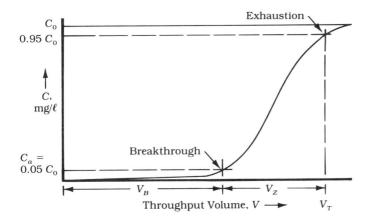

occurs, but it is negligible for illustrating the fundamentals. Once the sorption zone reaches the bottom of the column, the effluent solute concentration becomes a finite value and breakthrough begins, as shown in Figure 6.6. As the sorption zone disappears the effluent solute concentration increases to C_0 and the column is exhausted, resulting in a breakthrough curve as shown in Figure 6.6. In Figure 6.6 the allowable breakthrough concentration, C_a, is considered as $0.05\,C_0$. The allowable breakthrough concentration does not necessarily have to be this value, although 5 percent is commonly used. The exhaustion is considered as occurring at $C = 0.95\,C_0$. For the case of a symmetrical breakthrough curve (which is the usual case), it has been shown that the length of the sorption zone, Z_s, is related to the column height, Z, and the throughput volumes, V_B and V_T, by (Michaels, A.S., 1952),

$$Z_s = Z \left[\frac{V_Z}{V_T - 0.5\,V_Z} \right] \tag{6.3}$$

where $V_Z = V_T - V_B$. The area above the breakthrough curve represents the mass or amount of solute adsorbed by the column and is equal to $\int (C_0 - C)dV$ from $V = 0$ to $V =$ the allowable throughput volume under consideration. At the allowable breakthrough volume, V_B, the area above the breakthrough curve is equal to the mass or amount of solute adsorbed by the column. At complete exhaustion, $C = C_0$ and the area above the breakthrough curve is equal to the maximum amount of solute adsorbed. At complete exhaustion, the entire adsorption column is in equilibrium with the influent and effluent flows. Also, the solute concentration in the influent is equal to the concentration in the effluent.

It is not possible to accurately design a column without a test column breakthrough curve for the liquid of interest and the adsorbent solid to be used. In the subsequent text a scale-up and kinetic approach to design adsorption columns is presented. In both of the approaches a breakthrough curve from a test column, either laboratory or pilot scale, is required, and the column should be as large as possible to minimize side-wall effects. Neither of the design procedures requires the adsorption to be represented by an isotherm such as the Freundlich equation. A disadvantage to the scale-up and kinetic design approaches is that they do not take into account the effect of the unit hydraulic flow rate. This is incorporated in a third design approach using mass transfer fundamentals that has been developed and published by the author (Reynolds, T. D., & Pence, R. F., 1980).

Scale-Up Approach
This method was developed by Fornwalt and Hutchins (1966) for the design of carbon adsorption columns. The principal experimental data required is a breakthrough curve from a test column, either laboratory or pilot scale, which has been operated at the same liquid flow rate in terms of bed volumes per unit time, Q_b, as the design column. Since the contact time, T_c, is equal to ε/Q_b where ε is the pore fraction, the design column will have the same contact time as the test column. Since the contact times

are the same, it is assumed that the volume of liquid treated per unit mass of adsorbent, $\hat{V}_B$, for a given breakthrough in the test column is the same as for the design column. Before a breakthrough test can be performed, it is necessary to select a satisfactory liquid flow rate, Q_b, in bed volumes per unit time. This may be estimated from calculations using such information as the required breakthrough volume, the solute concentration, the maximum solid-phase concentration, and other pertinent data. Usually, the value of Q_b is from 0.2 to 3.0 bed volumes per hour.

The bed volume of the design column is given by

$$\text{Bed Volume } (BV) = \frac{Q}{Q_b} \qquad (6.4)$$

where Q is the design liquid flow rate. The mass or weight of the adsorbent, M, for the design column is determined from

$$M = (BV)(\rho_s) \qquad (6.5)$$

where ρ_s is the adsorbent bulk density. From the breakthrough curve for the laboratory- or pilot-scale column, the breakthrough volume, V_B, is determined for the allowable effluent solute concentration, C_a. The volume of liquid treated per unit mass of adsorbent, $\hat{V}_B$, is then determined by

$$\hat{V}_B = \frac{V_B}{M} \qquad (6.6)$$

where M is the mass of the adsorbent in the test column. The mass of adsorbent exhausted per hour, M_t, for the design column is computed from

$$M_t = \frac{Q}{\hat{V}_B} \qquad (6.7)$$

where Q is the design liquid flow rate. The breakthrough time, T, is

$$T = \frac{M}{M_t} \qquad (6.8)$$

where M is the mass of adsorbent in the design column. The calculated breakthrough volume, V_B, for the allowable breakthrough concentration, C_a, for the design column is

$$V_B = QT \qquad (6.9)$$

If the calculated breakthrough time, T, from Eq. (6.8) or the calculated breakthrough volume, V_B, from Eq. (6.9) is not acceptable, another liquid flow rate, Q_b, to give the required time or volume should be determined from the available breakthrough data. The breakthrough test on the laboratory- or pilot-scale column should be repeated using the new Q_b value. Then, the computations using Eqs. (6.4) through (6.9) should be repeated. Usually, one breakthrough test is adequate, although in some cases two tests must be done. The main advantages of this design procedure are its simplicity and the relatively small amount of experimental data required.

Example Problem 6.2
Fixed-Bed Column
Design by the Scale-Up
Approach

A phenolic wastewater having a TOC of 200 mg/ℓ is to be treated by a fixed-bed granular carbon adsorption column for a wastewater flow of 40,000 gal/day and the allowable effluent concentration, C_a, is 10 mg/ℓ as TOC. A breakthrough curve, shown in Figure 6.7, has been obtained from an experimental pilot column operated at 1.67 *BV*/hr. Other data concerning the pilot column are: inside diameter = 3.75 in., length = 41 in., mass of carbon = 2980 gm (6.564 lb), liquid flow rate = 12.39 ℓ/hr, unit liquid flow rate = 0.71 gpm/ft^2, and packed carbon density = 25 lb/ft^3 (401 gm/ℓ). Using the scale-up approach for design, determine:

1. The design bed volume, ft^3.
2. The design mass of carbon required, lb.
3. The breakthrough time, T, hr.
4. The breakthrough volume, V_B, gal.

Solution

The design bed volume is

$$BV = \frac{Q}{Q_b} = \frac{40,000\,\text{gal}}{24\,\text{hr}} \left| \frac{\text{hr}}{1.67\,BV} \right| \frac{\text{ft}^3}{7.48\,\text{gal}}$$

$$= \underline{133.42\,\text{ft}^3}$$

The mass of carbon required is

$$M = (BV)(\rho_s) = (133.42\,\text{ft}^3)(25\,\text{lb/ft}^3) = \underline{3336\,\text{lb}}$$

From the breakthrough curve, shown in Figure 6.7, the volume treated at the allowable breakthrough, C_a, of 10 mg/ℓ TOC is 2080 ℓ or 549.5 gal. The solution treated per pound of carbon, $\hat{V}_B$, is (549.5 gal/6.564 lb) or 83.71 gal/

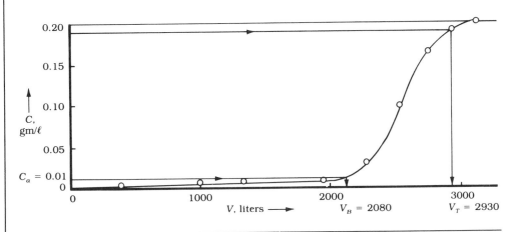

Figure 6.7. Breakthrough Curve for Example Problem 6.2

lb. The pounds of carbon exhausted per hour, M_t, is equal to the flow rate divided by the volume treated per pound of carbon or (40,000 gal/24 hr)(lb/83.71 gal), which is equal to 19.91 lb/hr. Thus, the breakthrough time, T, is

$$T = \frac{M}{M_t} = (3336\,\text{lb})(\text{hr}/19.91\,\text{lb})$$

$$= \underline{167.6\,\text{hr or }6.98\,\text{days}}$$

The breakthrough volume, V_B, for the design column is

$$V_B = QT = (40,000\,\text{gal}/24\,\text{hr})(167.6\,\text{hr})$$

$$= \underline{279,000\,\text{gal}}$$

Example Problem 6.2 SI Fixed-Bed Column Design by the Scale-Up Approach

A phenolic wastewater having a TOC of 200 mg/ℓ is to be treated by a fixed-bed granular carbon adsorption column for a wastewater flow of 150 m³/d and the allowable effluent concentration, C_a, is 10 mg/ℓ as TOC. A breakthrough curve, shown in Figure 6.7, has been obtained from an experimental pilot column operated at 1.67 BV/h. Other data concerning the pilot column are: inside diameter = 9.50 cm, length = 1.04 m, mass of carbon = 2.98 kg, liquid flow rate = 12.39 ℓ/h, unit liquid flow rate = 0.486 ℓ/s-m², and packed carbon density = 400 kg/m³. Using the scale-up approach for design, determine:

1. The design bed volume, m^3.
2. The design mass of carbon required, kg.
3. The breakthrough time, T, in hours and days.
4. The breakthrough volume, V_B, m^3.

Solution

The design bed volume is

$$BV = \frac{Q}{Q_b} = \frac{150\,m^3}{24\,hr} \left| \frac{h}{1.67\,BV} \right. = \underline{3.743\,m^3}$$

The mass of carbon required is

$$M = (BV)(\rho_s) = (3.743\,m^3)(400\,kg/m^3) = \underline{1497\,kg}$$

From the breakthrough curve, as shown in Figure 6.7, the volume treated at the allowable breakthrough, C_a, of 10 mg/ℓ TOC is 2080 ℓ. The solution treated per kg carbon, $\hat{V}_B$, is (2080 1/2.98 kg) or 698.0 ℓ/kg. The kilograms of carbon exhausted per hour, M_t, is equal to the flow rate divided by the volume treated per kilogram of carbon or (150 m^3/24 h)(kg/698.0 ℓ) (1000 ℓ/m^3) = 8.954 kg/h. The breakthrough time, T, is

$$T = \frac{M}{M_t} = (1497\,kg)(h/8.954\,kg) = \underline{167.2\,h\,or\,6.97\,d}$$

The breakthrough volume, V_B, for the design column is

$$V_B = QT = (150\,m^3/d)(6.97\,d) = \underline{1046\,m^3}$$

If the design procedure does not give the required breakthrough but gives a breakthrough time or volume near the required breakthrough, the design column volume can be computed from a rearrangement of Eq. (6.6), which is $M = V_B/\hat{V}_B$. The term $\hat{V}_B$ is the volume treated per unit weight of carbon in the breakthrough test. If, however, the design procedure gives a breakthrough time or volume which differs appreciably from that required, another test must be run. Using the data from the first test it is possible to estimate the flow rate, Q_b, in BV/hr to be used in the second test.

Kinetic Approach

This method utilizes a kinetic equation which is based on

the derivation by Thomas (1948). The kinetic equation may also be derived from an extension of the Bohart and Adams (1920) equation (Loebenstein, W. V., 1975). The principal experimental data required is a breakthrough curve from a test column, either laboratory or pilot scale.

The expression by Thomas for an adsorption column is as follows:

$$\frac{C}{C_0} \cong \frac{1}{1 + e^{\frac{k_1}{Q}(q_0M - C_0V)}} \tag{6.10}$$

where

C = effluent solute concentration;
C_0 = influent solute concentration;
k_1 = rate constant;
q_0 = maximum solid-phase concentration of the sorbed solute—for example, gm per gm*;
M = mass of the adsorbent—for example, gm*;
V = throughput volume—for example, liters;
Q = flow rate—for example, liters per hour.

Assuming the left side equals the right side, cross multiplying gives

$$1 + e^{\frac{k_1}{Q}(q_0M - C_0V)} = \frac{C_0}{C} \tag{6.11}$$

Rearranging and taking the natural logarithms of both sides yields the design equation

$$\ln\left(\frac{C_0}{C} - 1\right) = \frac{k_1 q_0 M}{Q} - \frac{k_1 C_0 V}{Q} \tag{6.12}$$

From Eq. (6.12), it can be seen that this is a straight-line equation of the type $y = b + mx$. The terms are $y = \ln(C_0/C - 1)$, $x = V$, $m = k_1 C_0/Q$, and $b = k_1 q_0 M/Q$.

The laboratory- or pilot-scale column used to obtain the breakthrough curve for the kinetic design approach should be operated at approximately the same flow rate in terms of bed volumes per hour as the design column. One advantage of the kinetic approach is that the breakthrough volume, V, may be selected in the design of a column.

---

*The solid-phase concentration could be expressed as lb per lb, the mass could be lb, and the flow rate could be gal per hr, and so on.

A phenolic wastewater having a TOC of 200 mg/ℓ is to be treated by a fixed-bed granular carbon adsorption column for a wastewater flow of 40,000 gal/day and the allowable effluent concentration, C_a, is 10 mg/ℓ as TOC. A breakthrough curve has been obtained from an experimental pilot column operated at 1.67 BV/hr. This is the same pilot column that was used in Example Problem 6.2 and the resulting breakthrough curve is shown in Figure 6.7. Other data concerning the pilot column are: inside diameter = 3.75 in., length = 41 in., mass of carbon = 2980 gm (6.564 lb), liquid flow rate = 12.39

Figure 6.8. Plot for Example Problem 6.3

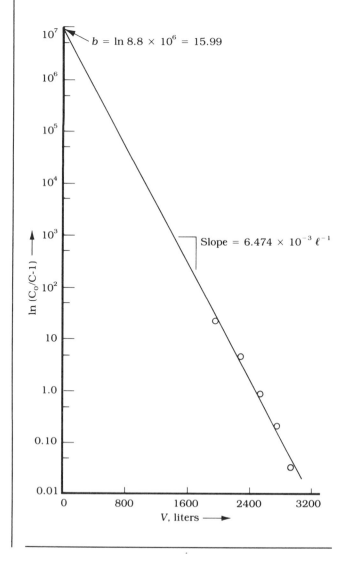

$b = \ln 8.8 \times 10^6 = 15.99$

Slope $= 6.474 \times 10^{-3}\ \ell^{-1}$

$\ln (C_o/C\text{-}1)$

V, liters $\longrightarrow$

ℓ/hr, unit liquid flow rate = 0.71 gpm/ft^2, and the packed carbon density = 25 lb/ft^3 (401 gm/ℓ). The design column is to have a unit liquid flow rate of 3 gpm/ft^2 and the allowable breakthrough volume is 280,000 gal. (This approximates the breakthrough volume in Example Problem 6.2 and it is used so that a comparison can be made between the design approach in Example Problem 6.2 and the approach in this problem.) Using the kinetic approach for design, determine:

1. The design reaction constant, k_1, gal/hr-lb.
2. The design maximum solid-phase concentration, q_0, lb/lb.
3. The pounds of carbon required for the design column.
4. The diameter and height of the design column.
5. The pounds of carbon required per 1000 gal treated.

Solution

The data from the breakthrough test are given in columns (1) and (2) of Table 6.2.

Table 6.2. Reduced Data from Breakthrough Test

(1) V (liters)	(2) C (gm/ℓ)	(3) C/C$_0$	(4) C$_0$/C	(5) C$_0$/C $-$ 1
378	0.009			
984	0.011			
1324	0.008			
1930	0.009	0.045	22.1	21.1
2272	0.030	0.150	6.67	5.67
2520	0.100	0.500	2.00	1.00
2740	0.165	0.825	1.21	0.21
2930	0.193	0.965	1.036	0.036
3126	0.200	1.000	1.00	0.00

The reduced data for the portion of the curve where breakthrough occurs is given in columns (3), (4), and (5). The plot of $\ln(C_0/C - 1)$ versus V is shown in Figure 6.8. The slope = $k_1 C_0/Q$ or $(\ln 8.8 \times 10^6 - \ln 1.0)/(0 - 2470)\ell = 6.474 \times 10^{-3} \ell^{-1}$. The value of k_1 = (slope)(Q/C$_0$) or k_1 = (6.474 × 10^{-3}/ℓ)(12.39 ℓ/hr)(ℓ/0.200 gm)(gal/3.785 ℓ)(454 gm/lb) = 48.11 gal/hr-lb. The y-axis intercept, b, equals $\ln 8.80 \times 10^6$ or 15.99. Since $b = k_1 q_0 M/Q$, rearranging gives $q_0 = bQ/k_1 M$ = (15.99)(12.39 ℓ/hr)(hr-lb/48.11 gal)(gal/3.785 ℓ)(1/6.564 lb) = 0.166 lb/lb. The mass of carbon, M, may be computed from

$$\ln\left(\frac{C_0}{C} - 1\right) = \frac{k_1 q_0 M}{Q} - \frac{k_1 C_0 V}{Q}$$

$$\ln\left(\frac{0.200}{0.010} - 1\right) = \frac{48.11\,\text{gal}}{\text{hr-lb}} \left| \frac{0.166\,\text{lb}}{\text{lb}} \right| \frac{\text{day}}{40,000\,\text{gal}}$$

$$\times \frac{24\,\text{hr}}{\text{day}} \left| M - \frac{48.11\,\text{gal}}{\text{hr-lb}} \right| \frac{0.200\,\text{gm}}{\ell} \left| \frac{3.785\,\ell}{\text{gal}} \right.$$

$$\times \frac{\text{lb}}{454\,\text{gm}} \left| \frac{\text{day}}{40,000\,\text{gal}} \right| \frac{280,000\,\text{gal}}{} \left| \frac{24\,\text{hr}}{\text{day}} \right.$$

From which, M = 3427 lb. The column diameter is D = $[(40,000$ gal/day)(day/1440 min)(min-ft^2/3.0 gal)(4/π)]$^{1/2}$ = 3.43 ft or 3 ft-6 in. Volume = (3427 lb)(1 ft^3/25 lb) = 137.08 ft^3. The carbon height is Z = (137.08 ft^3)(4/π)(1/3.5 ft)^2 = 14.25 ft. The pounds of carbon required per 1000 gal is (3427 lb/280,000 gal)(1000 gal/T gal) = 12.24 lb/1000 gal. The carbon required by the kinetic approach is 3427 lb and by the scale-up approach is 3336 lb.

Geometric Considerations

For a fixed-bed column it is desirable to have a large height to diameter ratio, Z/D, because the percent utilization of the maximum capacity of the adsorbent increases with this ratio. Usually, Z/D is from 3:1 to 5:1, and the unit liquid loading is from 4 to 10 gpm/ft^2. If head space is limited several short columns in series may be used to simulate a taller column.

Fixed-Bed Columns in Series

In most cases, it is advantageous to use two or more smaller fixed-bed columns operated in series rather than one large column containing the same amount of adsorbent. Fixed-bed columns are used in series when the headroom is limited and several small columns are used to simulate one large column. Also, fixed-bed columns in series are used when the increased utilization of the maximum adsorption capacity of the adsorbent will pay for the additional capital and operating costs of two or more smaller columns in lieu of one large column.

The operation of two columns in series may be illustrated by considering two granular activated carbon columns designated A and B, with A being the lead column. Once the sorption zone has passed from column A to column B, the carbon in column A is completely exhausted and the column is taken off-line, leaving column B on-line. After the carbon in column A is removed and replaced with fresh carbon, the column is placed back on-line but downstream from column B. Once the sorption zone has passed from column B to column A, column B is taken off-line and recharged with fresh carbon in the same manner as column A. This type of operation allows the carbon to become completely exhausted prior to regeneration. Thus, the carbon required per 1000 gal treated is less than that for a single, large fixed-bed column containing the same amount of carbon as the two smaller columns.

Moving-Bed Countercurrent Adsorption Columns

The moving-bed adsorption column (pulsed-bed) is a steady-state countercurrent operation since the adsorbent solid is moving downward through the column while the liquid is flowing upward. It is a common method of operation and approximately one-half the carbon columns used in advanced and tertiary wastewater treatment are of this type. A material balance on a moving-bed column at steady state gives

$$Q_L (C_1 - C_2) = L_S (X_1 - X_2) \tag{6.13}$$

where Q_L is liquid flow rate, L_S = adsorbent flow rate, C_1 and C_2 are the liquid-phase solute concentrations at the bottom and top of the column, and X_1 and X_2 are the solid-phase concentrations at the bottom and top of the column. The minimum theoretical height of a countercurrent adsorption column is equal to the height of the adsorption zone, Z_s. A column of the minimum theoretical height would require extreme care in operation to maintain the entire mass transfer front within the column. Therefore, in practice, the height of a countercurrent column is usually equal to the design height of a fixed-bed column for the same design situation. Once a column has been designed, V_T may be determined from the previous equations and the height of the adsorption zone may be estimated from Eq. (6.3).

Example Problem 6.4 Moving-Bed Countercurrent Adsorption Column Design

The phenolic wastewater in Example Problem 6.3 is to be treated by a moving-bed countercurrent granular carbon adsorption column and the allowable effluent concentration, C_a, is 10 mg/ℓ as TOC. Determine the pounds of carbon required per 1000 gal.

Solution

Equation (6.13) may be rearranged to give $L_S/Q_L = (C_1 - C_2)/(X_1 - X_2)$. $C_1 = 0.200$ gm/ℓ, $C_2 = 0.010$ gm/ℓ, $X_1 = 0.166$ gm/gm, and $X_2 = 0$ gm/gm. Thus, $L_S/Q_L = (0.200 - 0.010)(\text{gm}/\ell)/(0.166 - 0)(\text{gm/gm}) = 1.145$ gm/ℓ. The carbon per 1000 gal is (1.145 gm/ℓ)(lb/454 gm) × (3.785 ℓ/gal)(1000 gal/T gal) = 9.55 lb/1000 gal. The countercurrent system is more efficient than a single fixed-bed column since the carbon required in Example Problem 6.3 was 12.24 lb/1000 gal.

Example Problem 6.5
Sorption Zone Height

The adsorption column from Example Problem 6.3 contained 3427 lb carbon and was 3.50 ft in diameter and 14.25 ft in height. The flow was 40,000 gal/day and the allowable breakthrough was $C_a = 5$ percent C_0 or 0.010 gm/ℓ. Determine the height of the adsorption zone, Z_s, using the scale-up approach and the kinetic approach.

Solution

From the breakthrough curve, Figure 6.7, $V_B = 2080$ ℓ at $C = 5$ percent C_0 and $V_T = 2930$ ℓ at $C = 95$ percent C_0. The test column contained 6.564 lb carbon; thus, the volume treated per pound of carbon is $\hat{V}_B = (2080$ $\ell)(\text{gal}/3.785$ $\ell)(1/6.564$ lb) or 83.72 gal/lb and $\hat{V}_T = (2930$ $\ell)(\text{gal}/3.785$ $\ell)(1/6.564$ lb) or 117.93 gal/lb. Thus, V_B for the design column = (83.72 gal/lb)(3427 lb) = 286,908 gal and $V_T = (117.93$ gal/lb)(3427 lb) = 404,146 gal. The value $V_Z = 404,146 - 286,908 = 117,238$ gal. The sorption zone is now obtained from Eq. (6.3):

$$Z_s = 14.25 \text{ ft} \left[\frac{117,238}{404,146 - 0.5(117,238)} \right]$$

$$= 4.84 \text{ ft}$$

From the computations in Example Problem 6.3, $V_B = 280,000$ gal at $C = 5$ percent C_0. Using the equation in Example Problem 6.3 to determine V_T gives $\ln(1/0.95 - 1) = (48.11$ gal/hr-lb)(0.166 lb/lb)(day/40,000 gal) × (24 hr/day)(3427 lb) − (48.11 gal/hr-lb)(0.200 gm/ℓ)(3.785 ℓ/gal)(lb/454 gm)(day/40,000 gal)(24 hr/day)V_T. From the equation $V_T = 402,354$ gal. The value $V_Z = 402,354 - 280,000 = 122,354$ gal. The sorption zone height is now obtained from Eq. (6.3):

$$Z_s = 14.25 \, \text{ft} \left[\frac{122{,}354}{402{,}354 - 0.5(122{,}354)} \right]$$

$$= \underline{5.11 \, \text{ft}}$$

Fluidized Beds

Occasionally fluidized beds are used in water treatment and the advanced treatment of wastewaters. The fluidized bed consists of a bed of granular adsorbent solids. The liquid flows upward through the bed in the vertical direction. The upward liquid velocity is sufficient to suspend the adsorbent solids so that the solids do not have constant inter-particle contact. At the top of the solids there is a distinct interface between the solids and the effluent liquid. The principal advantage of the fluidized bed is that liquids with appreciable suspended solids content may be given adsorption treatment without clogging the bed, since the suspended solids pass through the bed and leave with the effluent. Usually they are operated in a continuous countercurrent fashion. The ratio of the adsorbent required per given amount of liquid, L_s/Q_L, is the same as for a countercurrent moving-bed adsorption column. For further reading the text by Weber (1972) is recommended.

Test Columns

The larger the laboratory or pilot column, the smaller will be the scale-up effects. A laboratory column should be at least 1 in. in diameter (ID) and the height should be at least 24 in. Carbon manufacturers have numerous good publications on laboratory techniques. It has been found that side-wall effects or the channelling of flow down the inside of a test column are of significant magnitude if the unit liquid flow rate is appreciable. The maximum unit liquid flow rates to prevent appreciable side-wall effects are as follows: $1\frac{3}{8}$ in. ID, 0.50 gpm/ft^2; $2\frac{3}{4}$ in. ID, 1.00 gpm/ft^2; and $3\frac{3}{4}$ in. ID, 1.50 gpm/ft^2.

Design Considerations

Carbon contactors may employ gravity-flow or pressurized-flow contactors. For pressurized contactors, which are usually fixed-bed columns, the carbon beds may be operated in a downflow or upflow fashion; however, the downflow fashion is the most common. For downflow fixed-bed columns, the carbon size is usually 8 × 30 mesh, whereas, for upflow fixed-bed columns, the carbon size is usually 12 × 40 mesh. Carbon beds have depths from 10 to 40 ft, with 15 to 20 ft being the most common. Column height must be sufficient to allow for 50 percent bed expansion for backwashing of downflow fixed beds or for 15 percent bed expansion for the backwashing of upflow fixed beds. For fixed-bed columns, it is desirable to have a large carbon bed depth to diameter ratio, Z/D, because the percent utilization of the adsorbent increases with this ratio. Usually Z/D is from 1.5:1 to 4:1. Prefabricated steel cylinders may be obtained in diameters up to 12 ft in 3-in. intervals; however, 6-in. intervals are more common, and diameters range from 2 ft–6 in. to 12 ft. The operating pressure drop through a carbon bed depends mainly upon the unit hydraulic loading and the carbon size. Usually it is less than 1.0 psi per ft of bed depth. For downflow or upflow fixed-bed columns, the filtration rate is from 2 to 5 gpm/ft^2, whereas for pulsed-bed columns, the filtration rate is from 2 to 10 gpm/ft^2. Backwash rates for downflow fixed beds using 8×30 mesh carbon are from 10 to 20 gpm/ft^2, which gives bed expansions of 10 to 50 percent. Backwashing of fixed-bed upflow columns using 12×40 mesh carbon merely consists of increasing the upflow rate to 10 to 12 gpm/ft^2, which gives a bed expansion of 10 to 15 percent. With upflow columns, a bed expansion of 10 percent is adequate for washing particulates from the bed. The backwash for downflow or upflow fixed-bed columns should last from 10 to 15 minutes. Backwashing of pulsed-bed columns consists of reversing the flow at a backwash rate of 10 to 12 gpm/ft^2 for about 10 to 15 minutes. Since the backwash is downwards, expansion of the bed does not occur.

Gravity-flow carbon beds usually are the downflow type using 8×30 mesh carbon and the filtration rates are from 2 to 4 gpm/ft^2. Most gravity-flow beds are open beds constructed using reinforced concrete and a bed depth of at least 10 ft is used. Backwash rates are from 15 to 20 gpm/ft^2, which will expand the bed depth from about 25 to 50 percent. Backwashing should be for a 10- to 15-min duration.

Expanded or fluidized beds are usually constructed using reinforced concrete and are open beds with a depth of at least 10 ft. The filtration rates are from 6 to 10 gpm/ft^2 and the carbon size is usually 12×40 mesh. Bed expansion is about 10 percent, which requires an upflow of 6 gpm/ft^2 for 12×40 mesh carbon or 10 gpm/ft^2 for 8×30 mesh carbon. Backwashing is not required since a 10 percent bed expansion allows the particulates to pass through the bed.

In order to have filter runs with a suitable duration, the feed flow to the carbon contactors, with the exception of the fluidized bed, should have less than 5 mg/ℓ suspended solids. For the usual flow schemes, this places the carbon contactors as one of the last treatment processes.

Due to the wide variety of industrial and municipal wastewaters treated by activated carbon, the numerous organic compounds present, and the different characteristics of the various activated carbons, it is usual practice to report results in terms of the most significant parameters. Usually, these are the unit hydraulic loading in gpm/ft^2, the empty-bed contact time, the pounds of COD removed per pound of carbon and the effluent COD. For municipal wastewaters treated by activated carbon in tertiary treatment or physical-chemical treatment, the unit hydraulic loadings are in the ranges that have been previously discussed for the various contacting methods. The empty-bed contact time is from 30 to 45 minutes and the organic removal is from 0.2 to 0.8 lb COD per pound of carbon. The effluent COD will be from 0.5 to 15 mg/ℓ.

References

Bohart, G. S., and Adams, E. Q. 1920. Some Aspects of the Behavior of Charcoal with Respect to Chlorine. *Jour. ACS* 42:523.

Burleson, N. K., Eckenfelder, W. W., Jr., and Malina, J. F., Jr. 1968. Tertiary Treatment of Secondary Industrial Effluents by Activated Carbon. *Proceedings of the 23rd Annual Purdue Industrial Waste Conference.* Part 1.

Conway, R. A., and Ross, R. D. 1980. *Handbook of Industrial Waste Disposal.* New York: Van Nostrand Reinhold.

Cookston, J. T., Jr. 1970. Design of Activated Carbon Adsorption Beds. *Jour. WPCF* 42, no. 12:2124.

Culp, R. L., and Culp, G. L. 1978. *Handbook of Advanced Wastewater Treatment.* 2nd ed. New York: Van Nostrand Reinhold.

Eckenfelder, W. W., Jr. 1980. *Principles of Water Quality Management.* Boston: CBI Publishing.

Environmental Protection Agency (EPA). 1973. *Carbon Adsorption.* EPA Process Design Manual, Washington, D.C.

Fornwalt, J. J., and Hutchins, R. A. 1966. Purifying Liquids with Activated Carbon. *Chem. Eng.* (11 April): 1979; (9 May): 155.

Keinath, T. M., and Weber, W. J., Jr. 1968. A Predictive Model for the Design of Fluid-Bed Adsorbers. *Jour. WPCF* 40, no. 5:741.

Loebenstein, W. V. 1975. Comparison of Column Decolorization Experiments with Theory. *Proceedings of the Fifth Technical Session on Bone Char.*

Lukchis, G. M. 1973. Adsorption Systems, Part I, II and III. *Chem. Eng.* (11 June):111; (9 July):83; (6 August):83.

Metcalf and Eddy, Inc. 1979. *Wastewater Engineering, Treatment, Disposal and Reuse.* New York: McGraw-Hill.

Michaels, A. S. 1952. Simplified Method of Interpreting Kinetics Data in Fixed-Bed Ion Exchange. *Ind. and Eng. Chem.* 44: 1922–1930.

Pittsburg Activated Carbon Division. 1966. The Laboratory Evaluation of Granular Activated Carbons for Liquid Phase Applications. Calgon Corp., Pittsburgh, Pa.

Ramalho, R. S. 1977. *Introduction to Wastewater Treatment Processes*. New York: Academic Press.

Reynolds, T. D., and Pence, R. F. 1980. Design of an Activated Carbon System for Wood Preserving Wastes. *Proceedings of the 35th Annual Purdue Industrial Waste Conference*, West Lafayette, Ind.

Sanks, R. L. 1978. *Water Treatment Plant Design*. Ann Arbor, Mich.: Ann Arbor Science Publishers.

Schroeder, E. D. 1977. *Waste and Wastewater Treatment*. New York: McGraw-Hill.

Sundstrom, D. W., and Klei, H. E. 1979. *Wastewater Treatment*. Englewood Cliffs, N.J.: Prentice-Hall.

Thomas, H. C. 1948. Chromatography: A Problem in Kinetics. *Annals of the New York Academy of Science* 49:161.

Water Pollution Control Federation (WPCF). 1977. *Wastewater Treatment Plant Design*. Manual of Practice no. 8, Washington, D.C.

Weber, W. J., Jr. 1972. *Physicochemical Processes for Water Quality Control*. New York: Wiley-Interscience.

Weber, W. J., Jr., and Ying, W. 1979. Bio-physicochemical Adsorption Model Systems for Wastewater Treatment. *Jour. WPCF* 51, no. 11:2661.

Witco Chemical Company. Column Evaluation Techniques. *Tech. Bull.* 5–6, New York.

Problems

1. A pilot column breakthrough test has been performed using the phenolic wastewater in Example Problem 6.1. Pertinent design data are: inside diameter = 3.75 in., length = 41 in., mass of carbon = 2980 gm (6.564 lb), liquid flow rate = 17.42 ℓ/hr, unit liquid flow rate = 1.00 gpm/ft^2, and packed carbon density = 25 lb/ft^3 (401 gm/ℓ). The breakthrough data are given in Table 6.3.

Table 6.3.

Throughput Volume, V (gal)	Effluent TOC, C (mg/ℓ)
4	12
18	16
42	24
72	16
100	16
180	20
255	28
292	32
321	103
340	211
372	350
409	400

Determine:

  a. The liquid flow rate in bed volumes per hour and the volume of liquid treated per unit mass of carbon—in other words, the gal/lb at an allowable breakthrough of 35 mg/ℓ TOC.

  b. The kinetic constants k_1 in gal/hr-lb and q_0 in lb/lb.

2. The phenolic wastewater in Problem 1 is to be treated by a fixed-bed granular carbon adsorption column for a wastewater flow of 60,000 gal/day. The allowable breakthrough concentration, C_a, is 35 mg/ℓ as TOC. The design column will have a unit liquid flow rate of 3.5 gpm/ft^2. The design procedure is to be the scale-up approach using the values from Problem 1. Determine the column dimensions.

3. The phenolic wastewater in Problem 1 is to be treated by a fixed-bed granular carbon adsorption column for a wastewater flow of 60,000 gal/day. The allowable breakthrough concentration, C_a, is 35 mg/ℓ as TOC. The design column will have a unit liquid flow rate of 3.5 gpm/ft^2 and the on-line time is seven days. The design procedure is to be the kinetic approach using the k_1 and q_0 values from Problem 1. Determine:

  a. The column dimensions.

  b. The number of columns in series to be used if the available headroom is 18 ft.

4. The phenolic wastewater in Problem 1 is to be treated by a countercurrent moving-bed column (pulsed-bed) and the allowable breakthrough concentration, C_a, is 35 mg/ℓ as TOC. The wastewater flow = 60,000 gal/day. Determine the pounds of carbon required per day.

5. A tertiary treatment plant treats a municipal secondary effluent and is to employ fixed-bed carbon adsorption columns. Parallel treatment using two rows of two columns in series (in other words, a total of four columns) will be used. The flowsheet consists of lime coagulation, settling, ammonia stripping, recarbonation, settling, multimedia filtration, carbon adsorption, and chlorination. From batch-type slurry tests it has been found that 0.42 lb of COD is adsorbed when $C_e = C_0$. Pertinent data are: flow = 1.5 MGD, COD in feed to carbon columns = 20 mg/ℓ, contact time function based on an empty carbon bed = 30 min, and unit liquid flow rate = 6.50 gpm/ft^2. Each column is kept on-line until the entire mass of carbon in the column is completely exhausted. Determine:

a. The volume of each column, ft³.
b. The diameter and height of each column, ft.
c. The mass of carbon in each column if the packed density is 25 lb/ft³.
d. The pounds of carbon exhausted per day if the COD removed is assumed to be 98 percent.
e. The on-line time for each of the columns.

6. A tertiary treatment plant treats a municipal secondary effluent and is to employ fixed-bed carbon adsorption columns. Parallel treatment using two rows of two columns in series (in other words, a total of four columns) will be used. The flowsheet consists of lime coagulation, settling, ammonia stripping, recarbonation, settling, multimedia filtration, carbon adsorption, and chlorination. From batch-type slurry tests it has been found that 0.42 g COD/g carbon is adsorbed when $C_e = C_0$. Pertinent data are: flow = 5700 m³/d, COD in feed to carbon columns = 20 mg/ℓ, contact time function based on an empty carbon bed = 30 min, and unit liquid flow rate = 4.4 ℓ/s-m². Each column is kept on-line until the entire mass of carbon in the column is completely exhausted. Determine:
a. The volume of each column, m³.
b. The diameter and height of each column, m.
c. The mass of carbon in each column if the packed density is 400 kg/m³.
d. The kilograms of carbon exhausted per day if the COD removed is assumed to be 98 percent.
e. The on-line time for each of the columns.

Chapter Seven

ION EXCHANGE

The ion exchange process consists of a chemical reaction between ions in a liquid phase and ions in a solid phase. Certain ions in the solution are preferentially sorbed by the ion exchanger solid and, because electroneutrality must be maintained, the exchanger solid releases replacement ions back into the solution. For instance, in the softening of water by the ion exchange process, the calcium and magnesium ions are removed from the solution and the exchanger solid releases sodium ions to replace the removed calcium and magnesium ions. The reactions are stoichiometric and reversible and obey the law of mass action.

The first commercially used ion exchange materials were naturally occurring porous sands that were commonly called zeolites. These minerals have a deficit of positive atoms within their crystalline structure and, as a result, they have a net negative charge which is balanced by exchangeable cations held within the pore capillaries. Zeolites were the first ion exchangers used to soften waters; however, they have been almost completely replaced in recent years by synthetic organic exchange resins which have a much higher ion exchange capacity. Synthetic cation exchange resins are polymeric materials that have reactive groups, such as the sulfonic, phenolic, and carboxylic, that are ionizable, and may be charged with exchangeable cations. Also, synthetic anion exchange resins are available that have ionizable groups, such as the quaternary ammonium or amine groups, which may be charged with exchangeable anions. Thus, synthetic resins are available that have both cation and anion exchange capabilities. Certain minerals, such as montmorillonite clays, have appreciable cation exchange capacity and are used for special ion exchange applications.

Ion exchange is used extensively in both water and wastewater treatment. Some of the common applications are water softening, demineralization, desalting, ammonia removal, treatment of heavy metal wastewaters, and treatment of some radioactive wastes. (1) One of the largest uses of ion exchange in environmental engineering is the softening of water by the exchange of sodium ions for calcium and magnesium ions. Since removal of all of the hardness is undesirable for a domestic water supply, a portion of the flow may bypass the exchangers to give a blended water of the desired hardness. This process is termed split-flow softening. (2) Ion exchange is used for the removal of all cations and anions from a water. In total demineralization, the cationic resins are charged with the hydrogen ion and the anionic resins are charged with the hydroxyl ion. The cationic resins exchange hydrogen ions for cations and the anionic resins exchange hydroxyl ions for anions. Thus the treated water has only hydrogen and hydroxyl ions, which makes it essentially pure water. Industries using high pressure boilers require demineralized water as boiler water. In addition, there are other industries that require demineralized water. (3) Ion exchange may be used for partial demineralization of wastewaters in tertiary treatment and of brackish waters for water supplies. Several ion exchange techniques may be employed, one of which is split-flow demineralization. (4) The natural zeolite, clinoptilolite, may be used to remove ammonia in advanced waste treatment plants in lieu of or in addition to other methods of ammonia removal. (5) Ion exchangers may be used to remove heavy metallic ions from certain wastewaters. The heavy metallic ions are thus concentrated in the spent regenerate. An example is the treatment of wastewaters from a metal plating industry that contains zinc, cadmium, copper, nickel, and chromium. (6) Clays and other minerals possessing ion exchange capacities are used to treat low- or moderate-level radioactive wastes to preferentially remove heavy metallic radionuclides such as Cs^{137}.

A number of texts and references are available on the theory and application of the ion exchange process and laboratory evaluation procedures (Betz, 1962; Dorfner, K., 1971; Dow, 1964; Helfferich, F., 1962; Kitchener, J. A., 1961; Kunin, R., 1960).

Theory

The softening of a hard water by an exchanger solid, either a zeolite or a synthetic resin, may be represented by the following reactions:

$$Ca^{+2} + 2\,Na \cdot Ex \rightleftharpoons Ca \cdot Ex_2 + 2\,Na^{+1} \qquad (7.1)$$

$$Mg^{+2} + 2\,Na \cdot Ex \rightleftharpoons Mg \cdot Ex_2 + 2\,Na^{+1} \qquad (7.2)$$

where Ex represents the exchanger solid. As shown by the reactions, a hard water may be softened by exchanging Na^{+1} from the exchanger solid for the Ca^{+2} and Mg^{+2} in the solution. After the solid is saturated with the Ca^{+2} and Mg^{+2}, it may be regenerated by a strong salt solution since the reactions are reversible. The regeneration reactions are

$$Ca \cdot Ex_2 + 2\,Na^{+1} \xrightarrow[\leftarrow]{strong\,brine} 2\,Na \cdot Ex + Ca^{+2} \quad (7.3)$$

$$Mg \cdot Ex_2 + 2\,Na^{+1} \xrightarrow[\leftarrow]{strong\,brine} 2\,Na \cdot Ex + Mg^{+2} \quad (7.4)$$

After regeneration, the exchanger solid is washed to remove the remaining brine and then is placed back on-line to soften more water.

In the demineralization of water, the water is first passed through cation exchange resins charged with the hydrogen ion. The cation removal may be represented by the reaction

$$M^{+x} + xH \cdot Re \rightleftharpoons M \cdot Re_x + xH^{+1} \qquad (7.5)$$

where M^+ represents the cationic species present and x is the valence number. After passing through the cation exchange resins, the water then passes through anion exchange resins charged with the hydroxyl ion. The anion removal may be represented by the reaction

$$A^{-z} + zRe \cdot OH \rightleftharpoons Re_z \cdot A + zOH^{-1} \qquad (7.6)$$

where A^- represents the anion species present and z is the valence number. After the resins become exhausted, the cation exchange resins are regenerated using a strong mineral acid such as sulfuric or hydrochloric acid. In a similar manner, the anion exchange resins are regenerated using a strong base such as sodium hydroxide.

Exchange resins are usually bead- or granular-shaped, having a size of about 0.1 to 1.0 mm. The ion exchange ability of resins may be classified as "strong" or "weak" according to the characteristics of the exchange capability (Dorfner, K., 1971). The strong acid cation resins have strong reactive sites such as the sulfonic group ($-SO_3H$), and the resins readily remove all cations. Conversely, the weak acid cation exchange resins have weak reactive sites such as the carboxylic group ($-COOH$), and these resins readily remove cations from the weaker bases such as Ca^{+2} and Mg^{+2} but have limited ability to remove cations from strong bases such as Na^{+1} and K^{+1}. The strong base anion exchange resins have reactive sites such as the quaternary ammonium group, and these resins readily remove all anions. The weak base anion exchange resins have reactive sites such as the amine group, and these resins remove mainly anions from strong acids such as SO_4^{-2}, Cl^{-1}, and NO_3^{-1}, with limited removal for HCO_3^{-1}, CO_3^{-2}, or SiO_4^{-4}.

Ion exchange resins or zeolites have a limited number of exchange sites available, and the total solid-phase concentration, $\hat{q}_0$, is termed the *ion exchange capacity*. For cation exchange resins it is usually between 200 to 500 meq per 100 gms. Since a cation exchanger must remain electrically neutral during the exchange reaction, all of the exchange sites must be occupied by sufficient cations to balance the negative charge of the exchanger. Thus, for a system involving Ca^{+2}, Mg^{+2}, and Na^{+1}, the sum of the solid-phase concentrations of these ions must, at any time, be equal to the cation exchange capacity, $\hat{q}_0$. Electroneutrality applies to anion exchangers as well as cation exchangers.

Since ion exchange is a chemical reaction, the law of mass action may be applied. The generalized equation for cation exchange by a resin may be represented by

$$M_1^+ + Re \cdot M_2 \rightleftarrows M_2^+ + Re \cdot M_1 \qquad (7.7)$$

where M_1^+, M_2^+ are cations of different species and Re is the resin. The mass action constant is as follows:

$$K\frac{M_1^+}{M_2^+} = \frac{[Re \cdot M_1][M_2^+]}{[Re \cdot M_2][M_1^+]} \qquad (7.8)$$

$$= \left[\frac{M_1}{M_2}\right]_{solid} \times \left[\frac{M_2}{M_1}\right]_{solution} \qquad (7.9)$$

where

$$K\frac{M_1^+}{M_2^+} = \text{mass action constant or selectivity coefficient.}$$

In Eqs. (7.8) and (7.9) the bracketed terms represent equilibrium concentrations expressed in an appropriate concentration unit. The magnitude of K represents the relative preference for ion exchange. Thus, it is the relative preference of the resin to sorb cation M_1^+ as compared to cation M_2^+.

The greater the selectivity coefficient, K, the greater is the preference for the ion by the exchanger. An ion exchanger tends to prefer: (1) ions of higher valence, (2) ions with a small solvated volume, (3) ions with greater ability to polarize, (4) ions that react strongly with the ion exchange sites of the exchanger solid, and (5) ions that participate least with other ions to form complexes. For the usual cation exchangers, the preference series for the most common cations is as follows (Helfferich, F., 1962):

$$Ba^{+2} > Pb^{+2} > Sr^{+2} > Ca^{+2} > Ni^{+2} > Cd^{+2} > Cu^{+2}$$

$$> Co^{+2} > Zn^{+2} > Mg^{+2} > Ag^{+1} > Cs^{+1} > K^{+1}$$

$$> NH_4^{+1} > Na^{+1} > H^{+1}$$

This series is for strong acid resins—that is, those having strong reactive sites such as the sulfonic group ($-SO_3H$). Weak acid resins—that is, those having weak reactive sites such as the carboxylic group ($-COOH$)—will have the H^{+1} position to the left of that shown above depending upon the strength of the reactive site. For very weak sites, the H^{+1} may fall to the left as far as Ag^{+1}. For the usual anion exchangers, the preference series for the most common anions is as follows (Helfferich, F., 1962):

$$SO_4^{-2} > I^{-1} > NO_3^{-1} > CrO_4^{-2} > Br^{-1} > Cl^{-1} > OH^{-1}$$

This series is for strong base resins—that is, those having strong reactive sites such as the quaternary ammonium group. For weak base resins—those having weak reactive sites such as the secondary or tertiary amine group—the OH^{-1} will fall farther to the left depending upon the strength of the reactive group. It should be understood that the previous series are to be used as a guides and that exceptions occur. Since equilibria are involved, the most suitable method to determine the uptake by a certain exchanger solid for a particular ion in a solution is to perform sorption tests. These may be done by (1) using a column to obtain the breakthrough curve for the particular ion of interest or (2) using batch-type slurry tests to obtain the uptake of the particular ion.

The rate of ion exchange depends upon the rates of the various transport mechanisms involved and the rate of the exchange reaction itself. These are as follows: (1) movement of the ions from the bulk solution to the film or boundary layer surrounding the exchanger solid, (2) diffusion of the ions through the film to the solid surface, (3) diffusion of the ions inward through the pores of the solid to the exchange sites, (4) exchange of the ions by the reaction, (5) diffusion of the exchanged ions outward through the pores to the solid surface, (6) diffusion of the exchanged ions through the liquid film or boundary layer surrounding the solid, and (7) movement of the exchanged ions into the bulk solution. Weber (1972) has shown that in a stirred batch process or in a continuous-flow process operated at the velocities used in water and waste treatment systems, the rate of exchange is usually controlled by step 2 or, in some cases, by step 3. Step 2, the diffusion of ions through the film or boundary layer, is frequently referred to as *film diffusion*, and step 3, the diffusion of ions through the pores, is termed *pore diffusion*.

Contacting Techniques and Equipment

All of the operating techniques ordinarily used for adsorption are also used for ion exchange. Thus, there are batch, column, or fluidized-bed processes. Column processes may consist of fixed-bed columns in which the beds are stationary or moving-bed columns in which both the liquid and the ion exchanger solid move countercurrently. Also, fluidized beds may be operated countercurrently. Of all the contacting systems, the fixed-bed columns are the most common and the second most popular are the countercurrent moving-bed columns. The popularity of these two contacting techniques is due mainly to their reduced labor costs. Frequently in ion exchange applications, there may be several columns or stages in series.

Figure 7.1 shows a schematic drawing of typical equipment required for a fixed-bed column used for water softening with a sodium-charged exchange resin. The equipment consists of the exchange column, a salt storage tank, a regeneration solution tank, and all required appurtenances. The exchange column is usually a steel cylindrical tank with an inlet distribution system and an underdrain collection system. The resins inside the vessel are supported on a graded gravel layer. The controller controls the flow to and from the column and contains the necessary valves to change the system from the softening cycle to the regeneration cycle and vice versa.

During the softening cycle, the water enters the top of the bed and flows downward at a constant flow rate of about 1 to 8 gpm/ft^2. Once the allowable breakthrough of hardness occurs in the effluent, the controller is activated so that backwashing is accomplished to remove any suspended material that may have accumulated by filtering action

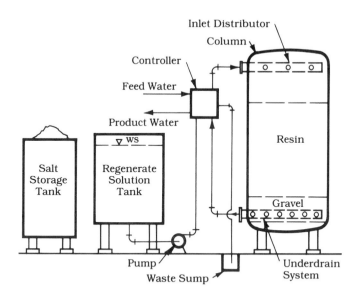

Figure 7.1. Typical Fixed-Bed Column Equipment for Softening Using a Sodium-Charged Resin

during the softening cycle. After backwashing, the salt solution, which is usually 5 to 10 percent salt, is passed through the exchanger bed at a controlled rate to regenerate the resins. Once regeneration is completed, a slow rinse flow is passed through the bed to displace the remaining regenerate solution to waste. This is followed by a fast, short rinse to remove the last traces of the regenerate solution from the bed. Once the fast rinse is completed, the column is placed back on-line to continue the softening process.

Design Procedures

The breakthrough curves for an ion exchange column and an adsorption column are similar, and the contacting techniques are almost identical. Consequently, the same procedures for the design of adsorption columns may be used for ion exchange columns. The scale-up approach by Fornwalt and Hutchins (1966) and the kinetic approach by Thomas (1948) also may be applied to the design of ion exchange columns by the same procedures presented in Chapter 6. A laboratory- or pilot-scale breakthrough curve is required for both procedures. The breakthrough curve for a column shows the solute or ion concentration in the

Figure 7.2.
Ion Exchange Softening

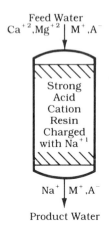

Feed Water
Ca^{+2}, Mg^{+2} | M^+, A^-

Strong
Acid
Cation
Resin
Charged
with Na^{+1}

Na^+ | M^+, A^-

Product Water

effluent on the y-axis versus the effluent throughput volume on the x-axis. The area above the breakthrough curve represents the amount of solute or ions taken up by the column and is $\int(C_0 - C)dV$ from $V = 0$ to $V =$ the allowable throughput volume under consideration. At the allowable breakthrough volume, V_B, the area above the breakthrough curve is equal to the amount of ions removed by the column. At complete exhaustion, $C = C_0$ and the area above the breakthrough curve is equal to the maximum amount of ions removed by the column. At complete exhaustion, the entire exchange column is in equilibrium with the influent and effluent flows. Also, the ion concentration in the influent is equal to the ion concentration in the effluent.

Another design approach is to determine the meqs or equivalent weights of the ions removed by a test column using the breakthrough curve. The throughput volume under consideration should be the allowable breakthrough volume, V_B, at the allowable breakthrough concentration, C_a. The ratio of the amount of ions removed per unit mass of exchanger is computed. Then, using this ratio, the mass of exchanger required is calculated from the allowable breakthrough volume for the design column and the concentration of the polyvalent metallic ions to be removed from the liquid flow. For this method to be valid, the flow rate used for the test column in terms of bed volumes per hour must be similar to the flow rate of the design column. A design procedure based on mass transfer fundamentals and the unit transfer concept is also available (Michaels, A. S., 1952).

Softening and Demineralization

Softening may be achieved by using a strong acid cation exchanger on the sodium cycle (that is, charged with Na^{+1}), as depicted in Figure 7.2. The Ca^{+2}, Mg^{+2}, and other divalent or polyvalent metallic ions are sorbed by the exchanger solid, and Na^{+1} ions are released into the solution. Since electroneutrality must be maintained, one meq of divalent or polyvalent cations removed causes one meq of sodium ions to be released. Once breakthrough occurs the bed is regenerated using a strong brine (NaCl) solution.

Demineralization where silica reduction is not required may be accomplished by the flowsheet shown in Figure 7.3. This consists of a strong acid cation exchanger on the hydrogen cycle, a weak base anion exchanger on the hydroxyl cycle, and a carbon dioxide stripping or degasification unit. Once the water passes through the cation ex-

Figure 7.3. Ion Exchange Demineralization

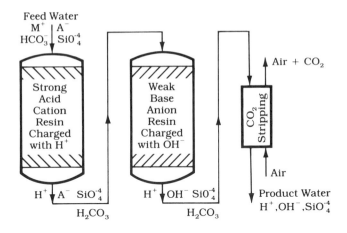

changer the cations have been removed and exchanged for hydrogen ions, resulting in an acidic water. The bicarbonates and carbonates present are converted to carbonic acid due to the low pH. The water then passes through a weak base anion exchanger and the anions, with the exception of silicates, are removed. The water then passes through a carbon dioxide stripper (degasifier), where excess carbon dioxide is removed. The product water will contain some silicate ions since weak base anion exchangers have limited removal for these. If the carbon dioxide content is not objectionable or the feed water has a low alkalinity, carbon dioxide stripping is not required. Regeneration is achieved using a solution of a strong mineral acid, such as sulfuric or hydrochloric, for the cation exchanger, and a sodium hydroxide solution for the anion exchanger.

Demineralization where silica removal is required may be accomplished by the flowsheet shown in Figure 7.4. This consists of a strong acid cation exchanger on the hydrogen cycle, a carbon dioxide stripper or degasifier, and a strong base anion exchanger on the hydroxyl cycle. Once the water passes through the cation exchanger, it goes through the carbon dioxide stripper where excess carbon dioxide is removed. The water then goes through the strong base anion exchanger, which removes all anions, including silicates. If the feed water has a low alkalinity the carbon dioxide stripper may be omitted. The product water will be essentially pure since the cations and anions have been removed, leaving only the hydrogen and hydroxyl ions, which form water.

Figure 7.4. Ion Exchange Demineralization

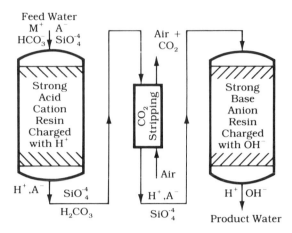

Figure 7.5. Mixed-Resin Demineralization

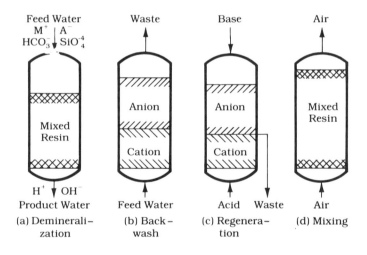

Demineralization may also be accomplished by a mixed-resin bed consisting of both strong acid cation exchange resins mixed with strong base anion exchange resins, as depicted in Figure 7.5 (a). The flowsheets shown in Figures 7.3 and 7.4 give up to 99 percent dissolved solids removal; however, a mixed-resin bed will give greater removal. Once breakthrough occurs in the mixed bed, it is backwashed to separate the resins. Since the anion resins are less dense than the cation resins, they will move to the top of the bed, as shown in Figure 7.5 (b). Regeneration is accomplished using a strong mineral acid solution and a strong base solution, as illustrated in Figure 7.5 (c). After regeneration, air is introduced into the bottom of the bed to remix the resins, as shown in Figure 7.5 (d). Frequently mixed-resin beds are used to polish waters demineralized by the flowsheets shown in Figures 7.3 and 7.4.

Other demineralization flowsheets employing more than two exchange columns may be used and are reported in the literature (Betz, 1962; Dorfner, K., 1971; Helfferich, F., 1962; Kitchener, J. A., 1961; Kunin, R., 1960; Nordell, E., 1961; Perry, R. H., & Chilton, C. H., 1973; Weber, W. J., 1972).

Example Problem 7.1
Ion Exchange in Waste
Treatment

An industrial wastewater with 107 mg/ℓ of Cu^{+2} (3.37 meq/ℓ) is to be treated by an exchange column. The allowable effluent concentration, C_a, is 5 percent C_0. A breakthrough curve, shown in Figure 7.6, has been obtained from an experimental laboratory column on the sodium cycle. Data concerning the column are: inside diameter = 0.5 in., length = 18 in., mass of resin = 41.50 gm on a moist basis (23.24 gm on a dry basis), moisture = 44 percent, bulk density of resin = 44.69 lb/ft^3 on moist basis, and liquid flow rate = 1.0428 ℓ/day. The design column will have a flow rate of 100,000 gal/day, the allowable breakthrough time is seven days of flow, and the resin depth is approximately two times the column diameter. Using the kinetic approach to column design determine:

1. The pounds of resin required.

Figure 7.6. Breakthrough Curve for Example Problem 7.1

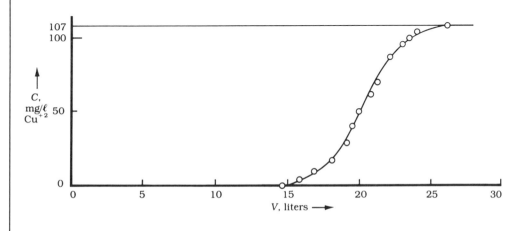

2. The diameter and depth.
3. The height of the sorption zone.

Solution

In this problem, the design equation to be used is Eq. (6.12). The data for the breakthrough test are given in columns (1) and (2) of Table 7.1.

Table 7.1. Reduced Data from Breakthrough Test

(1) V (liters)	(2) C (mg/ℓ)	(3) C (meq/ℓ)	(4) C/C_0	(5) C_0/C	(6) C_0/C-1
15.9	4.45	0.14	0.041	24.29	23.29
16.9	9.85	0.31	0.091	10.97	9.97
18.1	17.16	0.54	0.159	6.30	5.30
19.1	27.56	0.88	0.259	3.86	2.86
19.5	40.03	1.26	0.371	2.70	1.70
20.0	49.56	1.56	0.459	2.18	1.18
20.7	62.90	1.98	0.582	1.72	0.72
21.2	68.89	2.20	0.647	1.55	0.55
22.0	86.41	2.72	0.800	1.25	0.25
22.9	94.03	2.96	0.871	1.15	0.15
23.4	98.17	3.09	0.917	1.09	0.09
24.0	102.93	3.24	0.961	1.05	0.05
26.0	107.00	3.37	1.000	1.01	0.01

The plot of $\ln(C_0/C - 1)$ versus V is shown in Figure 7.7. The slope $= k_1C_0/Q$ or 0.7583 ℓ^{-1}. The value of k_1 = (slope)(Q/C_0) or k_1 = (0.7583/ℓ)(1.0428 ℓ/day)(ℓ/3.37 meq)(1000 meq/eq)(gal/3.785 ℓ) = 61.994 gal/day-eq. The y-axis intercept, b, equals $\ln 4.540 \times 10^6$ or 15.33. Since $b = k_1q_0M/Q$, rearranging gives $q_0 = bQ/k_1M$ = (15.33)(1.0428 ℓ/day)(day-eq/61.994 gal)(ℓ/23.24 gm)(454 gm/lb)(gal/3.785 ℓ) = 1.3309 eq/lb. The mass of resin may be computed from

$$\ln\left(\frac{C_0}{C} - 1\right) = \frac{k_1q_0M}{Q} - \frac{k_1C_0V}{Q}$$

or

$$\ln\left(\frac{C_0}{0.05\,C_0} - 1\right) = \frac{61.994\,\text{gal}}{\text{day-eq}} \left| \frac{1.3309\,\text{eq}}{\text{lb}} \right.$$

$$\times \frac{\text{day}}{100,000\,\text{gal}} \left| M - \frac{61.994\,\text{gal}}{\text{day-eq}} \right| \frac{3.37\,\text{meq}}{\ell}$$

$$\times \frac{\text{eq}}{1000\,\text{meq}} \left| \frac{3.785\,\ell}{\text{gal}} \right| \frac{\text{day}}{100,000\,\text{gal}} \left| 700,000\,\text{gal} \right.$$

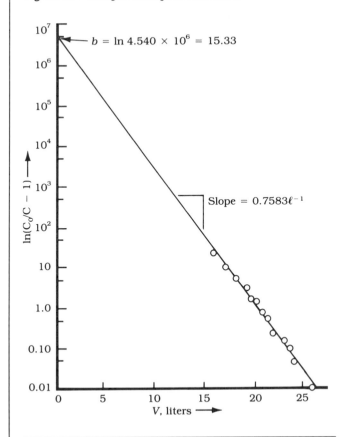

Figure 7.7. Plot for Example Problem 7.1

$b = \ln 4.540 \times 10^6 = 15.33$

$\ln(C_0/C - 1)$

Slope $= 0.7583\ell^{-1}$

V, liters

From which $M = \underline{10{,}278\ \text{lb dry weight}}$. $\text{Ft}^3 = (10{,}278$ lb$)(1/0.56)(\text{ft}^3/44.69\ \text{lb}) = 410.7\ \text{ft}^3$. Since $(\pi/4)(D^2)(2D)$ $= 410.7\ \text{ft}^3$, $D = 6.39\ \text{ft}$. Use $D = \underline{6.50\ \text{ft}}$ and $Z = (410.7$ $\text{ft}^3)(4/\pi)(1/6.50\ \text{ft})^2 = \underline{12.38\ \text{ft}}$. At 5 percent C_0 break-through, the breakthrough volume $= 700{,}000$ gal. Using 95 percent C_0 and the previous equation gives a throughput volume of 1,444,760 gal. Thus $V_T = 1{,}444{,}760$ gal, $V_B = 700{,}000$ gal, and $V_Z = 1{,}444{,}760 - 700{,}000 = 744{,}760$ gal. Substituting into the equation $Z_s = Z[V_Z/(V_T - 0.5\ V_Z)]$ gives

$$Z_s = 12.38\ \text{ft}\left[\frac{744{,}760}{1{,}444{,}760 - 0.5(744{,}760)}\right]$$

$$= \underline{8.60\ \text{ft}}$$

Example Problem 7.2

Ion Removal

For the test column and breakthrough curve given in Example Problem 7.1, determine the meq of Cu^{+2} ion removed per 100 gm resin on a dry weight basis at the allowable breakthrough volume, V_B, for $C_a = 0.05\ C_0$. Also, determine the meq of Cu^{+2} ion removed per 100 gm resin on a dry weight basis at complete exhaustion. The dry weight of resin used was 23.24 gm.

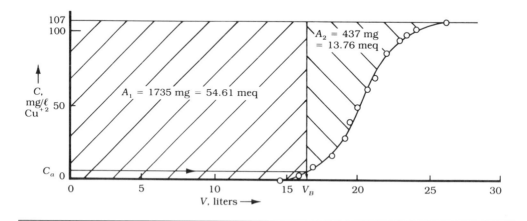

Figure 7.8. Breakthrough Curve for Example Problem 7.2

Solution

The meq weight of Cu^{+2} is 63.54/2 or 31.77 mg. The area above the breakthrough curve from V equal to zero to V equal to the volume under consideration is equal to the ion content removed by the exchanger. Figure 7.8 shows the breakthrough curve and the area above the breakthrough curve out to V_B, the allowable breakthrough volume for $C_a = 0.05\ C_0$. The area, A_1, is 1735 mg or 1735/31.77, which is equal to 54.61 meq. From the allowable breakthrough volume, V_B, to the volume at exhaustion, the area, A_2, is 437 mg or 437/31.77, which is equal to 13.76 meq. At exhaustion, the total area is 54.61 + 13.76 or 68.37 meq. The copper removed at $C_a = 0.05\ C_0$ is 54.61 meq/23.24 gm or 235.0 meq/ 100 gm dry weight. The copper removed at exhaustion is 68.37 meq/23.24 gm or 294.2 meq/100 gm dry weight.

Example Problem 7.3
Ion Exchange Softening

A well water is to be softened by split-flow ion exchange with the exchanger on the sodium cycle. The flow is 50 gpm, the hardness = 225 mg/ℓ as $CaCO_3$, the desired hardness is 50 mg/ℓ as $CaCO_3$, and the moisture content of the resin is 45 percent. A test column has been used in the laboratory to obtain a breakthrough curve. The computed hardness removed by the resin at the allowable breakthrough concentration $C_a = 0.05 \, C_0$ was 282 meq/100 gm resin on a dry weight basis. Determine the pounds of resin required if the allowable breakthrough is seven days.

Solution

A materials balance on the flow downstream from the exchanger is $Q_e(0) + (Q_b)(225) = (Q_e + Q_b)50$, where Q_e = flow through the exchanger and Q_b = flow bypassing the exchanger. From the balance, $Q_b = 0.28570 \, Q_e$. Since $Q_b + Q_e = 50$ gpm, $0.2857 \, Q_e + Q_e = 50$ gpm or $Q_e = 38.89$ gpm. The hardness of the well water in meq/ℓ is $(225 \, mg/\ell)(meq/50 \, mg)$ or 4.50 meq/ℓ. The hardness removed in seven days is $(38.89 \, gal/min)(1440 \, min/day)(7 \, day/week)(3.785 \, \ell/gal)(4.50 \, meq/\ell) = 6,676,930$ meq. Thus, the amount of resin $= (6,676,930 \, meq)(100 \, gm/282 \, meq)(lb/454 \, gm) = \underline{5215 \, lb \; dry \; weight}$. On a moist basis, this is $(5215 \, lb)[1/(1 - 0.45)] = \underline{9482 \, lb \; moist \; weight}$.

Example Problem 7.3 SI
Ion Exchange Softening

A well water is to be softened by split-flow ion exchange with the exchanger on the sodium cycle. The flow is 3.2 ℓ/s, the hardness is 225 mg/ℓ as $CaCO_3$, the desired hardness is 50 mg/ℓ as $CaCO_3$, and the moisture content of the resin is 45 percent. A test column has been used in the laboratory to obtain a breakthrough curve. The computed hardness removed by the resin at the allowable breakthrough concentration $C_a = 0.05 \, C_0$ was 282 meq/100 g resin on a dry weight basis. Determine the kilograms of resin required if the allowable breakthrough is seven days.

Solution

A materials balance on the flow downstream from the exchanger is $Q_e(0) + Q_b(225) = (Q_e + Q_b)50$, where Q_e = flow through the exchanger and Q_b = flow bypassing the exchanger. From the balance, $Q_b = 0.2857 \, Q_e$. Since $Q_b + Q_e = 3.2$ ℓ/s, $0.2857 \, Q_e + Q_e = 3.2$ ℓ/s or $Q_e = 2.49$ ℓ/s. The hardness of the well water in meq/ℓ is $(225 \, mg/\ell)(meq/50 \, mg)$ or 4.50 meq/ℓ. The hardness removed in seven days is $(2.49 \, \ell/s)(60 \, s/min)(1440 \, min/$

d)(7 d/week)(4.50 meq/ℓ) = 6,776,784 meq. Thus, the amount of resin = (6,776,784 meq)(100 g/282 meq)(kg/ 1000 g) = __2403 kg dry weight.__ On a moist basis, this is (2403 kg)[1/(1 − 0.45)] = __4369 kg moist weight.__

Design Considerations

Zeolite or resin beds used for softening usually have an exchanger bed of 2.0 to 8.5 ft in depth and operate at 1 to 8 gpm/ft^2. Since ion exchangers are regenerated without removing the exchanger solid from the vessel, the height to diameter ratio is not as critical as for carbon adsorption columns. Usually, the height of resin to diameter ratio is from 1.5:1 to 3:1. The column height should be sufficient to allow for expansion of the bed during the backwash. On backwashing, zeolites expand to about 25 percent of the bed depth, whereas synthetic resins expand to about 75 to 100 percent of the bed depth. The maximum column height is usually 12 ft. If a column height greater than 12 ft is required, two columns in series can be used. Prefabricated steel cylinders may be obtained in diameters up to 12 ft in 3-in. intervals; however, 6-in. intervals are more common and diameters range from 2 ft-6 in. to 12 ft. At an allowable breakthrough of 5 percent C_0, usually 65 to 85 percent of the ion exchange capacity will be used for removing hardness. Synthetic cation exchange resins with strong acid exchange sites have ion exchange capacities from about 350 to 520 meq/100 gm dry resin. Moist densities are from 43.0 lb/ft^3 (689 gm/ℓ) to 54.0 lb/ft^3 (865 gm/ℓ), and the moisture content may be from 40 to 60 percent. Dry resins may swell up to 55 percent of their original volume upon becoming moist. The ion exchange capacity, density, and moisture content depend upon the particular resin under consideration. Regeneration of synthetic resins on the sodium cycle is done with a 5 to 25 percent brine solution. Usually a 5 to 10 percent solution is used. In regeneration, the brine solution is passed through the bed in either a downward or upward direction. The unit liquid flow rate is 1 to 2 gpm/ft^2. After regeneration, a slow rinse flow is passed through the bed in the direction of the softening flow to rinse the brine from the bed. After the slow rinse, a short, fast rinse is used to flush any remaining brine from the bed. Usually 30 to 150 gal of rinse water are required per ft^3 resin. Salt requirements are from 5 to 20 lb salt per ft^3 resin bed, with 5 to 10 lb per ft^3 being typical.

Strong acid cation exchangers on the hydrogen cycle are regenerated using a H_2SO_4 or HCl solution. If H_2SO_4 is used, a 2 percent solution should first be passed through the exchanger followed by a 10 percent solution. The lower concentration used first is to avoid precipitation of $CaSO_4$ in the exchanger bed. The H_2SO_4 required is from 6 to 12 lb per ft^3 resin. If HCl is used, a 15 percent solution should be employed and the HCl required is from 5 to 10 lb per ft^3 resin. Strong base exchangers on the hydroxyl cycle are regenerated with a 2 to 10 percent NaOH solution. Regeneration requires from 3 to 6 lb NaOH per ft^3 resin.

Test columns should be regenerated in the same manner as a design column in order for the breakthrough curve to be representative. That is, the resin should be regenerated with the regenerate ion, then a test performed to complete exhaustion. Next, the resin should be regenerated, then the test should be rerun to obtain the design breakthrough curve.

References

American Water Works Association (AWWA) Research Foundation. 1973. *Desalting Techniques for Water Supply Quality Improvement.* Report for the Office of Saline Water, Department of the Interior.

Betz Laboratories. 1962. *Betz Handbook of Industrial Water Conditioning.* 6th ed. Trevose, Pa.: Betz Laboratories, Inc.

Conway, R. A., and Ross, R. D. 1980. *Handbook of Industrial Waste Disposal.* New York: Van Nostrand Reinhold.

Culp, R. L., and Culp, G. L. 1978. *Handbook of Advanced Wastewater Treatment.* 2nd ed. New York: Van Nostrand Reinhold.

Dorfner, K. 1971. *Ion Exchangers, Properties and Applications.* 3rd ed. Ann Arbor, Mich.: Ann Arbor Science Publishers.

Dow Chemical Co. 1964. DOWEX: *Ion Exchange.* Midland, Mich.: Dow Chemical Co.

Eckenfelder, W. W., Jr. 1980. *Principles of Water Quality Management.* Boston: CBI Publishing.

Environmental Protection Agency (EPA). 1971. Advanced Waste Treatment and Water Reuse Symposium. Sessions 1–5, Dallas, Tex.

Environmental Protection Agency (EPA). 1975. *Nitrogen Control.* EPA Process Design Manual, Washington, D.C.

Fornwalt, H. J., and Hutchins, R. A. 1966. Purifying Liquids with Activated Carbon. *Chem. Eng.* (11 April):179; (9 May):155.

Helfferich, F. 1962. *Ion Exchange.* New York: McGraw-Hill.

Kitchener, J. A. 1961. *Ion Exchange Resins.* London: Methuen & Co.; and New York: Wiley.

Kunin, R. 1960. *Elements of Ion Exchange.* New York: Reinhold.

Metcalf and Eddy, Inc. 1979. *Wastewater Engineering, Treatment, Disposal and Reuse.* New York: McGraw-Hill.

Michaels, A. S. 1952. Simplified Method of Interpreting Kinetic Data in Fixed-Bed Ion Exchange. *Ind. and Eng. Chem.* 44, no. 8:1922–1930.

Nordell, E. 1961. *Water Treatment for Industrial and Other Uses.* 2nd ed. New York: Reinhold.

Perry, R. H., and Chilton, C. H. 1973. *Chemical Engineers' Handbook.* 5th ed. New York: McGraw-Hill.

Ramalho, R. S. 1977. *Introduction to Wastewater Treatment Processes.* New York: Academic Press.

Reynolds, T. D., and Westervelt, R. 1969. Ion Exchange as a Tertiary Treatment. Paper presented at the American Society of Civil Engineers annual meeting and national meeting on Water Resources Engineering, New Orleans, La.

Rich, L. G. 1963. *Unit Processes of Sanitary Engineering*. New York: Wiley.

Sanks, R. L. 1978. *Water Treatment Plant Design*. Ann Arbor, Mich.: Ann Arbor Science Publishers.

Schroeder, E. D. 1977. *Water and Wastewater Treatment*. New York: McGraw-Hill.

Sundstrom, D. W., and Klei, H. E. 1979. *Wastewater Treatment*. Englewood Cliffs, N.J.: Prentice-Hall.

Thomas, H. C. 1948. Chromatography: A Problem in Kinetics. *Annals of the New York Academy of Science* 49:161.

Weber, W. J., Jr. 1972. *Physicochemical Processes for Water Quality Control*. New York: Wiley-Interscience.

Problems

1. A water is to be softened for an industrial water supply and the Ca^{+2} content is 107 mg/ℓ and the Mg^{+2} is 18 mg/ℓ. The allowable breakthrough, C_a, is 0.05 C_0, where C_0 is the hardness of the untreated water. A pilot column 4 in. in diameter and containing 3 ft of resin has been operated at a flow rate of 0.59 gal/hr to obtain breakthrough data. The following throughput volumes in gallons at the respective C_e/C_0 values (where C_e is the effluent hardness) were obtained: 620, 0.03; 768, 0.07; 820, 0.16; 860, 0.24; 895, 0.36; 922, 0.47; 940, 0.52; 980, 0.66; 1008, 0.74; 1066, 0.87; and 1220, 0.99. The resin has a specific weight of 44 lb/ft^3 and the design flow rate is 150,000 gal/day. Using the scale-up design approach, determine:
   a. The exchange column volume and pounds of resin.
   b. The rate at which the resins are expended per hour.
   c. The design column life in hours and days.
   d. The diameter and height if the resin height is twice the diameter.

2. A countercurrent ion exchange column removes copper ions (Cu^{+2}) from a wastewater stream. The copper ion concentration is 350 mg/ℓ and there is 99.9 percent removal. The wastewater has a flow upward at 4.5 gpm/ft^2 and the ion exchange resin flows downward. The ion exchange capacity of the resin is 4.50 meq/gm and the resin flow leaving the column is saturated with copper ions (that is, essentially all of the exchange sites are occupied by copper ions). Determine:
   a. The resin mass flow rate, lb/hr-ft^2.
   b. The gallons of wastewater treated by one pound of resin.
   c. The diameter of the column in feet if the flow rate is 100 gpm.

3. A countercurrent ion exchange column removes copper ions (Cu^{+2}) from a wastewater stream. The copper ion concentration is 350 mg/ℓ and there is 99.9 percent removal. The wastewater has a flow upward at 3.0 ℓ/s-m^2 and the ion exchange resin flows downward. The ion exchange capacity of the resin is 4.50 meq/g and the resin flow leaving the column is saturated with copper ions (that is, essentially all the exchange sites are occupied by copper ions). Determine:

a. The resin mass flow rate, kg/h-m^2.

b. The liters of wastewater treated by 1 kilogram of resin.

c. The diameter of the column in meters if the flow rate is 6.3 ℓ/s.

4. An industrial wastewater containing 205 mg/ℓ Cd^{+2} is to be treated by an ion exchange column, and the allowable effluent concentration is 5 percent C_0. A breakthrough curve has been obtained using a small laboratory column on the sodium cycle. Data concerning the column are: inside diameter = 0.5 in., mass of resin = 45.28 gm on a moist basis, moisture content = 48 percent, bulk density of the resin = 738 g/ℓ on a moist basis, and liquid flow rate = 1.120 ℓ/day. The following throughput volumes in liters at the respective effluent cadmium concentrations in mg/ℓ were obtained: 5.0, 0.0; 10.0, 0.0; 11.0, 14.06; 11.5, 26.43; 12.0, 42.17; 12.5, 68.60; 13.0, 104.59; 13.5, 141.70; 14.0, 163.07; 15.0, 196.81; 15.5, 202.43; 16.0, 205,00; and 16.5, 205.00. The design column will have a flow rate of 60,000 gal/day, the allowable breakthrough time is seven days of flow, and the resin depth is approximately twice the column diameter. Using the kinetic approach to design, determine:

a. The pounds of resin required.

b. The column diameter and depth.

c. The height of the adsorption zone.

5. The wastewater from a metal plating industry is to be treated by ion exchangers on the sodium cycle. Two columns in series are to be used and the lead column will reach 100 percent exhaustion before it is taken off-line. The flow is 310 m^3/day and the concentrations of the ions to be removed are 15 mg/ℓ zinc (Zn^{+2}), 20 mg/ℓ cadmium (Cd^{+2}), 25 mg/ℓ copper (Cu^{+2}), 20 mg/ℓ nickel (Ni^{+2}). The cation exchange resin to be used has a capacity of 375 meq/100 gm on a moist basis and 42 percent moisture. If the allowable breakthrough is seven days duration, determine the kilograms of dry resin

required and also the kilograms of moist resin required.

6. A well water is to be softened by split-flow ion exchange with the exchange resin on the sodium cycle. The flow is 40 gpm, the hardness is 195 mg/ℓ as $CaCO_3$, the desired hardness is 50 mg/ℓ as $CaCO_3$, the resin removes 296 meq/100 gm on a dry basis at an allowable breakthrough of 5 percent C_0, and the moisture content is 45 percent. The allowable breakthrough time is seven days. Determine the pounds of resin required on both a moist and dry basis.

Chapter Eight

MEMBRANE PROCESSES

I n the membrane processes, separation of a substance from a solution containing numerous substances is possible by the use of a selectively permeable membrane. The solution containing the components is separated from the solvent liquid by the membrane, which must first be differently permeable to the components. The main membrane processes are (1) dialysis, (2) electrodialysis, and (3) reverse osmosis. Figure 8.1 shows the effective separation ranges of the membrane processes and allows a comparison to other separation techniques. Each membrane process requires a driving force to cause mass transfer of the solute. The driving force is a difference in concentration for dialysis, in electric potential for electrodialysis, and in pressure for reverse osmosis. The main difficulty with the membrane processes is that the rate of mass transfer per unit area of the membrane (that is, mass flux) is relatively small.

Dialysis

Although dialysis is not used to an appreciable extent in environmental engineering, a knowledge of dialysis is necessary to understand the principles involved in electrodialysis and reverse osmosis.

Theory

Dialysis consists of separating solutes of different ionic or molecular size in a solution by means of a selectively permeable membrane. The driving force for dialysis is the difference in the solute concentration across the membrane. In

Figure 8.1. *Effective Ranges of Some Separation Techniques*

Adapted from "Membrane Separation Processes" by R. E. Lacey in *Chemical Engineering*, 4 September 1972. Copyright © 1972 by McGraw-Hill, Inc. Reprinted by permission.

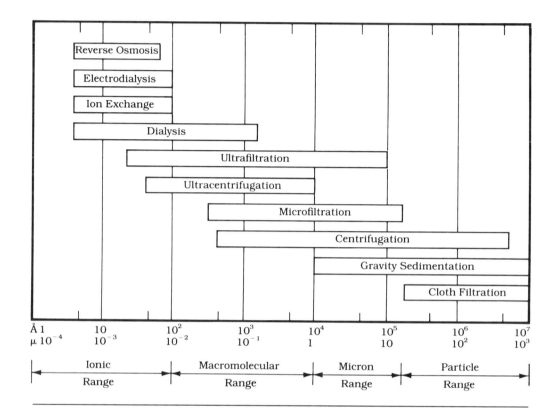

a batch dialysis cell, the solution to be dialyzed is separated from the solvent solution by the semipermeable membrane. The smaller ions or molecules pass through the membrane, whereas the larger ions and molecules do not pass through due to the relative size of the membrane pore openings. The passage of the smaller solutes is from the solution side to the solvent side because this is in the direction of a drop in concentration of the solutes.

The mass transfer of solute passing through the membrane in a batch dialysis cell at a given time is

$$M = KA\,\Delta C \tag{8.1}$$

where

M = mass transferred per unit time—for example, gm/hr;

K = mass transfer coefficient—for example, gm/(hr-cm^2)(gm/cm^3);

A = membrane area—for example, cm^2;

ΔC = difference in concentration of the solute passing through the membrane—for example, gm/cm^3.

Since a batch dialysis cell represents an unsteady-state condition, the term ΔC decreases with an increase in time. In continuous-flow dialysis cells, the flow of solvent and solution is countercurrent and, in application, numerous cells are pressed together in a stack with all cells being connected in parallel. The purpose of using a stack is to make the membrane area/stack volume as high as possible, thus resulting in a compact unit.

Application

Sodium hydroxide has been recovered from textile mill wastewaters by a continuous-flow dialysis stack (Nemerow, N.C., & Steel, W. R., 1955). The flow rate was from 420 to 475 gal per day and the recovery of sodium hydroxide was from 87.3 to 94.6 percent. Although good results were obtained, dialysis is limited to small flows because the mass transfer coefficient, K, is relatively small.

Electrodialysis

In environmental engineering, electrodialysis is presently used and has potential for more uses. In this process, the driving force for mass transfer is an electromotive force.

Theory

If dialysis is to be used to separate inorganic electrolytes from a solution, the presence of an electromotive force across the selectively permeable membrane will result in an increased rate of ion transfer. In this manner, the salt concentration of the treated solution is decreased. Since electrodialysis demineralizes, it has been used to produce fresh water from brackish water and seawater. Also, it has been used to demineralize effluents in tertiary treatment.

An electrodialysis stack consisting of three cells is shown in Figure 8.2. When a direct current is applied to the electrodes, all positively charged ions (cations) tend to migrate towards the cathode. Also, all negatively charged ions (anions) tend to migrate towards the anode. The cations can pass through the cation permeable membranes (designated C in Figure 8.2), but they are obstructed from passing through the anion permeable membranes (designated A). In a similar manner, the anions can pass through the anion permeable membranes but are obstructed from passing through the cation permeable membranes. As shown in Figure 8.2, alternate compartments are formed in which the ionic concentration is greater or less than the

Figure 8.2. Schematic of a Continuous-Flow Electrodialysis Stack
Adapted from "Membrane Separation Processes" by R. E. Lacey in *Chemical Engineering*, 4 September 1972, Copyright © 1972 by McGraw-Hill Inc. Reprinted by permission.

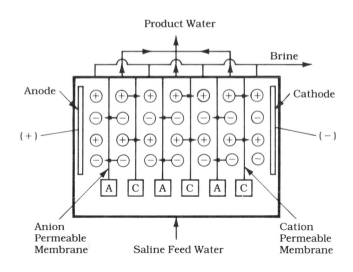

concentration in the feed solution. As a result, the flows from the stack consist of the product water, which has a low electrolyte concentration, and the brine solution, which has a high electrolyte concentration. The cells in the stack are connected for parallel flow. Gases are frequently formed at the electrodes, such as hydrogen at the cathode and oxygen and chlorine at the anode.

An electrodialysis membrane is a porous, sheetlike, structural matrix made of synthetic ion exchange resin. The matrix of a cation permeable membrane has a negative charge due to the ionization of its cation exchange sites. Exchangeable cations within the pore spaces balance the negative charge of the matrix. When a current flows, cations enter the pores and pass through the membrane since the electrical migratory forces acting upon the cations are greater than the attractive forces between the cations and the cation permeable membrane. Since the matrix is negatively charged it repels the negatively charged anions. Anion permeable membranes are made in a similar manner and allow anions to pass through but repel cations. Due to the nature of the membranes, some consider electrodialysis to be a specialized application of ion exchange.

The current required for an electrodialysis stack may be calculated using Faraday's laws of electrolysis. One Faraday (F) of electricity (96,500 ampere-seconds or coulombs) will cause one gram equivalent weight of a substance to migrate from one electrode to another. The number of gram equivalent weights removed per unit time in an electrolytic cell is

Number of gram equivalent weights/unit
time $= QNE_r$ (8.2)

where

Q = solution flow rate, liters/sec;
N = normality of the solution—that is, the number of gram equivalent weights/liter;
E_r = electrolyte removal as a fraction.

Since one Faraday is required per gram equivalent weight removed and a cell has a given current efficiency, the current required for a single cell can be derived from Eq. (8.2) as follows:

$$I = \frac{FQNE_r}{E_c}$$ (8.3)

where

I = current in amperes;
F = Faraday's constant (96,500 ampere-seconds per gram equivalent weight removed);
E_c = current efficiency as a fraction.

Since the same electrical current is passed through all of the cells in a stack, it is effectively used n number of times, where n is the number of cells. Thus, the current for the stack is given by a modification of Eq. (8.3), which is

$$I = \frac{FQNE_r}{nE_c}$$ (8.4)

where

n = number of cells.

The electrodialysis stacks used in desalting usually have from 100 to 250 cells (200 to 500 membranes). The current efficiency (E_c) for a particular electrodialysis stack and feed water must be determined experimentally. Usually, the current efficiency as a fraction is 0.90 or more. The electrolyte removal as a fraction (E_r) is usually from 0.25 to 0.50.

The capacity of an electrodialysis cell to pass an electric current is related to the current density and the normality of the feed water. The current density is defined as the current divided by the membrane area and is usually expressed as ma/cm^2. The normality of the feed water is, by definition, the number of gram equivalent weights per liter of solution. The parameter relating these variables is the current density/normality ratio. The ratio may vary from 400 to 700 when the current density is expressed as ma/cm^2. If the current density/normality ratio is too high, it will cause regions of low ionic concentrations near the membranes, resulting in polarization. This is to be avoided since high electrical resistance occurs, thus causing high electrical consumption.

The resistance of an electrodialysis stack treating a particular feed water must be determined experimentally. Once the electrical resistance, R, and the current flow, I, are known, the voltage required, E, is given by Ohm's law, which is $E = RI$, where E = volts, R = ohms, and I = amperes. The power required is given by $P = EI = RI^2$, where P = power in watts.

Example Problem 8.1
Electrodialysis

An electrodialysis stack having 200 cells is to be used to partially demineralize 90,000 gal per day of advanced treated wastewater so that it can be used by an industry. The salt content is 4000 mg/ℓ and the cation or anion content is 0.066 gram equivalent weights per liter. Pilot-scale studies using a multicellular stack have been made. It was found that the current efficiency, E_c, was 90 percent, the efficiency of salt removal, E_r, was 50 percent, the resistance was 4.5 ohms, and the current density/normality ratio was 400. The stack is to have 200 cells. Determine:

a. The current, I, required.
b. The area of the membranes.
c. The power required.

Solution

The current, I, is given by Eq. (8.4):

$$I = \frac{96,500\,\text{amp-sec}}{\text{gm-eq-wt}} \left| \frac{90,000\,\text{gal}}{\text{day}} \right| \frac{0.066\,\text{gm-eq-wt}}{\ell}$$
$$\times \frac{3.785\,\ell}{\text{gal}} \left| \frac{\text{day}}{86,400\,\text{sec}} \right| \frac{0.50}{} \left| \frac{}{200} \right| 0.90$$

$$= \underline{69.75\,\text{amps}}$$

Current density/normality ratio = 400. Since normality equals the number of gram equivalent weights per liter, normality = 0.066. The current density is therefore equal to (400)(normality) = (400)(0.066) = 26.4 ma/cm^2. Thus, the membrane area is

$$\text{Area} = (69.75\,\text{amps})(\text{cm}^2/26.4\,\text{ma})(1000\,\text{ma}/\text{amp})$$
$$= \underline{2642\,\text{cm}^2}$$

If the membranes are square in shape, the size is 51.4 cm × 51.4 cm. The power required is

$$\text{Power} = RI^2 = (4.5\,\text{ohms})(69.75\,\text{amp})^2$$

$$= \underline{21,900\,\text{watts}}$$

Application

The technical feasibility of demineralizing sea water and brackish waters by electrodialysis has been demonstrated by pilot installations and, in the case of brackish waters, by full-scale plants (AWWA, 1973; Browning, J. E., 1970; Collier, E. P., & Fulton, J. F., 1967). Electrodialysis stacks used for demineralization usually consist of several hundred cells (two membranes per cell). From Eq. (8.4), it can be seen that the electrical energy requirements are directly proportional to the amount of salt removed. Consequently, the electrical costs are governed by both the dissolved solids content of the feed water and the desired dissolved solids content of the product water. Also, energy consumption increases with deposition of scale upon the membranes and the amount of scale is related to the various dissolved salts in the feed water. Because of the scaling problem and high electrical consumption, electrodialysis is not well suited to the demineralization of sea water. However, electrodialysis is adaptable to the demineralization of brackish waters, and numerous desalination plants using electrodialysis have been constructed throughout the world for brackish water desalting. Studies have shown that electrodialysis is particularly adaptable for the deionization of brackish waters of 5000 mg/ℓ dissolved solids or less to produce a product water with a dissolved solids of about 500 mg/ℓ. Membrane replacement costs and electrical power costs are about 40 percent of the total cost.

Pilot plant studies have shown that electrodialysis is a technically feasible method of demineralizing secondary effluents; however, scaling problems and organic fouling of the membranes will occur (Brunner, C. A., 1967; Culp, G. L., 1975; EPA, 1971; HEW, 1962 & 1965; FWQA, 1970). It has been found that 25 to 50 percent of the dissolved salts can be removed in a single pass through an electrodialysis stack. In order to obtain greater salt removal, several electrodialysis stacks may be placed in series. Scaling and organic fouling may be reduced by proper pretreatments such as coagulation, settling, filtration, and activated carbon adsorption. Scaling may also be reduced by the addition of a small amount of acid to the feed stream, whereas organic fouling may be reduced by cleaning the membranes with an enzyme detergent solution. Scaling and membrane fouling must be controlled in order for electrodialysis to be economically feasible. Since reverse osmosis desalts at a comparable cost and gives other pollutant removals, it has a greater potential in tertiary treatment than electrodialysis.

Reverse Osmosis

In environmental engineering, reverse osmosis is presently used and has potential for more uses. For this process, the driving force for mass transfer is a hydrostatic pressure difference.

Theory

Reverse osmosis consists of separating a solvent, such as water, from a saline solution by the use of a semipermeable membrane and a hydrostatic pressure. Consider the batch dialysis cell shown in Figure 8.3 (a). The solvent flow is from the solvent side through the semipermeable membrane to the saline side because this is in the direction of a drop in solvent concentration. This transfer of solvent water through a semipermeable membrane from the solvent side to the saline side is referred to as osmosis. Eventually the system will reach equilibrium, as shown in Figure 8.3 (b), where the hydrostatic pressure is referred to as the *osmotic pressure*. If a force is applied to a piston to produce a pressure greater than the osmotic pressure, as shown in Figure 8.3 (c), there will be a transfer of the solvent in the reverse direction. This mass transfer of a solvent using a semipermeable membrane and a hydrostatic pressure is re-

---

Figure 8.3. Osmosis and Reverse Osmosis

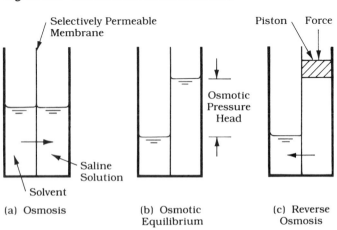

(a) Osmosis   (b) Osmotic Equilibrium   (c) Reverse Osmosis

ENVIRONMENTAL ENGINEERING

ferred to as *reverse osmosis*. Reverse osmosis is a useful separation method since it permits the passage of water and rejects the passage of molecules other than water and of ions. The reverse osmosis cell in Figure 8.3 (c) is a batch operation, whereas the cell in Figure 8.4 is a continuous-flow operation.

Reverse osmosis is similar to ultrafiltration and micro-filtration because all three techniques utilize semipermeable membranes and hydrostatic pressures to force the solvent through the membranes. In ultrafiltration and microfiltration, the separation is mainly due to a filtering action and not a reverse osmotic action; also, the size of the substances that are rejected by the ultrafiltration and microfiltration membranes is larger, as shown in Figure 8.1.

The osmotic pressure of solutions of electrolytes may be determined by

$$\pi = \phi v \frac{n}{V} RT \qquad (8.5)$$

where

π = osmotic pressure;
ϕ = osmotic coefficient;
v = number of ions formed from one molecule of electrolyte;
n = number of moles of electrolyte;
V = volume of solvent;
R = universal gas constant;
T = absolute temperature.

The osmotic coefficient, ϕ, depends upon the nature of the substance and its concentration. The osmotic pressure for seawater, which has 35,000 mg/ℓ dissolved solids, is 397 psi at 25°C. Since a salt solution of 35,000 mg/ℓ dissolved solids has an osmotic pressure of 397 psi, it may be assumed for practical purposes that an increase of 1000 mg/ℓ salt concentration results in an increase of 11.3 psi in osmotic pressure.

The schematic diagram of a continuous-flow reverse osmosis unit is shown in Figure 8.4. The saline feed is pressurized so that the differential pressure between the two compartments is greater than the difference in osmotic pressure. Although the transfer of solvent will begin when the pressure difference exceeds the osmotic pressure difference, the rate of solvent mass transfer increases as the pressure difference increases. In practice, the pressures used for the feed stream are from 250 to 800 psi. The design pressure depends mainly upon the osmotic pressure differential between the feed and product solution, the charac-

Figure 8.4. Schematic of a Continuous-Flow Reverse Osmosis Unit

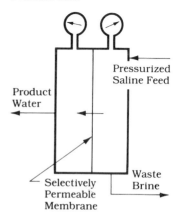

teristics of the membrane, and the temperature.

The main design parameters for a reverse osmosis unit are the production per unit area of membrane and the product water quality (Kaup, E. C., 1973). The production is measured by the flux of water through the membrane—for example, gal/day-ft^2. The flux of a membrane depends on the membrane characteristics (that is, thickness and porosity) and the system conditions (that is, temperature, differential pressure across the membrane, salt concentration, and flow velocity of the water through the membrane). In practice, the water flux is simply related to the pressures by (Kaup, E. C., 1973)

$$F_w = K(\Delta p - \Delta \pi) \tag{8.6}$$

where

F_w = water flux—that is, gal/day-ft^2;
K = mass transfer coefficient for a unit area of membrane;
Δp = pressure difference between the feed and product water;
$\Delta \pi$ = osmotic pressure difference between the feed and product water or the feed osmotic pressure minus the product water osmotic pressure.

The membrane flux value furnished by a manufacturer is usually for 25°C (77°F). As temperature varies, the diffusivity and viscosity vary also, and this in turn causes the flux to vary. Membrane area corrections (A_T/A_{25}) due to the respective temperatures are as follows: 10°C, 1.58; 15°C, 1.34; 20°C, 1.15; 25°C, 1.00; and 30°C, 0.84. The term A_T/A_{25} is the ratio of the areas required for temperatures of T (°C) and 25°C (Kaup, E. C., 1973).

The flux value will gradually decrease during the lifetime of a membrane due to a slow densification of the membrane structure, which results in a decrease in the pore passages. This gradual flux reduction occurs in all membranes, and it is permanent. The membrane must be replaced when the flux has reached the minimum acceptable value. Manufacturers can provide initial flux values and the expected rate of flux reduction for various operating pressures. The flux versus duration will plot a straight line on log-log paper. Usually, the life span of a membrane is from two months to two years.

The most common membrane used is made of cellulose acetate. These membranes are relatively tight; that is, they have low water permeability and can reject over 99 percent of the salts. However, the water flux is very low (about 10 gal/day-ft^2). The main types of mounting hardware employed in reverse osmosis equipment modules are classified as: (1) tubular, (2) hollow fiber, and (3) spiral wound. Figures 8.5 and 8.6 show schematic drawings of the tubular and the hollow-fiber types of mountings. In the tubular mounting the feed water enters the inside of the tubular membrane, which is encased by a porous support tube that gives the required tensile strength, as shown in Figure 8.5 (a). The portion of the water that passes through the membrane leaves as product water, while the remaining water leaves the end of the tube as brine. In the hollow-fiber mounting, a portion of the feed water passes from the outside of the tube to the inside, as shown in Figure 8.6 (a), and leaves as the product water, while the remaining water leaves as brine. In the spiral-wound mounting, a porous hollow tube is spirally wrapped with a porous sheet for the feed flow, and a membrane sheet and a porous sheet for the product water flow to give a spiral sandwich-type wrapping. The spiral module is encased in a pressure vessel, and the feed flow through the porous sheet is in an axial direction to the porous tube. As the feed flow passes through the porous sheet, a portion of the flow passes through the membrane into the porous sheet for the product water. From

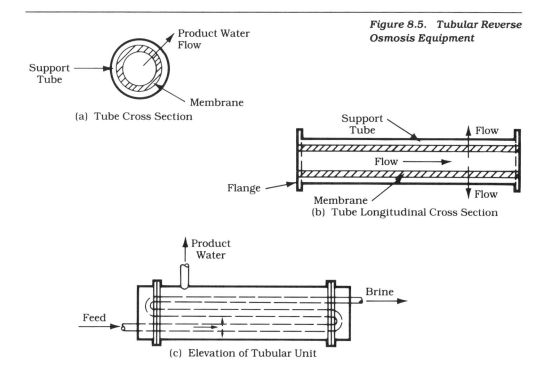

Figure 8.5. Tubular Reverse Osmosis Equipment

Product Water Flow

Support Tube

Membrane

(a) Tube Cross Section

Support Tube

Flow

Flow

Flow

Flange

Membrane

(b) Tube Longitudinal Cross Section

Product Water

Brine

Feed

(c) Elevation of Tubular Unit

Figure 8.6. Hollow-Fiber Reverse Osmosis Equipment

(a) Hollow-Fiber Cross Section

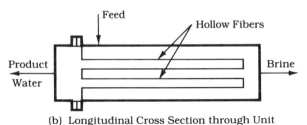

(b) Longitudinal Cross Section through Unit

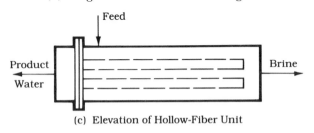

(c) Elevation of Hollow-Fiber Unit

there the product water flows spirally to the porous center tube and is discharged from the conduit in the tube. The brine is discharged from the downstream end of the porous sheet for the feed flow. Some of the characteristics of the various mountings are given in Table 8.1.

Table 8.1. Characteristics of Different Reverse Osmosis Modules

Module Type	Surface Area (ft² membrane area per ft³ of equipment)	Water Flux (gal/day-ft²)	Water Produced (gal/day per ft³ of equipment)
Tubular	20	32	640
Hollow fiber (cellulose acetate)	2,500	10	25,000
Hollow fiber (nylon)	5,400	1	5,400
Spiral wound	250	32	8000

Environmental Protection Agency, Advanced Waste Treatment and Water Reuse Symposium, sessions 1–5, Dallas, Texas, 1971.

From these data, it appears that hollow fibers are best; however, in practice their hydraulic inadequacies are serious.

A reverse osmosis unit is to demineralize 200,000 gal of tertiary treated effluent per day. Pertinent data are: mass transfer coefficient $= 0.0350$ gal/(day-ft^2)(psi) at 25°C, pressure difference between the feed and product water $= 350$ psi, osmotic pressure difference between the feed and product water $= 45$ psi, lowest operating temperature $= 10$°C, and $A_{10°} = 1.58\,A_{25°}$. Determine the membrane area required.

Solution

The water flux is given by Eq. (8.6),

$$F_w = [0.0350\,\text{gal}/(\text{day-ft}^2)(\text{psi})](350\,\text{psi} - 45\,\text{psi})$$

$$= 10.675\,\text{gal}/(\text{day-ft}^2)\,\text{at}\,25°C$$

The area at 10°C is given by

$$A = (200{,}000\,\text{gal}/\text{day})[(\text{day-ft}^2)/10.675\,\text{gal}](1.58)$$

$$= \underline{29{,}600\,\text{ft}^2}$$

A reverse osmosis unit is to demineralize 760,000 ℓ/d of tertiary treated effluent per day. Pertinent data are: mass transfer coefficient $= 0.2068$ ℓ/(d-m^2)(kPa) at 25°C, pressure difference between the feed and product water $= 2400$ kPa, osmotic pressure difference between the feed and product water $= 310$ kPa, lowest operating temperature $= 10$°C, and $A_{10°} = 1.58\,A_{25°}$. Determine the membrane area required.

Solution

The water flux is given by Eq. (8.6),

$$F_w = [0.2068\,\ell/(\text{d-m}^2)(\text{kPa})](2400\,\text{kPa} - 310\,\text{kPa})$$

$$= 432.21\,\ell/(\text{d-m}^2)\,\text{at}\,25°C$$

The area at 10°C is given by

$$A = (760{,}000\,\ell/\text{d})[(\text{d-m}^2)/(432.21\,\ell)](1.58)$$

$$= \underline{2778\,\text{m}^2}$$

Application

The technical feasibility of demineralizing sea water and brackish waters by reverse osmosis has been demonstrated by pilot plant installations (AWWA, 1973; Browning, J. E., 1970; Collier, E. P., & Fulton, J. F., 1967).

Pilot plant work in the United States has yielded a product water of 500 mg/ℓ dissolved solids from sea water and a product water of 250 mg/ℓ from a brackish water with 4500 mg/ℓ dissolved solids. Operating pressures up to 1500 psi for sea water and up to 750 psi for the brackish water were used in the pilot plant work. From pilot plant work it was found that the membranes have a short life and membrane replacement would represent about half the cost of desalting seawater. Because of the energy requirements for reverse osmosis and the short membrane life, reverse osmosis is best suited to the demineralization of brackish waters and not sea water. It has been shown that reverse osmosis is particularly adapted to desalting a feed water having a dissolved solids content from 3000 to 10,000 mg/ℓ. Membrane replacement expense is the largest single cost item, and this amounts to about 32 percent of the total cost.

Pilot plant studies have shown that reverse osmosis is a technically feasible method to demineralize secondary effluents; however, some organic fouling of the membranes will occur. The modified cellulose acetate membranes have been found to be the most applicable. These membranes will reject from 90 to 99 percent of the salts and 90 percent of the organic materials. Organic fouling of the membranes is usually not serious enough to require prior treatments such as activated carbon adsorption. Periodic cleaning of the membranes with an enzyme detergent solution usually controls the fouling. Reverse osmosis has a large potential for wastewater treatment since a high degree of removal for all contaminants can be accomplished by one unit operation. Pilot plants have shown typical removals from secondary effluents treated by reverse osmosis units operated at 450 psi and 8 gal/day-ft^2 are as follows: total organic carbon, 90 percent; total dissolved solids, 93 percent; phosphate, 94 percent; organic nitrogen, 86 percent; ammonia nitrogen, 85 percent; and nitrate nitrogen, 65 percent (Culp, G. L., 1975).

References

American Water Works Association (AWWA) Research Foundation. 1973. *Desalting Techniques for Water Supply Quality Improvement.* Report for Office of Saline Water, U.S. Department of the Interior.

Browning, J. E. 1970. Zeroing in on Desalting. *Chem. Eng.* (23 March):64.

Brunner, C. A. 1967. Pilot-Plant Experiences in Demineralization of Secondary Effluents Using Electrodialysis. *Jour. WPCF* 39, no. 10, part 2:R1.

Castellan, G. W. 1964. *Physical Chemistry*. Reading, Mass.: Addison-Wesley.

Collier, E. P., and Fulton, J. F. 1967. Water Desalination. Department of Energy, Mines and Resources, Ottawa, Canada.

Conway, R. A., and Ross, R. D. 1980. *Handbook of Industrial Waste Disposal*. New York: Van Nostrand Reinhold.

Culp, G. L. 1975. *Treatment Processes for Wastewater Reclamation for Groundwater Recharge*. Preliminary Report for the State of California, Department of Water Resources.

Culp G. L., and Hamann, C. L. 1974. Advanced Waste Treatment Process Selection: Parts 1, 2 and 3. *Public Works* (March, April, and May).

Culp, R. L., and Culp, G. L. 1978. *Handbook of Advanced Wastewater Treatment*. 2nd ed. New York: Van Nostrand Reinhold.

Eckenfelder, W. W., Jr. 1980. *Principles of Water Quality Management*. Boston: CBI Publishing.

Environmental Protection Agency (EPA). 1971. Advanced Waste Treatment and Water Reuse Symposium, sessions 1–5, Dallas, Tex.

Halper, B. M., and Olie, J. 1972. 1.25 MGD Electrodialysis Plant in Israel. *Jour. AWWA* 64, no. 11:735.

Katz, W. E. 1961. Preliminary Evaluation of Electric Membrane Processes for Chemical Processing Applications. In *Separation Processes in Practice*, edited by R. F. Chapman. New York: Reinhold.

Kaup, E. C. 1973. Design Factors in Reverse Osmosis. *Chem. Eng.* (2 April):48.

King, J. C. 1971. *Separation Processes*. New York: McGraw-Hill.

Lacy, R. E. 1972. Membrane Separation Processes. *Chem. Eng.* (4 September):56.

Mason, E. A., and Juda, W. 1959. Applications of Ion Exchange Membranes in Electrodialysis. *Chem. Eng. Progr. Symposium Series* 55, no. 24:155.

Mason, E. A., and Kirkham, T. A. 1959. Design of Electrodialysis Equipment. *Chem. Eng. Progr. Symposium Series* 55, no. 24:173.

Metcalf and Eddy, Inc. 1979. *Wastewater Engineering, Treatment, Disposal and Reuse*. New York: McGraw-Hill.

Nemerow, N. C., and Steel, W. R. 1955. *Proceedings of the 11th Annual Purdue Industrial Waste Conference*.

Perry, R. H., and Chilton, C. H. 1973. *Chemical Engineers Handbook*. 5th ed. New York: McGraw-Hill.

Ramalho, R. S. 1977. *Introduction to Wastewater Treatment Processes*. New York: Academic Press.

Rich, L. G. 1963. *Unit Processes of Sanitary Engineering*. New York: Wiley.

Robinson, R. A., and Stokes, R. H. 1959. *Electrolyte Solutions*. London: Butterworth.

Sanks, R. L. 1978. *Water Treatment Plant Design*. Ann Arbor, Mich.: Ann Arbor Science Publishers.

Schroeder, E. D. 1977. *Water and Wastewater Treatment*. New York: McGraw-Hill.

Sourirajan, S. 1970. *Reverse Osmosis*. New York: Academic Press.

Sundstrom, D. W., and Klei, H. E. 1979. *Wastewater Treatment*. Englewood Cliffs, N.J.: Prentice-Hall.

U.S. Department of Health, Education and Welfare (HEW). 1962. *Advanced Waste Treatment Research-1, Summary Report, June 1960 to Dec. 1961.*

U.S. Department of Health, Education and Welfare (HEW). 1965. *Advanced Waste Treatment Research, Summary Report, Jan. 1962 to June 1964,* AWTR-14.

U.S. Department of Interior (FWQA). 1970. Advanced Waste Treatment Seminar, sessions 1–4, San Francisco, Calif.

Weber, W. J., Jr. 1972. *Physicochemical Processes for Water Quality Control.* New York: Wiley-Interscience.

Problems

1. An electrodialysis stack is to be used to partially demineralize 100,000 gal/day of advanced treated wastewater so that it can be used for reuse by an industry. The membranes are 30 × 30 inches and the salt concentration is 4500 mg/ℓ. The cations and anions are 0.0742 eq/ℓ. The product water must not have more than 2250 mg/ℓ salt content. Pertinent data are: stack resistance = 4.5 ohms, current efficiency = 90 percent, maximum current density to normality ratio = 400 (with current density as ma/cm^2), power costs = 2½ ¢/kw-hr, brine density = 67 lb/ft^3, and brine = 12 percent salt. Determine:

   a. The removal efficiency, number of membranes, power consumption (watts), and power costs per 1000 gal.

   b. The product water and brine flows, gal/day.

2. A reverse osmosis unit is to demineralize 100,000 gal of tertiary treated effluent per day. Pertinent data are: mass transfer coefficient = 0.030 gal/(day-ft^2)(psi), pressure difference between the feed and product water = 380 psi, osmotic pressure difference between the feed and product water = 45 psi, lowest operating temperature = 10°C, and membrane area per unit volume of equipment = 2500 ft^2/ft^3. Determine:

   a. The membrane area required.

   b. The space required for the equipment, ft^3.

3. A reverse osmosis unit is to demineralize 760 m^3/d of tertiary treated effluent per day. Pertinent data are: mass transfer coefficient = 0.207 ℓ/(d-m^2)(kPa), pressure difference between the feed and product water = 2400 kPa, osmotic pressure difference between the feed and product water = 310 kPa, lowest operating temperature = 10°C, and membrane area per unit volume of equipment = 2500 m^2/m^3. Determine:

   a. The membrane area required.

   b. The space required for the equipment, m^3.

Chapter Nine

ACTIVATED SLUDGE

The activated sludge process utilizes a fluidized, mixed growth of microorganisms under aerobic conditions to use the organic materials in the wastewater as substrates, thus removing them by microbial respiration and synthesis. The main units of the system, as shown in Figure 9.1 (a), consist of a biological reactor with its oxygen supply (the aeration tank), a solid-liquid separator (the final clarifier), and the recycle sludge pumps.

The feed wastewater flow, Q, mixes with the recycled activated sludge flow, R, immediately prior to entering the biological reactor or immediately after entering. The activated sludge-wastewater mixture is termed the *mixed liquor,* and the *mixed liquor suspended solids (mLss)* usually range from 2000 to 4000 mg/ℓ by dry weight. Upon entering the reactor, the activated sludge rapidly adsorbs the suspended organic solids in the wastewater, this period lasting from about 20 to 45 minutes. After adsorption, the adsorbed organic solids are solubilized and oxidized by biological oxidation as the mixed liquor moves through the aeration tank. The soluble organic substances, on the other hand, are usually sorbed (that is, both adsorbed and absorbed) at the greatest rate at the upstream end of the tank. The rate of sorption gradually decreases as the mixed liquor passes through the tank. The sorbed soluble organic materials are oxidized by biological oxidation with a reaction time usually less than that required for the adsorbed suspended organic substances. The oxygen supply for the aer-

Figure 9.1. Activated Sludge Process

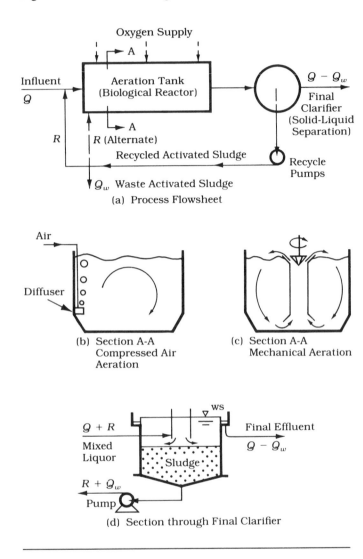

(a) Process Flowsheet

(b) Section A-A
Compressed Air
Aeration

(c) Section A-A
Mechanical Aeration

(d) Section through Final Clarifier

ation tank is usually furnished by diffused compressed air, as shown in Figure 9.1 (b), or by mechanical surface aeration, as shown in Figure 9.1 (c); however, pure oxygen has been used in some instances. Aeration by diffused compressed air or mechanical means has a dual purpose because it must supply the required oxygen for the aerobic bio-oxidation and it must provide sufficient mixing for adequate contact between the activated sludge and the organic substances in the wastewater. When pure oxygen is used as the oxygen source, it must be supplemented with mechanical mixing devices to furnish the required mixing.

At the downstream end of the reactor, the adsorbed and absorbed organic materials have been bio-oxidized and the mixed liquor flows to the solid-liquid separator (the final clarifier). There, the active biological solids settle to the bottom, as shown in Figure 9.1 (d), and the separated treated wastewater spills over the peripheral weirs into the effluent channels. Usually, effluent disinfection is required for municipal effluents prior to discharge into the receiving body of water; however, many industrial effluents do not require disinfection because there are no pathogens present. The activated sludge from the bottom of the final clarifier is pumped by recycle pumps, and the recycled activated sludge flow mixes with the incoming feed wastewater flow as previously described and as shown in Figure 9.1 (a).

The desired mixed liquor suspended solids (MLSS) concentration in the biological reactor is maintained at a constant level by recycling a fixed amount of the settled activated sludge from the solid-liquid separator. The recycle ratio, R/Q, is dependent upon the desired MLSS concentration and the concentration of the settled activated sludge in the recycle flow as measured by the sludge density index (SDI). In this control test, a volume of mixed liquor is taken from the downstream end of the aeration tank and 1 liter of the mixture is placed in a 1-liter graduate cylinder. At the same time, the suspended solids concentration in the mixed liquor is determined. After the sludge has settled in the cylinder for 30 minutes, the volume occupied by the settled sludge is read from the graduations. The concentration of the settled sludge in mg/ℓ is then computed; it represents the sludge density index (SDI). This test approximates the settling that occurs in the final clarifier. If the sludge density index is 10,000 mg/ℓ and the desired MLSS is 2500 mg/ℓ, a material balance for the activated sludge at the junction of the influent and recycle lines is

$$Q(0) + R(10{,}000) = (Q + R)(2500)$$

and, from this,

$$R/Q = 2500/7500 = \tfrac{1}{3} \text{ or } 33.3 \text{ percent}$$

Thus, the recycled sludge flow rate, R, must be 33.3 percent of the incoming flow rate, Q, in order to maintain the desired mixed liquor suspended solids at a constant concentration.

The reciprocal of the sludge density index, after appropriate unit conversions are made, is the sludge volume index (SVI). This is the volume in mℓ occupied by 1 gram of settled activated sludge, and it is a measure of the settling characteristics of the sludge. In a properly operating diffused air activated sludge treatment plant, the SVI is usually from 50 to 150 mℓ/gm.

Because in biological oxidation the substrate is used for respiration and for synthesis of new microbial cells, the net cell production (the waste activated sludge) must be removed from the system, as shown in Figure 9.1 (a), in order to maintain a constant mixed liquor suspended solids concentration in the reactor. Usually, the waste activated sludge flow rate amounts to 1 to 6 percent of the incoming feed wastewater flow rate.

The term *activated* stems from the sorptive properties of the biological solids. At the downstream end of the reactor, the activated sludge is in a low substrate environment and has utilized its sorbed organic substances; as a result, it has a relatively high sorption capacity for suspended and dissolved organic material. In the upstream region of the reactor, the sludge will have used most of its sorption capacity and will not be reactivated until it has biologically oxidized the sorbed organic material.

A simplified biochemical equation for the utilization of organic matter as a substrate for respiration and cell synthesis in the activated sludge process is

Rectangular Aeration Tank with Diffused Compressed Air

ENVIRONMENTAL ENGINEERING

$$\text{Organic matter} + O_2 \xrightarrow{\text{Aerobic microbes}} \text{New cells} + \text{Energy for cells}$$

$$+ CO_2 + H_2O + \text{Other end products} \qquad (9.1)$$

Some of the other end products include NH_4^{+1}, NO_2^{-1}, NO_3^{-1}, and PO_4^{-3}. The empirical equation that has usually been found to represent activated sludge is $C_5H_7O_2N$ (McKinney, R. E., 1962), which has a molecular weight of 113. The net mass of cells produced daily represents the mass of cells that must be disposed daily as waste activated sludge; it is equal to the total cell mass synthesized minus the cell mass that is endogenously decayed. The most common organic materials in municipal wastewaters not containing appreciable amounts of industrial wastes are carbohydrates, fats, proteins, urea, soaps, detergents, or their degradation or breakdown products. Carbohydrates and fats contain the elements carbon, hydrogen, and oxygen; proteins, in addition to these elements, contain nitrogen, sulfur, and phosphorus. Urea consists of carbon, hydrogen, oxygen, and nitrogen. Soaps consist mainly of carbon, hydrogen, and oxygen; detergents, in addition to these elements, contain phosphorus. Also, there may be trace amounts of other substances, such as pesticides, her-

Completely Mixed Aeration Tank with Mechanical Aerators

bicides, and other agricultural chemicals that enter by infiltration. The organic substances and compounds present in industrial wastewaters vary greatly from one type of industry to another. Also, there may be variation within an industry, such as the petrochemical industry. The organic materials remaining after biological treatment (that is, the nonbiodegradable fraction) include very slowly degraded and bioresistant substances.

Microorganisms

The mixed culture (a growth of two or more species) present in activated sludge is a dynamic system; the number of species and the particular species and their populations depend upon the specific wastewater being treated and the environmental conditions in the reactor-clarifier system. The microorganisms include bacteria (both single and multicellular), protozoa, fungi, rotifers, and sometimes nematodes. The bacteria, protozoa, and fungi are members of the Kingdom Protista, whereas the rotifers and nematodes are of the Animal Kingdom. The principal organisms involved in the bio-oxidation of organic substances in wastewaters are the single-celled bacteria.

Single-celled bacteria may be classified by the shape of the cells as *bacilli*, which are rod-shaped, *cocci*, which are spherical, and *spirilla*, which are spiral shaped. They may occur as single cells, clusters of cells, or in chains made up of cells, although some other groupings may occur. Cell shape and cell grouping are characteristics of each individual species. For example, *E. coli* cells are rod shaped and occur as single cells. Rod-shaped cells are usually from about 1.0 micron (μ) in diameter and about 3 to 5 μ long. Spherical-shaped cells are from about 0.2 to 2 μ in diameter. Spiral-shaped cells are about 0.3 to 5 μ in diameter and 6 to 15 μ long. Single-celled bacteria are microscopic organisms, usually nonphotosynthetic, that normally multiply by binary fission. In binary fission, the parent cell divides and breaks apart into two daughter cells. The division time, also called the *fission* or *generation time*, varies with the species and environmental conditions but usually it is about 20 minutes. Bacterial cells must have their food in soluble form and they may be capable of movement (motile) or incapable of movement (nonmotile), according to the species. Figure 9.2 shows a schematic drawing of a rod-shaped bacterial cell. The cytoplasm consists of a colloidal suspension of carbohydrates, proteins, and other complex organic compounds. Biochemical reactions, such as the synthesis of proteins, essential to the life processes of the cell take place in the cytoplasm. The nuclear area is well

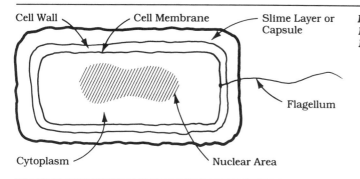

Figure 9.2. Schematic Drawing of a Rod-Shaped Bacterial Cell

Cell Wall

Cell Membrane

Slime Layer or Capsule

Flagellum

Cytoplasm

Nuclear Area

defined in some species and poorly defined in other species. The nuclear area is responsible for the hereditary characteristics of the cell that control reproduction of cell components. The cell membrane controls the flow of material into and out of the cell. The cell wall, which is semirigid, protects the cell interior, and also retains the cytoplasm in a definite volume and shape. The slime layer or capsule protects the cell from drying if it is placed in a dry environment and also serves as a reserve food. The flagella are hairlike organelles for movement and are present only in cells that are motile. Enzymes (not shown in Figure 9.2) are organic catalysts that catalyze necessary biochemical reactions. Some enzymes are within the cell (endocellular), whereas others are secreted to the outside of the cell (exocellular). Exocellular enzymes break down large molecules, such as proteins and starches, into smaller ones that can pass through the cell wall and cell membrane. When bacterial cells die, they lyse or break apart, thus releasing their cell contents.

Most of the single-celled bacteria used in wastewater treatment are soil microorganisms and very few are of enteric origin. The majority of the bacterial species, with the exception of some groups such as the nitrifying bacteria, are saprophytic heterotrophs because they require nonliving preformed organic materials as substrates. The nitrifying bacteria, which convert the ammonium ion to nitrite and the nitrite ion to nitrate, are autotrophs because they use carbon dioxide as their carbon source instead of preformed organic substances. The microbes present consist of both aerobes and facultative anaerobes since they use free molecular oxygen in their respiration process. The principal bacterial genera found in activated sludge when treating municipal wastewaters are: *Achromobacter, Arthrobacter, Cytophaga, Flavobacterium, Alkaligenes, Pseudomonas, Vibrio, Aeromonas, Bacillus, Zoogloea, Nitrosomonas,* and *Nitrobacter* (Lighthart, B., & Loew, G. A.,

1972; McKinney, R. E., 1962a). The cells of the first four genera are the most numerous (Lighthart, B., & Loew, G. A., 1972). In addition to the nitrifying bacteria, other specialized groups, such as some sulfur and iron bacteria, are present, although in small numbers. The enteric bacteria rapidly die off in the aeration tank because they cannot compete with the other microbes in the existing environment. Although some individual microbial cells are present in activated sludge, the majority are present as zoogloeal biomass particles, which consist of mixed species of cells embedded in masses of polysaccharide gums from the slime layers of living and lysed bacterial cells. The zoogloeal particles, frequently termed *floc*, are desirable because they have appreciable sorptive properties and settle quickly. The filamentous organisms, such as the bacterium *Sphaerotilis natans* and most fungi, are usually not numerous and are not desired because, in large numbers, they can create a sludge with poor settling characteristics. The protozoa do not use the organic substances in the wastewater themselves but instead feed on the bacterial population. The protozoa found include the stalked and free-swimming Ciliata and, to a lesser extent, the Suctoria. Rotifers are usually not numerous but are found in activated sludges that have

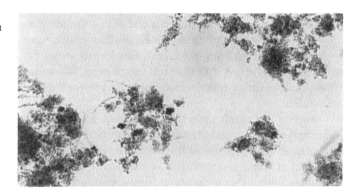

Photomicrograph of an Activated Sludge Culture with Excellent Settling Characteristics (100 ×). Note agglomerated floc.

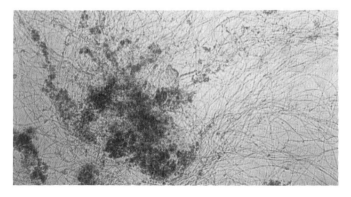

Photomicrograph of an Activated Sludge Culture with Extremely Poor Settling Characteristics (430 ×). Note wiry, filamentous growths.

undergone an appreciable aeration time. They feed on activated sludge fragments that are too large for protozoa. Nematodes are not numerous; however, they use organic materials not readily oxidized by other microorganisms.

The active biological solids in activated sludge are frequently assumed to be represented by the mixed liquor suspended solids (MLSS); however, the mixed liquor volatile suspended solids (MLVSS) are a more accurate representation.

Growth Phases

The growth of microorganisms is primarily related to the number of viable cells present and the amount of substrate or limiting nutrient present, in addition to other environmental factors. Many of the concepts involved in the continuous growth of microbes, as in the activated sludge process, are illustrated by growth relationships of batch cultures.

First, consider the case of a single species of bacteria inoculated in (that is, added to) a medium containing a substrate and all substances required for growth. The growth that will occur is shown in Figures 9.3 and 9.4. In Figure 9.3 the growth is measured by the number of viable cells, N. There are distinct phases to this curve, which are usually designated as (1) lag, (2) log growth, (3) declining growth, (4) maximum stationary, (5) increasing death, and (6) log death. Frobisher, Hinsdill, Crabtree, and Goodheart (1974) describe the following physiological changes that occur in the various phases.

Figure 9.3. Microbial Growth Phases Based on the Number of Viable Cells, N

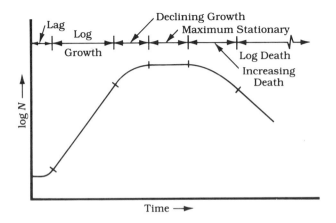

Figure 9.4. Microbial Growth Phases Based on the Mass of Viable Cells

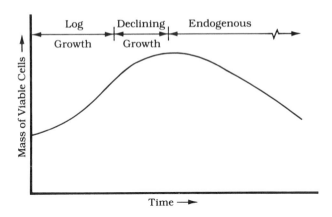

In the beginning of the *lag phase,* the microbes are becoming adjusted to their new environment. Although this phase is not completely understood, it is known that the organisms are recovering from transplant injuries and are absorbing water and substrate and secreting exocellular enzymes to break down large substrate molecules. There is no increase in the number of viable cells at the start of the lag phase; however, towards the end of the phase, metabolic activity begins and the cells begin to increase. During the latter part of the lag phase, the division time gradually decreases, cell size diminishes, and the fission rate reaches a maximum determined by the species and growth conditions. The duration of the lag phase is greatly dependent upon the age of the inoculum culture and the amount of inoculum. If the parent culture is young and biologically active, the lag phase will be extremely short. Also, if the inoculum is relatively large, the lag phase will be minimized. A large inoculum will introduce minute but effective amounts of enzymes, certain essential nutrients, and growth factors from the parent culture or it will quickly synthesize or release these due to the large number of cells in the inoculum.

During the *log growth phase,* the rate of fission is the maximum possible and the average size of the cells is at its minimum for the species. The cell wall and membranes are the thinnest during this phase and the metabolic activities are at the maximum rate. The cells are physiologically young, biologically active, and more vulnerable to deleterious influences than mature, less active, cells. It is a phase of exponential growth, and the log of the number of viable cells versus time is a straight-line relationship. The fission time or generation time depends upon the species and the nutrient and environmental conditions and may vary from a few minutes to as much as several days. Usually, however, it is from 10 to 60 minutes.

Towards the end of the log growth phase the cells begin to encounter difficulties such as the depletion of the substrate or an essential nutrient and the accumulation of toxic end products, which may have reached an inhibitory level. This is the beginning of the *declining growth phase* in which the rate of fission begins to decline and the microorganisms die in increasing numbers so that the increase in the number of viable cells is at a slower rate. In most cases in wastewater treatment, the declining growth phase is due to a depletion of substrate and not due to an accumulation of toxic end products.

Eventually, the number of cells dying equals the number of cells being produced and the culture is in the *maximum stationary phase*, in which the population of viable cells is at a relatively constant value. The time required to reach this phase depends primarily on the species, the concentration of the microbes, the composition of the medium, and the temperature. As the environment becomes more and more adverse to microbial growth, the *increasing death phase* begins, in which the cells reproduce more slowly and the rate at which the cells die exceeds the growth rate. Finally, the increasing death phase progresses into the *log death phase.*

If the cell growth is measured as the total mass of viable cells produced instead of the number of viable cells, the growth curve will be as shown in Figure 9.4. The slope of a tangent to the curve at any time represents the rate of cell production based upon mass. Immediately after inoculation, the cells begin to absorb water and substrate and the cell mass starts to increase. Thus, the *log phase* based on mass includes both the lag and log phase based on viable cell numbers. As the cells begin to multiply, the log growth phase progresses and the rate of cell mass production increases.

When the substrate or an essential nutrient or other factor becomes limiting, or if there is an inhibitory level of accumulated toxic end products, the rate of cell mass production begins to decrease. This condition represents the end of the log growth phase and the beginning of the *declining growth phase.* As environmental conditions become more unfavorable for cell growth, the decrease in growth becomes more pronounced. Finally, when the rate at which the cell mass being produced equals the rate at which the mass is decreasing, the curve reaches a maximum value. This represents the end of the declining growth phase and the beginning of the endogenous phase.

In the *endogenous phase*, the microbes utilize as food their own stored food materials and protoplasm in addition to a portion of the dead cells in the environment. The net effect of the endogenous phase is a decrease in total cell mass with respect to time. The rate at which the cell mass is endogenously decayed is a relatively constant value of about 5 to 20 percent per day. The endogenous degradation rate is continuous; it exists in the log growth and declining growth phases but is masked by the much larger rate of growth (Wilner, B., & Clifton, C. E., 1954; Santer, M., & Ajl, S., 1954).

Consider an aerated laboratory vessel containing a substrate and inoculated with an acclimated mixed culture of microbes (one developed to use the particular substrate). Figure 9.5 shows the cells and substrate concentrations versus time after inoculation for an initial substrate to microbe ratio of 680/495 or 1.374 mg COD/mg activated sludge. The rate of substrate conversion at any time, t, is equal to the slope of a tangent to the substrate curve at that time. A lag phase, based on the number of cells, exists be-

Figure 9.5. Batch Activated Sludge Data: Initial S/X Ratio = 680/495 = 1.374 mg COD/ mg cells

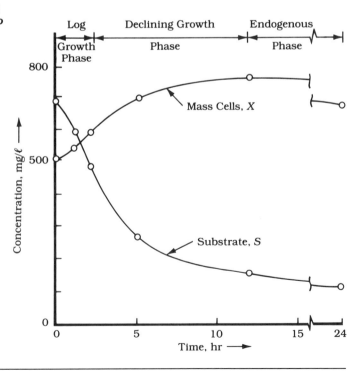

ENVIRONMENTAL ENGINEERING

cause the maximum substrate utilization rate does not occur at time zero, but instead is at about 1.5 hours. The log growth phase ends at about 2.5 hours because the inflection point on the cell curve is at that time and, also, the rate of substrate utilization begins to decrease. The declining growth phase exists from about 2.5 to 12 hours, because at 12 hours the cell mass concentration has reached a maximum value. Beyond that time the endogenous decay rate predominates, the endogenous phase exists, and the mass of microorganisms has begun to decrease.

Figure 9.6 shows the cell and substrate concentrations versus time for the same substrate but with a much larger inoculum. The initial substrate to microorganism ratio is 680/1920 or 0.354 mg COD/mg activated sludge. Since the maximum substrate utilization rate is at time zero, there

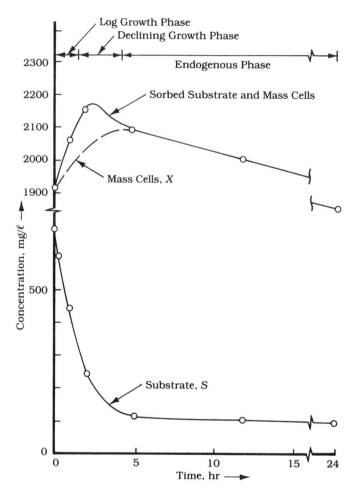

Figure 9.6. Batch Activated Sludge Data: Initial S/X Ratio = 680/1920 = 0.354 mg COD/ mg cells

is no lag phase; this is due to the extremely large inoculum. The log growth phase ends at approximately 1.5 hours, and it can be seen from the graph that the cell growth rate of the microorganisms starts decreasing at that time. The declining growth phase exists from about 1.5 to 4.5 hours and, after 4.5 hours, the endogenous phase exists and the cell mass decreases with respect to time. Also depicted in the graph is the mass of substrate sorbed by the activated sludge from time zero to about 5 hours.

If the biological reactor in Figure 9.1 is a long, narrow tank, a quantity of the MLSS will pass through the reactor with a travel time equal to the theoretical detention time. Curves similar to those shown in Figure 9.6 will represent the cell-substrate relationship for the biological reactor. The maximum substrate removal is at the upstream end of the reactor, and the rate is equal to the slope of a tangent to the substrate curve at time zero. If the reactor has a reaction time greater than about 4.5 hours, the growth phases that will be present are the log growth, declining growth, and endogenous phases. The log growth phase will exist in the upstream reach in the reactor, the declining growth phase will be in the middle reach, and the endogenous phase will be present in the downstream reach. Although endogenous degradation is the principal means by which cell mass changes occur in the endogenous phase, it is continuous through all phases; the endogenous decay rate is merely concealed by the larger log growth and declining growth rates.

Nutrition and Respiration
Microbes may be classified according to the source of their energy and carbon requirements (Frobisher, et al., 1974). Microbes that chemically oxidize exogenic inorganic substances for their energy needs are termed *chemolithotrophs*, whereas those that oxidize exogenic organic substances are termed *chemoorganotrophs*. Those that use light as an energy source but require some inorganic substances for respiration are termed *photolithotrophs*, and those using light but requiring some organic substances for respiration are *photoorganotrophs*. Microbes that use inorganic substances as their carbon source, such as carbon dioxide, are termed *autotrophs*, and those using organic substances as a carbon source are *heterotrophs*. Most microbes used in the activated sludge process use organic materials for both energy (that is, respiration) and cell synthesis; thus, they are chemoorganotrophic and are heterotrophs. The nitrifying bacteria, *Nitrosomonas* and *Nitrobacter*, use the chemical oxidation of an inorganic

nitrogen compound for their energy source and use carbon dioxide for synthesis; thus, they are chemolithotrophic and are autotrophs.

The essential elements and substances required for nutrition are classified as (1) major elements, (2) minor elements, (3) trace elements, and (4) growth factors. The major elements are carbon, hydrogen, oxygen, nitrogen, and phosphorus. The minor elements are principally sulfur, potassium, sodium, magnesium, calcium, and chlorine. The trace elements mainly consist of iron, manganese, cobalt, copper, boron, zinc, molybdenum, and aluminum. The growth factors are essential substances required in trace amounts that the cells themselves cannot synthesize. These principally include vitamins, essential amino acids, and precursors for the synthesis of essential amino acids or other required synthesized compounds. It should be understood that not all the minor elements, trace elements, and growth factors previously listed are required for all microbes because the requirements vary for each species.

In municipal wastewaters, all the essential elements and growth factors are present in adequate amounts if no significant industrial wastewaters enter the collection system. In certain industrial wastewaters, there may be a deficiency of nitrogen and phosphorus, and these must be added so that the $N:BOD_5$ ratio is at least 1:32 and the $P:BOD_5$ ratio is at least 1:150 (Sawyer, C. N., 1956). For industrial wastewaters, the minor elements, trace elements, and growth factors are usually present in sufficient amounts in the carriage water.

Microorganisms may be classified according to their respiration requirements as *aerobes, anaerobes,* and *facultative anaerobes* (Frobisher, et al., 1974). Respiration produces the energy for their life processes and essentially consists of enzymatically removing hydrogen atoms from a hydrogen donor substance which, as a result, is oxidized. The hydrogen atoms are united with a hydrogen acceptor, which is constantly reduced. In aerobic respiration, the hydrogen acceptor is molecular oxygen and the end product is water. In anaerobic respiration, the hydrogen acceptor may be combined oxygen in the form of radicals, such as the carbonate, nitrate, and the sulfate ion or it may be an organic compound. The respective end products for these acceptors are methane, ammonia, hydrogen sulfide, or a reduced organic compound. Facultative anaerobes use aerobic respiration if molecular oxygen is in the environment; if it is absent, they employ anaerobic respiration. In the activated sludge process, the microbes consist of both aerobes and facultative anaerobes.

Environmental Factors Affecting Microbial Activity

Environmental factors may be broadly classified as physical, chemical, and biological according to their nature. The effects of environmental factors upon microbial activity are an important consideration because: (1) it is desirable to maintain an activated sludge culture at its optimum activity, (2) environmental factors are important in evaluating the feasibility of treating certain organic wastewaters by the activated sludge process, and (3) environmental factors are important in disinfecting effluent from biological treatment processes. The environmental factors to be discussed in this section are those that affect the activated sludge process. It should be understood that there are other factors that are important in other applications in environmental engineering. For example, microorganisms must have liquid water available for their feeding, and in the activated sludge process this is always the case. However, oxidation or holding ponds for seasonal wastewaters, such as cannery wastes, may freeze throughout their depth in the winter in severe climates and, if this occurs, microbial action ceases.

Some of the major physical factors affecting the activated sludge process are (1) temperature, (2) osmotic pressure, and (3) the presence of molecular oxygen. Since the temperature within a microbial cell is virtually equal to the temperature of the environment, an increase in temperature increases the microbial activity up to that point where the cells are killed. A temperature increase of 10°C approximately doubles the microbial activity. Microbes may be classified according to their optimum temperature range as psychrophils, mesophils, and thermophils, which have the respective optimum temperature ranges of 0 to 10°C, 10 to 45°C, and 45 to 75°C. Above 75°C microbes are rapidly killed if the contact time is sufficient. The majority of microbes utilized in the activated sludge process are psychrophils and mesophils, although there are always some thermophils present. The osmotic pressure, which is dependent upon the salt concentration in the environment, must be within a certain range because microbes feed by osmosis. Most microbes are not affected by the salt content if it is between 500 to 35,000 mg/ℓ. Hydrostatic pressure within the range encountered in the activated sludge process does not affect the microbial activity. Molecular or dissolved oxygen must be present for aerobes and facultative anaerobes when using aerobic respiration. Usually, a concentration of about 2.0 mg/ℓ is used as a design value for an aerobic biological reactor.

The major chemical factors affecting the activated sludge process are (1) pH, (2) the presence of certain acids and bases, (3) the presence of oxidizing and reducing agents, (4) the presence of heavy metal salts and ions, and (5) the presence of certain chemicals. The microbes utilized in the activated sludge process, like the majority of all microbes, thrive best at a neutral pH range of 6.5 to 9.0. Since carbon dioxide is one of the end products from aerobic bio-oxidation, the carbonate-bicarbonate buffering system will be established in the biological reactor and this will assist in maintaining a neutral pH. However, certain industrial wastewaters may have such a low or high pH that neutralization is required prior to treatment.

Certain acids, such as benzoic acid, and certain bases, such as ammonium hydroxide, are toxic to microbes if present in sufficient concentrations; however, at high concentrations all acids and bases are toxic. Strong oxidizing or reducing reagents are toxic to microorganisms at relatively low concentrations. All of the halogens (chlorine, fluorine, bromine, and iodine) and their salts are very toxic if the halogen has a valence above its lowest valence state because, in this form, it is a strong oxidizing agent.

Heavy metal salts and heavy metal ions are toxic in relatively low concentrations. The toxicity of a metallic ion, in general, increases with an increase in the atomic weight. The heavy metals that may be encountered in wastewaters are usually mercury, arsenic, lead, chromium, zinc, cadmium, copper, barium, and nickel. Silver sometimes is found because it is used in photographic operations; however, due to its value it is usually recovered. Mercury, arsenic, and lead are the most toxic of the heavy metals.

Certain industrial chemicals such as organic acids, alcohols, ethers, aldehydes, phenol, chlorophenol, cresols, and dyes are toxic if present in sufficient concentrations. Soaps and detergents are toxic in relatively high concentrations; however, in the concentrations usually found in wastewaters, they are usually not toxic. Antibiotics produced by pharmaceutical fermentations are toxic in minute concentrations. Greases, if present in significant amounts, will coat the microbes in activated sludge and interfere with their aerobic respiration.

The effect of all the previously discussed chemical agents and compounds is primarily a function of the concentration, temperature, and contact time. As the contact time increases and usually as the temperature increases, the relative toxicity increases. At a constant contact time and temperature, the effect of concentration upon microbial activity is as depicted below:

Concentration →

| No Effect | Stimulates | Inhibits | Kills |

At relatively low concentrations there is no effect; however, as the concentration increases, the chemical agent or compound stimulates microbial growth. A further increase in concentration causes inhibition, and an increase beyond this range results in a killing effect or toxicity to the microorganisms. For example, phenol in sufficient concentrations is a good disinfectant. However, if a phenolic wastewater has a phenol concentration less than about 500 mg/ℓ, it readily stimulates microbial activity and is bio-oxidized by the activated sludge process, providing it is an acclimated culture. Also, petrochemical wastewaters containing organic compounds such as organic acids, alcohols, ethers, and aldehydes are readily bio-oxidized if the concentrations are dilute. Even pharmaceutical wastes from the production of antibiotics may be treated by the activated sludge process if the concentrations are extremely dilute.

Biological factors affecting the activated sludge process are usually a result of undesirable microorganisms in the mixed culture. For example, excessive growth of the filamentous genus, *Sphaerotilis*, creates a sludge having poor settling characteristics, making proper final clarification difficult.

Kinetics and Types of Reactors

The study of chemical kinetics is concerned with the rates of reactions by which the reactants are converted into the products. The rate of a reaction at a given temperature is usually expressed as the change in concentration with respect to time, dC/dt. The rate equation for a reaction is the differential equation, which expresses the rate as a function of the concentration of each species that affects the rate. Consider the simple conversion of reactant A to products. The generalized equation for the rate is

$$-\frac{dC_A}{dt} = kC_A^n \tag{9.2}$$

where

k = rate constant;
C_A = concentration of A, mass per unit volume;
n = exponential power.

The reaction order is equal to the exponential power in the rate equation. For example, for a *zero-order* reaction, $n = 0$, whereas for a *first-order* reaction, $n = 1$. The rates of many but not all reactions may be expressed by such a simple differential rate equation as previously shown.

Reactors may be classifed according to their mode of operation as *continuous-flow* and *batch* reactors. A continuous-flow reactor has a continuous stream of reactants entering and a continuous stream of products leaving. A batch reactor, however, does not have continuous streams. The reactants are added, the reaction occurs, and then the products are discharged.

Continuous-flow reactors, such as the reactor basins or aeration tanks used for activated sludge, may be classified according to their flow regime as plug-flow, dispersed plug-flow, and completely mixed reactors. In a *plug-flow reactor*, shown schematically in Figure 9.7, the elements of the fluid that enter the reactor at the same time flow through it with the same velocity and leave at the same time. The flow through the reactor is similar to a piston or plug moving through it. The travel time of the fluid elements is equal to the theoretical detention time, and there is no longitudinal mixing. Long, narrow tanks approach a plug-flow regime. In a *completely mixed reactor*, shown schematically in Figure 9.8, the fluid elements, upon entering, are immediately dispersed throughout the reactor volume; the reactor contents are uniform and are identical with the effluent stream. Completely mixed reactor basins are usually circular or square tanks. Plug-flow and com-

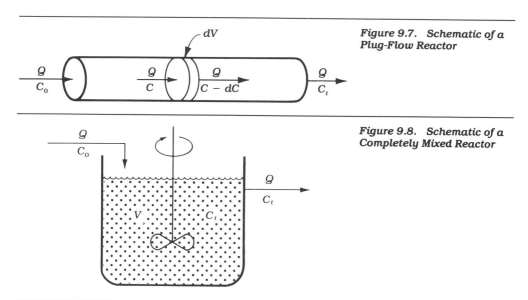

Figure 9.7. Schematic of a Plug-Flow Reactor

Figure 9.8. Schematic of a Completely Mixed Reactor

pletely mixed reactors are ideal reactors; however, with proper design the flow regime of plug-flow and completely mixed flow may be approached. The *dispersed plug-flow reactor* has longitudinal dispersion of the fluid elements as they pass through the reactor. The flow regime of the dispersed plug-flow reactor is between plug-flow and completely mixed flow. Dispersed plug-flow reactor basins are usually rectangular tanks with significant longitudinal mixing. For rectangular activated sludge tanks with a dispersed plug-flow regime, the longitudinal dispersion is mainly due to their relatively small length-to-width ratios.

For the case of a plug-flow reactor, consider the reactor shown in Figure 9.7 along with the element of volume, dV. Assume that the element of volume, dV, is located at a fixed position anywhere along the length of the reactor. The material balance is

$$[\text{Accumulation}] = [\text{Input}] - \begin{bmatrix} \text{Decrease} \\ \text{due to} \\ \text{reaction} \end{bmatrix}$$

$$- [\text{Output}] \tag{9.3}$$

For steady state, the accumulation term is zero; thus, the material balance becomes

$$[\text{Input}] = [\text{Output}] + \begin{bmatrix} \text{Decrease} \\ \text{due to} \\ \text{reaction} \end{bmatrix} \tag{9.4}$$

The rate reaction for a simple first-order chemical reaction is

$$-\frac{dC}{dt} = kC = -r \tag{9.5}$$

where

dC/dt = rate of reaction, mass/(volume)(time);
k = rate constant, time^{-1};
C = reactant concentration, mass/volume;
r = rate of reaction, mass/(volume)(time).

Substituting values into the material balance for the element of volume, dV, gives

$$QC = Q(C - dC) + r\,dV \tag{9.6}$$

From which

$$Q\,dC = r\,dV \tag{9.7}$$

From Eq. (9.5), $r = -kC$. Substituting this into Eq. (9.7) gives the following:

$$Q dC = - kC dV \qquad (9.8)$$

Rearranging and setting the limits for integration gives

$$\int_{C_0}^{C_t} \frac{dC}{C} = -\frac{k}{Q} \int_0^V dV \qquad (9.9)$$

where

    C_0 = reactant concentration entering the reactor;
    C_t = reactant concentration leaving the reactor;
    V = volume of the reactor.

Integrating and inserting the limits gives

$$\ln C_t - \ln C_0 = -k \frac{V}{Q} \qquad (9.10)$$

Since the detention time or reaction time $\theta = V/Q$, Eq. (9.10) becomes

$$\ln C_t - \ln C_0 = -k\theta \qquad (9.11)$$

Rearranging gives

$$\frac{C_t}{C_0} = e^{-k\theta} \qquad (9.12)$$

For the case of the completely mixed reactor, consider the reactor shown in Figure 9.8. The rate equation for a simple first-order chemical reaction is

$$-\frac{dC_t}{dt} = kC_t \qquad (9.13)$$

where

    dC_t/dt = rate of reaction, mass/(volume)(time);
    k = rate constant, time^{-1};
    C_t = concentration of reactant in the reactor and in the effluent, mass/volume.

The material balance is [Accumulation] = [Input] − [Decrease due to reaction] − [Output]. From Eq. (9.13), $- dC_t = kC_t dt$. Substituting this and other values into the material balance gives

$$dC_t \cdot V = Q C_0 dt - V k C_t dt - Q C_t dt \qquad (9.14)$$

Dividing by Vdt produces

$$\frac{dC_t}{dt} = \frac{Q}{V}C_0 - kC_t - \frac{Q}{V}C_t \qquad (9.15)$$

Since the accumulation term $dC_t/dt = 0$ for steady state and $\theta = V/Q$, Eq. (9.15) becomes

$$0 = \frac{C_0}{\theta} - kC_t - \frac{C_t}{\theta} \qquad (9.16)$$

Rearranging gives

$$kC_t = \frac{C_0 - C_t}{\theta} \qquad (9.17)$$

or

$$\theta = \frac{C_0 - C_t}{kC_t} \qquad (9.18)$$

For a plug-flow reactor where $k = 0.40 \text{ hr}^{-1}$, $C_0 = 200$ mg/ℓ and $C_t = 20$ mg/ℓ, Eq. (9.12) gives $20/200 = \exp[(-0.40)\theta]$. From this equation, $\theta = 5.76$ hr. For a completely mixed reactor producing the same effluent, the reaction time from Eq. (9.18) is $\theta = (200 - 20)/[(0.40) \times (20)]$, from which $\theta = 22.5$ hr. Thus, for the same effluent concentration or the same degree of removal, the completely mixed reactor requires a much longer reaction time or detention time. Also, it can be shown that for the same effluent concentration or the same degree of removal, a dispersed plug-flow reactor has a reaction time or detention time between that of a plug-flow reactor and that of a completely mixed reactor.

For a batch reactor, there are no continuous flows into or out of the reactor; thus the terms [Input] and [Output] in the material balance are zero. The material balance, [Accumulation] = [Input] − [Decrease due to reaction] − [Output], becomes [Accumulation] = − [Decrease due to reaction]. Therefore, for a simple first-order chemical reaction, the material balance is

$$\frac{dC}{dt} = -kC \qquad (9.19)$$

Rearranging for integration gives

$$\int_{C_0}^{C_t} \frac{dC}{C} = -k \int_0^t dt \qquad (9.20)$$

Integration and substituting limits gives

$$\ln C_t - \ln C_0 = -kt \qquad (9.21)$$

or

$$\frac{C_t}{C_0} = e^{-kt} \qquad (9.22)$$

Eq. (9.22) is identical with Eq. (9.12) for a plug-flow reactor if the reaction time, t, for the batch reactor equals the detention time or reaction time, θ, for the plug-flow reactor. Thus, for the same degree of removal, the plug-flow and the batch reactor require the same reaction time.

Two basic approaches are used to determine the rate of substrate removal that occurs in the activated sludge process. The first approach, which uses the Michaelis-Menten, or Monod, equation, utilizes relationships applied in the fermentation industries (McCarty, P. L., & Lawrence, A. W., 1970). The second approach uses a modification of chemical kinetics (Eckenfelder, W. W., 1966). If a batch activated sludge reactor is inoculated and the values of the cell mass and substrate mass concentrations versus time of reaction are obtained, the data may be analyzed to give the kinetic constants required for design by either of the two approaches. The inoculum must come from an acclimated parent culture developed in a continuous-flow reactor in order to obtain meaningful data.

Biochemical Kinetics

The first approach employs formulations that have been developed for industrial fermentations. It has been found that most enzyme-catalyzed reactions involving a single substrate are zero order with respect to the substrate at relatively high substrate concentrations, and are first order with respect to the substrate at relatively low substrate concentrations. When this occurs in a batch reactor, the substrate utilization is zero order for a period after the inoculation and, later, the substrate utilization becomes first order. An explanation of this occurrence is that at high substrate concentrations the surface of the enzymes is saturated with substrate; thus, the reaction is independent of the substrate concentration. At relatively low substrate concentrations, the portion of the surface of the enzymes that is covered with the substrate is proportional to the substrate concentration. This phenomenon is explained by the Michaelis-Menten concept.

The Michaelis-Menten equation gives the relationship

between the rate of fermentation product production, dP/dt, and the substrate concentration, S. The rate of product production may also be expressed as the specific rate of product production, $(1/X)(dP/dt)$, where X is the cell mass concentration (Aiba, S., Humphery, A. E., & Millis, N. F., 1965). Since the specific rate of substrate utilization, $(1/X)(dS/dt)$, is proportional to the specific rate of product production (Aiba, et al., 1965), it may be incorporated into the Michaelis-Menten formulation as follows:

$$\frac{1}{X}\frac{dS}{dt} = k_s\left(\frac{S}{K_m + S}\right) \tag{9.23}$$

where

$(1/X)(dS/dt)$ = specific rate of substrate utilization, mass/(mass microbes)(time);

dS/dt = rate of substrate utilization, mass/(volume)(time);

k_s = maximum rate of substrate utilization, mass/(mass microbes)(time);

K_m = substrate concentration when the rate of utilization is half the maximum rate, mass/volume;

S = substrate concentration, mass/volume.

It can be shown that $K_m = S$ when the specific rate of substrate utilization is half its maximum value, k_s. Eq. (9.23) is similar to the formulation developed by Monod (1949), giving the relationship between the growth rate of microorganisms and the substrate concentration. The relationship of Eq. (9.23) may be used to relate the rate of substrate utilization to the substrate and cell concentrations in the activated sludge process.

In examining Eq. (9.23), it can be noted that there are two limiting cases. If S is relatively large, K_m may be neglected. S cancels out in the numerator and denominator and the reaction is zero order in substrate (Brey, W. S., 1958); thus,

$$\frac{1}{X}\frac{dS}{dt} = k_s = K_0 \tag{9.24}$$

where K_0 is the rate constant for a zero-order reaction. If, on the other hand, S is relatively small, it may be neglected in the denominator and the reaction is first order in substrate (Brey, W. S., 1958) or

$$\frac{1}{X}\frac{dS}{dt} = \frac{k_s}{K_m}(S) = K_1 S \tag{9.25}$$

where K_1 is the rate constant for a first-order reaction in substrate.

In summarizing the approach using the Michaelis-Menten concept, high substrate concentrations yield zero order in substrate utilization, and low substrate concentrations yield first order in substrate utilization. This relationship has worked extremely well for industrial fermentations where the mass of inoculum is relatively small compared to the maximum cell mass synthesized and where endogenous degradation is negligible compared to cell growth. In order for a cell growth of several magnitudes times the inoculum to occur, the initial substrate to cell ratio must be relatively large.

The second approach to biochemical kinetics applied to the activated sludge process utilizes chemical kinetic theory. Since the Michaelis-Menten approach indicated that for high substrate concentrations the reaction is zero order in substrate, the rate equation for a pseudo–zero–order reaction is

$$-\frac{1}{X}\frac{dS}{dt} = K \tag{9.26}$$

where K is the rate constant. The negative sign indicates that the substrate is decreasing with respect to time. Rearranging Eq. (9.26) for integration gives

$$\int_{S_0}^{S_t} dS = -K\overline{X}\int_0^t dt \tag{9.27}$$

where

K = rate constant, time^{-1};
$\overline{X}$ = average cell mass concentration during the biochemical reaction—that is, $\overline{X} = (X_0 + X_t)\frac{1}{2}$, where X_0 and X_t are the cell mass concentrations at the respective times, $t = 0$ and $t = t$, mass/volume;
S_t = substrate concentration at time, t, mass/volume;
S_0 = substrate concentration at time, $t = 0$, mass/volume.

Integration of Eq. (9.27) yields

$$S_t - S_0 = -K\overline{X}t \tag{9.28}$$

or

$$S_t = S_0 - K\overline{X}t \tag{9.29}$$

Equation (9.29) is of the form $y = mx + b$; thus, plotting S_t on the y-axis versus $\overline{X}t$ on the x-axis on arithmetical paper will result in a straight line if the reaction is pseudo–zero order and the slope will equal $-K$.

A first-order reaction at low substrate concentrations was indicated by the Michaelis-Menten concept; thus, the rate equation for a pseudo–first-order reaction is

$$-\frac{1}{X}\frac{dS}{dt} = KS \tag{9.30}$$

where K is the rate constant, volume/(mass microbes)(time). Rearranging for integration gives

$$\int_{S_0}^{S_t}\frac{dS}{S} = -K\overline{X}\int_0^t dt \tag{9.31}$$

Integration results in

$$\ln S\bigg]_{S_0}^{S_t} = -K\overline{X}t \tag{9.32}$$

Equation (9.32) may be simplifed to

$$\ln S_t = \ln S_0 - K\overline{X}t \tag{9.33}$$

Equation (9.33) is of the form $y = mx + b$; thus, plotting S_t on the y-axis versus $\overline{X}t$ on the x-axis on semilog paper will result in a straight line if the reaction is pseudo–first order, and the line slope will equal $-K$.

Since in wastewaters there are numerous substrates present, the rate constant, K, for the kinetic equations and the constants for the Michaelis-Menten equation represent overall average values.

The value of the rate constant, K, for the kinetic equations depends primarily on the specific wastewater of interest because the species of organic substrates will vary for different wastewaters and, in particular, for industrial wastewaters and municipal wastewaters containing significant amounts of industrial wastes. Because only certain microbial species in an acclimated sludge can bio-oxidize a particular organic compound and each species has its own particular rate of utilization, it follows that the overall rate constant, K, will vary for different type wastewaters. Table 9.1 gives some typical wastewaters along with the reaction order and rate constant based on the biodegradable TOC by acclimated activated sludges. It can be seen that

Table 9.1. Reaction Orders and Rate Constants for Some Selected Wastewaters

Type of Wastewater	Pseudoreaction Order	Reaction Rate Constant K[a] ℓ/(gm MLVSS)(hr) at 25°C
Pulp and paper mill	First	0.375
Pulp and paper mill	First	0.528
Chemical manufacture	First	0.479
Chemical manufacture	First	0.601
Oil refinery	First	0.504
Oil refinery	First	0.660
Petrochemical manufacture	First	0.592
Petrochemical manufacture	First	0.686
Petrochemical manufacture	First	0.713
Petrochemical manufacture	First	0.911
Petrochemical manufacture	First	1.221
Petrochemical manufacture	First	1.333
Municipal (domestic)	First	1.717

a. Based on biodegradable total organic carbon (TOC).

the variation of K is widespread and even for a certain type of industry, such as the petrochemical industry, there is a significant variation in the rate constant. The petrochemical wastewaters having the lower K values are from plants manufacturing organic compounds relatively difficult to bio-oxidize, such as insecticides, herbicides, fungicides, organic compounds for production of hard plastics, and certain organic solvents. The petrochemical wastewaters having the higher K values are from plants producing compounds more easily biodegraded, such as unsaturated hydrocarbons for synthetic rubber and flexible plastic manufacture. It can be noted that the municipal wastewater has substrates more easily degraded than any of the industrial wastewaters; however, this is not always true. Based on limited data, the rate constant, K, for municipal wastewaters ranges from 0.10 to 1.25 ℓ/(gm MLSS)(hr) using total BOD_5 as a measure of the organic content—that is, both soluble and insoluble BOD_5. Due to the wide range in K values, care should be exercised in arbitrarily assuming a K value for design when no bench-scale or pilot-scale

Module of Four Batch Activated Sludge Reactors (left)

Module of Four Batch Activated Sludge Reactors in Operation (right)

Continuous-Flow Completely Mixed Activated Sludge Reactor. Baffle between mixing and settling chambers may be lowered or raised to adjust recycle.

Continuous-Flow Completely Mixed Activated Sludge Reactor in Operation. Effluent leaves by vertical tube (not visible) passing through bench top.

studies are done. In the absence of such studies, a K value in the range of 0.10 to 0.40 ℓ/(gm MLSS)(hr) is recommended.

The rate constant, K, may be determined from batch reactor studies, as shown in Example Problem 9.1, or from three or more continuous-flow, completely mixed activated sludge reactors, as shown later in this chapter. If possible, pilot plant studies should be used since they simulate field conditions better than laboratory studies.

Example Problem 9.1
Biochemical Kinetics

A kinetic study of a soluble organic wastewater has been done in the laboratory using a batch reactor inoculated from a parent acclimated culture developed in a continuous-flow activated sludge reactor. The COD and MLSS concentrations for the various reaction times were as shown in Table 9.2.

Table 9.2. Kinetic Data from Batch Test

Reaction time (hr)	COD (mg/ℓ)	MLSS (mg/ℓ)
0	680	1910
1	440	2180
2	240	2210
3	165	2190
4	128	2130
5	115	2090
24	102	1860

At 24 hours, the BOD_5 = 4.2 mg/ℓ, the BOD_5 = 0.35 BOD_u (ultimate first-stage BOD), and the BOD_u is equal to the degradable COD. The MLVSS = 88 percent of the MLSS. Determine the reaction order and the reaction rate constant in terms of MLSS and MLVSS.

Solution

The nondegradable COD is equal to the COD at 24 hours minus the BOD_u or $102 - 4.2/0.35 = 102 - 12 = 90$ mg/ℓ. Assuming a pseudo–first-order reaction, the data are shown in Table 9.3 for a time up to 5 hours.

Table 9.3. Reduced Data from Batch Test

Time, t (hr)	Degradable COD (mg/ℓ)	X MLSS (mg/ℓ)	$\overline{X}$ MLSS (mg/ℓ)	$\overline{X}t$ (mg/ℓ)(hr)
0	590	1910		
1	350	2180	2045	2045
2	150	2210	2060	4120
3	75	2190	2050	6150
4	38	2130	2020	8080
5	25	2090	2000	10,000

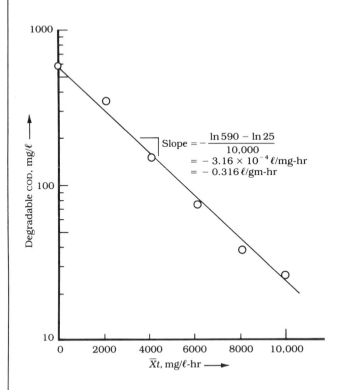

Figure 9.9. Graph for Example Problem 9.1

The plot of $1n \, S_t$ versus $\overline{X}t$ is shown in Figure 9.9. Since it is a straight line, the reaction is pseudo–first order and from the slope, K (MLSS) = 0.316 ℓ/(gm MLSS-hr). K (MLVSS) = K (MLSS)/0.88; therefore, K (MLVSS) = 0.316/0.88 or K (MLVSS) = 0.359 ℓ/(gm MLVSS-hr).

Food to Microbe Ratio (F/M) and Mean Cell Residence Time (θ_c)

The instantaneous food to microorganism ratio is equal to the specific rate of substrate utilization, $(1/X)(dS/dt)$, and integration of this equation yields

$$\frac{F}{M} = \frac{\Delta S}{\overline{X}\Delta t} \tag{9.34}$$

where $\Delta S/\Delta t$ is the rate of substrate removal, mass/time. The units for the F/M ratio are mass substrate/(mass microbes)(time)—that is, 1b BOD_5/(lb MLVSS)(day). The mean cell residence time, θ_c, is

$$\theta_c = \frac{\overline{X}}{X_w} \qquad (9.35)$$

where the active biological solids in the reactor and in the waste activated sludge flow are $\overline{X}$ and X_w, respectively. The units for the mean cell residence time are days and, frequently, the mean cell residence time is referred to as the sludge age. Both the food to microorganism ratio and the mean cell residence time are used to characterize the performance of an activated sludge process. For example, a high food to microorganism ratio and a low mean cell residence time usually produces filamentous growths that have poor settling characteristics. On the other hand, a low food to microorganism ratio and a large mean cell residence time can cause the biological solids to undergo excessive endogenous degradation and cell dispersion. For municipal wastewaters, the mean cell residence time must be at least three to four days to attain proper settling in the final clarifier. If significant nitrification is desired for a municipal wastewater, the mean cell residence time must be at least ten days, as shown in Figure 9.10, and the dissolved oxygen (DO) at least 2.0 mg/ℓ.

The relationship between the mean cell residence time, θ_c, and the food to microbe ratio, F/M, can be derived by starting with the equation for cell production, which is

$$\frac{\Delta X}{\Delta t} = Y\frac{\Delta S}{\Delta t} - k_e\overline{X} \qquad (9.36)$$

Figure 9.10. Nitrification versus Mean Cell Residence Time
Adapted from "Nitrification in Wastewater Treatment Plants (Activated Sludge)" by A. C. Petrasek. Dissertation, Civil Engineering Department, Texas A&M University, 1975. Reprinted by permission.

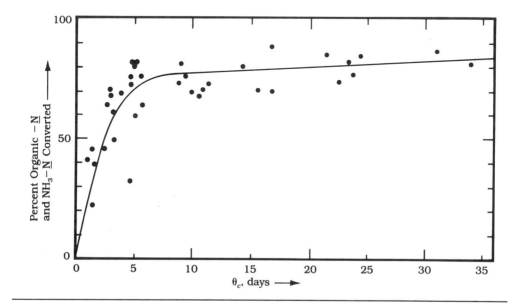

where

$$\Delta X/\Delta t = \text{rate of cell production, mass/time;}$$
$$Y = \text{cell yield coefficient, mass cells created/mass substrate removed;}$$
$$k_e = \text{endogenous decay coefficient, mass cells/(total mass cells)(time);}$$
$$\overline{X} = \text{average cell concentration, mass.}$$

Dividing by $\overline{X}$ gives

$$\frac{\Delta X/\Delta t}{\overline{X}} = Y \frac{\Delta S/\Delta t}{\overline{X}} - k_e \tag{9.37}$$

The mean cell residence time, θ_c, is the average time a cell remains in the system; thus,

$$\theta_c = \frac{\overline{X}}{\Delta X/\Delta t} \tag{9.38}$$

The food to microbe ratio, F/M, is the rate of substrate removal per unit weight of the cells, or

$$\frac{F}{M} = \frac{\Delta S/\Delta t}{\overline{X}} \tag{9.39}$$

Substituting Eqs. (9.38) and (9.39) into Eq. (9.37) gives

$$\frac{1}{\theta_c} = Y \frac{F}{M} - k_e \tag{9.40}$$

The F/M ratio is also given by Eq. (9.34), which is another form of Eq. (9.39). Substituting Eq. (9.34) into Eq. (9.40) gives

$$\frac{1}{\theta_c} = Y \frac{\Delta S}{\overline{X}\Delta t} - k_e \tag{9.41}$$

In this expression, $F/M = \Delta S/(\overline{X}\Delta t)$. In many cases, Eqs. (9.40) and (9.41) are approximations because the coefficients Y and k_e frequently have been found to decrease as θ_c increases (WPCF, 1977). An example using Eq. (9.41) is as follows: If the influent to an aeration tank has a BOD_5 = 150 mg/ℓ, the effluent BOD_5 = 10 mg/ℓ, the MLSS = 2500 mg/ℓ, Y = 0.7 lb MLSS/lb BOD_5, k_e = 0.05 day^{-1}, and the aeration time (Δt) = 6 hr, then θ_c^{-1} = (0.7)(150 − 10)/ (2500)(6/24) − 0.05 or θ_c = 9.4 days.

The *conventional* activated sludge process, Figure 9.11, uses a rectangular aeration tank as the biological reactor; the food to microbe ratio, F/M, is usually from 0.15 to 0.4 lb BOD$_5$/lb MLSS-day. The space loading is usually less than 25 lb BOD$_5$/day-1000 ft^3 of aeration volume.

The wastewater is preliminary treated to remove screened materials and grit, such as sand and silt; then the wastewater is primary treated to remove as much settleable suspended solids as possible. The primary sludge must be treated by some method such as anaerobic digestion to bio-oxidize the organic solids and thus stabilize the primary sludge solids. The clarified wastewater, Q, is mixed with the return activated sludge flow, R, to inoculate the wastewater with active biological solids, and the mixed liquor, $Q + R$, enters the biological reactor. As the mixed liquor passes through the reactor, the activated sludge solids sorb the organic matter, both the insoluble and soluble portions, and bio-oxidize the materials to produce carbon dioxide, water, and other end products and to synthesize new cells. As the bio-oxidation proceeds in the reactor, endogenous degradation is also occurring simultaneously. Aerobic conditions are maintained by compressed air or mechanical aeration devices, spaced uniformly along the length of the reactor, and the minimum operating dissolved oxygen level is 2.0 mg/ℓ (Great Lakes Board, 1978).

The mixed liquor passes from the reactor to the final clarifier (solid-liquid separator), where the activated sludge solids settle by gravity. The effluent spills over the clarifier weirs and is usually disinfected by chlorine prior to dis-

Plug-Flow and Dispersed Plug-Flow Reactors

Figure 9.11. Conventional Activated Sludge Process

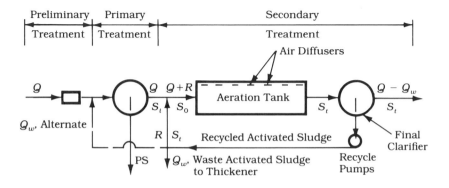

charge. The settled activated sludge is pumped by recycle pumps and the major portion, R, of the recycled sludge is mixed with the feed wastewater to inoculate the flow, Q, to the reactor. The recycle ratio, R/Q, is from 30 to 100 percent, depending on the desired MLSS and the sludge density index (SDI).

A portion of the recycled activated sludge, Q_w, is wasted to remove the net mass synthesized in the system. In plants having a capacity less than several MGD, the waste activated sludge, Q_w, is usually mixed with the feed to the primary clarifier, as shown in Figure 9.11. It is subsequently settled and removed with the primary sludge, PS. In this manner, the sludge is thickened and a large portion of its water is removed prior to sludge treatment. In large plants, the waste activated sludge flow, Q_w, is usually thickened by separate sludge thickeners. The primary sludge and waste activated sludge solids must be oxidized or stabilized, usually by anaerobic or aerobic digestion, prior to disposal. Usually the waste activated sludge flow is from about 1 to 6 percent of Q by volumetric flow rate.

The *tapered aeration* activated sludge process, shown in Figure 9.12, is only a modification of the conventional process. It is identical with the conventional process except that the aeration devices are spaced along the length of the reactor in accordance to the oxygen demand. At the upstream end of the reactor, as shown in Figure 9.13, the oxygen demand is high because the BOD_5 is the maximum in the reactor. As the mixed liquor passes through the reactor, the BOD_5 gradually reduces; thus, the oxygen demand also gradually reduces. Both the BOD_5 and the oxygen demand become minimum at the downstream end. The oxygen required for each quarter for treating municipal wastewater with a reaction time less than eight hours is shown in Table 9.4.

Figure 9.12. Tapered Aeration Activated Sludge Process

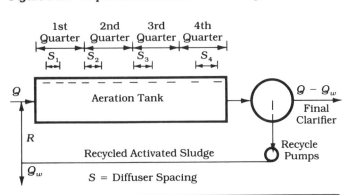

Figure 9.13. *Relative Oxygen Demand for the Conventional or Tapered Aeration Activated Sludge Process (Municipal Wastewater)*

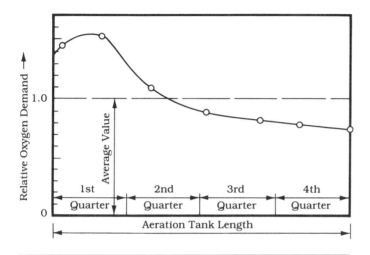

Table 9.4 Oxygen Demand versus Reactor Length for Municipal Wastewaters

Quarter Length of Reactor	Quarter Oxygen Demand/Total Oxygen Demand for the Entire Reactor (%)
1st	35
2nd	26
3rd	20
4th	19

The flow regime in biological reactors used in the conventional or tapered aeration process is dispersed plug flow; however, it may approach plug flow in some cases. In a plug-flow reactor, as previously mentioned, there is negligible induced mixing between elements of the fluid along the axial direction of flow. Actually, in the conventional or tapered aeration reactor, there is some longitudinal or axial mixing but, in long, narrow tanks, the flow regime may approach plug flow. The longitudinal mixing is characterized by the dispersion number, D/vL, where D is the dispersion coefficient, v is the axial velocity, and L is the tank length. Usually, the dispersion number is from 0 to 0.20 for rectangular activated sludge tanks (Metcalf and Eddy, 1979).

The performance of a plug-flow biological reactor exhibiting a pseudo–first-order reaction is the same as for a batch reactor, as given by Eq. (9.33), which may be rearranged to yield (Eckenfelder, W. W., 1966)

$$\frac{S_t}{S_0} = e^{-K\overline{X}\theta} \tag{9.42}$$

where the detention time, θ, for the plug-flow reactor is equal to the reaction time, t, for the batch reactor. The reaction or detention time, θ, for a plug-flow reactor may be determined from Eq. (9.42). For a dispersed plug-flow reactor, the detention time, θ, in the previous plug-flow equation must be increased to maintain the same S_t value. The ratio of the detention times of a dispersed plug-flow and plug-flow reactor is the same as the ratio of their volumes. Thus, the volume ratio may be determined from Figure 9.14, and then the detention time required for a dispersed plug-flow reactor may be determined. The volume of the reactor, V, for either plug-flow or dispersed plug flow is given by

$$V = (Q + R)\theta \tag{9.43}$$

Figure 9.14. Comparison between Dispersed Plug-Flow and Ideal Plug-Flow Reactors for a First-Order Reaction of the Type, A→Products
Adapted with permission from "Backmixing in the Design of Chemical Reactors" by O. Levenspiel and K. B. Bischoff in *Industrial and Engineering Chemistry* 51, no. 12 (December 1959):1431; and from "Reaction Rate Constant May Modify the Effects of Backmixing" by O. Levenspiel and K. B. Bischoff in *Industrial and Engineering Chemistry* 53, no. 4 (April 1961):313. Copyright 1959, 1961 American Chemical Society.

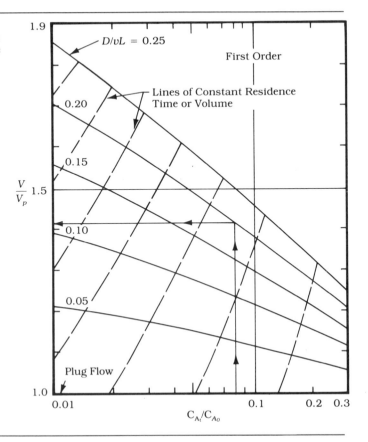

where V is the reactor volume.

For the conventional or tapered aeration activated sludge process, the aeration time (based on the flow, Q) is usually from 4 to 8 hours, the mean cell residence time is from 5 to 18 days, the recycle ratio is usually from 30 to 100 percent, and the MLSS concentration is usually from 1500 to 4000 mg/ℓ (WPCF, 1977). The performançe is from 90 to 95 percent BOD_5 removal and 85 to 95 percent suspended solids removal. Either process will give good nitrification if the mean cell residence time is greater than 10 days and the DO is adequate.

Example Problem 9.2
Plug-Flow and
Dispersed Plug-Flow
Biological Reactors

A municipal wastewater having an influent BOD_5 of 250 mg/ℓ is to be treated by the tapered aeration activated sludge process. The flow diagram is as shown in Figure 9.12; the primary clarifier removes 35 percent of the influent BOD_5. The effluent must have a BOD_5 of 10 mg/ℓ or less. The design MLSS is 2500 mg/ℓ and the SDI is 10,000 mg/ℓ. The reaction is pseudo–first order with a reaction constant of 0.195 ℓ/gm-hr based on MLSS. Determine:

1. The reaction or detention time if the flow regime in the reactor is plug flow.
2. The reaction or detention time if the flow in the reactor is dispersed plug flow with a dispersion number, $D/\upsilon L$, of 0.20.

Solution

The BOD_5 in the primary effluent is $(250\ mg/\ell)(1-0.35)$ $= 162.5\ mg/\ell$. A material balance on the biological solids at the junction of the primary effluent and the return sludge line is

$$(Q)(0) + (R)(10,000) = (Q + R)(2500)$$

The recycle ratio, R/Q, from the above equation is 33.3 percent. A material balance on the BOD_5 at the junction is

$$(Q)(162.5) + (R)(10) = (Q + R)(S_0)$$

or

$$(Q)(162.5) + (0.333\ Q)(10) = (Q + 0.333\ Q)(S_0)$$

The BOD$_5$ in the mixed liquor, S_0, from the above equation is 124.4 mg/ℓ. The performance equation for a plug-flow reactor is

$$\frac{S_t}{S_0} = e^{-K\bar{X}\theta}$$

or

$$\frac{10}{124.4} = e^{-(0.195)(2.50)\theta}$$

Therefore,

θ = 5.17 hours for a plug-flow reactor.

The fraction of BOD$_5$ remaining is

$$\frac{S_t}{S_0} = \frac{10}{124.4} = 0.0804$$

As shown in Figure 9.14, the ratio of the dispersed plug-flow volume to the plug-flow volume, V/V_p, is 1.42 for a dispersion number of 0.20 and a fraction remaining equal to 0.0804. Since the volume ratio is the same as the detention time ratio,

$$\theta = (5.17\,\text{hours})(1.42)$$

or

θ = 7.34 hours for a dispersed plug-flow reactor

Completely Mixed Reactors

The completely mixed activated sludge process, as shown in Figures 9.15 and 9.16, generally uses circular or square aeration tanks as the biological reactor; however, rectangular tanks have also been employed. The food to microorganism ratio, F/M, is generally from 0.05 to 0.6 lb BOD$_5$/lb MLSS-day. The space loading is usually less than 60 lb BOD$_5$/day-1000 ft^3.

After preliminary and primary treatment, as shown in Figure 9.15, the clarified wastewater flow, Q, mixes with the return activated sludge flow, R, to inoculate the incoming wastewater with the active biological solids. The mixed liquor flow, $Q + R$, then enters the reactor where it is quickly dispersed throughout the reactor volume. An alternate method of introducing the return activated sludge flow, as

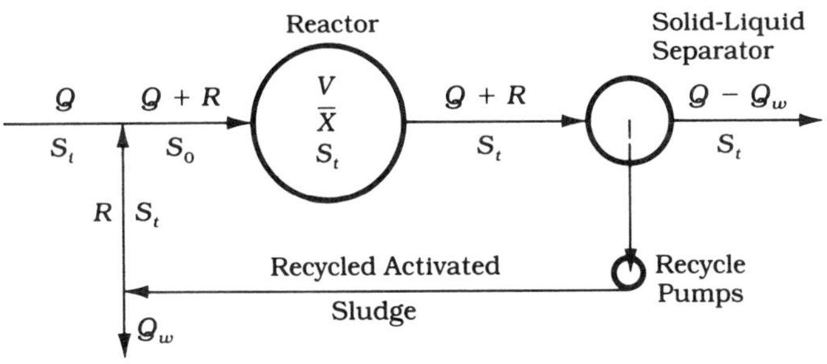

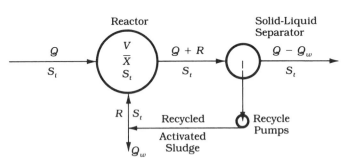

Figure 9.15. Flowsheet for the Completely Mixed Activated Sludge Process

Figure 9.16. Flowsheet for the Completely Mixed Activated Sludge Process

shown in Figure 9.16, is to discharge the recycled flow, R, directly to the reactor in the vicinity of the aerators so that it is rapidly mixed throughout the tank volume. Introducing the recycled sludge directly in the reactor assists in minimizing the effect of a toxic slug of material in the feed wastewater. In this respect, it is preferable to discharge the recycled sludge directly to the reactor instead of mixing it with the influent feed flow. The contents in the reactor are uniform throughout the volume and have the same characteristics as the reactor effluent. In the reactor, the active biological solids sorb the organic matter, both insoluble and soluble fractions, and bio-oxidize these materials to produce aerobic end products and synthesize new microbial cells. As the bio-oxidation proceeds, endogenous degradation is also occurring. Aerobic conditions are usually maintained by mechanical aeration devices and the minimum operating dissolved oxygen level is 2.0 mg/ℓ (Great Lakes–Upper Mississippi Board of State Sanitary Engineers, 1978). The solid-liquid separation in the final clarifier, the recycled sludge system, and the method of disposal of the waste activated sludge are carried out in the same manner as in the conventional or tapered aeration activated sludge processes.

The completely mixed activated sludge process is unique because the contents throughout the reactor volume have the same characteristics as the mixed-liquor flow leaving the reactor. Thus, the soluble substrate concentration in the reactor is the same as in the discharged mixed-liquor flow and in the final effluent. Also, the total substrate concentration in the reactor, both soluble and insoluble fractions, has essentially the same concentration as in the flow to the final clarifier and in the final effluent.

The completely mixed activated sludge process has several advantages over the conventional and tapered aeration processes, as well as other activated sludge modifications, because it has (1) maximum equalization in the oxygen uptake rate, (2) maximum dampening of slug or shock loads because they are quickly dispersed throughout the reactor volume, (3) maximum neutralization of the carbon dioxide produced during the aerobic bio-oxidation, (4) maximum reduction in the toxicity of a slug of toxic substance because it is quickly mixed throughout the reactor volume (consequently, it will be of lower concentration than at the upstream end of a conventional or tapered aeration activated sludge reactor), (5) relatively uniform environmental conditions for the active biological mass, and (6) greater flexibility than the other activated sludge processes. The principal disadvantage of the completely mixed process is that the reactor volume for a given organic removal for a soluble organic wastewater must be larger than the volume for a conventional process or most of the other process modifications. It is particularly applicable for treating industrial wastewaters having a high organic content. If a 20- to 24-hour reaction time is used, the rate of oxygen demand in mg/ℓ-hour will be relatively low and the large endogenous decay of the sludge minimizes the amount of waste activated sludge to be disposed.

Numerous researchers have developed design equations for the completely mixed activated sludge process (Stack, V. T., & Conway, R. A., 1959; Busch, A. W., 1961; McKinney, R. E., 1962b; Rich, L. G., 1963; Pipes, W. O., Grieves, R. B., & Milbury, W. F., 1964; Reynolds, T. D., & Yang, J. T., 1966; Smith, H. S., & Paulson, W. L., 1966; McCarty, P. L., & Lawrence, A. W., 1970; Eckenfelder, W. W., 1970; Eckenfelder, W. W., Adams, C. E., & Hovious, J. C., 1975) and they have usually been based on kinetic equations, growth equations, and material balances on the substrate or biological solids or a combination of these. In 1966 the author presented a model based on growth relationships and material balances on the substrate and the biological cell mass (Reynolds, T. D., & Yang, J. T., 1966). All these approaches have shown similar agreement.

For a kinetic derivation, consider the flowsheet of a completely mixed reactor shown in Figure 9.15. Although the subsequent kinetic derivation is based on the flowsheet shown in Figure 9.15, it is also applicable to the flowsheet in Figure 9.16. It is assumed that the reaction is pseudo–first order and, thus, the rate of substrate utilization is

$$-\frac{1}{X}\left(\frac{dS_t}{dt}\right) = KS_t \tag{9.44}$$

and the material balance on the substrate is given by

$$[\text{Accumulation}] = [\text{Input}] - \begin{bmatrix} \text{Decrease} \\ \text{due to} \\ \text{reaction} \end{bmatrix}$$
$$- [\text{Output}] \tag{9.45}$$

Thus,

$$dS_t \cdot V = (Q + R)S_0 dt - V[dS_t]_{\text{Growth}}$$
$$- (Q + R)S_t dt \tag{9.46}$$

From Eq. (9.44), $[dS_t]_{\text{Growth}} = K\overline{X}S_t dt$; thus, substituting this in Eq. (9.46) gives

$$dS_t \cdot V = (Q + R)S_0 dt - VK\overline{X}S_t dt - (Q + R)S_t dt \tag{9.47}$$

Dividing Eq. (9.47) by Vdt yields

$$\frac{dS_t}{dt} = \left(\frac{Q+R}{V}\right) S_0 - K\overline{X}S_t - \left(\frac{Q+R}{V}\right) S_t \tag{9.48}$$

Since the accumulation term $(dS_t/dt) = 0$ for steady state and since $V/(Q+R) = \theta$, Eq. (9.48) may be rearranged to produce

$$\theta = \frac{S_0 - S_t}{K\overline{X}S_t} \tag{9.49}$$

This may be rearranged to (Eckenfelder, W. W., 1970)

$$\frac{S_0 - S_t}{\overline{X}\theta} = KS_t \tag{9.50}$$

Thus, if $(S_0 - S_t)/\overline{X}\theta$ is plotted on the y-axis and S_t on the x-axis, the data will plot a straight line with a slope $= K$ if the reaction is pseudo–first order.

ACTIVATED SLUDGE   291

In applying the previous equations to the flowsheet shown in Figure 9.16, it would be imagined that the flows, Q and R, join prior to entering the reactor to produce a flow, $Q + R$, as shown in Figure 9.15. A material balance would be made to determine S_0 for the imaginary flow. Thus, the flowsheet shown in Figure 9.16 is solved in the same fashion as the flowsheet in Figure 9.15.

For the completely mixed activated sludge process, the reaction or aeration time (based on the flow, Q) is usually from 4 to 36 hours (4 to 8 hours for municipal wastes), the mean cell residence time is from 3 to 30 days, the recycle ratio is from 50 to 150 percent, and the MLSS concentration is from about 3000 to 6000 mg/ℓ (WPCF, 1977). Since the process is very flexible, it is possible to design a completely mixed system to operate as a high-rate aeration process at one extreme or, at the other extreme, it may be designed to operate as an extended aeration process. The performance in terms of BOD$_5$ removal and suspended solids removal will depend upon the particular design; however, BOD$_5$ and suspended solids removals are usually from 85 to 95 percent. For municipal wastewaters, nitrification will occur if the mean cell residence time is at least 10 days and the DO is adequate.

Example Problem 9.3
Completely Mixed
Biological Reactor

A completely mixed activated sludge reactor is to be designed to treat a petrochemical industrial wastewater having a negligible suspended solids content and a COD of 1460 mg/ℓ. The design MLSS is 3000 mg/ℓ and the design SDI is 6000 mg/ℓ. The effluent COD must be less than 180 mg/ℓ. Pilot-scale studies have been conducted and it was found that the biochemical reaction is pseudo–first order. The rate constant based on MLVSS is 0.532 ℓ/(gm)(hr) at 18°C. The MLVSS is 70 percent of the MLSS, and the nonbiodegradable COD is 155 mg/ℓ. The flowsheet is the same as shown in Figure 9.15. Determine the reaction time, θ.

Solution

A material balance on the biological solids at the junction of the influent flow, Q, and the recycle flow, R, is

$$(Q)(0) + (R)(6000) = (Q + R)(3000)$$

The recycle ratio, R/Q, from the above equation is 100.0 percent. A material balance on the COD at the junction is

$$(Q)(1460) + (R)(180) = (Q + R)(S_0)$$

or

$$(Q)(1460) + (1.0\,Q)(180) = (Q + 1.0\,Q)(S_0)$$

The COD in the mixed liquor, S_0, from the above equation is 820 mg/ℓ. The reaction time is given by the equation

$$\theta = \frac{S_0 - S_t}{K\overline{X}S_t}$$

where S_0 and S_t are expressed as biodegradable COD. Thus $S_0 = 820 - 155$ or 665 mg/ℓ and $S_t = 180 - 155$ or 25 mg/ℓ. The value $\overline{X}$ is (3.00) 0.70 or 2.10 gm volatile suspended solids per liter. The reaction time, θ, is

$$\theta = \frac{665 - 25}{(0.532)(2.10)(25)}$$

from which

$$\underline{\theta = 22.9 \text{ hours}}$$

Note: If the flowsheet were as shown in Figure 9.16, it would be imagined that the flows, Q and R, join prior to entering the reactor to produce a flow, $Q + R$, as shown in Figure 9.15. A material balance would be made to determine S_0 and the problem would be solved in the same manner as above, giving the same answer.

Other Activated Sludge Process Modifications

Several other modifications of the activated sludge process have been made and each was developed to attain a particular operational or design objective. With the exception of the modified aeration process, all subsequent activated sludge processes will achieve nitrification if the mean cell residence time is sufficient and the environmental conditions are favorable, particularly the dissolved oxygen level.

Step Aeration
The step aeration process, shown in Figure 9.17, has a food to microorganism ratio, F/M, of 0.2 to 0.5 lb BOD$_5$/lb MLSS-day. The space loading is from 25 to 35 lb BOD$_5$/day-1000 ft^3. This particular activated sludge process modification was developed to even out the oxygen demand of the mixed

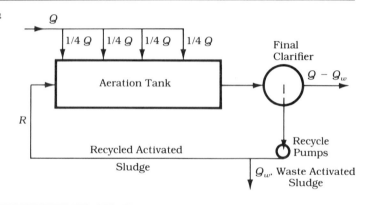

Figure 9.17. Step Aeration Process

liquor throughout the length of the reactor. This is accomplished by making each of the step loadings to the aeration tank equal to one-fourth the total wastewater inflow. Although the oxygen demand at the upstream end of the reactor is greater than at the downstream end, it has been evened out considerably compared to the conventional activated sludge process. The aeration tank is a dispersed plug-flow reactor with step inputs of the feed flow.

In the step aeration process, the reaction or aeration time is normally 4 to 8 hours (based on the flow, Q), the mean cell residence time is 4 to 12 days, the recycle ratio is from 20 to 80 percent, and the MLSS concentration is from about 2000 to 3500 mg/ℓ (WPCF, 1977). The performance is usually from 85 to 95 percent BOD$_5$ and suspended solids removal.

Modified Aeration

The modified aeration process operates with a food to microbe ratio, F/M, of 1.5 to 2 lb BOD$_5$/lb MLSS-day and was designed to provide a lower degree of treatment than the other activated sludge processes. The space loading is usually less than 100 lb BOD$_5$/day-1000 ft^3. Usually, the aeration time is from 1 to 3 hours (based on the flow, Q) and the BOD$_5$ removal is from 60 to 75 percent. The mean cell residence time is less than 1.0 day, the recycle ratio is from 10 to 30 percent, and the MLSS concentration is from 500 to 1500 mg/ℓ (WPCF, 1977). Since the mean cell residence time is so small, the suspended solids concentration in the effluent is high in relation to the other activated sludge processes. With municipal wastewaters virtually no nitrification occurs.

Contact Stabilization or Biosorption

The sorption (that is, both adsorption and absorption)

characteristics of a wastewater with soluble organic materials and a wastewater with mainly suspended and colloidal organic materials are illustrated in Figure 9.18. Particulate organic solids exhibit rapid sorption by activated sludge flocs, whereas soluble organic materials are sorbed more slowly. Municipal wastewaters usually have about 80 to 85 percent of the organic content in the form of particulate matter, and an aeration time less than one hour usually removes from 85 to 95 percent of the BOD_5. The contact stabilization or biosorption process, shown in Figure 9.19, was designed to provide two reactors, one for the sorption of organic materials and one for bio-oxidation of the sorbed

Figure 9.18. Sorption Characteristics of Soluble and Particulate Organic Materials

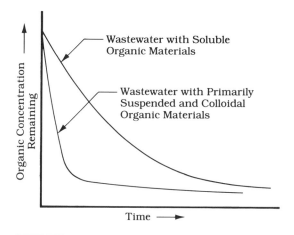

Figure 9.19. The Contact Stabilization or Biosorption Process

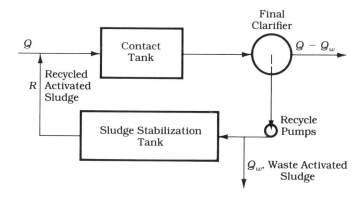

materials. Usually, this type of plant does not have a primary clarifier system. The food to microbe ratio, F/M, is from 0.2 to 0.6 lb BOD_5/lb MLSS-day. The space loading is usually from 25 to 50 lb/day-1000 ft^3.

In the contact tank, which has a contact time of 20 to 60 minutes (based on Q), the active biological solids sorb the suspended organic matter and much of the dissolved organic substances, and the active biological solids are then separated from the treated wastewater in the final clarifier. The separated solids flow or recycled sludge flow is from about 25 to 75 percent of the incoming wastewater flow rate to the plant. The biological solids are aerated in a sludge bio-oxidation or stabilization tank for a 3- to 6-hour reaction time (based on R) and, in the reactor, the sorbed organic materials are bio-oxidized to yield end products and new microbial cells. Separating the sorption process from the bio-oxidation process by providing two reactors requires a total aeration volume of only 50 to 60 percent of that for a conventional plant. The contact tank is usually 30 to 35 percent of total tank volume at a plant.

The contact stabilization process was developed for municipal wastewaters that have an appreciable amount of the organic matter in the form of particulate solids that are readily sorbed. For other wastewaters, bench-scale or pilot-scale studies should be made to determine the feasibility of the process for that particular wastewater because, in many cases, sufficient sorption of the organic substances in the wastewater does not occur in the short time provided in the contact tank. This may preclude the use of this type of activated sludge process. The flow regime in the contact tank is usually completely mixed, and in the bio-oxidation tank it is dispersed plug flow.

In the contact stabilization process, the mean cell residence time is from 4 to 18 days, the MLSS concentration in the contact tank is from 2000 to 4000 mg/ℓ, and the MLSS concentration in the bio-oxidation tank is about 6000 to 10,000 mg/ℓ (WPCF, 1977). The performance is usually from 85 to 95 percent BOD_5 and suspended solids removal.

Extended Aeration
The extended aeration process, shown in Figures 9.20 and 9.21, has a food to microorganism ratio, F/M, from 0.05 to 0.2 lb BOD_5/lb MLSS-day and was developed to minimize waste activated sludge production by providing a large endogenous decay of the sludge mass. The space loading is usually less than 15 lb BOD_5/day-1000 ft^3. The process is designed so that the mass of cells synthesized per day equals the mass of cells endogenously degraded per day.

Figure 9.20. Extended Aeration Process (Rectangular Tank Layout)

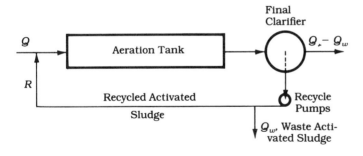

Figure 9.21. Extended Aeration Process (Racetrack Layout, Commonly Called an Oxidation Ditch)

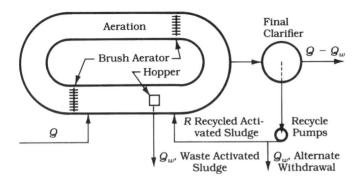

Thus, there is theoretically no net production of cell mass. The wastewater usually receives only preliminary treatment prior to entering the reactor because most of these plants do not have a primary clarifier system. The reaction time is from about 16 to 36 hours (based on Q) and the reactors are either square, round, or rectangular in plan view or a racetrack design. The racetrack design, commonly called an *oxidation ditch*, uses brush-type aerators that have a rotating axle with radiating steel bristles or some other axle type aeration device.

To make the net sludge production equal zero, it is necessary that the synthesized degradable cell mass equal the endogenously degraded cell mass. Some of the cell mass synthesized, such as peptidoglycans or mucopeptides used in the microbial cell walls, is nonbiodegradable in the reaction times employed. Zero net cell production is given by

$$X_w = Y_b S_r - k_e f \overline{X} = 0 \qquad (9.51)$$

where Y_b is the biodegradable yield coefficient, S_r is the substrate removed per day, and f is the fraction of degradable solids in the system, which is usually 0.7 to 0.8. The total mass of cells in the process, $\overline{X}$, is much larger than that in a conventional plant having the same capacity. Rearranging Eq. (9.51) gives (Eckenfelder, W. W., 1970)

$$\overline{X} = \frac{Y_b S_r}{k_e f} \qquad (9.52)$$

The above equation gives the total cell mass that is required in the system and, knowing the design MLSS value, it is possible to determine the reactor volume required. Another design procedure is to assume a reaction time and a recycle ratio and, after computing X and the reactor volume, the minimum required MLSS concentration may be determined. Although there is no net cell production, there is an accumulation of lysed cell fragments that are very slowly degraded. This necessitates the periodic wasting of some of the sludge. The fraction of viable cells in the mixed liquor is so low that usually the waste sludge may be dewatered without requiring digestion.

The circular or square reactors have a completely mixed flow regime, whereas the rectangular reactors are dispersed plug flow and the racetrack type (oxidation ditch) approaches plug flow. In the oxidation ditch, the velocity must be from 0.8 to 1.2 fps to maintain the solids in suspension. Two sets of rotors should be provided and each should have the capability of furnishing the required oxygen demand

A Large Oxidation Ditch Facility
Courtesy Lakeside Equipment Corporation

ENVIRONMENTAL ENGINEERING

with one rotor out of service. The ditch should be lined with reinforced concrete. Provisions to vary the rotor depth of submergence should be provided. In the extended aeration process, the mean cell residence time is from 12 to 30 days, the recycle ratio is from 50 to 300 percent, and the MLSS concentration is from about 3000 to 6000 mg/ℓ (WPCF, 1977). Performance in terms of BOD$_5$ removal is from 75 to 95 percent. Due to the long aeration time and mean cell residence time, considerable amounts of cell fragments usually will be present in the final effluent. Consequently, the effluent suspended solids are relatively high compared to the other modifications of the activated sludge process. The extended aeration process is very applicable for small installations, such as small communities and isolated institutions.

Oxidation Ditch with Two Rotating Brush Aerators. The ditch is concrete lined to prevent erosion.

Rotating Brush Aerators for an Oxidation Ditch. The aerators furnish the required oxygen and impart sufficient velocity to the mixed liquor to maintain the solids in suspension.

An oxidation ditch is to be designed for a community of 6000 persons. The flow is 100 gal/cap-day and the influent BOD$_5$ is 225 mg/ℓ. The plant has 90 percent BOD$_5$ removal, the yield coefficient Y_b is 0.65 lb MLVSS per lb BOD$_5$ oxidized, the endogenous coefficient is 0.06 day^{-1}, the biodegradable fraction of the total solids is 0.8, and the MLVSS is 50 percent of the MLSS. Determine:

1. The reactor volume if the reaction time is one day and the expected recycle ratio is 1.00.
2. The operating MLSS concentration.
3. The layout dimensions if the ditch has a trapezoidal cross section and the distance across the racetrack layout is 150 ft from ditch centerline to centerline. The design cross section is 15 ft wide at the base, has 2:1 side slopes, and has a 6-ft water depth.

Solution

The flow is (6000 persons)(100 gal/cap-day) or 600,000 gal per day. Since $V = (Q + R)\theta$, the volume is

$$V = \frac{600,000 \text{ gal}}{\text{day}} \left| \frac{1 + 1}{} \right| \frac{1 \text{ day}}{} \left| \frac{\text{ft}^3}{7.48 \text{ gal}} \right.$$

$$= \underline{160,427 \text{ ft}^3}$$

The BOD$_5$ removed per day is

$$S_r = \frac{600,000 \text{ gal}}{\text{day}} \left| \frac{8.34 \text{ lb}}{\text{gal}} \right| \frac{225}{10^6} \left| 0.90 \right.$$

$$= 1013 \text{ lb BOD}_5/\text{day}$$

The volatile solids are

$$\overline{X} = \frac{Y_b S_r}{k_e f} = \frac{0.65 \text{ lb MLVSS}}{\text{lb BOD}_5} \left| \frac{1013 \text{ lb BOD}_5}{\text{day}} \right| \frac{\text{day}}{0.06}$$

$$\times \frac{1}{0.8}$$

$$= 13,718 \text{ lb MLVSS}$$

Operating MLSS is

$$\overline{X} = \frac{13,718 \text{ lb MLVSS}}{160,427 \text{ ft}^3} \left| \frac{1.0 \text{ lb MLSS}}{0.50 \text{ lb MLVSS}} \right| \frac{1 \text{ ft}^3}{62.4 \text{ lb}} \left| 10^6 \right.$$

$$= 2740 \, \text{ppm} = \underline{2740 \, \text{mg}/\ell}$$

The cross sectional area of the ditch is (15 ft + 39 ft) × (½) (6 ft) or 162 ft². The ditch length, L, is L = (160,427 ft³) (1/162 ft²) = $\underline{990.4 \, \text{ft}}$. The length of the curved part of the racetrack is (π) 150 ft or 471.2 ft. The straight length of the ditch, L_s, is L_s = (990.3 ft − 471.2 ft) (½) = 259.6 ft or $\underline{260 \, \text{ft}}$.

An oxidation ditch is to be designed for a community of 6000 persons. The flow is 380ℓ/cap-day and the influent BOD$_5$ is 225 mg/ℓ. The plant has 90 percent BOD$_5$ removal, the yield coefficient Y_b is 0.65 g MLVSS per g BOD$_5$ oxidized, the endogenous coefficient is 0.06 day^{-1}, the biodegradable fraction of the total solids is 0.8, and the MLVSS is 50 percent of the MLSS. Determine:

1. The reactor volume if the reaction time is one day and the expected recycle ratio is 1.00.
2. The operating MLSS concentration.
3. The layout dimensions if the ditch has a trapezoidal cross section and the distance across the racetrack layout is 45 m from ditch centerline to centerline. The design cross section is 4.5 m wide at the base, has 2:1 side slopes, and has a 1.8-m water depth.

Solution

The flow is (6000 persons)(380 ℓ/cap-day)(m³/1000 ℓ) = 2280 m³/day. Since $V = (Q + R)\theta$, the volume is

$$V = \frac{2280 \, \text{m}^3}{\text{day}} \left| \frac{1+1}{} \right| \frac{1 \, \text{day}}{}$$

$$= \underline{4560 \, \text{m}^3}$$

The BOD$_5$ removed per day is

$$S_r = \frac{2280 \, \text{m}^3}{\text{day}} \left| \frac{1000 \, \ell}{\text{m}^3} \right| \frac{225 \, \text{mg}}{\ell} \left| \frac{\text{kg}}{10^6 \, \text{mg}} \right| 0.9$$

$$= 461.7 \, \text{kg}$$

The volatile solids are

$$\overline{X} = \frac{Y_b S_r}{k_e f} = \frac{0.65 \, \text{g MLVSS}}{\text{g BOD}_5} \left| \frac{461.7 \, \text{kg BOD}_5}{\text{day}} \right| \frac{\text{day}}{0.06}$$

$$\times \frac{1}{0.8}$$

$$= 6252 \, \text{kg MLVSS}$$

Operating MLSS is

$$\overline{X} = \frac{6252 \text{ kg MLVSS}}{4560 \text{ m}^3} \left| \frac{1.0 \text{ kg MLSS}}{0.5 \text{ kg MLVSS}} \right| \frac{10^6 \text{ mg}}{\text{kg}} \left| \frac{\text{m}^3}{1000 \, \ell} \right.$$

$$= \underline{2742 \text{ mg}/\ell}$$

The cross sectional area of the ditch is $(4.5 \text{ m} + 11.7 \text{ m})(\frac{1}{2})(1.8 \text{ m}) = 14.58 \text{ m}^2$. The ditch length, L, is $L = (4560 \text{ m}^3)(1/14.58 \text{ m}^2) = \underline{312.8 \text{ m}}$. The length of the curved part of the racetrack is (π) 45 m or 141.4 m. The straight length of the ditch, L_s, is $L_s = (312.8 \text{ m} - 141.4 \text{ m})(\frac{1}{2}) = \underline{85.7 \text{ m}}$.

Pure Oxygen Activated Sludge

The pure oxygen activated sludge process uses covered aeration tanks under a slight pressure and pure oxygen as the oxygen supply instead of air. The tanks are compartmented to give a series of completely-mixed biological reactors and rotary devices, such as turbines, are employed for mixing. Usually, at least 3 stages are provided. The partial pressure of the oxygen in the atmosphere above the mixed liquor is about 0.8 atmosphere, thus, the oxygen saturation concentration in the mixed liquor is about 4 times the concentration normally encountered. In pure oxygen activated sludge systems, the food to microbe ratio, F/M, is from 0.4 to 1.0 lb BOD_5/day-lb MLSS, the aeration time (based on Q) is 1 to 3 hours, the mean cell residence times is 8 to 20 days, the recycle ratio is from 25 to 50 percent, the MLSS concentration is 6,000 to 10,000 mg/ℓ and the operating dissolved oxygen level is 4 to 10 mg/ℓ. The space loading is usually less than 120 lb BOD_5/day-1000 ft^3. The advantages reported for pure oxygen systems are reduced reaction times, decreased amounts of waste activated sludge, increased sludge settling characteristics and reduced land requirements.

Temperature Effects

The biochemical reactions employed in microbial metabolism for substrate utilization and endogenous degradation are enzyme-catalyzed reactions. Reactions catalyzed by enzymes are temperature dependent and, in general, the reaction rate or reaction velocity approximately doubles for each 10°C rise in temperature up to a temperature at which the enzymes are denatured. In the activated sludge process, both the reaction rate constant for substrate utilization, K, and the rate constant for endogenous degradation, k_e, are temperature dependent.

The relationship between the reaction rate constant for substrate utilization and the temperature is

$$K_2 = K_1 \cdot \theta^{(T_2 - T_1)} \tag{9.53}$$

where

K_1, K_2 = reaction rate constants at the respective temperatures, T_1 and T_2, °C;

θ = temperature correction coefficient;

T_1 = temperature of the mixed liquor, °C, for K_1;

T_2 = temperature of the mixed liquor, °C, for K_2.

Usually θ is from 1.03 to 1.09 (Eckenfelder, W. W., 1970). The variation of θ is due to the difference between various activated sludge systems, and this is dependent upon the specific wastewater of interest. Also, it is indicated that the θ value is dependent upon the food to microorganism ratio, F/M, with high F/M ratios resulting in high θ values and low F/M ratios resulting in low θ values.

The effect of temperature upon the endogenous degradation rate constant, k_e, may be estimated from a relationship similar to Eq. (9.33), which is

$$k_{e_2} = k_{e_1} \cdot \theta^{(T_2 - T_1)} \tag{9.54}$$

where

$k_{e_1} k_{e_2}$ = rate constants for endogenous degradation at the respective temperatures, T_1 and T_2, °C;

θ = temperature correction coefficient;

T_1 = temperature of the mixed liquor, °C, for k_{e_1};

T_2 = temperature of the mixed liquor, °C, for k_{e_2}.

The value of θ in this equation depends upon the specific activated sludge system and wastewater of interest and the θ value may range from 1.065 to 1.085 (Wuhrmann, K., 1964).

Temperature affects the rate of microbial growth because it is also dependent upon enzyme-catalyzed reactions. For example, one study of a particular microbial species showed that the generation time (cell division or fission time) at 4°C was 180 minutes, while at 42°C it decreased to about 20 minutes (Clifton, C. E., 1957).

The temperature of the mixed liquor in Example Problem 9.3 was 18°C (64.4°F). In this problem assume that the temperature correction coefficient θ is 1.05 for a mixed liquor temperature of 28°C (82.4°F). Determine:

1. The effluent COD.
2. The percent COD removal or conversion at 28°C.
3. The percent COD removal or conversion at 18°C.

Solution

From Example Problem 9.3, the reaction constant at 18°C is 0.532 ℓ/(gm)(hr), the recycle ratio is 100.0 percent, the total COD in the feed flow is 1460 mg/ℓ, and the nonbiodegradable COD is 155 mg/ℓ. The flowsheet for the process is shown in Figure 9.15. The reaction constant at 28°C is

$$K_2 = (0.532)\,1.05^{(28-18)} = 0.866\ \ell/(gm)(hr)$$

The biodegradable COD in the feed flow is $1460 - 155$ or 1305 mg/ℓ. A material balance on the substrate at the junction of the feed flow, Q, and the recycled flow, R, is

$$(Q)(1305) + (R)S_t = (Q + R)S_0$$

Since $R = 1.00\,Q$,

$$(Q)(1305) + (1.0\,Q)(S_t) = (2.0\,Q)(S_0)$$

From the previous equation,

$$S_0 = 652.5 + 0.500\,S_t \tag{9.55}$$

The detention time, θ, for the completely mixed biological reactor is given by

$$\theta = \frac{S_0 - S_t}{K\overline{X}S_t}$$

This may be rearranged to

$$S_0 = S_t(1 + K\overline{X}\theta) = S_t[1 + (0.866)(2.10)(22.91)]$$

or

$$S_0 = 42.644\,S_t \tag{9.56}$$

Combining Eq. (9.56) with Eq. (9.55) gives

$$42.664\,S_t = 652.5 + 0.500\,S_t$$

from which $S_t = 15.5$ mg/ℓ COD (biodegradable). This is the biodegradable COD; thus, the total COD in the effluent is $15.5 + 155$ or

$$\text{Effluent COD} = 170.5\,\text{mg}/\ell = \underline{171\,\text{mg}/\ell\ \text{total COD}}$$

The fraction removal at 28°C is $(1460 - 171) \div 1460$ or 0.883. Therefore, percent COD removal at 28°C = $\underline{88.3\,\text{percent}}$. The fraction removal at 18°C is $(1460 - 180) \div 1460 = 0.877$. Thus, percent COD removal at 18°C = $\underline{87.7\,\text{percent}}$.

Other Kinetic Relationships

Another kinetic approach to biochemical reactions is that based on the Michaelis-Menten relationship, Eq. (9.3), or the Monod relationship. The fundamental expressions for the Monod approach are

$$\frac{dX}{dt} = \mu X = Y\frac{dS}{dt} \tag{9.57}$$

where

dX/dt = rate of cell growth, mass/(volume)(time);
μ = growth coefficient, time^{-1}.

The growth coefficient is given by Monod's expression, which is

$$\mu = \mu_{max}\left(\frac{S}{K_s + S}\right) \tag{9.58}$$

where

μ_{max} = maximum value of the growth coefficient, time^{-1};
K_s = substrate concentration with $\mu = \frac{1}{2}\mu_{max}$.

The rate of endogenous decay is given by

$$\frac{dX}{dt} = k_e X \tag{9.59}$$

Consider the completely mixed reactors shown in Figures 9.22 and 9.23. For the *completely mixed system without recycle*, a material balance on the cells gives [Accumulation]

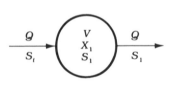

Figure 9.22. Completely Mixed Activated Sludge without Recycle

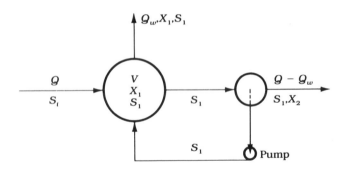

Figure 9.23. Completely Mixed Activated Sludge with Recycle

= [Increase due to growth] − [Decrease due to endogenous decay] − [Output] or

$$dX_1 \cdot V = V\mu X_1 dt - V k_e X_1 dt - Q X_1 dt \tag{9.60}$$

Dividing by $V dt$ gives

$$\frac{dX_1}{dt} = \mu X_1 - k_e X_1 - \frac{Q}{V} X_1 \tag{9.61}$$

Since $dX_1/dt = 0$ for steady state and the detention time, θ_t, based on the influent flow is $\theta_t = V/Q$, Eq. (9.61) can be rearranged to give

$$\mu = \frac{1}{\theta_t} + k_e \tag{9.62}$$

A material balance on the substrate gives [Accumulation] = [Input] − [Output] − [Decrease due to growth] or

$$dS_1 \cdot V = Q S_t dt - Q S_1 dt - V[dS_1]_{\text{Growth}} \tag{9.63}$$

The decrease due to growth is given by a rearrangement of Eq. (9.57), which is $[dS_1]_{\text{Growth}} = (\mu/Y)(X_1)\, dt$. Substituting this into Eq. (9.63) gives

$$dS_1 \cdot V = Q S_t dt - Q S_1 dt - V \frac{\mu}{Y} X_1 dt \tag{9.64}$$

Dividing by $V dt$ gives an expression for unsteady-state conditions that is as follows:

$$\frac{dS_1}{dt} = \frac{Q}{V} S_t - \frac{Q}{V} S_1 - \frac{\mu}{Y} X_1 \tag{9.65}$$

Since $dS_1/dt = 0$ for steady state and $\theta_t = V/Q$, Eq. (9.65) can be rearranged to give

$$\mu = \frac{Y}{X_1\theta_t}(S_t - S_1) \tag{9.66}$$

For the nonrecycle system, $\theta_t = \theta_c$; thus, equating Eqs. (9.66) and (9.62) and rearranging gives the design equation

$$X_1 = \frac{Y(S_t - S_1)}{1 + k_e\theta_c} \tag{9.67}$$

Using Eq. (9.67) and knowing the other parameters, it is possible to determine the MLSS required, X_1. Another useful equation is given by equating Eqs. (9.62) and (9.58) and substituting $\theta_c = \theta_t$ and $S_1 = S$ to give

$$\frac{1}{\theta_c} = \mu_{max}\left(\frac{S_1}{K_s + S_1}\right) - k_e \tag{9.68}$$

Equation (9.68) may be rearranged to give the design equation

$$S_1 = \frac{K_s(1 + k_e\theta_c)}{\theta_c(\mu_{max} - k_e) - 1} \tag{9.69}$$

To determine the values of Y and k_e, Eqs. (9.66) and (9.62) may be equated and rearranged to give

$$\frac{S_t - S_1}{X_1\theta_t} = \frac{k_e}{Y} + \frac{1}{Y}\frac{1}{\theta_t} \tag{9.70}$$

This is of the form $y = b + mx$. Using a continuous-flow reactor or several continuous-flow reactors operated at several flow rates will give the data to plot $(S_t - S_1)/X_1\theta_t$ on the y-axis and $1/\theta_t$ on the x-axis. The slope of the line equals $1/Y$ and the y-intercept is k_e/Y. To determine K_s and μ_{max}, Eqs. (9.62) and (9.58) can be equated and rearranged to give the following expression relating K_s and μ_{max} to the terms S_1, k_e, and θ_t:

$$\left(\frac{\theta_t}{1 + k_e\theta_t}\right)S_1 = \frac{K_s}{\mu_{max}} + \frac{1}{\mu_{max}}S_1 \tag{9.71}$$

This is of the form $y = b + mx$; thus, plotting $[\theta_t/(1 + k_e\theta_t)]S_1$ values on the y-axis and S_1 values on the x-axis will give a line with a slope equal to $1/\mu_{max}$ and a y-axis intercept of K_s/μ_{max}.

For the *completely mixed system with recycle*, shown in Figure 9.23, a material balance on the cells gives [Accumulation] = [Increase due to growth] − [Decrease due to decay] − [Output] or

$$dX_1 \cdot V = V\mu X_1 dt - Vk_e X_1 dt - Q_w X_1 dt$$
$$- (Q - Q_w)X_2 dt \tag{9.72}$$

Dividing by Vdt gives

$$\frac{dX_1}{dt} = \mu X_1 - k_e X_1 - \frac{Q_w X_1}{V} - \frac{(Q - Q_w)X_2}{V} \tag{9.73}$$

Since $dX_1/dt = 0$ for steady state, setting Eq. (9.73) equal to zero and dividing by X_1 and rearranging gives

$$\mu = \frac{Q_w X_1 + (Q - Q_w)X_2}{VX_1} + k_e \tag{9.74}$$

Since $\theta_c = (VX_1)/[Q_w X_1 + (Q - Q_w)X_2]$, Eq. (9.74) yields

$$\mu = \frac{1}{\theta_c} + k_e \tag{9.75}$$

A material balance on the substrate gives [Accumulation] = [Input] − [Output] − [Decrease due to growth] or

$$dS_1 \cdot V = QS_t dt - Q_w S_1 dt - (Q - Q_w)S_1 dt$$
$$- V[dS_1]_{\text{Growth}} \tag{9.76}$$

Inserting the value $[dS_1]_{\text{Growth}} = (\mu/Y)X_1 dt$ yields

$$dS_1 \cdot V = QS_t dt - Q_w S_1 dt - (Q - Q_w)S_1 dt$$
$$- V\frac{\mu}{Y}X_1 dt \tag{9.77}$$

Dividing by Vdt and simplifying gives

$$\frac{dS_1}{dt} = \frac{Q}{V}S_t - \frac{Q}{V}S_1 - \frac{\mu}{Y}X_1 \tag{9.78}$$

Since $dS_1/dt = 0$ for steady state and $\theta_t = V/Q$, Eq. (9.78) can be rearranged to give

$$\mu = \frac{Y}{X_1}\left(\frac{S_t - S_1}{\theta_t}\right) \tag{9.79}$$

Equating Eqs. (9.79) and (9.75) and simplifying gives the design equation

$$X_1 = \frac{\theta_c}{\theta_t} \cdot \frac{Y(S_t - S_1)}{1 + k_e \theta_c} \tag{9.80}$$

Another useful equation is given by equating Eqs. (9.75) and (9.58) and substituting $S_1 = S$ to give

$$\frac{1}{\theta_c} = \mu_{max} \left(\frac{S_1}{K_s + S_1} \right) - k_e \tag{9.81}$$

which is identical to Eq. (9.68) for the nonrecycle system. Equation (9.81) can be rearranged to give the design equation

$$S_1 = \frac{K_s(1 + k_e \theta_c)}{\theta_c(\mu_{max} - k_e) - 1} \tag{9.82}$$

which is identical to Eq. (9.69) for the nonrecycle system. To determine the values of Y and k_e, Eqs. (9.79) and (9.75) can be equated and simplified to give

$$\frac{S_t - S_1}{X_1 \theta_t} = \frac{k_e}{Y} + \frac{1}{Y} \frac{1}{\theta_c} \tag{9.83}$$

This is the form $y = b + mx$. Using one or more continuous-flow reactors operated at different θ_c values will give the data to plot $(S_t - S_1)/X_1 \theta_t$ on the y-axis and $1/\theta_c$ on the x-axis. The slope of the resulting straight line is $1/Y$ and its y-axis intercept is k_e/Y. To determine K_s and μ_{max}, Eqs. (9.75) and (9.58) can be equated, S_1 substituted for S, and the equation can be rearranged to give

$$\left(\frac{\theta_c}{1 + k_e \theta_c} \right) S_1 = \frac{K_s}{\mu_{max}} + \frac{1}{\mu_{max}} S_1 \tag{9.84}$$

This is of the form $y = b + mx$. Using one or more continuous-flow reactors operated at several θ_c values will give the data to plot $[\theta_c/(1 + k_e \theta_c)] S_1$ on the y-axis and S_1 on the x-axis. The slope of the resulting straight line is $1/\mu_{max}$ and the y-axis intercept is K_s/μ_{max}.

For the case of a *plug-flow reactor*, the performance equation that can be derived using the Monod relationship is

$$\frac{1}{\theta_c} = \frac{\mu_{max}(S_t - S_1)}{(S_t - S_1) + CK_s} - k_e \tag{9.85}$$

where $C = (1 + \alpha)\ln[(\alpha S_1 + S_t)/(1 + \alpha)S_1]$ and $\alpha = R/Q$. In these expressions, S_t is the substrate concentration in the influent flow prior to mixing with the recycle. The limit of C as $\alpha \to 0$ is equal to $\ln(S_t/S_1)$. If $\alpha < 1$, this approximation may be made and Eq. (9.85) reduces to

$$\frac{1}{\theta_c} = \frac{\mu_{max}(S_t - S_1)}{(S_t - S_1) + K_s \ln(S_t/S_1)} - k_e \tag{9.86}$$

The relationship for the cell concentration is

$$X_1 = \frac{\theta_c}{\theta_t} \cdot \frac{Y(S_t - S_1)}{1 + k_e\theta_c} \tag{9.87}$$

From the previous relationships the hydraulic detention time, θ_t, may be determined for a given set of conditions. The volume for a plug-flow reactor, V_p, may be determined from $V_p = Q\theta_t$.

For the case of the *dispersed plug-flow reactor*, the previous equations may be used to get the volume required, V_p, for a plug-flow reactor. Then this volume can be used to obtain the volume, V, for a dispersed plug-flow reactor using Figure 9.14. This method was illustrated in Example Problem 9.2.

Some investigators (McCarty, P. L., & Lawrence, A. W., 1970) in their research work have used a modification of the Monod and the Michaelis-Menten equations as a basis for their derivations. The modified equation in the form usually used is

$$\frac{dS}{dt} = kX\left(\frac{S}{K_s + S}\right) \tag{9.88}$$

where

k = maximum rate constant for substrate utilization;
K_s = substrate concentration when the rate of substrate utilization is half the maximum rate.

The design equations that have been derived are similar to Eqs. (9.67), (9.68), (9.69), (9.80), (9.81), (9.82), (9.85), (9.86), and (9.87) except that the value Yk is equal to μ_{max} in these equations.

Inspection of the previous derivations shows that Eqs. (9.68) and (9.69) for the nonrecycle reactor are identical to Eqs. (9.81) and (9.82) for the recycle reactor. For both systems there will be a minimum mean cell residence time, θ_w, at which washout of the cells from the reactors will occur. The washout mean cell residence time, θ_w, can be determined from Eqs. (9.68) and (9.81) by substituting S_t for S_1 because this will occur for a washout condition.

Several empirical relationships have been developed to give the performance of activated sludge plants when used for treating municipal wastewaters. One of the most widely used is the expression by Fair and Thomas (1950), commonly termed the *National Research Council* (NRC) *equation*, which is,

$$E_s = \frac{100}{1 + 0.03\left(\dfrac{y_0}{W\theta_t}\right)^{0.42}} \qquad (9.89)$$

where

E_S = efficiency of the secondary stage—that is, the aeration tank and the final clarifier, percent;

y_0 = pounds of BOD$_5$ applied to the aeration tank per day;

W = thousands of pounds of activated sludge in the aeration tank;

θ_t = aeration time based on the influent flow, hours.

In the derivation of this expression, the aeration time, θ_t, is assumed to be V/Q. Another convenient form of Eq. (9.89) is

$$W\theta_t = \frac{y_0}{4200}\left(\frac{E_S}{100 - E_S}\right)^{2.38} \qquad (9.90)$$

The use of these expressions is given in Example Problem 9.6.

Example Problem 9.6
Activated Sludge
Performance by the NRC
Equation

A tapered aeration activated sludge plant is to treat the wastewater from a population of 150,000 persons, and the effluent BOD$_5$ is to be 20 mg/ℓ. Other pertinent data are: flow = 100 gal/cap-day, BOD$_5$ of the primary clarifier effluent = 162.5 mg/ℓ, and MLSS = 2500 mg/ℓ. Determine the required aeration time in hours.

Solution

The efficiency of the secondary stage is [(162.5 − 20)/(162.5)] × 100 = 87.69 percent. The value of y_0 is (150,000 persons)(100 gal/cap-day)(8.34 lb/gal)(162.5/10^6) = 20,329 lb BOD$_5$ per day. The value of $W\theta_t$ is given by Eq. (9.90) or

$$W\theta_t = \left(\frac{20,329}{4200}\right)\left(\frac{87.69}{100 - 87.69}\right)^{2.38}$$

$$= 517.9 \tag{9.91}$$

Since $V = Q\theta_t$ and $Q = (150,000 \text{ persons})(100 \text{ gal/cap-day}) = 15$ MGD,

$$W = \left(\frac{15,000,000 \text{ gal}}{24 \text{ hr}}\right)\left(\frac{8.34 \text{ lb}}{\text{gal}}\right)\left(\frac{2500}{10^6}\right)\left(\frac{\theta_t}{1000}\right)$$

$$= 13.03\,\theta_t \tag{9.92}$$

Combining Eqs. (9.91) and (9.92) gives $\theta_t = \underline{6.30 \text{ hours}}$.

Reactor Basins

Usually, plug-flow or dispersed plug-flow reactor basins are rectangular tanks constructed of reinforced concrete and employ diffused compressed air. Rectangular basins can use common wall construction when multiple basins are provided. The walls are vertical and usually have 1 to 2 ft freeboard, as shown in Figure 9.24. The mixed liquor depth

Figure 9.24. Section through an Aeration Tank

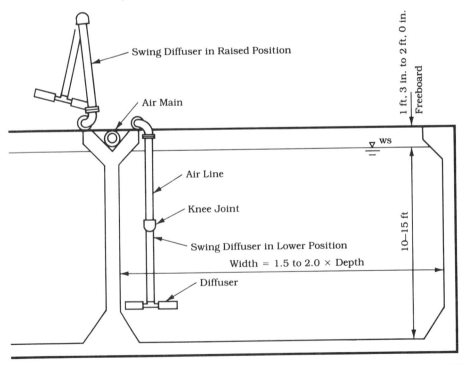

ENVIRONMENTAL ENGINEERING

is usually 10 to 15 ft and the tank width is from 1.5 to 2.0 times the depth. The length:width ratio is 5:1 or more and the tank lengths may be up to 500 ft or more. The air diffusers are usually mounted from 1.5 to 2.5 ft above the tank bottom. Fillets are provided at the bottom of the walls and the tops of the walls flare out at 45° to promote a spiral roll to the mixed liquor. Multiple basins should be provided when the flow exceeds 1.0 MGD. It is desirable to have a diffused air system that allows the diffusers to be raised above the mixed liquor surface for cleaning. If mechanical surface aerators are used, the tank width must be great enough to contain the spray and mist from the units; also, high water velocities must be avoided to minimize splashing against the basin walls. Three types of mechanical aerators have been successfully used in rectangular tanks; these are the slow-speed vertical turbine or radial-flow turbine, the submerged turbine with a sparger ring, and the horizontal rotating brush aerator. The slow-speed vertical turbine has a gear reduction box to obtain low rotational speeds (30 to 60 rpm), and the turbine is barely submerged below the still water surface. As the turbine rotates, the flow moves radially outward from the unit. In rectangular basins, these units usually have a draft tube to obtain satisfactory mixing. The submerged turbine has a slow-speed turbine mounted slightly above the tank bottom with a sparger ring beneath the turbine. The sparger ring allows compressed air to be released below the turbine and results in a relatively high oxygen transfer. When turbines are used in rectangular tanks, they usually are located halfway between the walls. The basin width is usually larger than the width when compressed air is used. The rotating brush aerator, which is also used in oxidation ditches, consists of a horizontal axle with radiating steel bristles. As the axle rotates, the bristles spray the water outward giving mixing and also entraining air. In rectangular tanks, the rotating axles are mounted along the same wall.

Usually, completely mixed reactor basins for small installations are square or circular concrete tanks with vertical walls and may use diffused compressed air or mechanical aeration. For large basins, square or rectangular tanks with a length:width ratio of 3:1 or less are used. The sides of square or rectangular basins may be vertical reinforced concrete walls or sloped embankments having a reinforced concrete liner on the slope to avoid erosion. The floor of a basin is usually a reinforced concrete slab. Sometimes, gunite or asphalt liners are used instead of reinforced concrete because they are less expensive. Due to their relatively large areas, widths, and lengths, completely mixed basins usually employ mechanical aeration. Widths and lengths of several hundred feet are common. In particular, the slow-speed vertical turbine type aerator mounted on piers has been very popular. Sometimes these units are float mounted for flexibility. For slow-speed vertical turbines, tank depths are from 6 to 30 ft, with 12 to 15 being most common. Tanks from 15 to 30 ft deep usually require draft tubes or a second turbine mounted at about mid-depth to obtain adequate mixing. Aeration basins are usually designed so that the area of influence around each aerator is approximately a square.

Biochemical Equations

Biochemical equations are useful in illustrating the principles of respiration and synthesis involved in bio-oxidation. A compound or substance is oxidized if (1) it is united with oxygen, (2) hydrogen atoms are removed, or (3) electrons are removed. Conversely, a compound or substance is reduced if (1) oxygen is removed from the substance, (2) hydrogen atoms are added, or (3) electrons are added. If the empirical equation for the substrate and the yield coefficient for the cells (that is, mass of cells produced per mass substrate used) are known, it is possible to develop respiration and synthesis equations, and also an overall biochemical equation representing the bio-oxidation (Hoover, S. R., & Porges, N., 1952). Since biological oxidation involves oxidation-reduction reactions, these are utilized in the development of equations for biochemical reactions. For chemoorganotrophs, the organic substrate is oxidized and molecular oxygen is reduced; thus, the steps necessary to develop the respiration equation require an oxidation reaction for the substrate and a reduction reaction for the molecular oxygen.

If the empirical equation for the organic portion of a wastewater is $C_6H_{14}O_2N$ and the nitrogen end product is the ammonium ion, the development of the respiration equation for a slightly basic condition is as follows.

Step 1. Balance all the key elements except oxygen and hydrogen to form the end products of CO_2 and NH_4^+.

$$C_6H_{14}O_2N \rightarrow 6\,CO_2 + NH_4^+$$
$$O_2 \rightarrow 2\,H_2O$$

Step 2. Balance the oxygen with the oxygen of water.

$$C_6H_{14}O_2N + 10\,H_2O \rightarrow 6\,CO_2 + NH_4^+$$
$$O_2 \rightarrow 2\,H_2O$$

Step 3. Balance the hydrogen with the hydrogen ion.

$$C_6H_{14}O_2N + 10\,H_2O \rightarrow 6\,CO_2 + NH_4^+ + 30\,H^+$$
$$4\,H^+ + O_2 \rightarrow 2\,H_2O$$

Step 4. Balance the charges with electrons.

$$C_6H_{14}O_2N + 10\,H_2O \rightarrow 6\,CO_2 + NH_4^+ + 30\,H^+ + 31\,e^-$$
$$4\,H^+ + 4\,e^- + O_2 \rightarrow 2\,H_2O$$

Step 5. Add the two reactions to eliminate the electrons.

$$C_6H_{14}O_2N + 10\,H_2O \rightarrow 6\,CO_2 + NH_4^+ + 30\,H^+$$
$$+ 31\,e^-$$
$$31\,H^+ + 31\,e^- + 7.75\,O_2 \rightarrow 15.5\,H_2O$$

$$C_6H_{14}O_2N + 7.75\,O_2 + H^+ \rightarrow 6\,CO_2 + NH_4^+$$
$$+ 5.50\,H_2O \qquad (9.93)$$

Equation (9.93) represents the respiration in a slightly acidic solution.

Step 6. For a slightly basic solution, add hydroxyl ions to unite with the hydrogens to form water. Adding one hydroxyl ion to both sides of Eq. (9.93) yields

$$C_6H_{14}O_2N + 7.75\,O_2 \rightarrow 6\,CO_2 + NH_4^+ + 4.50\,H_2O$$
$$+ OH^- \qquad (9.94)$$

Thus, Eq. (9.94) represents the respiration equation in a slightly basic solution.

If the empirical equation for the activated sludge produced is $C_5H_7O_2N$, the synthesis equation for a slightly basic solution is developed in a similar manner as the respiration equation and is as follows:

Step 1. Balance all the key elements except oxygen and hydrogen. Since the nitrogen in the respiration equation terminated as the ammonium ion, use the ammonium ion in the balance.

$$5\,C_6H_{14}O_2N + NH_4^+ \rightarrow 6\,C_5H_7O_2N$$
$$O_2 \rightarrow 2\,H_2O$$

Step 2. Balance the oxygen with the oxygen of water.

$$5\,C_6H_{14}O_2N + NH_4^+ + 2\,H_2O \rightarrow 6\,C_5H_7O_2N$$

$$O_2 \rightarrow 2\,H_2O$$

Step 3. Balance the hydrogen with the hydrogen ion.

$$5C_6H_{14}O_2N + NH_4^+ + 2\,H_2O \rightarrow 6\,C_5H_7O_2N + 36\,H^+$$

$$4\,H^+ + O_2 \rightarrow 2\,H_2O$$

Step 4. Balance the charges with electrons.

$$5C_6H_{14}O_2N + NH_4^+ + 2\,H_2O \rightarrow 6\,C_5H_7O_2N + 36\,H^+ + 35\,e^-$$

$$4\,H^+ + 4\,e^- + O_2 \rightarrow 2\,H_2O$$

Step 5. Add the two reactions to eliminate the electrons.

$$5C_6H_{14}O_2N + NH_4^+ + 2\,H_2O \rightarrow 6\,C_5H_7O_2N + 36\,H^+ + 35\,e^-$$

$$35\,H^+ + 35\,e^- + 8.75\,O_2 \rightarrow 17.5\,H_2O$$

$$\overline{\phantom{5C_6H_{14}O_2N + NH_4^+ + 8.75\,O_2}}$$

$$5\,C_6H_{14}O_2N + NH_4^+ + 8.75\,O_2 \rightarrow 6\,C_5H_7O_2N + H^+$$

$$+ 15.5\,H_2O \qquad (9.95)$$

Equation (9.95) represents the synthesis equation for a slightly acidic solution.

Step 6. For a slightly basic solution, add hydroxyl ions to each side to unite with the hydrogen ion to form water. Adding one hydroxyl to each side yields

$$5\,C_6H_{14}O_2N + NH_4^+ + 8.75\,O_2 + OH^- \rightarrow 6\,C_5H_7O_2N$$

$$+ 16.5\,H_2O \qquad (9.96)$$

Equation (9.96) represents the synthesis equation for a slightly basic solution. In some cases, no oxygen will be required in the synthesis equation.

To combine the respiration equation with the synthesis equation to give the overall biochemical reaction requires the yield coefficient, Y, which is defined as the mass of cells produced per unit mass of substrate utilized. The theoretical oxygen demand (TOD) may be determined from the respiration equation and, since the pseudomolecular weight of the substrate is 132, the TOD is (7.75)(32)/132 or

1.879 mg oxygen per mg substrate. If the yield coefficient, Y, is assumed to be 0.40 mg cells per mg TOD, the yield coefficient in terms of milligrams of substrate is

$$Y = (0.40 \text{ mg cells/mg TOD})(1.879 \text{ mg TOD/mg substrate})$$

$$= 0.752 \text{ mg cells/mg substrate}$$

Assuming 100 mg of cells are produced, the substrate used is

$$\text{Substrate} = (100 \text{ mg cells})(1 \text{ mg substrate/} 0.752 \text{ mg cells})$$

$$= 132.98 \text{ mg}$$

Letting x be the substrate required for respiration and y the substrate required for synthesis gives

$$x + y = 132.98 \text{ mg}$$

The pseudomolecular weight for the cells is 113; thus, from Eq. (9.96),

$$\frac{y}{5(132)} = \frac{100}{6(113)} \text{ or } y = 97.35 \text{ mg for synthesis}$$

Thus, $x = 132.98 - 97.35$ or $x = 35.63$ mg of substrate used for respiration. The fraction of substrate used for respiration is $35.63/132.98$ or 0.268 and the fraction of substrate used for synthesis is $97.35/132.98$ or 0.732. To combine the respiration equation, Eq. (9.94), with the synthesis equation, Eq. (9.96), requires multiplying Eq. (9.94) by 0.268 and Eq. (9.96) by $0.732(\frac{1}{5})$. This yields the following equations, which may be combined:

$$0.268 \, C_6H_{14}O_2N + 2.077 \, O_2 \rightarrow 1.608 \, CO_2 + 0.268 \, NH_4^+$$
$$+ 1.206 \, H_2O + 0.268 \, OH^-$$

$$0.732 \, C_6H_{14}O_2N + 0.146 \, NH_4^+ + 1.281 \, O_2$$
$$+ 0.146 \, OH^- \rightarrow 0.878 \, C_5H_7O_2N + 2.416 \, H_2O$$

$$\overline{C_6H_{14}O_2N + 3.358 \, O_2 \rightarrow 0.878 \, C_5H_7O_2N + 1.608 \, CO_2}$$
$$+ 0.122 \, NH_4^+ + 3.622 \, H_2O$$
$$+ 0.122 \, OH^- \qquad\qquad (9.97)$$

Thus, Eq. (9.97) represents the overall biochemical equation representing the bio-oxidation of the organic wastewater having an empirical equation of $C_6H_{14}O_2N$ for the organic fraction.

The theoretical oxygen demand of the cells may be determined from the following equation for the oxidation:

$$C_5H_7O_2N + H^+ + 5\,O_2 \rightarrow 5\,CO_2 + NH_4^+ + 2\,H_2O \quad (9.98)$$

Thus, the TOD is $(5)(32)/113$ or 1.42 mg oxygen per mg cells. If the cell yield coefficient, Y, is expressed in terms of cell TOD produced per unit substrate TOD bio-oxidized, Y may be determined using values from Eq. (9.97) as

$$Y = \frac{Cell\,TOD}{Substrate\,TOD} = \frac{(0.878)(113)(1.42)}{(132)(1.879)}$$

$$= 0.567 \quad (9.99)$$

The oxygen coefficient, Y', is defined as the mass of oxygen used per unit mass of substrate utilized. If the substrate is expressed as TOD, the oxygen coefficient, Y', may be determined using values from Eq. (9.97) as

$$Y' = \frac{Oxygen\,used}{Substrate\,TOD} = \frac{(3.358)(32)}{(132)(1.879)} = 0.433 \quad (9.100)$$

Thus, the sum of the yield coefficient (Y) and the oxygen coefficient (Y') is

$$Y + Y' = 0.567 + 0.433 \quad (9.101)$$

or

$$Y + Y' = 1.00 \quad (9.102)$$

In summary, the previous biochemical equation for the aerobic bio-oxidation of the substrate, Eq. (9.97), shows that when the cells produced are expressed as TOD and the substrate utilized is expressed as TOD, the sum of the yield coefficient, Y, and the oxygen coefficient, Y', is unity (Stack, V. T., & Conway, R. A., 1959). Thus, if the yield coefficient, Y, is known, then Y' may be computed. If Y is in terms of cell mass produced per unit TOD utilized, then the yield coefficient in terms of cell TOD produced is $1.42\,Y$. Therefore,

$$1.42\,Y + Y' = 1.00 \quad (9.103)$$

where Y is the mass of cells produced per unit substrate TOD oxidized.

Frequently BOD_5 is used instead of the TOD; for this case, the units may be converted from BOD_5 knowing the BOD_5:TOD ratio. For example, if the BOD_5 is 0.40 TOD, then

$$1.42\,Y\left(\frac{cells}{BOD_5}\right)\left(\frac{0.40\,BOD_5}{1.00\,TOD}\right)$$

$$+ \; Y'\left(\frac{oxygen}{BOD_5}\right)\left(\frac{0.40\,BOD_5}{1.00\,TOD}\right) = 1.00 \qquad (9.104)$$

where the units in italics represent the units to be utilized in the calculations. From Eq. (9.104), it can be seen that the BOD_5:TOD ratio merely converts the units to TOD. From Eq. (9.104) it follows that

$$0.568\,Y + 0.40\,Y' = 1.00 \qquad (9.105)$$

where

Y = mass of cells produced per unit BOD_5 oxidized— for example, mg cells/mg BOD_5;

Y' = mass of oxygen required per unit BOD_5 oxidized— for example, mg oxygen/mg BOD_5.

Equations (9.102) and (9.103) will not hold true if: (1) the wastewater has an appreciable volatile fraction that will be stripped from the water, (2) a significant amount of organic materials exert an immediate chemical oxygen demand, and (3) a sizeable amount of substrate is stored as food material. In general, though, the previous expressions are useful in determining the oxygen requirements for substrate bio-oxidation.

An overall biochemical equation developed for the conversion of the ammonium ion to nitrate is (McCarty, P. L., 1970)

$$22\,NH_4^+ + 37\,O_2 + 4\,CO_2 + HCO_3^- \rightarrow C_5H_7O_2N$$
$$+ \; 21\,NO_3^- + 20\,H_2O + 42\,H^+ \qquad (9.106)$$

From this equation, the oxygen requirements for nitrification are (37)(32)/(22)(18) or 2.99 mg oxygen per mg ammonium ion. In terms of ammonia nitrogen the oxygen required is (37)(32)/(22)(14) or 3.84 mg oxygen per mg ammonia nitrogen.

Sludge Production

As shown in the previous section on biochemical reactions, the bio-oxidation of an organic substrate produces a certain number of cells due to synthesis. At the same time, sludge is being endogenously decayed and, because endogenous degradation occurs during all growth phases, the entire mass of cells in the biological reactor is being endogenously degraded. The bio-oxidation of the ammonium ion to the nitrate ion synthesizes a certain mass of cells; however, this cell mass is usually very small in relation to the cells synthesized from the organic substrate, and it is usually neglected in determining sludge production. A material balance on the cells in the reactor-clarifier system is

$$[\text{Accumulation}] = [\text{Input}] - \begin{bmatrix} \text{Output due} \\ \text{to effluent} \end{bmatrix}$$

$$- \begin{bmatrix} \text{Output due} \\ \text{to wastage} \end{bmatrix} + \begin{bmatrix} \text{Increase due} \\ \text{to synthesis} \end{bmatrix}$$

$$- \begin{bmatrix} \text{Decrease due} \\ \text{to decay} \end{bmatrix} \qquad (9.107)$$

Because the accumulation term is zero for steady state, the input of active biological solids in the feed wastewater is negligible, and the output of active biological solids in the effluent is negligible, the material balance becomes

$$\begin{bmatrix} \text{Output due} \\ \text{to wastage} \end{bmatrix} = \begin{bmatrix} \text{Increase due} \\ \text{to synthesis} \end{bmatrix}$$

$$- \begin{bmatrix} \text{Decrease due} \\ \text{to decay} \end{bmatrix} \qquad (9.108)$$

The mathematical representation for the net cell production rate in Eq. (9.108) is

$$\frac{dX}{dt} = Y \frac{dS}{dt} - k_e X \qquad (9.109)$$

where

dX/dt = rate of net cell production, mass/time;
Y = yield coefficient, mass microbes produced/ mass substrate utilized;
dS/dt = rate of substrate removal, mass/time;
k_e = endogenous degradation coefficient, mass cells/(total mass cells)(time);
X = total cell mass in the biological reactor, mass.

Equation (9.109) may be rearranged to give

$$dX = YdS - k_e Xdt \qquad (9.110)$$

Preparing for integration produces

$$\int_0^{X_w} dX = Y \int_0^{S_r} dS - k_e \overline{X} \int_0^1 dt \qquad (9.111)$$

Integration yields the equation by Eckenfelder and Weston (1956),

$$X_w = YS_r - k_e \overline{X} \qquad (9.112)$$

where

X_w = waste cells produced, mass/day;
S_r = substrate removed, mass/day;
k_e = endogenous coefficient, mass cells/(total mass cells) (day);
$\overline{X}$ = average cell concentration in the biological reactor, mass.

The nature of the wastewater determines the values of Y and k_e. They may be determined from three or more batch reactors, as shown in Example Problem 9.8, or from three or more completely mixed, continuous-flow reactors, as shown in Example Problem 9.10.

Oxygen Requirements

As depicted in the previous section on biochemical reactions, the bio-oxidation of a substrate requires a certain amount of oxygen for respiration and, in most cases, for synthesis. Also, oxygen is required for the endogenous degradation of the cell mass and for nitrification. Thus, the total oxygen required is given by (AWARE, 1974)

$$O_r = Y'S_r + k'_e \overline{X} + O_n \qquad (9.113)$$

where

O_r = total oxygen required, mass/day;
Y' = oxygen coefficient, mass oxygen/mass substrate utilized;
k'_e = endogenous respiration coefficient, mass oxygen/(mass cells) (day);
O_n = oxygen required for nitrification, mass/day.

It was shown from Eq. (9.106) that 3.84 parts of oxygen are required per part ammonia nitrogen bio-oxidized. Thus, the oxygen required for nitrification, O_n, may be determined once the amount of organic and ammonia nitrogen to be converted has been computed.

Frequently, the amount of oxygen required for nitrification is relatively small compared to the oxygen required for carbonaceous substrate removal and endogenous respiration. For this case, the term O_n in Eq. (9.113) is omitted.

Example Problem 9.7
Sludge Production and
Oxygen Requirements

An activated sludge plant treats 5.0 MGD of municipal wastewater. The recycle ratio, R/Q, is $\frac{1}{3}$, the aeration time is 8 hours, the MLSS concentration is 2500 mg/ℓ, and the MLVSS is 75 percent of the MLSS. The yield coefficient, Y, is 0.39 mg MLVSS/mg COD removed and the endogenous coefficient, k_e, is 0.08 day^{-1}. The oxygen coefficient, Y', is 0.45 mg oxygen/mg COD removed and the endogenous oxygen coefficient, k'_e, is 0.18 mg oxygen/mg MLVSS-day. The sum of the organic and ammonia nitrogen in the primary effluent is 20 mg/ℓ and, excluding the nitrogen synthesized, 90 percent is converted to nitrate. The COD may be assumed to be approximately equal to the TOD. Assume that the flow from the primary clarifier has a COD of 350 mg/ℓ and the effluent from the plant has a COD of 50 mg/ℓ. Determine:

1. The net sludge produced due to COD removal and endogenous decay.
2. The oxygen required.
3. The sludge produced due to nitrification.

Solution

The reactor volume is

$$V = \frac{5.0 \times 10^6 \, \text{gal}}{\text{day}} \left| \frac{\text{day}}{24 \, \text{hr}} \right| \frac{1 + 0.333}{} \left| 8 \, \text{hr} \right.$$

$$= 2.222 \times 10^6 \, \text{gal}$$

The total mass of MLVSS in the reactor is

$$\overline{X} = \frac{2.222 \times 10^6 \, \text{gal}}{} \left| \frac{8.34 \, \text{lb}}{\text{gal}} \right| \frac{2500}{10^6} \left| 0.75 \right.$$

$$= 34,747 \, \text{lb}$$

The substrate removed per day is

$$S_r = \frac{350\text{-}50}{10^6} \left| \frac{5.0 \times 10^6 \text{gal}}{\text{day}} \right| \frac{8.34 \, \text{lb}}{\text{gal}}$$

$$= 12,510 \, \text{lb}$$

322

ENVIRONMENTAL ENGINEERING

The volatile suspended solids produced is

$$X_w = YS_r - k_e \overline{X}$$

$$= \left(\frac{0.39 \text{ lb vss}}{\text{lb COD}} \right) \left(\frac{12,510 \text{ lb COD}}{\text{day}} \right)$$

$$- \left(\frac{0.08}{\text{day}} \right) (34,747 \text{ lb})$$

$$= \underline{2099 \text{ lb vss/day}}$$

A material balance for the nitrogen is

$$[\text{Input}] = [\text{Output}] + \left[\begin{array}{c} \text{Decrease due} \\ \text{to synthesis} \end{array} \right]$$

$$+ \left[\begin{array}{c} \text{Decrease due} \\ \text{to nitrification} \end{array} \right]$$

The input is $(5 \times 10^6 \text{ gal/day})(20/10^6)(8.34 \text{ lb/gal})$ or 834 lb N/day. Since the cells are represented by $C_5H_7O_2N$, the percent nitrogen is $(14/113)(100)$ or 12.39 percent. Thus, the decrease due to synthesis is (2099 lb vss/ day)(0.1239 lb N/lb vss) or 260.1 lb N/day. Therefore,

$$[\text{Output}] + \left[\begin{array}{c} \text{Decrease due} \\ \text{to nitrification} \end{array} \right] = 834 - 260.1$$

$$= 573.9 \text{ lb N/day}$$

Since 90 percent of the organic and ammonia nitrogen is nitrified, the mass N converted is $(0.90)(573.9) = 516.5$ lb N/day. The oxygen required for nitrification is 3.84 lb/lb N. Thus, the total oxygen required is

$$O_r = Y'S_r + k_e'\overline{X} + O_n$$

$$= \left(\frac{0.45 \text{ lb oxygen}}{\text{lb COD}} \right) \left(\frac{12,510 \text{ lb COD}}{\text{day}} \right)$$

$$+ \left(\frac{0.18 \text{ lb oxygen}}{\text{lb MLVSS-day}} \right) (34,747 \text{ lb MLVSS})$$

$$+ \left(\frac{516.5 \text{ lb N}}{\text{day}} \right) \left(\frac{3.84 \text{ lb oxygen}}{\text{lb N}} \right)$$

$$= 5630 + 6254 + 1983$$

$$= \underline{13,867 \text{ lb oxygen/day}}$$

From the overall biochemical equation due to nitrification, Eq. (9.106), the cells synthesized due to nitrification are $113/(22)(14) = 0.367$ lb cells per lb $\underline{N}$. Thus, the mass of nitrifying microorganisms produced is

$$\left(\frac{516.5\,\text{lb}\,\underline{N}}{\text{day}}\right)\left(\frac{0.367\,\text{lb cells}}{1\,\text{lb}\,\underline{N}}\right)$$

$$= \underline{189.6\,\text{lb cells/day}}$$

The cells produced due to COD removal equal $(0.39)(12,510)$ or 4878.9 lb/day. The nitrifying bacteria equal $(189.6)(100/4878.9)$ or 3.89 percent of the cell production due to synthesis of the organic substrate. *Note:* This example problem shows that the amount of cells synthesized in nitrification is negligible compared to the amount of cells synthesized in carbonaceous removal. Usually, sludge production computations only consider carbonaceous removal.

Activated Sludge Coefficients

The yield coefficient, Y, and the endogenous decay or degradation constant, k_e, may be determined using a series of bench-scale batch reactors inoculated with a parent activated sludge culture developed from a continuous-flow activated sludge reactor. The acclimated parent culture from a continuous-flow reactor will be biologically active and, as a result, the growths in the series of batch reactors will exhibit little, if any, lag phase. If the acclimated parent culture is from a fill and draw reactor, a lag phase usually will occur. In general, the test for activated sludge coefficients is for a 24-hour period because a duration of this length will give more accurate values of Y and k_e than a shorter duration. The proof for the test is based on cell-substrate relationships. The net rate of cell production is given by

$$\frac{dX}{dt} = Y\frac{dS}{dt} - k_e X \tag{9.114}$$

where

dX/dt = rate of cell mass production, mass/(volume)(time);

dS/dt = rate of substrate bio-oxidation, mass/(volume)(time);

X = cell mass concentration in the reactor, mass/volume.

Dividing by X, the mass concentration of cells in the reactor, gives

$$\frac{1}{X}\frac{dX}{dt} = \frac{Y}{X}\frac{dS}{dt} - k_e \qquad (9.115)$$

If measurable increments are used, Eq. (9.115) becomes

$$\frac{1}{\overline{X}}\frac{\Delta X}{\Delta t} = \frac{Y}{\overline{X}}\frac{\Delta S}{\Delta t} - k_e \qquad (9.116)$$

where

ΔX = change in cell mass concentration, mass/volume;

ΔS = change in substrate mass concentration, mass/volume;

Δt = time increment;

$\overline{X}$ = average cell mass concentration during the time increment, mass/volume—that is, $\overline{X} = (X_0 + X_t)(\frac{1}{2})$, where X_0 and X_t are the cell mass concentrations at times $t = 0$ and $t = t$.

If the test is for one day, Δt is unity and Eq. (9.116) reduces to

$$\frac{\Delta X}{\overline{X}} = Y\frac{\Delta S}{\overline{X}} - k_e \qquad (9.117)$$

Equation (9.117) is of the straight-line form $y = mx + b$; thus, if $\Delta X/\overline{X}$ for each reactor is plotted on the y-axis and $\Delta S/\overline{X}$ is plotted on the x-axis, the data result in a straight line with a slope equal to Y and a y-axis intercept equal to k_e. The oxygen coefficient, Y', may usually be computed from the yield coefficient, Y, and Eqs. (9.102), (9.103), or a modification of these using the appropriate unit conversions. The endogenous oxygen coefficient k'_e theoretically equals $1.42\ k_e$.

It is also possible to determine Y' and k'_e by measuring the oxygen uptake rates in each reactor throughout the test duration. The mass of oxygen used, the average cell mass, and mass of substrate removed for each reactor may be correlated to give Y' and k'_e. This is desirable because it gives a check on the Y' and k'_e values computed from the previous theoretical relationships. Usually, the measured value of k'_e is larger than that computed from $k'_e = 1.42\ k_e$. The equation for determining the oxygen coefficients from batch reactor data is

$$\frac{O_r}{X} = Y' \frac{\Delta S}{X} + k'_e \qquad (9.118)$$

where O_r is the oxygen used, mass/volume. The k'_e value at a temperature other than the test temperature may be obtained by multiplying the test k'_e value by the ratio of the k_e values at the two respective temperatures.

The nature of the wastewater determines the values of Y, Y', k_e, and k'_e. Y ranges from about 0.3 to 0.8 lb vss/lb BOD$_5$ removed and k_e from about 0.04 to 0.25 day^{-1}. For municipal wastewaters, Y is about 0.5 to 0.7 and k_e is about 0.04 to 0.10. Y' ranges from about 0.3 to 0.8 lb oxygen/lb BOD$_5$ removed, and k'_e ranges from about 0.05 to 0.35 lb oxygen/lb MLVSS-day. For municipal wastewaters, Y' is about 0.5 to 0.7 and k'_e is about 0.05 to 0.15.

The values of Y, k_e, Y', and k'_e may be determined from three or more batch reactors or from three or more completely mixed, continuous-flow reactors.

Example Problem 9.8
Activated Sludge
Coefficients

An acclimated parent culture of activated sludge was produced using a continuous-flow activated sludge reactor for a soluble organic wastewater. Four batch reactors were inoculated using the parent culture and aerated for 24 hours. The MLVSS was 70 percent of the MLSS and the COD was approximately equal to the TOD. A summary of the data is given in Table 9.5. Determine Y, k_e, and the theoretical values of Y' and k'_e.

Table 9.5. Data from Batch Reactor Tests

Reactor No.	X_0 MLSS (mg/ℓ)	X_0 MLVSS (mg/ℓ)	X_t MLSS (mg/ℓ)	X_t MLVSS (mg/ℓ)	S_0 COD (mg/ℓ)	S_t COD (mg/ℓ)
1	420	294	737	516	706	95
2	830	581	1115	781	706	89
3	1620	1134	1910	1337	706	77
4	3670	2569	3820	2674	706	70

Solution

A summary of the reduced data is given in Table 9.6, where $\overline{X}$ is MLVSS. A plot of $\Delta X/\overline{X}$ versus $\Delta S/\overline{X}$ is shown in Figure 9.25. From the slope and y-axis intercept, Y = 0.40 mg MLVSS/mg COD and k_e = 0.06 mg MLVSS/mg MLVSS-day. The theoretical value Y' = 1 − 1.42 Y = 1

− 1.42(0.40) or $Y' = 0.43$ mg oxygen/mg COD. The theoretical value $k_e' = 1.42 \, k_e = 1.42(0.06)$ or $\underline{k_e' = 0.09}$ mg oxygen/mg MLVSS-day.

Table 9.6. Reduced Data from Batch Reactor Tests

Reactor No.	$\overline{X}$	ΔX	ΔS	$\Delta X/\overline{X}$	$\Delta S/\overline{X}$
1	405	222	611	0.548	1.509
2	681	200	617	0.294	0.906
3	1236	203	629	0.164	0.509
4	2622	105	636	0.040	0.243

Figure 9.25. *Graph for Example Problem 9.8*

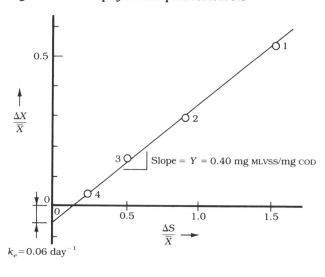

Example Problem 9.9
Activated Sludge
Coefficients

An acclimated parent culture of activated sludge was produced using a continuous-flow activated sludge reactor for a soluble organic wastewater. Four batch reactors were inoculated using the parent culture and aerated for 24 hours. The oxygen uptake rates in mg/ℓ-hr were measured for each reactor at periodic intervals during the 24 hours. The MLVSS was 72 percent of the MLSS. A summary of the data is given in Table 9.7. It is desired to determine the Y' and k_e' values.

Table 9.7. Data from Batch Reactor Tests

Reactor No.	X_0 MLSS (mg/ℓ)	X_0 MLVSS (mg/ℓ)	X_t MLSS (mg/ℓ)	X_t MLVSS (mg/ℓ)	S_0 COD (mg/ℓ)	S_t COD (mg/ℓ)
1	817	588	1020	734	712	101
2	1636	1178	1780	1282	712	87
3	3263	2349	3342	2406	712	80
4	5161	3716	5120	3686	712	73

Solution

Figure 9.26 shows a plot of the oxygen uptake rate versus time for reactor no. 1, and the area under the curve, 459 mg/ℓ, represents the total oxygen used. Similar plots for the other reactors showed the oxygen used to be 552, 780, and 1125 mg/ℓ for reactors no. 2, 3, and 4, respectively. Using X as MLVSS and correlating the oxygen and other data results in the reduced data in Table 9.8.

Figure 9.26. Graph for Example Problem 9.9

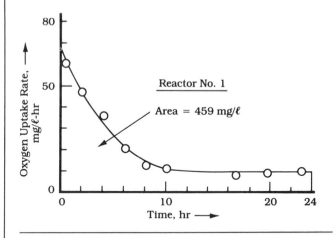

Table 9.8. Reduced Data from Batch Reactor Tests

Reactor No.	$\bar{X}$	ΔS	O_2	$\Delta S/\bar{X}$	$O_r/\bar{X}$
1	661	611	459	0.924	0.694
2	1230	625	552	0.508	0.449
3	2378	632	780	0.266	0.328
4	3701	639	1125	0.173	0.304

A plot of $\Delta S/\overline{X}$ versus $O_r/\overline{X}$ is shown in Figure 9.27. From the graph, $\underline{Y' = 0.53}$ mg oxygen/mg COD and $\underline{k'_e = 0.20}$ mg oxygen/mg MLVSS-day.

Figure 9.27. Graph for Example Problem 9.9

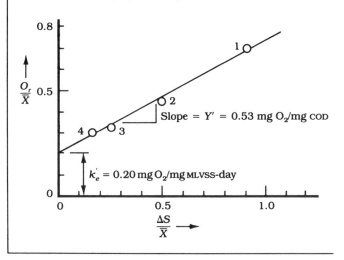

It is also possible to determine the reaction rate constant, K, and the coefficients Y, k_e, Y', and k'_e from three or more continuous-flow, completely mixed lab or pilot reactors operated at different F/M ratios or one reactor operated at three or more F/M ratios. If possible, it is desirable to use pilot-scale reactors. These simulate field conditions better than laboratory reactors because the quality of the wastewater stream may vary with time. It is important that the reactor or reactors have a completely mixed flow regime in order for this procedure to be valid. Usually three or more reactors are operated simultaneously to conserve time. The reactors are operated with increasing F/M values until acclimated sludges are developed and the reactors are at the desired F/M ratios. The end of the acclimation period can best be determined by measuring the organic substrate concentrations in the effluents from the reactors. Once the substrate concentration in the effluent from a reactor has reached the minimum value, the activated sludge in the reactor has become completely acclimated. After acclimation has been attained, the reactors are operated at steady-state for a week or more. The influent and effluent substrate concentrations are measured daily, along with the oxygen uptake rates and the amounts of sludge that must be re-

Rate Constants and Coefficients from Continuous-Flow Biological Reactors

moved from the reactors to keep the MLSS values at the desired concentrations. The rate constant, K, can be obtained by modification of Eq. (9.50), which is

$$\frac{S_t - S_t}{\overline{X}\theta_t} = KS_t \tag{9.119}$$

where S_t is the substrate concentration in the influent to the reactors. Since in most model reactors it is not possible to measure the recycle flow (R), $S_t = S_0$ and the reaction or detention time is given by $\theta_t = V/Q$. Plotting the $(S_t - S_t)/(\overline{X}\theta_t)$ values on the y-axis and the S_t values on the x-axis will give a straight line if the biochemical reaction is pseudo–first order. The slope of the line is equal to the rate constant, K.

The coefficients Y and k_e may be determined by a modification of Eq. (9.116), which is

$$\frac{\Delta X/\Delta t}{\overline{X}} = Y\left(\frac{S_t - S_t}{\overline{X}\theta_t}\right) - k_e \tag{9.120}$$

Plotting the $(\Delta X/\Delta t)/\overline{X}$ values on the y-axis versus the $(S_t - S_t)/(\overline{X}\theta_t)$ values on the x-axis will give a straight line with a slope equal to Y and a y-axis intercept equal to k_e. The coefficients Y' and k'_e may be determined by a modification of Eq. (9.118), which is

$$\frac{O_r}{\overline{X}} = Y'\left(\frac{S_t - S_t}{\overline{X}\theta_t}\right) + k'_e \tag{9.121}$$

where O_r is the oxygen used, mass/(volume)(day). Plotting the $O_r/\overline{X}$ values on the y-axis versus the $(S_t - S_t)/(\overline{X}\theta_t)$ values on the x-axis will give a straight line with a slope equal to Y' and a y-axis intercept equal to k'_e. In the previous equation, Eq. (9.121), the term for nitrification, O_n, has been omitted; thus the Y' value includes the oxygen required for carbonaceous removal and also nitrification. By determining the nitrogen values (all forms) in the influent and effluent flows and computing the nitrogen synthesized into cells, the nitrogen converted to NO_3^{-1} can be determined. Using the amount of nitrogen converted to NO_3^{-1}, the amount of oxygen required for nitrification can be computed. Subtracting the oxygen for nitrification from the O_r values gives the net oxygen required for carbonaceous removal. When this is done, the plot will give a Y' value only for carbonaceous removal. Frequently, the nitrification that occurs in a model reactor is not sufficient to require an oxygen adjustment to the O_r values and, when this occurs, the Y' value from the plot is essentially that required for carbonaceous removal alone.

The previous concepts are illustrated by the data in Table 9.9, which is from four completely mixed, continuous-flow activated sludge reactors. The acclimated cultures were developed for a soluble organic wastewater at F/M ratios of about 0.2, 0.4, 0.6, and 0.8 lb BOD_5/lb MLVSS-day. The MLVSS was 72 percent of the MLSS. After a week of steady-state operation, the data in Table 9.9 were obtained.

Table 9.9. Data from Continuous-Flow Reactor Tests

Reactor No.	$\overline{X}$ MLSS (mg/ℓ)	$\overline{X}$ MLVSS (mg/ℓ)	θ_t (hr)	S_i BOD$_5$ (mg/ℓ)	S_t BOD$_5$ (mg/ℓ)	O_2 Uptake (mg/ℓ-hr)	$\Delta X/\Delta t$ MLVSS/ day (mg/ℓ- day)
1	1276	919	25.3	885	61	29	308.8
2	1875	1350	23.7	885	47	35	340.2
3	2868	2065	24.4	885	31	39	312.2
4	5394	3884	24.9	885	12	58	135.9

Determine K, Y, k_e, Y', and k_e'.

Solution

A summary of the reduced data is given in Table 9.10, where X = MLVSS.

Table 9.10. Reduced Data from Continuous-Flow Reactor Tests

Reactor No.	$\dfrac{S_i - S_t}{\overline{X}\theta_t}$ (mg/mg-day)	$\dfrac{\Delta X/\Delta t}{\overline{X}}$ (mg/mg-day)	$\dfrac{O_r}{\overline{X}}$ (mg/mg-day)	S_t BOD$_5$ (mg/ℓ)
1	0.851	0.336	0.760	61
2	0.629	0.252	0.630	47
3	0.407	0.151	0.450	31
4	0.217	0.035	0.360	12

A plot of the $(S_i - S_t)/(\overline{X}\theta_t)$ values versus the S_t values is shown in Figure 9.28. The slope from the graph gives

$$K = 0.588 \, \ell/(\text{gm MLVSS})(\text{hr})$$

Since the MLVSS is equal to 72 percent of the MLSS, the K based on MLSS is (0.588)(0.72) or

$$K = 0.423 \, \ell/(\text{gm MLSS})(\text{hr})$$

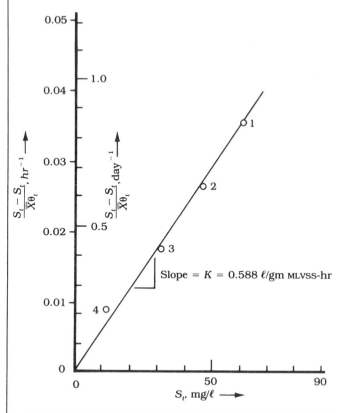

Figure 9.28. Graph for Example Problem 9.10

The line in Figure 9.28 passes through the origin because the organic concentration is the BOD_5, which represents the biodegradable substrate. A plot of the $(\Delta X/\Delta t)/\overline{X}$ versus the $(S_t - S_t)/(\overline{X}\theta_t)$ values is shown in Figure 9.29. From the graph, $Y = 0.47$ mg MLVSS/mg BOD_5 and $k_e = 0.05$ day^{-1}. A plot of the $O_r/\overline{X}$ values versus the $(S_t - S_t)/(\overline{X}\theta_t)$ values is shown in Figure 9.30. From the graph, $Y' = 0.66$ mg O_2/mg BOD_5 and $k_e' = 0.20$ mg O_2/MLVSS-day.

Note: In the previous problem, the biodegradable COD or the total COD may be used instead of BOD_5. All graphs will be similar to those shown except for Figure 9.28 for the case of total COD. For this case, the straight line will not pass through the origin, but will intersect the x-axis at a value equal to the nondegradable COD, and the slope of the line will be equal to the reaction rate constant, K. A plot of this type is shown in Figure 9.31.

It is also possible to use three or more continuous-flow activated sludge reactors to determine the rate con-

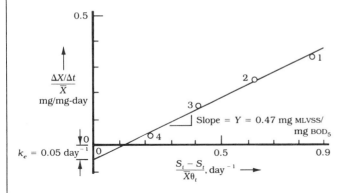

Figure 9.29. Graph for Example Problem 9.10

Slope = Y = 0.47 mg MLVSS/ mg BOD₅

$\frac{\Delta X/\Delta t}{\overline{X}}$ mg/mg-day

$k_e = 0.05$ day⁻¹

$\frac{S_i - S_t}{\overline{X}\theta_i}$, day⁻¹ ⟶

Figure 9.30. Graph for Example Problem 9.10

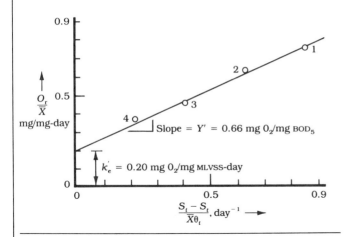

Slope = Y' = 0.66 mg O₂/mg BOD₅

$\frac{O_r}{\overline{X}}$ mg/mg-day

$k_e' = 0.20$ mg O₂/mg MLVSS-day

$\frac{S_i - S_t}{\overline{X}\theta_i}$, day⁻¹ ⟶

Figure 9.31. Plot for Using Total COD for Determining the Reaction Constant K

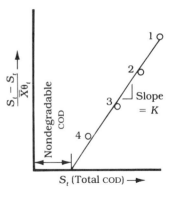

$\frac{S_i - S_t}{\overline{X}\theta_i}$

Nondegradable COD

Slope = K

S_t (Total COD) ⟶

stants, μ_{max} and K_m, and the coefficients Y and k_e for the relationships derived from the Monod equation. For reactors without recycle, Eqs. (9.70) and (9.71) are used and, for reactors with recycle, Eqs. (9.83) and (9.84) are used. The procedure for analyzing the data was previously discussed in the text where these equations were presented.

Operational Problems

The most common operating problem in the activated sludge process is inadequate solid-liquid separation in the final clarifier. This results in appreciable amounts of biological solids spilling over the effluent weirs and unacceptable amounts of solids leaving with the effluent. This condition, generally termed *sludge bulking*, may be caused by the occurrence of a biological sludge that has poor settling characteristics, hydraulic and solids overloading, or inadequate recycle flow rate. For diffused air activated sludge processes treating municipal wastewaters, the sludge density index is usually from about 6500 mg/ℓ (svi $\approx$ 150) to 20,000 mg/ℓ (svi $\approx$ 50). For mechanical aeration activated sludge processes treating municipal wastewaters, the sludge density index is usually about 3500 mg/ℓ (svi $\approx$ 300) to 5000 mg/ℓ (svi $\approx$ 200) (Culp, R. L., 1970; Steel, E. W., & McGhee, T. J., 1979). Thus, in mechanical aeration tanks, the added agitation is sufficient to cause some shearing of the biological floc. Slow-speed turbines impart less shear forces to the water than high-speed turbines; consequently, the sludge density index is greater for the slow-speed units. If the svi values are greater than given above, a poor effluent in terms of suspended solids and BOD_5 can be expected. For municipal wastewaters, a sludge with excellent settling characteristics will be produced when the environmental conditions are favorable, the mean cell residence time is from 5 to 15 days, and the food to microbe ratio (F/M) is from 0.2 to 0.7 lb BOD_5/lb mlss-day.

The flocculating capability of an activated sludge culture is lowest during the log growth phase (high F/M ratio), and it increases during the declining growth phase and is maximum during the beginning of the endogenous phase (low F/M ratio). The flocculating capacity is partly due to the presence of gummy slime layers or capsules surrounding the viable cells and the accumulation of this material from the attrition of the slime layers. In the log growth phase, the slime layers or capsules are the thinnest and have reached the maximum thickness at the beginning of the endogenous phase. The slime layers are usually polysaccharide gums sometimes having proteinaceous or lipoidal material associated with them (Frobisher, et al., 1974). From the attrition of the slime layers comes a gelatinous agglomerating substance that is relatively bioresistant and assists in floc aggregation. The flocculating capacity of an activated sludge is also partly due to the inactivity of the microbial cells and their behavior as colloids. In the log growth phase (high F/M ratio), the cells are very active and their motility prevents their flocculation (McKinney, R. E., 1962a). In the declining growth phase and, in particular, the endogenous phase (low F/M ratio), the motility of the cells is appreciably decreased and, if the cells come in close contact due to mixing, the forces of attraction become greater than the forces of repulsion and the cells flocculate.

If the aeration time is inadequate, relatively small amounts of the polysaccharide gums will be present, thus contributing to poor settling characteristics. Also, if an excessive slug of organic material enters the reactor, the amount of agglomerating gum may not be sufficient and, as a result, poor biological flocculation will occur. In addition, an excessive slug of organic matter will cause cell activity and motility to increase, thus significantly decreasing the flocculating capacity of the biological suspension.

When filamentous microbes (such as the *Sphaerotilus* group) outgrow the desirable microorganisms, they become predominant and, due to their wiry nature, they prevent proper sludge settling and compaction in the final clarifier. Also, in a similar manner, excessive growths of fungi cause a poor settling sludge. Conditions favoring filamentous growths are F/M ratios greater than about 0.8 lbs BOD_5/lb MLSS-day for several days, wastewaters containing a high carbohydrate content, lack of nitrogen and phosphorus nutrients, and low operating dissolved oxygen concentrations. Fungi growths are also promoted by wastewaters having a high carbohydrate content and inadequate amounts of nitrogen and phosphorus nutrients.

Overaeration creates shear forces that prevent proper agglomeration of the biological floc and, also, shear apart floc particles. Usually, if the power level is not greater than about 0.30 hp/1000 gal or 35 scfm/1000 ft^3, overaeration will not occur. Also, if the aeration time is too long, shearing will break apart the floc, large amounts of lysed cell fragments will be present, and poor settling will result.

Underaeration usually occurs when the oxygen level is much less than 1.0 mg/ℓ and, as a result, the inner portions of the biological particles may become anaerobic and produce gaseous end products. Also, anaerobic action causes the metabolism of carbohydrates and some other materials to be shunted, resulting in formation of intracellular products such as glycerol, which has specific gravity less than one. The overall effect of the entrained gaseous end products in the floc and the lower specific gravity substances in the cells is a reduction in the specific gravity of the floc particles which, as a result, causes poor settling characteristics.

When the dissolved oxygen level in the mixed liquor is low and the sludge is retained for numerous hours in the final clarifier, anaerobic action will occur and, consequently, floating sludge masses will result that may contribute to a poor effluent. If the dissolved oxygen level in the mixed liquor is at least 2.0 mg/ℓ and the sludge retention time in the final clarifier is less than about one hour, this problem is usually avoided.

If any of the chemical ions, compounds, or substances given in the previous section, titled "Environmental Factors Affecting Microbial Activity," are present in inhibitory or toxic concentrations for sufficient time, they will interfere with the microbial metabolic functions and result in a decreasing rate of substrate or BOD_5 utilization. As a result, the BOD_5 removal efficiency will decrease and, if the condition persists for enough time or if the chemical concentration is great enough, killing of the microbial population will occur. When the cells die they lyse or break apart into fragments, and this will result in a sludge having poor settling characteristics.

Excessive grease can coat the biological floc particles and cause them to become anaerobic. This coating will interfere with the sorption of organic substances, the attrition of slime layers, the agglutination of the biological floc particles, and the transfer of oxygen. If the grease content in the sludge is less than 5 to 7 percent, this usually will not be a problem (Steel, E. W., & McGhee, T. J., 1979).

In bulking, where the cause is excessive or shock hydraulic loads, the influent flow rate reaches a point where the overflow velocity in the final clarifier exceeds the settling velocity of the mixed liquor. As a result, the biological solids accumulate in the final clarifier at a concentration near that of the mixed liquor. The sludge blanket rate of rise will be approximately equal to the difference between the overflow velocity and the sludge settling velocity (Texas Water Utilities Association, 1971). For example, if the overflow rate is 1600 gal/sq ft-day, the overflow velocity is 8.9 ft/hr and, if the mixed liquor suspended solids settling velocity is 4.5 ft/hr, the sludge blanket will rise at a rate of about 8.9 − 4.5 or 4.4 ft/hr. If the high hydraulic loading persists, the tank will become filled with activated sludge solids and the biological solids will overflow with the effluent. When this occurs, the solids concentration in the effluent will approach that of the mixed liquor in the reactor, which is usually several thousand mg/ℓ. When the overflow velocity exceeds the sludge settling velocity, the bulking cannot be controlled by increasing the underflow or recycled sludge flow rate. The only remedy is the proper reduction in the mixed liquor flow rate to the clarifier or the addition of another clarifier.

If the solids loading to the final clarifier is excessive, the solids will not thicken and be removed as the underflow at the rate the solids enter the clarifier. As a result, the sludge blanket will rise and, if the time duration is sufficient, excessive solids will spill over the effluent weirs. When this occurs, the solids concentration in the effluent will approach that of the mixed liquor. Increasing the recycle flow rate will not eliminate the situation. The only remedy is a reduction in the mixed liquor flow rate to the final clarifier, a reduction in the mixed liquor suspended solids concentration, or the addition of another final clarifier. The design solids loading based on the average hourly flow is usually from about 0.60 to 1.20 lb/hr-ft^2, and the design solids loading based on the peak hourly flow is from about 1.25 to 2.0 lb/hr-ft^2. The larger solids loading values are for plants located in warm climates because the sludge settling rate increases with an increase in temperature.

In the operation of an activated sludge plant, it is essential to recycle the appropriate amount of activated sludge to maintain the desired MLSS concentration and, also, to determine daily the solids concentration in the reactor and the sludge blanket depth in the final clarifier. Normally, the top of the sludge blanket is less than half the final clarifier tank depth. If the recycle is inadequate, the top of the blanket will rise and eventually spill over the effluent weirs. This will result in an excessive suspended solids concentration in the effluent, and the solids concentration may approach or exceed that of the mixed liquor. If desired, a plant may be operated with a sludge blanket depth as small as about 2 feet. Final clarifiers with suction-type sludge removal usually have sludge blankets less than 2 feet deep.

References

Aiba, S., Humphery, A. E., and Millis, N. F. 1965. *Biochemical Engineering.* New York: Academic Press.

AWARE, Inc. 1974. *Process Design Techniques for Industrial Waste Treatment,* edited by C. E. Adams, Jr. and W. W. Eckenfelder, Jr. Nashville, Tenn.: Enviro Press.

Benefield, L. D., and Randall, C. W. 1980. *Biological Process Design for Wastewater Treatment.* Englewood Cliffs, N.J.: Prentice-Hall.

Blakebrough, N., ed. 1967. *Biochemical and Biological Engineering Science.* Vol. 1. London: Academic Press.

Boyle, W. C., Crabtree, K., Rohlich, G. A., Iaccarino, E. P., and Lightfoot, E. N. 1969. Flocculation Phenomena in Biological Systems. In *Advances in Water Quality Improvement,* edited by E. F. Gloyna and W. W. Eckenfelder, Jr. Austin, Tex.: University of Texas Press.

Brey, W. S., Jr. 1958. *Principles of Physical Chemistry.* New York: Appleton-Century-Crofts.

Busch, A. W. 1961. Treatability vs. Oxidizability of Industrial Wastes and the Formulation of Process Design Criteria. In

Proceedings of the 16th Annual Purdue Industrial Waste Conference.

Clifton, C. E. 1957. Introduction to Bacterial Physiology. New York: McGraw-Hill.

Conway, R. A., and Ross, R. D. 1980. Handbook of Industrial Waste Disposal. New York: Van Nostrand Reinhold.

Culp, R. L. 1970. The Operation of Wastewater Treatment Plants. Public Works Magazine (October, November, and December).

Downing, A. L., Painter, H. A., and Knowles, G. 1964. Nitrification in the Activated Sludge Process. Jour. and Proc. Inst. Sewage Purif., part 2:130.

Eckenfelder, W. W., Jr. 1966. Industrial Water Pollution Control. New York: McGraw-Hill.

Eckenfelder, W. W., Jr. 1970. Water Quality Engineering for Practicing Engineers. New York: Barnes and Noble.

Eckenfelder, W. W., Jr. 1980. Principles of Water Quality Management. Boston: CBI Publishing.

Eckenfelder, W. W., Jr., Adams, C. E., and Hovious, J. C. 1975. A Kinetic Model for Design of Completely Mixed Activated Sludge Treating Variable Strength Industrial Wastewaters. Water Research 9, no. 1:37.

Eckenfelder, W. W., Jr., and Ford, D. L. 1970. Water Pollution Control. Austin, Tex.: Pemberton Press.

Eckenfelder, W. W., Jr., Irvine, R. L., and Reynolds, T. D. 1972. Wastewater Treatment Cost Models and Required Parameters. Technical Report to Galveston Bay Study. Texas Water Quality Board.

Eckenfelder, W. W., Jr., and O'Connor, D. J. 1961. Biological Waste Treatment. London: Pergamon Press.

Eckenfelder, W. W., Jr., and Weston, R. F. 1956. Kinetics of Biological Oxidation. In Biological Treatment of Sewage and Industrial Wastes. Vol. 1, edited by J. McCabe and W. W. Eckenfelder. New York: Reinhold.

Fair, G. M., Geyer, J. C., and Okun, D. A. 1968. Water and Wastewater Engineering. Vol. 2. Water Purification and Wastewater Treatment and Disposal. New York: Wiley.

Fair, G. M., and Thomas, H. A., Jr. 1950. The Concept of Interface and Loading in Submerged, Aerobic, Biological Sewage-Treatment Systems. Jour. and Proc. Inst. Sewage Purif., part 3:235.

Farquhar, G. J., and Boyle, W. C. 1971. Identification and Occurrence of Filamentous Microorganisms in Activated Sludge. Jour. WPCF 43, no. 4:604; and 43, no. 5:779.

Ford, D. L., Gloyna, E. F., and Yang, Y. T. 1967. Development of Biological Treatment Data for Chemical Wastes. Proceedings of the 22nd Annual Purdue Industrial Waste Conference. Part 1.

Ford, D. L., and Reynolds, T. D. 1965. Aerobic Oxidation in the Thermophilic Range. Proceedings of the 5th Industrial Water and Waste Conference, Dallas, Tex.

Frobisher, M., Hinsdill, R. D., Crabtree, K. T., and Goodheart, C. R. 1974. Fundamentals of Microbiology. 9th ed. Philadelphia, Pa.: Saunders.

Great Lakes—Upper Mississippi Board of State Sanitary Engineers. 1978. Recommended Standards for Sewage Works. Ten state standards, Albany, N.Y.

Hoover, S. R., and Porges, N. 1952. Assimilation of Dairy Wastes by Activated Sludge. *Sew. and Ind. Wastes* 24, no. 3:306.

Imhoff, K., Muller, W. J., and Thistlethwayte, D. K. B. 1971. *Disposal of Sewage and Other Water-borne Wastes.* 2nd ed. Ann Arbor, Mich.: Ann Arbor Science Publishers.

Levenspiel, O., and Bischoff, K. B. 1959. Backmixing in the Design of Chemical Reactors. *Ind. and Eng. Chem.* 51, no. 12:1431.

Levenspiel, O., and Bischoff, K. B. 1961. Reaction Rate Constant May Modify the Effects of Backmixing. *Ind. and Eng. Chem.* 53, no. 4:313.

Lighthart, B., and Loew, G. A. 1972. Identification Key for Bacteria Clusters from an Activated Sludge Plant. *Jour. WPCF* 44, no. 11:2078.

McCarty, P. L. 1970. Biological Processes for Nitrogen Removal: Theory and Application. *Proceedings of the Twelfth Sanitary Engineering Conference.* Urbana, Ill.: University of Illinois.

McCarty, P. L., and Lawrence, A. W. 1970. Unified Basis for Biological Treatment Design and Operation. *Jour. SED* 96, no. SA3:757.

McKinney, R. E. 1962a. *Microbiology for Sanitary Engineers.* New York: McGraw-Hill.

McKinney, R. E. 1962b. Mathematics of Complete Mixing Activated Sludge. *Jour. SED* 88, no. SA3:87.

Metcalf and Eddy, Inc. 1979. *Wastewater Engineering, Treatment, Disposal and Reuse.* New York: McGraw-Hill.

Middlebrooks, E. J., and Garland, C. F. 1968. Kinetics of Model and Field Extended-Aeration Wastewater Treatment Units. *Jour. WPCF* 40, no. 4:586.

Monod, J. 1949. The Growth of Bacterial Cultures. *Annual Review of Microbiology* 3:371.

Pearson, E. A. 1968. Kinetics of Biological Treatment. In *Advances in Water Quality Improvement,* edited by E. F. Gloyna and W. W. Eckenfelder, Jr. Austin, Tex.: University of Texas Press.

Petrasek, A. C. 1975. Nitrification in Wastewater Treatment Plants (Activated Sludge). Dissertation, Civil Engineering Department, Texas A&M University.

Pipes, W. O., Jr., Grieves, R. B., and Milbury, W. F. 1964. A Mixing Model for Activated Sludge. *Jour. WPCF* 36, no. 5:619.

Ramalho, R. S. 1977. *Introduction to Wastewater Treatment Processes.* New York: Academic Press.

Reynolds, T. D., and Yang, J. T. 1966. Model of the Completely Mixed Activated Sludge Process. In *Proceedings of the 21st Annual Purdue Industrial Waste Conference.* Part 2.

Rich, L. G. 1963. *Unit Processes of Sanitary Engineering.* Clemson, S.C.: L. G. Rich.

Rich, L. G. 1973. *Environmental Systems Engineering.* New York: McGraw-Hill.

Santer, M., and Ajl, S. 1954. Metabolic Reactions of *Pasteurella pestis,* L. Terminal Oxidation. *J. Bacteriology* 67:379.

Sawyer, C. N. 1956. Bacterial Nutrition and Synthesis. In *Biological Treatment of Sewage and Industrial Wastes.* Vol. 1, edited by J. McCabe and W. W. Eckenfelder. New York: Reinhold.

Schroeder, E. D. 1977. *Waste and Wastewater Treatment.* New York: McGraw-Hill.

Servizi, J. A., and Bogan, R. H. 1964. Thermodynamic Aspects of Biological Oxidation and Synthesis. *Jour. WPCF* 38, no. 5:607.

Smith, H. S., and Paulson, W. L. 1966. Homogeneous Activated

Sludge-Wastewater Treatment at Lower Cost. *Civil Engineering* (May):56.

Stack, V. T., Jr., and Conway, R. A. 1959. Design Data for Completely Mixed Activated Sludge Treatment. *Sew. and Ind. Wastes* 31:1181.

Steel, E. W., and McGhee, T. J. 1979. *Water Supply and Sewerage.* New York: McGraw-Hill.

Stewart, M. J. 1964. Activated Sludge Process Variations: The Complete Spectrum. Part 1: Basic Concepts; Part 2: Process Descriptions; Part 3: Effluent Quality-Process Loading Relationships. *Water and Sewage Works* 111, no. 4:153; 111, no. 5:246; 111, no. 6:295.

Sundstrom, D. W., and Klei, H. E. 1979. *Wastewater Treatment.* Englewood Cliffs, N.J.: Prentice-Hall.

Texas Water Utilities Association. 1971. *Manual of Wastewater Operations.* 4th ed. Austin, Tex.

Water Pollution Control Federation (wpcf). 1977. *Wastewater Treatment Plant Design.* wpcf Manual of Practice no. 8, Washington, D.C.

Water Resources Symposium no. 6. 1973. *Application of Commercial Oxygen to Water and Wastewater Systems,* edited by R. E. Speece and J. F. Malina, Jr. Austin, Tex.: University of Texas Press.

Weston, R. F., and Eckenfelder, W. W. 1955. Applications of Biological Treatment to Industrial Wastes: I. Kinetics and Equilibria of Oxidative Treatment. *Sew. and Ind. Wastes* 27, no. 7:802.

Wilner, B., and Clifton, C. E. 1954. Oxidative Assimilation by *Bacillus subtilis. J. Bacteriology* 67:571.

Wu, Y. C., and Kao, D. F. 1976. Yeast Plant Wastewater Treatment. *Jour. SED* 102, no. EE5:969.

Wuhrmann, K. 1964. Die Grundlagen Der Dimensionierung der Berlüftung bei Belebschlammanlagen Vortrag. Swiss Federal Institute of Technology, Zurich, Switzerland.

Problems

1. An industrial wastewater from a large milk processing plant has been treated using a continuous-flow activated sludge reactor to develop an acclimated culture of activated sludge. Using the acclimated culture, three batch reactors (each containing the same amount of wastewater) were inoculated with the same amount of activated sludge. The reactors were operated over a 24-hour duration to obtain data for kinetic evaluations. The laboratory data from the three batch tests (average values) are shown in Table 9.11.

At 24 hours, the BOD_5 = 5.2 mg/ℓ, the BOD_5 = 0.35 BOD_u, and the BOD_u = COD (biodegradable). Determine the nonbiodegradable COD, the reaction order, and the rate constant based on the biodegradable COD and both the MLVSS and the MLSS values.

Table 9.11.

Time (hr)	COD (mg/ℓ)	MLSS (mg/ℓ)	MLVSS (mg/ℓ)	% Volatile Solids
0	918	1688	1418	84
0.5	821	1652	1437	87
1.0	740	1716	1459	85
1.5	675	1689	1486	88
2.0	528	1718	1512	88
2.5	428	1812	1540	85
3.0	363	1875	1575	84
3.5	297	2106	1622	77
4.0	265	1948	1695	87
5.0	220	2023	1740	86
6.0	162	1982	1744	88
8.0	134	1948	1714	88
24.0	107	1626	1415	87

2. A soluble organic industrial wastewater has been treated using a continuous-flow, activated sludge reactor to develop an acclimated culture of activated sludge. The desired amount of acclimated sludge was placed in a batch reactor and the reactor was operated over a 24-hour duration to obtain data for kinetic evaluations. The observed COD and MLSS concentrations are in Table 9.12.

Table 9.12.

Time (hr)	MLSS X (mg/ℓ)	COD C (mg/ℓ)
0	3670	706
0.017	3672	604
0.5	3675	372
1.0	3690	276
2.0	3760	147
4.0	3860	93
8.0	3880	93
24.0	3820	89

The MLVSS = 0.85 MLSS, BOD_5 = 7 mg/ℓ, and BOD_5 = 0.36 BOD_u at 24 hr.

a. Determine the nondegradable COD and plot the concentration of mixed liquor suspended solids, X, and the biodegradable COD, C, versus time, t, on arithmetic paper. Is a lag phase evident?

b. Plot the rate, dC/dt, versus the biodegradable COD, C, on arithmetic paper and show the progress of the reaction versus C.

c. Determine the reaction order and rate constant if the substrate concentration is represented by the biodegradable COD.

3. A soluble organic industrial wastewater has a BOD_5 of 420 mg/ℓ and is to be treated by the activated sludge process to produce an effluent BOD_5 of 20 mg/ℓ. The rate constant K is 0.12 ℓ/(gm MLSS)(hr), the MLSS concentration is 4000 mg/ℓ, and the sludge density index is 10,000 mg/ℓ. Determine:

a. The reaction time for a plug-flow reactor.

b. The reaction time for a dispersed plug-flow reactor if the dispersion number is 0.2.

c. The reaction time for a completely mixed reactor.

4. Four laboratory-scale, continuous-flow, completely mixed, activated sludge reactors have been operated using a wastewater from a yeast production plant. The data found (Wu, Y. C., & Kao, C. F., 1976) are shown in Table 9.13.

Table 9.13.

Reactor No.	θ_t (hr)	X_1 MLSS (mg/ℓ)	S_t BOD_5 (mg/ℓ)	S_1 BOD_5 (mg/ℓ)	$\Delta X_1/\Delta t$ (mg/ℓ-day)
1	24	4,165	2,462	546	1,417
2	24	6,566	2,406	305	1,279
3	24	8,996	2,383	224	1,259
4	24	10,781	2,436	172	1,025

a. Determine the parameters Y, k_e, μ_{max}, and K_s.

b. A completely mixed activated sludge plant is to treat the wastewater, and the influent BOD_5 is the average found in the studies. Determine:

    (1) The mean cell residence time, θ_c, if the effluent BOD_5 is 240 mg/ℓ.

    (2) The hydraulic detention time if the MLSS is 7500 mg/ℓ.

    (3) The aeration tank volume if the flow is 250,000 gal/day.

    (4) The pounds of sludge produced per day.

5. Using the parameters from Problem 4, determine for a dispersed plug-flow reactor:

a. The aeration time if the dispersion number is 0.2, the MLSS concentration = 7500 mg/ℓ, the sludge density index = 12,500 mg/ℓ, and the effluent BOD$_5$ is 240 mg/ℓ.

b. The aeration tank volume if the flow is 250,000 gal/day.

c. The pounds of sludge produced per day.

6. For the data in Problem 4, determine:
   a. The parameters K, Y, and k_e. These are to be used in b, c, and d.
   b. The hydraulic detention time for a completely mixed activated sludge plant with an MLSS of 7500 mg/ℓ, the sludge density index = 12,500 mg/ℓ, and an effluent BOD$_5$ of 240 mg/ℓ.
   c. The aeration tank volume if the flow is 250,000 gal/day.
   d. The pounds of sludge produced per day.

7. Using the parameters from Problem 6, determine for a dispersed plug-flow reactor:
   a. The aeration time if the dispersion number is 0.2, the MLSS is 7500 mg/ℓ, the sludge density index = 12,500 mg/ℓ, and the effluent BOD$_5$ is 240 mg/ℓ.
   b. The aeration tank volume if the flow is 250,000 gal/day.
   c. The pounds of sludge produced per day.

8. A design is needed for a tapered aeration activated sludge plant having a design or projected population of 12,000 persons. The influent BOD$_5$ is 200 mg/ℓ and the average annual per capita flow is 100 gal/day. The effluent BOD$_5$ is to be 10 mg/ℓ. The aeration tank is to have a width, W, to depth, H, ratio of 1.75:1. The length, L, to width, W, ratio is to be 5:1. The design is to be by the NRC equation. Determine:
   a. The required BOD$_5$ removal as a percent for the secondary treatment. (Assume a removal efficiency of 35 percent for the primary treatment and a negligible removal for the chlorination facilities.)
   b. The recirculation ratio, R/Q, if the MLSS concentration in the aeration tanks is to be maintained at 2500 mg/ℓ and the sludge density index is 10,000 mg/ℓ.
   c. The required detention time in hours for the aeration tank and the required volume for the aeration tank in ft^3.
   d. The dimensions of the aeration tank.
   e. The blower capacity in cfm if the air required is 1000 ft^3 per lb BOD$_5$ applied to the aeration tank (not including BOD$_5$ in the return sludge).
   f. The space loading—that is, the lb BOD$_5$/1000 ft^3.

9. A tapered aeration activated sludge plant is to be designed for a municipality having a population of 10,000 persons and the BOD_5 of the effluent is to be equal to 10 mg/ℓ. Pertinent design data are: $Y = 0.81$ lb vss/lb BOD_5, $k_e = 0.07$ day^{-1}, $Y' = 0.73$ lb oxygen/lb BOD_5, $k_e' = 0.16$ lb O_2/lb MLVSS-day, design MLSS $= 2500$ mg/ℓ, MLVSS $= 0.70$ MLSS, SDI $= 10,000$ mg/ℓ, dispersion number $(D/vL) = 0.15$ for the dispersed plug-flow reactor, flow $= 100$ gal/cap-day, BOD_5 in effluent from the primary clarifier $= 155$ mg/ℓ, organic and ammonia nitrogen in the effluent from the primary clarifier $= 25$ mg/ℓ, 90 percent of the organic and ammonia nitrogen remaining after synthesis is converted to nitrate, 3.84 lb oxygen/lb $\underline{\text{N}}$ converted to nitrate, formulation for the cells in the volatile suspended solids $= C_5H_7O_2N$, plant elevation $= 3000$ ft above sea level, and design temperature $= 68°F$ (20°C). The biochemical reaction is pseudo–first order and the rate constant based on MLSS is 0.200 ℓ/gm-hr. Determine:

a. The required aeration time and volume of the reactor.

b. The net sludge production in pounds of sludge per day (that is, the pounds of solids in the waste activated sludge flow per day).

c. The waste activated sludge flow in gallons per day if the specific gravity is 1.01.

d. The oxygen demand in pounds of oxygen per day and pounds per hour, and also the average oxygen uptake rate in mg/ℓ-hr.

e. The average oxygen demand in pounds of oxygen per hour for the first, second, third, and fourth quarters of the aeration tank and the oxygen uptake in mg/ℓ-hr for each quarter of the tank.

f. The standard cubic feet of air required per minute for each quarter of the tank length if the oxygen diffusers have an efficiency of 6 percent at an operating DO of 2.0 mg/ℓ.

g. The cubic feet of compressed air per gallon of wastewater and per pound of BOD_5 removed.

Note: From the graph in the Appendix, the oxygen content at El. 3000 at 20°C is 0.01562 lb oxygen per cubic foot of air.

10. A completely mixed activated sludge plant is to be designed for a soluble organic industrial waste having a flow rate of 2.0 MGD, and the total COD of the effluent during winter conditions is to be 120 mg/ℓ. Pertinent design data are: influent COD (total) $= 1020$ mg/ℓ, non-

degradable COD = 90 mg/ℓ, Y = 0.40 lb VSS/lb COD, k_e = 0.06 day^{-1} (at 25°C), Y' = 0.63 lb oxygen/lb COD, k'_e = 0.14 lb O_2/lb MLVSS-day (at 25°C), K = 0.370 ℓ/gm-hr based on MLVSS (at 25°C), MLVSS = 90 percent of MLSS, design MLSS = 3000 mg/ℓ, sludge density index (SDI) = 6000 mg/ℓ, minimum operating temperature of the mixed liquor = 18°C, maximum operating temperature of the mixed liquor = 28°C, temperature correction coefficient (θ) for the rate constant = 1.05, temperature correction coefficient (θ) for the endogenous degradation coefficient = 1.07, depth of the basin = 16 ft, basin geometry is square in plan view, specific gravity of waste activated sludge flow = 1.01, and four mechanical aerators are to be used. The biochemical reaction is pseudo–first order based on the biodegradable COD. For the winter operating conditions, determine:

a. The required aeration time, the reactor volume, and the reactor dimensions.
b. The net sludge production in pounds of sludge per day (the pounds of solids in the waste activated sludge flow per day).
c. The waste activated sludge flow in gallons per day and as a percent of the influent flow.
d. The oxygen demand in pounds of oxygen per day and pounds per hour, and also the pounds per hour per aerator.

Next, for summer operating conditions, determine:

e. The biodegradable COD and total COD in the effluent.
f. The net sludge production in pounds of sludge per day.
g. The waste activated sludge flow in gallons per day and as a percent of the influent flow.
h. The oxygen demand in pounds of oxygen per day and pounds per hour, and also the pounds per hour per aerator.

Which condition (summer or winter) will control in the design of a reactor size for the maximum effluent COD? In this problem, which condition controlled the design of the waste activated sludge handling facilities? Which condition controlled the oxygen requirements?

11. An organic industrial wastewater has been treated using a continuous-flow activated sludge reactor to develop an acclimated culture of activated sludge. Using the acclimated culture, four batch reactors were operated to obtain data to determine the coefficients Y, Y', k_e, and k'_e. The test was 24-hours duration and the laboratory data are shown in Table 9.14.

Table 9.14.

Reactor No.	X_0 MLSS (mg/ℓ)	X_t MLSS (mg/ℓ)	S_0 COD (mg/ℓ)	S_t COD (mg/ℓ)
1	500	665	680	111
2	946	1050	680	83
3	1920	1850	680	95
4	3650	3310	680	88

The MLVSS was 88 percent of the MLSS. Determine the coefficients Y and k_e and the theoretical coefficients Y' and k'_e.

12. The empirical equation for the organic fraction of an organic industrial wastewater is $C_5H_{14}O_2N$.
   a. Develop the equation for the respiration of the organic material if the nitrogen is oxidized to the ammonium ion, NH_4^+, and the oxidation products for respiration are CO_2, H_2O, and NH_4^+.
   b. Develop the equation for the synthesis of microbial cells if the empirical equation for the cells is $C_5H_7O_2N$.
   c. Develop the overall biochemical equation for respiration and synthesis if the cell yield is 0.39 mg TOD.
   d. What is the theoretical oxygen demand per 1.0 lb organic matter?
   e. What is the ratio of the ultimate first-stage BOD to the theoretical oxygen demand?
   f. What is the ratio of the five-day BOD to the theoretical oxygen demand if the five-day BOD is 70 percent of the ultimate first-stage BOD?

13. An industrial wastewater from a sugar refining plant has sucrose, $C_{12}H_{22}O_{11}$, as essentially all of its organic content and the nitrogen source to be added is in the form of the ammonium ion, NH_4^+.
   a. Develop the equation for the respiration of sucrose.
   b. Develop the equation for the synthesis of microbial cells from sucrose if the empirical equation for the cell mass is $C_5H_7O_2N$.
   c. Develop the overall biochemical equation for respiration and synthesis if the cell yield is 0.39 mg cells per mg TOD.
   d. What is the theoretical oxygen demand per 1.0 lb organic matter?
   e. What is the ratio of the ultimate first-stage BOD to the theoretical oxygen demand?

14. A completely mixed activated sludge plant is to be designed for a soluble organic industrial waste having a flow rate of 7600 m^3/d, and the total COD of the effluent during winter conditions is to be 120 mg/ℓ. Pertinent design data are: influent COD (total) = 1020 mg/ℓ, nondegradable COD = 90 mg/ℓ, Y = 0.40 g VSS/g COD, k_e = 0.06 day^{-1} (at 25°C), Y' = 0.63 g oxygen/g COD, k'_e = 0.14 g O_2/g MLVSS-day (at 25°C), K = 0.370 ℓ/g-hr based on MLVSS (at 25°C), MLVSS = 90 percent of MLSS, design MLSS = 3000 mg/ℓ, sludge density index (SDI) = 6000 mg/ℓ, minimum operating temperature of the mixed liquor = 18°C, maximum operating temperature of the mixed liquor = 28°C, temperature correction coefficient, θ, for the rate constant = 1.05, temperature correction coefficient, θ, for the endogenous degradation coefficient = 1.07, depth of the basin = 4.9 m, basin geometry is square in plan view, specific gravity of waste activated sludge flow = 1.01, and four mechanical aerators are to be used. The biochemical reaction is pseudo–first order based on the biodegradable COD. For the winter operating conditions, determine:

a. The required aeration time, the reactor volume, and the reactor dimensions.

b. The net sludge production in kilograms of sludge per day (the kilograms of solids in the waste activated sludge flow per day).

c. The waste activated sludge flow in m^3 per day and as a percent of the influent flow.

d. The oxygen demand in kilograms of oxygen per day and kilograms per hour, and also the kilograms per hour per aerator.

Next, for summer operating conditions, determine:

e. The biodegradable COD and total COD in the effluent.

f. The net sludge production in kilograms of sludge per day.

g. The waste activated sludge flow in m^3 per day and as a percent of the influent flow.

h. The oxygen demand in kilograms of oxygen per day and kilograms per hour, and also the kilograms per hour per aerator.

Which condition (summer or winter) will control in the design of a reactor size for the maximum effluent COD? In this problem, which condition controlled the design of the waste activated sludge handling facilities? Which condition controlled the oxygen requirements?

Chapter Ten

OXYGEN TRANSFER
AND MIXING

I n activated sludge, aerated lagoon, and aerobic digestion processes, oxygen must be supplied to the biological solids for aerobic respiration, and mixing
must be sufficient to maintain the solids in suspension. Oxygen transfer and mixing are provided by diffused
compressed air, mechanical aeration, a combination of diffused compressed air and mechanical aeration (such as the
submerged-turbine type of aerator), or by pure oxygen with
mechanical agitation for mixing. This chapter covers primarily diffused compressed air aeration and mechanical
aeration because these are the most common systems used.
However, the principles presented also apply to pure oxygen
systems.

The transfer of a solute gas from a gas mixture into a liquid
that is in contact with the mixture can be described by the
two-film theory of Lewis and Whitman (1924). Figure 10.1
shows a schematic drawing of the two phases in contact
with each other. The partial pressures of the solute gas in
the bulk gas and at the gas interface are p_G and p_{G_i}, respectively. The concentrations of the solute gas at the liquid
interface and in the bulk liquid are C_{L_i} and C_L, respectively.
The solute gas must diffuse through the gas film (laminar
layer), pass through the interface, and then diffuse through
the liquid film (laminar layer). The interface offers no resistance to the solute gas transfer. For gases that are very
soluble in the liquid, the rate limiting step is the diffusion
of the solute gas through the gas film. For gases that are

Oxygen Transfer

349

Figure 10.1. Schematic Drawing Illustrating the Two-Film Theory

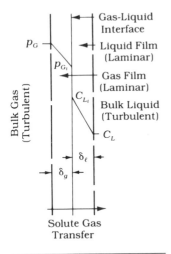

Bulk Gas (Turbulent)

p_G

p_{G_i}

C_{L_i}

C_L

Gas-Liquid Interface

Liquid Film (Laminar)

Gas Film (Laminar)

Bulk Liquid (Turbulent)

δ_ℓ

δ_g

Solute Gas Transfer

slightly soluble in the liquid, such as oxygen in water, the rate limiting step is the diffusion of the solute gas through the liquid film. The diffusion transfer coefficient, K_L, for oxygen diffusing through the water film is given by $K_L = D/\delta_\ell$, where D is the diffusivity coefficient of oxygen in water and δ_ℓ is the film thickness. Multiplying K_L by a, the interfacial bubble area per unit volume of water, gives the overall mass transfer coefficient, $K_L a$. Since the liquid resistance is controlling, $p_G = p_{G_i}$ and $C_{L_i} = C_s$, C_s being the saturation dissolved oxygen concentration in equilibrium with the oxygen partial pressure in air bubbles, p_G. The saturation dissolved oxygen concentration, C_s, is in equilibrium with the partial pressure, p_G, in accordance with Henry's law, one form of this law being $p_G = $ (constant) (C_s). The driving force for mass transfer is $C_{L_i} - C_L$ but, since $C_s = C_{L_i}$, the driving force is $C_s - C_L$. Since the rate of oxygen mass transfer is equal to the mass transfer coefficient times the driving force, the mass transfer is expressed by

$$\frac{dC}{dt} = K_L a (C_s - C_L) \qquad (10.1)$$

where

dC/dt = rate of oxygen transfer, mass/(volume)(time)— for example, mg/ℓ-hr;

$K_L a$ = overall liquid mass transfer coefficient, time^{-1}—for example, hour^{-1};

C_s = saturation dissolved oxygen concentration, mass/volume—for example, mg/ℓ;

C_L = dissolved oxygen concentration in the liquid, mass/volume—for example, mg/ℓ.

The mass transfer per unit time is given by

$$\frac{dM}{dt} = N = K_L a V (C_s - C_L) \qquad (10.2)$$

where

dM/dt = rate of oxygen transfer, mass/time;

N = rate of oxygen transfer, mass/time;

V = liquid volume.

Aeration devices for diffused compressed air may be classified as coarse or fine bubble diffusers. The efficiency of oxygen transfer depends primarily upon the design of the diffuser, the size of the bubbles produced, and the depth of submergence. The coarse bubble diffusers usually consist of a short vertical tube and/or an orifice device as

Figure 10.2. Coarse Bubble Diffusers

Figure 10.3. Fine Bubble Diffusers

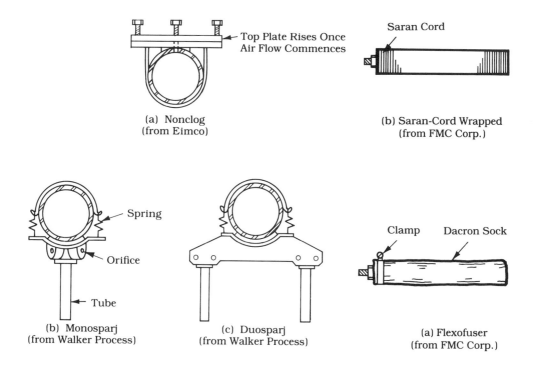

(a) Nonclog
(from Eimco)

(b) Saran-Cord Wrapped
(from FMC Corp.)

(b) Monosparj
(from Walker Process)

(c) Duosparj
(from Walker Process)

(a) Flexofuser
(from FMC Corp.)

in Figure 10.2. The diffusers must be spaced far enough apart so that the upward flowing bubble plumes do not significantly interfere with each other. The transfer efficiency for coarse bubble diffusers is usually from 4 to 8 percent, with a value of 6 percent being typical. The coarse bubble type has a lower efficiency than the fine bubble type; however, it is less susceptible to clogging. The fine bubble types, some of which are shown in Figure 10.3, are usually made of porous ceramic tubes or cylindrical metal frames covered with a wrapping such as a Dacron cloth or Saran cord. Sometimes porous ceramic plates covering an air channel cast in the tank bottom are used. The transfer efficiency of the fine bubble type is from 8 to 12 percent, with a value of 9 percent being typical. Usually, the air flow per unit for either the fine or coarse bubble type is from 4 to 16 scfm (at 20°C and one atmosphere).

For diffused aeration, Eckenfelder and Ford (1968)

Air Compressors at a Diffused Compressed Air Type Activated Sludge Plant

Activated Sludge Aeration Tank with Swing-Type Air Diffusers
Courtesy FMC Corporation, Material Handling Systems Division

have presented the following formulation for the rate of oxygen mass transfer:

$$N = C G_a^{1-n} D^{0.67} (C_m - C_L) \cdot 1.02^{(T-20)} \cdot \alpha \qquad (10.3)$$

where

N = rate of oxygen transfer, lb/hr;

C and n = constants;

G_a = air flow, standard cubic feet per minute (at 20°C and one atmosphere);

D = depth to diffusers, ft;

C_m = saturation dissolved oxygen in the wastewater at the mid-depth of the tank at operating conditions;

T = operating temperature, °C;

α = $K_L a$ (wastewater)/$K_L a$ (water).

It can be seen in Eq. (10.3) that the terms, $C G_a^{1-n} D^{0.67} \times 1.02^{(T-20)} \alpha$, represent $K_L a V$ in Eq. (10.2). Usually the values of the experimental constants C and n are furnished by the manufacturer.

Since the solubility of oxygen varies with pressure, the saturation dissolved oxygen, C_m, is determined at the tank mid-depth and is given by (Oldshue, J. Y., 1956)

$$C_m = C_w \left(\frac{P_r}{29.4} + \frac{O_e}{42} \right) \qquad (10.4)$$

where

C_m = saturation dissolved oxygen concentration of wastewater at the tank mid-depth at operating conditions;

C_w = saturation dissolved oxygen concentration of the wastewater at operating conditions;

P_r = absolute pressure in psi at the depth of air release;

O_e = percent oxygen content in the exit air flow.

The saturation dissolved oxygen of the wastewater at the operating atmospheric pressure, C_w, is given by $C_w = \beta C_s$, where β is the ratio of the saturation dissolved oxygen for the wastewater divided by the saturation dissolved oxygen of tap water at the same temperature.

Mechanical aerators usually employ impellers and, as shown in Figure 10.4, they may be float or fixed mounted. They may be classified as the high-speed axial-flow pump type, the slow-speed vertical turbine, the submerged slow-speed vertical turbine with a sparger ring, and the rotating brush aerator. The pump-type aerator, as shown in Figure 10.4 (a), is a high-speed axial-flow pump with a propeller; it is mainly used for aerated lagoons. Motor sizes are from 1 to 150 horsepower, and they have speeds from 900 to 1800 rpm, with 900 to 1200 being typical. The operating water depth is from 3 to 18 ft and draft tubes are usually required when the depth is greater than about 10 to 12 ft. Oxygen transfer occurs as the spray passes through the air, and also at the impingement area because a considerable amount of turbulence is created. These aerators are not suited for severely cold climates where freezing of the water

Figure 10.4. Some Impellers Employed by Mechanical Aerators

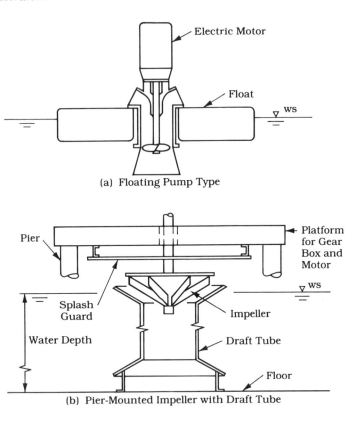

(a) Floating Pump Type

(b) Pier-Mounted Impeller with Draft Tube

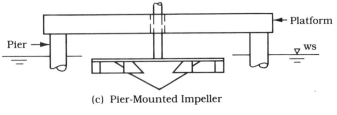

(c) Pier-Mounted Impeller

might occur. The oxygen transfer rate is from 2.0 to 3.9 lb/hp-hr(nameplate hp), with the smaller machines having the highest transfer.

The slow-speed vertical turbine, shown in Figure 10.4 (b) and (c), may be used for activated sludge, aerobic digesters, and aerated lagoons. Motor sizes are from 3 to 150 hp, and turbine sizes range from about 3 to 12 ft in diameter. The speeds are from about 30 to 60 rpm; thus, they always require a gear reduction box. They may be fixed mounted, as shown in Figure 10.4 (b) and (c), or float mounted for flexibility. The operating water depth is from 3 to 30 ft, with 12 to 15 ft being the most common. Depths from about 15 to 30 ft require draft tubes, as shown in Figure 10.4 (b). The turbines are submerged only a few inches below the still water surface and, when operating, the flow goes radially outward. Air entrainment occurs in the immediate vicinity of the turbine, and also oxygen absorption occurs

Mechanical Surface Aerator. A drawing of the impeller and draft tube for this type of unit is shown in Figure 10.4(b).

Mechanical Surface Aerator. A drawing of the impeller for this type of unit is shown in Figure 10.4(c).

due to turbulence. The proper depth of submergence is critical for good oxygen transfer. Transfer rates are from about 2.0 to 3.9 lb/hp-hr(nameplate), with the smaller machines having the greatest transfer.

For mechanical aeration, Eckenfelder and Ford (1968) have presented the following formulation for the rate of oxygen mass transfer:

$$N = N_0 \left(\frac{C_w - C_L}{9.17} \right) 1.02^{(T-20)} \alpha \qquad (10.5)$$

where

N = rate of oxygen transferred at operating conditions, lb/hr or lb/(hp)(hr);

N_0 = oxygen rating of the aerator, lb/hr or lb/(hp)(hr).

The rating of a mechanical aerator, N_0, is for 20°C and a zero dissolved oxygen concentration; thus, the parenthetical term in Eq. (10.5) is the ratio of the actual driving force to the rated driving force.

The submerged slow-speed vertical turbine with a sparger ring is located about 1.5 ft above the bottom of a tank; the sparger ring allows compressed air to be released below the turbine, thus creating oxygen transfer. Oxygen transfer is from about 2.0 to 3.0 lb/hp-hr(nameplate). Usually, the turbine diameter is 0.1 to 0.2 times the tank width. In a circular tank, four baffles mounted on the wall at 90-degree intervals in plan view are required to minimize rotational flow. In a square tank, two baffles mounted on opposite walls usually is satisfactory. In rectangular tanks, baffles are usually not required if the length to width is 1.5:1 or more. One advantage of these units is that they can be used in deep, narrow tanks. A disadvantage is that they require a source of compressed air.

The rotating brush aerator is mainly used for oxidation ditches; however, in Europe they are often used for aeration in long, narrow tanks used for the activated sludge process. The aerator consists of a long horizontal axle with radiating steel bristles partly submerged in the still water surface. As the axle rotates, the bristles spray the liquid outwards and oxygen transfer occurs due to air entrainment in the immediate vicinity of the bristles, and also due to the spray and impingement area. Oxygen transfer rates are from 3.0 to 3.5 lb/hp-hr(nameplate).

The value of α may be determined in the laboratory using a vessel equipped with a diffused aeration stone and employing a constant liquid depth in the vessel and a constant air flow rate. The vessel is filled with tap water and

Submerged Slow-Speed Vertical Turbine Aerator. Note the upper turbine used to provide additional mixing and the compressed air release below the lower turbine.

Rotating Brush Aerator. The radiating steel bristles furnish the required agitation.

deoxygenated by adding sodium sulfite and cobalt chloride, the cobalt serving as a catalyst. The aeration is begun and the dissolved oxygen concentration versus time is recorded. This represents an unsteady state, and, for this condition,

$$\frac{dC}{dt} = K_L a (C_s - C) \tag{10.6}$$

where C = dissolved oxygen concentration, mg/ℓ. Rearranging for integration gives

$$\int_{C_0}^{C_t} \frac{-dC}{C_s - C} = K_L a \int_0^t - dt \tag{10.7}$$

where

C_0 = dissolved oxygen concentration at time zero;
C_t = dissolved oxygen concentration at time, t.

Integrating and rearranging gives

$$\ln(C_s - C_t) = \ln(C_s - C_0) - K_L a t \tag{10.8}$$

Equation (10.8) is of the straight-line form $y = mx + b$; thus, plotting $\ln(C_s - C_t)$ versus time, t, gives a straight line of the slope equal to $K_L a$. This gives $K_L a$ for tap water. Now the vessel is emptied and refilled to the same depth with the untreated wastewater and, using the same air flow rate, the experiment is repeated to give $K_L a$ for the untreated wastewater. Similarly, using treated wastewater, the value of $K_L a$ for treated wastewater is obtained. The value for $K_L a$ for the mixed liquor is the average of the two values. The value α is given by $\alpha = K_L a(\text{wastewater})/K_L a(\text{water})$. Here, $K_L a(\text{wastewater})$ is equal to the mean value of $K_L a(\text{untreated}$ wastewater) and $K_L a$ (treated wastewater). The $K_L a$ (wastewater) is equal to $K_L a$ for the mixed liquor. The saturation dissolved oxygen concentration for the waste may be obtained by filling the vessel with wastewater and aerating the waste until the dissolved oxygen reaches a maximum, thus giving the saturation dissolved oxygen concentration for the wastewater. This is repeated for the treated wastewater, giving a dissolved oxygen saturation concentration for the treated waste. The value C_w is equal to the average of the two saturation values for the wastewater—that is, the values for the untreated and treated wastewater. The β value is the average C_w divided by the saturated oxygen concentration, C_s, for tap water at the same temperature.

In the previous aeration equations, the values of C_s

and, in turn, C_w, are for the operating temperature of the mixed liquor and the operating atmospheric pressure. If the elevation of the plant is much greater than sea level, the operating pressure is less than one atmosphere (760 mm Hg), and the saturated dissolved oxygen concentration is less than that given in standard solubility tables. The atmospheric pressures for various elevations are shown in Table 10.1.

Table 10.1. Atmospheric Pressures for Various Elevations

Elevation (ft)	Atmospheric Pressure (mm Hg)
0	760
1000	733
2000	706
3000	680
4000	656
5000	632
6000	609
7000	586
8000	564
9000	543

The saturation dissolved oxygen concentration at an elevation above sea level is equal to the C_s value given in a standard table at 760 mm multiplied by the ratio of the atmospheric pressure at that elevation divided by 760. For instance, at an elevation of 5500 ft, the atmospheric pressure using interpolation of the above values is 620 mm; thus, the solubility of the dissolved oxygen at 20°C is (620/760)(9.17) or 7.48 mg/ℓ. If an activated sludge plant for this elevation were designed using a saturated dissolved oxygen value at 760 mm and an operating dissolved oxygen value of 2.0 mg/ℓ, the aeration system would be underdesigned by (9.17 − 2.0)/(7.48 − 2.0) minus 1 or 31 percent. Consequently, in some instances a correction for elevation must be made.

Mixing

Mixing has not been rationally analyzed to a great extent and it is common to express mixing requirements in terms of power intensity or air flow rate for a given volume. For instance, Metcalf and Eddy (1979) give aeration intensities required to keep activated sludge in complete suspension as 20 to 30 scfm/1000 ft^3 of tank volume. For mechanical aerators they recommend 0.5 to 1.0 hp/1000 ft^3 or 0.067 to 0.134 hp/1000 gal. For aerobic digesters, Metcalf and Eddy recommend 20 to 30 scfm/1000 ft^3 of diffused air for mixing or 0.067 to 0.134 hp/1000 gal of mechanical aeration for mixing. Eckenfelder (1966) has given power intensities of 0.10 to 0.13 hp per 1000 gal as the range required for surface aerators for activated sludge. For diffused aeration, the degree of mixing is related to the depth of the diffusers; thus, the lower requirements, such as 20 scfm/1000 ft^3, are for deep tanks and the higher requirements, such as 30 scfm/1000 ft^3, are for shallow tanks. Normally activated sludge tanks are from about 10 to 16 feet deep. For mechanical surface aeration, the degree of mixing is inversely related to the tank depth; thus, the higher values, such as 0.13 hp/1000 gal, are for deep tanks and the lower values, such as 0.10 hp/1000 gal, are for shallow tanks.

Compressor Requirements

The power required by a compressor to furnish diffused compressed air may be determined from the following equation, which gives brake (shaft) horsepower for the adiabatic compression of a gas (Perry, R. H., & Chilton, C. H., 1973):

$$\text{hp (brake or shaft)} = \frac{FRT_1}{33,000 \, nE} \left[\left(\frac{p_2}{p_1} \right)^n - 1 \right] \quad (10.9)$$

where

F = mass air flow, lb/min;
R = gas constant = 53.5;
T_1 = inlet absolute temperature, °R;
p_1 = absolute inlet pressure, lb per sq. in. absolute (psia);
p_2 = absolute outlet pressure, (psia);
n = 0.283 for air;
E = efficiency of the compressors as a fraction (usually 0.70 to 0.80).

The mass air flow in lb/min is given by the formulation

$$F\text{(lb/min)} = G_a\text{(cfm)} \times \rho_{\text{Air}} \text{(lb/ft}^3) \quad (10.10)$$

The specific weight or density of air depends upon the pressure and temperature and is given by Figure 10.5. In compressor work, a standard cubic foot is at one atmosphere (14.7 psia), 20°C (68°F), and 36 percent relative humidity. The term *free air* refers to air at conditions prevailing at the compressor inlet. In using Eq. (10.9), the power requirements should be determined by the free air flow in lb/min and the actual inlet temperature. The two types of com-

Figure 10.5. Density or
Specific Weight of Air (36%
relative humidity)

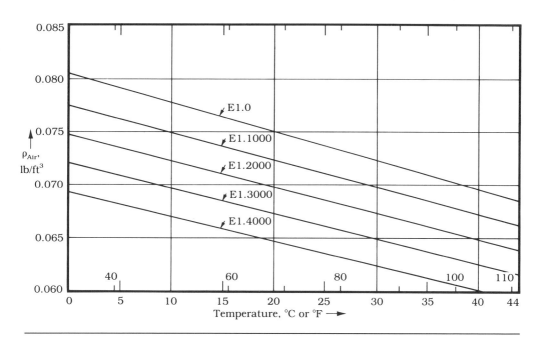

pressors (frequently called *blowers*) used in wastewater
treatment are the centrifugal and the rotary type. The centrifugal type has a split impeller and operates on the same
principle as a centrifugal pump. The rotary type is a positive
displacement compressor that uses two high-speed rotors
or impellers. Eq. (10.9) gives the power that must be applied to the shaft of the compressor; therefore, to get the
power required by the motor, the shaft horsepower of the
compressor must be divided by the efficiency of the motor.
Electric motors used in compressor work have about 95 to
98 percent efficiency. The operating pressure is a function
of the head loss through the piping and diffusers and the
depth of submergence of the diffusers. Most of the pressure
is due to the depth of submergence, and operating pressures are usually 5 to 10 psi.

References

Dobbins, W. E. 1956. The Nature of the Oxygen Transfer Coefficient in Aeration Systems. In *Biological Treatment of Sewage
and Industrial Wastes.* Vol. 1, edited by J. McCabe and W. W.
Eckenfelder, Jr. New York: Reinhold.
Dreier, D. E. 1956. Theory and Development of Aeration Equipment. In *Biological Treatment of Sewage and Industrial*

Wastes. Vol. 1, edited by J. McCabe and W. W. Eckenfelder, Jr. New York: Reinhold.

Eckenfelder, W. W., Jr. 1966. *Industrial Water Pollution Control.* New York: McGraw-Hill.

Eckenfelder, W. W., Jr. 1980. *Principles of Water Quality Management.* Boston: CBI Publishing.

Eckenfelder, W. W., Jr., and Ford, D. L. 1966. Engineering Aspects of Surface Aeration Design. In *Proceedings of the 22nd Annual Purdue Industrial Waste Conference.*

Eckenfelder, W. W., Jr., and Ford, D. L. 1968. New Concepts in Oxygen Transfer and Aeration. In *Advances in Water Quality Improvement,* edited by E. F. Gloyna and W. W. Eckenfelder, Jr. Austin, Tex.: University of Texas Press.

Eckenfelder, W. W., Jr., and Ford, D. L. 1970. *Water Pollution Control.* Austin, Tex.: Pemberton Press.

Gaden, E. L., Jr. 1956. Aeration and Oxygen Transport in Biological Systems—Basic Considerations. In *Biological Treatment of Sewage and Industrial Wastes.* Vol. 1, edited by J. McCabe and W. W. Eckenfelder. New York: Reinhold.

Lewis, W. K., and Whitman, W. G. 1924. Principles of Gas Absorption. *Ind. and Eng. Chem.* 16, no. 12:1215.

Metcalf and Eddy, Inc. 1979. *Wastewater Engineering, Treatment, Disposal and Reuse.* New York: McGraw-Hill.

Oldshue, J. Y. 1956. Aeration of Biological Systems Using Impellers. In *Biological Treatment of Sewage and Industrial Wastes.* Vol. 1, edited by J. McCabe and W. W. Eckenfelder, Jr. New York: Reinhold.

Perry, R. H., and Chilton, C. H. 1973. *Chemical Engineer's Handbook.* 5th ed. New York: McGraw-Hill.

Ramalho, R. S. 1977. *Introduction to Wastewater Treatment Processes.* New York: Academic Press.

Schroeder, E. D. 1977. *Water and Wastewater Treatment.* New York: McGraw-Hill.

Sundstrom, D. W., and Klei, H. E. 1979. *Wastewater Treatment.* Englewood Cliffs, N.J.: Prentice-Hall.

Von der Emde, W. 1968. Aeration and Developments in Europe. In *Advances in Water Quality Improvement,* edited by E. F. Gloyna and W. W. Eckenfelder, Jr. Austin, Tex.: University of Texas Press.

Problems

1. A tapered aeration activated sludge plant located 1500 ft above sea level treats a municipal wastewater. The aeration tank is 15 ft deep, 24 ft wide, and 140 ft long, and the oxygen demands for the first, second, third, and fourth quarters are 31.0, 23.0, 17.7, and 16.8 lb per hour, respectively. The diffuser units are located 12 in. above the tank bottom, and the center to center spacings for the units are 12 in., 15 in., 17.5 in., and 17.5 in. for the respective quarters. Other pertinent data are: operating mixed liquor temperature = 23°C, α = 0.85, β = 0.95, and operating dissolved oxygen

(C_L) = 2.0 mg/ℓ. The diffusers are dual-tube coarse bubble type and the performance relationship is

$$N = 0.00170\, G_{\hat{a}}^{0.9} D^{0.67} (C_m - C_L) 1.02^{T\text{-}20} \alpha$$

where

N = pounds of oxygen transferred per hour per diffuser unit;
G_a = air flow, scfm per diffuser unit;
D = depth to diffusers, ft;
C_m = saturation dissolved oxygen concentration of the wastewater at mid-depth, mg/ℓ;
C_L = operating dissolved oxygen concentration, mg/ℓ;
T = temperature of mixed liquor, °C;
α = $K_L a$(wastewater)/$K_L a$(tap water).

Determine the air flow required for each of the diffusers in the respective quarters and the total air flow required for each quarter.

2. A pump-type floating surface aerator with a 65-hp (nameplate) electric motor is to be rated at standard conditions, which are: water temperature = 20°C, atmospheric pressure = 760 mm Hg, water quality equal to tap water, and the dissolved oxygen concentration = 0.0 mg/ℓ. A test has been performed in which the pump-type aerator was placed in a test tank that was 95 feet in diameter and had an 8.5 foot depth of water. The dissolved oxygen in the water was removed by adding Na_2SO_3 and $CoCl_2$ as a catalyst. Once the dissolved oxygen had reached approximately zero and there was a negligible amount of $SO_3^=$ left, the aerator was started and the dissolved oxygen was measured at four points versus time. The DO concentrations measured are given in Table 10.2.

   Points 1 and 3 were 1.0 ft below the surface, whereas points 2 and 4 were 1.0 ft above the bottom. Other pertinent data are: water temperature = 28°C, saturation dissolved oxygen in the water = 7.80 mg/ℓ, $K_L a$ (20°C) = $K_L a(T) \div 1.02^{(T\text{-}20)}$, saturation dissolved oxygen concentration in clear tap water at 20°C = 9.17 mg/ℓ, and the line hp drawn by the electric motor = 62.0. Determine:
   a. The pounds of oxygen transferred per hp-hr (nameplate) at 20°C and at a DO equal zero.
   b. The pounds of oxygen transferred per hp-hr (line hp) at 20°C and at a DO equal zero.

Table 10.2.

| Time (min) | 5 ft from Spray | | 1 ft from Wall | |
	Point 1 Top	Point 2 Bottom	Point 3 Top	Point 4 Bottom
0	0.30	0.30	0.20	0.20
3	1.60	1.60	1.50	1.70
6	2.40	2.50	2.70	2.70
8	3.40	3.40	3.40	3.40
11	4.30	4.30	4.30	4.30
14	5.20	5.20	5.10	4.90
17	5.65	5.60	5.70	5.50
20	6.10	6.20	6.00	5.90
23	6.60	6.60	6.50	6.35
26	6.80	6.80	6.70	6.60
29	7.10	7.10	7.05	6.90

3. For the equation for diffused air aeration units in Problem 1, determine the constant equivalent to 0.00170 so that the equation gives kg oxygen transferred per hour, with G_a equal to m^3 air/1000 m^3 volume and D equal to depth in m.

4. A completely mixed activated sludge plant treating an industrial wastewater has four mechanical aerators, and the plant elevation is 1200 ft. The oxygen demand during the summer is 242.0 lb/hr, and the wastewater temperature is 28°C. The oxygen demand during the winter is 170.4 lb/hr, and the wastewater temperature is 18°C. Aeration studies have been performed in the laboratory using an aeration vessel with a diffuser stone that used compressed air. The K_La values found were 0.388 hr^{-1} for wastewater, 0.526 hr^{-1} for effluent, and 0.554 hr^{-1} for water. All K_La values were at the same temperature. The saturation value of oxygen in water, wastewater, and effluent was 8.68, 8.2, and 8.3 mg/ℓ, respectively. The mechanical aerators to be used transfer 2.1 lb O_2/hp-hr (nameplate) at one atmosphere, 20°C, and zero dissolved oxygen. The operating DO level = 2.0 mg/ℓ. Determine:
   a. The α value for the mixed liquor.
   b. The β value.
   c. The nameplate horsepower required for summer conditions.
   d. The nameplate horsepower required for winter conditions.
   e. Which condition summer or winter controlled aerator requirements? What horsepower should be used?

5. A completely mixed activated sludge plant treats the wastewater from a refinery and petrochemical plant. The reactor basin is 120 ft × 120 ft and has a depth of 13 ft. Four 150-hp slow-speed turbines furnish the oxygen for the process. The α value is 0.9 and the β value is 0.85. A field test was made that showed the oxygen uptake rate to be 61.2 mg/ℓ-hr, and the average operating DO was 2.2 mg/ℓ. The mixed liquor temperature was 18°C, and the saturation dissolved oxygen value was 9.3 mg/ℓ. Determine the oxygen transfer of the aerators in lb O_2/hp-hr.

6. An activated sludge plant has diffusers that have a 14-ft depth of submergence. The air head loss through the diffusers is 21 in. of water column (the usual range is from 14 to 28 in.), and the air piping has a head loss of 6 in. of water column (the usual range is from 4 to 8 in. for properly designed piping). Determine the operating pressure of the compressors.

7. An activated sludge plant is located at El. 1800. The design air flow using Eq. (10.3) is 1130 scfm, and the operating pressure is 7.0 psi. Pertinent data are: maximum summer temperature = 100°F, efficiency of the compressors = 75 percent, and electric motor efficiency = 95 percent. Determine:
   a. The free air flow in cfm.
   b. The required motor horsepower.

Chapter Eleven

TRICKLING FILTERS AND ROTARY BIOCONTACTORS

These processes are similar because both employ cultures of microorganisms that are attached to a media surface. For trickling filters the media is stationary, whereas for rotary biocontactors the media discs are rotating.

Trickling Filters

As a wastewater passes through a filter bed, the microbial growths sorb an appreciable amount of the organic materials for use as food substances, mainly by aerobic bio-oxidation. The final clarifier, which is immediately downstream of the filter, serves to remove microbial growths that periodically slough from the filter media. Usually large gravel or crushed stone is used as the media; however, synthetic plastic media or other packing is also used. In the past, the trickling filter process has been commonly used for small- to medium-sized cities, primarily because of its simplicity and dependability. For very small towns, a trickling filter plant may require an operator for only a few hours each day. Because of higher effluent standards, activated sludge processes have largely displaced trickling filters in plants presently being constructed, even for very small communities. A flowsheet for a trickling filter plant treating a municipal wastewater is shown in Figure 11.1. The wastewater receives primary treatment prior to the secondary treatment, which consists of the trickling filter and its final clarifier.

Figure 11.1. Flowsheet for a Trickling Filter

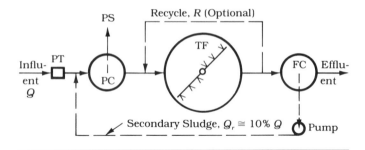

Filters have been classified as low or standard rate and high rate according to the organic loading, the unit liquid loading, and the recycle employed. Low-rate filters, which were developed first, have organic loadings from 300 to 800 pounds of BOD_5/ac-ft-day, unit liquid loadings from 1 to 6 million gal/ac-day, and bed depths of about 5 to 9 feet, and they employ recycle only during times, such as the night period, when the wastewater flow is inadequate to turn the rotary distributor. Low-rate trickling filters attain high BOD_5 removals, such as 90 to 95 percent, when treating municipal wastewaters, and better nitrification than high-rate filters; however, the volume of media required is much greater. When treating municipal wastewaters, a low-rate trickling filter plant will produce an effluent having about 12 to 25 mg/ℓ BOD_5. High-rate trickling filters have organic loadings of 800 to 8000 lb of BOD_5/ac-ft-day, unit liquid loadings of 10 to 40 million gal/ac-day, and bed depths of about 3 to 8 feet, and they employ continuous recycle. When treating municipal wastewaters, high-rate trickling filters employing two stages in series will produce BOD_5 removals of 85 to 90 percent; however, very little nitrification occurs. When treating municipal wastewaters, a two-stage high-rate trickling filter plant will give an effluent having about 15 to 35 mg/ℓ of BOD_5, whereas a single-stage high-rate trickling filter plant will produce an effluent having about 35 to 50 mg/ℓ. A current use of the trickling filter is to employ a single high-rate filter with a high organic loading as a roughing unit prior to an activated sludge treatment for wastewaters having extremely high BOD_5. A roughing filter usually employs plastic packing as a media and has an organic loading of 2000 to 15,000 lb of BOD_5/ac-ft-day. The bed depth may be 10 to 40 ft and the hydraulic load is from 40 to 200 million gal/ac-day. Trickling filters are usually constructed using reinforced concrete tanks and are from 10 to 250 ft in diameter. Standard diameters are in 5-ft intervals; however, any diameter may be specially made.

*Rotary Distributor Trickling
Filter. The jet action from the
nozzles drives the rotary
distributor.*

*Trickling Filter with a Rotary
Distributor*
Courtesy of Dorr-Oliver, Inc.

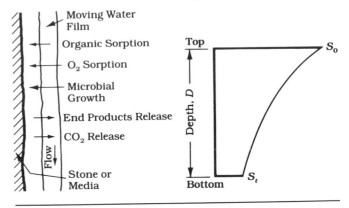

Figure 11.2. Schematic Showing the Biological Action of a Trickling Filter

Moving Water Film

Organic Sorption

O_2 Sorption

Microbial Growth

End Products Release

CO_2 Release

Flow

Stone or Media

Figure 11.3. Profile Showing Organic Concentration in the Wastewater versus Filter Depth

Top

Depth, D

Bottom

S_o

S_t

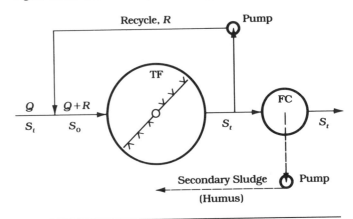

Figure 11.4.   Flowsheet of a Trickling Filter

Recycle, R

Pump

TF

Q

S_i

$Q+R$

S_o

S_t

FC

S_t

Secondary Sludge
(Humus)

Pump

Biological Filtration

As wastewater passes over microbial growths, Figure 11.2, an appreciable amount of the organic material is sorbed by the growths along with molecular oxygen. Aerobic bio-oxidation occurs and the oxidized organic and inorganic end products are released into the moving water film. The wastewater passes through a filter in a matter of minutes; however, the sorbed organic materials are retained for several hours as they undergo bio-oxidation. The microbial action is essentially biosorption, similar to that described in Chapter 9.

Most of the microbial growth is aerobic; however, the growth immediate to the media surface is usually anaerobic. As time progresses, the thickness of the microbial slime increases due to the new cell mass synthesized; finally the growth attains a thickness that causes it to slough off the media. The amount of organic material removed per foot of bed depth is greatest at the top of the filter, Figure 11.3, and smallest at the bottom. The nitrifying microbial growths will be located in the lower depths of the filter bed. The BOD_5 in the effluent from a filter is the same as in the effluent from its final clarifier, as indicated in Figure 11.4 by the designation S_t. Recycling the effluent by spraying it back on the top of the filter allows the microbial growths to have successive opportunities for sorption and biological oxidation of the organic matter, thus increasing the filter efficiency.

The growths that have sloughed from the filter media (that is, the humus) are subsequently settled in the final clarifier, and the secondary sludge is usually sent to the head of the plant, where it is resettled in the primary clarifier and thickened prior to sludge treatment.

Filter Performance

The first performance equations for trickling filters were empirical (Fairall, J. M., 1956; National Research Council, 1946; Rankin, R. S., 1955); however, later developments have produced equations based on biochemical kinetics (Eckenfelder, W. W., 1961 & 1970; Velz, C. J., 1948).

Kinetic Equations

One of the earliest performance equations based on kinetics is that by Velz (1948). He observed that the rate of organic removal per interval of depth is proportional to the remaining concentration of removable organic matter. His basic equation, which is a modification of a first-order rate equation, is $-dL/dD = kL$, where dL is an increment of remaining organic concentration, dD is an increment of depth, k is the rate constant, and L is the remaining removable organic concentration expressed as the ultimate first-stage BOD. Velz made an analogy between his rate equation and a first-order type rate equation since an increment of depth, dD, is proportional to an increment of time, dt. His integrated equation in terms of base 10 logarithms is

$$\frac{L_D}{L} = 10^{-kD} \tag{11.1}$$

where

L_D = removable ultimate first-stage BOD concentration at depth, D;

L = removable ultimate first-stage BOD concentration applied to the bed;

$$k = \text{rate constant;}$$
$$D = \text{depth of the bed, ft.}$$

For low-rate filters treating municipal wastewaters and operated 2 to 6 MG/acre-day, Velz found the k value to be 0.175 and the removable fraction to be 90 percent. For high-rate filters operated at approximately 20 MG/acre-day, he found the k value to be 0.1505 and the removable fraction to be 78.4 percent. Velz also developed the equation, $k_2 = k_{20} \cdot 1.047^{T_2-20}$, where $k_2 =$ rate constant at $T_2°C$ and $k_{20} =$ rate constant at 20°C, for correcting the rate constant for a change in temperature.

Eckenfelder (1970) has also developed a performance equation based on the specific rate of substrate removal for a pseudo–first-order reaction, which is

$$-\frac{1}{X}\frac{dS}{dt} = kS \tag{11.2}$$

where

$$(1/X)(dS/dt) = \text{specific rate of substrate utilization, mass/(microbes)(time);}$$
$$dS/dt = \text{rate of substrate utilization, mass/(volume)(time);}$$
$$k = \text{rate constant, volume/(mass microbes)} \times \text{(time);}$$
$$S = \text{substrate concentration, mass/volume.}$$

Rearranging Eq. (11.2) for integration gives

$$\int_{S_0}^{S_t} \frac{dS}{S} = -k\overline{X} \int_0^t dt \tag{11.3}$$

where

$$k = \text{rate constant;}$$
$$\overline{X} = \text{average cell mass concentration, mass/volume;}$$
$$S_t = \text{substrate concentration after the contact time, } t, \text{ mass/volume;}$$
$$S_0 = \text{substrate concentration applied to the filter, mass/volume.}$$

Integration of Eq. (11.3) gives

$$\frac{S_t}{S_0} = e^{-k\overline{X}t} \tag{11.4}$$

The average cell mass concentration, $\overline{X}$, is proportional to the specific surface area of the packing, A_s—in other words, the ft^2 of surface per bulk ft^3 of volume. Thus,

$$\overline{X} \sim A_s \qquad (11.5)$$

where A_s is the specific surface area. The mean contact time, t, for a filter is represented by (Howland, W. E., 1957)

$$t = CD/Q_L^n \qquad (11.6)$$

where

t = mean contact time;
D = depth of the bed;
Q_L = unit liquid loading or surface loading;
C, n = experimental constants.

Substituting Eqs. (11.5) and (11.6) into Eq. (11.4) and combining the constants k and C gives the performance equation by Eckenfelder (1970):

$$\frac{S_t}{S_0} = e^{-KA_s^m D/Q_L^n} \qquad (11.7)$$

where

S_t = substrate concentration in the filter effluent, mass/volume;
S_0 = substrate concentration applied to the filter, mass/volume;
K = rate constant;
A_s = specific surface area of the packing, area/volume;
D = filter depth;
Q_L = unit liquid loading or surface loading;
m, n = experimental constants.

The value of n depends on the flow characteristics through the packing and is usually about 0.50 to 0.67. The specific surface areas (ft^2/ft^3) of some typical filter media are as follows (Balakrishnan, S., Eckenfelder, W. W., & Brown, C., 1969): $2\frac{1}{4}$- to 4-in. rock, 15.0; 1- to 3-in. crushed granite, 29.0; Surfpac, 28.0; $1\frac{1}{2}$-in. Flexxirings, 40.0; and Flexipac, 70.0. For a particular wastewater and media of interest, Eq. (11.7) may be simplified by combining KA_s^m to give

$$\frac{S_t}{S_0} = e^{-KD/Q_L^n} \qquad (11.8)$$

where K is the new rate constant.

The following equation has been developed for correcting trickling filter rate constants for temperature changes (Eckenfelder, W. W., 1970):

$$K_T = K_{20} \cdot 1.035^{(T-20)} \tag{11.9}$$

where

K_T = rate constant at temperature, T, °C;
K_{20} = rate constant at 20°C;
T = temperature, °C.

One of the most common kinetic equations for filter performance when treating municipal wastewaters on stone media has been developed by Eckenfelder (1961) and is as follows:

$$\frac{S_t}{S_0} = \frac{1}{1 + 2.5\left(\dfrac{D^{0.67}}{Q_L^{0.50}}\right)} \tag{11.10}$$

where

S_t = BOD$_5$ in the filter effluent, mg/ℓ;
S_0 = BOD$_5$ in the wastewater discharged on the filter bed, mg/ℓ;
D = filter depth, ft;
Q_L = unit liquid loading, MG/acre-day.

The values of S_t and S_0 are depicted in the flowsheet shown in Figure 11.4 and, if no recycle is employed, $S_t = S_0$. Equation (11.10) may be derived from the second-order kinetic equation, $(1/X)(dS/dt) = KS^2$, as follows. The integration of the second-order equation gives $S_t/S_0 = 1/(1 + S_0 K\overline{X}t)$. The mean contact time is represented by $t = C\,D^{0.67}/Q_L^{0.50}$. Substituting the expression for t in the integrated equation and combining the constants S_0, K, $\overline{X}$, and C gives $S_t/S_0 = 1/(1 + \text{Constant}\,D^{0.67}/Q_L^{0.50})$. For municipal wastewaters, Eckenfelder has found the constant to be 2.5.

Example Problem 11.1
Low-Rate Trickling
Filter

A city has a population of 5000 persons and is served by a low-rate trickling filter with a polishing pond downstream from the final clarifier. Pertinent data for the plant are: wastewater flow = 100 gal/cap-day, BOD$_5$ after primary clarification = 150 mg/ℓ, BOD$_5$ after final clarification = 25 mg/ℓ, and filter bed depth = 4 to 6 ft. Determine a suitable diameter of the filter and its depth.

Solution

The flowsheet for a low-rate trickling filter is shown in Figure 11.4 and, since the recycle, R, is only used at night, $S_t = S_o$. Assume $D = 5$ ft. Substituting $S_o = 150$, $S_t = 25$, and $D = 5$ in Eq. (11.10) yields

$$\frac{25}{150} = \frac{1}{1 + 2.5\left(\dfrac{5^{0.67}}{Q_L^{0.50}}\right)}$$

The above equation gives Q_L as 2.161 MG/ac-day. The wastewater flow is (5000 persons)(100 gal/cap-day)(1 MG/10^6 gal) or 0.50 MG/day. Thus, the plan area is (0.50 MG/day)(ac-day/2.161 MG)(43,560 ft^2/ac) or 10,079 ft^2. The diameter, D, is

$$D = \left[(10,079\,\text{ft}^2)\frac{4}{\pi}\right]^{1/2}$$

$$= 113.3\,\text{ft or }\underline{115\,\text{ft for standard size}}$$

Determining the area for 115 ft diameter gives 0.2385 acre. This results in a hydraulic load, Q_L, of 2.0964 MG/ac-day. Substituting 2.0964 MG/ac-day, $S_t = 25$ and $S_o = 150$ into Eq. (11.10) gives

$$\text{Depth}(D) = 4.89\,\text{ft} \qquad \underline{\text{Use 5 ft–0. in.}}$$

Example Problem 11.1 SI
Low-Rate Trickling
Filter

A city has a population of 5000 persons and is served by a low-rate trickling filter with a polishing pond downstream from the final clarifier. Pertinent data for the plant are: wastewater flow = 380 ℓ/cap-day, BOD$_5$ after primary clarification = 150 mg/ℓ, BOD$_5$ after final clarification = 25 mg/ℓ, and filter bed depth = 1.5 m. Determine the diameter of the filter.

Solution

The flowsheet for a low-rate trickling filter is shown in Figure 11.4 and, since the recycle, R, is only used at night, $S_t = S_o$. For the SI system, the constant 2.5 in Eq. (11.10) is replaced by 5.358 if the depth is meters and the hydraulic loading is m^3/ day-m^2. Substituting $S_o = 150$, $S_t = 25$, and $D = 1.5$ m in Eq. (11.10) yields

$$\frac{25}{150} = \frac{1}{1 + 5.358 \left(\dfrac{1.5^{0.67}}{Q_L^{0.50}} \right)}$$

The above equation gives Q_L as 1.977 m³/day-m². The wastewater flow is (5000 persons)(380 ℓ/cap-day)(m³/1000ℓ) or 1900 m³/day. The plan area is (1900 m³/day)(day-m²/1.977 m³) or 961.1 m². The diameter, D, is

$$D = \left[(961.1 \text{ m}^2) \frac{4}{\pi} \right]^{1/2} = \underline{35.0 \text{ m}}$$

Example Problem 11.2
High-Rate Trickling
Filter

Assume the city in Example Problem 11.1 is served by a high-rate trickling filter having a recycle ratio of 2 with a polishing pond downstream from the final clarifier. The depth is to be 5 to 7 ft. Determine a suitable filter diameter and depth.

Solution

The pertinent data for the plant are the same as in Example Problem 11.1. A materials balance on the BOD₅ at the junction of the recycle flow and the primary effluent yields (Q) 150 + $(2Q)$ 25 = $(1 + 2)$ $(Q)S_0$ or S_0 = 66.67 mg/ℓ. Assume a depth of 6 ft. Substituting the values in Eq. (11.10) yields

$$\frac{25}{66.67} = \frac{1}{1 + 2.5 \left(\dfrac{6^{0.67}}{Q_L^{0.50}} \right)}$$

The above equation gives Q_L as 24.82 MG/ac-day. The wastewater flow, Q, is 0.50 MGD; thus, the recycle flow, R, is (2)0.50 or 1.0 MGD. The total flow applied to the filter is 0.50 + 1.0 or 1.5 MGD. The plan area is (1.5 MGD)(ac-day/24.82 MG)(43,560 ft²/ac) or 2633 ft². The diameter, D, is

$$D = \left[(2633 \text{ ft}^2) \frac{4}{\pi} \right]^{1/2}$$

$$= 57.9 \text{ ft or } \underline{60 \text{ ft for standard size}}$$

Determining the area for 60-ft diameter gives 0.06491 acre. This results in a hydraulic load, Q_L, of 23.11 MG/ac-day. Substituting 23.11 MG/ac-day, $S_t = 25$, and $S_0 = 66.67$ into Eq. (11.10) gives

Depth $(D) = 5.69$ ft   Use 5 ft–9 in.

NRC Equations

The National Research Council (NRC) made a study of the data collected from numerous wastewater treatment plants operated at military bases during World War II. Their study included low-rate, single-stage high-rate, and two-stage high-rate trickling filter plants. The empirical equation developed in the NRC study (1946) for a single trickling filter and its final clarifier is

$$E_S = \frac{100}{1 + 0.0085 \sqrt{\dfrac{y_0}{VF}}} \qquad (11.11)$$

where

E_S = filter efficiency, percent;
y_0 = lb of BOD$_5$ applied to the filter per day;
V = stone media volume, ac-ft;
F = recycle factor or the effective number of passes through the filter.

The effective number of passes of the organic matter through the filter, F, is given by

$$F = \frac{1 + R/Q}{(1 + 0.1 R/Q)^2} \qquad (11.12)$$

where R/Q is the recycle ratio. In the NRC study, it was found that the maximum feasible recycle ratio is 8. If the filter is a low-rate filter, the value of F in Eq. (11.11) is 1.

If the trickling filter is the second-stage filter of a two-stage plant, the filter must degrade materials more resistant to bio-oxidation than the first-stage filter. Due to this, the equation developed in the NRC study (1946) for a second-stage filter is

$$E_{S_2} = \frac{100}{1 + \dfrac{0.0085}{1 - E_{S1}} \sqrt{\dfrac{y_0'}{V_2 F_2}}} \qquad (11.13)$$

where

E_{S_2} = second-stage filter efficiency, percent;
E_{S_1} = first-stage filter efficiency as a fraction;
y_0' = lb of BOD$_5$ applied to the second-stage filter per day, $y_0' = y_0(1 - E_{S_1})$
V_2 = volume of the stone media, ac-ft;
F_2 = recycle factor or the effective number of passes through the second-stage filter.

The term $1 - E_{s_1}$ accounts for the more bioresistant materials treated by the second-stage filter.

Example Problem 11.3
NRC Equations

A two-stage high-rate trickling filter plant treats 2.0 MGD of municipal wastewater that has a BOD$_5$ of 200 mg/ℓ. The filters are of equal diameter and depth and have equal recycle ratios. Pertinent data for the plant are: BOD$_5$ removal by the primary clarifier = 35 percent, filter diameter = 70 ft, filter depth = 5.5 ft, and the recycle ratio = 1.0. Determine the BOD$_5$ in the final effluent.

Solution

The BOD$_5$ applied to the first-stage filter, y_0, is $(2.0 \times 10^6$ gal/day)(8.34 lb/gal)(200/10^6)(1 − 0.35) or 2168.4 lb/day. The stone volume, V, for each filter is $(70 \text{ ft})^2 \times (\pi/4)(5.5 \text{ ft})(\text{ac-ft}/43{,}560 \text{ ft}^3)$ or 0.4859 ac-ft. The recycle factor, F, is $(1 + 1)/(1 + 0.1)^2$ or 1.65. Substituting these values into Eq. (11.11) yields

$$E_{S_1} = \frac{100}{1 + 0.0085 \sqrt{\dfrac{2168.4}{(0.4859)(1.65)}}}$$

The above equation gives the efficiency of the first-stage filter, E_{S_1}, as 69.35 percent. The BOD$_5$ applied to the second-stage filter is therefore (2168.4)(1 − 0.6935) or 664.6 lb/day. Substituting the appropriate values into Eq. (11.13) yields

$$E_{S_2} = \frac{100}{1 + \dfrac{0.0085}{(1 - 0.6935)} \sqrt{\dfrac{664.6}{(0.4859)(1.65)}}}$$

The above equation gives the efficiency of the second-stage filter, E_{S_2}, as 55.60 percent. The effluent BOD$_5$ is therefore

$$\text{Effluent BOD}_5 = (200)(1 - 0.35)(1 - 0.6935)$$
$$\times (1 - 0.5560)$$
$$= \underline{17.7 \text{ mg}/\ell}$$

Figure 11.5 shows the usual flowsheets that have been employed for single-stage and two-stage high-rate trickling filter plants. The single-stage plant shown in Figure 11.5 (b) has the recycled flow, R, clarified prior to discharge on the filter. Also, the two-stage high-rate trickling filter plant shown in Figure 11.5 (d) has the recycled flows, R_1 and R_2, clarified prior to discharge on the filters. It has been found that clarification of the recycle flows has a negligible effect upon filter performance. Because clarification of the recy-

Flowsheets for High-Rate Trickling Filter Plants

Figure 11.5. Flowsheets for High-Rate Trickling-Filter Plants

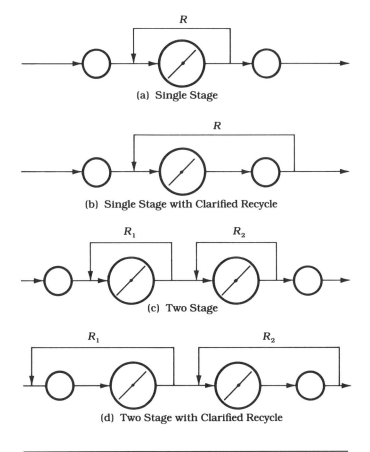

(a) Single Stage

(b) Single Stage with Clarified Recycle

(c) Two Stage

(d) Two Stage with Clarified Recycle

---

cled flow requires a clarifier to be larger than otherwise, most plants do not use clarified recycle and, as a result, the flowsheets shown in Figures 11.5 (a) and (c) are the most commonly used. Two-stage high-rate trickling filter plants usually have filters of equal size and the recycled flows, R_1 and R_2, are usually equal.

Filter Details

The media used in trickling filters is usually gravel, crushed stone, or slag having a size of $2\frac{1}{2}$ to $3\frac{1}{2}$ inches. If crushed stone is employed, the stone should be of a hard nature such as trap rock or granite. Soft stones, such as limestone, should not be used because they will gradually break apart due to the end products from microbial action and weathering. Synthetic media such as plastics are usually used when stone is not locally available. In general, the synthetic media have a larger specific area (that is, the surface area per unit bulk volume) than stone media. Surfpac is a PVC medium made in corrugated sheets similar to industrial grating. It is assembled in stacks within the filter. Actifil and Flexxirings are randomly dumped media similar to that used in packed towers for gas absorption and stripping.

Figure 11.6 (a) shows a typical underdrain block made of vitrified clay. It serves both to support the media and also to provide small channels through which the collected effluent flows. Figure 11.6 (b) shows the plan of a typical trickling filter floor and, as indicated, it slopes gradually to the effluent channel that crosses the center of the filter. The underdrain blocks are placed so that their channels align and the collected effluent flows to the effluent channel as shown in Figure 11.6 (b). Figure 11.6 (c) shows a section through the filter at right angles to the underdrain block channels. Figure 11.6 (d) shows a profile through the filter bed illustrating the collected effluent discharging from the underdrain blocks into the effluent channel. The underdrain channels and the effluent channel should be designed to flow not more than half full at the maximum hydraulic flow during the day in order to provide proper ventilation.

The most common distributor is the rotary type that is mounted on a center pier in the middle of the filter. The distributor is usually driven by the jet action of the wastewater discharging from the filter arms onto the bed. Extremely small filters, less than about 10 feet in diameter, are usually electrically driven. Also, small filters may have a single spray in the center of the filter instead of a rotary distributor. Formerly, rectangular trickling filters were used that were dosed by fixed nozzles located above the filter

Figure 11.6. Trickling Filter Details

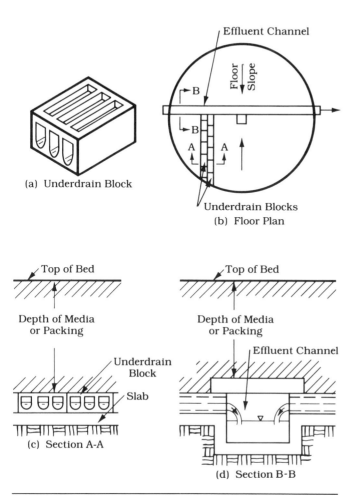

(a) Underdrain Block

Effluent Channel

Floor Slope

B

B

A   A

Underdrain Blocks
(b) Floor Plan

Top of Bed

Depth of Media
or Packing

Underdrain
Block

Slab

(c) Section A-A

Top of Bed

Depth of Media
or Packing

Effluent Channel

(d) Section B-B

bed; however, frequent clogging of the nozzles has caused this type of filter to be almost totally discontinued. The hydraulic head required to drive a rotary distributor depends on the ratio of the maximum to minimum flow and whether the distributor is equipped with overflow arms or overflow jets used for high flows. If the distributor has four arms with two being used for high flows, or has double compartment arms with one compartment and its jets being used for high flows, the head required will be minimized. If the ratio of the maximum to minimum flow is less than 4.0 and the distributor has overflow arms or overflow jets, the head required at the center of the distributor is usually about 3.50 ft above the top of the filter bed.

Operational Problems

In general, trickling filters have fewer operational problems than the activated sludge process; however, some problems may be encountered. Trickling filter humus settles rapidly in a final clarifier and rarely will poor settling result if the scraper mechanism removes the sludge as it collects at the bottom of the tank. Floating sludge masses may be encountered if anaerobic conditions occur within the settled sludge.

The most common problem with trickling filters is the presence of the *Psychoda* fly that breeds in filter beds. Ordinarily they remain in the immediate vicinity of the filters; however, they may be a nuisance to the operators. The most effective control is to provide for flooding the filter and, if a filter is flooded and the media allowed to stay submerged for one day of each summer month, the fly larvae will usually be killed.

Odors may be a problem when low-rate filters are employed, and the wastewater is stale when it reaches the plant. However, odors are rarely a problem for high-rate trickling filters if the recycle flow is continuous throughout the day and the wastewater is not stale.

Rotating Biological Contactors

A single-stage rotary biological contactor consists of a vat through which the wastewater flows and multiple plastic discs mounted on a horizontal shaft that passes through the center of the discs. The shaft is mounted at right angles to the wastewater flow, with about 40 percent of the total disc area being submerged. As the shaft rotates, the biological growths on the discs pass through the wastewater and sorb organic materials in addition to losing excess growths, which slough off. As the biological growths pass through the air, oxygen is adsorbed to keep the growths as aerobic as possible. The biosorption and bio-oxidation that occurs is similar to that of a trickling filter. A multistage rotary biological contactor consists of two or more stages connected in series to achieve greater BOD_5 removal than occurs for a single stage. Recycle is not employed and the growths sloughing from a single- or multistage contactor are removed by a final clarifier. The secondary sludge is mixed with the raw wastewater and resettled in the primary clarifier in order to thicken the sludge. The primary sludge, being a mixture of incoming settleable solids and secondary settlings, usually is 4.2 to 6.5 percent dry solids (Antonie, R. L, Kluge, D. L, & Mielke, J. H., 1974). Biological action is significantly reduced when the wastewater temperature drops below 55°F (13°C). In cold climates, the discs must be covered to avoid heat loss from the wastewater and to protect against freezing temperatures. In warm climates,

Rotating Biological Contactors
Courtesy of Bio-Systems Division, Autotrol Corporation

the units do not need to be enclosed except for aesthetic purposes. In warm climates, a simple sun roof is frequently used.

The principal design parameter for rotary biological discs is the wastewater flow rate per unit surface area of the discs—that is, the hydraulic loading in gal/day-ft^2 area (gpd/ft^2). This is indirectly a food to microbe ratio since the flow rate is related to the mass of substrate per unit time and the disc area is related to the mass of microbes in the system. Usually, four stages are provided for municipal wastewaters and the peripheral speed of the discs is 60 ft/min. In treating municipal wastewaters with a BOD_5 of 200 mg/ℓ and using four stages, a 20-mg/ℓ BOD_5 effluent can be obtained if the hydraulic loading is less than 3.8 gpd/ft^2 and the wastewater temperature is greater than 55°F. This is assuming 35 percent BOD_5 removal by the primary clarifier. For the same conditions, a 10-mg/ℓ BOD_5 effluent can be obtained if the hydraulic loading is less than 1.9 gpd/ft^2. The main advantages of the rotary biological contactors are the low energy requirements compared to the activated sludge process, the ability to handle shock loads, and the ability of multistage units to achieve a high degree of nitrification.

References

Antonie, R. L., Kluge, D. L., and Mielke, J. H. 1974. Evaluation of a Rotary Disk Wastewater Treatment Plant. *Jour. WPCF* 46, no. 3:498.

Balakrishnan, S., Eckenfelder, W. W., Jr., and Brown, C. 1969. Organics Removal by a Selected Trickling Filter Media. *Water and Wastes Eng.* 6, no. 1:A-22.

Bio-Systems Division, Autotrol Corp. 1978. *Autotrol Wastewater Treatment Systems.* Design Manual. Milwaukee, Wis.

Conway, R. A., and Ross, R. D. 1980. *Handbook of Industrial Waste Disposal.* New York: Van Nostrand Reinhold.

Eckenfelder, W. W., Jr. 1980. *Principles of Water Quality Management.* Boston: CBI Publishing.

Eckenfelder, W. W., Jr. 1961. Trickling Filter Design and Performance. *Jour. SED* 87, no. SA4, part 1:33.

Eckenfelder, W. W., Jr. 1970. *Water Quality Engineering for Practicing Engineers.* New York: Barnes and Noble.

Fairall, J. M. 1956. Correlation of Trickling Filter Data. *Sew. and Ind. Wastes* 28, no. 9:1056.

Galler, W. S., and Gotaas, H. B. 1966. Optimization Analysis for Biological Filter Design. *Jour. SED* 92, no. SA1:163.

Great Lakes–Upper Mississippi Board of State Sanitary Engineers. Recommended Standards for Sewage Works. Ten state standards, Albany, N.Y.

Howland, W. E. 1957. Flow Over Porous Media as in a Trickling Filter. *Proc. of the 12th Annual Purdue Ind. Wastes Conference.*

McKinney, R. E. 1962. *Microbiology for Sanitary Engineers.* New York: McGraw-Hill.

Metcalf and Eddy, Inc. 1979. *Wastewater Engineering, Treatment, Disposal and Reuse.* New York: McGraw-Hill.

National Research Council. 1946. Trickling Filters in Sewage Treatment at Military Installations. *Sew. Works Jour.* 18, no. 5:417.

Ramalho, R. S. 1977. *Introduction to Wastewater Treatment Processes.* New York: Academic Press.

Rankin, R. S. 1955. Evaluation of the Performance of Biofiltration Plants. *Trans. ASCE* 120:823.

Schroeder, E. D. 1977. *Water and Wastewater Treatment.* New York: McGraw-Hill.

Stack, V. T. 1957. Theoretical Performance of Trickling Filter Processes. *Sew. and Ind. Wastes* 29, no. 9:987.

Sundstrom, D. W., and Klei, H. E. 1979. *Wastewater Treatment.* Englewood Cliffs, N.J.: Prentice-Hall.

Velz, C. J. 1948. A Basic Law for the Performance of Biological Beds. *Sew. Works Jour.* 20, no. 4:607.

Water Pollution Control Federation (WPCF). 1977. *Wastewater Treatment Plant Design.* WPCF Manual of Practice no. 8, Washington, D.C.

Problems

1. A city of 6500 persons is to be served by a low-rate trickling filter with a polishing pond downstream from the final clarifier. Pertinent data for the plant are: wastewater flow = 80 gal/cap-day, BOD_5 after primary clarification = 165 mg/ℓ, BOD_5 after final clarification = 25 mg/ℓ, and filter bed depth = 5 to 8 ft. Determine a suitable diameter and depth of the filter using Eq. (11.10).

2. Assume the city in Problem 1 is served by a high-rate trickling filter having a recycle ratio of 2.5, with a polishing pond downstream from the final clarifier. The pertinent data for the plant are the same as in Problem 1. Determine a suitable diameter and depth of the filter using Eq. (11.10).

3. The data for a two-stage high-rate trickling filter plant are the same as for Example Problem 11.3. Determine the BOD_5 in the final effluent using Eq. (11.10).

4. A single-stage high-rate trickling filter plant treats a municipal wastewater from a population of 10,000 persons. Pertinent data are: wastewater flow = 100 gal/cap-day, BOD_5 = 0.17 lb/cap-day, recycle ratio = 2.5, filter depth = 5 to 8 ft, BOD_5 removal by the primary clarifier = 35 percent, and effluent BOD_5 = 25 mg/ℓ. Determine a suitable filter diameter and depth using the NRC equation.

5. A two-stage high-rate trickling filter plant is to treat a municipal wastewater. Pertinent plant data are: flow = 2.0 MGD, influent BOD_5 = 250 mg/ℓ, BOD_5 removal by primary clarifiers = 35 percent, filter depth = 5 to 8 ft, first- and second-stage filters are equal diameter, recycle ratio = 1.0 for each filter, and final effluent BOD_5 = 20 mg/ℓ. Determine a suitable diameter and depth of the filters using the NRC equations.

6. The data for a two-stage high-rate trickling filter plant are the same as for problem 5. Determine a suitable diameter and depth of the filters using Eq. (11.10).

7. A two-stage high-rate trickling filter plant treats 1.2 MGD of domestic waste having a BOD_5 of 250 mg/ℓ. Both filters have recycle directly around each filter and have the same recycle ratio. The primary clarifier removes 35 percent of the BOD_5, and the filters are to be 5 to 6 ft deep and have a recycle ratio of 1.0. The effluent BOD_5 is 20 mg/ℓ and the NRC formulas are to be used for design. Determine:
   a. The volume of each filter, ac-ft.
   b. The suitable depth and diameter of each filter, ft.

8. A city of 8000 persons is served by a high-rate trickling filter with a polishing pond downstream from the final clarifier. Pertinent data for the plant are: wastewater flow = 300 ℓ/cap-day, BOD_5 after primary clarification = 160 mg/ℓ, BOD_5 after final clarification = 25 mg/ℓ, filter bed depth = 1.6 m, and recycle ratio = 2.5. Determine the diameter of the filter using Eq. (11.10). For Eq. (11.10) the constant 2.5 is replaced by 5.358 if the depth = m and the hydraulic load is m^3/day-m^2.

9. A low-rate trickling filter plant gives 95 percent BOD_5 removal. The primary clarifier removes 35 percent of the BOD_5 applied to it. The secondary sludge amounts to 0.029 lb dry solid/cap-day. If the raw wastewater has

a BOD$_5$ of 0.17 lb/cap-day, determine the secondary sludge produced per day per pound of BOD$_5$ removed.

10. A high-rate trickling filter plant gives 90 percent BOD$_5$ removal. The primary clarifier removes 35 percent of the BOD$_5$ applied to it. The secondary sludge amounts to 0.044 lb dry solid/cap-day. If the raw wastewater has a BOD$_5$ of 0.17 lb/cap-day, determine the secondary sludge produced per day per pound of BOD$_5$ removed.

11. An industrial wastewater having a flow of 0.8 MGD and a BOD$_5$ of 320 mg/ℓ is to be treated using a high-rate trickling filter with plastic packing. The recycle ratio is 2:1 and the filter depth is to be from 18 to 20 ft. A pilot plant study has been made and the values K and n in Eq. (11.7) have been found to be equal to 0.073 and 0.55, respectively. The hydraulic loading is expressed as gpm/ft^2 and the depth is in ft. The effluent BOD$_5$ is to be 20 mg/ℓ. Determine:
   a. The filter diameter.
   b. The filter depth.

12. A four-stage rotating biological disc facility is to be designed for a city of 4000 persons. The flow is 100 gal/cap-day and the hydraulic loading is 1.9 gpd/ft^2. Determine the disc area in ft^2 and m^2.

Chapter Twelve

STABILIZATION PONDS
AND AERATED LAGOONS

T he stabilization or oxidation pond, shown in Figure 12.1, consists of a quiescent diked pond in which the wastewater enters and, due to the microbial action, the organic materials are bio-oxidized giving CO_2, NH_3, inorganic radicals such as SO_4^{-2} and PO_4^{-3}, and new microbial cells as end products. The algal population uses the CO_2, inorganic radicals, and sunlight to produce dissolved oxygen and new algal cells as end products. Thus, the microbial and algal populations have a synergistic relationship in which both groups benefit from each other. Although most stabilization ponds have effluents discharging directly into the receiving body of water, the future trend in design is to use polishing treatments such as intermittent sand beds to remove algal growths from the effluent, thus giving a better degree of treatment. Stabilization ponds are used for both municipal and industrial wastewater treatment, particularly for small municipalities and seasonal industrial wastewaters.

The aerated lagoon, shown in Figure 12.2, consists of a diked pond with artificial aeration, usually by floating pump-type aerators, and a downline settling tank or facultative stabilization pond that serves as the final clarifier. The biological solids developed in the aerated lagoon are removed from the effluent by the settling tank or stabilization pond. If a settling tank is used, it is usually a poured in place concrete clarifier with mechanical sludge rakes for continuous sludge removal. If a stabilization pond is used, the biological solids settle to the pond bottom and undergo anaerobic decomposition. The pond is drained periodically,

**Figure 12.1. Details of a
Stabilization Pond System**

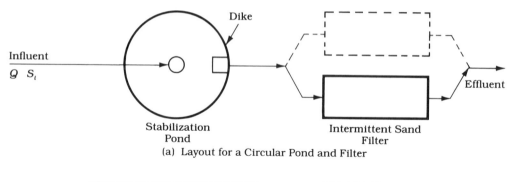

Dike

Influent
Q S_i

Stabilization
Pond

Intermittent Sand
Filter

Effluent

(a) Layout for a Circular Pond and Filter

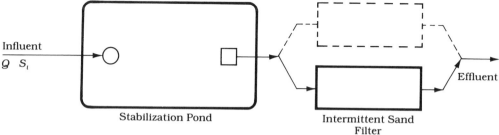

Influent
Q S_i

Stabilization Pond

Intermittent Sand
Filter

Effluent

(b) Layout for a Rectangular Pond and Filter

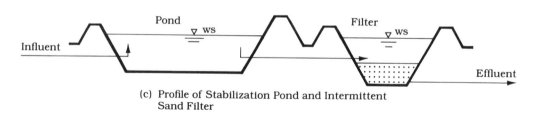

Pond

∇ ws

Influent

Filter

∇ ws

Effluent

(c) Profile of Stabilization Pond and Intermittent
Sand Filter

usually every two or three years, and the digested solids are removed and disposed by sanitary landfill. The aerated lagoon process is essentially a nonrecycle activated sludge process that usually has biological solids at a concentration of about 200 to 500 mg/ℓ. Aerated lagoons are widely used in industrial wastewater treatment because they are less expensive than the activated sludge process. However, they have much greater land requirements, which must be considered. Frequently, an industrial wastewater treatment facility is initially a stabilization pond system and, as the waste load increases, artificial aeration is added to convert some of the ponds into aerated lagoons.

Stabilization ponds are widely used in the southern and southwestern United States because the climate is

Figure 12.2. Details of an Aerated Lagoon System

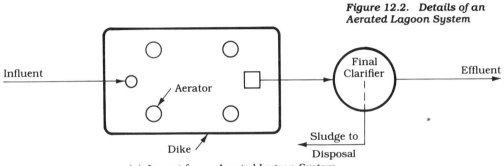

(a) Layout for an Aerated Lagoon System

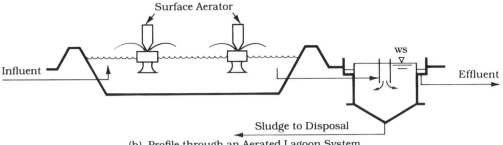

(b) Profile through an Aerated Lagoon System

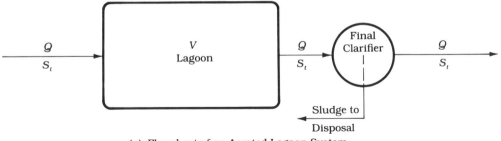

(c) Flowsheet of an Aerated Lagoon System

Large Aerated Lagoon. Each floating, pump-type aerator has two pilings to moor the aerator and prevent sinking due to leaky floats.

warm and there is a high sunlight intensity. Aerated lagoons are also used in these regions because the lagoon temperature is relatively high due to the climate and sunlight intensity.

Stabilization Ponds

Stabilization ponds may be classified as aerobic, facultative, or anaerobic according to their oxygen profile. An aerobic pond has dissolved oxygen throughout its depth, whereas an anaerobic pond has no dissolved oxygen at any depth and the facultative pond has dissolved oxygen in the upper zone of water and no dissolved oxygen in the lower zone. Most ponds are facultative since these are loaded higher than aerobic ponds, yet few odors are produced because the upper pond depth is aerobic.

The amount of oxygen present in a facultative pond depends on the organic loading and the sunlight intensity. There will be a diurnal variation in the dissolved oxygen concentration in the upper zone, and the variation will be greatest during the summer. During the night, the dissolved oxygen will be low because the microbial and some of the algal population uses dissolved oxygen. After the sun rises, photosynthesis starts to occur and the dissolved oxygen, which is an algal end product, begins to increase, with the maximum concentration occurring during the middle of the day. Shortly after sunset, photosynthesis ceases, dissolved oxygen production stops, and the dissolved oxygen concentration decreases due to the oxygen demand of the microbial and algal population. In the summer in the southern and southwestern United States, the variation in dissolved oxygen may be from about 2 to 3 mg/ℓ during the night to above 14 to 16 mg/ℓ during the day. Even for facultative ponds, anaerobic conditions may occur throughout the pond depth during overcast days. Odors, however, are usually not a problem unless windy conditions cause waves on the surface.

As the algal and microbial cells die, they lyse and the cell fragments settle to the bottom of the pond, where they undergo anaerobic decomposition. For the usual facultative pond loadings, the sludge accumulation is very slow and may be only a fraction of an inch per year.

Approximately one-third to one-half the influent organic carbon in the raw wastewater is synthesized into microbial and algal cells that leave with the pond effluent unless an effluent polishing treatment is used. Many times in the past, the BOD_5 reported for an effluent was for a filtered effluent sample, which was misleading because the algal and microbial growths exerted an oxygen demand in the receiving body of water. The BOD_5 removal based on filtered effluent samples will be from 80 to 90 percent, whereas the BOD_5 removal for unfiltered effluent samples will be from about 45 to 65 percent. Coliform removal on unfiltered samples is usually from 85 to 95 percent.

Facultative ponds may range from 3 to 8 feet deep, with the usual water depth being from 3 to 6 feet. When future conversion to an aerobic lagoon system is anticipated, the pond should be deep to accommodate surface aerators. Ponds are usually circular or rectangular in plan view, as shown in Figures 12.1 (a) and (b), unless the terrain necessitates some other geometry. Usually, two or more ponds are provided and the piping is arranged so that series or parallel operation is possible. For single ponds or ponds operated in parallel when treating municipal wastewaters, an organic loading of 25 to 35 lb BOD_5/day-acre is frequently used. For series operation when treating municipal wastewaters, the first pond is usually loaded as high as 75 to 80 lb BOD_5/day-acre, and the downstream pond at a loading of 25 to 35 lb BOD_5/day-acre.

Gloyna (1976) recommends the following equation for determining the volume of a facultative stabilization pond treating municipal wastewaters:

$$V = 3.5 \times 10^{-5} QS_t [\theta^{(35-T)}] ff' \tag{12.1}$$

where

V = pond volume, m^3;

Q = flow, ℓ/day;

S_t = ultimate influent BOD or COD, mg/ℓ;

f = algal toxicity factor, $f = 1$ for municipal wastewaters and many industrial wastewaters;

f' = sulfide or other immediate chemical oxygen demand, $f' = 1$ for SO_4^{-2} equivalent ion concentration of less than 500 mg/ℓ;

θ = temperature coefficient;

T = average water temperature for the pond during winter months, °C.

The value of θ ranges from 1.036 to 1.085. The value of 1.085 is recommended because it is conservative and field data supports this analysis (Gloyna, E. F., 1976). For the case where $\theta = 1.085$ and an effective depth = 5 ft, the area of the pond may be computed using Eq. (12.1). The area in English units is

$$A = 2.148 \times 10^{-2} QS_t [1.085^{(35-T)}] ff' \tag{12.2}$$

where

A = area of pond (acres) for a depth of 5 ft + 1 ft of sludge storage;

Q = flow, MGD.

The BOD$_5$ removal efficiency can be expected to be 80 to 90 percent based on unfiltered influent samples and filtered effluent samples. The efficiency based on unfiltered effluent samples can be expected to vary unless a maturation pond is used as a followup unit.

The recommended minimum depth of a facultative pond is one meter. Additional depth to compensate for sludge storage is desirable. The minimum depth of about one meter is required to control potential growth of emergent vegetation. If the depth is too great, there will be inadequate surface area to support photosynthetic action; also, deep ponds tend to stratify during hot periods. The suggested design guidelines for depth are shown in Table 12.1.

Table 12.1. Design Guidelines for Depth

Case	Depth	Related Conditions
1	1 meter	Generally ideal condition, very uniform temperature, tropical to subtropical, minimum settleable solids.
2	1.25 meters	Same as above but with modest amounts of settleable solids. Surface design based on one-meter depth and 0.25 meters used for reserve volume.
3	1.5 meters	Same as case 2 except for significant seasonal variation in temperature, major fluctuations in daily flow. Surface design based on one meter of depth.
4	2 meters and greater	For soluble wastewaters that are slowly biodegradable and retention is controlling.

For wastewaters containing considerable amounts of biodegradable settleable solids, Gloyna (1976) recommends that an anaerobic pond precede the facultative stabilization pond. The high temperature coefficient, 1.085, indicates that pond performance is very sensitive to temperature changes. To determine the design pond loading for an industrial wastewater for which no field data exists, Gloyna (1968) and Eckenfelder and Ford (1970) have developed laboratory procedures using bench-scale stabilization ponds.

Ponds should be located so that the prevailing winds passing over the ponds are not directed towards populated areas where possible odors would create a problem. The pond bottom and sides should be relatively impervious and, if the soil is porous, a clay liner should be used. The inlet to a pond should be designed so that the influent is distributed outward into the pond, and outlet structures should be baffled to prevent floating materials from leaving with the effluent. The earthen dikes usually have side slopes of 1:3 or less, and the inside slope is protected against wave erosion by riprap. The outside slopes should be sodded to prevent erosion due to surface runoff. Surface drainage must be excluded by dikes or ditching. At least 3 feet of freeboard should be provided above the maximum water surface, and the ponds should be designed so that the water level may be varied by 6 inches to assist in mosquito control.

Aerated Lagoons

Aerated lagoons may be classified as aerobic or facultative lagoons according to their dissolved oxygen profile. In aerobic lagoons, the oxygen furnished is sufficient to maintain dissolved oxygen throughout the pond depth, and the mixing is sufficient to keep the biological solids in suspension. In the aerobic lagoon, final clarification is provided by a settling tank or a facultative stabilization pond. The facultative lagoon has dissolved oxygen in the upper depths of the lagoon; however, no dissolved oxygen is present in the lower depths. In the facultative lagoon, the mixing is insufficient to maintain all of the biological solids in suspension, and solids deposition occurs on the lagoon bottom. These solids undergo anaerobic decomposition and are removed at infrequent intervals, such as every few years. The facultative lagoon usually has no final clarification except that which occurs in the aerated lagoon itself; consequently, the solids content in the effluent often precludes their use. In the following discussion, the fundamentals and application of the aerobic aerated lagoon are presented.

Aerated lagoons are widely used in the treatment of biodegradable organic industrial wastewaters because they occupy less land area than stabilization ponds and have less construction and operating costs than activated sludge plants. However, they have appreciable land area requirements when compared to activated sludge plants. The cost of an aerated lagoon system is between that of a stabilization pond system and an activated sludge plant. The oxygen requirements for an aerobic lagoon are furnished by artificial aeration and, in particular, mechanical surface aerators such as the floating pump-type are used. Occasionally, perforated pipes laid on the lagoon bottom to disperse compressed air are employed. Earthen dikes are used to form a lagoon and to exclude surface runoff.

The aerated lagoon is essentially an activated sludge system without recycle; thus, biochemical kinetics may be used for design formulations. Because the biological solids in lagoons do not vary appreciably, the pseudo–first-order reaction representing the rate of removal is

$$-\frac{dS_t}{dt} = KS_t \qquad (12.3)$$

where

dS_t/dt = rate of substrate utilization, mass/(volume) × (time);
K = reaction rate constant;
S_t = substrate concentration at any time, mass/volume.

A material balance on the substrate is given by

$$[\text{Accumulation}] = [\text{Input}] - \begin{bmatrix} \text{Decrease} \\ \text{due to} \\ \text{reaction} \end{bmatrix} - [\text{Output}] \qquad (12.4)$$

Assuming the system is completely mixed and using the designations shown in Figure 12.2 gives

$$V(dS_t) = QS_t dt - V[dS_t]_{\text{Growth}} - QdS_t dt \qquad (12.5)$$

From Eq. (12.3), $[dS_t]_{\text{Growth}} = KS_t dt$; thus, substituting this expression in Eq. (12.5) gives

$$V(dS_t) = QS_t dt - VKS_t dt - QS_t dt \qquad (12.6)$$

Dividing Eq. (12.6) by Vdt yields

$$\frac{dS_t}{dt} = \left(\frac{Q}{V}\right)S_t - KS_t - \left(\frac{Q}{V}\right)S_t \qquad (12.7)$$

Since the accumulation term $(dS_t/dt) = 0$ for steady state and $(V/Q) = \theta_t$, Eq. (12.7) may be rearranged to give the following formulation (Eckenfelder, W. W., 1970; Eckenfelder, W. W., & Ford, D. L., 1970):

$$\frac{S_t}{S_i} = \frac{1}{1 + K\theta_t} \qquad (12.8)$$

which is the design equation for an aerated lagoon. Experimental studies can be performed using a completely mixed activated sludge unit without recycle to obtain the reaction rate constant, K. The unit should be operated at a detention time of several days until acclimation is attained and the substrate and biological solids concentrations in the effluent are measured. The flow rate is then increased and, once steady state occurs, the substrate and biological solids in the effluent are again measured. Four or five increases in the flow rate should be made to obtain sufficient data for plotting. To evaluate the data, Eq. (12.8) may be rearranged to give

$$K = \frac{S_i - S_t}{S_t\theta_t} \qquad (12.9)$$

Thus, plotting $S_i - S_t$ on the y-axis versus $S_t\theta_t$ on the x-axis, as in Figure 12.3, will give a straight line with a slope $= K$. To evaluate the data to determine the yield coefficient, Y, and the endogenous decay constant, k_e, as shown in Figure 12.4, the following equation developed by Reynolds and Yang (1966) as well as other researchers (Metcalf and Eddy, 1979; Wu, Y. C., & Kao, D. F., 1976) may be used:

$$\frac{S_i - S_t}{\overline{X}\theta_t} = \frac{k_e}{Y} + \left(\frac{1}{Y}\right)\frac{1}{\theta_c} \qquad (12.10)$$

Figure 12.3. Graph for Determining K

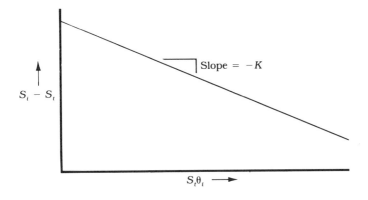

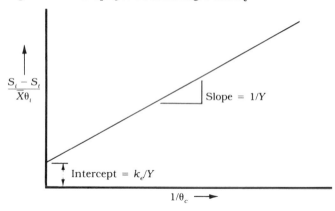

Figure 12.4. Graph for Determining Y and k_e

where

$\overline{X}$ = average cell mass concentration during the reaction;

θ_t = detention time based on the influent flow;

Y = yield coefficient, mass of cells produced per mass of substrate used;

k_e = endogenous decay constant, time^{-1};

θ_c = mean cell residence time, time.

The mean cell residence time, θ_c, is equal to the detention time for a nonrecycle reactor. A plot of $(S_i - S_t)/(\overline{X}\theta_t)$ on the y-axis versus $1/\theta_c$ on the x-axis will yield a straight line with a slope of $1/Y$ and a y-axis intercept of k_e/Y, as shown in Figure 12.4.

Since the biological solid concentration in an aerated lagoon is relatively low, the water temperature has a significant effect upon lagoon performance. The variation of the rate constant over a temperature range of 10 to 30°C has been found to be represented by

$$K_{T_2} = K_{T_1} \cdot \theta^{(T_2 - T_1)} \tag{12.11}$$

where

K_{T_2} = rate constant at temperature, T_2 (°C);

K_{T_1} = rate constant at temperature, T_1 (°C);

θ = temperature correction coefficient, θ = 1.06 to 1.10 (Eckenfelder, W. W., & Ford, D. L., 1970).

The operating temperature of a pond depends on the temperature of the influent wastewater, the heat loss by convection, radiation, and evaporation, and the heat gain by solar radiation. Mancini and Barnhart (1968) have found the heat loss by evaporation and the heat gain by solar radiation to be small compared to other terms. Their findings show that the temperature of a lagoon is given by

$$T_i - T_w = \frac{fA(T_w - T_a)}{Q} \qquad (12.12)$$

where

T_i = influent wastewater temperature, °F;
T_w = temperature of water in the lagoon, °F;
T_a = temperature of air, °F;
A = lagoon area, ft^2;
Q = influent wastewater flow, MGD;
f = experimental factor.

They found the value of the factor f to be 12×10^{-6} for the central and eastern part of the United States and 20×10^{-6} for the Gulf states. Equation (12.12) may also be used to obtain the temperature of stabilization ponds.

The total oxygen requirements of a lagoon are given by the following equation:

$$O_r = Y'S_r + k_e'\overline{X} + O_n \qquad (12.13)$$

where

O_r = total oxygen required, lb/day;
Y' = oxygen coefficient, pounds of oxygen required per pound of BOD$_5$ or COD removed;
S_r = substrate removed per day, pounds of BOD$_5$ or COD removed;
k_e' = endogenous oxygen coefficient, pounds of O$_2$ required per pound of cells decayed-day;
$\overline{X}$ = mass of biological solids in the lagoon, lb;
O_n = oxygen required for nitrification, lb.

The term k_e' may be assumed to equal 1.42 k_e, where k_e is the endogenous decay coefficient as time^{-1}. Frequently, the term $k_e'\overline{X}$ is very small compared to $Y'S_r$ and, in such cases it may be neglected. The oxygen requirements should be at least 1.6 lb oxygen per lb BOD$_5$ applied to the lagoon. The oxygen furnished by artificial aeration may be determined by equations presented in Chapter 10. The power level supplied by artificial aeration should be at least 0.03 hp per 1000 gal to insure maintaining the biological solids suspension.

Since aerated lagoons are completely mixed activated

sludge systems without recycle, they may be also designed by the appropriate equations presented in Chapter 9. These kinetic equations utilized the Monod expression in their derivation.

Aerated lagoons usually have a rectangular layout except where the terrain precludes the use of this geometry. If rectangular in layout, the length to width ratio is usually 2:1. At least two lagoons should be provided for flexibility and piping should be arranged so that parallel or series operation is possible. Surface drainage must be excluded by dikes and ditching. The earthen dikes forming the lagoon usually have side slopes of 1:3 or less and riprap or other suitable protection should be used to avoid wave erosion. The outside slopes should be sodded to avoid erosion due to rainfall runoff. The lagoon bottom and sides should be provided with a clay liner as a sealer if the soil is pervious. Inlets must distribute the influent outward into the lagoon and outlets should be baffled to avoid floating matter from leaving the effluent. A freeboard of at least 3 feet should be provided. Concrete pads must be furnished under each aerator to protect the bottom from erosion. Final clarifiers or stabilization ponds should be provided to remove biological solids from the final effluent. Lagoon depths are from 8 to 16 ft.

References

AWARE, Inc. 1974. *Process Design Techniques for Industrial Waste Treatment*, edited by C. E. Adams and W. W. Eckenfelder, Jr. Nashville, Tenn.: Enviro Press.

Conway, R. A., and Ross, R. D. 1980. *Handbook of Industrial Waste Disposal*. New York: Van Nostrand Reinhold.

Eckenfelder, W. W., Jr. 1980. *Principles of Water Quality Management*. Boston: CBI Publishing.

Eckenfelder, W. W., Jr. 1970. *Water Quality Engineering for Practicing Engineers*. New York: Barnes and Noble.

Eckenfelder, W. W., Jr., and Ford, D. L. 1970. *Water Pollution Control*. Austin, Tex.: Pemberton Press.

Gloyna, E. F. 1968. Basis for Waste Stabilization Pond Designs. In *Advances in Water Quality Improvement*, edited by E. F. Gloyna and W. W. Eckenfelder, Jr. Austin, Tex.: University of Texas Press.

Gloyna, E. F. 1976. Facultative Waste Stabilization Pond Design. In *Ponds as a Wastewater Treatment Alternative*, edited by E. F. Gloyna, J. F. Malina, Jr., and E. M. Davis. Austin, Tex.: University of Texas Press.

Gloyna, E. F. 1965. Waste Stabilization Pond Concepts and Experiences. Lecture presented at the Tenth Summer Institute in Water Pollution Control, Manhattan College, Bronx, N.Y.

Gloyna, E. F., and Hermann, E. R. 1958. Waste Stabilization Ponds. Reprint no. 74, Bureau of Engineering Research, University of Texas, Austin, Tex.

Great Lakes–Upper Mississippi Board of State Sanitary Engineers. Recommended Standards for Sewage Works. Ten state standards, Albany, N.Y.

Mancini, J. L., and Barnhart, E. L. 1968. Industrial Waste Treatment in Aerated Lagoons. In *Advances in Water Quality Improvement*, edited by E. F. Gloyna and W. W. Eckenfelder, Jr. Austin, Tex.: University of Texas Press.

Metcalf and Eddy, Inc. 1979. *Wastewater Engineering, Treatment, Disposal and Reuse*. New York: McGraw-Hill.

Ramalho, R. S. 1977. *Introduction to Wastewater Treatment Processes*. New York: Academic Press.

Reynolds, T. D., and Yang, J. T. 1966. Model of the Completely Mixed Activated Sludge Process. *Proceedings of the 21st Annual Purdue Industrial Waste Conference*. Part 2.

Sawyer, C. N. 1968. New Concepts in Aerated Lagoon Design and Operation. In *Advances in Water Quality Improvement*, edited by E. F. Gloyna and W. W. Eckenfelder, Jr. Austin, Tex.: University of Texas Press.

Schroeder, E. D. 1977. *Water and Wastewater Treatment*. New York: McGraw-Hill.

Sundstrom, D. W., and Klei, H. E. 1979. *Wastewater Treatment*. Englewood Cliffs, N.J.: Prentice-Hall.

Texas Department of Health Resources. 1976. *Design Criteria for Sewerage Systems*. Austin, Tex.

Wu, Y. C., and Kao, D. F. 1976. Yeast Plant Wastewater Treatment. *Jour. SED* 102, no. EE5:969.

Problems

1. A facultative oxidation pond facility is to be designed for a population of 20,000 persons and is to have primary treatment prior to the ponds. The facility is to have two ponds in parallel and the city is located in the Gulf Coast area. Pertinent data are: BOD_5 of raw wastewater $= 200$ mg/ℓ, 35 percent BOD_5 removal by primary treatment, winter wastewater temperature $= 69°F$, average January air temperature $= 55°F$, wastewater flow $= 100$ gal/cap-day, pond depth $= 3.5$ ft, length, L, to width, W, ratio $= 2:1$, and $BOD_5 = 0.68\ BOD_u$. Determine:
   a. The operating temperature of the ponds.
   b. The total pond area, acres.
   c. The detention time, days.
   d. The surface loading, lb BOD_5/acre-day.
   e. The dimensions of the ponds.

2. An aerobic aerated lagoon system is to be designed to treat a soluble organic industrial wastewater having a flow of 1.5 MGD. The COD of the wastewater is 1020 mg/ℓ and the plant effluent is to have a COD not more than 40 mg/ℓ. A bench-scale laboratory study has been performed using a completely mixed reactor without recycle and the data obtained at 23°C are shown in Table 12.2.

Table 12.2.

Detention time (hr) θ	Influent COD (mg/ℓ) S_i	Effluent COD (mg/ℓ) S_t	Biological Solids (mg/ℓ) X_t
11.20	1003	47	409
9.00	1003	58	401
6.00	1003	80	388
3.75	1003	162	358
3.40	1003	220	359
2.80	1003	303	324

The COD after ten days of aeration was 14 mg/ℓ and may be assumed to be the nondegradable COD. Other pertinent data are: wastewater temperature = 82°F, January average air temperature = 55°F, July average air temperature = 83°F, plant location is in the Gulf states, aerated lagoon depth = 8.0 ft, number of lagoons = 2, the lagoons are rectangular in layout, length:width = 2:1, final clarification is to be furnished by settling tanks, organic nitrogen in wastewater = 32.2 mg/ℓ, ammonia nitrogen in wastewater = 2.9 mg/ℓ, oxygen furnished by surface aerators = 1.5 lb O$_2$/hp-hr, BOD$_5$ of the untreated wastewater = 42 percent of the COD and 3.84 lb O$_2$ required per pound of nitrogen nitrified, and MLVSS = 88 percent of MLSS. Determine:

a. The reaction constant, K, based on biodegradable COD.

b. The yield coefficient, Y, and the endogenous decay constant, k_e.

c. The oxygen coefficient, Y', and the endogenous oxygen coefficient, k'_e.

d. The detention time in days for winter conditions.

e. The length and width of each lagoon.

f. The soluble COD in the effluent for summer conditions.

g. The percent nitrification based on the mean cell residence time.

h. The pounds of oxygen required per day for each lagoon based on summer conditions.

i. The horsepower required for mechanical surface aerators for each lagoon.

j. The horsepower required per 1000 gal volume.

k. The pounds of oxygen required per pound of BOD$_5$ applied.

3. A facultative oxidation pond facility is to be designed for a population of 10,000 persons and is to have primary treatment prior to the pond. The facility will have

two ponds operated in parallel after primary treatment. Pertinent data are: BOD_5 of the raw wastewater = 200 mg/ℓ, 35 percent BOD_5 removal by the primary clarifier, wastewater flow = 100 gal/cap-day, pond depth = 3.5 ft, surface loading = 35 lb BOD_5/acre-day, and length, L, to width, W, ratio = 2:1. Determine:

a. The total pond area, acres.

b. The dimensions of the ponds.

c. The detention time of the ponds, days.

4. An aerated lagoon is to be designed for a pulp and paper mill wastewater. After coagulation, flocculation, and settling, the waste has a BOD_5 of 160 mg/ℓ. Pertinent data are: flow = 38,000 m^3/day, BOD_5 after treatment = 10 mg/ℓ, K = 2.4 day^{-1} at 20°C, minimum operating temperature = 12.8°C, θ = 1.080, and pond depth = 3.65 m. Determine:

a The detention time, days.

b. The area of the ponds, km^2.

5. A municipality of 4500 persons has a wastewater flow of 360 ℓ/cap-day and the raw wastewater has a COD of 380 mg/ℓ. A wastewater stabilization pond is to treat the wastewater after preliminary treatment. The average January temperature is 8.8°C. Determine the pond area in m^2 and the depth in m.

Chapter Thirteen

ANAEROBIC DIGESTION

Anaerobic digestion, which is the most common sludge treatment, may be defined as the biological oxidation of degradable organic sludges by microbes under anaerobic conditions. Most of the microbes used in the process are facultative anaerobes; however, some, such as the methane-producing bacteria, are obligate anaerobes. The term *sludge* refers to the solids that settle and are removed when a liquid with suspended solids is passed through a settling tank. Organic sludges may originate from several sources in a wastewater treatment plant, and each sludge has its own characteristics, such as solids concentration and ease of biodegradation. The various sludges from wastewater treatment plants are as follows: (1) *raw* or *primary sludge* is sludge from the primary settling of untreated wastewater, (2) *waste activated sludge* is excess sludge that is produced from the activated sludge process, (3) *trickling filter secondary sludge* or *humus* is sludge that originates from the secondary settling of trickling filter effluent, (4) *secondary sludge* is sludge from the secondary clarifier of an activated sludge or trickling filter plant, (5) *fresh sludge* refers to untreated organic sludges, (6) *digested sludge* is sludge that has undergone biological oxidation, and (7) *dewatered sludge* is sludge that has had the major portion of its water removed.

All organic sludges undergo considerable changes in their physical, chemical, and biological properties during anaerobic digestion. Before digestion the volatile solids will be quite high, depending on the type of sludge. Primary sludge and waste activated sludge usually have 65 to 75 percent volatile solids, whereas trickling filter humus has 45 to 70 percent. The dry solids are usually 4 to 6 percent for a mixture of primary and secondary sludge, and the specific gravity is about 1.01. It is difficult to separate the water from the solids and the solids will readily decompose under anaerobic conditions. The color is usually tan. The fuel value of undigested sludges is from about 6500 to 8000 Btu per lb dry solid.

After digestion the volatile solids are usually reduced to 32 to 48 percent. The dry solids are usually from 8 to 13 percent and the specific gravity is from 1.03 to 1.05. The water will readily separate from the solids and the solids are stable and not degradable. The color will be blackish

Two-Stage Anaerobic Digester System. Small building between the digesters houses sludge heaters, recirculation pumps, piping, valves, and other appurtences. Mixing of the first stage is provided by gas recirculation.
Courtesy of Walker Process Corporation

and the odor will be tarlike. The fuel value of digested sludges is from about 3500 to 4000 Btu per lb dry solid. Approximately 99.8 percent of the coliforms in municipal primary sludges are destroyed during digestion.

Chemically precipitated sludges, such as from an advanced wastewater treatment plant (chemical-physical treatment), usually will digest anaerobically as well as primary, waste activated sludge, and trickling filter sludges if the pH range is satisfactory.

Frequently, sludge is thickened prior to digestion because it reduces the daily fresh sludge volume, thus decreasing the digester size and the amount of supernatant liquor. For instance, if a fresh sludge with 4 percent solids content is thickened to 8 percent, then the thickened sludge will be half the original volume.

Anaerobic digesters are usually constructed using cylindrical reinforced concrete tanks from 12 to 36 ft deep and from 15 to 125 ft in diameter. The most common depth is from 18 to 24 ft. The reinforced concrete floor is shaped like an inverted cone and slopes to the center of the tank where the discharge pipe for digested sludge is located.

Process Fundamentals

Anaerobic digestion employs microbes that thrive in an environment in which there is no molecular oxygen and there is a substantial amount of organic matter. The organic material is a food source for the microbes, and they convert it into oxidized materials, new cells, energy for their life processes, and some gaseous end products, such as methane and carbon dioxide. The generalized equation for anaerobic action is

$$\begin{array}{l} \dfrac{\text{Organic}}{\text{matter}} + \dfrac{\text{Combined}}{\text{oxygen}} \xrightarrow[\text{microbes}]{\text{Anaerobic}} \dfrac{\text{New}}{\text{cells}} \\ + \dfrac{\text{Energy for}}{\text{cells}} + CH_4 + CO_2 + \dfrac{\text{Other end}}{\text{products}} \quad (13.1) \end{array}$$

The sources of combined oxygen include the radicals of CO_3^{-2}, SO_4^{-2}, NO_3^{-1}, and PO_4^{-3}. Some of the other end products include the gases H_2S, H_2, and N_2.

The microbial action during anaerobic digestion consists of three stages: (1) liquefaction of solids, (2) digestion of the soluble solids, and (3) gas production. Digestion is accomplished by two groups of microorganisms: (1) the organic-acid–forming heterotrophs and (2) the methane-producing heterotrophs. The organic-acid–forming heterotrophs use the complex organic substrates, such as carbohydrates, proteins, fats, oils, or their degradation products and produce organic fatty acids, primarily acetic and propionic with some butyric and valeric acids. Carbohydrates include sugars, starches, and cellulose or plant fiber. Proteins occur in all raw animal and plant food materials—in particular, animal tissue has a high protein content. Fats and oils are similar in structure; however, fats are solids at ordinary temperatures, while oils are liquids. Fats and oils include animal and vegetable oils and grease from animal tissue. The breakdown of carbohydrates is

$$\text{Carbohydrates} \rightarrow \frac{\text{Simple}}{\text{sugars}} \rightarrow \frac{\text{Alcohol}}{\text{Aldehydes}} \\ \rightarrow \frac{\text{Organic}}{\text{acids}} \qquad (13.2)$$

The breakdown of proteins is

$$\text{Proteins} \rightarrow \text{Amino acids} \rightarrow \text{Organic acids} + NH_3 \quad (13.3)$$

The breakdown of animal or vegetable fats and oils is

$$\begin{matrix}\text{Fats} \\ \text{and} \\ \text{Oils}\end{matrix} \rightarrow \text{Organic acids} \qquad (13.4)$$

In Eqs. (13.2), (13.3), and (13.4), the final breakdown products of carbohydrates, proteins, and fats and oils are organic fatty acids frequently called *volatile acids.* Most of the organic-acid–forming bacteria are soil microorganisms and are facultative anaerobes. McKinney (1962) lists the following species as being common in anaerobic digestion of municipal sludges: *Pseudomonas, Flavobacterium, Alcaligenes, Escherichia,* and *Aerobacter.* These microbes thrive in a relatively wide pH range. The methane-producing heterotrophs use the organic acids produced by the acid formers as substrates and produce methane and carbon dioxide. The methane producers grow more slowly than the acid formers and require a rather narrow pH range of about 6.7 to 7.4. The action of the methane bacteria is

$$\text{Organic acids} \rightarrow CH_4 + CO_2 \qquad (13.5)$$

The gas produced in a properly operating digester is about 55 to 75 percent methane, 25 to 45 percent carbon dioxide, and has trace amounts of such gases as hydrogen sulfide, hydrogen, and nitrogen. Some of the carbon dioxide from the microbial action reacts with the water present to establish the bicarbonate buffering system, which is important in the digestion process. In a properly operating digester, the organic acids are utilized as rapidly as they are produced. The bicarbonate buffering system gives flexibility in operation because, up to a certain limit, the acids may be temporarily produced faster than they are broken down and the buffering system will maintain the pH in the proper range. If, however, excessive amounts of the acids are produced, the buffering system will be overcome and an unwanted pH drop will result which will inhibit the methane producers. The methane producers are strict anaerobes; McKinney (1962) lists the principal species as *Methanococcus, Methanobacterium,* and *Methanoscarcina.* Figure 13.1 shows the pathway for methane production from organic solids found in municipal wastewaters. The figure shows that about 72 percent of the methane is produced from acetic acid. Figure 13.2 depicts the relationship be-

Figure 13.1. Pathways for Methane Production from Complex Wastes such as Municipal Wastewater Sludges. (Percentages are COD conversion.)
Adapted from "Anaerobic Waste Fundamentals, Part 1" by P. L. McCarty in *Public Works Magazine* 95, no. 9 (September 1964):107. Reprinted by permission.

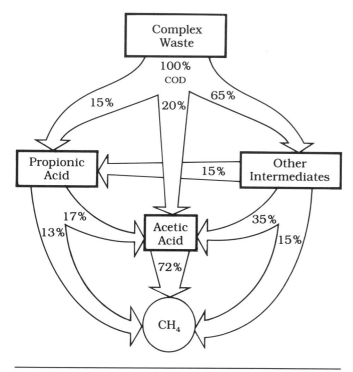

Figure 13.2. Relationship between pH, Bicarbonate Concentration, and CO₂ Content near 95°F
Adapted from "Anaerobic Waste Fundamentals, Part 2" by P. L. McCarty in *Public Works Magazine* 95, no. 10 (October 1964):123. Reprinted by permission.

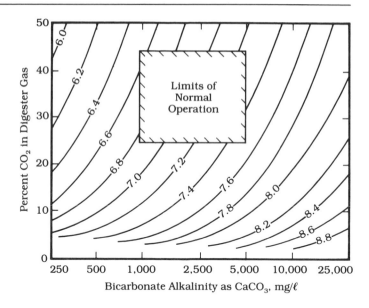

Figure 13.3. Schematic Diagrams of Low-Rate and High-Rate Anaerobic Digesters

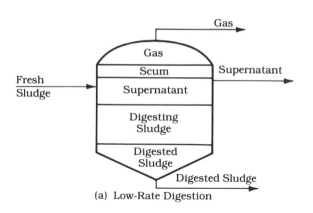

(a) Low-Rate Digestion

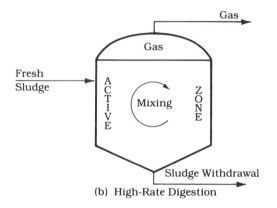

(b) High-Rate Digestion

tween the bicarbonate alkalinity, pH, and the amount of carbon dioxide in the digester gas, and also shows the range for normal operation.

A schematic diagram of a *conventional* or *low-rate digester* and a *high-rate digester* is shown in Figure 13.3 (a) and (b). Conventional or low-rate digesters have intermittent mixing, intermittent sludge feeding, and intermittent sludge withdrawal. When mixing is not being done, the digester contents are stratified, as shown in Figure 13.3 (a). High rate digesters have continuous mixing, as shown in Figure 13.3 (b), and continuous or intermittent sludge feeding and sludge withdrawal. If a high-rate digester is a single-stage digester, usually mixing is stopped and the contents are allowed to stratify before digested sludge and supernatant are withdrawn. Typical design and operational parameters for low- and high-rate digesters are shown in Table 13.1. Most digesters, both low and high rate, operate in the mesophilic temperature range. Table 13.2 shows parameters for mesophilic anaerobic digestion. Low-rate digesters are semicontinuous-flow biological reactors without recycle. High-rate digesters, however, are continuous-flow biological reactors without any recycle.

Table 13.1. Typical Design and Operational Parameters for Standard-Rate and High-Rate Digesters

Parameter	Low Rate	High Rate
Digestion Time, days	30 to 60	10 to 20
Organic Solid Loading, lb vss/ft^3-day	0.04 to 0.10	0.15 to 0.40
Volume Criteria, ft^3/ capita		
a. Primary Sludge	2 to 3	1⅓ to 2
b. Primary Sludge and Trickling Filter Sludge	4 to 5	2⅔ to 3⅓
c. Primary Sludge and Waste Activated Sludge	4 to 6	2⅔ to 4
Mixture of Primary and Secondary Sludge Feed Concentration, Percent Solids (dry basis)	2 to 5	4 to 6
Digester Underflow Concentration, Percent Solids (dry basis)	4 to 8	4 to 6

Adapted from Environmental Protection Agency, *Sludge Treatment and Disposal*. EPA Process Design Manual, Washington, D.C.

Table 13.2. Conditions for Mesophilic Anaerobic Digestion

Temperature	
Optimum	98°F(35°C)
Usual range of operation	85°–98°F
	(29.4°–35.0°C)
pH	
Optimum	7.0–7.2
Usual limits	6.7–7.4
Solids Reduction	
Volatile solids	50–75 percent
Suspended solids	35–50 percent
Gas Production	
Per pound of volatile solids added	6–12 ft^3
Per pound of volatile solids destroyed	12–18 ft^3
Per pound of volatile solids destroyed	1.05–1.75 lb
Per capita served (primary sludge only)	0.6–0.8 ft^3/day
Per capita served (primary plus secondary sludge)	1.0–1.2 ft^3/day
Gas Composition	
Methane	55–75 percent
Carbon dioxide	25–45 percent
Hydrogen sulfide	trace
Hydrogen	trace
Nitrogen	trace
Gas fuel value	530–730 Btu/ft^3
Volatile Acids Concentration as Acetic Acid	
Normal operation	50–250 mg/ℓ
Maximum	approx. 2000 mg/ℓ
Alkalinity Concentrations as CaCO$_3$	
Normal operation	1000–5000 mg/ℓ

Low-Rate Digesters

Low-rate or conventional digesters have digestion times of 30 to 60 days, organic solid loadings of 0.04 to 0.10 lb vss/ft^3-day, intermittent mixing, intermittent feeding and

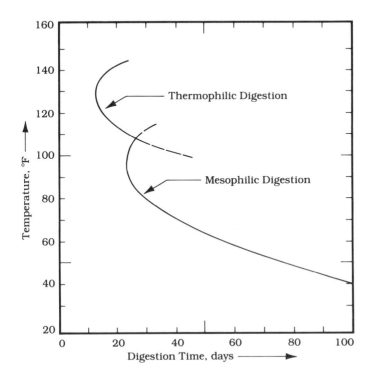

Figure 13.4. Digestion Time versus Temperature for Conventional Digesters. (Time required for 90 percent digestion of municipal wastewater sludges.)
Adapted from *Water Purification and Wastewater Treatment and Disposal*, Vol. 2, *Water and Wastewater Engineering* by G. M. Fair, J. C. Geyer, and D. A. Okun. Copyright © 1968 by John Wiley and Sons, Inc. Reprinted by permission.

sludge withdrawal, and are in a stratified state except when mixing is being done.

The digestion time required to digest 90 percent of the degradable solids in primary sludge as a function of the digestion temperature is shown in Figure 13.4. The mesophilic range extends up to about 108°F (42.2°C), whereas the thermophilic range is above 108°F. The digestion time for mesophilic digestion decreases with an increase in temperature up to the optimum temperature of 98°F (35°C). This increase in digestion rate with an increase in temperature should be expected because it is a microbial process. Above 98°F the mesophilic digestion time increases. In cold weather, digesters are heated to near the optimum temperature; the heated range is usually from 85°F to 100°F (29.4

to 37.8°C). For the thermophilic range, the optimum temperature is 130°F (54.4°C). Attempts have been made to use thermophilic digestion; however, it has not been successful because the thermophiles are very sensitive to environmental changes.

Conventional or low-rate digesters are single stage and may have either floating covers or fixed covers. Standard diameters for floating covers are from 15 to 125 ft and are available in 5-ft intervals. Figure 13.5 shows a section through a digester with a floating cover, along with details showing the piping required. For a floating cover digester,

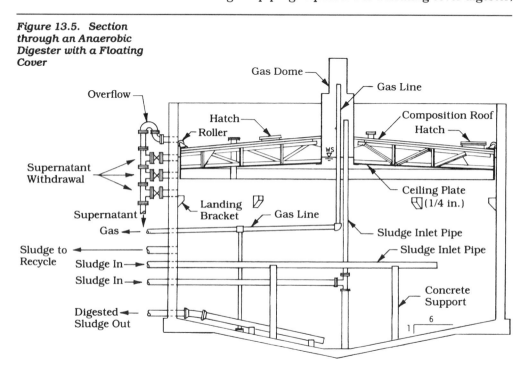

Figure 13.5. Section through an Anaerobic Digester with a Floating Cover

Figure 13.6. Floating Cover Digester: Sludge Being Added

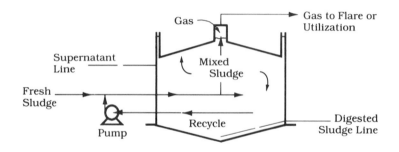

ENVIRONMENTAL ENGINEERING

when sludge is being added, recycle is done to seed the incoming fresh sludge with digesting sludge, as shown in Figure 13.6. The fresh and recycled sludge is usually added in the middle of the tank and in the gas dome of the cover. The discharge in the gas dome breaks up any scum or grease that has floated to the gas-water interface. While sludge is added, no supernatant or digested sludge is withdrawn, and fresh sludge is usually added daily. After the addition of fresh sludge into the tank, the recycle is stopped and the mixture stratifies, with the digesting sludge in the bottom portion of the digester and the supernatant in the top portion (see Figure 13.7). The water given up during digestion, the supernatant liquor, is withdrawn from the digester every few days. The digested sludge is usually withdrawn about every two weeks. If open drying beds are used and it is rainy weather, the sludge is not withdrawn, but instead is kept in the digester until suitable weather occurs. The grease that rises to the top of the supernatant is kept submerged by the cover to assist in breakdown by bacterial action. The gas produced during digestion is collected in the dome at the center of the floating cover. If sludge is added over a period of time and no digested sludge or supernatant is withdrawn, the floating cover rises. In a similar manner, when digested sludge or supernatant is withdrawn, the cover will slowly drop.

The fixed cover digester is not as flexible in operation as the floating cover type because there is a limit to the amount of fresh sludge that may be added over a given period of time. Also, there is a limit to the amount of digested sludge or supernatant that may be removed at one time. When fresh sludge is added and no supernatant or digested sludge is withdrawn, the gas will be compressed and the maximum allowable pressure is about 8 inches of water column. When sludge or supernatant is withdrawn, the gas

Figure 13.7. Floating Cover Digester: Sludge Digesting

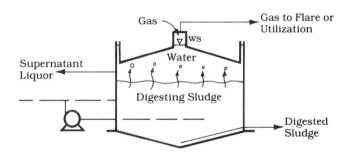

expands and the pressure reduces, with the minimum allowable pressure being about 3 inches of water column. Fixed cover digesters have more problems with grease floating to the top of the supernatant and drying than the floating cover type. When the grease dries, there is very little biochemical breakdown and the grease accumulates and frequently causes operational problems, such as plugging of the supernatant outlets. Due to their operational problems, many states do not allow fixed cover digesters to be used except when the design population is less than about 10,000 persons.

The capacity of the recirculation or recycle pump used to provide mixing should be sufficient to pump one tank volume in 30 minutes. Although many conventional digesters do not have any method for mixing, they are subject to frequent acidification and are not considered good engineering practice. At least three supernatant withdrawal pipes should be provided along with supernatant sampling petcocks. The withdrawal pipes should be at least 2.0 ft apart in elevation in order to provide selective withdrawal.

High-Rate Digesters

High rate digesters have digestion times of 10 to 20 days, organic solid loadings of 0.15 to 0.40 lb vss/day-ft^3, continuous mixing, and continuous or intermittent sludge feeding and sludge withdrawal, and the contents are in a homogeneous state.

In conventional digestion the mixing or recirculation system is usually operated only when fresh sludge solids are added, which ordinarily is less than a one-hour period during each day. Thus, most of the time the digester contents are stratified and the digesting sludge occupies approximately the bottom half of the digester. In high-rate digestion the mixing is continuous; thus, the entire digester volume is available for digesting sludge and the mixing provides better contact between the seeded sludge and the fresh solids that have been added. This allows a high-rate digester to operate at organic loadings (that is, the pounds of volatile solids added/day-ft^3) of several magnitudes greater than conventional digesters. Also, the detention times are much shorter than for conventional digesters. The digestion times from Figure 13.4 are for conventional or low-rate digesters; therefore, the curves are not applicable for high-rate digestion. As fresh sludge is added to a high-rate digester, the digested sludge may be displaced into a holding tank where the supernatant liquor is separated from the sludge. An alternate method for removing sludge or supernatant liquor is to stop the mixing and allow the contents to stratify. Once stratification has occurred, the withdrawals can be made. Most high-rate digesters have an additional mixing system other than a sludge recycle pump. Typical mixing systems are shown in Figure 13.8. Mixing may be provided, as shown in Figure 13.8 (a), by recycling gas to a draft tube mounted in the center of the digester. As the gas rises, it causes sludge to be lifted through the draft tube and discharged from the top of the tube. Mixing may also be provided by recycling gas, as shown in Figure 13.8 (b), and discharging it near the bottom of the digester by gas diffusers. Also, the gas may be discharged at about one-half the tank depth by pipes suspended from the ceiling. Mixing may be accomplished, as shown in Figure 13.8 (c), by an impeller mounted with a draft tube in the center of the digester. If the tank diameter is greater than about 50 ft, three or more draft tubes and turbine impellers are required. If three draft tubes and impellers are used, they

Figure 13.8. Types of Mixing Systems for Anaerobic Digesters

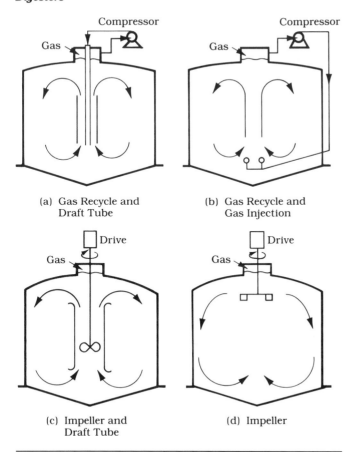

(a) Gas Recycle and Draft Tube

(b) Gas Recycle and Gas Injection

(c) Impeller and Draft Tube

(d) Impeller

*Anaerobic Digester with
Impeller and Draft Tube for
Mixing*
Courtesy Dorr-Oliver, Inc.

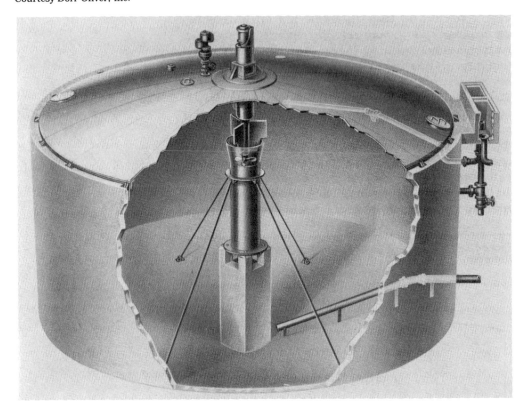

are placed the same distance from the center of the digester
and are located 120 degrees apart in plan view. Mixing has
also been done using turbine impellers mounted in the upper depths of a digester, as shown in Figure 13.8 (d). Mixing
by gas recycle has an advantage over mechanical mixing
because there are no moving mechanical parts inside the
digester where the environment is very corrosive.

Two-Stage Digesters   A two-stage digester system such as shown in Figure 13.9
is usually provided when the design population is over
about 30,000 to 50,000 persons. In the first stage, the main
biochemical action is liquification of organic solids, digestion of soluble organic materials, and gasification. In the
second stage some gasification occurs; however, its main
use is supernatant separation, gas storage, and digested
sludge storage. The first stage is usually a high-rate digester
employing a fixed cover and continuous mixing, while the
second stage is usually a conventional digester with a float-

Figure 13.9. Two-Stage Digestion

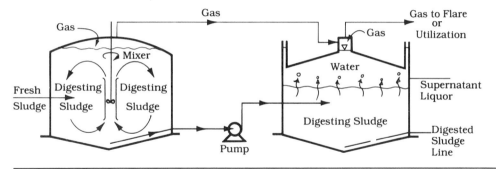

ing cover and intermittent mixing. The organic loading applied to the first stage is usually several magnitudes greater than the loading to the second stage.

Digester Operation

Most digesters are heated to 85°F to 100°F during cold weather in order to give a rapid digestion time. The digester gas produced can easily be used for heating purposes.

The optimum pH range of 7.0 to 7.2 may usually be maintained if the daily fresh sludge added is properly seeded and the sludge additions and withdrawals are not excessive. Usually, acidification will not occur if the dry solids added or withdrawn daily do not exceed 3 to 5 percent of the dry solids in the digester. Acidification is characterized by a drop in pH, inhibition of the methane bacteria, a decrease in gas production, and a decrease in the methane content of the sludge gas. Also, there may be bad odors, foaming, and floating sludge. Acidification may be temporarily controlled by adding lime to raise the pH; however, a permanent solution requires changing the environmental conditions so that the methane producers are not inhibited and proper digestion can occur.

If the daily additions of fresh sludge contain inhibitory substances, such as heavy metals, they can interfere with the digestion process. When this occurs the source of the inhibitory material must be eliminated prior to emptying the digester and restarting it.

The supernatant liquor is the water released during digestion; it may have a BOD_5 as high as 2000 mg/ℓ and a suspended solids concentration as high as 1000 mg/ℓ. Usually it is gradually fed back to the influent to the primary clarifiers.

The degree of digestion that is accomplished may be measured by the volatile solids reduction and the amount of sludge gas produced. Table 13.2 shows a summary of the general conditions for mesophilic sludge digestion and also gives the gas analyses and the amount of gas production in a properly operated anaerobic digester.

Digester Volume

The volume of digesting sludge in a digester is a function of the volume of fresh sludge added daily, the volume of digested sludge produced daily, and the required digestion time in days. Additional volume must be provided for the supernatant liquor, gas storage, and storage of digested sludge. The volume required for gas storage is usually relatively small compared to total digester volume.

Batch digestion experiments have shown that, if the supernatant is removed from a batch of digesting sludge as it is produced, the volume of the remaining digesting sludge versus digestion time is a parabolic function. For a parabolic function, the average volume is the initial volume minus two-thirds the difference between the initial and final volumes. Thus, the volume of digesting sludge, V_s, is given by (Fair, G. M., Geyer, J. C., & Okun, D. A., 1968)

$$V_s = [V_1 - \tfrac{2}{3} (V_1 - V_2)]t \tag{13.6}$$

where

V_s = volume of digesting sludge, ft^3;
V_1 = volume of fresh sludge added daily, ft^3/day;
V_2 = volume of digested sludge produced daily, ft^3/day;
t = digestion time, days.

The digestion time is a function of the tank operating temperature and can be estimated from Figure 13.4 for municipal sludges treated in conventional or low-rate digesters. The reduction in sludge volume during digestion is mainly due to the release of water from the sludge solids. The volume from Eq. (13.6) is for the digesting sludge; therefore, an additional volume must be provided for supernatant liquor, gas, and digested sludge storage.

The digester requirements of most state health departments are based on digestion volume per capita served. These criteria are for the total volume required, which includes digesting sludge, supernatant liquor, gas, and digested sludge storage. Typical values for low- and high-rate digesters for various types of municipal sludges are shown in Table 13.1.

Digesters may also be designed on the basis of the organic loading—that is, the pounds of volatile solids added per ft^3-day or the solids retention time. Typical organic solid loadings for low- and high-rate digesters are shown in Table 13.1. The solid retention time, θ_s, based on the solids produced per day in the digested sludge is given by

$$\theta_s = \frac{X}{\Delta X} \qquad (13.7)$$

where

θ_s = solid retention time, days;
X = pounds of dry solids in the digester;
ΔX = pounds of dry solids produced per day in the digested sludge.

The design of high-rate digesters is frequently done using the mean cell residence time, θ_c. For high-rate digesters, mixing is continuous; thus, the system is a completely mixed, continuous-flow biological reactor without recycle. Since the number of cells in the feed to the digester is negligible compared to the cells in the digester and the digested sludge flow, the mean cell residence time, θ_c, is approximately equal to the hydraulic residence time, θ_h, and also the solids retention time, θ_s. As the mean cell residence time decreases, a minimum time, θ_c^{Min}, will be reached at which the cells are washed from the system faster than they can multiply. Suggested values of the minimum mean cell residence time are given in Table 13.3 (McCarty, P. L., 1964).

Table 13.3. Suggested Values of the Minimum Mean Cell Residence Time

Temperature, °C	θ_c^{Min} (days)
18	11
24	8
30	6
35	4
40	4

Since the minimum cell residence time, θ_c^{Min}, is a critical condition, the design mean cell residence time, θ_c, is much longer than the minimum time. Usually, the design mean cell residence time is 2.5 times the minimum.

Example Problem 13.1
Low-Rate Digester

A heated, low-rate anaerobic digester is to be designed for an activated sludge plant treating the wastewater from 25,000 persons. The fresh sludge has 0.25 lb dry solids/cap-day, volatile solids are 70 percent of the dry solids, the dry solids are 5 percent of the sludge, and the wet specific gravity is 1.01. Sixty percent of the volatile solids are destroyed by digestion and the fixed solids remain unchanged. The digested sludge has 8 percent dry solids and a wet specific gravity of 1.03. The operating temperature is 95°F and the sludge storage time is 30 days. The sludge occupies the lower half of the tank depth and the supernatant liquor and the gas occupy the upper half of the tank depth. Determine the digester volume.

Solution

The fresh sludge solids represent an Input. Fresh solids = (25,000 person)(0.25 lb/cap-day) = 6250 lb/day. vss = (6250)(0.70) = 4375 lb/day. fss = (6250)(0.30) = 1875 lb/day. vss destroyed = (4375) (0.60) = 2625 lb/day. The digested sludge solids represent an Output. For steady state, [Output] = [Input] − [Decrease due to reaction]. Thus, vss in the digested sludge = 4375 − 2625 = 1750 lb/day. fss in the digested sludge = 1875 − 0 = 1875 lb/day. The total solids in the digested sludge = 1750 + 1875 = 3625 lb/day. The fresh sludge volume = (6250 lb/day)(ft^3/62.4 lb)(100 lb/5 lb)(1/1.01) or 1983 ft^3/day. The digested sludge volume = (3625 lb/day)(ft^3/62.4 lb)(100 lb/8 lb)(1/1.03) or 705 ft^3/day. $V_{avg} = V_1 - \frac{2}{3}(V_1 - V_2) = 1983 - \frac{2}{3}(1983 - 705) = 1131$ ft^3/day. From Figure 13.4 the digestion time is 23 days. Thus, the total sludge volume is

$$\text{Volume of sludge} = (1131)(23) + (705)(30)$$

$$= 47,163 \text{ ft}^3$$

Therefore, the total digester volume is

$$V = (47,163)(2) = \underline{94,325 \text{ ft}^3}$$

Note: Some sources assume that the fixed solids do not change during the digestion process, as was assumed in this problem. Other sources (Fair, G. M., Geyer, J. C., & Okun, D. A., 1968) assume that 25 percent of the volatile solids destroyed is converted to fixed solids.

Example Problem 13.1 SI
Low-Rate Digester

A heated, low-rate anaerobic digester is to be designed for an activated sludge plant treating the wastewater from 25,000 persons. The fresh sludge has 0.11 kg dry solids/cap-day, volatile solids are 70 percent of the dry solids, the dry solids are 5 percent of the sludge, and the wet specific gravity is 1.01. Sixty percent of the volatile solids is destroyed by digestion and the fixed solids remain unchanged. The digested sludge has 8 percent dry solids and a wet specific gravity of 1.03. The operating temperature is 35°C and the sludge storage time is 30 days. The sludge occupies the lower half of the tank depth and the supernatant liquor and the

ENVIRONMENTAL ENGINEERING

gas occupies the upper half of the tank depth. Determine the digester volume.

Solution

The fresh sludge solids represent an Input. Fresh solids = (25,000 persons)(0.11 kg/cap-day) = 2750 kg/day. vss = (2750)(0.70) = 1925 kg/day. fss = (2750)(0.30) = 825 kg/day. vss destroyed = (1925) (0.60) = 1155 kg/day. The digested sludge solids represent an Output. For steady state, [Output] = [Input] − [Decrease due to reaction]. Thus, vss in the digested sludge = 1925 − 1155 = 770 kg/day. fss in the digested sludge = 825 − 0 = 825 kg/day. The total solids in the digested sludge = 770 + 825 = 1595 kg/day. The fresh sludge volume = (2750 kg/day)(m^3/1000 kg)(100 kg/5 kg)(1/1.01) or 54.46 m^3/day. The digested sludge volume = (1595 kg/day)(m^3/1000 kg)(100 kg/8 kg)(1/1.03) or 19.36 m^3/day. $V_{avg} = V_1 − \frac{2}{3}(V_1 − V_2)$ = 54.46 − $\frac{2}{3}$ (55.46 − 19.36) = 30.39 m^3/day. From Figure 13.4 the digestion time at 35°C (95°F) is 23 days. Thus, the total sludge volume is

$$\text{Volume of sludge} = (30.39)(23) + (19.36)(30)$$
$$= 1279.8 \, \text{m}^3$$

Therefore, the total digester volume is

$$V = (1279.8)(2) = \underline{2559.6 \, \text{m}^3}$$

Example Problem 13.2
High-Rate Digester

A heated high-rate anaerobic digester is to be designed for an activated sludge plant treating the wastewater from 25,000 persons. The feed to the digester (both primary and secondary sludge) is 1983 ft^3/day and the operating temperature is 95°F. Determine the digester volume.

Solution

From the previous table, the minimum cell residence time, θ_c^{Min}, for 95°F is 4 days. The design mean cell residence time, θ_c, is 2.5 × 4 days or 10 days. For a completely mixed biological reactor without recycle, the θ_c value is equal to the hydraulic detention time, θ_h. Thus, the digester volume, V, is

$$V = Q\theta_h = (1983 \, \text{ft}^3/\text{day})(10 \, \text{days}) = \underline{19,830 \, \text{ft}^3}$$

Moisture-Weight Relationships

The specific gravity of a wet or dried sludge, s, depends upon the water content, the solids content, and the specific gravity of the dried solids, s_s. The percent water or moisture content, p_w, is given by

$$p_w = \frac{100\,W_w}{(W_w + W_s)} \tag{13.8}$$

where

p_w = percent water;
W_w = weight of the water;
W_s = weight of the dry solids.

The percent solids, p_s, is given by the formulation

$$p_s = (100 - p_w) = \frac{100\,W_s}{W_w + W_s} \tag{13.9}$$

The specific gravity of dried sludge solids, s_s, is a function of the specific gravities of the volatile and fixed fractions, which are designated s_v and s_f, respectively. If the percent volatile material is p_v and the percent fixed material is p_f, the following may be written:

$$\frac{100}{s_s} = \frac{p_v}{s_v} + \frac{100 - p_v}{s_f} \tag{13.10}$$

Rearranging Eq. (13.10) yields

$$s_s = \frac{100\,s_f s_v}{100\,s_v + p_v(s_f - s_v)} \tag{13.11}$$

For engineering purposes, the specific gravity of the volatile fraction, s_v, may be considered as 1.0 and the fixed fraction, s_f, as 2.5. Substituting these into Eq. (13.11) and rearranging gives

$$s_s = \frac{250}{100 + 1.5\,p_v} \tag{13.12}$$

The specific gravity of wet sludge, s, may be determined from the equation

$$s = \frac{p_w + (100 - p_w)}{p_w + (100 - p_w)/s_s} = \frac{100\,s_s}{p_w s_s + (100 - p_w)} \tag{13.13}$$

It has been possible, from evaluations of vast amounts of operating data and records, to make estimates of the amounts of sludge that are produced at municipal wastewater treatment plants, and also to estimate the solids concentrations, volatile and fixed fractions, and other pertinent characteristics. Table 13.4 shows typical solids contributions per capita and solids concentrations for various sludges from different types of treatment plants. In design, the percent solids shown in the table may be used without much reservation because they have been found to be relatively consistent. However, the dry solids contributions in pounds per capita per day should be used with caution because they vary from plant to plant.

Sludge Quantities and Solids Concentrations

Table 13.4. Sludge Solids Contributions and Solids Contents

Type of Treatment and Type of Sludge	Dry Solids (lb/cap-day)	Solids Content (percent)
Plain sedimentation		
Fresh, wet	0.119	2.5–5.0
Digested, wet	0.119	10–15
Digested, air dried	0.075	45
Activated Sludge (Conventional)		
Excess activated sludge, wet	0.068	0.5–1.5
Primary and excess activated sludge, wet	0.187	4–5
Digested primary and excess activated sludge, wet	0.121	6–8
Digested primary and excess activated sludge, air dried	0.121	45
Thickened excess activated sludge, wet	0.068	1–2
Digested thickened excess activated sludge, wet	0.041	2–3
Trickling Filter Plant (High rate)		
Secondary, wet	0.044	5
Primary and secondary, wet	0.163	5
Digested primary and secondary, wet	0.106	10
Digested primary and secondary, air dried	0.106	45
Trickling Filter Plant (Standard rate)		
Secondary, wet	0.029	5–10
Primary and secondary, wet	0.148	3–6
Digested primary and secondary, wet	0.095	10
Digested primary and secondary, air dried	0.095	45
Chemically Precipitated Raw Wastewater		
Fresh wet	0.198	2–5
Digested	0.125	10

Adapted from *Sewage Treatment* by K. Imhoff and G. M. Fair. Copyright © 1956 by John Wiley and Sons, Inc. Reprinted by permission.

If sludge pumping records are available giving the flows of primary and secondary sludge per day and also the solids contents, these data may be used to determine the pounds of primary and secondary sludge produced per day. If these records are not available, it is necessary to use the following procedure to determine the amount of primary and secondary sludge solids. Determining the pounds of primary sludge solids produced per day requires the influent suspended solids concentration, the fraction of suspended solids removed (usually 0.60 to 0.65), and the influent flow rate of the wastewater. The suspended solids concentration should come from long-term suspended solids analyses to be representative. Determining the pounds of secondary solids from the activated sludge process requires the pounds of BOD_5 or COD removed per day and requires the use of the equation (Eckenfelder, W. W., & Weston, R. F., 1956)

$$\text{lb solids/day} = YS_r - k_e \overline{X} \qquad (13.14)$$

where

Y = yield coefficient, pounds of total suspended or volatile suspended solids per pound of BOD_5 or COD removed per day;

S_r = pounds of BOD_5 or COD removed per day;

k_e = endogeneous coefficient, day^{-1};

$\overline{X}$ = pounds of total suspended or volatile suspended solids in the aeration tank.

Determining the pounds of secondary sludge solids from the trickling filter process requires the pounds of BOD_5 or COD removed per day by the secondary units and the pounds of biological solids produced per pound of BOD_5 or COD removed. Low-rate trickling filters produce from 0.25 to 0.35 lb of biological solids per lb of BOD_5 removed, whereas high-rate trickling filters produce from 0.40 to 0.50 lb biological solids per pound of BOD_5 removed. If the BOD_5/COD ratio is known, these values may be expressed in terms of COD. Determining the pounds of suspended solids produced per day by chemical coagulation of raw municipal wastewater requires the influent suspended solids concentration, the fraction of suspended solids removed (usually from 0.95 to 0.99), and the influent flow rate of the wastewater. Frequently, estimating the amount of sludge produced can be facilitated by the use of population equivalents. The population equivalents of BOD_5, suspended solids, and flow are 0.17 lb, 0.20 lb, and 100 gal per capita per day, respectively. However, the increasing popularity of garbage grinders may cause these values to increase in the future.

A low-rate trickling filter plant treats the wastewater from 5000 persons. Short-term analyses using composite samples of the influent wastewater show the BOD_5 and suspended solids concentrations to be that expected from average values for municipal wastewaters in the United States. The plant gets 95 percent BOD_5 removal, 35 percent BOD_5 removal by the primary clarifier, and 65 percent suspended solids removal by the primary clarifier. Assume that the primary sludge has 4 percent dry solids, the primary sludge specific gravity = 1.01, the secondary sludge has 5 percent dry solids, the specific gravity = 1.02, and the biological solids produced is 0.35 lb/lb BOD_5 removed. Determine the primary and secondary sludge produced per day.

Solution

The primary solids = (5000)(0.20)(0.65) = 650 lb/day. The primary sludge flow = (650 lb/day)(100 lb/4.0 lb) × (gal/8.34 lb)(1/1.01) = 1929 gal/day. Thus,

Primary sludge = 1929 gal/day

The BOD_5 to the secondary units = (0.17)(5000)(1 − 0.35) = 552.5 lb/day. The BOD_5 from the plant = (0.17)(5000)(1 − 0.95) = 42.5 lb/day. The BOD_5 removed by the secondary units = 552.5 − 42.5 = 510 lb/day. Thus, the biological solids produced per day = (510 lb/day)(0.35 lb/lb) = 178.5 lb/day. The secondary sludge flow = (178.5 lb/day)(100 lb/5 lb)(gal/8.34 lb) (1/1.02) = 420 gal/day. Thus,

Secondary sludge = 420 gal/day

In moderate and cold climates, it is necessary to heat digesters during the winter to maintain the digester temperature within the desired range. The heat furnished must be sufficient to (1) increase the temperature of the incoming fresh sludge to the temperature in the digester, and (2) make up for heat losses from the digester through the walls, bottom, and cover. Usually, the heat required to raise the incoming sludge temperature to the digester temperature is larger than the heat required to make up for heat losses.

The heat required to raise the temperature of the incoming fresh sludge to the digester temperature is given by

Digester Heat Requirements

$$Q_s = P \times \frac{100}{p_s} \times (T_d - T_s) \times \frac{1}{24} \times c_p \qquad (13.15)$$

where

Q_s = Btu/hr required;
P = pounds of fresh dry sludge solids added per day;
p_s = percent dry solids in the fresh sludge;
T_d = temperature in the digester, °F;
T_s = temperature of the fresh sludge, °F;
c_p = specific heat constant, equal to 0.97 to 1.00.

The heat required to make up for losses through the top, walls, and bottom is given by

$$Q_d = CA \cdot \Delta T \qquad (13.16)$$

where

Q_d = Btu/hr required;
C = coefficient of heat flow, Btu/ft^2-hr-°F;
A = surface area, ft^2;
ΔT = difference between the tank temperature and the outside material being considered, °F.

Figure 13.10 Externally Heated Digester

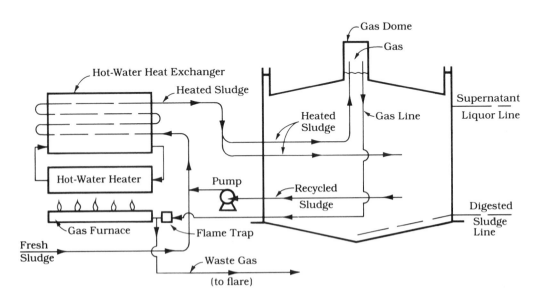

ENVIRONMENTAL ENGINEERING

The C values for concrete walls and slabs against other materials are 0.08 for dry earth, 0.30 for air, and 0.26 for wet earth. The C value for a floating cover is 0.16 (Imhoff, K., Mueller, W. J., & Thistlethwayte, D. K. B., 1971).

The most common method to heat a digester is to use a hot-water heater and heat exchanger outside the digester and heat the incoming and recycled sludge, as shown in Figure 13.10. The first heated digesters used hot-water heaters or boilers external to the digesters, and the hot water was circulated through coils of water pipes mounted inside the digesters. It was found, however, that sludge caked around the pipes and greatly reduced the thermal efficiency. Since digesters are rarely emptied, it was not feasible to clean the cake from the pipes. The use of an external heater, which heats the pumped sludge outside the digester as shown in Figure 13.10, controls the caking problem since any sludge caking will occur in the pipes in the heat exchanger. The elbows may be removed from the heat exchanger and the cake may be rodded loose from the pipes, then the elbows may be replaced and heating resumed. Digesters with a mixing system consisting of recycled gas discharging in a draft tube in the center of the digester frequently are heated by internal heat exchangers mounted inside the draft tube. These heat exchangers can be removed through the access hatch in the center of the dome to clean off any caking that may have occurred.

Example Problem 13.4
Sludge Heat
Requirements

An activated sludge treatment plant has an anaerobic digester and serves a population of 25,000 persons. The mixture of primary and secondary sludge amounts to 0.28 lb/cap-day, the fresh sludge has 4.5 percent solids on a dry weight basis, and the digester operating temperature is 95°F. During the coldest month, January, the sludge temperature is 55°F. If the specific heat constant is 1.00, determine the heat required to raise the fresh sludge temperature to that of the digester.

Solution

The number of pounds of fresh dry sludge solids added per day is (25,000 capita) (0.28 lb/cap-day) = 7000 lb/day. The heat required, Q_s, is given by Eq. (13.15), which is Q_s = (7000 lb/day)(100 lb/4.5 lb)(95°F − 55°F) (day/24 hr)(1.00) or

$$Q_s = \underline{259,259\,Btu/hr}$$

The anaerobic digester for the municipality in Example Problem 13.3 is 75 ft in diameter, has a wall height of 30 ft, and has a floating cover. The digester is recessed 2 ft in the earth and has a freeboard of 3 ft. For insulation, earth is mounded around the digester up to a height of 15 ft. A wall height of 30 ft − (2 ft + 3 ft + 15 ft), or 10 ft, is exposed to air on the outside and digester contents on the inside. The average monthly temperature in January is 45°F. The digester operating temperature = 95°F and the dry earth mounded around the digester has a temperature equal to the average of the digester operating temperature and the average monthly temperature. The earth temperature below the digester floor is 95°F, and the temperature of the earth around the wall at the 2-ft recess is 95°F. Determine the heat required to make up heat losses from the digester.

Solution

The area of the floating cover is $(\pi/4)(75 \text{ ft})^2 = 4418 \text{ ft}^2$. The area of the wall where air is on the outside and digester contents are on the inside is $(\pi)(75 \text{ ft})(10 \text{ ft}) = 2356 \text{ ft}^2$. The area of the wall where dry earth is on the outside and digester contents are on the inside is $(\pi)(75 \text{ ft})(15 \text{ ft}) = 3534 \text{ ft}^2$. The temperature of the dry earth mounded around the digester is $(95°F + 45°F)(\frac{1}{2})$ or 70°F. The heat required, Q_d, to make up digester heat losses is Q_d = (4418 ft^2)(0.16 Btu/ft^2-hr-°F)(95°F − 45°F) + (2356 ft^2)(0.30 Btu/ft^2-hr-°F)(95°F − 45°F) + (3534 ft^2)(0.08 Btu/ft^2-hr-°F)(95°F − 70°F) = 35,344 Btu/hr + 35,340 Btu/hr + 7068 Btu/hr or

$$Q_d = \underline{77,752 \text{ Btu/hr}}$$

Digester Gas Utilization

Digester gas consists of methane (about 55 to 75 percent), carbon dioxide (about 25 to 45 percent), water vapor, and trace amounts of such gases as hydrogen sulfide, hydrogen, and nitrogen. Sludge gas is explosive, and the hydrogen sulfide is a corrosive and toxic gas. The usual methods for using sludge gas are for heating purposes and for power generation. Sludge heaters will burn the gas without scrubbing; however, for use as a fuel for internal combustion engines, the gas must be scrubbed. Scrubbing will remove hydrogen sulfide and some carbon dioxide, thus increasing

External Sludge Heater
Courtesy of Walker
Process Corporation

the fuel value in terms of Btu per ft^3. Internal combustion engines may be used to drive air compressors, pumps, and generators for production of electricity. If the gas utilization facilities are extensive, the gas should be scrubbed to make it less corrosive. The amount of sludge gas produced will meet all the energy requirements for a municipal wastewater treatment plant if the digesters are operating properly. Excess sludge gas is usually burned in a flare. The gas collection and circulation system must have safety equipment such as condensate traps, pressure regulation valves, flame traps on lines to burners, engines and flares, and a waste gas flare.

Example Problem 13.6
Sludge Gas Utilization

The sludge, for the municipality in Example Problems 13.4 and 13.5, has 70 percent volatile matter; 60 percent of the volatile matter is destroyed during digestion. The sludge gas produced amounts to 14 ft^3 per lb volatile solids destroyed and is 70 percent methane. The fuel value of methane is 963 Btu/ft^3 at standard conditions. The sludge heater that heats the incoming sludge and makes up the digester heat losses has a thermal efficiency of 60 percent. Determine what percent of the fuel value produced is required for the sludge heater.

Solution

The volume of sludge gas produced per day = (7000 lb/day)(0.70) (0.60)(14 ft^3/lb) = 41,160 ft^3/day. The fuel value available is (41,160 ft^3/24 hr)(0.70)(963 Btu/ft^3) = 1,156,082 Btu/hr. The heat required = (259,259 Btu/hr + 77,752 Btu/hr)(1/0.60) = 561,685 Btu/hr. The percent of the fuel value used = (561,685/1,156,082)(100 percent) or

Fuel value used = <u>48.6 percent</u>

Fertilizer Value of Dried Sludge

Fresh municipal sludges are not suitable as fertilizers because of hygienic and aesthetic problems. Air-dried digested sludges should not be used on crops that are eaten raw by humans, such as vegetables. Heat-dried sludges are sterile and may be used on any crops. The fertilizer value of sludges is about the same as barnyard manure and, in addition to their fertilizer value, sludges also serve as excellent soil conditioners.

Fluid Properties of Sludge

The frictional head losses for organic municipal wastewater sludges may be estimated using hydraulic formulas, such as the Hazen-Williams equation. The head loss depends mainly upon the nature of the sludge, the type of flow (that is, laminar, transitional, or turbulent), the solids content, and the temperature. Sludges exhibit laminar, transitional, and turbulent flow at much higher Reynolds numbers than water. Laminar and transitional flow occur up to about 3.5 fps in pipes 4 to 8 inches in diameter and turbulent flow occurs above about 3.5 fps (WPCF, 1977). Hazen-Williams coefficients for laminar and transitional flow for fresh and digested primary sludges at various solid contents are shown in Table 13.5.

Table 13.5. Hazen-Williams Coefficients for Sludge versus Solids Content

Percent Solids	Raw Primary[a] C_{HW}	Digested C_{HW}
0	100	100
2	81	90
4	61	85
6	45	75
8.5	32	60
10	25	55

a. Data taken from S. G. Brisbin, "Flow of Concentrated Raw Sewage Sludges in Pipes," *Proc. SED*, ASCE 83 (1957):SA3.

Activated sludges usually have solid contents from 0.5 to 1.5 percent and, for these sludges, the Hazen-Williams coefficients for digested sludges may be used.

The Hazen-Williams equation for pipe flow is

$$V = 1.318\,C_{HW}R^{0.63}S^{0.54} \qquad (13.17)$$

where

V = velocity in fps;
C_{HW} = Hazen-Williams friction coefficient;
R = hydraulic radius in feet;
S = slope of the energy gradient.

Example Problem 13.7
Sludge Pumping

A centrifugal sludge pump with a capacity of 300 gpm pumps a mixture of primary and secondary sludge from the primary clarifier to the digester at a municipal treatment plant. The secondary sludge is mixed with the influent to the primary clarifier and allowed to resettle with the primary settlings in order to thicken the secondary sludge. The piping from the primary clarifier to the sludge pump and from the sludge pump to the digester is 8 in. in diameter. The discharge pipe in the digester runs vertically up to the gas dome of the floating cover, and the elevation difference between the water surface of the primary clarifier and the end of the vertical discharge pipe is 18.60 ft. The gas is under a pressure of 8 in. of water column. The sludge mixture has 4.5 percent dry solids and the specific gravity is 1.01. The length of 8-in. pipe is 325 ft and the fittings or form losses are equivalent to 150 ft of pipe. The pump is operated by a time clock and it runs at a preset time interval once every hour. Determine the head at which the pump operates.

Solution

Writing Bernoulli's equation between point 1 at the primary clarifier water surface and point 2 at the discharge end of the piping gives

$$\frac{V_1^2}{2g} + \frac{p_1}{\gamma} + Z_1 + H_p = \frac{V_2^2}{2g} + \frac{p_2}{\gamma} + Z_2 + 0.5\frac{V_2^2}{2g} + H_L$$

where

$$\begin{aligned}
V_1, V_2 &= \text{velocities;} \\
p_1, p_2 &= \text{pressures;} \\
Z_1, Z_2 &= \text{elevation datums;} \\
H_p &= \text{head on pump;} \\
0.5\ V_2^2/2g &= \text{entrance loss;} \\
&= \text{total head losses.}
\end{aligned}$$

Assume a datum that makes Z_1 = El. 0.0 ft. Since V_1 = 0, p_1 = 0, and Z_1 = 0, Bernoulli's equation becomes

$$H_p = 1.5\frac{V_2^2}{2g} + \frac{8}{12}\text{ft} + 18.60\text{ ft} + H_L$$

$$= 1.5\frac{V_2^2}{2g} + 19.27\text{ ft} + H_L$$

Table 13.5 gives C_{HW} = 61 for 4 percent solids and C_{HW} = 45 for 6 percent solids. Interpolation for 4.5 percent solids gives C_{HW} = 57. The flow of 300 gpm is equal to 0.668 ft³/sec. The area for an 8-in. pipe is 0.349 ft²; thus, the velocity is (0.668 ft³/sec)(1/0.349 ft²) = 1.91 fps. The hydraulic radius = (8/12 ft)($\frac{1}{4}$) = 0.1667 ft. Inserting the known values into the Hazen-Williams equation gives

$$1.91 = (1.318)(57)(0.1667)^{0.63} S^{0.54}$$

which yields

$$S = 0.0090\text{ ft/ft or } 9.00\text{ ft/1000 ft}$$

Substituting into Bernoulli's equation gives

$$H_p = (1.5)\left[\frac{(1.91)^2}{2g}\right] + 19.27$$

$$+ \left(\frac{9.00}{1000}\right)(325 + 150)$$

or

$$H_p = 0.08 + 19.27 + 4.28 = \underline{23.63\text{ ft}}$$

At many wastewater treatment plants, particularly large ones, the fresh sludge is thickened to increase its solids content prior to digestion. Thickening prior to digestion is becoming more common because it reduces the daily fresh sludge volume, thus decreasing the required size of the digester and also the amount of supernatant liquor to be disposed. Thickening may be accomplished by gravity thickeners, which are the most widely used, or by centrifuges. Gravity thickeners are similar to circular clarifiers; the most common type has vertical pickets mounted on the trusswork for the bottom scraper blades. The pickets extend up to about half the depth of the tank, and as they rake through the sludge they break up sludge arching and release much of the entrained water. Gravity thickeners usually thicken a sludge to about twice the original solids content, thus decreasing the volume of the fresh sludge to about half the original volume. Surface loadings are usually 600 to 800 gal/day-ft^2 based on the supernatant flow. The allowable solids loading in lb/day-ft^2 depends upon the nature of the sludge. The thickening of various sludges has given the following percent solids in the thickened flow: (1) raw primary sludges at 20 to 30 lb/day-ft^2 gave 8 to 10 percent solids; (2) mixtures of raw primary and waste activated sludge at 6 to 10 lb/day-ft^2 gave 5 to 8 percent solids; (3) mixtures of raw primary and trickling filter humus at 10 to 12 lb/day-ft^2 gave 7 to 9 percent solids; (4) waste activated sludge at 5 to 6 lb/day-ft^2 gave 2.5 to 3.0 percent solids; and (5) trickling filter humus at 8 to 10 lb/day-ft^2 gave 7 to 9 percent (EPA, 1979).

Sludge Thickening

Anaerobically digested sludge may be dewatered by air drying, vacuum filters, centrifuges, filter presses, and sludge lagoons. Sludge air dried on sand beds may have a solids content as high as 30 to 45 percent, and it can usually be removed once the solids are about 25 percent. The solids content obtained by dewatering using mechanical means depends primarily upon the nature of the sludge, its solids content, and whether or not a conditioner is used.

Vacuum filtration of a chemically conditioned digested primary sludge at filter loadings of 5 to 8 lb/hr-ft^2 has produced a cake with 25 to 32 percent solids. Vacuum filtration of a chemically conditioned digested mixture of primary and secondary sludge at filter loadings of 3.5 to 6 lb/hr-ft^2 has produced a cake with 14 to 22 percent solids (EPA, 1979).

Centrifugation of chemically conditioned digested primary sludge has produced a cake of 28 to 35 percent solids and centrifugation of a chemically conditioned digested

Sludge Dewatering

mixture of primary and secondary sludge has produced a cake of 15 to 30 percent solids. Filter presses dewatering a chemically conditioned digested mixture of primary and secondary sludge has produced a cake of 45 to 50 percent solids (EPA, 1979).

When sludge lagoons are used for dewatering, the digested sludge is added to about a 2.0-ft depth and allowed to dry; then the filling and drying is repeated. Once the lagoon is filled with dry sludge, the sludge is removed. The lagoon loading is usually from 2.2 to 2.4 lb/yr-ft^3 volume (EPA, 1979).

References

Brisbin, S. G. 1957. Flow of Concentrated Raw Sewage Sludges in Pipes. *Proc. SED,* ASCE 83:SA3.

Conway, R. A., and Ross, R. D. 1980. *Handbook of Industrial Waste Disposal.* New York: Van Nostrand Reinhold.

Eckenfelder, W. W., Jr. 1980. *Principles of Water Quality Management.* Boston: CBI Publishing.

Eckenfelder, W. W., Jr., and O'Connor, D. J. 1961. *Biological Waste Treatment.* London: Pergamon Press.

Eckenfelder, W. W., Jr., and Weston, R. F. 1956. Kinetics of Biological Oxidation. In *Biological Treatment of Sewage and Industrial Wastes.* Vol. 1, edited by J. McCabe and W. W. Eckenfelder, Jr. New York: Reinhold.

Environmental Protection Agency (EPA). 1974. *Upgrading Existing Wastewater Treatment Plants.* EPA Process Design Manual, Washington, D.C.

Environmental Protection Agency (EPA). 1979. *Sludge Treatment and Disposal.* EPA Process Design Manual, Washington, D.C.

Fair, G. M., Geyer, J. C., and Okun, D. A. 1968. *Water Purification and Wastewater Treatment, and Disposal.* Vol. 2. *Water and Wastewater Engineering.* New York: Wiley.

Fair, G. M., and Moore, E. W. 1932. Heat and Energy Relations in the Digestion of Sewage Solids. *Sewage Works Jour.* 4 (February):242; (May):248; (July):589; (September):728.

Fair, G. M, and Moore, E. W. 1937. Observations on the Digestion of Sewage Sludge over a Wide Range of Temperatures. *Sewage Works Jour.* 9, no. 1:3.

Ford, D. L. 1970. General Sludge Characteristics. In *Advances in Water Quality Improvement by Physical and Chemical Processes,* edited by E. F. Gloyna and W. W. Eckenfelder, Jr. Austin, Tex.: University of Texas Press.

Great Lakes–Upper Mississippi Board of State Sanitary Engineers. 1978. Recommended Standards for Sewage Works. Ten state standards, Albany, N.Y.

Hardenbergh, W. A., and Rodie, E. B. 1963. *Water Supply and Waste Disposal.* Scranton, Pa.: International Textbook Co.

Heukelekian, H. 1956. Basic Principles of Sludge Digestion. In *Biological Treatment of Sewage and Industrial Wastes.* Vol. 2, edited by J. McCabe and W. W. Eckenfelder, Jr. New York: Reinhold.

Imhoff, K., and Fair, G. M. 1956. *Sewage Treatment.* New York: Wiley.

Imhoff, K., Mueller, W. J., and Thistlethwayte, D. K. B. 1971. *Disposal of Sewage and Other Waterborne Wastes*. Ann Arbor, Mich.: Ann Arbor Science Publishers.

Kampelmacher, E. H., and van Noorde, J. L. M. 1972. Reduction of Bacteria in Sludge Treatment. *Jour. WPCF* 44, no. 2:309.

Langford, L. L. 1956. Mesophilic Anaerobic Digestion: Design and Operating Considerations. In *Biological Treatment of Sewage and Industrial Wastes*. Vol. 2, edited by J. McCabe and W. W. Eckenfelder, Jr. New York: Reinhold.

Lawrence, A. W., and McCarty, P. L. 1969. Kinetics of Methane Fermentation in Anaerobic Treatment. *Jour. WPCF* 41, no. 2, part 2:R1.

Malina, J. F., Jr., and Mihotits, E. M. 1968. New Developments in the Anaerobic Digestion of Sludges. In *Advances in Water Quality Improvement*, edited by E. F. Gloyna and W. W. Eckenfelder, Jr. Austin, Tex.: University of Texas Press.

McCarty, P. L. 1968. Anaerobic Treatment of Soluble Wastes. In *Advances in Water Quality Improvement*, edited by E. F. Gloyna and W. W. Eckenfelder, Jr. Austin, Tex.: University of Texas Press.

McCarty, P. L. 1964. Anaerobic Waste Treatment Fundamentals. *Public Works Magazine* 95, no. 9:107; 95, no. 10:123; 95, no. 11:91; and 95, no. 12:95.

McKinney, R. E. 1962. *Microbiology for Sanitary Engineers*. New York: McGraw-Hill.

Metcalf and Eddy, Inc. 1979. *Wastewater Engineering, Treatment, Disposal and Reuse*. New York: McGraw-Hill.

Ramalho, R. S. 1977. *Introduction to Wastewater Treatment Processes*. New York: Academic Press.

Sawyer, C. N. 1956. An Evaluation of High-Rate Digestion. In *Biological Treatment of Sewage and Industrial Wastes*. Vol. 2, edited by J. McCabe and W. W. Eckenfelder, Jr. New York: Reinhold.

Schroeder, E. D. 1977. *Water and Wastewater Treatment*. New York: McGraw-Hill.

Sundstrom, D. W., and Klei, H. E. 1979. *Wastewater Treatment*. Englewood Cliffs, N.J.: Prentice-Hall.

Texas State Department of Health. 1976. Design Criteria for Sewage Systems. Austin, Tex.

Water Pollution Control Federation (WPCF). 1968. *Anaerobic Sludge Digestion*. WPCF Manual of Practice no. 16, Washington, D.C.

Water Pollution Control Federation (WPCF). 1977. *Wastewater Treatment Plant Design*. WPCF Manual of Practice no. 8, Washington, D.C.

Problems

1. A conventional anaerobic digester is to be designed to treat the primary and waste activated sludge from a municipal wastewater treatment plant serving 18,000 persons. The waste activated sludge is added to the plant influent flow ahead of the primary clarifier to thicken the solids and, as a result, the waste activated sludge solids leave with the underflow from the primary clarifier. Pertinent data are: influent flow = 100 gal/cap-day, influent solids = 0.25 lb/cap-day, influent

suspended solids removal by the primary clarifier = 65 percent, waste activated sludge flow = 2.0 percent of plant influent, sludge volume index = 80 mℓ/gm, solids in underflow from primary clarifier = 4.5 percent, volatile solids in underflow = 70 percent, design temperature = 95°F, storage time for digested solids = 30 days, volatile solids destroyed by digestion = 60 percent of feed volatile solids, 7.0 percent solids in digested sludge, 12 ft^3 of sludge gas at STP are produced per pound of volatile solids destroyed, 65 percent of sludge gas by volume is methane, depth of sludge when stratified = $\frac{1}{2}$ swd, 2.5 ft of freeboard for floating cover, and depth of supernatant and sludge ≈ 23 ft. The specific gravity for the fresh sludge is 1.01 and for the digested sludge is 1.03. Assume that the fixed solids remain unchanged during digestion. Determine:

a. The digestion time, days.

b. The required digester diameter and wall height, ft.

c. The fuel value of the sludge gas, Btu/day.

2. A heated conventional anaerobic digester is to be designed for treating a primary and waste activated sludge from a wastewater treatment plant serving 70,000 persons. The digester is recessed 3 ft below finished grade. Pertinent data are: solids contribution (primary plus secondary) = 0.25 lb/cap-day, fresh sludge solids content = 4.0 percent, volatile solids in fresh sludge = 70 percent, digested sludge solids = 7.0 percent, digester operating temperature = 95°F, fresh sludge temperature during January = 60°F, average air temperature during January = 55°, storage time for digested sludge = 30 days, depth of sludge when stratified = $\frac{1}{2}$ of swd, freeboard for floating cover = 2.5 ft, volatile matter destroyed by digestion = 60 percent of total feed volatile solids, height:diameter ratio = 1:3 to 1:4, maximum wall height = 34 ft, exposed concrete wall height = 10 ft, temperature of earth adjacent to walls = average of the design digestion temperature and outside air temperature, temperature of earth along the 3-ft recess and below the floor = digester operating temperature of 95°F, specific heat of wet sludge solids = 1.0 Btu/lb-mass-°F, 12 ft^3 of sludge gas at STP are produced per pound of volatile solids destroyed, and gas composition = 70 percent methane and 30 percent carbon dioxide. Assume that the fixed solids remain unchanged during digestion. The specific gravity for the fresh sludge is 1.01 and for the digested sludge is 1.03. Determine:

a. The size of the digester—that is, both diameter and wall height.
b. The material balance for the digester giving all flows.
c. The volatile fraction of the digested sludge.
d. The heat requirements.
e. The pseudomolecular weight of the sludge gas and its unit volume (ft^3/lb) at STP.
f. The percent of the gas required for digester heating if the heat exchangers have an efficiency of 75 percent.
g. The percent of the gas to burn in a waste-gas flare.

3. A conventional anaerobic digester is to be designed to treat the primary and waste activated sludge from a municipal wastewater treatment plant serving 18,000 persons. The waste activated sludge is added to the plant influent flow ahead of the primary clarifier to thicken the solids and, as a result, the waste activated sludge solids leave with the underflow from the primary clarifier. Pertinent data are: influent flow = 380 ℓ/cap-day, influent solids = 0.12 kg/cap-day, influent suspended solids removal by the primary clarifier = 65 percent, waste activated sludge flow = 2.0 percent of plant influent, sludge volume index = 80 mℓ/gm, solids in underflow from primary clarifier = 4.5 percent, volatile solids in underflow = 70 percent, design temperature = 35°C, storage time for digested solids = 60 days, volatile solids destroyed by digestion = 60 percent of feed volatile solids, 7.0 percent solids in digested sludge, 150 ℓ of sludge gas at STP are produced per kg of volatile solids destroyed, 65 percent of sludge gas by volume is methane, depth of sludge when stratified = $\frac{1}{2}$ swd, 0.75 m of freeboard for floating cover, and depth of supernatant and sludge = 6.1 m. The specific gravity for the fresh sludge is 1.01 and for the digested sludge is 1.03. Assume that the fixed solids remain unchanged during digestion. Determine:
a. The digestion time, days.
b. The required digester diameter and wall height, meters.
c. The fuel value of the sludge gas, watts.

4. A two-stage high-rate trickling filter plant gives 90 percent BOD$_5$ removal with the primary clarifier removing 35 percent. The secondary sludge amounts to 0.05 lb/cap-day. If the raw wastewater has a BOD$_5$ of 0.17 lb/cap-day, determine the pounds of secondary sludge produced per pound of BOD$_5$ removed.

5. A heated high-rate anaerobic digester is to be designed for an activated sludge plant treating the wastewater from 45,000 persons. The thickened feed sludge to the digester (both primary and secondary sludge) is 8.0 percent dry solids, 68 percent volatile solids, and amounts to 0.30 lb dry solid/cap-day. Determine the volume of the digester in ft^3 if the operating temperature is 95°F.

6. A digester gas is 72 percent methane, 28 percent carbon dioxide, and has trace amounts of other gases. Determine the pseudomolecular weight and the specific volume in ft^3/lb at STP.

7. A low-rate anaerobic digester is to be designed to treat the primary and waste activated sludge from a municipal wastewater treatment plant serving 20,000 persons. The waste activated sludge is added to the plant influent flow ahead of the primary clarifier to thicken the solids and, as a result, the waste activated sludge solids leave with the underflow from the primary clarifier. Pertinent data are: primary and secondary sludge mixture is 0.32 lb dry solid/cap-day, solids in the underflow from the primary clarifier = 4.5 percent, volatile solids in underflow = 70 percent, design temperature = 95°F, storage time for digested solids = 30 days, volatile solids destroyed by digestion = 65 percent of feed volatile solids, 25 percent of volatile solids destroyed are converted to fixed solids, 7.0 percent solids in digested sludge, and the depth of sludge when stratified = half the side water depth. Determine:
   a. The volume of the digester, ft^3.
   b. The volume per capita, ft^3/cap.

Chapter Fourteen

AEROBIC DIGESTION

Aerobic digestion may be defined as the biological oxidation of organic sludges under aerobic conditions. It closely resembles the activated sludge process since the aeration equipment and tanks are similar. Most of the microbes used in the process are facultative; however, some are obligate aerobes, such as the nitrifying bacteria. The following types of sludge have been successfully treated by aerobic digestion: (1) waste activated sludge, (2) primary and waste activated sludge, (3) trickling filter secondary sludge (humus), and (4) primary and trickling filter secondary sludge (humus). The most widespread use of aerobic digestion has been the treatment of waste activated sludge. Most of the aerobic digesters built up to the present have been for small- to medium-size plants and most package plants have aerobic digestion systems.

Some advantages of aerobic digestion compared to anaerobic digestion are fewer operational problems, thus less laboratory control and daily maintenance required, much lower BOD concentrations in the supernatant liquor, and lower capital costs. The disadvantages of aerobic digestion compared to anaerobic digestion are higher energy requirements because an appreciable amount of aeration and mixing is required, methane that is a usable by-product is not produced, and the digested sludge has a lower solids content, and thus the volume of sludge to be dewatered is larger. For both aerobic and anaerobic digestion, the volatile solids loading and percent solids destruction are about the same.

Aerobic digesters can be operated in either a batchwise or continuous-flow manner; however, the majority of the systems are continuous flow. Not only do continuous-flow systems have lower operational costs, but they also provide relatively constant environmental conditions that aid in rapid digestion. If a batch digester is used, it is operated on a fill and draw basis. The continuous-flow systems may be operated with or without a thickener but, in nearly all cases, a thickener-clarifier is provided. If a digester does not have a thickener, the solids concentration in the digester will be the same as in the feed sludge and the solids retention time will be equal to the hydraulic detention time. When a thickener-clarifier is used, the solids concentration in the digester will be several magnitudes larger than in the feed sludge; this has the desirable result of a long solids retention time.

Two types of digester facilities employing thickener-clarifiers are shown in Figures 14.1 and 14.2. The system in Figure 14.1, which is the most common arrangement, has the thickener-clarifier placed downline from the digester. The thickener-clarifier serves to remove supernatant liquor, which is released during the digestion process, from the digester system and to thicken the sludge so the recycled sludge flow rate will be a minimum. The digested sludge, which is periodically withdrawn, may be removed from the recycled sludge flow or directly from the digester. If withdrawal is from the digester, the digested sludge may be removed while the digester is operating or while the aeration system is shut down. Usually, aeration is discontin-

Figure 14.1. Aerobic Digester System with Thickener-Clarifier Downstream from Digester

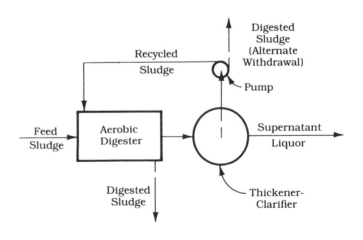

ued so that the sludge will settle and thicken prior to removal. This will reduce the volume of the digested sludge withdrawn and decrease the costs of sludge dewatering. The system in Figure 14.2 has the thickener placed prior to the digester and the recycled sludge is mixed with the feed sludge ahead of the thickener. In the thickener, as much supernatant is removed as possible from the feed and recycled sludges. Usually, the digested sludge is removed directly from the digester and aeration is temporarily discontinued in order for the sludge to settle and thicken prior to removal.

It is essential to accurately estimate the sludge flows and their solids concentrations that are to be treated, because these significantly affect the design of the digester, the thickener-clarifier, and other digester facilities. If the digester system is inadequate in size, the solids will not be properly digested and the thickener-clarifier will not give the proper solid-liquid separation.

Aerobic Biochemical Equations

When primary sludge is mixed with waste activated sludge or trickling filter humus and the combination is aerobically digested, there will be both oxidation of the organic matter in the primary sludge and endogenous oxidation of the cell mass produced from the biological oxidation and from the activated sludge or filter humus.

The generalized biochemical equation for the aerobic digestion of primary sludge solids is

---

Figure 14.2. Aerobic Digester System with Thickener-Clarifier Prior to Digester

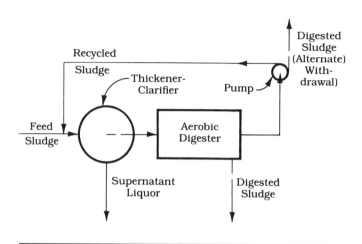

$$\begin{array}{c}\dfrac{\text{Organic}}{\text{matter}} + O_2 \xrightarrow[\text{microbes}]{\text{Aerobic}} \dfrac{\text{New-}}{\text{cells}} + \dfrac{\text{Energy for}}{\text{cells}} \\[2mm] + CO_2 + H_2O + \dfrac{\text{Other end}}{\text{products}} \hspace{1cm} (14.1)\end{array}$$

Some of the other end products include NH_4^{+1}, NO_2^{-1}, NO_3^{-1}, and PO_4^{-3}. During the biological oxidation there is an appreciable amount of nitrogen converted to the nitrate ion. The computed oxygen for the oxidation of primary BOD_5 is about 1.45 lb O_2 per lb of BOD_5. However, studies have shown that up to 1.9 lb O_2 per lb BOD_5 are usually required. The excess above 1.45 lb is the amount of oxygen required for nitrification.

The amount of living cell mass originally present from waste activated sludge or trickling filter humus, and the living cell mass produced by the oxidation of the primary solids is gradually reduced by endogenous decay until it is almost completely stabilized. The biochemical equation for the endogenous decay of cells with the nitrogen terminating in ammonia is

$$C_5H_7NO_2 + 5O_2 \rightarrow 5CO_2 + 2H_2O + NH_3 \hspace{1cm} (14.2)$$

For this equation the amount of oxygen required per pound of cell mass auto-oxidized (MW = 113) is $5 \times 32/113$ or 1.42 lb of oxygen per pound of cells. An appreciable amount of ammonia produced from this auto-oxidation is subsequently oxidized to the nitrate ion as the digestion proceeds. The biochemical equation for the overall endogenous decay of cell mass to form NO_3^{-1} as one of the end products is given by

$$C_5H_7NO_2 + 7O_2 \rightarrow 5CO_2 + 3H_2O + H^{+1} + NO_3^{-1} \hspace{0.5cm} (14.3)$$

In the above reaction, the cell nitrogen terminates as NO_3^{-1}. For the previous reaction the amount of oxygen required per pound of cell mass is $7 \times 32/113$ or 1.98 lb of oxygen per pound of cell mass oxidized. Research on the aerobic digestion of waste activated sludge has shown that from 1.79 to 1.86 lb (average = 1.84) of oxygen are required per pound of volatile solids destroyed. Approximately two-thirds of the cell mass is auto-oxidized during aerobic digestion. The remaining cell material consists of organic compounds, such as peptidoglycans, that are not readily biodegradable.

The pH of the digesting sludge is dependent upon the buffering capacity of the system and may drop as low as 5 to 6 at long hydraulic detention times; however, this does not inhibit the microbial action. The pH drop is possibly due to carbon dioxide production, which reduces the pH of the system, and possibly due to the increase in the nitrate ion concentration.

The rate of aerobic oxidation of solid organic materials frequently has been found to be represented by the following pseudo–first-order biochemical equation:

Kinetics of Aerobic Biological Oxidation

$$-\frac{dX}{dt} = K_a X \qquad (14.4)$$

where

dX = change in the biodegradable organic matter;
dt = time interval;
K_d = reaction rate or degradation constant;
X = concentration of biodegradable materials at any time t.

Equation (14.4) may be arranged for integration between definite limits as follows:

$$\int_{X_0}^{X_t} \frac{dX}{X} = -K_d \int_0^t dt \qquad (14.5)$$

where X_t and X_0 represent the biodegradable matter at the respective times, $t = t$ and $t = 0$. Integration gives

$$\ln \frac{X_t}{X_0} = -K_d t \qquad (14.6)$$

The above equation may be rearranged to give

$$\frac{X_t}{X_0} = e^{-K_d t} \qquad (14.7)$$

Equation (14.6) will plot a straight line on semilog paper with the X_t/X_0 values on the y-axis and the t values on the x-axis. The slope will be $-K_d/2.303$.

The most important design and operational parameters for aerobic digesters are shown in Tables 14.1 and 14.2. The hydraulic detention time is equal to the digester volume divided by the feed sludge flow rate. If sludge recycle is used, the recycled sludge flow rate is not incorporated in the computations. The hydraulic detention time depends upon the nature of the sludge and the operational temperature. Waste activated sludge is more readily degradable than a

Design Considerations

Table 14.1. Aerobic Digester Design Parameters

Parameter	Value
Hydraulic Detention Time, days at 20°C	
Primary and waste activated sludge or trickling filter sludge	18–22[l]
Waste activated sludge from a biosorption or contact stabilization plant (no primary settling)	16–18[l]
Waste activated sludge	12–16[l]
Minimum Solids Retention Time (SRT), days	
Primary and waste activated sludge	15–20[a,h,k,l]
Waste activated sludge	10–15[c,j,n]
Maximum Solids Retention Time (SRT), days	45–60
Solids Concentration, mg/ℓ	Up to 50,000
Organic Loading, lb volatile solids per ft^3-day	0.04 to 0.20[c,g,i,k,l,n]
Volume Loading, ft^3 per capita	1.5–4[d,i,n,q]
Operating Temperature	$> 15°$ C[f]
Volatile Solids Destruction, percent	40–75[a,n,o]
Solids Destruction, percent	35–55[a,n,o]
Oxygen Requirements	
Primary sludge, lb O_2/lb BOD_5 destroyed	1.9[l]
Waste activated sludge, lb O_2/lb solids destroyed	2.0[l,n]
Trickling filter humus, lb O_2/lb solids destroyed	2.0[l]
Minimum Dissolved Oxygen, mg/ℓ	1.0–2.0[l,m,n,o]
Mixing Requirements	
Diffused aeration for primary and waste activated sludge, scfm/1000 ft^3	> 60[c,n,p]
Diffused aeration for waste activated sludge, scfm/1000 ft^3	20–35[b,n,p]
Mechanical aeration for primary and waste activated sludge or waste activated sludge, hp/1000 ft^3	0.5–1.25[e]

a. E. Barnhart, "Application of Aerobic Digestion to Industrial Waste Treatment," *Proceedings of the 16th Annual Purdue Industrial Waste Conference*, May 1961, p. 612.

b. R. S. Burd, "A Study of Sludge Handling and Disposal," FWPCA Publication no. WP-20-4, May 1968.

c. D. E. Dreier, "Aerobic Digestion of Solids," *Proceedings of the 18th Annual Purdue Industrial Waste Conference*, May 1963, p. 123.

d. D. E. Dreier, "Discussion on Aerobic Sludge Digestion," by N. Jaworski, G. W. Lawton, & G. A. Rohlich, paper presented at the Conference on Biological Waste Treatment, Manhattan College, New York City, April 1960.

e. Environmental Protection Agency, *Sludge Treatment and Disposal*, EPA Process Design Manual, September 1979.

f. Environmental Protection Agency, *Upgrading Existing Wastewater Treatment Plants*, EPA Process Design Manual, October 1974.

g. R. H. L. Howe, "What to Do with Supernatant," *Waste Engineering* 30, no. 1 (January 1959): 12.

h. N. Jaworski, G. W. Lawton, & G. A. Rohlich, "Aerobic Sludge Digestion," paper presented at the Conference on Biological Waste Treatment, Manhattan College, New York City, April 1960.

i. C. E. Levis, C. R. Miller, & L. R. Bosburg, "Design and Operating Experiences Using Turbine Dispersion for Aerobic Ingestions," *Jour. WPCF* 43, no. 3 (March 1971):417.

mixture of primary and waste activated sludge; thus, the hydraulic detention time will be less. Table 14.1 gives the usual detention times for various sludges at a 20°C operating temperature. The required hydraulic detention time at temperatures other than 20°C is given by the equation

$$\theta_{h_2} = \theta_{h_{20}} \cdot \theta^{20-T_2} \qquad (14.8)$$

where

θ_{h_2} = hydraulic detention time in days at T_2 (°C);
$\theta_{h_{20}}$ = hydraulic detention time at 20°C;
T_2 = temperature, °C.

Table 14.2. Reaction Rate or Degradation Constants (K_d)

Type Sludge	Temperature (°C)	Solids Concentration (mg/ℓ)	K_d(days^{-1})
Primary and Waste Activated Sludges	15	32,000	0.017[h]
Primary and Waste Activated Sludges	20	32,000	0.180[h]
Primary and Waste Activated Sludges	35	32,000	0.177[h]
Primary and Waste Activated Sludges	—	—	0.30[a]
Waste Activated Sludges			
Municipal Wastes	25	7,800	0.71[n]
Municipal Wastes	25	12,400	0.62[n]
Municipal Wastes	25	15,050	0.51[n]
Municipal Wastes	25	21,260	0.44[n]
Municipal Wastes	25	22,700	0.34[n]
Municipal Wastes	—	—	0.28[a]
Municipal and Textile Waste	—	—	0.43[a]
Pharmaceutical Waste	—	—	0.46[a]
Spent Sulfite Liquor Waste	—	—	0.19[a]
Primary and Waste Activated Sludge, Pulp and Paper Waste	—	—	0.14[a]

j. R. C. Loehr, "Aerobic Digestion: Factors Affecting Design," *Water and Sewage Works* 112 (30 November 1965):R169.

k. J. F. Malina & H. M. Burton, "Aerobic Stabilization of Primary Wastewater Sludge," *Proceedings of the 19th Annual Purdue Industrial Waste Conference*, May 1964, p. 123.

l. Metcalf and Eddy, Inc., *Wastewater Engineering, Treatment, Disposal and Reuse* (New York: McGraw-Hill, 1979).

m. C. N. Randall & C. T. Koch, "Dewatering Characteristics of Aerobically Digested Sludge," *Jour. WPCF* 41, no. 5, part 2 (May 1969):R215.

n. T. D. Reynolds, "Aerobic Digestion of Thickened Waste Activated Sludge," *Proceedings of the 28th Annual Purdue Industrial Waste Conference*, Part 1, May 1973, p. 12.

o. L. Ritter, "Design and Operating Experiences Using Diffused Aeration for Sludge Digestion," *Jour. WPCF* 42, no. 10 (October 1970):1782.

p. R. Smith, R. G. Eilers, & E. D. Hall, "A Mathematical Model for Aerobic Digestion," EPA, Office of Research and Monitoring, Advanced Waste Treatment Research Laboratory, Cincinnati, Ohio, February 1973.

q. Texas State Department of Health, "Design Criteria for Sewage Systems," Austin, Tex., 1976.

The value of θ ranges from 1.02 to 1.11; however, most θ values are in the upper half of this range. This high θ range shows that the process is extremely temperature dependent at usual detention times. Because appreciable air-water contact occurs during aeration, it may be assumed that the temperature of the digester contents could approach the average monthly temperature.

The solids retention time is given by

$$\theta_s = \frac{X}{\Delta X} \qquad (14.9)$$

where

$\quad X \quad$ = pounds of solids in the digester;
$\quad \Delta X$ = pounds of solids produced per day in the digested sludge.

In Eq. (14.9) the solids may be total or volatile solids. If the digester is operated without a thickener, the solids in the digester will be at the same concentration as the solids in the feed sludge and the solids retention time will be equal to the hydraulic detention time. However, if a thickener is used, the solids concentration in the digester will be greater than in the feed sludge and the solids retention time will be larger than the hydraulic detention time.

The reaction or degradation constant (K_d), Table 14.2, depends on the nature of the sludge, its solids concentration, and the temperature. Waste activated sludge, which is readily degraded, will have a high K_d value. A mixture of primary and waste activated sludge is not as easily degraded and, as a result, will have a lower K_d value. As the concentration of a sludge is increased, the K_d value decreases. The K_d value and the fraction of the sludge that is biodegradable may be determined by laboratory or pilot plant studies.

The solids concentration may be up to 50,000 mg/ℓ; however, the usual range is from 25,000 to 35,000 mg/ℓ. It is essential to have a proper thickener-clarifier design in order to insure that solids are maintained at these levels. Mixing requirements to keep the solids in suspension must be determined in addition to oxygen requirements.

The organic loading, based on limited data, should be in the range of 0.04 to 0.20 lb of volatile solids per day-ft^3, which is similar to anaerobic digester loadings. Usually, aerobic digester design is based on hydraulic detention time and solids retention time; however, the organic loading should be checked. The volumetric loading expressed in terms of population equivalents is usually from 1.5 to 4 ft^3 per capita equivalent.

The degree of volatile solids reduction or destruction

depends on the nature of the sludge, the hydraulic detention time, the solids retention time, and the operating temperature, provided that no toxic substances are present in inhibitory concentrations. For mixtures of primary and waste activated sludge at operating temperatures of 15 to 35°C, the volatile solids destruction has approximated a pseudo–first-order reaction. The required time and degree of volatile solids destruction will vary with the characteristics of the sludge and the operating temperature. The maximum volatile solids destruction is usually from 40 to 75 percent and the maximum total solids destruction is usually from 35 to 55 percent.

The maximum oxygen required for aerobic digestion of primary sludge is 1.9 lb of O_2 per pound of BOD_5 destroyed. Normally BOD_5 destruction in aerobic digestion is from 80 to 90 percent. For waste activated sludge and trickling filter humus, a maximum of 2.0 lb of O_2 are required per pound of solids destroyed. The dissolved oxygen concentration in a digester should be at least 1 to 2 mg/ℓ so that aerobic conditions will be maintained and digestion will proceed normally. If the oxygen concentration drops below 1 mg/ℓ, the inner portions of the solids will become anaerobic and poor digestion will occur. This, in turn, results in a sludge that has poor dewatering characteristics.

The mixing requirements depend primarily upon the nature of the sludge, the solids concentration, the sludge temperature, and the tank depth. Usually, the air required for oxygen requirements is sufficient to attain adequate mixing; however, mixing requirements should be evaluated. A rational approach to mixing is needed; however, data collected so far indicate the following. For diffused compressed air used for the aerobic digestion of waste activated sludge, the air required for mixing is from 20 to 35 scfm per 1000 ft^3. The higher air requirements are needed for high suspended solids concentrations, relatively low sludge temperatures, and relatively shallow tanks. The lower air requirements are for low suspended solids concentrations, relatively high sludge temperatures, and relatively deep tanks. For diffused compressed air used in the aerobic digestion of mixtures of primary and waste activated sludge, the air required for mixing is about 60 scfm per 1000 ft^3 or more. For mechanical aeration, mixing requirements are usually from 0.5 to 1.25 hp per 1000 ft^3, and bottom mixers are required if the solids are greater than about 8000 mg/ℓ. Since the degree of mixing by a mechanical aerator decreases with tank depth, high power levels are required for relatively deep tanks and lower power levels are used for relatively shallow tanks.

The tanks or basins used for aerobic digestion are similar to those used for the activated sludge process, and both diffused compressed air and mechanical aeration are used. The criteria for aeration tanks and aeration systems, presented in Chapter 9, may also be used for aerobic digester design.

Thickener-Clarifiers

The majority of aerobic digestion systems have the thickener-clarifier placed downline from the digester. The thickener-clarifier should be designed for the feed sludge plus the recycled sludge flow. It should have the capacity to clarify the supernatant liquor leaving the digester system and also to thicken the settled sludge so that it may be recycled to the head of the digester. Inadequate thickener-clarifier capacity is a common problem that results in high suspended solids concentrations in the supernatant liquor and high recycled sludge flow rates. A high solids concentration in the supernatant liquor is undesirable because it will have a pronounced effect on units that receive the flow. Surface skimmers should be provided to remove any floating materials, such as grease, from the system.

Example Problem 14.1
Aerobic Digestion

An aerobic digester is to be designed to treat the waste activated sludge flow from an industrial wastewater treatment plant. The waste sludge flow is 35,000 gal per day and has a suspended solids content of 10,000 mg/ℓ (specific gravity = 1.00). The volatile suspended solids are 73 percent of the suspended solids. The aerobic digester will operate at 25,000 mg/ℓ suspended solids (specific gravity = 1.01). The diffused air system to be used has an oxygen transfer efficiency of 6 percent. The operating conditions are 20°C and one atmosphere pressure.

Batch aerobic digestion tests have been performed using the waste activated sludge at a concentration of 25,000 mg/ℓ and at 20°C. The results obtained are shown in Table 14.3.

Consider that the suspended solids in the supernatant liquor are negligible and may be assumed to be zero. Assume that the digested sludge is withdrawn daily. Batch settling tests on the digesting sludge gave an allowable overflow rate of 400 gal/day-ft^2 and an allowable solids flux of 6 lb/day-ft^2. The flow diagram for the aerobic digester system is shown in Figure 14.3. The process design is required for the digester and the thickener-clarifier. The design is to be based on the hydraulic detention time using the fresh sludge flow and the solid

ENVIRONMENTAL ENGINEERING

retention time. The digester should have 95 percent biodegradable solids destruction.

Table 14.3. Results of Batch Aerobic Digestion Tests

vss at $t = 0$	73.0 percent
FSS at $t = 0$	27.0 percent
Digestion time, t_d	12 days
vss at $t = t_d$	48.0 percent
FSS at $t = t_d$	52.0 percent
Fraction of ss destroyed	48.1 percent
Fraction of vss destroyed	65.9 percent
K_d	$0.325 \ \text{day}^{-1}$
Settled suspended solids concentration (specific gravity $= 1.01$)	$37{,}000 \ \text{mg}/\ell$

Figure 14.3. Schematic Diagram of an Aerobic Digester System
Note: The units for the flows and so on are Q_S, Q_D, Q_E = gal/day; x_1, x_2, x_3, x_4, x_R = mass or weight fractions, ppm.

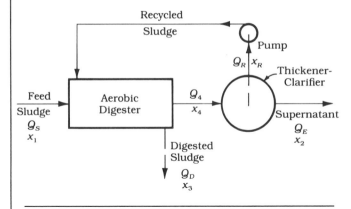

Solution

The fresh sludge solids represent an Input. Fresh solids $= (35{,}000 \ \text{gal/day})(8.34 \ \text{lb/gal})(10{,}000/10^6) = 2919$ lb/day. vss $= (2919)(0.73) = 2131$ lb/day. FSS $= (2919)(0.27) = 788$ lb/day. The digested sludge solids represent an Output. vss destroyed $= (2131)(0.659) = 1404$ lb/day. For steady state, [Output] $=$ [Input] $-$ [Decrease due to reaction]. Thus, vss in digested sludge $= 2131 - 1404 = 727$ lb/day. FSS in the digested sludge $= 788 - 0 = 788$ lb/day. The total solids in the digested sludge $= 727 + 788 = 1515$ lb/day. Percent vss in the digested sludge $= (727/1515)(100 \ \text{percent}) = 48.0$ percent. Assume a digestion time of 12 days based on flow. The volume of the digester, V, is

$$V = (35,000 \text{ gal/day})(12 \text{ days})(\text{ft}^3/7.48 \text{ gal})$$

$$= \underline{56,150 \text{ ft}^3}$$

The digested sludge flow, Q_d, is

$$Q_D = (1515 \text{ lb/day})(\text{gal/8.34 lb})(1/1.01)(10^6/37,000)$$

$$= \underline{4,861 \text{ gal/day}}$$

The supernatant flow, Q_E, when sludge is withdrawn, is given by a material balance on flows in Figure 14.3. The material balance is

$$Q_S = Q_D + Q_E$$

or

$$Q_E = Q_S - Q_D = 35,000 - 4861 = \underline{30,139 \text{ gal/day}}$$

The supernatant flow when no sludge is withdrawn is given by

$$Q_E = 35,000 - 0 = \underline{35,000 \text{ gal/day}}$$

The solids retention time for 95 percent stabilization is given by the equation $X_t/X_0 = e^{-K_d t}$ or $t = -1/K_d \ln X_t / X_0$. Since $X_t = 0.05 X_0$, the time is

$$t = [-1/(0.325/\text{day})]\ln(0.05/1) = \underline{9.22 \text{ days}}$$

The solids in the digester $= (56,150 \text{ ft}^3)(62.4 \text{ lb/ft}^3)(25.000/10^6)(1.01) = 88,470 \text{ lb}$. The solids retention time based on the solids produced per day in the digested sludge is

$$\theta_s = 88,470 \text{ lb}/(1515 \text{ lb/day}) = \underline{58.4 \text{ days}}$$

Since $\theta_s > t$, 95 percent stabilization will occur. A solids balance on the thickener-clarifier gives $(Q_4)(25,000) = (Q_R)(37,000)$ or $Q_4 = 1.48 Q_R$. A material balance on the flows to and from the digester will give

$$Q_S + Q_R = Q_4 + Q_D$$

Substituting for Q_4 and rearranging gives

$$Q_R = (1/0.48)(Q_S - Q_D)$$

The maximum value of Q_R occurs when $Q_D = 0$. Or

$$Q_{R(max)} = (35,000 \, \text{gal/day})(1/0.48)(1 \, \text{day}/1440 \, \text{min})$$
$$= \underline{50.6 \, \text{gpm}}$$

The oxygen required based on 2.0 lb O_2/lb solid destroyed is $(1404 \, \text{lb/day})(2.0 \, \text{lb/lb}) = 2808 \, \text{lb/day}$. At 20°C and one atmosphere air contains $0.0175 \, \text{lb} \, O_2/\text{ft}^3$. Thus, the air required for bio-oxidation is

$$\text{Scfm} = (2808 \, \text{lb/day})(\text{day}/1440 \, \text{min})$$
$$\times (\text{scf}/0.0175 \, \text{lb})(1/0.06)$$
$$= \underline{1857 \, \text{scfm}}$$

The air required for mixing using 20 scfm per $1000 \, \text{ft}^3$,

$$\text{Scfm} = (20/1000)(56,250) = \underline{1125 \, \text{scfm}}$$

Therefore, the air for bio-oxidation controls. The area required for clarification in the thickener-clarifier is

$$\text{Area} = (35,000 \, \text{gal/day})(\text{day-ft}^2/400 \, \text{gal}) = \underline{87.5 \, \text{ft}^2}$$

To determine the area for thickening requires the value of Q_4. $Q_4 = 1.48 \, Q_R = (1.48)(50.6) = 74.9 \, \text{gpm}$. The area for thickening is

$$\text{Area} = (74.9 \, \text{gal/min})(1440 \, \text{min/day})(8.34 \, \text{lb/gal})$$
$$\times (25,000/10^6)(1.01)(\text{day-ft}^2/6 \, \text{lb})$$
$$= \underline{3785 \, \text{ft}^2}$$

Thickening controls, thus $(\pi/4)(D^2) = 3785 \, \text{ft}^2$,

$$D = 69.4 \, \text{ft} \quad \underline{\text{Use 70.0 ft for standard size.}}$$

An aerobic digester is to be designed to treat the waste activated sludge flow from an industrial wastewater treatment plant. The waste sludge flow is 133,000 ℓ/d and has a suspended solids content of 10,000 mg/ℓ (specific gravity = 1.00). The volatile suspended solids are 73 percent of the suspended solids. The aerobic digester will operate at 25,000 mg/ℓ suspended solids (specific gravity = 1.01). The diffused air system to be used has an oxygen transfer efficiency of 6 percent. The operating conditions are 20°C and one atmosphere pressure.

Batch aerobic digestion tests have been performed using the waste activated sludge at a concentration of 25,000 mg/ℓ and at 20°C. The results obtained are shown in Table 14.4.

Table 14.4. Results of Batch Aerobic Digestion Tests

VSS at $t = 0$	73.0 percent
FSS at $t = 0$	27.0 percent
Digestion time, t_d	12 days
VSS at $t = t_d$	48.0 percent
FSS at $t = t_d$	52.0 percent
Fraction of SS destroyed	48.1 percent
Fraction of VSS destroyed	65.9 percent
K_d	0.325 day^{-1}
Settled suspended solids concentration (specific gravity = 1.01)	37,000 mg/ℓ

Consider that the suspended solids in the supernatant liquor are negligible and may be assumed to be zero. Assume that the digested sludge is withdrawn daily. Batch settling tests on the digesting sludge gave an allowable overflow rate of 16 m³/d-m² and an allowable solids flux of 30 kg/d-m². The flow diagram for the aerobic digester system is shown in Figure 14.3. The process design is required for the digester and the thickener-clarifier. The design is to be based on the hydraulic detention time using the fresh sludge flow and the solid retention time. The digester should have 95 percent biodegradable solids destruction.

Solution

The fresh sludge solids represent an Input. Fresh solids = (133,000 ℓ/d)(10,000 mg/ℓ)(kg/10⁶ mg) = 1330 kg/d. VSS = (1330)(0.73) = 970.9 kg/d. FSS = (1330)(0.27) = 359.1 kg/d. The digested sludge solids represent an Output. VSS destroyed = (970.9)(0.659) = 639.8 kg/d. For steady state, [Output] = [Input] − [Decrease due to reaction]. Thus, VSS in digested sludge = 970.9 − 639.8 = 331.1 kg/d. FSS in the digested sludge = 359.1

$- 0 = 359.1$ kg/d. The total solids in the digested sludge $= 331.1 + 359.1 = 690.2$ kg/d. Percent vss in the digested sludge $= (331.1/690.2)(100$ percent$) = 48.0$ percent. Assume a digestion time of 12 days based on flow. The volume of the digester, V, is

$$V = (133{,}000 \, \ell/d)(12 \, d)(m^3/1000 \, \ell) = \underline{1596 \, m^3}$$

The digested sludge flow, Q_D, is

$$Q_D = (690.2 \, kg/d)(\ell/37{,}000 \, mg)(10^6 \, mg/kg)(1/1.01)$$
$$= \underline{18{,}469 \, \ell/d}$$

The supernatant flow, Q_E, when sludge is withdrawn, is given by a material balance on flows in Figure 14.3. The material balance is

$$Q_S = Q_D + Q_E$$

or

$$Q_E = Q_S - Q_D = 133{,}000 - 18{,}469 = \underline{114{,}531 \, \ell/d}$$

The supernatant flow when no sludge is withdrawn is given by

$$Q_E = 133{,}000 - 0 = \underline{133{,}000 \, \ell/d}$$

The solids retention time for 95 percent stabilization is given by the equation $X_t/X_0 = e^{-K_d t}$ or $t = -1/K_d \ln X_t/X_0$. Since $X_t = 0.05 \, X_0$, the time is

$$t = [- 1/(0.325/day)] \ln (0.05/1) = \underline{9.22 \, days}$$

The solids in the digester $= (1596 \, m^3)(1000 \, \ell/m^3)(25{,}000 \, mg/\ell)(kg/10^6 mg) \, (1.01) = 40{,}299$ kg. The solids retention time based on the solids in the digested sludge is

$$\theta_s = (40{,}299 \, kg)/(690.2 \, kg/d) = \underline{58.4 \, days}$$

Since $\theta_s > t$, 95 percent stabilization will occur. A solids balance on the thickener-clarifier gives $(Q_4)(25{,}000) = (Q_R)(37{,}000)$ or $Q_4 = 1.48 \, Q_R$. A material balance on the flows to and from the digester will give

$$Q_S + Q_R = Q_4 + Q_D$$

Substituting for Q_4 and rearranging gives

$$Q_R = (1/0.48)(Q_S - Q_D)$$

The maximum value of Q_R occurs when $Q_D = 0$. Or

$$Q_{R(max)} = (133{,}000 \, \ell/d)(1/0.48)(d/1440 \, min)$$

$$= 192.4 \, \ell/min$$

The oxygen required based on 2.0 kg O_2/kg solid destroyed is $(639.8 \, kg/d)(2.0 \, kg/kg) = 1280 \, kg/d$. At 20°C and one atmosphere air contains 0.2806 kg O_2/m^3. Thus, the air required for bio-oxidation is

$$m^3/h = (1280 \, kg/d)(m^3/0.2806 \, kg)(d/24 \, h)(1/0.06)$$

$$= 3168 \, m^3/h$$

The air required for mixing using 1200 m^3/h per 1000 m^3 is

$$m^3/h = (1200 \, m^3/h\text{-}1000 \, m^3)(1596 \, m^3)$$

$$= 1915 \, m^3/h$$

Therefore, the air for bio-oxidation controls. The area required for clarification in the thickener-clarifier is

$$\text{Area} = (133{,}000 \, \ell/d)(m^3/1000 \, \ell)(d\text{-}m^2/16 \, m^3)$$

$$= 8.31 \, m^2$$

To determine the area for thickening requires the value of Q_4. $Q_4 = 1.48 \, Q_R = (1.48)(192.4) = 284.8 \, \ell/min$. The area for thickening is

$$\text{Area} = (284.8 \, \ell/min)(25{,}000 \, mg/\ell)(1 \, kg/10^6 \, mg)$$

$$\times (1440 \, min/d)(d\text{-}m^2/30 \, kg)$$

$$= 341.8 \, m^2$$

Thickening controls, thus $(\pi/4)(D^2) = 341.8 \, m^2$,

$$D = 20.86 \, m$$

References

Ahlberg, N. R., and Boyko, B. I. 1972. Evaluation and Design of Aerobic Digesters. *Jour. WPCF* 44, no. 4:634.

AWARE, Inc. 1974. *Process Design Techniques for Industrial Waste Treatment*, edited by C. E. Adams and W. W. Eckenfelder, Jr. Nashville, Tenn.: Enviro Press.

Barnhart, E. 1961. Application of Aerobic Digestion to Industrial Waste Treatment. *Proceedings of the 16th Annual Purdue Industrial Waste Conference.*

Burd, R. S. 1968. A Study of Sludge Handling and Disposal. FWPCA Publication no. WP-20-4.

Cameron, J. W. 1972. Aerobic Digestion of Activated Sludge to Reduce Sludge Handling Costs. Paper presented at 45th Annual Conference, Water Pollution Control Federation, Atlanta, Ga.

Conway, R. A., and Ross, R. D. 1980. *Handbook of Industrial Waste Disposal.* New York: Van Nostrand Reinhold.

Dreier, D. E. 1960. Discussion on Aerobic Sludge Digestion, by Jaworski, N., Lawton, G. W., and Rohlich, G. A. Paper presented at the Conference on Biological Waste Treatment, Manhattan College, New York City.

Dreier, D. E. 1963. Aerobic Digestion of Solids. *Proceedings of the 18th Annual Purdue Industrial Waste Conference.*

Dreier, D. E., and Obma, C. A. 1963. *Aerobic Digestion of Solids.* Walker Process Equipment Co., bulletin no. 26-S-18194, Aurora, Ill.

Eckenfelder, W. W., Jr. 1980. *Principles of Water Quality Management.* Boston: CBI Publishing.

Eckenfelder, W. W., Jr. and O'Connor, D. J. 1961. *Biological Waste Treatment.* New York: Pergamon Press.

Environmental Protection Agency (EPA). 1979. *Sludge Treatment and Disposal.* EPA Process Design Manual, Washington, D.C.

Environmental Protection Agency (EPA). 1974. *Upgrading Existing Wastewater Treatment Plants.* EPA Process Design Manual, Washington, D.C.

Great Lakes–Upper Mississippi Board of State Sanitary Engineers. 1978. Recommended Standards for Sewage Works. Ten state standards, Albany, N.Y.

Howe, R. H. L. 1959. What to Do with Supernatant. *Waste Engineering* 30, no. 1:12.

Jaworski, N., Lawton, G. W., and Rohlich, G. A. 1960. Aerobic Sludge Digestion. Paper presented at the Conference on Biological Waste Treatment, Manhattan College, New York City.

Lawton, G. W., and Norman, J. D. 1964. Aerobic Digestion Studies. *Jour. WPCF* 36, no. 4:495.

Levis, C. E., Miller, C. R., and Bosburg, L. E. 1971. Design and Operating Experiences Using Turbine Dispersion for Aerobic Digestion. *Jour. WPCF* 43, no. 3:417.

Loehr, R. C. 1965. Aerobic Digestion: Factors Affecting Design. *Water and Sewage Works* 112:R169.

Malina, J. F., and Burton, H. M. 1964. Aerobic Stabilization of Primary Wastewater Sludge. *Proceedings of the 19th Purdue Industrial Waste Conference.*

Metcalf and Eddy, Inc. 1979. *Wastewater Engineering, Treatment, Disposal and Reuse.* New York: McGraw-Hill.

Ramalho, R. S. 1977. *Introduction to Wastewater Treatment Processes.* New York: Academic Press.

Randall, C. N., and Koch, C. T. 1969. Dewatering Characteristics of Aerobically Digested Sludge. *Jour. WPCF* 41, no. 5, part 2:R215. (Research Supplement)

Reynolds, T. D. 1967. Aerobic Digestion of Waste Activated Sludge. *Water and Sewage Works* 114, no. 22:37.

Reynolds, T. D. 1973. Aerobic Digestion of Thickened Waste Activated Sludge. *Proceedings of the 28th Annual Purdue Indus-*

trial *Waste Conference.*

Ritter, L. 1970. Design and Operating Experiences Using Diffused Aeration for Sludge Digestion. *Jour. WPCF* 42, no. 10:1782.

Schroeder, E. D. 1977. *Water and Wastewater Treatment.* New York: McGraw-Hill.

Smith, A. R. 1971. Aerobic Digestion Gains Favor. *Water and Waste Engineering* 8, no. 2:24.

Smith, R., Eilers, R. G., and Hall, E. D. 1973. A Mathematical Model for Aerobic Digestion. EPA, Office of Research and Monitoring, Advanced Waste Treatment Research Laboratory, Cincinnati, Ohio.

Sundstrom, D. W., and Klei, H. E. 1979. *Wastewater Treatment.* Englewood Cliffs, N.J.: Prentice-Hall.

Texas State Department of Health. 1976. Design Criteria for Sewage Systems. Austin, Tex.

Water Pollution Control Federation (WPCF). 1977. *Wastewater Treatment Plant Design.* WPCF Manual of Practice no. 8, Washington, D.C.

Problems

1. An aerobic digester having a flowsheet as shown in Figure 14.4 is to be designed to treat the waste activated sludge from an industrial wastewater treatment plant. The feed sludge flow is 35,000 gal per day, the solids content is 10,000 ppm, the fraction of feed solids destroyed per day is 45 percent, the design suspended solids concentration in the digester is 25,000 ppm, and laboratory tests of the digesting sludge show that it will settle to 35,000 mg/ℓ. For Figure 14.4, Q = flow rate and x = solids in ppm.

Figure 14.4.

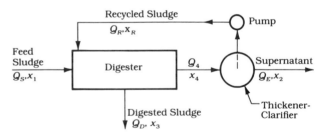

Other pertinent data are: digester size = 15 ft × 24 ft × 115 ft, volatile solids content in the feed waste activated sludge = 70 percent, $x_2 \cong 0$, oxygen requirements = 2.0 lb per pound of solids destroyed, operating air and water temperature = 20°C (68°F), operating pressure = 1 atmosphere, oxygen transfer efficiency = 6.0 percent, oxygen concentration = 0.0175 lb oxygen per ft^3 of air at 20°C and 1 atmosphere, thickener-clarifier overflow rate = 400 gal/day-ft^2, and the thick-

ener-clarifier solids loading = 6 lb/day-ft^2. The specific gravity of the digesting and digested sludge = 1.01. Determine:
a. The oxygen requirements, lb per day.
b. The air required for oxygen mass transfer, scfm and scfm per 1000 ft^3.
c. The solids loading, lb volatile solids/day-ft^3.
d. The flows, Q_E and Q_D, gal/day.
e. The minimum sludge recycle flow, Q_R, gal/day, to maintain 25,000 ppm of solids in the digester. Also, maximum recycle.
f. The area and diameter of the thickener-clarifier.

2. An aerobic digester having a flowsheet the same as Problem 1 is to treat waste activated sludge from a bio-sorption (contact stabilization) plant. Pertinent data are: waste activated sludge flow = 20,000 gal/day, solids content in waste activated sludge = 10,000 ppm, solids content in the digester = 35,000 ppm, recycled sludge solids content = 47,000 ppm, specific gravity of the wet sludge in recycle line = 1.01, solid loading for the thickener-clarifier = 5 lb/day-ft^2, overflow rate for the thickener-clarifier = 400 gal/day-ft^2, and solids in the supernatant equal zero. If the solids destruction in the digester and the digested sludge flow are ignored, determine:
a. The recycled sludge rate of flow, gal/day.
b. The thickener-clarifier area required for clarification.
c. The thickener-clarifier area required for thickening.
d. The diameter of the thickener-clarifier if it is circular in plan view.

3. An aerobic digester having a flowsheet the same as Problem 1 is to be designed to treat the primary and waste activated sludge from a municipal wastewater treatment plant serving 45,000 persons. Pertinent data are: wastewater flow = 380 ℓ/d-cap, influent BOD$_5$ = 225 mg/ℓ, BOD$_5$ removal by primary clarifier = 35 percent, influent suspended solids = 320 mg/ℓ, suspended solids removal by primary clarifier = 65 percent, waste activated sludge flow = 2.0 percent of plant influent flow, solids in waste activated sludge flow = 1.0 percent, solids in primary and secondary sludge = 3.0 percent, volatile solids in primary and secondary sludge = 68 percent, specific gravity of primary and secondary sludge = 1.01, hydraulic detention time of digester = 20 days, length:width ratio of digester = 6:1, width:depth ratio of digester = 2:1, solids destroyed in

digester = 55 percent of feed solids, primary BOD$_5$ destroyed in digester = 90 percent, operating solids concentration = 45,000 ppm, 1.9 kg of oxygen required per kg BOD$_5$ destroyed, 2.0 kg of oxygen required per kg waste activated sludge solids destroyed, aeration efficiency = 6.0 percent, plant elevation = 1000 ft, 0.2806 kg O$_2$/m^3 at 20°C and 1 atmosphere, settled sludge solids = 65,000 ppm, specific gravity of the settled sludge = 1.01, operating temperature = 20°C and K_d = 0.180 day^{-1}. Determine:

a. The size of the digester.
b. The material balance on the system giving all flows and solids content.
c. The minimum and maximum recycle flow rate.
d. The air required, m^3/h and m^3/h-1000 m^3.
e. The solids retention time for 95 percent stabilization.
f. The actual solids retention time.

Chapter Fifteen

SOLIDS HANDLING

I n water treatment plants, such as the coagulation and softening type, and in wastewater treatment plants, sludges are produced that require disposal. Solids handling consists of the satisfactory processing of sludges for ultimate disposal to the environment, which is usually land disposal. In some cases, ultimate disposal may be by the air or water environment. In conventional wastewater treatment plants, the sludges are principally of an organic nature, such as primary sludge, excess (waste) activated sludge, or trickling filter humus; however, in advanced wastewater treatment, the sludges are mainly of a chemical nature because they result from coagulation or precipitation. Although the sludges in advanced wastewater treatment are considered chemical sludges, they do have some organic matter associated with them. In a wastewater treatment plant, which has both organic and chemical sludges, it is usually advantageous not to mix the two but, instead, process each sludge separately. Mixing of the two types of sludges frequently results in a mixture that is difficult to process. In water treatment, the sludges from coagulation or precipitation are primarily of a chemical nature, although some organic material will be present.

Wastewater Treatment Plant Sludges (Organic)

Organic sludges are those produced by primary, secondary, or sludge treatment in biological wastewater treatment plants. Fresh or undigested sludges, such as primary or secondary sludges, are rather difficult to dewater as compared to digested sludges. The operations and processes used in solids handling in biological wastewater treatment plants may be classified as thickening, stabilization, conditioning, dewatering, incineration, and disposal. Primary sludges and secondary sludges, such as excess activated sludge and trickling filter humus, are frequently mixed by adding the secondary sludge to the incoming raw wastewater and resettling it in the primary clarifiers. This thickens, to a certain degree, the primary sludge and, in the case of excess activated sludge, the secondary sludge. The solids content of various municipal wastewater sludges is given in Table 13.4.

Typical solids handling systems for primary and secondary sludge mixtures mixed in this manner are (1) anaerobic digestion, sand-bed dewatering, and landfill disposal of the solids; (2) anaerobic digestion, chemical conditioning, vacuum filtration or centrifugation, and landfill disposal of the solids; (3) gravity thickening, chemical conditioning, vacuum filtration or centrifugation, multiple-hearth or fluidized-bed incineration, and landfill disposal of the ash; and (4) gravity thickening, chemical conditioning, vacuum filtration, flash drying, and land disposal of the solids as a fertilizer.

When primary sludge and excess activated sludge are thickened separately, typical solids handling systems are (1) gravity thickening of the primary sludge, air-flotation thickening of the excess activated sludge, anaerobic digestion, chemical conditioning, vacuum filtration or centrifugation, and landfill disposal of the solids, and (2) gravity thickening of the primary sludge, air-flotation thickening of the excess activated sludge, chemical conditioning, vacuum filtration or centrifugation, multiple-hearth or fluidized-bed incineration, and landfill disposal of the ash.

Excess activated sludge from activated sludge plants without primary clarification (extended aeration and contact stabilization plants) is frequently handled by the following system: aerobic digestion, sand-bed dewatering, and landfill disposal of the solids.

In general, the larger the treatment plant, the more complex will be the solids handling system because the most favorable disposal of the solids is desired.

Thickening

Thickening consists of increasing the solids content of a sludge which, as a result, reduces the volume of sludge to be processed by subsequent units. If, for example, a sludge having 3 percent solids content is thickened to 6 percent solids, the volume of sludge leaving the thickener is $\frac{3}{6}$ or $\frac{1}{2}$

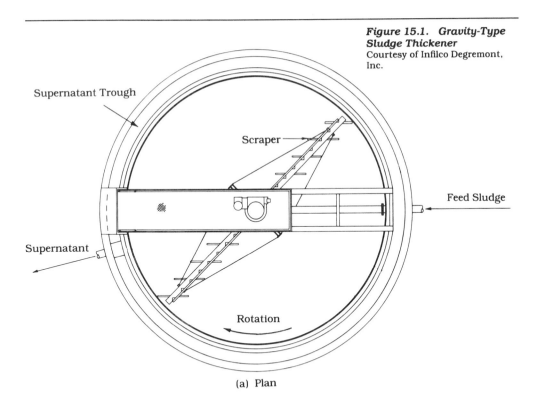

Figure 15.1. Gravity-Type Sludge Thickener
Courtesy of Infilco Degremont, Inc.

Supernatant Trough

Scraper

Feed Sludge

Supernatant

Rotation

(a) Plan

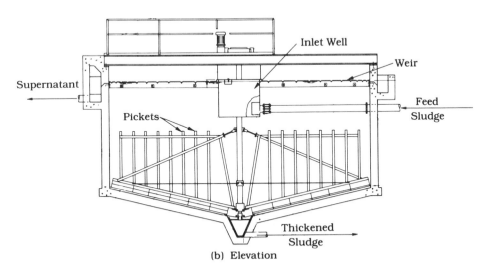

Inlet Well

Weir

Supernatant

Feed Sludge

Pickets

Thickened Sludge

(b) Elevation

the volume of the feed sludge.

Gravity thickening, which is the most common thickening method, can be used when the specific gravity of the solids is greater than one. Figure 15.1 shows a gravity-type

Figure 15.2. Schematic Diagram of an Air-Flotation System with Recycle

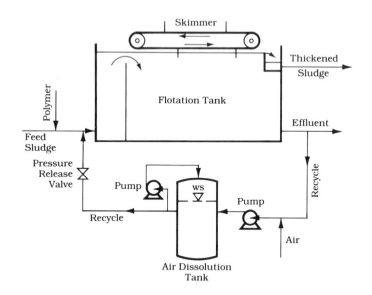

thickener, and it can be seen that it is very similar to a clarifier except that the floor slope is much greater and vertical pickets are mounted on the trusswork. The feed sludge enters the thickener at the inlet well in the center of the unit. As the flow goes radially outward from the inlet well, the solids settle into the sludge blanket that occupies the lower depths of the thickener. The thickened sludge leaves as the underflow and the supernatant liquid leaves by discharging over the effluent weirs. The pickets mounted on the moving trusswork slowly rake through the sludge mass and break up sludge arching or bridging, thus releasing entrained water that rises to the surface, where it leaves as the supernatant. Also, the slow agitation will release gas bubbles that have been formed and the scraper blades move the thickened sludge to the center of the unit where it is removed as the underflow. The thickener must serve two purposes: the thickening of the sludge solids and the clarification of the supernatant liquid. The design parameters for these two requirements are the solid loading in lb solids/day-ft^2 and the overflow rate in gal/day-ft^2.

These may be determined by the same methods of analysis as for activated sludge final clarifiers, presented in Chapter 3. The degree of thickening that may be accomplished, as expressed by the thickened sludge solids content, is a function of the nature of the sludge. Typical performance of gravity thickeners processing municipal sludges is shown in Table 15.1. The overflow rate is usually 400 to 800 gal/day-ft^2 based on the supernatant flow. It is important that the solids retention time in the thickener is not great enough to cause anaerobic conditions that will result in odors. The solids retention time is usually taken as the volume of the sludge blanket divided by the thickened sludge flow rate and is usually from 0.5 to 2 days (EPA, 1979). Low values should be used in warm climates because microbial action will be greater. The sidewater depth of a thickener is usually about 10 to 12 ft.

Table 15.1. Gravity Thickener Performance for Processing Fresh Municipal Sludges

Type of Sludge	Solids Loading (lb/day-ft^2)	Percent Solids Unthick- ened	Thickened
Primary[a]	20–30	2.5–5	8–10
Waste Activated Sludge[a]	5–6	0.5–1	2.5–3
Waste Activated Sludge, Pure Oxygen	7–34	1.5–3.5	4–8
Trickling Filter Humus[a]	8–10	5–10	7–10
Primary and Waste Activated Sludge[a]	6–10	4–5	5–10
Primary and Trickling Filter Humus[a]	10–12	3–6	7–9

a. D. Newton, "Thickening by Gravity and Mechanical Means," in *Sludge Concentration, Filtration and Incineration*, University of Michigan School of Health Continuing Education Series 113, no. 4 (1964).

Thickening using dissolved *air flotation* may be used whenever the specific gravity of the solids is near unity. It has been very successfully used for thickening excess activated sludges. Figure 15.2 shows a schematic diagram of a typical flotation system, consisting of the flotation tank, the air dissolution tank, and the necessary pumps. Flotation uses the formation of air bubbles on the solid particles to buoy them to the surface, where they are skimmed from the flotation tank. It usually is accomplished by three methods. The first method, which is the most commonly used, is depicted in Figure 15.2. It consists of pressurizing the recycled flow in an air tank to dissolve air gases in the flow. The recycled flow then mixes with the feed sludge flow and the mixture enters the flotation tank. Due to the release in pressure, the air gases come out of solution and form air bubbles around the sludge solids, which buoys them to the surface. They are skimmed off and leave as the thickened sludge flow. The air pressure in the dissolution tank is usually 40 to 70 psi and the recycle flow is usually from 30 to 150 percent of the feed flow. The second method, depicted in Figure 15.3, consists of pressurizing the entire feed flow to dissolve air gases. The pressure is released just prior to the flow entering the flotation tank. The flotation action and removal of solids is the same as for the first method. Usually, pressurizing the entire flow is limited to small installations. The third method consists of aerating the feed sludge flow so that the liquid is saturated with dissolved air gases. The flow then goes to a covered flotation tank, where a vacuum is drawn upon the system. This causes the air gases to come out of solution and buoy the solids to the surface, where they are removed by skimming. Air-flotation thickening of waste-activated sludge has an advantage over gravity thickening because the solids content in the thickened sludge is usually higher and the cost of the flotation system is usually less. Flotation aids, such as polymers, aid significantly and nearly all installations employ them. Air-flotation thickening of waste activated sludge with about 10,000 mg/ℓ suspended solids will result

Figure 15.3. Schematic Diagram of an Air-Flotation System without Recycle

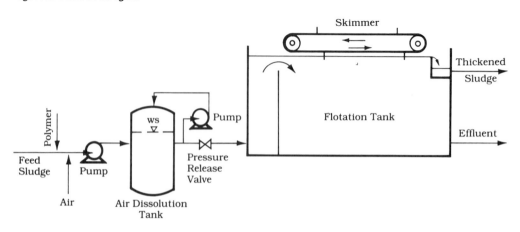

in a thickened sludge of 4 to 6 percent solids when flotation aids are used. Without flotation aids, the thickened sludge will be about 3 to 5 percent solids (Smith, J. E., Hathaway, S. W., Farrell, J. B., & Dean, R. B., 1972).

Although waste activated sludges from industrial wastewater treatment plants may flotate with more or less ease than waste activated sludges from municipal plants, nearly all waste activated sludges may be thickened by flotation. In the use of dissolved air flotation for the thickening of waste activated sludges, the air/solids ratio is usually from 0.005 to 0.06. The design procedure for dissolved air-flotation units, both with and without recycle, is presented in Chapter 16.

In *centrifugal thickening*, solid bowl, disc-nozzle, and basket-type centrifuges have been used to thicken waste activated sludges. The disc-nozzle type, shown in Figure 15.4, has been the most popular. In this centrifuge the feed sludge enters as shown in Figure 15.4 and passes upward through the discs with each disc acting as a single centrifuge. The centrifuged solids slide down each disc due to the centrifugal force and the angle of the discs, and leave through the outlet port. The centrifugation of waste activated sludge with 6500 to 20,000 mg/ℓ suspended solids

---

Figure 15.4. Section through a Disc-Nozzle Centrifuge

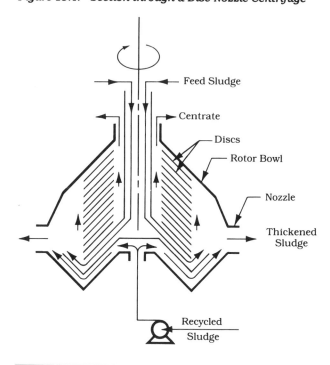

has thickened the sludges to 3 to 6 percent solids (WPCF, 1977). If the waste activated sludge is from a plant not having primary clarification (for example, the biosorption process), the sludge should be screened prior to centrifugal thickening. Since centrifuges have appreciable maintenance and energy costs, centrifugation is usually used when gravity or flotation thickening are not feasible due to such factors as space limitations.

Stabilization

Stabilization consists of treating the sludge so that future decomposition by biological action does not occur. It results in a sludge that will not undergo bacterial decomposition, has good dewatering characteristics, and has very little odor. In the case of municipal sludges, it also results in a low pathogen content.

Anaerobic digestion, which has been discussed in detail in Chapter 13, consists of the bio-oxidation of organic sludges under anaerobic conditions. One advantage that it has over aerobic digestion is the production of methane gas, which is a usable by-product.

Aerobic digestion, which has been presented in Chapter 14, consists of the bio-oxidation of organic sludges under aerobic conditions. One advantage it has over anaerobic digestion is that less technical skill is required in operation since the process avoids many of the operational problems of anaerobic digestion.

Lagoons have found limited use in the digestion and disposal of waste activated sludges. In order to function as oxidation ponds and not merely as sludge storage lagoons, they must be designed properly. Oxidation ponds consist of earthen embankments with a pond depth of 3 to 5 feet. In their operation, the sludge settles to the bottom, where it undergoes anaerobic digestion. The top portion of the water depth remains aerobic, which prevents odors. If the organic loadings are too great, anaerobic conditions will occur throughout the water depth and odors will be produced. Oxidation ponds are limited to warm climates having high sunlight intensities and where land is relatively cheap and occasional odors are not objectionable.

Wet combustion uses chemical combustion of sludge with oxygen under wet conditions. One manufactured process uses pressures from 1000 to 1800 psi. Injected steam raises the temperature of the sludge to a self-sustaining level, which is usually about 500°F (260°C). The chemical oxidation or combustion that occurs is not complete, however; the treated sludge is sterile and dewaters readily.

In *chemical treatment,* lime has been used to stabilize primary and excess activated sludges, temporarily preventing odors. If sufficient slaked lime is added to raise the pH above 11, almost all biological action ceases and a calcium carbonate precipitate is formed that greatly assists in the dewatering characteristics. Essentially all *E. coli* and *Salmonella typhosa* have been killed by high lime treatment at a pH of 11 to 11.5 and a 4-hour contact time at 15°C (59°F) (Riehl, M. L., Weiser, H. H., & Rheims, B. T., 1952). It has been found that lime-stabilized sludges disposed in lagoons have a gradual pH reduction and a gradual increase in biological action. Thus, lime treatment should be considered as a temporary sludge stabilization method because bio-oxidation will eventually occur. Chlorine has also been used for sludge stabilization. It differs from lime in that chemical oxidation occurs because chlorine is a strong oxidizing agent. Usually, about 2000 mg/ℓ chlorine concentration is employed, which produces a treated sludge that is stable and dewaters well on drying beds. The pH, however, will be about 2, and usually the sludge must be neutralized prior to dewatering by mechanical methods. Neutralization is necessary due to the corrosive nature of the sludge, and also because the low pH interferes with any chemical conditioners that are required. The high concentration of chloramines in the water removed during dewatering must be considered in the treatment of such flows.

The *composting* of thickened and dewatered undigested primary and secondary sludges has been applied to a limited extent in the United States. Both soil and mechanical aeration systems have been employed, usually with the sludge being mixed with solid wastes. The stabilization is essentially an aerobic digestion process. The lack of use is primarily due to the lack of demand for the compost product.

Conditioning

Conditioning consists of treating a sludge prior to dewatering and sometimes thickening to enhance its dewatering characteristics.

In *chemical treatment,* both organic polymers and inorganic coagulants have been used for conditioning; however, the present trend is to use organic polymers (polyelectrolytes) due to their effectiveness. Sludges are stable suspensions, and the addition of polyelectrolytes causes coagulation and aggregation of the sludge solids and a release of the entrained water. Polyelectrolytes may have positive charges (cationic type), negative charges (anionic type), or neutral charges (nonionic type). The cationic type is the most commonly used in sludge conditioning because the sludge particles are slightly negatively charged. The coagulation of sludge particles occurs mainly due to bridging action between reactive groups on the polymer and the sludge particles; however, some coagulation by charge reduction does occur. Polymers may be obtained in powder, pellet, or liquid form; however, the liquid form is the easiest to mix with water. If the sludge is to be incinerated, polymers have an advantage over inorganic coagulant salts because inorganic coagulants decrease the fuel value of the treated sludge. Polymer requirements depend on the nature of the sludge and may be from 3 to 25 lb per ton of dry solids. Inorganic coagulants, particularly ferric chloride, have been used in sludge conditioning. Because of the high coagulant dosages used, the pH drop can be significant because ferric chloride is an acidic salt. The pH is maintained in the optimum coagulation range by the addition of hydrated lime, which also assists in the formation of the ferric hydroxide precipitate. Coagulation by inorganic salts is caused by charge reduction, enmeshment in the hydroxide precipitate, and also by chemical bridging.

Elutriation consists of washing a sludge with water to remove components such as bicarbonates and fine solids that result in high chemical dosages when subsequent conditioning is done using inorganic salts. In particular, it has been used for anaerobically digested sludges that are to be chemically coagulated prior to dewatering. However, due to the widespread use of polymers, elutriation is rapidly decreasing in popularity.

Heat treatment has, in particular, been used to condition waste activated sludges. Since the solids are mainly microbial cells, they contain significant amounts of water that are not removed by the usual dewatering methods. Heat treatment, by such processes as the Porteous process (EPA, 1979) causes the cells to rupture or lyse; the intercellular material, which has a high water content, is released. The temperature, pressure, and treatment durations are usually from 350 to 400°F (177 to 204°C), 250 to 400 psi, and 15 to 60 minutes, respectively. The dewatering characteristics of excess activated sludge are enhanced, and dewatering without chemical conditioning using vacuum filtration will produce a cake of 30 to 50 percent solids. The principal disadvantages of the process are the disposal of the supernatant liquor produced and the treatment of odorous sidestream gases. The supernatant is odorous, has a BOD_5 of 5000 to 15,000 mg/ℓ, and has a high nitrogen and phosphorus content. The supernatant has a pH of 4 to 5 and, after neutralization, it may be treated by biological methods.

Ash obtained by sludge incineration may be added to sludges to increase their dewatering characteristics. The addition of ash to fresh primary and secondary sludges at a ratio of 0.25 to 0.5 lb ash per pound of dry solids has resulted in a vacuum-filtered cake containing 33 percent solids (EPA, 1979).

Dewatering

Dewatering consists of removing as much water from a sludge as possible so that the dewatered sludge volume to be subsequently processed is minimized. Fresh waste activated sludge is very difficult to dewater and prior conditioning is nearly always required.

The most commonly used mechanical-type dewatering device is the *rotary vacuum filter.* It consists of a cylindrical drum having a filter medium, which may be cloth, wire mesh, or coil springs. The filters may be classified as drum, belt, and coil-spring type. Figure 15.5 shows a schematic section of the drum type which has a filter medium that is maintained on the drum throughout the filter cycle. The drum slowly rotates through the sludge vat and the vacuum within the drum causes the sludge cake to form. Much of the water in the cake is drawn through the medium into the drum. The cake formed undergoes further dewatering in the sector shown in Figure 15.5. The piping inside the drum is arranged so that a vacuum exists in the sector shown for cake formation and dewatering. In the discharge sector, the cake is released by a flow of compressed air blowing through the medium, and a scraper knife peels it from

---

Figure 15.5. Schematic of a Rotary Vacuum Filter (Drum Type)

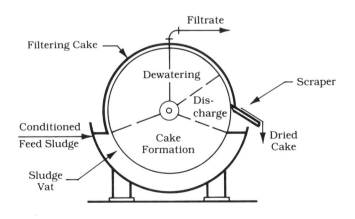

the drum. Then, the medium may be spray washed prior to submergence in the sludge vat. Figure 15.6 shows a schematic cross section of the belt-type filter. In the discharge sector, the medium or belt and the cake leave the drum and, as the belt passes over the first rollers, the cake falls from the belt. The belt is then washed and returns to the drum. The coil-spring type has a belt made of springs and operates in a manner similar to the belt type. The rotary vacuum filter requires appurtenances, such as a vacuum pump, filtrate receiver and pump, and a sludge feed pump. For vacuum filters, the performance is measured by the pounds of dry solids filtered per hour per square foot filter area, and also by the percent dry solids in the filter cake. The performance of vacuum filters processing municipal sludges is shown in Table 15.2. Generally, rotary vacuum filters have shown good results when filtering primary sludge or mixtures of primary and secondary sludges.

Table 15.2. Vacuum Filter Performance for Processing Municipal Sludges Conditioned By Ferric Chloride and Lime

Type of Sludge	Solids Loading (lb/hr-ft^2)	Percent Solids of Cake
Fresh Primary	6–8	25–38
Fresh Primary and Waste Activated Sludge	4–5	16–25
Fresh Primary and Waste Activated Sludge (Pure Oxygen)	5–6	20–28
Fresh Primary and Trickling Filter Humus	4–6	20–30
Digested Primary (Anaerobic)	5–8	25–32
Digested Primary and Waste Activated Sludge (Anaerobic)	4–5	14–22

From the Environmental Protection Agency, *Sludge Treatment and Disposal*, EPA Process Design Manual, Washington, D.C., 1979.

The basic equation for evaluating vacuum filtration is (Coackley, P., 1956)

$$\frac{dV}{dt} = \frac{\Delta p A^2}{\mu(wVR + AR_f)} \tag{15.1}$$

where

V = volume of filtrate;

t = time;

Vacuum Filter at a
Wastewater Treatment Plant
Courtesy Envirex, Inc., a Rexnord
Co.

Figure 15.6. Schematic of a Rotary Vacuum Filter
(Belt Type)

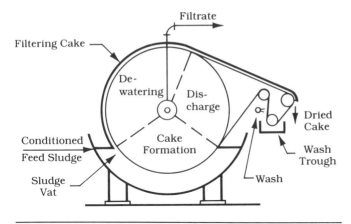

Δp = vacuum pressure differential;

A = filter area;

μ = absolute or dynamic viscosity of the filtrate;

w = weight of the dry sludge solids per unit volume
     of filtrate;

R = specific resistance of the sludge cake;

R_f = specific resistance of the filter medium.

The specific resistance, R, of the sludge cake is the principal
design parameter for vacuum filtration. The specific resistance is the reciprocal of the permeability. The specific resistance may be obtained in the laboratory using the

apparatus shown in Figure 15.7. For a constant vacuum pressure, Eq. (15.1) can be integrated and rearranged to give

$$\frac{t}{V} = \left(\frac{\mu\, wR}{2\Delta pA^2}\right) V + \frac{\mu R_f}{\Delta pA} \tag{15.2}$$

Using the laboratory apparatus, a batch of sludge can be dewatered and the volume of the filtrate, V, versus filtration time, t, can be obtained. Equation (15.2) is of the straight-line form, $y = mx + b$; thus, plotting the t/V values on the y-axis and the V values on the x-axis will result in a straight line. The slope of the line, m, will be equal to the term ($\mu wR/2\Delta pA^2$) in Eq. (15.2). The specific resistance may then be obtained from

$$R = \left(\frac{2\Delta pA^2}{\mu w}\right) m \tag{15.3}$$

From the vacuum filtration test, the value of w may be obtained by drying the solids on the filter paper and dividing this weight by the volume of filtrate obtained during the

---

Figure 15.7. Laboratory Apparatus for Vacuum Filtration Testing

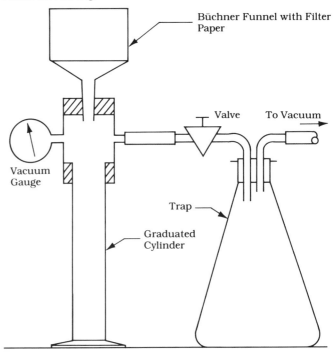

Büchner Funnel with Filter Paper

Valve

To Vacuum

Vacuum Gauge

Trap

Graduated Cylinder

test. The w value may also be computed using the dry solids concentration in the unfiltered sludge and in the sludge cake. The equation for w is

$$w = \frac{\gamma}{[(1 - x)/x] - [(1 - x_c)/x_c]}$$

(15.4)

where

w = weight of solids per unit volume of filtrate;
γ = specific weight of water;
x = dry solids content in the unfiltered sludge expressed as a fraction;
x_c = dry solids content in the cake expressed as a fraction.

If sludge conditioning with chemicals is to be done, the sludge must be conditioned prior to the vacuum filtration test. The effects of various chemical conditioners and their concentrations on vacuum filtration can be evaluated by a series of vacuum filtration tests. Conditioners will reduce the specific resistance of a sludge. The plot of specific resistance versus chemical dosage, shown in Figure 15.8, will give the optimum chemical dosage. Some organic sludges

---

Figure 15.8. Specific Resistance versus Chemical Dosage Used for Conditioning

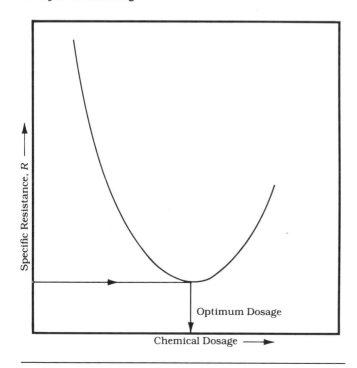

produce compressible sludge cakes causing the specific resistance to vary with the vacuum level used. For these sludges, it is necessary to obtain a relationship giving specific resistance as a function of the vacuum level. A relationship for this is

$$R = r'\Delta p^s \tag{15.5}$$

where r' is the cake compressibility constant and s is the coefficient of compressibility. These constants are a characteristic of each sludge cake. To evaluate these constants, a series of vacuum filter tests may be conducted at different vacuum levels. Taking the log of Eq. (15.5) gives $\log R = \log r' + s \log \Delta p$. Thus, plotting the specific resistance, R, on the y-axis of log-log paper and the vacuum levels, Δp, on the x-axis will result in a straight line with a slope equal to s. The y-axis value when $\Delta p = 1$ equals the constant r'. If $s = 0$, the sludge cake is incompressible and the specific resistance is independent of the vacuum level. Beginning with Eq. (15.1), a derivation can be made to give the following performance equation for vacuum filters:

$$Y = \left(\frac{2\Delta p w \alpha}{\mu R \theta}\right)^{1/2} \tag{15.6}$$

where

Y = filter yield;
Δp = vacuum pressure differential;
w = weight of dry sludge solids per unit volume of filtrate;
α = ratio of form time/cycle time;
μ = absolute or dynamic viscosity of the filtrate;
R = specific resistance of the sludge cake;
θ = cycle time of the rotating drum—that is, time for one revolution.

In the derivation of Eq. (15.6), it is assumed that the specific resistance of the filter medium, R_f, is negligible. The filter yield is expressed as mass/(area)(time), such as lb/hr-ft^2. Proper dimensional conversions must be made in using Eq. (15.6) to obtain the yield in the desired terms. In inspecting Eq. (15.6), it can be seen that the filter yield can be increased by increasing Δp and α or by decreasing θ. Usually, the highest possible vacuum is maintained and operational control is done by varying the drum rotational speed, which changes θ.

A conditioned digested sludge is to be dewatered on a rotary vacuum filter under a vacuum of 25 in. Hg. A vacuum filtration test has been done in the laboratory; it gave a specific resistance of 2.4×10^7 sec^2/gm, the unfiltered sludge solids were 4.5 percent by dry weight, and the sludge cake had 32 percent dry solids. The filtration temperature is 70°F, the cycle time is 6 min, and the form time is 40 percent of the cycle time. Determine the filter yield in lb/hr-ft^2.

Solution

The dry solids per unit volume of filtrate is given by

$$w = \frac{62.4 \text{ lb/ft}^3}{[(1 - 0.045)/0.045] - [(1 - 0.32)/0.32]}$$

$$= 3.27 \text{ lb/ft}^3$$

The specific resistance is

$$R = (2.4 \times 10^7 \text{ sec}^2/\text{gm})(454 \text{ gm/lb})$$

$$= 1.0896 \times 10^{10} \text{ sec}^2/\text{lb}$$

The vacuum differential is

$$\Delta p = \left(\frac{14.7 \text{ psi}}{29.92 \text{ in. Hg}} \right) (25 \text{ in. Hg}) = 12.28 \text{ lb/in}^2$$

The operating temperature of 70°F is 21.1°C. From a table giving the viscosity of water, a viscosity, μ, of 0.9980 centipoise is obtained by interpolating between 21°C and 22°C. The viscosity is

$$\mu = (0.9980 \text{ centipoise})(6.72 \times 10^{-4})$$

$$= 6.707 \times 10^{-4} \text{ lb/sec-ft}$$

The cycle time, θ, is (6 min)(60 sec/min) = 360 sec. The α value is 0.40. Thus, the yield is given by

$$Y = \left[\frac{2(12.28) \text{ lb}}{\text{in}^2} \middle| \frac{144 \text{ in}^2}{\text{ft}^2} \middle| \frac{3.27 \text{ lb}}{\text{ft}^3} \middle| 0.40 \right.$$

$$\left. \times \frac{10^4 \text{ sec-ft}}{6.707 \text{ lb}} \middle| \frac{\text{lb}}{1.0896 \times 10^{10} \text{ sec}^2} \middle| \frac{}{360 \text{ sec}} \right]^{1/2}$$

$$= 1.326 \times 10^{-3} \, \text{lb/sec-ft}^2$$

$$= (1.326 \times 10^{-3} \, \text{lb/sec-ft}^2)(3600 \, \text{sec/hr})$$

$$= \underline{4.77 \, \text{lb/hr-ft}^2}$$

Example Problem 15.2
Filter Drum Area

If the sludge in Example Problem 15.1 has a flow of 85,000 gal/day, determine the required filter drum area.

Solution

The pounds of solids per hour, assuming that the specific gravity is approximately 1.0, is

$$\text{lb/hr} = (85,000 \, \text{gal}/24 \, \text{hr})(8.34 \, \text{lb/gal})(4.5 \, \text{lb}/100 \, \text{lb})$$

$$= 1329.2$$

Thus, the required area is

$$A = (1329.2 \, \text{lb/hr})(\text{hr-ft}^2/4.77 \, \text{lb}) = \underline{279 \, \text{ft}^2}$$

The *pressure filter* is a batch operated filter consisting of numerous vertical filter plates mounted on a horizontal shaft, as depicted in Figure 15.9. The filter plates have recesses covered with filter cloth and filtrate drain holes that discharge into an outlet port. The vertical plates are movable and are mounted on a horizontal shaft, as shown in Figure 15.10. When the press is closed, the plates are pressed together with either a mechanical screw-type ram or a hydraulic ram. The sealing pressure must be sufficient

Figure 15.9. Section through a Filter Press

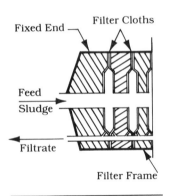

Figure 15.10. Elevation of a Filter Press

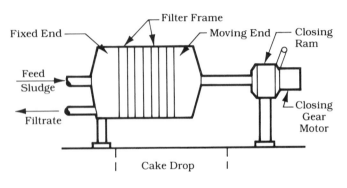

to withstand the hydraulic pressure that exists during the filtration cycle. During the filtration cycle, the sludge is pumped under pressure into the space between the plates, as shown in Figure 15.9. The filtrate passes through the filter cloth into the filtrate drain holes and passes from the press by an outlet port that is in the axial direction. The pressures used during the filtration cycle are usually from 100 to 250 psi. The sludge flow is maintained until all the spaces within the press are filled with sludge cake, which usually is less than one hour. The fluid pressure is maintained until all the filterable water has passed from the cake. To remove the cake, the back pressure is released and the moving end is slid to the closing gear mechanism. Each individual plate then moves over the gap between the plates and the moving end, thus allowing the cake to fall from the press onto a conveyor or into a trailer. Frequently, presses are located at a sufficient height above ground level to allow the filtered cake to fall directly into trailers. Once the press has been emptied, it is closed again and the filtration cycle is repeated. The complete filtration cycle time is from about 1.5 to 2.5 hours. Filter presses are one of the most successful dewatering means used to dewater waste activated sludges. The following percentages of dry solids in the processed filter cake have been obtained for conditioned municipal sludges dewatered by filter presses: (1) raw primary sludge, 45 to 50 percent, (2) raw primary and fresh waste activated sludge, 45 to 50 percent, (3) fresh waste activated sludge, 50 percent, and (4) digested primary and waste activated sludge, 45 to 50 percent (Forster, H. W., 1972).

Belt presses have numerous designs; however, the design shown in Figure 15.11 is frequently used. The press consists of two converging belts mounted on rollers. The lower belt is made of fine wire mesh and is very porous. As conditioned sludge moves onto the belt, some of the en-

---

Figure 15.11. Schematic of a Belt Press

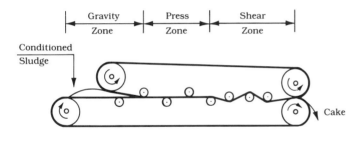

Sludge Filter Press at a Water Treatment Plant. The press dewaters sludge from alum coagulation.

trained water drains through the belt in the gravity drain zone. The sludge passes through the press zone where the pressure provided by the converging belts and rollers causes dewatering. Finally, in the shear zone, the sludge must zigzag at changing directions, thus giving further dewatering. The following percentages of dry solids in the processed sludge cake have been obtained for municipal sludges chemically conditioned by polymers and dewatered by belt filter presses: (1) raw primary, 28 to 44 percent, (2) raw primary and waste activated sludge, 20 to 35 percent, (3) raw primary and trickling filter humus, 20 to 40 percent, (4) digested primary (anaerobic digestion), 26 to 36 percent, (5) digested primary and waste activated sludge (anaerobic digestion), 12 to 18 percent, and (6) digested primary and waste activated sludge (aerobic digestion), 12 to 18 percent (EPA, 1979). The belt press has low energy requirements and does not require a vacuum or pressure pump.

The main types of *centrifuges* that have been successfully used in dewatering sludges are the countercurrent solid bowl, shown in Figure 15.12, the concurrent solid-bowl, and the basket type centrifuge shown in Figure 15.13.

In a countercurrent solid-bowl centrifuge, shown in Figure 15.12, both the bowl and the scroll conveyer rotate, with the speed of rotation of the scroll slightly less than that of the bowl. The feed sludge, upon entering, is slung outward from the shaft and, due to the centrifugal force, the solids collect on the inner wall of the bowl. The scroll conveyer, which spins at a slightly slower speed than the bowl, moves the collected solids down the bowl and up the

Figure 15.12. Continuous Countercurrent Solid-Bowl Centrifuge
Courtesy of Sharples-Stokes Division, Pennwalt Corp.

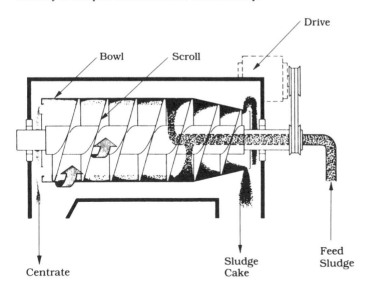

Figure 15.13. Schematic of the Operation of a Basket Centrifuge

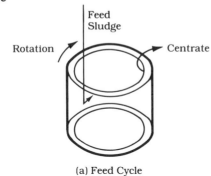

(a) Feed Cycle

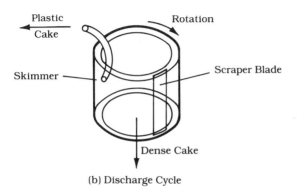

(b) Discharge Cycle

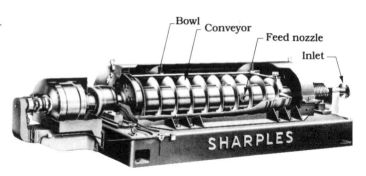

inclined beach, where the cake is discharged. The clarified liquid moves countercurrent to the solids, and the centrate and the cake discharge at the opposite ends of the machine. The concurrent solid-bowl centrifuge has a rotating bowl and scroll conveyer like the countercurrent type; however, the liquid and the collected solids move in the same direction. The centrate and the cake leave the same end of the machine. The main advantage of the concurrent-type centrifuge is the concurrent movement of both the liquid and the collected solids. This tends to cause less disturbance to the solids than the countercurrent type. As a result, the centrate has less suspended solids content.

The following percentages of dry solids in the processed sludge cake have been obtained for municipal sludges chemically conditioned and dewatered by solid-bowl centrifugation: (1) raw primary sludge, 25 to 35 percent, (2) raw primary and fresh waste activated sludge, 15 to 23 percent, (3) raw primary and fresh trickling filter humus, 24 to 30 percent, (4) digested primary sludge (anaerobic digestion), 25 to 35 percent, (5) digested primary and waste activated sludge (anaerobic digestion), 14 to 18 percent, (6) digested primary and waste activated sludge (anaerobic digestion), 10 to 14 percent, and (7) digested primary and trickling filter humus (anaerobic digestion), 20 to 28 percent (Bird Machine Co., 1971). In all of the above cases, the solids capture was about 90 percent.

The basket-type centrifuge, shown in Figure 15.13, has a semi–batch-type operation. The sludge is introduced into the vertical mounted spinning bowl and the solids are collected at the sides. The centrate overflows the bowl rim and is discharged. Once the solids have built up to a maximum thickness, the feed sludge is stopped, the rotation of the bowl is decreased, and a scraper blade peels the sludge cake from the walls and it discharges out of the bottom of the bowl. Once the cake is removed, the bowl rotation is increased and the feed sludge flow is resumed. The following percentages of dry solids in the processed sludge cake have been obtained for municipal sludges chemically conditioned by polymers and dewatered by basket-type centrifugation: (1) raw primary sludge, 25 to 30 percent, (2) raw primary and fresh waste activated sludge, 12 to 14 percent, and (3) digested primary and waste activated sludge, 12 to 14 percent (EPA, 1979). The solids capture was from 75 to 97 percent.

Example Problem 15.3
Sludge Centrifugation

A chemically conditioned digested sludge is to be dewatered using a centrifuge that has 90 percent solids capture. The digested primary and secondary sludge has 8 percent dry solids and a specific gravity of 1.03. The digested sludge flow is 20,000 gal/day and the centrifuged cake has 18 percent dry solids. Determine the lb/day of cake and the gal/day of centrate.

Solution

Use one day of operation as a basis for calculation. The pounds of dry solids are (20,000 gal/day)(8.34 lb/gal)(1.03)(8 lb/100 lb)(1 day) or 13,744 lb. The pounds of dry solids in the cake are (13,744 lb)(0.90) or 12,370 lb. The pounds of cake are (12,370 lb)(100 lb/18 lb) or 68,722 lb. A material balance on the total flows gives [Input] = [Output] or

(20,000 gal)(8.34 lb/gal)(1.03)
= 68, 722 lb + (C)(8.34 lb/gal)

where C is the gallons of centrate. From the previous equation, C = 12,360 gal. Thus, the cake is 68,700 lb/day and the centrate is 12,400 gal/day.

Example Problem 15.3 SI
Sludge Centrifugation

A chemically conditioned digested sldge is to be dewatered using a centrifuge that has 90 percent solids capture. The digested primary and secondary sludge has 8 percent dry solids and a specific gravity of 1.03. The digested sludge flow is 76,000 ℓ/d and the centrifuged cake has 18 percent dry solids. Determine the kg/d of cake and the ℓ/d of centrate.

Solution

Use one day of operation as a basis for calculation. The

kilograms of dry solids are (76,000 ℓ/d)(kg/ℓ)(1.03)(8 kg/100 kg)(1d) or 6262.4 kg. The kilograms of dry solids in the cake are (6242.4 kg)(0.90) or 5636.2 kg. The kilograms of cake are (5636.2 kg)(100 kg/18 kg) or 31,312 kg. A material balance on the total flows gives [Input] = [Output] or

$$(76,000\ \ell/\text{d})(\text{kg}/\ell)(1.03) = 31,312\ \text{kg} + (C)(\text{kg}/\ell)$$

where C is the liters of centrate. From the previous equation, C = 46,968 ℓ. Thus, the cake is 31,300 kg/d and the centrate is 47,000 ℓ/d.

Basket Centrifuge
Courtesy of Sharples-Stokes
Division. Pennwalt Corporation.

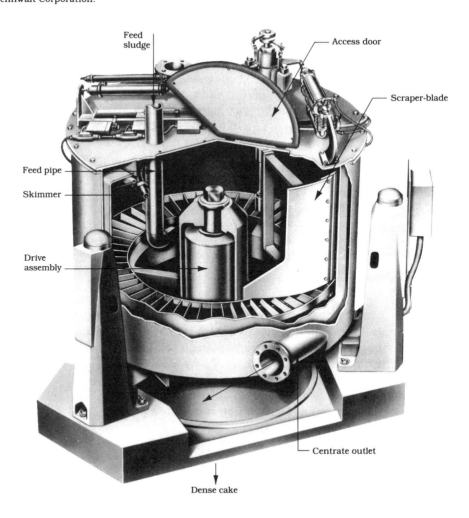

Feed sludge

Access door

Scraper-blade

Feed pipe

Skimmer

Drive assembly

Centrate outlet

Dense cake

Sludge-drying beds, shown in Figure 15.14, are the most commonly used dewatering method for digested sludges from small- to medium-size installations. If the population is greater than about 25,000 persons, other methods of dewatering should be considered because of the space required and labor costs involved. Drying beds usually are 20 to 30 feet wide, 25 to 125 feet long, and consist of a 6- to 10-in. coarse sand layer above a 6- to 12-in. graded gravel layer. The subgrade is sloped to a drain tile line that is beneath the center of the bed and extends the full length of the bed. At least two beds must be provided. Sludge is usually applied in an 8- to 12-in. depth. Dewatering occurs by both drainage from the bed and evaporation of water to the atmosphere. Sludge drainage occurs up to about two days after application of the sludge to the bed. Usually, the drainage is returned to the head of the plant. The final dewatering is by evaporation to the atmosphere. In dry weather, sludge may dewater sufficiently for removal after about

---

Figure 15.14. Sludge-Drying Bed

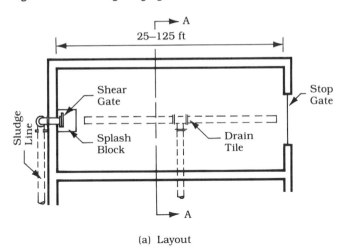

(a) Layout

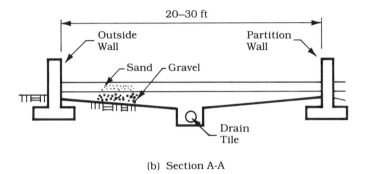

(b) Section A-A

---

two to four weeks, with the solids content of the dried sludge being about 30 to 40 percent. The amount of bed area required per population equivalent is dependent primarily on the type of digestion and the rainfall. The area required per capita for digested primary and secondary sludge mixtures is given by the equation (Texas State Department of Health, 1976)

$$\hat{A} = K(0.01R + 1.0) \tag{15.7}$$

where

$\hat{A}$ = area per capita, ft^2/cap;

K = factor depending upon type of digestion, K = 1.0 for anaerobic digestion, K = 1.6 for aerobic digestion;

R = annual rainfall, in.

Only digested sludges may be dewatered on drying beds because fresh sludges will decompose anaerobically and create severe odor problems.

The main dewatering action in *drying lagoons* is atmospheric evaporation. Lagoons are a simple, low-cost method of dewatering; however, they can occupy considerable land space in humid and rainy climates. Lagoons are usually used in pairs and must be surrounded by dikes to prevent surface runoff from entering. In operation, a lagoon is filled with about 2 feet of digested sludge and allowed to dry, then the filling and drying are repeated. Once the lagoon is filled with dried sludge, the sludge is removed. The

Sludge Drying Beds at a Wastewater Treatment Plant. Note concrete splash block at shear gate inlet and concrete pads for truck wheels.

ENVIRONMENTAL ENGINEERING

solids loading is usually from 2.2 to 2.4 lb/yr-ft^3 (Great Lakes–Upper Mississippi Board of State Sanitary Engineers, 1978). Only digested sludges should be dewatered by lagoons if severe odor problems are to be avoided.

Heat Drying

Heat drying is used when fresh sludge is to be processed to produce a fertilizer. Heat drying is accomplished by flash drying, kiln drying, or multiple-hearth furnaces. In flash drying, the fresh sludge is mixed with some previously dried sludge, then the mixture is dried by a stream of hot combustion gases from a fuel-fired furnace. The dried sludge is separated from the gases by a cyclone. A portion of the dried material is mixed with incoming fresh sludge, while the remainder is removed as the dried product. The vapors from the cyclone are returned to the furnace for deodorization. In the kiln dryer, the fresh sludge enters a sloped rotating kiln and moves through it countercurrent to hot combustion gases that dry the material. The dried sludge is discharged from the lower end of the kiln. In the multiple-hearth furnace, the sludge is dried as it passes downward through the hearths. The temperature used is usually about 700 to 900°F (371 to 482°C) and is not sufficient to cause sludge combustion. In all the drying methods, a fuel must be employed that involves considerable operational costs. The dried sludge has about 10 percent moisture, is sterile, and is a good fertilizer and soil conditioner. Although fertilizer production from sludge is not profit making, it does reduce handling costs by about 30 to 40 percent. It is limited to large operations due to the complexity of the process and the fact that the product demands are for large quantities.

Incineration

Incineration consists of dry combustion of a sludge to produce an inert ash. The ash is usually disposed in a sanitary landfill. The fuel requirements depend on the fuel value of the sludge solids and the water content. Raw primary and undigested secondary sludges will have fuel values from about 7000 to 8000 Btu/lb dry solids and, if they are dewatered to about 25 percent solids, incineration will be self-sustaining. Auxiliary fuel is required only during the start-up of the incinerator. Digested sludge nearly always requires an auxiliary fuel because the fuel value is only about half that of fresh sludge. The two types of incinerators used are the multiple-hearth type, shown in Figure 15.15, and the fluidized-bed incinerator, shown in Figure 15.16.

Figure 15.15. Multiple-Hearth Incinerator
Courtesy of Nichols Engineering and Research Corp.

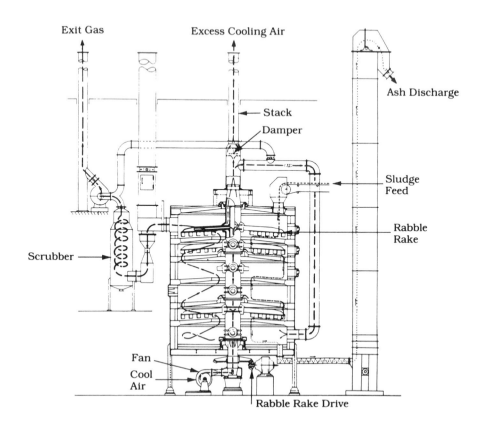

Figure 15.16. Fluidized-Bed Incinerator

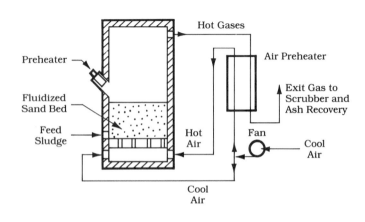

The *multiple-hearth incinerator,* depicted in Figure 15.15, consists of a hearth-type furnace in which the dewatered sludge is fed to the first hearth at the top. In the incinerator, the solids move downward by the raking action of the rabble arms. In the upper hearths, the water content is vaporized and the sludge solids are dried at a temperature of about 900 to 1200°F (482 to 649°C). In the middle hearths, the sludge solids are ignited and burned at a temperature ranging from about 1200 to 1500°F (649 to 816°C). In the lower hearths, the slow burning material is burned and the ash undergoes cooling at a temperature of about 600°F (316°C). The ash produced leaves at the bottom of the incinerator. The excess air required by this type of incinerator is from 50 to 100 percent of the theoretical amount. The efficiency of multiple-hearth furnaces is about 55 percent.

Example Problem 15.4
Sludge Incineration

A fresh sludge having a fuel value of 7000 Btu/lb dry solids is to be dewatered and then incinerated in a multiple-hearth furnace that has an efficiency of 55 percent—that is, 55 percent of the fuel value is available to evaporate water. The sludge has a winter temperature of 45°F. Determine the required dewatering if the furnace is to operate without auxiliary fuel.

Solution

The heat required to raise the temperature of the sludge from 45°F to 212°F is $212 - 45$ or 167 Btu/lb. From steam tables, the heat required to evaporate one pound of water at 212°F is 970.3 Btu/lb. Thus,the heat required to raise the temperature to 212°F and then to evaporate the water is equal to $167 + 970.3 = 1137.3$ Btu/lb. If x is the lb dry solids/lb wet sludge, the heat available is $(x)(7000)$(furnace efficiency as a fraction) and the heat required is $(1 - x)(1137.3)$. Thus,

$$(x)(7000)(0.55) = (1 - x)(1137.3)$$

From this equation, $x = 0.228$. Therefore, the sludge must be dewatered to obtain at least 22.8 percent dry solids in order to incinerate without auxiliary fuel.

The *fluidized-bed incinerator,* shown in Figure 15.16, consists of a combustion reactor containing a bed of sand above a grid. To start the furnace, the preheater is ignited and the fluidizing air is passed upward through the bed to suspend the sand. Once the sand temperature has reached about 1400 to 1500°F (760 to 816°C), the sludge feed to the incinerator is begun. The water is vaporized and the sludge solids are burned in the fluidized sand bed. The intense agitation caused by the water vaporization and the solids combustion allows the furnace to be operated with only about 20 to 25 percent excess air. The ash created is carried from the reactor by the exit combustion gases and is subsequently removed by a cyclone or a scrubber.

Ultimate Disposal
The solids produced from sludge dewatering or the ash from incineration are usually disposed on land. Digested, air-dried municipal sludge may be spread on agricultural land and plowed under, thus serving as both a fertilizer and a soil conditioner. Wet digested sludge may also be spread on land and, once dried, it may be plowed under. Air-dried digested sludge may be spread on lawns for both fertilization and soil conditioning. Heat-dried sludge is sold as a fertilizer because it is sterile and has fertilizer value. Dewatered sludge and incinerator ash frequently are disposed in sanitary landfills.

Wastewater Treatment Plant Sludges (Chemical)

The chemical coagulants generally used for coagulation and precipitation in tertiary treatment of secondary effluents are lime and alum, with lime being the most common. The coagulant generally used for the coagulation of raw municipal wastewaters is usually lime. Both alum and lime coagulants are effective in removing suspended solids and phosphorous.

Lime Sludges
Coagulation with lime requires large chemical dosages and, as a result, the amount of sludge produced is relatively large. Typical lime dosages are 100 to 500 mg/ℓ calcium oxide. Although the sludge may be disposed on land, it is generally recalcined to recover the lime because this greatly reduces the amount of new lime required and also minimizes the problem of sludge disposal. The operations and processes used for solids handling of lime sludges consist of thickening, dewatering, and lime recalcining.

Lime sludges are very dense and may be easily *thickened* using a gravity-type thickener. For a feed sludge of 5000 to 10,000 gm/ℓ solids, a thickener with a solids loading of about 200 lb/day-ft^2 and an overflow rate of about

1000 gal/day-ft^2 has produced a thickened sludge with 8 to 20 percent solids (Culp, R. L., & Culp, G. L., 1978; South Tahoe Public Utility District, 1971). Pumps for lime sludges are usually recessed-impeller centrifugals and should be located near the thickener to minimize the length of suction lines. Provisions should be made for rodding the lines.

Thickened lime sludges *dewater* readily. Concurrent flow centrifuges have dewatered a sludge with 8 to 20 percent solids to a cake of about 50 to 55 percent solids and a solids capture of about 90 percent (Culp, R. L., & Culp, G. L., 1978; South Tahoe Public Utility District, 1971). Lime sludges will contain both calcium carbonate and calcium phosphate. Since the two precipitates have different densities, it is possible to operate a centrifuge so that the calcium phosphate is separated from the sludge to be recalcined. When this is done, two centrifuges are required. The first is operated to produce a cake of about 40 percent solids for recalcining, with the calcium phosphate leaving in the centrate. The centrate is then sent to the second centrifuge, which produces a cake containing the calcium phosphate precipitate. The calcium phosphate cake from the second centrifuge is usually disposed in a sanitary landfill.

Lime recalcining or lime recovery is usually accomplished by multiple-hearth furnaces similar to those used for sludge incineration. Usually six hearths are employed and the hearth temperatures are about 800°F (427°C), 1250°F (677°C), 1850°F (1010°C), 1850°F (1010°C), 1850°F (1010°C), and 750°F (399°C) for the respective hearths from the top of the furnace to the bottom. As the lime sludge passes down through the furnace, the water is evaporated and the decomposition of calcium carbonate occurs as follows:

$$CaCO_3 \xrightarrow{\Delta} CaO + CO_2 \qquad (15.8)$$

The solids from the bottom hearth are passed through a grinder to break up large particles, and the recalcined lime is then cooled and conveyed to storage. Since some inerts will be present in lime, it is necessary to periodically waste some lime sludge to remove the inerts from the system. Lime recalcining may also be accomplished by fluidized-bed furnaces and rotary kilns. In lime recalcining, the organic solids in the sludge are incinerated during the recovery process.

Alum Sludges

Alum coagulation removes both suspended solids and phosphorus, requiring an alum dosage much less than that required for lime. Alum sludges are difficult to dewater and, where possible, they are frequently mixed with sewage sludges and sent to anaerobic digestion. In the digestion process, the aluminum hydroxide and the aluminum phosphate precipitates remain as solids and are disposed along with the digested sludge. In some cases, alum in sludges has been recovered either by an alkaline or acid treatment (Culp, R. L., & Culp, G. L., 1978).

Water Treatment Plant Sludges

In water coagulation, sludges are produced from the clarifier operations and the backwashing of the filters. The aluminum or iron salt coagulants create a gelatinous sludge that will contain the organic and inorganic materials that are coagulated, and also the hydroxide precipitate. Dewatering of these sludges is very difficult. In the past, at most plants the clarifier sludge and the filter backwash water were returned to the water supply source, which was usually a river or lake. The present trend is to process the sludge from clarifiers for ultimate disposal and to catch the backwash water in basins and gradually return it to the treatment plant for reprocessing.

In water softening, the sludges produced are mainly calcium carbonate and magnesium hydroxide precipitates, although some organic and inorganic materials are present that have been removed from the water. These sludges dewater rather easily, and it is common to process them for ultimate disposal. The backwash water is usually caught in basins and slowly returned to the treatment plant for reprocessing.

The unit operations and processes used for solids handling in water treatment are similar to those employed in wastewater treatment. The main operations and processes include thickening, conditioning, dewatering, ultimate disposal, and, in some cases, coagulant or lime recovery.

Discharge to Sanitary Sewers

The gelatinous sludge from the clarifiers in a coagulation plant may be discharged to the sanitary sewage system if the primary clarifiers and solids handling facilities at the wastewater treatment plant have adequate capacity to process this additional sludge. Where this is practiced, the sludge is frequently discharged to the sewer system during the night when the wastewater flow is relatively low. Softening sludges should not be disposed in this manner because they are of a large quantity, may readily fill the digesters, and will produce encrustations on weirs, channels, piping, and so on.

Thickening

Gravity thickening of lime softening sludges has increased the solids content from 1 to about 30 percent when thickener loadings of 12.5 lb/day-ft^2 have been used (AWWA, 1969). Gravity thickening of alum sludges from water coagulation has increased the solids content from about 1 to 2 percent when thickener loadings of 4.0 lb/day-ft^2 have been used (AWWA, 1969).

Conditioning

The sludge from alum or iron salt coagulation may be conditioned to improve its dewatering characteristics.

Gelatinous hydroxide sludges may be *heated* in reactors at elevated temperatures and pressures to cause the bound water to be released. One study showed that vacuum filtration of heat-treated sludges at 10 to 20 lb/hr-ft^2 produced a cake having about 21 percent solids (Schroeder, R. P., 1970).

The *freezing* and *thawing* of gelatinous hydroxide sludges causes the bound water to be released, thus improving the dewatering characteristics. An alum sludge thickened to about 2 percent solids, then frozen and thawed, created a sludge with about 20 percent solids that was subsequently dewatered by vacuum filtration to 34 percent solids (AWWA, 1969).

Dewatering

The sludges from water softening plants dewater readily; however, the metallic hydroxide sludges from water coagulation are difficult to dewater.

Vacuum filtration of gravity-thickened lime-softening sludges containing about 30 percent solids has produced a cake of about 65 percent solids at a vacuum filter loading of 40 lb/hr-ft^2 (AWWA, 1969). Vacuum filtration of coagulation sludges requires the use of a precoated *rotary vacuum filter*. In this operation, a filter aid is added to create a 2- to 4-inch thick precoat on the rotary filter drum prior to sludge addition. Tests have shown that a cake of 29 to 32 percent solids could be obtained (Mahoney, P. F., & Duensing, W. J., 1972).

Centrifugation has been widely used to dewater lime softening sludges. Most installations use a gravity thickener to thicken the feed sludge to about 15 to 25 percent solids. Subsequent centrifugation produces a cake having about 55 to 60 percent solids (AWWA, 1969). In one installation, the underflow from the clarifier had 5 to 10 percent solids, and centrifugation of the underflow produced a cake of 55 to 60 percent solids (AWWA, 1969). Most centrifuges are the countercurrent solid-bowl type; however, the recently designed concurrent solid-bowl type gives a centrate with less solids. Recent installations usually employ this type. Centrifuges may also be used to classify softening sludges where recalcining is practiced. A lime softening sludge contains both calcium carbonate and magnesium hydroxide and, since these precipitates have different densities, centrifuges can be operated to separate them. The calcium carbonate will leave in the cake and the magnesium hydroxide in the centrate of the first centrifuge. A second centrifuge handling the centrate from the first can be used to remove the magnesium hydroxide. The solids capture by centrifuges handling softening sludges is usually from 70 to 90 percent without polymer aids and up to 95 percent when polymer aids are used. The centrifugation of alum sludges has not been very successful.

Filter presses have been used to process lime softening sludges to yield a cake having 60 to 65 percent solids. The filter press is the only mechanical dewatering device that is highly successful in dewatering coagulation sludges. Untreated alum sludges containing 1.5 to 2 percent solids have been processed to produce a cake of 15 to 20 percent solids (AWWA, 1969). If alum sludge is conditioned with lime or polymers, cakes are produced having much higher percent solids. Filter precoating with diatomaceous earth and lime conditioning of an alum sludge at one installation has produced cakes containing 40 to 50 percent solids (Weir, P., 1972).

Lagoons are the most widely used method to handle softening and coagulation sludges and, although the operating costs are rather low, the land requirements are appreciable. They should be used in pairs and have dike embankments to exclude surface runoff. The sludge is added until the lagoon is full; then it is taken out of service and allowed to dry. Once drying is sufficient, the sludge is removed; thus, the lagoons serve for both thickening and dewatering. Lime softening sludges are effectively handled by lagoons and the solids content of the dewatered sludge is usually about 50 percent (AWWA, 1969). Lagoons have not been nearly as successful in dewatering alum sludges. Usually alum sludges can be dewatered to about 1 to 10 percent solids (AWWA, 1969), and it is common to remove the partially dewatered sludge by drag lines and place it in trucks, then spread it on land for ultimate disposal. A climatic condition of alternate freezing and thawing assists in releasing the bound water in coagulation sludges.

Ultimate Disposal

The final disposal of dewatering sludges from water treatment is land disposal either in a sanitary landfill or spreading it on land; the sanitary landfill is the most widely used.

Lime or Coagulant Recovery

The lime sludge produced from water softening may be calcified by centrifuges to separate the calcium carbonate from the magnesium hydroxide. The lime can then be recalcined and the quicklime produced reused.

Alum recovery, although it is not widely used, can be accomplished using acidification with sulfuric acid as follows:

$$\underline{2Al(OH)}_3 + 3H_2SO_4 \rightarrow Al_2(SO_4)_3 + 6H_2O \qquad (15.8)$$

After acidification, the supernatant, which contains the alum, is separated from the solids, and the recovered solution is used as a liquid coagulant.

A magnesium carbonate and lime coagulation system has been developed which allows recovery of one or both of the coagulants. The coagulation reaction is

$$MgCO_3 + Ca(OH)_2 \rightarrow \underline{Mg(OH)}_2 + \underline{CaCO}_3 \qquad (15.9)$$

The sludge will contain calcium carbonate, magnesium hydroxide, and the coagulated matter. The sludge is carbonated to dissolve the magnesium hydroxide as follows:

$$\underline{Mg(OH)}_2 + 2CO_2 \rightarrow Mg(HCO_3)_2 \qquad (15.10)$$

The sludge is filtered and the filtrate, which contains the magnesium bicarbonate, is returned as a liquid coagulant solution. The solution when added with lime coagulates as follows:

$$Mg(HCO_3)_2 + 2Ca(OH)_2 \rightarrow \underline{Mg(OH)}_2$$
$$+ \underline{2CaCO}_3 + 2H_2O \qquad (15.11)$$

It is also possible to separate the calcium carbonate in the sludge and recalcine it to produce quicklime, which may be reused.

References

American Water Works Association (AWWA). 1971. *Water Quality and Treatment.* New York: McGraw-Hill.
American Water Works Association (AWWA) Research Foundation. 1969. Disposal of Wastes from Water Treatment Plants, Parts

1, 2, 3, and 4. *Jour. AWWA* 61, no. 10:541; 61, no. 11:619; 61, no. 12:681.

Bird Machine Co. 1971. *Bird Machine Company Product Manual.*

Coackley, P. 1956. Laboratory Scale Filtration Experiments and Their Application to Sewage Sludge Dewatering. In *Biological Treatment of Sewage and Industrial Wastes.* Vol. 2, edited by J. McCabe and W. W. Eckenfelder, Jr. New York: Reinhold.

Conway, R. A., and Ross, R. D. 1980. *Handbook of Industrial Waste Disposal.* New York: Van Nostrand Reinhold.

Culp, R. L., and Culp, G. L. 1978. *Handbook of Advanced Wastewater Treatment.* 2nd ed. New York: Van Nostrand Reinhold.

Culp, G. L., and Culp, R. L. 1974. *New Concepts in Water Purification.* New York: Van Nostrand Reinhold.

Dick, R. I., and Ewing, B. B. 1967. Evaluation of Activated Sludge Thickening Theories. *Jour. SED* 93, SA4:9.

Eckenfelder, W. W., Jr. 1980. *Principles of Water Quality Management.* Boston: CBI Publishing.

Environmental Protection Agency (EPA). 1979. *Sludge Treatment and Disposal.* EPA Process Design Manual, Washington, D.C.

Environmental Protection Agency (EPA). 1974. *Upgrading Existing Wastewater Treatment Plants.* EPA Process Design Manual, Washington, D.C.

Fair, G. M., Geyer, J. C., and Okun, D. A. 1968. *Water and Wastewater Engineering.* Vol. 2. *Water Purification and Wastewater Treatment and Disposal.* New York: Wiley.

Forster, H. W. 1972. Sludge Dewatering by Pressure Filtration. Paper presented at the AIChE Annual Meeting, New York City.

Great Lakes–Upper Mississippi Board of State Sanitary Engineers. 1978. Recommended Standards for Sewage Works. Ten state standards, Albany, N.Y.

Krasausakas, J. W. 1969. Review of Sludge Disposal Practices. *Jour. AWWA* 61, no. 5:225.

Mahoney, P. F., and Duensing, W. J. 1972. Precoat Vacuum Filtration and Natural-Freeze Dewatering of Alum Sludge. *Jour. AWWA* 64, no. 10:655.

Metcalf and Eddy, Inc. 1979. *Wastewater Engineering, Treatment, Disposal, and Reuse.* New York: McGraw-Hill.

Newton, D. 1964. Thickening by Gravity and Mechanical Means. In *Sludge Concentration, Filtration and Incineration.* University of Michigan, School of Public Health, Continuing Education Series 113, no. 4.

Ramalho, R. S. 1977. *Introduction to Wastewater Treatment Processes.* New York: Academic Press.

Riehl, M. L., Weiser, H. H., and Rheims, B. T. 1952. Effect of Lime-Treated Water on Survival of Bacteria. *Jour. AWWA* 44, no. 5:466.

Sanks, R. L. 1978. *Water Treatment Plant Design.* Ann Arbor, Mich.: Ann Arbor Science Publishers.

Schroeder, E. D. 1977. *Water and Wastewater Treatment.* New York: McGraw-Hill.

Schroeder, R. P. 1970. *Alum Sludge Disposal—Report no. 1.* R&D Project no. DP-6551, Eimco Corporation.

Smith, J. E., Hathaway, S. W., Farrell, J. B., and Dean, R. B. 1972. Sludge Conditioning with Incinerator Ash. Paper presented at 27th Annual Purdue Industrial Waste Conference.

South Tahoe Public Utility District. 1971. *Advanced Wastewater Treatment as Practiced at South Tahoe.* Technical Report for the EPA, Project 17010 ELQ (WPRD 52-01-67).

Sundstrom, D. W., and Klei, H. E. 1979. *Wastewater Treatment.* Englewood Cliffs, N.J.: Prentice-Hall.

Texas State Dept. of Health. 1976. Design Criteria for Sewage Systems. Austin, Tex.

Thompson, C. G., Singley, J. E., and Black, A. P. 1972. Magnesium Carbonate: A Recycled Coagulant. *Jour. AWWA* 64, no. 1:11; 64, no. 2:93.

Veislind, P. A. 1974. *Treatment and Disposal of Wastewater Sludges.* Ann Arbor, Mich.: Ann Arbor Science Publishers.

Water Pollution Control Federation (WPCF). 1969. *Sludge Dewatering.* WPCF Manual of Practice no. 20, Washington, D.C.

Water Pollution Control Federation (WPCF). 1977. *Wastewater Treatment Plant Design.* WPCF Manual of Practice no. 8, Washington, D.C.

Weir, P. 1972. Research Activities by Water Utilities—Atlanta Water Dept. *Jour. AWWA* 64, no. 10:634.

Yoshioka, N., Hotta, Y., Tanaka, S., Naito, S., and Tsugami, S. 1957. Continuous Thickening of Homogeneous Flocculated Slurries. *Chem. Eng.* 21, Tokyo, Japan (in Japanese with English abstract).

Problems

1. A thickener is to thicken waste activated sludge from an industrial wastewater treatment plant. Pertinent data are: sludge flow = 250,000 gal/day, solids in the waste activated sludge = 10,000 mg/ℓ, specific gravity of the sludge = 1.002, solids in the thickened sludge = 25,000 mg/ℓ, specific gravity of the thickened sludge = 1.005, solids in the supernatant = 800 mg/ℓ, overflow rate = 600 gal/day-ft^2, and solids loading = 5 lb/day-ft^2. Determine:

   a. The thickened sludge flow, gal/day.

   b. The diameter of the thickener.

2. Vacuum filters are to be used to dewater a raw primary and waste activated sludge mixture from a municipal wastewater treatment plant. Pertinent data are: sludge flow = 200,000 gal/day, solids in the sludge = 4.5 percent, specific gravity of the sludge = 1.008, cake solids = 20 percent, solids in the filtrate = 1000 gm/ℓ, and solids loading = 4.5 lb/hr-ft^2. Determine:

   a. The filtrate flow, gal/day.

   b. The cake produced, tons/day.

   c. The area of the filters, ft^2.

3. The solids in the primary and waste activated sludge from an industrial wastewater treatment plant treating a pulp and paper mill wastewater are to be incinerated using a fluidized-bed incinerator. The solids have a fuel value of 16,260 kJ/kg and the efficiency of the incinerator is 55 percent. That is, 55 percent of the heat released during combustion is available to evaporate the water in the sludge. The remaining heat is lost, mainly through the stack. Assume the heat required to

evaporate the water is 2610 kJ/kg water. Determine the minimum solids content in the dewatered sludge if the incinerator is to operate without auxiliary fuel.

4. A centrifuge is to be used to dewater a sludge from a municipal lime-soda softening plant. The maximum plant capacity is 31 MGD and 800 lb of dry sludge solids are produced per million gal treated. The solids content in the sludge is 10 percent, the cake from the centrifuge is 50 percent solids, and the solids capture is 90 percent. The specific gravity of the dry solids (s_s) is 2.50, and the specific gravity of an aqueous slurry is given by

$$s = \frac{p + (100 - p)}{p + (100 - p)/s_s}$$

where

s = specific gravity of the aqueous slurry;
p = percent water;
s_s = specific gravity of the dry solids.

Determine:
a. The gallons of wet sludge and centrate produced per day.
b. The pounds of cake produced per day.
c. The solids content of the centrate, mg/ℓ.

5. A rotary vacuum filter is to dewater a digested sludge from a plant treating a wastewater from 200,000 persons. The digested sludge produced is 0.10 kg solids per person per day, and the solids content of the digested sludge is 10 percent. Six percent of the solids are lost in the filtrate, the sludge cake is 25 percent solids, and the filter is loaded at 12 kg/h-m^2 on a dry solids basis. Determine:
a. The kilograms of cake produced per day.
b. The liters of filtrate produced per day and its solids content.
c. The size of the filter if the length is twice the diameter.

6. A centrifuge is to dewater digested sludge from an anaerobic digester in an activated sludge plant. The sludge solids in the feed sludge to the digester amount to 60,000 lb/day and the digested sludge solids produced from the digester are 45 percent of the sludge solids that enter. The feed sludge to the centrifuge contains 5.0 percent solids. The cake from the centrifuge is 30 percent solids and the solids recovery is 90 percent. Determine:

a. The pounds of cake produced per day.

b. The gallons of centrate produced per day.

c. The solids content of the centrate, mg/ℓ.

7. A conditioned digested sludge is to be dewatered on a rotary vacuum filter under a vacuum of 25 in. Hg. A vacuum filtration test has been done in the laboratory giving a specific resistance of 3.9×10^7 sec²/gm. The unfiltered solids were 4.9 percent by dry weight and the sludge cake had 22 percent dry solids. The filtration temperature is 65°F, the cycle time is 5 min, and the form time is 42 percent of the cycle time. Determine the filter yield in lb/hr-ft².

8. For the conditions described in Problem 7, determine the filter drum area if the sludge flow is 62,000 gal/day and the volatile solids are 35 percent of the dry solids.

9. A laboratory vacuum filtration test was performed using a conditioned, anaerobically digested sludge. The filtrate volumes shown in Table 15.3 were obtained at the indicated filtration times.

Table 15.3.

t (sec)	V (mℓ)
20	68
40	91
60	118
80	129
100	148

Other pertinent data from the test are: vacuum = 24.5 in. Hg, dry solids content in the conditioned sludge = 8.2 percent, dry solids in the cake = 21 percent, temperature = 23°C, and the diameter of the Buchner funnel = 9.5 cm. Determine the specific resistance in sec²/gm.

Chapter Sixteen

MISCELLANEOUS UNIT OPERATIONS AND UNIT PROCESSES

T his chapter presents a brief coverage of some miscellaneous unit operations and processes that are used in both water and wastewater treatment. This chapter is intended to be introductory and only some of the most common treatments are presented.

Water Treatment

Screening, aeration, defluoridation, and fluoridation are some of the unit operations and processes used in water treatment that have not been presented in previous chapters.

Screening

Coarse bar racks and fine traveling screens are employed at intake structures on reservoirs and rivers. Coarse bar screens usually have clear spaces up to 3 in. between the bars and are used to prevent the entry of large debris, such as logs, into the intake structure. A trolley and hoist must be provided to remove any logs that may hang against the bar rack. Fine traveling screens located behind the bar racks usually have square openings of about $\frac{3}{8}$ to $\frac{1}{2}$ in. and are used to prevent the entry of small debris, such as sticks, bark, leaves, and fish.

Aeration

Aeration may be used for gas stripping (degasification) to remove unwanted gases, such as carbon dioxide and hy-

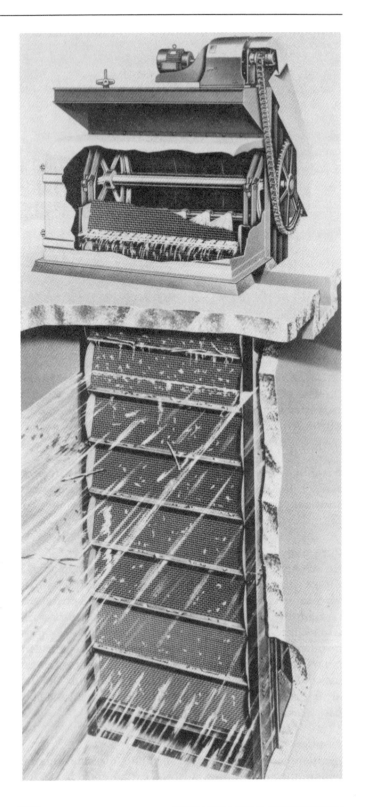

drogen sulfide, and iron and manganese. Groundwaters, in particular, may require aeration to remove these contaminants. If present in the free ion form, iron and frequently manganese may be oxidized to insoluble compounds that may be removed by coagulation (if required), sedimentation, and filtration. Usually, aeration is accomplished by cascades, multiple tray aerators, or spray nozzles.

Fluoridation

A fluoride content of 0.7 to 1.2 mg/ℓ in drinking water has been found to be beneficial because it results in the development of a hard decay-resistant enamel during the formation of permanent teeth. Fluoride is usually added in the form of sodium fluoride salt, although hydrofluorosilicic acid and sodium silico-fluoride salt have been used.

Defluoridation

Many waters, in particular well waters, may have excessive amounts of the fluoride ion, which will result in the mottling of tooth enamel during the formation of permanent teeth. Although several methods have been tried for fluoride removal, the most successful has been the adsorption by calcined alumina. Regeneration of the adsorbent requires the use of a strong base, such as sodium hydroxide, followed by a sulfuric acid rinse. This process has resulted in the reduction of the fluoride ion from about 8 to 1 mg/ℓ at Bartlett, Texas (Steel, E. W., & McGhee, T. J., 1979).

Wastewater Treatment

Screening, shredding, sand and silt (grit) removal, flow equalization, quality equalization, neutralization, and flotation are some of the unit operations and processes used in wastewater treatment that have not been presented in previous chapters.

Screening and Shredding

Screens and shredders are used in municipal wastewater treatment and sometimes in industrial wastewater treatment for the removal or shredding of coarse solids. Typical coarse solids in municipal wastewaters are pieces of wood, plastic materials, and rags. Although hand cleaned screens were formerly used, almost all screens presently installed are mechanically cleaned and the screened solids are dumped into receptacles for disposal. Usually, disposal is by sanitary landfill or incineration. For manually cleaned bar screens, the bar spacing is usually from 1 to 2 in. and the bars are mounted at a 30- to 75-degree angle to the horizontal, 30 to 45 degrees being typical. For mechanically cleaned screens, the bar spacing is usually $\frac{1}{2}$ to $1\frac{1}{2}$ in., and

Mechanically Cleaned Bar Screen
Courtesy, Envirex, Inc., a Rexnord Co.

the bars are at a 45- to 90-degree angle to the horizontal, 60 degrees being typical. Usually, the mechanical rakes that clean a screen are operated by a time clock or floats that activate the drive motor when the head loss across the screen is greater than about 2 in. The approach channel for a manual or mechanically cleaned bar screen should be straight for several feet ahead of the screen to give uniform flow across the screen; the approach velocity should be at least 1.5 fps to avoid solids deposition. The velocity through the bars should be about 2 fps at minimum flow and not more than 3 fps at maximum flow. At least two screens should be provided. At small plants, it is common to have one mechanical screen and one manual screen, which is used only when the mechanical screen is inoperative. At medium- to large-size plants, two or more mechanical screens are provided. Stop gate slots should be provided ahead of and behind a screen so that the unit may be dewatered for maintenance. The head loss through bar screens may be determined from

$$h_L = \frac{(V_b^2 - V_a^2)}{2g} \times \frac{1}{0.7} \qquad (16.1)$$

where

h_L = head loss, ft;
V_a = approach velocity, fps;
V_b = velocity through the bar openings, fps;
g = acceleration due to gravity.

Screenings usually amount to 0.5 to 6.0 ft^3 per million gallons treated.

The purpose of shredding coarse solids is to reduce their size so they will be removed by subsequent treatment operations, such as primary clarification, where both floating and settling solids are removed. Shredding may be accomplished by grinders, barminutors, and comminutors. Grinders shred the solids removed by a mechanically cleaned bar screen, and the shredded solids are returned to the wastewater flow downstream from the screen. The barminutor is a bar screen with a shredder that moves up and down the screen; rotating cutter blades pass through the bar spaces, thus cutting the accumulated coarse solids. The comminutor, Figure 16.1, consists of a rotating slotted cylinder through which the entire wastewater flow passes. Solids too large to pass through the slots are cut by blades as the cylinder rotates, thus reducing their size until they pass through the slot openings.

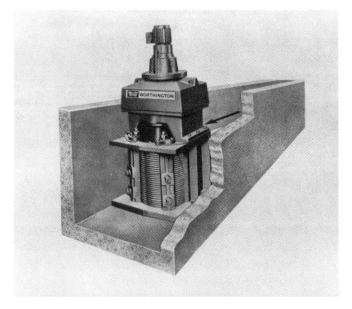

Comminutor Unit.
Courtesy Worthington Group,
McGraw-Edison Company.

Figure 16.1. **Comminutor**
Installation
Courtesy of Chicago Pump
Products, Clow Corporation

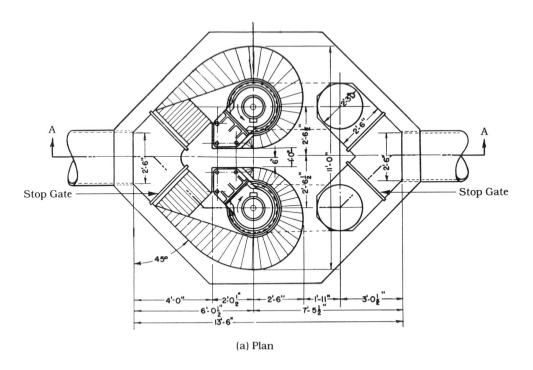

(a) Plan

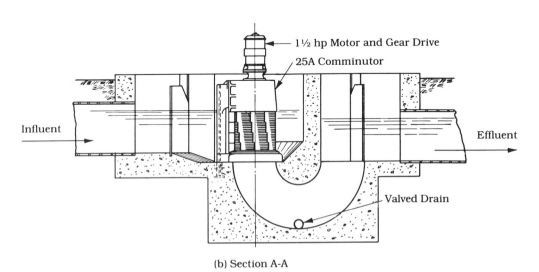

(b) Section A-A

Grit Removal

Grit (sand and silt) removal is usually limited to municipal wastewaters and is used to remove as much sand and silt as possible to prevent wear on pumps, accumulations in aeration tanks, clarifiers, and digesters, and clogging of sludge piping. Grit removal is usually by square or longitudinal settling chambers or diffused air tanks. In settling chambers, the sand and silt settles and the horizontal velocity is sufficient to prevent the settling of organic solids. The horizontal velocity in longitudinal grit chambers is from 0.75 to 1.25 fps, and the velocity is controlled by a proportional weir or a Parshall flume at the end of the chamber. The overflow rate is usually from 45,000 to 50,000 gal/day-ft^2 at maximum flow. This range in overflow rates will remove quartz sand particles 0.2 mm in diameter with a specific gravity of 2.65 and yet allow organic solids to remain in suspension. The grit is usually removed by mechanical rakes, discarded into receptacles, and disposed in sanitary landfills. If longitudinal chambers are used, at least two are provided, one of which is mechanically cleaned. It is desirable to have all chambers mechanically cleaned.

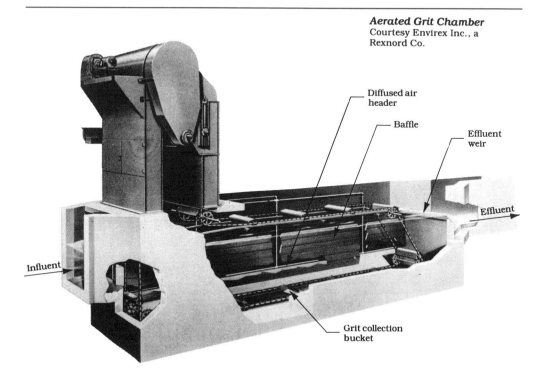

Aerated Grit Chamber
Courtesy Envirex Inc., a
Rexnord Co.

Diffused air header

Baffle

Effluent weir

Effluent

Influent

Grit collection bucket

Figure 16.2 Aerated Grit Removal Chamber
Courtesy of Walker Process Corporation

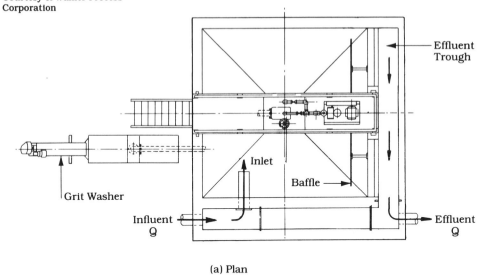

(a) Plan

(b) Elevation

The aerated grit removal tanks, Figure 16.2, allow the sand and silt to settle, whereas the organic solids tend to float. The bottom is sloped and terminates in a hopper or sump where the grit is removed by mechanical means or air lifts. Disposal is by sanitary landfills. The diffused air type grit removal chambers are very popular because the tanks are compact, the grit is cleaner, and the dissolved oxygen imparted creates aerobic conditions, thus reducing odor problems. For the diffused air type grit chamber, the velocity of roll of the wastewater controls the density of the particles removed. If the velocity is excessive, the heavier grit will leave with the effluent, and, if the velocity is insufficient, the light organic solids will be removed with the grit. This type of chamber has an airflow regulating valve that can be adjusted to give proper operation. At small plants, one chamber may be used; however, medium- to large-size plants usually have two or more tanks. The detention time at maximum flow is from 1 to 3 minutes, with 3 minutes being typical. The grit may be removed by air lifts, screw conveyors, chain and bucket conveyors, or grab buckets. The grit removed is from 0.3 to 24 ft^3 per million gallons treated.

Flow Equalization

The equalization of flow to give a relatively constant flow rate to a wastewater treatment plant is applicable to both municipal and industrial wastewaters. Although the following discussion is directed towards municipal wastewater treatment plants, the principles employed are also applicable to industrial treatment plants.

The use of flow-equalization basins after preliminary treatment (that is, screening and grit removal) provides a relatively constant flow rate to the subsequent treatment operations and processes; thus, it enhances the degree of treatment. Not only does equalization dampen the daily variation in the flow rate, but it also dampens the variation in the concentration of BOD_5, suspended solids, and so on, throughout the day. Flow equalization can significantly improve the performance of an existing plant and, in the case of a new plant design, it will reduce the required size of the downstream treatment facilities. Equalization is feasible for dry weather flows in separate sanitary sewers and sometimes for storm infiltration flows.

The equalization basins, as shown in Figure 16.3, may be either in-line facilities or side-line facilities. For the in-line equalization basin, Figure 16.3 (a), the entire wastewater flow is pumped at a relatively constant rate to the

Figure 16.3. Equalization Basins

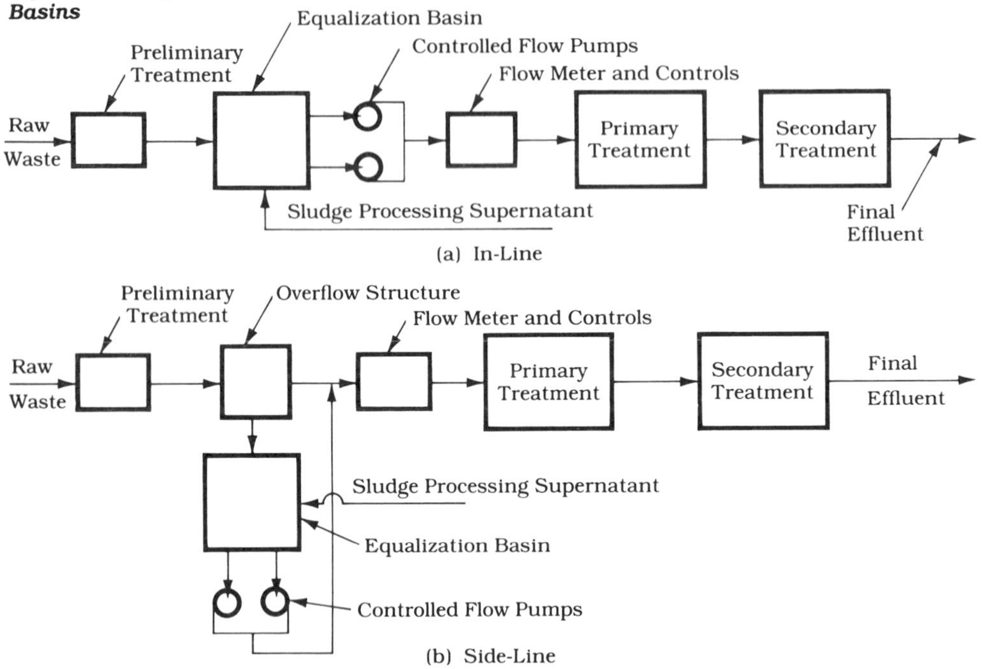

(a) In-Line

(b) Side-Line

Figure 16.4. Section through Equalization Basin

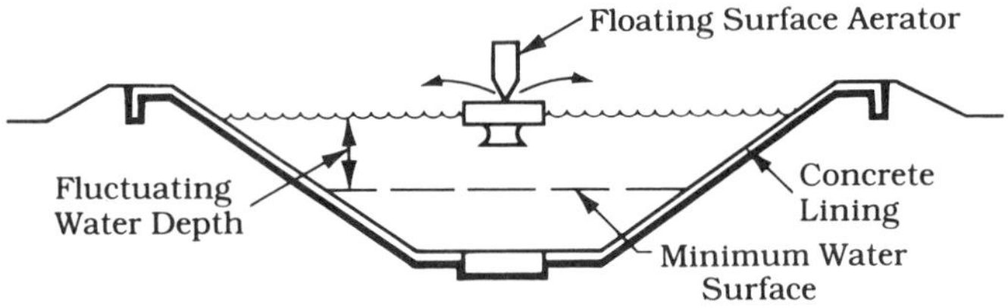

Figure 16.5. Fluctuating Volume Determined by Hydrograph

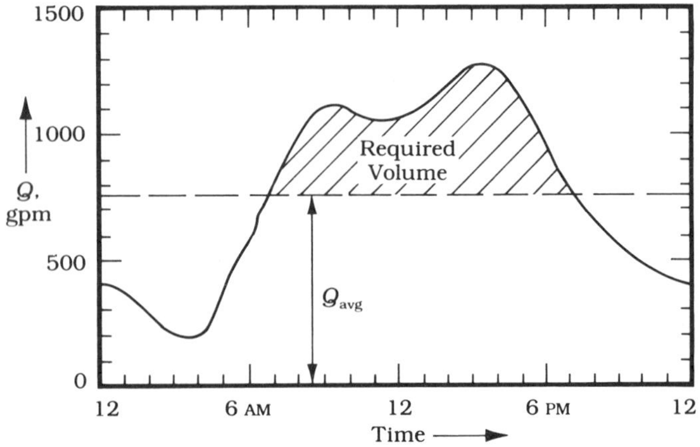

downstream wastewater treatment facilities. In the side-line equalization basin, Figure 16.3 (b), the flow during the day that is in excess of the average hourly flow overflows to the equalization basin and, when the influent flow rate becomes less than the average flow, the wastewater is pumped from the basin to the downstream treatments. The in-line system provides greater dampening of the concentration of BOD_5, suspended solids, and so on, than attained in the side-line system. The equalization basin will have a fluctuating water level, as shown in Figure 16.4, and aeration must be provided to keep the solids in suspension and maintain aerobic conditions. Usually, the fluctuating volume is from 10 to 25 percent of the average daily dry weather flow and may be determined from a flow hydrograph of the influent flow to the plant, as shown in Figure 16.5. Although basins may be constructed with earthen sides and bottoms, concrete-lined sides and bottoms are better engineering practice because erosion is eliminated.

For in-line basins, approximately 10 to 20 percent of the BOD_5 entering is stabilized in the basin (EPA, 1974a). At an existing wastewater treatment plant, the use of flow equalization has increased the suspended solids removal in the primary settling units from 23 to 47 percent (EPA, 1974a). Also, biological treatment performance significantly benefits from equalization because shock loads are minimized and the flow rate approaches steady state, which is beneficial to biological units and final clarifiers. At a new activated sludge plant employing side-line flow equalization and multimedia filtration as tertiary treatment, the final effluent is reported to have a BOD_5 less than 4 mg/ℓ and suspended solids less than 5 mg/ℓ. The volume required for the fluctuating water depth depicted in Figure 16.4 may be determined using a hydrograph of the hourly flow rate throughout the day, as shown in Figure 16.5. The average hourly flow rate is determined and a horizontal line representing this flow rate is drawn on the hydrograph. The area between the horizontal line depicting the average flow and the hourly flow exceeding the average represents the volume required for the fluctuating water depth once proper unit conversions have been made. Details for flow equalization basin design are given in appropriate publications (EPA, 1974a, 1974b).

Example Problem 16.1
Equalization Basin

An in-line flow equalization basin is to be designed for a wastewater treatment plant. From plant records a compilation has been made that gives the average flow rate versus hour as shown in Table 16.1.

Table 16.1. Average Flow Rate versus Hour

Hour	Gpm	Hour	Gpm
12 AM	1740	1 PM	2330
1	1630	2	2290
2	1390	3	2220
3	1180	4	2220
4	1040	5	2150
5	910	6	2010
6	910	7	1940
7	920	8	1940
8	1530	9	1880
9	2080	10	1630
10	2270	11	1560
11	2330	12 AM	1740
12 Noon	2330		

Determine the fluctuating volume required for the basin.

Solution

A plot of the hourly flow rate versus hour is shown in Figure 16.6. From the previous data, the average hourly

Figure 16.6. Graph for Example Problem 16.1.

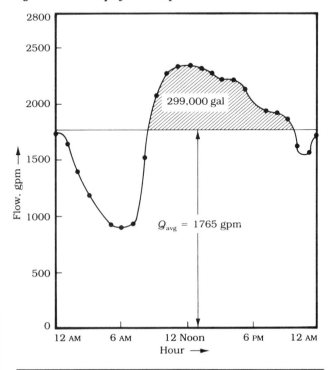

flow for the day is 1765 gpm. Plotting the average flow on Figure 16.6 and determining the areas above the average flow gives the required fluctuating volume as

$$\text{Fluctuating volume} = \underline{299,000\,\text{gal}}$$

Quality Equalization

Quality equalization consists of using a constant level equalization basin ahead of a municipal or industrial wastewater treatment plant to dampen the variation in the concentration of BOD_5, suspended solids, and so on throughout the day. Usually, quality equalization basins are employed when the variation in the organic concentration of 4-hour composite samples exceeds 4:1 during a day (EPA, 1974a). The use of quality equalization not only will increase the performance of an existing plant but also will give better treatment by a proposed plant. Equalization tanks are similar to the one shown in Figure 16.4 except that the wastewater is at a constant level. Aeration must be provided to keep suspended solids in suspension and to maintain aerobic conditions in the basin.

Neutralization

Frequently, industrial wastewaters may be acidic or basic and may require neutralization prior to subsequent treatments or release to a municipal sanitary sewer system. If the downstream treatment is a biological process, the wastewater should have a pH between about 6.5 and 9.0 to avoid inhibition. Sometimes it is feasible to mix an acidic waste stream with a basic waste stream and then use a constant level equalization basin as a neutralization tank.

Acidic wastewaters may be neutralized by passage through limestone beds, by the addition of slaked lime, $Ca(OH)_2$, caustic soda, NaOH, or soda ash, Na_2CO_3. Limestone beds may be of the upflow or downflow type; however, the upflow type is the most common. Limestone beds should not be used if the sulfuric acid content is greater than 0.6 percent because the $CaSO_4$ produced will be deposited on the crushed limestone and, as a result, effective neutralization ceases. Also, metallic ions such as Al^{+3} and Fe^{+3}, if present in sufficient amounts, form hydroxide precipitates that will coat the crushed limestone and reduce neutralization. Upflow beds are the most common because the products of the reaction, such as CO_2, are removed more effectively than in downflow beds. Laboratory- or pilot-scale studies should be made before limestone beds are used. Acidic wastewaters may be neutralized with slaked lime, $Ca(OH)_2$, and usually two or three agitated vessels are used in series, as shown in Figure 16.7. Each vessel has a pH sensor that controls the slaked lime feed rate. Since slaked lime is less expensive than other bases or soda ash, it is the most commonly used chemical for acidic neutralization. Caustic soda, NaOH, or soda ash, Na_2CO_3, may be used in similar manner as slaked lime.

Alkaline wastewaters may be neutralized with a strong mineral acid, such as H_2SO_4 or HCl, or with CO_2. Usually, if a source of CO_2 is not available, neutralization is done with H_2SO_4 because it is cheaper than HCl. The reaction with mineral acids is rapid, and agitated vessels are used with pH sensors that control the acid feed rate. Neutralization of alkaline wastewaters with CO_2 usually consists of bubbling CO_2 from a perforated pipe grid in the bottom of the neutralization tank, thus creating carbonic acid, H_2CO_3, which reacts with the alkaline substances. Frequently, flue gas is available as a source of CO_2, which makes the neutralization process more economical.

Flotation
Low density solid or liquid particles may be separated from a liquid by flotation. In this operation, fine air bubbles are introduced into the liquid, resulting in the attachment of the bubbles to the particles. The attached bubbles cause the particles to rise to the liquid surface, where they are removed by skimming.

Flotation may be accomplished by three methods of air introduction. The first is the injection of compressed air

Figure 16.7. Three-Stage Lime Neutralization

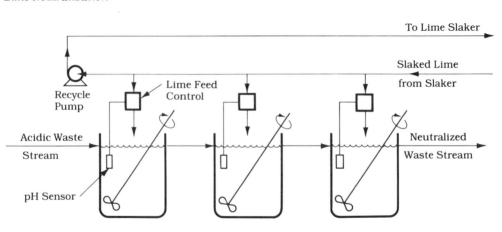

ENVIRONMENTAL ENGINEERING

through diffusers in the flotation tank to buoy the particles to the surface. This has not been highly successful except for flotating grease in wastewaters having an unusually high grease content. The second method consists of super-saturating the wastewater flow with air gases, then sending the flow to an open flotation tank, thus releasing the pressure. The air gases come out of solution and form minute bubbles around the solid and liquid particles, thus causing them to float to the surface where they are removed by skimming. For small installations, the entire flow is injected with air, as shown in Figure 16.8 (a), then pumped under a pressure of 30 to 60 psi to a closed retention tank with several minutes detention time. From there the flow goes to the flotation tank. For large installations, 30 to 150 percent of the effluent is recycled and pressurized to become supersaturated with air gases, as in Figure 16.8 (b). The recycle is then mixed with the main wastewater flow and

Figure 16.8. Air Flotation Systems

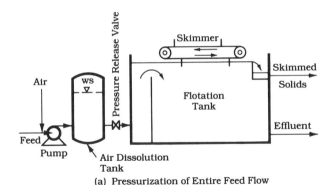

(a) Pressurization of Entire Feed Flow

(b) Pressurization of Recycled Flow

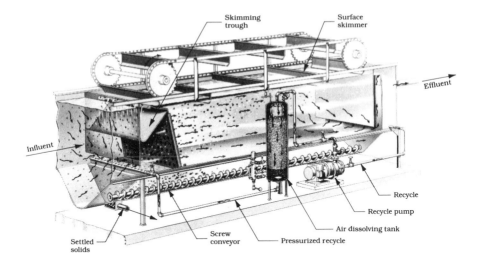

the mixture goes to the flotation tank. The third method consists of saturating the main wastewater flow with air gases by air injection as the flow goes to a covered tank where a small vacuum is drawn by a vacuum pump. This causes the air gases to come out of solution, flotation occurs, and particle removal is by skimming. Frequently, chemical coagulants are added ahead of the flotation operation to enhance the efficiency.

Flotation has been used to remove oil emulsions from wastewaters, thicken biological sludges, and remove suspended solids from secondary effluents, in addition to numerous applications in industrial wastewater treatment. The use of flotation to thicken sludges is presented in Chapter 15.

The principal parameter in the design of a dissolved air-flotation system is the air to solids ratio, A/S, which is expressed as mass/mass. To determine the optimum A/S ratio, bench-scale batch flotation experiments are used; the procedures are outlined by Eckenfelder and Ford (1970). The air to solids ratio for a system in which all the wastewater flow is pressurized is as follows:

$$\frac{A}{S} = \frac{1.3\,a_s(fP-1)}{S_s} \tag{16.2}$$

where

A/S = air to solids ratio, mg/mg;
a_s = air solubility, mℓ/ℓ;

f = fraction of air dissolved at a given pressure, usually 0.5 to 0.8;

P = absolute pressure in atmospheres;

S_s = suspended solids concentration, mg/ℓ.

For a system where pressurized recycle is used, Eq. (16.2)
is modified as follows:

$$\frac{A}{S} = \frac{1.3\, a_s(fP - 1)R}{S_s Q} \tag{16.3}$$

where

R = recycle flow rate;

Q = wastewater flow rate.

The solubility of air at 0, 10, 20, and 30°C is 29.2, 22.8,
18.7, and 15.7 mℓ/ℓ, respectively. Figure 16.9 shows a schematic of the laboratory equipment required. For a nonrecycle system, the air dissolving chamber is partially filled
with the wastewater, and the vessel is pressurized with
compressed air, shaken, then allowed to set to absorb air.
The pressure is then released, causing flotation of the solids. The floated sludge and the clarified liquid are withdrawn, the suspended solids are determined, and the air to
solids ratio is computed. The test is repeated at several different air pressures to give a curve of A/S versus effluent
suspended solids concentration. For a recycle system, the
air dissolving chamber is filled partly with effluent, pres-

Figure 16.9. Flotation Test Equipment

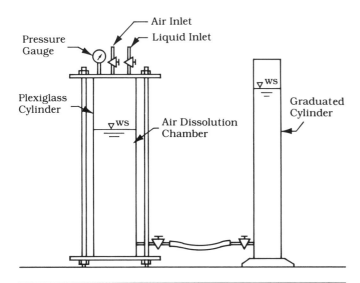

surized with air, shaken, and allowed to set to dissolve the air. The saturated liquid is released by a tube in the bottom of a graduated cylinder containing the wastewater. Flotation occurs and the floated sludge and clarified liquid are withdrawn and the suspended solids are determined. The experiment is repeated at several pressures to obtain a curve of A/S versus effluent suspended solids. The amount of pressurized effluent used depends on the recycle ratio. If R/Q is $\frac{1}{4}$ and 800 mℓ of wastewater are used, then the saturated effluent volume to be used is $(800)\frac{1}{4}$ or 200 mℓ. In design, the area of the flotation tank is based on an overflow rate of 0.2 to 4.0 gal/min-ft^2.

Example Problem 16.2
Air Flotation without
Recycle

A wastewater flow of 0.50 MGD has 200 mg/ℓ suspended solids. Air-flotation tests show that 0.05 mg air/mg solids gives optimum flotation. The design temperature is 20°C and a_s = 18.7 mℓ/ℓ at this temperature. The fraction of absorption is 0.5 and the overflow rate is 2.0 gal/min-ft^2. Determine the required pressure and flotation tank area.

Solution

Substituting into Eq. (16.2) gives

$$0.05 = \frac{(1.3)(18.7)(0.5P - 1)}{200}$$

or

$$\underline{P = 2.07\,\text{atmos}}$$

Plan area = (500,000 gal/day)(day/1440 min)
    × (min-ft^2/2.0 gal)

$$= \underline{174\ \text{ft}^2}$$

Example Problem 16.3
Air Flotation with
Recycle

For the data in Example Problem 16.2, determine the recycle ratio if the operating pressure is 3.00 atmospheres. Also determine the flotation tank area.

$$0.05 = \frac{(1.3)(18.7)(0.5 \times 3.00 - 1)R}{200\,Q}$$

or

$$\underline{R/Q = 0.82}$$

$$\text{Plan area} = (500{,}000 \text{ gal/day})(\text{day}/1440 \text{ min})$$
$$\times (1 + 0.82)(\text{min-ft}^2/2.0 \text{ gal})$$

$$= \underline{316 \text{ ft}^2}$$

Oil Separation

Since oils are relatively insoluble in water, they frequently may be separated from a wastewater by gravity separators, such as the American Petroleum Institute (API) separator. If oils are present as an emulsion, they usually require coagulation followed by gravity separation or air flotation for effective removal.

Anaerobic Contact Process

This is essentially an anaerobic activated sludge treatment process. The influent wastewater mixes with the recycled biological solids and goes to a covered, continuously mixed anaerobic reactor where most of the organic matter is biologically oxidized. The mixed liquor goes to a covered clarifier where the biological solids are separated from the supernatant, which leaves as the effluent. The settled biological solids are recycled and mixed with the influent flow. The off-gases of CH_4 and CO_2 are vented from the system. Becaues of the low cell synthesis in anaerobic systems, the waste sludge volume is much smaller than from an activated sludge system. It has been mainly used for high BOD_5 wastewaters.

Submerged Anaerobic Filter

This consists of an anaerobic filter with a packing, usually of stone. The filter is totally enclosed to maintain anaerobic conditions. The wastewater enters the bottom of the filter and passes upward through the media, which has fixed-microbial growths. The organic matter is biologically oxidized by anaerobic biochemical action by the fixed growths. The effluent leaves from the top of the unit and the off-gases of CH_4 and CO_2 are vented from the filter. This has been used on low BOD_5 wastewaters.

Land Application

This has been feasible, in some cases, for the disposal of primary and secondary effluents from municipal wastewater treatment plants. The most common systems that have been used are irrigation, infiltration-percolation, and overland flow. The irrigation system is suitable for soils with moderate permeability, such as sandy clays or clayey sands. The effluent is spread on the land by sprinkler or ridge and furrow irrigation. The water is disposed mainly by evapotranspiration and deep percolation. Frequently, the crop is grass that is used for animal fodder. Infiltration-percolation requires a very permeable soil, such as sand. The effluent is usually distributed by a ridge and furrow system. A vegetative crop is not required because the water is disposed by evaporation and deep percolation. The overland flow system is used on soils with low permeability, such as clays and silts, and with a surface slope of 2 to 4 percent. The effluent is applied by sprinkler or is distributed by ditches at the upper reaches of the sloped land. A vegetative crop is required and the water is disposed by evapotranspiration and percolation. The excess treated water flows overland and is collected by intercepting ditches. In land application, the organic matter is biologically oxidized by soil microorganisms. Phosphorus removal is mainly by soil uptake by the clay fraction in a soil along with plant uptake. Nitrogen removal is by plant uptake, nitrification, denitrification, soil uptake, and by leaching in the form of nitrates. Land application systems must be properly designed to avoid health hazards, nuisance conditions, and contamination of groundwater. Design details are available in an EPA publication (1977).

References

AWARE, Inc. 1974. *Process Design Techniques for Industrial Waste Treatment*, edited by C. E. Adams, Jr., and W. W. Eckenfelder, Jr. Nashville, Tenn.: Enviro Press.

Eckenfelder, W. W., Jr. 1966. *Industrial Water Pollution Control*. New York: McGraw-Hill.

Eckenfelder, W. W., Jr. 1980. *Principles of Water Quality Management*. Boston: CBI Publishing.

Eckenfelder, W. W., Jr. 1970. *Water Quality Engineering for Practicing Engineers*. New York: Barnes and Noble.

Eckenfelder, W. W., Jr., and Ford, D. L. 1970. *Water Pollution Control*. Austin, Tex.: Pemberton Press.

Environmental Protection Agency (EPA). 1974a. Flow Equalization. EPA Technology Transfer Seminar Publication, Washington, D.C.

Environmental Protection Agency (EPA). 1977. *Land Treatment of Municipal Wastewaters*. EPA Process Design Manual, Washington, D.C.

Environmental Protection Agency (EPA). 1974b. *Upgrading Existing Wastewater Treatment Plants*. EPA Process Design Manual, Washington, D.C.

Fair, G. M., Geyer, J. C., and Okun, D. A. 1968. *Water and Wastewater Engineering*. Vol. 2. *Water Purification and Wastewater Treatment and Disposal*. New York: Wiley.

Great Lakes—Upper Mississippi Board of State Sanitary Engineers. 1978. Recommended Standards for Sewage Works. Ten state standards, Albany, N.Y.

Metcalf and Eddy, Inc. 1979. *Wastewater Engineering, Treatment, Disposal and Reuse*. New York: McGraw-Hill.

Ramalho, R. S. 1977. *Introduction to Wastewater Treatment Processes.* New York: Academic Press.

Sanks, R. L. 1978. *Water Treatment Plant Design.* Ann Arbor, Mich.: Ann Arbor Science Publishers.

Schroeder, E. D. 1977. *Water and Wastewater Treatment.* New York: McGraw-Hill.

Steel, E. W., and McGhee, T. J. 1979. *Water Supply and Sewerage.* New York: McGraw-Hill.

Sundstrom, D. W., and Klei, H. E. 1979. *Wastewater Treatment.* Englewood Cliffs, N.J.: Prentice-Hall.

Water Pollution Control Federation (WPCF). 1977. *Wastewater Treatment Plant Design.* WPCF Manual of Practice no. 8, Washington, D.C.

Problems

1. A mechanically cleaned bar screen has $\frac{3}{8}$-in. thick bars and $1\frac{1}{4}$-in. clear space between the bars. The approach velocity is 1.8 fps at maximum flow. Determine the head loss through the screen.

2. An in-line equalization basin is to be designed for a wastewater treatment plant. From plant records, a compilation has been made that gives the hourly flow rate versus hour as shown in Table 16.2.

Table 16.2.

Hour	MGD	Hour	MGD
12 AM	3.0	1 PM	6.6
1	2.2	2	6.6
2	2.1	3	7.1
3	2.0	4	7.6
4	2.6	5	8.7
5	3.3	6	9.9
6	3.6	7	9.9
7	5.5	8	8.4
8	6.6	9	6.1
9	7.4	10	4.2
10	7.6	11	3.1
11	7.1	12 AM	3.0
12 Noon	6.8		

Determine the fluctuating volume required for the equalization basin.

3. An industrial wastewater treatment plant has a flow rate of 195 gpm and a pH of 11. It is to be neutralized to pH 8.5 before it enters the municipal sanitary sewer system. Titration using 0.1 $\underline{N}$ H_2SO_4 gives the pH versus acid used as shown in Table 16.3.

Table 16.3.

pH	mℓ acid
11.0	0
10.8	2
10.4	4
10.2	6
10.0	8
9.8	10
9.5	12
9.1	14
8.2	16
7.1	18
6.7	20

Determine the pounds of commercial grade sulfuric acid (93.2 percent H_2SO_4) required per day.

4. The maintenance shops at a large international airport have a process wastewater flow of 0.4 MGD and a free and emulsified oil content of 180 mg/ℓ as oil and grease. Air flotation using pressurized recycle and a polymer coagulant is to be employed to lower the oil and grease to 50 mg/ℓ. The flotation effluent is to be mixed with the sanitary wastewater and treated at an activated sludge plant owned by the airport authority. Laboratory experiments have been done to find the optimum coagulant dosage. Using this dosage, laboratory batch air-flotation studies have been performed to determine the air required. The data obtained are shown in Table 16.4.

Table 16.4.

Air/Solids (gm/gm)	Effluent Oil and Grease (mg/ℓ)
0.056	12
0.042	27
0.028	47
0.015	77
0.008	113
0.003	155

The fraction of air dissolved in a retention tank having a one-minute detention time is 0.5. Other pertinent data are: average operating temperature = 20°C, air pressure = 45 psia, and overflow rate based on $Q + R$ = 3.0 gpm/ft^2. Determine:

a. The recycle flow R, MGD.

b. The area of the flotation tank.

c. The pounds of air required per day.

d. The ft^3 of air required per day.

5. A design of a dissolved air flotation unit without recycle is to be made for the airport and conditions stated in Problem 4. Determine:

a. The area of the flotation tank.

b. The air pressure required in atmospheres and psia.

Chapter Seventeen

DISINFECTION

Disinfection is defined as the destruction of pathogenic microorganisms. It does not apply to nonpathogenic microorganisms or to pathogens that might be in the spore state. The term that applies to the destruction of all living organisms and especially to microorganisms including spores is *sterilization* (McCarthy, J. J., & Smith, C. H., 1974).

In the United States, water and wastewater disinfection is accomplished almost solely by chlorination. Very little use is made of other disinfection techniques. Findings of recent years, however, relating to the generation of undesirable trihalomethanes (Rook, J. J., 1977) and other chlorinated organic products (Jolley, R. L., 1975) by the chlorination process suggest that future disinfection practice in the U.S. might well be modified to lessen the magnitude of these undesirable consequences.

The chlorination of drinking water supplies was introduced in the United States in 1908, and by 1914 "the greater part of the water supplied in cities in the U.S. [was] treated in this [or an equivalent] way . . . (National Academy of Sciences, 1977). As a result, a marked decrease in the incidence of water-borne diseases also occurred. The Mills-Reincke theorem holds that for every death from water-borne typhoid, there are several deaths from other diseases for which the causal agents are transmitted by water. The National Academy of Sciences Safe Drinking Water Committee (1977) observes that the theorem appears to have considerable merit, but although disinfection to levels based on coliform criteria has ensured freedom from typhoid fever, it does not give a similar guarantee to freedom from other infections.

Disinfection is described by Chang (1971) as a complex rate process dependent upon:

1. physico-chemistry of the disinfectant;
2. cyto-chemical nature and physical state of the pathogens;
3. the interaction of 1 and 2 above; and
4. quantitative effects of factors in the reaction medium, such as:
   a. temperature;
   b. pH;
   c. electrolytes;
   d. interfering substances.

He also classified disinfectants as follows:

1. oxidizing agents (ozone, halogens, halogen compounds);
2. cations of heavy metals (silver, gold, mercury);
3. organic compounds;
4. gaseous agents;
5. physical agents (heat, uv and ionizing radiation, pH).

The cyto-chemical nature and physical state of the pathogens encompasses a broad range of organisms categorized by Englebrecht as including bacteria, viruses, protozoa, and helminths. Bacterial (and fungal) spores are much more resistant to disinfection than are vegetative forms. Spore-formers, however, are usually not important in water and wastewater disinfection. The same might be said of some bacterial pathogens, such as the tuberculosis organism, which are more resistant than the gram-negative coliform cells that serve as our criterion of water and wastewater disinfection. Finally, cysts and ova of protozoa and helminths are also more resistant to disinfection than vegetative bacterial cells, but other treatment processes, notably flocculation, sedimentation, and filtration, are usually quite effective in their removal when they are working properly.

Among the more important differences that exist between various microorganisms and their resistance to disinfection processes are those relating to various viruses on the one hand and coliforms on the other. The former, in general, are much more resistant to chloramines, and are also more resistant under many naturally occurring circumstances, such as when embedded or enmeshed in tissues or suspended material, or when aggregated in clumps. A technique often used to measure the effectiveness of various disinfection processes has been to seed with substantial numbers of viruses, but these seeded viruses have been found to be considerably more vulnerable than the viruses that naturally occur.

The rate of disinfection by a chemical agent may in many cases conform to Chick's law of disinfection (1908), which is presented by the following pseudo–first-order reaction:

$$-\frac{dN}{dt} = kN \tag{17.1}$$

where

dN/dt = rate of cell destruction, number/time;
  k = rate constant;
  N = number of living cells remaining at time t.

The constant, k, depends on the species and form of the microorganisms being destroyed, the disinfectant and its nature, the concentration of the disinfectant, and environmental factors such as pH and temperature. Chick's law is limited in application since, in many cases, it has been found that the rate of kill in the latter duration of the disinfection process may be more or less than indicated by Eq. (17.1).

An empirical equation that relates the concentration of a disinfectant and the contact time is

$$C^n t_c = K \tag{17.2}$$

where

  C = concentration of disinfectant at time, $t = 0$;
  t_c = time of contact required to kill a given percentage of the microbes;
  K,n = experimental constants.

The value of n depends on the nature of the disinfectant used. If n is greater than one, the disinfecting action is greatly dependent on the concentration of the disinfectant. Conversely, if n is less than one, the disinfectant action is primarily dependent on the time of contact. The value of the constant, K, depends on the types of microorganisms being destroyed and environmental factors such as pH and temperature. For the usual range of microbe concentration encountered in water treatment, there is usually very little effect on Eq. (17.2) due to microbial concentration. Equation (17.2) plots as a straight line on log-log paper, as shown in Figures 17.1 and 17.2. Figure 17.1 illustrates the effectiveness of the various forms of chlorine for *Escherichia coli* destruction. It can be seen that the free chlorine forms

Figure 17.1. Concentration versus Contact Time for 99 Percent Kill of E. coli by Various Forms of Chlorine at 2 to 6° C
Adapted from "Influence of pH and Temperature on the Survival of Coliforms and Enteric Pathogens When Exposed to Free Chlorine" by C. T. Butterfield, E. Wattie, S. Megregian, and C. W. Chambers in *U.S. Public Health Reports* 58 (1943):1837.

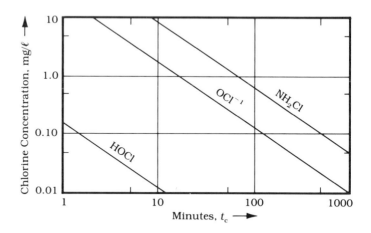

Figure 17.2. Concentration versus Contact Time for 99 Percent Kill of E. coli and Three Enteric Viruses by HOCl at 0 to 6° C
Adapted from "The Virus Hazard in Water Supplies" by G. Berg in *Journal of New England Water Works Association* 78 (1964):79. Reprinted by permission.

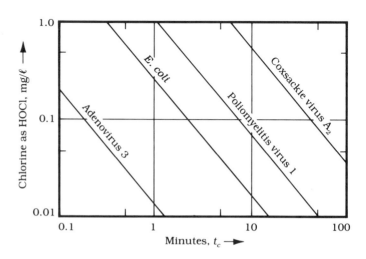

are much more effective than combined chlorine in the form of monochloramine. Figure 17.2 shows the effectiveness of free chlorine for the destruction of *E. coli* and three enteric viruses. As depicted, two of the viruses were more resistant than *E. coli*.

Temperature affects disinfecting action because an increase in temperature results in a more rapid kill. If organic matter is present, chemical disinfecting reagents may react with it, thus reducing the effective concentration of the disinfectant. In chlorination particularly, pH is important because it influences the relative distribution of agents of varying effectiveness, as will be discussed. The currently used disinfection models are particularly deficient when applied to long contact periods (Engelbrecht, R. S.).

The generation of undesirable organo-chlorines in the chlorination process and the increasing need to treat used waters (that is, wastewaters) for subsequent reuse is resulting in study and consideration of other disinfection methods. Most prominent among these are ozonation, chlorine dioxide, uv irradiation, high pH treatment, and the use of other halogens, such as iodine and bromine. The latter two are not discussed in the following text.

Chlorination

Chlorine is the most widely used disinfectant because it is effective at low concentration, cheap, and forms a residual if applied in sufficient dosage. It may be applied as a gas or as a hypochlorite, the gas form being most common. The gas is liquified at five to ten atmospheres and shipped in steel cylinders. Pressurized liquid chlorine (99.8 percent Cl_2) is available in cylinders containing 100, 150, or 2000 lb of the liquified gas. The disinfecting ability of chlorine is due to its powerful oxidizing properties, which oxidize those enzymes of microbial cells that are essential to the cells' metabolic processes (Butterfield, C. T., Wattie, E., Megregian, S., & Chambers, C. W. 1943).

Reaction

Chlorine gas reacts readily with water to form hypochlorous acid, HOCl, and hydrochloric acid:

$$Cl_2 + H_2O \rightarrow HOCl + HCl \qquad (17.3)$$

In dilute solution and with pH greater than 3, the reaction is appreciably displaced to the right and very little molecular chlorine gas will remain dissolved and unreacted. The hypochlorous acid produced then dissociates to yield hypochlorite ion:

$$HOCl \rightleftarrows H^+ + OCl^-$$ (17.4)

The relative distribution of HOCl and OCl^- is a function of pH, as shown in Figure 17.3.

Hypochlorite salts are available in dry (calcium hypochlorite) or liquid (sodium hypochlorite) form. The dry form is cheaper but must be dissolved in water:

$$Ca(OCl)_2 \xrightarrow{H_2O} Ca^{+2} + 2OCl^{-1}$$ (17.5)

The OCl^{-1} will then seek an equilibrium with the hydrogen ions as indicated in Eq. (17.4), and therefore when hypochlorites are used in such applications as swimming pools, it is often necessary to add acid.

Although both hypochlorous acid and hypochlorite ion are excellent disinfecting agents, the acid form is the more effective (Engelbrecht, R. S.). They also react with certain inorganic and organic materials in water. One of the important reactions is with ammonia:

Figure 17.3. Relative Amounts of Chlorine as HOCl and OCl^{-1} at 20°C versus pH
Adapted from "Behavior of Chlorine as a Disinfectant" by G. M. Fair, et al., *Journal of the American Water Works Association* 40, no. 10 (October 1948):1051. By permission. Copyright 1948, the American Water Works Association.

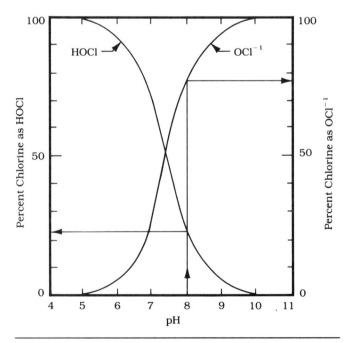

$$NH_3 + HOCl \rightarrow NH_2Cl + H_2O$$
monochloramine $\qquad$ (17.6)

$$NH_2Cl + 2HOCl \rightarrow NHCl_2 + 2H_2O$$
dichloramine $\qquad$ (17.7)

$$NHCl_2 + 3HOCl \rightarrow NCl_3 + 3H_2O$$
nitrogen trichloride $\qquad$ (17.8)

The relative amounts of the various chloramines that are formed are mainly a function of the hypochlorous acid present and the pH. The monochloramine form predominates at a pH greater than about 6.0, and the dichloramine at about pH 5.

Reactions also occur with reduced substances and with organic materials. Dissolved chlorine gas will react with hydrogen sulfide to produce sulfuric and hydrochloric acids. It will react with other inorganic reducing ions or substances such as Fe^{+2}, Mn^{+2}, and NO_2^{-1}. Dissolved chlorine will also react with organic compounds, particularly unsaturated compounds. Two of the organo reactions are particularly important—those that result in chlorophenols and those that produce trihalomethanes. Chlorophenols, formed from the reaction of chlorine with phenols, impart undesirable tastes and odors to water that are detectable at phenol concentrations less than one microgram per liter. Reaction of chlorine with innocuous humic substances results in the formation of trihalomethanes including:

$CHCl_3$	chloroform
$CHCl_2Br$	bromodichloromethane
$CHClBr_2$	dibromochloromethane

These compounds are limited by drinking water regulations to an in toto 0.1 milligram per liter because of tumorigenic properties.

Chloramines are effective compounds against bacteria but are not nearly so effective against viruses. The difference in effectiveness of chloramines was illustrated in a bench study (Figure 17.4) by Kruse, Hsu, Griffiths, and Stringer (1970), who utilized a synthetic waste, *Escherichia coli*, and F2 coliphage. A plot of the data of Durham and Wolf (1973) shows the same type of results on effluents from two trickling filter plants using total coliforms and any phages accepted by *E. coli* K12(f^+) cells. The extrapolation of these findings on coliphages to all animal viruses is not warranted, but the results are meant to convey some inherent differences that can exist between bacteria and viruses in their susceptibility to chloramines.

Figure 17.4. Chloramine Effects on Bacteria and Virus
a. C. W. Kruse, Y-C, Hsu, A. C. Griffiths, and R. Stringer, "Halogen Action on Bacteria, Viruses, and Protozoa," Proceedings of the National Specialty Conference on Disinfection, American Society of Civil Engineers, 1970, p. 113.
b.H. W. Wolf, R. S. Safferman, A. R. Mixson, and C. E. Stringer, "Virus Inactivation during Tertiary Treatment," *Jour. AWWA* 66, no. 9 (September 1974):526.

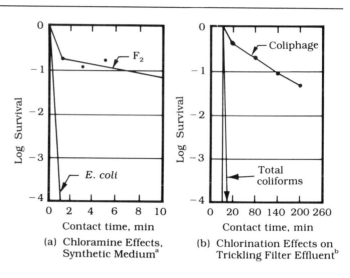

(a) Chloramine Effects, Synthetic Medium[a]

(b) Chlorination Effects on Trickling Filter Effluent[b]

Figure 17.5. Chlorine Dosages, Demands, and Residuals
Adapted from *Chemistry for Environmental Engineers* by C. N. Sawyer and P. L. McCarty. Copyright © 1978 by McGraw-Hill Book Co., Inc. Reprinted by permission.

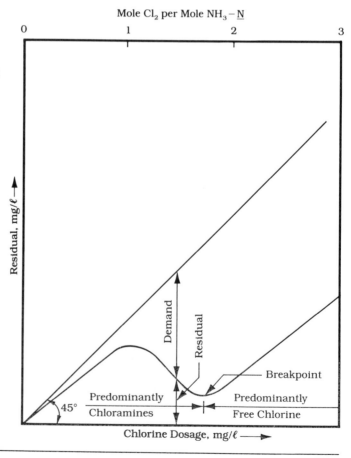

ENVIRONMENTAL ENGINEERING

Dosages, Demands, and Residuals

The dosage is the amount of chlorine added, the demand is the amount used for oxidation of materials present, and the residual is the amount remaining after oxidation. The relationships between these are shown in Figure 17.5. The residual equals the dosage minus the demand.

Contact time is very important in the disinfection process. In chlorination, increased time of contact results not only in greater destruction of microorganisms but in an increased demand and, if appropriate precursors and free chlorine are present, in an increased amount of various chlorinated by-products.

Chlorine gas, hypochlorous acid, and the hypochlorite ion remaining after the demand is satisfied are collectively termed *free chlorine residuals.* The chloramines and other reactive chlorine forms remaining after the demand is satisfied are referred to as *combined chlorine residuals.* Free chlorine residuals are faster acting than combined residuals, and, for the same concentration and time, the free chlorine residuals have much greater disinfecting capacity than combined residuals, especially for viruses.

As depicted in Figure 17.5, an increase in chlorine dosage results in an equivalent increase in the residual up to a molar ratio of chlorine to ammonia nitrogen of 1:1. The residual formed is predominantly mono- and dichloramine. If the chlorine dosage is increased above this ratio, some nitrogen trichloride will be formed; however, as the dosage is increased, most of the chloramines will be oxidized to nitrogen gases. The oxidation reaction is essentially complete, for any particular point in time, at the minimum dip in the residual curve, which is termed the *breakpoint.* The breakpoint occurs at a chlorine dosage of about $1\frac{1}{2}$ to 2 moles of chlorine per 1 mole of ammonia nitrogen and represents the dosage when the chloramines have been converted to the nitrogen gases. Some of the gases have been identified as free nitrogen, nitrous oxide, and nitrogen trichloride, with free nitrogen being the most predominant. Continued addition of chlorine beyond the breakpoint gives a residual that is predominantly free chlorine.

The breakpoint dosage is very much dependent upon water quality, but for many drinking water supplies it ranges over 4 to 10 mg/ℓ. The desirable residual to be maintained at the farthest tap on the distribution system is at least 0.2 mg/ℓ, all free chlorine. The National Primary Drinking Water Regulations promote such a residual by allowing a reduction in the number of bacteriological samplings if that residual is maintained.

Wastewater chlorination practice will vary with states'

policies. Some states utilize a fecal coliform criterion such as 200/100 mℓ. Other states specify a type of residual after a specific contact period. Texas specifies 1.0 mg/ℓ total residual after 20 minutes of contact. Breakpoint chlorination of wastewater is seldom practiced. Dosages of 50–70 mg/ℓ may be necessary to reach breakpoint in many wastewaters, and this renders the effluents highly toxic to much of the aquatic life (Brungs, W. A., 1973).

Chlorine Application

Gaseous chlorine may be dissolved in water using any one of a variety of proprietary chlorinators, and the concentrated solution is then piped to the water stream to be disinfected. Hypochlorites can be added using solution-type feeders. Dry hypochlorites are first dissolved in water in a plastic or clay vessel; the liquid is then decanted off by a solution-type feeder. Some hypochlorites are packaged in polymers to render them amenable to dry feed equipment, but for the usual size of installation these cannot compete economically with gaseous chlorine.

In water and wastewater treatment, prechlorination is the application of chlorine prior to any treatments, whereas postchlorination is chlorination after all treatments. Prechlorination is practiced to control undesirable growth such as might occur in a pipeline aqueduct. Similarly, prechlorination in wastewater treatment might be applied to sewers to control odors that develop as a result of undesirable growth. Postchlorination is sometimes called *terminal disinfection*.

Ozonation

Ozone is an allotrope of oxygen. It is a powerful oxidant and is more powerful than hypochlorous acid. In aqueous solution it is relatively unstable, having a half-life of 20–30 minutes in distilled water at 20°C. The presence of oxidant demanding materials in solution will render the half-life even shorter (Rice, R. G., Miller, G. W., Robson, C. M., & Hill, A. G., 1979).

Ozone is widely used in drinking water treatment practice in Europe. Its first application was in 1893 at Oudshoorn, Netherlands, where it was used for settled and filtered Rhine River water. Today more than 1000 plants throughout the world use ozone. Canada has 22 plants, with Montreal probably having the world's largest (Rice, et al., 1979). Ozone use in wastewater treatment is very limited but has considerable potential when the wastewater is well treated.

Ozone must be produced on-site because it cannot be

stored as can chlorine. This is not necessarily bad since serious accidents have happened with chlorine because of breaks in storage systems. Ozone is produced by passing air between oppositely charged plates or through tubes in which a core and the tube walls serve as the oppositely charged surfaces. Air is refrigerated to below the dew point to remove much of the atmospheric humidity, then passes through desiccants, such as silica gel, activated alumina, to dry the air to minus 40–60°C dew point. The use of dry, clean air results in less frequent ozone generator maintenance, long-life units, and more ozone production per unit of power used (Jolley, R. L., 1975).

Gomella and coworkers observed complete destruction of poliovirus samples in distilled water at a residual of 0.3 mg/ℓ at the end of 3 minutes of exposure. They then observed the same effectiveness when the viruses were suspended in Seine River water and recommended the use of 0.4 mg/ℓ after a contact of 4 minutes (Coin, L., Gomella, C., & Hannoun, C., 1964). It is important that ozone demand be completely satisfied before this disinfecting step.

Usual French practice uses two contactors. In the first, with a contact time of 8–12 minutes, the ozone demand is satisfied and a residual of 0.4 mg/ℓ obtained. In the second, with a contact time of 4–8 minutes, the 0.4 mg/ℓ residual is maintained (Rice, et al., 1979).

German use of ozone often exploits its ability to render some refractory organics biodegradable. Sontheimer uses a 20-minute contact period in this application. The dose of ozone to be used is determined from pilot studies that are conducted in Europe for one-year periods and even longer. Ozone-treated water is then passed through granular activated carbon, which serves as a fixed-bed biological contactor allowing saprophytic organisms to decompose the biodegradable materials. Such a contactor biologically regenerates itself and is called a BAC, or biological activated carbon, process. Ozone is never used as a terminal treatment because experience has shown that organisms can under certain circumstances proliferate in distribution systems, causing all types of problems. Hence, many European plants utilize the desirable residual action of chlorine as a terminal disinfectant, but the dose is very low, such as 0.1 to 0.3 mg/ℓ.

It is the capability of ozone to render some refractory organics biodegradable that indicates its future use in wastewater treatment. McCarthy and Smith (1974) observed a marked affinity of ozone for suspended matter. Thus, for good results, the suspended solids should not be too high. Buys (1980) observed that for a petrochemical wastewater, a dose of 12 mg O_3/ℓ had no effect on the raw wastewater, but on the treated wastewater, BOD_5 values were substantially increased by about 50 percent. This observation has been reported for municipal wastewater effluents. He also noted that the nonadsorbable COD was decreased by about 50 percent. The suggestion is that ozone is likely to have an increased application in future wastewater treatment.

It should be pointed out that there are uncertainties regarding health effects of some oxidized synthetic organics, a topic of current research.

Chlorine Dioxide

Chlorine dioxide (ClO_2) was discovered in 1811 but was not used in water treatment until 1944 at Niagara Falls, New York. Its application was primarily to alleviate taste and odor problems that arose from chlorination of phenolic contaminants in the water. A 1958 survey indicated 56 water treatment plants were using ClO_2, most for taste and odor control, some for iron or manganese removal, and others for disinfection.

ClO_2 is a more powerful oxidant than chlorine. It does not react with water as does chlorine gas, is easily removed from water by aeration, is readily decomposed by exposure to ultraviolet irradiation, does not react with ammonia as does chlorine, and persists to maintain a stable residual.

There are at least four general methods of preparing ClO_2:

1. Acid and sodium chlorite

$$NaClO_2 + HCl \rightarrow ClO_2 + NaCl + H^+ \qquad (17.9)$$

2. Chlorine gas and sodium chlorite (excess chlorine)

$$(Cl_2 + H_2O \rightarrow HOCl + HCl)$$

$$HOCl + HCl + 2\ NaClO_2 \rightarrow 2ClO_2$$
$$+ 2\ NaCl + H_2O \qquad (17.10)$$

3. Sodium hypochlorite and sodium chlorite

$$2\ NaClO_2 + NaOCl + 2\ HCl \rightarrow 2\ ClO_2$$
$$+ 3\ NaCl + H_2 \qquad (17.11)$$

4. Sodium chlorate
   Four different processes are used to produce ClO_2 from sodium chlorate but none are used in water treatment.

The process most widely used in water treatment is the excess chlorine method. The excess chlorine is to ensure that all of the chlorite ions will produce chlorine dioxide. Some uncertainties exist regarding the health effects of the chlorite ion, ClO_2^-, which might also be introduced after the oxidizing reaction.

The ClO_2 residual is longer lasting than is HOCl. Terminal disinfection is practiced using only 0.10–3.0 mg/ℓ ClO_2. However, taste and odor control applications may see dosages up to 10 mg/ℓ (Miller, G. W., Rice, R. G., Robson, C. M., & Wolf, H., 1978).

UV Irradiation

Ultraviolet irradiation has been utilized for the disinfection of drinking water supplies aboard ships for many years. Its utility in the disinfection of secondary effluents has been only recently investigated, but the pilot studies performed indicate it to be a most viable process (Johnson, J. D., Aldrich, J., Francisco, D. E., Wolff, T., & Elliott, M., 1978; Petrasek, A. C., Wolf, H. W., Esmond, S. E., & Andrews, D. C., 1980).

UV can effectively disinfect both water and wastewater. The lack of a residual is a major disadvantage in water treatment, and this is recognized in the Center for Disease Control regulations, which recommend the use of chlorine aboard ship even when UV systems are installed and functioning. If chlorine must be used, there does not appear to be any justification to also use UV. The lack of a residual, however, becomes a virtue in wastewater treatment application.

Two different types of UV installations were tested in Dallas, where it was found that the submergence of the UV bulb encased in a quartz tube was superior to an overhead bulb radiating downward through a shallow 1–2-inch depth of water. The major factor in achieving good microorganism kill is the ability of the radiation to pass through the water and get to the target organism. Interestingly, this transmittance was found to be markedly less a function of turbidity or suspended solids, but very much a function of the COD or TOC. Since lower COD or TOC values are associated with long mean cell residence times accompanying highly nitrified operation, an association of good transmittance, hence good kill, with lower NH_3-N values was observed.

Comparative dose-response equations for three organisms showed only a little spread and similar slopes:

$$y = 1.48x - 3.21 \quad \text{(fecal coliforms)} \tag{17.12}$$

$$y = 1.62x - 4.48 \quad \text{(polioviruses)} \tag{17.13}$$

$$y = 1.59x - 4.68 \quad \text{(coliphages)} \tag{17.14}$$

where

y = log reduction; and

x = log uv dose.

Unlike chloramines, uv performs very well against viruses.

A slime buildup occurs on the quartz sleeves housing the uv lamps. A proprietary cleaning solution was effective in cleaning the sleeves. A cleaning frequency of once each two weeks, or perhaps three weeks, is necessary to keep the problem under control.

The mean intensity of radiation, measured by a radiometer in microwatts per square centimeter (μw/cm^2), was multiplied by the detention time (in seconds) of the flow in the irradiation chamber to yield the indicated dose, D_I, in watt-sec/cm^2. For total and fecal coliforms, the relationship of D_I to log reduction was determined to be

$$D_I = 24800\ x\ -\ 60000 \quad \text{(fecal coliforms)} \qquad (17.15)$$

$$D_I = 20000\ x\ -\ 48200 \quad \text{(total coliforms)} \qquad (17.16)$$

where

x = log reduction.

Figure 17.6 is a plot of the log of the fecal coliform density whose probability of exceedence is equal to or greater than five percent versus the log of the D_I. The curve indicates that a fecal coliform density of 200/100 mℓ can be expected to be exceeded five percent of the time if the indicated dose is 24,400 μwatt-sec/cm^2 (Petrasek, et al., 1980).

High-pH Treatment

The use of lime in treating the contents of privies, dead animals, and battlefield mortalities to alleviate nuisance conditions is historic. Its effectiveness in destroying bacteria in water treatment application at high pH has been known since the 1920s. Even so, it is not relied on as the sole disinfectant and chlorine is always additionally employed because after neutralization of the high pH, there remains no residual to protect the water.

Berg, Dean, and Dahling (1968) studied virus removal from secondary effluents by lime flocculation and observed higher removals at higher pH values. Dallas studies using seeded polio I and coliphages showed very high removals of both viruses (neither was detected in the effluents) and of vegetative bacteria from secondary effluents, an observation reinforced by South African studies (Nupen, E. M., Bateman, B. W., & McKenny, N.C., 1974). Prior work in

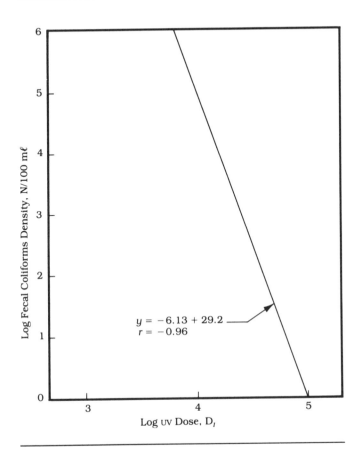

$$y = -6.13 + 29.2$$
$$r = -0.96$$

Dallas on coliphages alone had indicated the critical pH to contact time relationships to be in the range pH 11.2–11.3 and a time of 1.56 to 2.40 hours. In the studies showing no surviving viruses, a pH of 11.0 or greater and a theoretical contact time of 5 hours and 10 minutes were employed. It should be pointed out that a residence time distribution function study showed peak dye concentrations in the effluent occurring only two hours after the addition of a slug load. The pH was neutralized as water left the high-pH tank and growth of surviving or contaminating bacteria was observed.

High-pH treatment for disinfection of drinking water supplies, like ozone or UV, would be highly effective against viruses but still requires the use of an additional residual disinfectant. However, the additional disinfectant need only be added in minimal amount. In wastewater treatment, high-pH treatment for disinfection could be coupled with the additional objectives of ammonia and phosphorus removal. Recarbonation prior to discharge would also be necessary unless discharge is to sodic or acidic soils.

High-pH treatment of wastewater sludges, called *lime stabilization*, is a recommended procedure. Lime is added to pH 12 and maintained at that value for 30 minutes. Noland, Edwards, and Kipp (1978) observed on sludge what Berg, Dean, and Dahling (1968) had observed on wastewater—that is, the higher the pH, the greater the bacterial reduction. However, a pH of 12 or greater appears to be required by sludges. Unfortunately, virus evaluations were not undertaken and parasite data indicate little effect of the high pH; however, the time of exposure to the high pH in the study appears to have been limited to only 30 minutes (Wolf, H. W., Safferman, R. S., Mixson, A. R., & Stringer, C. E., 1974).

References

Berg, G. 1964. The Virus Hazard in Water Supplies. *Jour. New England Water Works Assoc.* 78:79.

Berg, G., Dean, R. B., and Dahling, D. R. 1968. Removal of Poliovirus I from Secondary Effluents by Lime Flocculation and Rapid Sand Filtration. *Jour. AWWA* 60, no. 2:193.

Brungs, W. A. 1973. Effects of Residual Chlorine on Aquatic Life. *Jour. WPCF* 45, no. 10:2180.

Butterfield, C. T., Wattie, E., Megregian, S., and Chambers, C. W. 1943. Influence of pH and Temperature on the Survival of Coliforms and Enteric Pathogens When Exposed to Free Chlorine. *U.S. Pub. Health Repts.* 58:1837.

Buys, R. E. 1980. The Effect of Solids Retention Time on Tertiary Ozonation and Carbon Adsorption of Petrochemical Wastewaters. M.S. thesis, Texas A&M University.

Chang, S. L. 1971. Modern Concept of Disinfection. *Jour. SED* 97, no. SA5:689.

Chick, H. 1908. Investigation of the Laws of Disinfection. *Jour. Hygiene* 8:92.

Coin, L., Gomella, C., and Hannoun, C. 1964. Inactivation of Poliomyelitis Virus by Ozone in the Presence of Water. *la Presse Med.* 72 (37):2153.

Durham, D., and Wolf, H. W. 1973. Wastewater Chlorination: Panacea or Placebo? *Water and Sewage Works* 120:67.

Engelbrecht, R. S. In press. Microbiological Criteria and Standards for Potable Reuse of Water.

Fair, G. M., Morris, J. C., Chang, S. L., Weil, I., and Burden, R. P. 1948. Behavior of Chlorine as a Disinfectant. *Jour. AWWA* 40, no. 10:1051.

Gomella, C. Personal communication.

Green, D. E., and Stumpf, P. K. 1946. The Mode of Action of Chlorine. *Jour. AWWA* 38, no. 11:1301.

Johnson, J. D., Aldrich, J., Francisco, D. E., Wolff, T., and Elliott, M. 1978. UV Disinfection of Secondary Effluent. *Progress in Wastewater Disinfection Technology*, edited by A. D. Venosa. Environmental Protection Agency, Washington, D.C.

Jolley, R. L. 1975. Chlorine-Containing Organic Constituents in Chlorinated Effluents. *Jour. WPCF* 47, no. 3:601.

Kruse, C. W., Hsu, Y.-C., Griffiths, A. C., and Stringer, R., 1970. Halogen Action on Bacteria, Viruses, and Protozoa. *Proceedings of the National Specialty Conference on Disinfection.* American Society of Civil Engineers.

McCarthy, J. J., and Smith, C. H. 1974. The Use of Ozone in Advanced Wastewater Treatment. *Jour. AWWA* 66, no. 12:718.

Miller, G. W., Rice, R. G., and Robson, C. M., Kuhr, W., and Wolf, H. 1978. *An Assessment of Ozone and Chlorine Dioxide Technologies for Treatment of Municipal Water Supplies.* Environmental Protection Agency Grant R 804385-01, Washington, D.C.

National Academy of Sciences. 1977. *Drinking Water and Health.* Washington, D.C.

Noland, R. F., Edwards, J. D., and Kipp, M. 1978. Full Scale Demonstration of Lime Stabilization. Environmental Protection Agency Report 600/2-68-171, Washington, D.C.

Nupen, E. M., Bateman, B. W., and McKenny, N. C. 1974. The Reduction of Virus by the Various Unit Processes Used in the Reclamation of Sewage to Potable Waters. *Virus Survival in Water and Wastewater Systems.* Austin, Tex.: University of Texas Press.

Petrasek, A. C., Jr., Wolf, H. W., Esmond, S. E., and Andrews, D. C. 1980. *Ultraviolet Disinfection of Municipal Wastewater Effluents.* Environmental Protection Agency Report 600/2-80-102, Washington, D.C.

Rice, R. G., Miller, G. W., Robson, C. M., and Hill, A. G. 1979. Ozone Utilization in Europe. AIChE, 8th Annual Meeting, Houston, Tex.

Rook, J. J. 1977. Chlorination Reactions of Fulvic Acids in Natural Waters. *Environmental Sci. and Tech.* 11:478.

Sawyer, C. N., and McCarty, P. L. 1978. *Chemistry for Environmental Engineers.* 3rd ed. New York: McGraw-Hill.

Sontheimer, H. Personal communication.

Webster's New Collegiate Dictionary. 1956. Springfield, Mass.: G. & C. Merriam.

Wolf, H. W., Safferman, R. S., Mixson, A. R., and Stringer, C. E. 1974. Virus Inactivation during Tertiary Treatment. *Jour. AWWA* 66, no. 9:526.

Appendixes

APPENDIX A

Conversions and Measures

1. Length

mi	yd	ft	in.	mm	cm	m	km
1	1760	5280	—	—	—	—	1.609
—	1	3	36	—	91.44	0.9144	—
—	—	1	12	—	30.48	—	—
—	—	—	1	25.40	2.54	—	—
—	—	—	—	10	1	—	—
—	1.094	3.281	—	—	100	1	—
0.6215	—	—	—	—	—	1000	1

2. Area

mi^2	ac	yd^2	ft^2	$in.^2$	cm^2	m^2	km^2
1	640	—	—	—	—	—	—
—	1	—	43,560	—	—	—	—
—	—	1	9	—	—	—	—
—	—	—	1	144	—	—	—
—	—	—	—	1	6.45	—	—
—	—	1.196	10.764	—	10^2	1	—
0.3863	247	—	—	—	—	10^6	1

1 hectare (ha) = 10,000 m^2 = 2.47 ac

3. Volume

ft^3	Imp. gal	U.S. gal	U.S. qt	$in.^3$	ℓ	cm^3	m^3
1	6.23	7.481	29.92	1728	28.32	—	—
—	1	1.2	4.8	277.4	4.536	—	—
—	—	1	4	231	3.785	—	—
—	—	—	1	57.75	0.946	—	—
—	—	0.264	1.057	61.02	1	1000	—
35.31	—	264.2	—	—	1000	10^6	1

4. Weight or Mass

ton	long ton	lb	grains	mg	gm	kgm	tonne
1	—	2000	—	—	—	—	—
—	1	2240	—	—	—	—	—
—	—	1	7000	—	454	—	—
—	—	—	15.43	1000	1	—	—
—	—	2.203	—	—	1000	1	—
—	—	—	—	—	—	1000	1

5. Pressure

psi	ft of water	in. of Hg	mm of Hg	atm.	kilopascal
1	2.307	2.036	51.714	–	6.8957
0.4335	1	0.8825	22.416	–	–
0.4912	1.33	1	25.4	–	–
14.70	33.93	29.92	760	1	101.37

6. Velocity

mi/hr	ft/sec	in./min	cm/s	km/h
1	1.467	1056	–	1.609
–	–	1	0.423	–
0.6215	–	–	–	1

7. Discharge

ft³/sec	MGD	gpm	ℓ/s	ℓ/min	m³/sec
1	0.6463	448.8	28.32	1699	–
1.547	1	694.4	–	–	–
–	–	15.85	1	–	–
35.32	22.82	15,850	1000	60,000	1

8. Power

kw	hp	ft-lb/sec	joule/sec (watt)
1	1.341	737.6	1000
0.7457	1	550	745.7

9. Work, Energy, Heat

kw-hr	hp-hr	ft-lb	Btu	gm-cal	joule
1	1.341	–	3412	–	–
0.7457	1	–	2544	–	–
–	–	1	–	–	1.356
–	–	777.5	1	252	1054
–	–	–	–	1	4.184
–	–	0.7376	–	–	1

10. Miscellaneous

1 U.S. gal = 8.34 lb of water

1 Imp. gal = 10 lb of water

1 ft^3 = 62.43 lb of water

1 m^3 = 2283 lb of water

1 newton (N) = 0.2248 lb force = 1 kgm-m/s^2

1 pascal (Pa) = 1 N/m^2

1 joule (J) = 1 N-m

1 joule/sec (J/sec) = 1 N-m/sec

1 watt (W) = 1 joule/sec = 1 N-m/sec

1 ppm = 1 mg/ℓ if the specific gravity = 1.00

°F = 9/5 (°C) + 32

°C = 5/9 (°F − 32)

°R = °F + 460

°K = °C + 273

g = 32.17 ft/sec^2 = 9.806 m/s^2

APPENDIX B

International Atomic Weights of Numerous Elements (1961), Based on Carbon 12

Elements	Symbol	Atomic Number	Atomic Weight
Aluminum	Al	13	26.98
Antimony	Sb	51	121.75
Arsenic	As	33	74.92
Barium	Ba	56	137.34
Boron	B	5	10.81
Bromine	Br	35	79.91
Cadmium	Cd	48	112.40
Calcium	Ca	20	40.08
Carbon	C	6	12.01
Chlorine	Cl	17	35.45
Chromium	Cr	24	52.00
Cobalt	Co	27	58.93
Copper	Cu	29	63.54
Fluorine	F	9	19.00
Gold	Au	79	196.97
Helium	He	2	4.00
Hydrogen	H	1	1.008
Iodine	I	53	126.90
Iron	Fe	26	55.85
Lead	Pb	82	207.19
Magnesium	Mg	12	24.31
Manganese	Mn	25	54.94
Mercury	Hg	80	200.59
Molybdenum	Mo	42	95.94
Nickel	Ni	28	58.71
Nitrogen	N	7	14.01
Oxygen	O	8	16.00
Phosphorous	P	15	30.97
Platinum	Pt	78	195.09
Potassium	K	19	39.10
Selenium	Se	34	78.96
Silicon	Si	14	28.09
Silver	Ag	47	107.87
Sodium	Na	11	22.99
Strontium	Sr	38	87.62
Sulfur	S	16	32.06
Tin	Sn	50	118.69
Zinc	Zn	30	65.37

The above is a partial listing. In particular, the radioactive elements are not included. For a complete listing see *Lange's Handbook of Chemistry* (New York: McGraw-Hill, 1961).

APPENDIX C

Density and Viscosity of Water

Temperature (°C)	Density (gms/cm³, γ)	Absolute Viscosity (centipoise[a], μ)	Kinematic Viscosity (centistokes[b], ν)
0	0.99987	1.7921	1.7923
1	0.99993	1.7320	1.7321
2	0.99997	1.6740	1.6741
3	0.99999	1.6193	1.6193
4	1.00000	1.5676	1.5676
5	0.99999	1.5188	1.5188
6	0.99997	1.4726	1.4726
7	0.99993	1.4288	1.4288
8	0.99988	1.3872	1.3874
9	0.99981	1.3476	1.3479
10	0.99973	1.3097	1.3101
11	0.99963	1.2735	1.2740
12	0.99952	1.2390	1.2396
13	0.99940	1.2061	1.2068
14	0.99927	1.1748	1.1756
15	0.99913	1.1447	1.1457
16	0.99897	1.1156	1.1168
17	0.99880	1.0876	1.0888
18	0.99862	1.0603	1.0618
19	0.99843	1.0340	1.0356
20	0.99823	1.0087	1.0105
21	0.99802	0.9843	0.9863
22	0.99780	0.9608	0.9629
23	0.99757	0.9380	0.9403
24	0.99733	0.9161	0.9186
25	0.99707	0.8949	0.8975
26	0.99681	0.8746	0.8774
27	0.99654	0.8551	0.8581
28	0.99626	0.8363	0.8394
29	0.99597	0.8181	0.8214
30	0.99568	0.8004	0.8039
31	0.99537	0.7834	0.7870
32	0.99505	0.7670	0.7708
33	0.99473	0.7511	0.7551
34	0.99440	0.7357	0.7398
35	0.99406	0.7208	0.7251
36	0.99371	0.7064	0.7109
37	0.99336	0.6925	0.6971
38	0.99299	0.6791	0.6839
39	0.99262	0.6661	0.6711

a. 1 centipoise = 10^{-2} (gram mass)/(cm)(sec). To express the absolute viscosity (μ) as (lb mass)/(ft)(sec), multiply centipoise by 6.72×10^{-4}. To express the absolute viscosity (μ) as (lb force)(sec)/(ft²), multiply centipoise by 2.088×10^{-5}.
b. 1 centistoke = 10^{-2} cm²/sec. To express the kinematic viscosity (ν) as ft²/sec, multiply centistokes by 1.075×10^{-5}.
From *International Critical Tables*, 1928 and 1929.

APPENDIX D

Dissolved Oxygen Saturation Values in Fresh and Sea Water Exposed to an Atmosphere Containing 20.9 Percent Oxygen under a Pressure of 760 mm of Mercury

Tempera-ture (C)	Dissolved oxygen (mg/ℓ) for stated concentrations of chloride (mg/ℓ)				
	0	5,000	10,000	15,000	20,000
0	14.62	13.79	12.97	12.14	11.32
1	14.23	13.41	12.61	11.82	11.03
2	13.84	13.05	12.28	11.52	10.76
3	13.48	12.72	11.98	11.24	10.50
4	13.13	12.41	11.69	10.97	10.25
5	12.80	12.09	11.39	10.70	10.01
6	12.48	11.79	11.12	10.45	9.78
7	12.17	11.51	10.85	10.21	9.57
8	11.87	11.24	10.61	9.98	9.36
9	11.59	10.97	10.36	9.76	9.17
10	11.33	10.73	10.13	9.55	8.98
11	11.08	10.49	9.92	9.35	8.80
12	10.83	10.28	9.72	9.17	8.62
13	10.60	10.05	9.52	8.98	8.46
14	10.37	9.85	9.32	8.80	8.30
15	10.15	9.65	9.14	8.63	8.14
16	9.95	9.46	8.96	8.47	7.99
17	9.74	9.26	8.78	8.30	7.84
18	9.54	9.07	8.62	8.15	7.70
19	9.35	8.89	8.45	8.00	7.56
20	9.17	8.73	8.30	7.86	7.42
21	8.99	8.57	8.14	7.71	7.28
22	8.83	8.42	7.99	7.57	7.14
23	8.68	8.27	7.85	7.43	7.00
24	8.53	8.12	7.71	7.30	6.87
25	8.38	7.96	7.56	7.15	6.74
26	8.22	7.81	7.42	7.02	6.61
27	8.07	7.67	7.28	6.88	6.49
28	7.92	7.53	7.14	6.75	6.37
29	7.77	7.39	7.00	6.62	6.25
30	7.63	7.25	6.86	6.49	6.13

Under any other barometric pressure, P(mm), the solubility, DO'_s (mg/ℓ) can be computed from the corresponding value in the table by the equation

$$DO'_s = DO_s \frac{P - p}{760 - p}$$

where DO_s is the solubility at 760 mm (29.92 in.) and p is the pressure (mm) of saturated water vapor at the temperature of the water. For elevations less than 3000 ft and temperatures less than 25°C, p can be ignored. The equation then becomes

$$DO'_s = DO_s \frac{P}{760}$$

From Whipple and Whipple, *Jour. Am. Chem. Soc.* 33, (1911): 362.

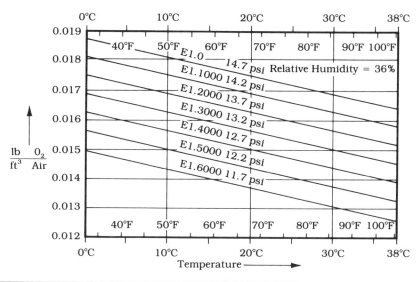

547

*Figure A.2.   Construction
Details for a Batch Reactor
Module of Four 2-ℓ Reactors*

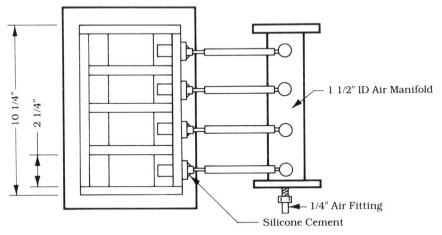

10 1/4"

2 1/4"

1 1/2" ID Air Manifold

1/4" Air Fitting

Silicone Cement

Plan View

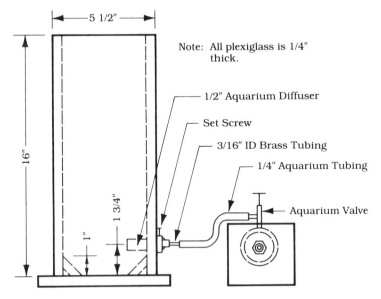

5 1/2"

Note: All plexiglass is 1/4"
thick.

1/2" Aquarium Diffuser

Set Screw

3/16" ID Brass Tubing

1/4" Aquarium Tubing

Aquarium Valve

16"

1 3/4"

1"

Elevation View

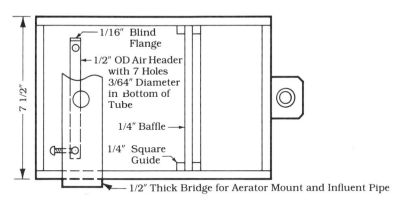

Plan View

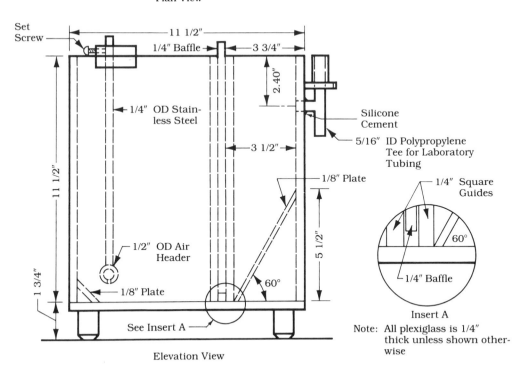

Elevation View

Insert A

Note: All plexiglass is 1/4" thick unless shown otherwise

Figure A.4. Schematic Drawing of a 10-ℓ Completely Mixed Activated Sludge Reactor in Operation

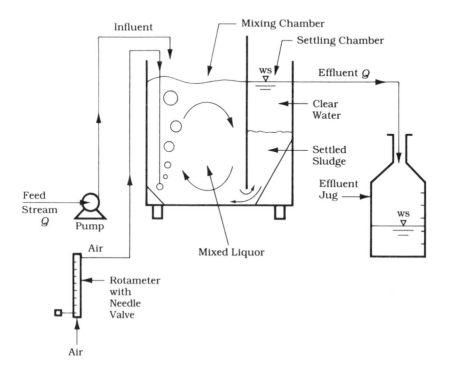

Operational Notes:

1. To acclimate a sludge for a particular wastewater, fill the reactor to the desired MLSS concentration using activated sludge from a municipal plant. Start aeration and feed the reactor in a stepwise manner using 0.05 lb BOD$_5$/lb MLSS-day increments for 2 to 3 days until the desired F/M ratio is attained. Waste sludge during the period to keep the desired MLSS concentration. If filamentous organisms occur, restart the reactor and use longer feed intervals.
2. For proper recycle, the space below the baffle should be from 1/4 to 3/8 in.
3. The baffle should be pulled and the contents allowed to mix when withdrawing samples for MLSS analyses, oxygen uptake rates, settling tests, and for wasting sludge and withdrawing sludge for batch reactor studies. The effluent pipe should be plugged prior to pulling the baffle.
4. The oxygen uptake rate can be determined with a BOD bottle fitted with a DO probe and stirrer. A mixed liquor sample larger than 300 mℓ is aerated by shaking in a stoppered flask. The sample is poured into the BOD bottle, and the oxygen concentration versus time is obtained. The slope of the plot is the oxygen uptake rate.
5. The detention time, θ, is based on θ = V/Q where V is the total reactor volume (both aeration and settling chambers, that is, 10 ℓ).

Glossary

Some of the terms in the Glossary are reprinted from the publication titled, *Glossary—Water and Wastewater Engineering,* by permission of the Water Pollution Control Federation.

acclimated culture or acclimated activated sludge A culture of microbes that has been developed to use a particular substrate or wastewater. To develop a culture, activated sludge from a municipal plant is gradually fed increasing amounts of the substrate or wastewater until the microbes that use these food substances have grown in abundance.

activated carbon Carbon particles made by carbonization of a cellulosic material in the absence of air. It has a high adsorptive capacity.

activated sludge The active biological solids in an activated sludge wastewater treatment plant.

activated sludge process A biological wastewater treatment process in which a mixture of the wastewater and activated sludge is aerated in a reactor basin or aeration tank. The active biological solids bio-oxidize the waste matter and the biological solids are then removed by secondary clarification or final settling.

advanced waste treatment Consists of: (1) The use of physical and chemical means for treating raw wastewaters, such as municipal wastewaters, in lieu of primary and secondary treatment. (2) The use of physical, chemical, and biological means to upgrade the quality of a secondary effluent. (3) The use of flowsheets, treatment operations, and/or processes not routinely used in wastewater treatment.

aerated lagoon A wastewater treatment pond in which mechanical or diffused-air aeration is used to artificially supplement the oxygen supply.

aeration tank A tank or basin in which mixed liquor, wastewater, sludge, or other liquid is aerated.

Aerobacter aerogenes One of the species of bacteria in the coliform group.

aerobic Requiring the presence of free molecular oxygen.

aerobic bacteria Bacteria that require free molecular oxygen for their life processes.

aerobic digestion Digestion of suspended organic matter by aerobic microbial action.

algae Primitive plantlike organisms, single or multicellular, usually aquatic and capable of utilizing food materials by photosynthesis.

alkalinity The ability of a water to neutralize an acid. It is due to the presence of bicarbonate, carbonate, and hydroxide ions, although occasionally it is caused by the presence of borate, silicate, and phosphate ions. It is expressed as mg/ℓ of equivalent $CaCO_3$ or meq/ℓ.

allowable breakthrough concentration The maximum acceptable concentration of a solute in the effluent from an adsorption or ion exchange column.

amino acid Organic acid containing one or more amino groups ($-NH_2$). They are the building blocks for proteins.

ammonia stripping The passage of a high-pH wastewater downward through a packed tower countercurrent to an induced air flow passing upwards. Due to the favorable equilibrium, the ammonia is stripped from the water and leaves with the air.

anaerobic Requiring combined oxygen, such as SO_4^{-2}, PO_4^{-3}, NO_3^{-1}, and so on, and requiring the absence of free molecular oxygen.

anaerobic bacteria Bacteria that require combined oxygen and the absence of free molecular oxygen.

anaerobic digestion Digestion of suspended organic matter by anaerobic microbial action.

anion A negatively charged ion in an electrolyte solution that migrates to the anode when an electrical potential is applied.

autotrophic bacteria Bacteria that use inorganic materials for energy and growth.

available chlorine A measure of the oxidizing power of hypochlorous acid and the hypochlorite ion.

bacilli Rod-shaped or cylindrical bacterial cells.

bacteria A group of universally distributed, rigid, essentially unicellular microscopic organisms lacking chlorophyll. Usually, they have a spheriod, rodlike, or spiral shape. Some use organic matter as a foodstuff, while others use inorganic matter.

bar rack or bar screen A screen consisting of parallel bars, either vertical or inclined, that is placed in a waterway to remove debris. The screenings are raked from the screen by manual or mechanical means.

batch activated carbon process The use of activated carbon adsorption in a batchwise manner. That is, the activated carbon and water or wastewater are added to a vessel, mixing is provided, transfer of the organic material from the liquid to the carbon occurs, the contents are allowed to settle, then the treated liquid is withdrawn.

batch activated sludge process The use of the activated sludge process in a batchwise manner. That is, the wastewater and activated sludge are placed in a reaction vessel, aeration is provided until the biochemical reaction is essentially complete, the contents are allowed to settle, and then the treated wastewater is withdrawn.

batch operation or process An operating technique that is batchwise in manner—that is, there are no continuous flows in or out of the operation or process.

batch reactor A reactor that does not have continuous streams entering or leaving. The reactants are added, the reaction occurs, then the products are discharged.

bench-scale A laboratory scale operation, process, or combination thereof.

binary fission The manner in which most bacteria multiply. The parent cell divides into two daughter cells.

biochemical action A chemical change resulting from the metabolism of living organisms.

biochemical oxidation An oxidation caused by biological activity resulting in a chemical combination of oxygen with organic matter to produce relatively stable end products. Same as bio-oxidation and biological oxidation.

biochemical oxygen demand (BOD) The amount of oxygen required by microbes in the stabilization of a decomposable waste under aerobic conditions.

biodegradation (biodegradability) The biological oxidation of natural or synthetic organic materials by soil microorganisms, either in soils, waterbodies, or wastewater treatment plants.

biological oxidation An oxidation caused by biological activity resulting in a chemical combination of oxygen with organic matter to produce relatively stable end products. Same as biochemical oxidation and bio-oxidation.

biomass Living biological matter.

bio-oxidation Same as biochemical oxidation or biological oxidation.

biosorption activated sludge See contact stabilization activated sludge process.

blowdown (1) Water discharged from a boiler or cooling tower to dispose of accumulated salts. (2) Removal of a portion of flow to maintain constituents of flow within desired levels.

BOD₅ See five-day biochemical oxygen demand.

BOD_u See ultimate biochemical oxygen demand.

boiler feedwater Water to be used in a boiler.

brackish water Water having a dissolved solids content between fresh water and seawater.

breakpoint chlorination Addition of chlorine to a water or wastewater until the chlorine demand has been satisfied. Further additions result in a residual that is directly proportional to the amount added beyond the breakpoint.

breakthrough curve A performance curve for a test or design column used for carbon adsorption and ion exchange. The solute concentration in the effluent is plotted on the y-axis versus the effluent throughput volume on the x-axis.

Brownian movement The random movement of microscopic particles suspended in a liquid medium.

brush aerator A surface aerator consisting of a rotating horizontal axle with protruding steel bristles partially submerged in the still water surface. Oxygen is transferred by air entrainment in the vicinity of the rotating bristles and also by the spray and impingement area. Known as the Kessener brush in Europe.

buffer action The action of certain ions in solution to oppose a change in pH.

bulking sludge An activated sludge that settles poorly because of a floc with a low bulk density.

cake Sludge that has been dewatered.

carbohydrates Organic compounds containing carbon, hydrogen, and oxygen. Common carbohydrates are sugars, starches, and cellulose.

carbon adsorption The use of activated carbon to remove dissolved organic matter from a water or wastewater.

carbonate hardness (CH) That part of the total hardness that is chemically equivalent to the bicarbonate plus carbonate alkalinity present in the water. Usually expressed as mg/ℓ of equivalent $CaCO_3$ or meq/ℓ. See hardness.

cathode A negatively charged electrode in an electrolytic cell.

cation A positively charged ion in an electrolyte solution that migrates to the cathode when an electrical potential is applied.

centrate The liquid flow leaving a centrifuge.

centrifuge A mechanical device that uses a centrifugal force to separate solids from a liquid.

chemical analysis Analysis by chemical methods to determine the composition and concentration of substances present.

chemical coagulation The destabilization and initial aggregation of colloidal and finely suspended matter by the addition of a floc-forming chemical coagulant.

chemical oxygen demand (COD) The amount of oxygen required to chemically oxidize the organic and sometimes inorganic matter in a water or wastewater. Usually expressed in mg/ℓ. The COD test does not measure the oxygen required to convert ammonia to nitrites and nitrites to nitrates; thus, COD is frequently assumed to be equal to the ultimate first-stage biochemical oxygen demand.

chemical precipitation Precipitation resulting from the addition of a chemical.

chemical sludge Sludge produced by chemical coagulation or chemical precipitation.

chemical treatment A process involving the addition of chemicals to obtain a desired result.

chemically coagulated raw wastewater A raw wastewater that has been chemically coagulated. Usually, it has been settled and filtered.

chemically conditioned sludge A sludge that has had chemicals added to enhance its dewatering characteristics.

chemically treated secondary effluent A secondary effluent that has been chemically treated, usually by coagulation along with other processes or operations.

chloramines Compounds of organic or inorganic nitrogen and chlorine.

chlorination The addition of chlorine to a water or wastewater usually for disinfection purposes; however, sometimes it is for accomplishing other biological or chemical results.

chlorine contact tank A tank that allows added chlorine sufficient time to disinfect or accomplish other desired results.

chlorine demand The difference between the amount of chlorine added to a water or wastewater and the amount of residual chlorine remaining after a specific contact duration.

clarification The removal of settleable suspended solids from a water or wastewater by gravity settling in a quiescent tank or basin. Also called *sedimentation* or *settling*.

clarified wastewater A wastewater that has had most of the settleable solids removed by clarification. Also called *settled wastewater*.

clarifier Same as sedimentation basin.

clear well A ground storage reservoir for filtered water that is sufficient volume to allow the filtration plant to operate at a constant flow rate on the day of maximum demand.

clinoptilolite A naturally occurring zeolite that has ion exchange ability. It is used in removing the ammonium ion.

coagulant A compound that causes coagulation—that is, a floc-forming agent.

coagulant aid A chemical or substance used to assist in coagulation.

coagulation In water or wastewater treatment, the destabilization and initial aggregation of colloidal and finely divided suspended solids by the addition of a floc-forming chemical.

coarse rack A rack with relatively wide spaces between the bars, usually one inch or more.

coarse screen A mesh or bar screen with openings of one inch or more.

cocci Spherical bacterial cells.

COD See chemical oxygen demand.

coliform bacteria A group of bacteria predominately living in the intestines of man and warm-blooded animals but also found elsewhere, such as in soils. It includes all aerobic and facultative anaerobic, gram-negative, non–spore forming bacilli that ferment lactose with gas production.

coliphage A virus pathogenic to coliforms.

colloids Fine suspended solids that will not settle by gravity but may be removed by coagulation, biological action, or membrane filtration.

combined chlorine The concentration of chlorine that is combined with ammonia as chloramines or as other chloroderivatives, yet is still available for chemical oxidation. Same as combined available chlorine.

combined chlorine residual That part of the total residual chlorine remaining in a water or wastewater at the end of a specific contact duration. It will react chemically or biologically as chloramines or organic chloramines. Same as combined available chlorine residual.

comminution The cutting and screening of solids contained in a wastewater prior to pumping or any treatments at a wastewater treatment plant.

complete treatment A wastewater treatment that uses both primary and secondary treatment.

completely mixed activated sludge An activated sludge process with a completely mixed reactor basin. The usual basin is square, circular, or slightly rectangular in plan view and the influent, on entering, is almost immediately dispersed throughout the reactor basin.

completely mixed reactor A reactor where the fluid elements, on entering, are dispersed almost immediately throughout the reactor volume. Usually, it is circular, square, or slightly rectangular in plan view.

contact stabilization activated sludge process An activated sludge process with a contact tank where sorption of the organic materials occurs and a sludge stabilization tank where the sludge bio-oxidizes the sorbed organic matter. Same as biosorption.

continuous-flow operation or process An operating technique that is continuous in manner; that is, there are continuous flows in and out of the operation or process.

continuous-flow reactor A reactor that has a continuous stream of reactants entering and a continuous stream of products leaving.

conventional activated sludge process An activated sludge plant with rectangular reactor basin and air diffusers or aerators spaced uniformly along the basin length.

conventional digester A low-rate anaerobic digester.

conventional wastewater treatment The use of primary and secondary treatment.

cooling water Water used to reduce the temperature of fluids and gases by use of industrial condensers.

countercurrent An operation or process that has two streams of materials in contact with each other and moving in opposite directions. For example, in countercurrent ion exchange, the ion exchange resin moves continuously down the column or bed, while the water or wastewater continuously moves upwards.

Crenothrix A genus of bacteria that occur as filaments having a sheath of deposited iron. They cause color, odor, and objectionable tastes in a water.

culture Microbial growth that has been developed by provision of suitable nutrients and environment.

decomposition of wastewater The breakdown of organic matter in wastewater by microbial action. It may be under aerobic or anaerobic conditions.

degradation The breakdown of substances by biological oxidation.

demineralization The removal of all salts from a water.

desalting Removal of salts from a brackish water or a tertiary effluent.

dewater To remove part of the water present in a sludge.

dewatered sludge A sludge that has had some of its water content removed.

diatomaceous-earth filter A filter usually used in water treatment that utilizes a builtup layer of diatomaceous earth as a filter medium.

diffused-air aeration
Aeration produced in a liquid by the use of compressed air passed through air diffusers.

digested sludge Sludge digested by aerobic or anaerobic action to the degree that the volatile content is low enough for the sludge to be stable.

digester A tank used for sludge digestion. See sludge digestion.

digestion The biological oxidation of organic matter in sludge, resulting in stabilization. See sludge digestion.

disinfection The killing or inactivation of most of the microorganisms in or on a substance with the probability that all pathogenic bacteria are killed by the disinfecting agent used.

dispersed plug-flow activated sludge An activated sludge process with a dispersed plug-flow reactor basin. The basin is rectangular in plan view and has significant longitudinal or axial dispersion of fluid elements throughout its length.

dispersed plug-flow reactor A reactor that is rectangular in plan view and has significant longitudinal mixing of fluid elements throughout its length.

dissolved oxygen The oxygen dissolved in a liquid, usually expressed in mg/ℓ. Abbreviated DO.

dissolved solids Dissolved substances in a water or wastewater.

DO See dissolved oxygen.

domestic wastewater Wastewater mainly from dwellings, business buildings, institutions, and so on.

dry suspended solids The suspended matter in water and, in particular, wastewater, which is removed by laboratory filtration and is dried for one hour at 103°C.

dry-weather flow The flow of wastewater in a sanitary sewer during dry weather. It is wastewater and dry weather infiltration.

effluent Wastewater or other liquid, partially or completely treated, or in its natural state flowing out of a basin, reservoir, treatment plant, or industrial treatment plant or parts thereof.

electrolyte A substance that dissociates into ions when it is dissolved in water.

endocellular Inside a microbial cell—for example, endocellular enzymes.

endogenous decay The continuous process in which microbial cell tissue decays.

end products Substances created by microbial metabolism and growth.

enteric bacteria Bacteria that inhabit the intestines of man and animals.

enzymes Organic catalysts that are proteins and are produced by living cells.

equalization basin A holding basin where variations in flow rate and liquid composition are averaged. An equalization basin provides an effluent of reasonably uniform composition and flow rate to subsequent treatment operations or processes.

Escherichia coli (E. coli) A species of bacteria in the coliform group. Its presence is considered indicative of fresh fecal contamination.

essential amino acid An amino acid required by a living cell that it cannot synthesize.

excess activated sludge Same as waste activated sludge.

exocellular Outside a microbial cell—for example, exocellular enzymes.

extended aeration activated sludge process An activated sludge process with a detention time long enough to allow the amount of cells synthesized to be endogenously decayed.

facultative anaerobic bacteria Bacteria that use either free molecular oxygen, if available, or combined oxygen. Also called *facultative bacteria.*

facultative stabilization pond A basin for retention of a wastewater, in which biological oxidation of the organic matter occurs by bacterial action. A significant amount of molecular oxygen used by the microbes is supplied by algal photosynthesis.

fats Triglyceride esters of fatty acids. Mistakenly used as synonymous with grease.

fermentation (1) A biochemical change caused by a ferment, such as yeast enzymes. (2) A biochemical change in organic matter or organic wastes caused by anaerobic biological action.

fill and draw reactor A batch operated activated sludge reactor. See batch activated sludge process.

filter A tanklike structure with a granular-bed and underdrain system that is used to remove fine suspended solids and colloids from a water or wastewater. The separation occurs as the liquid passes through the bed.

filter bed A tank used for water or wastewater filtration that has an underdrain system covered by a granular filter medium.

filter operating table A table set on the operating floor of a rapid sand filtration plant. It is placed in front of the filter that it operates and supports all the equipment for control and operation of the filter. It has the controls for all valves, the rate-of-flow gauge, the loss-of-head gauge, and so on.

filter plant In water treatment works, it consists of all operations and processes, structures, and appurtenances required for the filtration of water. Also called rapid sand filter plant.

filter press A mechanical press used to dewater sludges.

filter run The time interval between the backwashings of a rapid sand filter.

filter underdrains The system of underdrains for a granular medium filter. It is used for collecting the filtrate and for distributing the washwater.

filtration The unit operation that consists of passing a liquid through a granular medium for the removal of suspended and colloidal matter.

final clarifier The last settling basin or settling tank at a wastewater treatment plant. In the activated sludge process, it separates the biological solids from the final effluent. In the trickling filter process, it separates the trickling filter humus (that is, sloughed growths) from the final effluent.

final effluent The effluent from the final clarifier, final sedimentation basin, or final settling tank at a wastewater treatment plant.

fine rack A bar rack with clear spaces of about one inch or less between the bars.

fine screen In water treatment, a screen with openings less than 1 in. In wastewater treatment, a screen with openings less than 1/16 in.

first-stage biochemical oxygen demand That part of the biochemical oxygen demand that results from the biological oxidation of carbonaceous materials, as distinct from nitrogenous materials. Generally, the major portion of the carbonaceous materials are bio-oxidized before the bio-oxidation of nitrogenous materials (the second-stage biochemical oxygen demand) begins.

five-day biochemical oxygen demand (BOD_5) The oxygen required by microbes in the stabilization of a decomposable waste under aerobic conditions for a period of five days at 20°C and under specified conditions. It represents the breakdown of carbonaceous materials as distinct from nitrogenous materials.

fixed-bed In carbon adsorption or ion exchange treatments using columns or open beds, this refers to a bed that is stationary in the column or in the structure for the open bed.

fixed solids The residue remaining after the ignition of suspended or dissolved solids at 600°C.

floc In water treatment, the small, gelatinous masses formed in the water by the adding of a coagulant. In wastewater treatment, the small, gelatinous biological solids formed at an activated sludge treatment plant.

flocculation In water and wastewater treatment, the slow stirring of a coagulated water or wastewater to aggregate the destabilized particles and form a rapid settling floc. In biological wastewater treatment where a coagulant is not used, the aggregation may be accomplished biologically.

flocculator A basin in which flocculation is done or a mechanical device to enhance the formation of floc in a liquid.

flotation The raising of suspended matter to the surface of a liquid, where it is removed by skimming.

flowsheet A diagrammatic representation of unit operations and processes used in a water or wastewater treatment scheme.

fluidized bed (1) In carbon adsorption or ion exchange treatment, this refers to a bed in which the particles are not in continuous contact due to the upward flow of the water or wastewater. (2) In sludge incineration, this refers to an incineration bed of sand particles that are not in continuous contact due to the upward flow of the combustion air.

free chlorine The amount of chlorine available as dissolved gas, hypochlorous acid, or hypochlorite ion.

free chlorine residual The portion of the total residual chlorine remaining in a water or wastewater at the end of a specific contact duration, which will react chemically as hypochlorous acid or hypochlorite ion. Same as free available chlorine residual.

fresh sludge Undigested organic sludge.

fresh wastewater Raw wastewater containing some dissolved oxygen.

fungi Small, multicellular, nonphotosynthetic, plantlike organisms lacking chlorophyll, roots, stems, or leaves and that feed on organic matter. Their decomposition after death may cause disagreeable tastes and odors in a water. They are found in water, wastewater, wastewater effluents, and soil.

gas stripping An operation in which a solute gas is stripped from a liquid.

gasification The conversion of soluble organic matter into gas during anaerobic bio-oxidation.

gram-negative See Gram stain.

gram-positive See Gram stain.

Gram stain A staining procedure used in the identification of a bacterial species. A microscopic slide is made of the culture and a purple stain, such as crystal violet, is added to the slide. Then the slide is washed using special solutions and a red stain, such as safranin, is added. The slide is then washed with water. Once dry, the slide is viewed under a microscope. A species stained purple is gram-positive and a species stained red is gram-negative. A species always has the same staining characteristics—for example, *E. coli* is always gram-negative.

granular medium A granular material, such as sand or crushed anthracite coal, that serves as the filter bed.

granular medium filtration Filtration through a bed of granular material.

gravity filters Filters that have gravity flow of the water through the filter bed.

gravity thickening Thickening of a sludge using gravity settling in a tank. Pickets, usually mounted on the trusswork for the sludge scrapers, rake through the sludge releasing entrained water. This allows the sludge to subside and concentrate.

grease In wastewater treatment, a group of substances such as fats, waxes, free fatty acids, calcium and magnesium soaps, mineral oils, and other fatlike materials.

grit The dense, mineral, suspended matter present in a water or wastewater, such as silt and sand.

grit chamber In wastewater treatment, a settling chamber to remove grit from organic solids.

grit removal The removal of heavy suspended mineral matter present in a wastewater, such as sand and silt.

groundwater Subsurface water from which springs and wells are fed.

hard water A water with significant hardness. Waters with hardness greater than 75 mg/ℓ as $CaCO_3$ are usually considered hard. See hardness.

hardness The ability of a water to consume excessive amounts of soap prior to forming a lather and to produce scale in hot water heaters, boilers, or other units in which the temperature of the water is significantly increased. It is due to the presence of polyvalent metallic ions, mainly calcium and magnesium. Occasionally strontium, ferrous, and manganous ions are present. Usually expressed as mg/ℓ of equivalent $CaCO_3$ or meq/ℓ.

heavy metals Metals that can be precipitated by hydrogen sulfide in an acid solution—for example, lead, mercury, cadmium, zinc, silver, gold, bismuth, and copper. Above certain concentrations, they can inhibit microbial action in wastewater treatment plants and are toxic to humans.

helminth An intestinal worm.

herbicide An agent used to inhibit or kill plant growth.

heterotrophs Bacteria that feed on preformed organic matter.

high-rate digester An anaerobic digester with continuous mixing, continuous feeding, and digester heating.

high-rate trickling filter A trickling filter with continuous recycle, an organic loading greater than 800 lb BOD_5/ac-ft-day, and a hydraulic loading greater than 10 MGD/ac.

high service pumps Pumps that pump from a clear well to a distribution system.

hydrocarbons Organic compounds consisting of carbon and hydrogen.

Imhoff tank A deep, two-storied wastewater treatment tank. The upper story is a continuous-flow primary settling chamber and the lower story is an anaerobic digestion chamber. The sludge from the upper chamber passes through trapped slots to the lower chamber.

incineration The combustion of organic sludges to produce CO_2, H_2O, and other stable forms. The material remaining will be ash.

industrial wastewater The liquid wastes from industrial processes.

industrial water Water used in an industrial water system for process water, cooling water, and boiler water.

infiltration water Water that has migrated from the ground into a sewer system.

influent Water, wastewater, or other liquid flowing into a reservoir, treatment plant, or any unit thereof.

inoculate Introduce viable cells in a medium.

inorganic industrial wastewater An industrial wastewater that has inorganic ions or compounds as the objectionable constituents.

insecticide An agent used to kill insects.

institutional wastewater Wastewater from institutions such as hospitals, prisons, or charitable institutions.

ion A charged atom, molecule, or radical.

ion exchange A chemical process involving the reversible exchange of ions between a liquid and a solid.

ion exchange softening A process in which calcium and magnesium ions are removed from water by ion exchange. The exchanger exchanges sodium ions for the calcium and magnesium ions that are removed.

ion exchange treatment The use of ion exchangers such as resins to remove unwanted ions from a liquid and substitute more desirable ions.

iron bacteria Bacteria that assimilate iron and excrete its compounds in their life processes.

isoelectric point The pH at which electroneutrality occurs.

lime or slaked lime Calcium hydroxide, $Ca(OH)_2$.

lime recalcination The heat treatment of a sludge resulting from lime coagulation. It converts the calcium carbonate precipitate to calcium oxide.

lime slaking Reacting quicklime, CaO, with water to produce calcium hydroxide, $Ca(OH)_2$.

lime-soda softening A process in which calcium and magnesium ions are precipitated from a water by reaction with lime and soda ash.

liquefaction The changing of organic solids to soluble forms.

low-rate digester An anaerobic digester with intermittent mixing and intermittent feeding. It may or may not be heated. Same as conventional digester.

low-rate trickling filter A trickling filter with recycle only during periods of low flow, an organic loading less than 800 lb BOD_5/ac-ft-day, and a hydraulic load less than 6 MGD/ac. Same as standard-rate trickling filter.

make-up water Water added to a circulating water system to replace water lost by leakage, evaporation, or blowdown.

manganese bacteria Bacteria that utilize dissolved manganese and deposit it as hydrated manganic hydroxide.

mass transfer The transfer of a substance from one phase to another. For example, in diffused air aeration, there is transfer of oxygen from the diffused compressed air into the liquid.

mean cell residence time (θ_c) The average time a microbial cell remains in an activated sludge system. It is equal to the mass of cells divided by the rate of cell wastage from the system.

mechanical aeration (1) The transfer of oxygen from the atmosphere into a liquid by the mechanical action of a turbine or other mechanisms. (2) The mixing by mechanical means of the mixed liquor in the reactor basin or aeration tank of an activated sludge treatment plant.

media For trickling filters, this is the packing, such as stone, in the filter bed.

medium (1) A porous material used in filters. (2) A nutrient material used to cultivate bacteria.

mesophilic digestion Anaerobic digestion by biological oxidation by anaerobic action at or below 45°C (110°F).

microbe Same as microorganism.

microbial activity Chemical changes resulting from biochemical action—that is, the metabolism of living organisms.

microorganism Minute organisms, some being plantlike, others animallike, visible only by means of a microscope. Same as microbe.

microscopic Visible only by magnification with an optical microscope. The size range is from 0.5 to 100 µ.

microscreen A rotating drum with a screen mesh surrounding the drum. The wastewater enters the interior of the drum and flows outward. A backwash at the top of the drum continuously removes the screenings.

milliequivalent (meq) The weight in milligrams of a substance that combines with or displaces one milligram of hydrogen. It is equal to the formula weight in mg divided by its valence.

mixed culture A microbial culture consisting of two or more species.

mixed liquor A mixture of wastewater and activated sludge undergoing aeration in a reactor basin or aeration tank in an activated sludge wastewater treatment plant.

mixed liquor suspended solids (MLSS) The suspended solids in the mixed liquor—that is, the mixture of wastewater and activated sludge undergoing aeration at an activated sludge wastewater treatment plant.

mixed liquor volatile suspended solids (MLVSS) The volatile fraction of the mixed liquor suspended solids. The MLVSS is usually considered to be more representative of the active biological solids than the MLSS.

mixing tank A tank to provide thorough mixing of chemicals added to a liquid. Also called *mixing basin*, *mixing chamber*, or *rapid-mix tank*.

MLSS See mixed liquor suspended solids.

MLVSS See mixed liquor volatile suspended solids.

moisture content The amount of water present in wastewater sludge, industrial wastewater sludge, water treatment sludge, or soil. Usually expressed as a percentage of the wet weight.

morphology The shape and grouping of bacterial cells. The three basic shapes are bacilli (rod or cylindrical shapes), cocci (spherical shape), and spirilli (spiral shape). Cells usually appear as single cells, bunches, or chains, although some other groupings do occur. It is a characteristic of a species to have the same morphology. For example, *E. coli* is always rod-shaped and occurs as single cells.

most probable number (MPN) The statistical number of organisms in a 100-mℓ sample.

moving-bed In carbon adsorption or ion exchange treatments using columns or open beds, this refers to a bed that moves downward through the column or the structure for the open bed. The bed depth is a constant value because the carbon or exchanger material continuously enters above the bed.

MPN See most probable number.

multimedia Two or three granular media used for filter beds, such as crushed coal and sand, or crushed coal, sand, and garnet.

multimedia filtration The filtration of a water or wastewater through a granular bed containing two or more filter media.

multiple-hearth furnace A furnace consisting of numerous hearths which is used to incinerate organic sludges or recalcinate lime.

municipal wastewater Wastewater derived principally from dwellings, business buildings, institutions, and the like. It may or may not contain groundwater, surface water, storm water, and industrial wastewater.

nematode A worm that feeds on minute organisms.

nitrification The conversion of nitrogenous matter into nitrates by bacterial action.

Nitrobacter A genus of bacteria that oxidizes nitrites to nitrates.

Nitrosomonas A genus of bacteria that oxidizes ammonia to nitrites.

noncarbonate hardness (NCH) That amount of the hardness equal to the total hardness minus the carbonate hardness. Usually expressed as mg/ℓ of equivalent $CaCO_3$ or meq/ℓ. See hardness.

nonsetteable solids Suspended solids in a wastewater that will not settle by gravity means within a given duration of time, usually considered one hour for laboratory testing.

operating floor The floor in a rapid sand filter building on which the operating and indicating devices are generally installed.

organic industrial wastewater An industrial wastewater that has organic compounds as the objectionable constituents.

organic matter Chemical substances of animal and vegetable origin that are carbon compounds.

organic-matter degradation The conversion of organic matter to inorganic forms by biological action.

organic nitrogen Organic compounds or molecules containing nitrogen, such as proteins, amines, and amino acids.

organic sludges Sludges that have a high organic content. Usually, these are primary or secondary sludges at a wastewater treatment plant.

oxidation ditch An extended aeration process with a race-track-shaped aeration basin.

oxidation pond A pond where the organic matter in a wastewater is biologically oxidized. Same as a waste stabilization pond or, if artificial aeration is provided, an aerated lagoon.

ozone Oxygen in molecular form consisting of three atoms of oxygen forming each molecule (O_3).

parasitic bacteria Bacteria that require a living host organism but do not harm the host.

pathogen Pathogenic or disease-producing organism.

pathogenic bacteria Bacteria that require a living host organism and harm the host by causing disease.

pesticide Any substance or chemical applied to kill or control pests, such as weeds, insects, algae, rodents, and so on.

petrochemicals Products or compounds produced by the chemical processing of petroleum and natural gas hydrocarbons.

pH The reciprocal of the logarithm of the hydrogen ion concentration in gram moles per liter. Neutral water has a pH of 7.

phosphorous The phosphorous radicals found in a wastewater. Usually, the major portion is orthophate, PO_4^{-3}.

photosynthesis The synthesis of complex organic materials from carbon dioxide, water, and inorganic salts using sunlight as the energy source and a catalyst such as chlorophyll.

physical analysis The examination of a water or wastewater to determine physical characteristics, such as temperature, turbidity, color, odor, and so on.

physical-chemical treatment The use of physical and chemical means for treating a raw wastewater or a secondary effluent.

pilot scale A pilot-scale operation, process, or combination thereof. It is operated in the field.

pipe gallery A gallery provided in a rapid sand filtration plant that houses the pipes going to and from the filters, along with their valves and other appurtenances. It also serves as an access passageway.

plain sedimentation The gravity settling of suspended solids in a water or wastewater without the aid of chemical coagulants.

plug-flow reactor A reactor in which all fluid elements that enter the reactor at the same time flow through it with the same velocity and leave at the same time. The travel time of the fluid elements is equal to the theoretical detention time and there is no longitudinal mixing. It is approached in long, narrow tanks.

polishing ponds Stabilization lagoons or oxidation ponds with detention times less than 24 hours and that are used as a finishing treatment. They are frequently used for trickling filter effluents.

pollution A condition caused by the presence of harmful or objectional material in a water.

polyelectrolytes Organic polymers used as coagulant aids or coagulants.

polymers Organic polyelectrolytes used as coagulant aids or coagulants.

precipitate To separate from solution as a precipitate.

precipitation The phenomenon that occurs when a substance held in solution in a liquid passes out of solution into solid form.

precursor A substance from which another substance is formed.

preliminary treatment In a wastewater treatment plant, this refers to unit operations like screening, comminution, or grit removal that prepare the wastewater for subsequent major operations.

pressure filters Filters that operate under pressure. Usually the flow is pumped to and through the filter.

primary clarifier In a wastewater treatment plant, this is the first clarifier used. It is for removal of settleable suspended solids. Also, called *primary settling tank.*

primary sludge The sludge from a primary clarifier.

primary treatment The treatment of a wastewater by sedimentation to remove a substantial amount of the suspended solids.

process water Industrial water used in making or manufacturing a product.

proteins Complex nitrogenous compounds formed in living organisms. The compounds consist of amino acids bound together by the peptide linkage.

protozoa Small, single-celled, animallike microbes including amoebae, ciliates, and flagellants.

pulsed-bed A countercurrent bed with intermittent bed movement.

pure culture A microbial culture consisting of only one species.

pure oxygen activated sludge process An activated sludge process that uses pure molecular oxygen for microbial respiration instead of atmospheric oxygen.

purification The removal of objectionable material from a water or wastewater by natural or artificial means.

quicklime A calcined material that is mainly calcium oxide, CaO.

rack A device consisting of parallel bars evenly spaced that is placed in a waterway to remove suspended or floating solids from a wastewater.

rapid-mix basins Basins or tanks with agitation that are used to disperse and dissolve one or more chemicals in a water or wastewater.

rapid sand filter A granular-medium filter used in water treatment in which the water is passed downward through the filter medium. Typical media are sand and/or crushed anthracite coal. The water is pretreated by coagulation, flocculation, and sedimentation.

rapid sand filtration plant A water treatment plant consisting of coagulation, flocculation, sedimentation, rapid sand filtration, and disinfection. Rapid sand filters used contain a granular medium or multimedia.

rate-of-flow controller An automatic device that controls the rate of flow of a fluid, usually flowing in a pipe.

rate-of-flow indicator A device that indicates the rate of flow of a fluid at any time. It may or may not have a recorder.

raw sludge A settled sludge that is undigested. Also called *undigested sludge* and *fresh sludge*.

raw wastewater A wastewater before it receives any treatments.

raw water Untreated water. Usually the water entering the first treatment unit at a water treatment plant.

reactor basin An aeration tank at an activated sludge plant.

recarbonation The diffusion of carbon dioxide in a water or wastewater after lime coagulation to lower the pH.

receiving body of water A natural watercourse, lake, or ocean into which a treated or untreated wastewater is discharged.

recycle ratio For an activated sludge or trickling filter plant, it is equal to the rate of recycled flow divided by the rate of influent flow.

residual chlorine Chlorine remaining in a water or wastewater at the end of a specified contact time as free or combined chlorine. Same as chlorine residual.

resins Synthetic organic resins that have ion exchange ability.

respiration The furnishing of energy for microbial metabolism and growth.

returned sludge Settled activated sludge that is returned to mix with the raw or primary settled wastewater.

rotary distributor A revolving distributor with long arms with jets that are used to spray wastewater on a trickling filter bed.

rotary biological filter A biological filter consisting of circular discs mounted on a horizontal rotating axle. Fixed biological growths are on the discs and the discs are partially submerged in a vat containing the wastewater. As the discs rotate, the fixed biological growths sorb the organic matter and bio-oxidize the materials. Oxygen is supplied by absorption from the atmosphere as the discs are partially exposed during their rotation.

rotifer A minute, usually microscopic, multicellular invertebrate animal that feeds on organic matter.

sanitary landfill A landfill for disposing of solid wastes.

sanitary wastewater (1) Domestic wastewater without storm and surface runoff. (2) Wastewater from the sanitary conveniences in dwellings, office buildings, industrial plants, and institutions. (3) The water supply of a community after it has been used and discharged to a sewer.

saprophytic bacteria Bacteria that feed on dead or nonliving organic matter.

SDI See sludge density index.

secondary clarifier The final settling basin at a wastewater treatment plant. Also called *secondary settling tank, secondary settling basin,* or *final clarifier.*

secondary effluent The effluent leaving the secondary or final clarifier at a wastewater treatment plant.

secondary sludge The sludge from the final clarifier at a wastewater treatment plant. For the activated sludge process, it is the sludge to be recycled. For the trickling filter process, it is the trickling filter growths that have sloughed off—that is, the trickling filter humus.

secondary treatment The treatment of a wastewater by biological oxidation after primary treatment by sedimentation.

second-stage biochemical oxygen demand That part of the biochemical oxygen demand that results from the biological oxidation of nitrogenous materials. This includes the bio-oxidation of ammonia to nitrites and nitrites to nitrates. The oxidation of nitrogenous materials usually does not begin until a significant portion of the carbonaceous material has been bio-oxidized in the first stage.

sedimentation The removal of settleable suspended solids from a water or wastewater by gravity settling in a quiescent tank or basin. Also called *clarification* or *settling.*

sedimentation basin A basin or tank through which a water or wastewater is passed to remove settleable suspended solids by gravity settling. Also called *sedimentation tank, settling tank, settling basin,* or *clarifier.*

sedimentation tank Same as sedimentation basin.

settleable solids Suspended solids removed by gravity settling.

settled wastewater Wastewater that has been treated by sedimentation. Also called a *clarified wastewater.*

settling Same as sedimentation and clarification.

settling basin Same as sedimentation basin.

settling tank Same as sedimentation basin.

sewage See wastewater.

shock load See slug load.

side water depth The depth of water in a tank measured from the bottom of the tank to the water surface at an exterior wall.

skimmings The grease, solids, and scum skimmed from a wastewater settling tank or basin.

slake To mix with water so that a chemical combination takes place, such as the slaking of lime where CaO reacts with H_2O to produce $Ca(OH)_2$.

slow sand filter A filter for water purification in which water without previous treatment is passed through a sand bed. It is characterized by a slow rate of filtration, such as 3 to 6 MGD/acre of area.

sludge (1) The accumulated solids removed from a sedimentation basin, settling tank, or clarifier in a water or wastewater treatment plant. (2) The precipitate resulting from the chemical coagulation, flocculation, and sedimentation of a water or wastewater.

sludge age Same as mean cell residence time.

sludge bed An area consisting of natural or artificial layers of porous materials on which digested wastewater sludge is dewatered by drainage and evaporation.

sludge blanket (1) The depth of activated sludge in a final clarifier or thickener. (2) The accumulation of sludge hydrodynamically suspended in a solids-contact unit.

sludge bulking A phenomenon that occurs in activated sludge treatment plants in which the sludge does not settle and concentrate readily.

sludge cake Sludge that has been dewatered to increase the solids content to 15 to 45 percent based on dry solids.

sludge collector A mechanical device for scraping the sludge along the bottom of a settling tank to a hopper where it can be withdrawn.

sludge conditioning Treatment of a sludge usually by chemical means to enhance its dewatering characteristics.

sludge density index (SDI) The concentration of an activated sludge after settling one liter of the mixed liquor for 30 minutes in a one-liter graduated cylinder. It is the reciprocal of the sludge volume index with appropriate conversions. Usually expressed as mg/ℓ.

sludge dewatering The removal of part of the water in a sludge by any method such as centrifugation, filter pressing, vacuum filtration, or passing through a belt press.

sludge digestion The biological oxidation of organic or volatile matter in sludges to produce more stable substances. See anaerobic and aerobic digestion.

sludge digestion gas Gas produced from the biological oxidation of organic matter by anaerobic digestion.

sludge digestion tank Tank used for the aneraobic digestion of organic sludges.

sludge dryer A device that utilizes heat for the removal of a large portion of the water in a sludge.

sludge drying The removal of a large portion of the moisture in a sludge by drainage or evaporation.

sludge-gas utilization The use of digester gas from anaerobic digesters for beneficial purposes such as heating and fueling engines.

sludge moisture content The amount of water in a sludge, usually expressed by percentage of dry weight.

sludge processing The collection, treatment, and disposal of sludge.

sludge rakes Rakes used in the collection of settled sludge from a clarifier used in water or wastewater treatment.

sludge thickener A tank or a piece of equipment that concentrates the solids in a sludge.

sludge thickening To increase the solid content of a sludge.

sludge treatment The processing of wastewater sludges to render them innocuous. Common methods are anaerobic or aerobic digestion followed by sludge dewatering.

sludge volume index (SVI) The volume in $m\ell$ occupied by one gram of mixed liquor after settling for 30 minutes in a liter graduate cylinder. Expressed as $m\ell/gm$.

slug load A sudden load, either organic or hydraulic, to a wastewater treatment plant. Same as shock load.

soft water Water having a low concentration of calcium and magnesium ions. Usually considered to have less than 75 mg/ℓ as equivalent $CaCO_3$.

softening The removal of hardness—that is, the divalent metallic ions such as calcium and magnesium—from a water.

solids handling system In water or wastewater treatment, it is the system for the collection, treatment, and dewatering of sludges.

solute A substance dissolved in a fluid.

Sphaerotilus **bulking** Sludge bulking caused by the presence of the filamentous bacteria *Sphaerotilus* in large numbers.

spirilla Spiral-shaped bacterial cells.

spore A reproductive body created by some bacteria and capable of development into a new bacterial cell under proper environmental conditions.

spore-formers Bacteria capable of spore formation.

stabilization In lime-soda softening or lime coagulation, any process that minimizes or eliminates scale-forming tendencies.

stabilization pond An oxidation pond where the organic matter in a wastewater is biologically oxidized without artificial aeration.

stale wastewater Raw wastewater that does not contain any dissolved oxygen.

standard biochemical oxygen demand The biochemical oxygen demand determined by standard laboratory procedure for five days at 20°C. Usually expressed in mg/ℓ. See five-day biochemical oxygen demand.

Standard Methods Methods for the Examination of Water and Wastewater, which are jointly published by the American Public Health Association, the American Water Works Association, and the Water Pollution Control Association.

standard-rate trickling filter Same as low-rate trickling filter.

substrate The food substances for microorganisms in a liquid solution.

sulfur bacteria Bacteria that use dissolved sulfur or dissolved sulfur compounds for their metabolism or growth.

supernatant liquor The liquid released during anaerobic digestion. In a stratified tank, it lies above the sludge and beneath the scum layer.

surface water Water that appears on the surface of the earth as distinguished from groundwater.

suspended matter Solids in suspension in a water or wastewater that can be removed by laboratory filtration techniques, such as membrane filtration.

suspended solids Solids in suspension in a water or wastewater that can be removed by laboratory filtration techniques, such as membrane filtration. Same as suspended matter.

svi See sludge volume index.

synthesis The forming of new cell tissue in microbial metabolism and growth.

tapered aeration activated sludge process An activated sludge plant with a rectangular reactor basin and air diffusers or aerators spaced along the reactor length in accordance to the oxygen demand.

tertiary treatment The use of physical, chemical, or biological means to upgrade a secondary effluent.

theoretical oxygen demand (TOD) The amount of oxygen stoichiometrically required to convert organic matter to stabilized substances such as CO_2, H_2O, NO_3^{-1}, and so on.

thermophilic digestion Anaerobic digestion at a temperature within the thermophilic range usually considered to be between 110° and 145°F.

θ_c See mean cell residence time.

thickened sludge Sludge that has had some of its water removed.

thickener-clarifier A clarifier that settles and thickens the solids in the influent, while at the same time it clarifies the effluent leaving the tank.

thickening of sludge See sludge thickening.

TOC See total organic carbon.

TOD See theoretical oxygen demand.

total hardness (TH) The sum of the polyvalent metallic ion concentration in a water. It is principally due to calcium and magnesium ions in the water, although occasionally strontium, ferrous, and manganous ions are present. Usually expressed as mg/ℓ of equivalent $CaCO_3$ or meq/ℓ. See hardness.

total organic carbon (TOC) A measure of the organic matter in a water or wastewater in terms of the organic carbon content. Usually reported as mg/ℓ.

total solids The sum of the dissolved and suspended solids in a water or wastewater. Usually expressed in mg/ℓ.

traveling screen A rotating trash screen.

treated wastewater A wastewater that has undergone complete treatment or partial treatment.

trickling filter A biological filter consisting of a bed of coarse material, such as stone, over which wastewater is distributed by a spray from a moving distributor or other device. The wastewater trickles through the bed to the underdrains, giving the microbial slimes an opportunity to absorb the organic material and clarify the wastewater.

trickling filter media The packing in a trickling filter.

turbidity Suspended matter in a water or wastewater that causes the scattering or absorption of light rays.

ultimate biochemical oxygen demand (BOD_u) (1) Generally, the amount of oxygen required to completely satisfy the first-stage BOD. (2) More strictly, the oxygen required to completely satisfy the first- and second-stage BOD.

ultimate first-stage biochemical oxygen demand (BOD$_u$) The total amount of oxygen required to satisfy the first-stage biochemical oxygen demand. The relationship between the BOD(y) at any day during the first-stage and the BOD$_u$ is $y = $ BOD$_u$ [1-exp $(-kt)$], where k is the rate constant to base e and t is the time in days.

undigested sludge Untreated, settled sludge that has been removed from a sedimentation basin or tank. Also called *fresh* or *raw sludge.*

unsaturated hydrocarbons Hydrocarbons having at least one double bond between two carbon atoms.

vacuum filter A sludge dewatering filter consisting of a cylindrical drum mounted on a horizontal axle, covered with a filter cloth, and rotating partly submerged in a sludge vat. A vacuum is maintained within the drum for the major part of the revolution to form the sludge cake on the drum and to dewater the cake. The cake is continuously scraped off or removed by other means.

viable cells Living cells.

virus The smallest form capable of producing disease in man, animals, and plants, being 10 to 300 mμ in diameter.

volatile acids Fatty acids with six or fewer carbon atoms that are water soluble. Usually expressed as equivalent acetic acid in mg/ℓ.

volatile matter Matter within a residue that is lost at 600°C ignition temperature. The ignition time must be sufficient to reach a constant weight of residue, usually 15 minutes. See volatile solids.

volatile solids The quantity of solids, either suspended or dissolved, lost in ignition at 600°C.

waste activated sludge The excess activated sludge produced by the microbial solids in an activated sludge plant. This amount has to be wasted from the system at the rate it is produced. Same as excess activated sludge.

waste treatment Any operation or process that removes objectionable constituents from a wastewater and renders it less offensive or dangerous. Same as wastewater treatment.

wastewater The spent water that consists of a combination of liquid and water-carried wastes from dwellings, business buildings, industrial plants, and institutions, along with any surface or groundwater infiltration. In recent years, the term *wastewater* has taken precedence over the word *sewage.*

wastewater analysis The determination of the physical, chemical, and biological characteristics of a wastewater or treatment plant effluent.

wastewater composition (1) The relative quantities of the various solids, liquids, and gases in a wastewater. (2) The physical, chemical, and biological constituents of a wastewater.

wastewater renovation The treatment of a wastewater for reuse.

wastewater treatment Any operation or process that removes objectionable constituents from a wastewater and renders it less offensive or dangerous. Same as waste treatment.

water analysis The determination of the physical, chemical, and biological characteristics of a water.

water-borne disease A disease caused by organisms or toxic materials transported by water. The most common water-borne diseases are typhoid fever, cholera, dysentery, and other intestinal disturbances.

water quality The physical, chemical and biological characteristics of a water in regards to its suitability for a particular use.

water softening The process of removing calcium and magnesium ions from a water. The removal may be partial or total.

water treatment The treatment of a water by operations and processes to make it acceptable for a specific use.

water treatment plant A plant for removal of objectionable constituents from a water to make it satisfactory for a particular use.

zeolite Natural minerals or synthetic resins that have ion exchange capabilities.

zooglea The gelatinous material resulting from the attrition of bacterial slime layers. An important constituent of activated sludge floc and trickling filter growths.

GLOSSARY REFERENCES

Glossary—Water and Wastewater Engineering. 1969. American Water Works Association, Water Pollution Control Federation, American Society of Civil Engineers, and American Public Health Association.

McKinney, R. E. 1962. *Microbiology for Sanitary Engineers.* New York: McGraw-Hill.

Sawyer, C. N., and McCarty, P. L. 1978. *Chemistry for Environmental Engineers.* 3rd ed. New York: McGraw-Hill.

Standard Methods for the Examination of Water and Wastewater. 1980. 15th ed. American Water Works Association, Water Pollution Control Federation, and American Public Health Association.

Webster's Third New International Dictionary. 1968. Vols. I, II, and III. Copyright 1966, G. & C. Merriam Co. Encyclopaedia Brittannica, Inc., London, England.

Answers to Odd-Numbered Problems

Chapter 2

1. a. 5 ft-6 in. × 5 ft-6 in. × 7 ft-0 in.
   b. 8.51 hp
   c. 88.9 rpm
3. a. 3.32 m × 3.32 m × 6.64 m
   b. 54.9 m³/h
   c. 8 diffusers
5. CaO = 2566 lb, Na_2CO_3 = 1080 lb

Chapter 3

1. a. Diameter = 55 ft, depth = 8.91 ft
   b. Head = 1.85 in.
   c. Depth = 2.85 in.
   d. Depth (y_c) = 0.57 ft
   e. Depth (H_o) = 1.07 ft
   f. Depth = 18 in.
3. a. Area for clarification = 4,320 ft²
   b. Area for thickening = 4,430 ft²
   c. Diameter = 80 ft
5. Total head loss = 7.63 ft
   Primary ws = El. 328.48
   Aeration tank ws = El. 326.20
   Final ws = El. 325.81
7. a. Overflow rate = 29.3 m³/(m²)(day); detention time = 2.03 h
   b. Diameter = 20.3 m; depth = 2.48 m
9. Existing capacity = 778,000 gal/day
   Increased capacity = 5,180,000 gal/day

Chapter 4

1. Backwash rate = 7.15 gpm/ft^2
3. a. Depth = 0.0968 ft or 1.16 in.
   b. Depth (y_c) = 0.64 ft
   c. Depth (H_0) = 1.17 ft
   d. Depth = 18 in.
5. Head loss = 1.51 ft
7. a. Velocity = 0.00914 ft/sec or 0.548 ft/min
   b. Backwash flow = 4.10 gpm/ft^2
   c. Head loss = 2.43 ft
   d. Depth = 2.87 ft

Chapter 5

1. a. Theoretical air flow = 2480 lb/(hr)(ft²)
   b. Design air = 4340 lb/(hr)(ft²)
   c. Design air = 1070 cfm/ft²;air/gal = 537 ft³/gal
   d. Size = 32.3 ft × 32.3 ft
3. a. Theoretical demand = 105,000 lb/month
   b. Actual demand = 112,000 lb/month

Chapter 6

1. a. Flow rate = 2.35 BV/hr and gal/lb = 44.6
   b. Constant = 55.62 gal/lb-hr and solid phase concentration = 0.171 lb/lb
3. a. Diameter = 4 ft-0 in.; height = 14 ft-3 in.
   b. Two columns in series
5. a. Volume = 2090 ft³
   b. Diameter = 10 ft-6 in.; height = 24 ft-3 in.
   c. Mass = 52,500 lb
   d. Pounds/day = 584
   e. On-line time = 89.9 days

Chapter 7

1. a. Volume = 2780 ft³; pounds = 122,000
   b. Expended rate = 97.3 lb/hr
   c. Column life = 1230 hr and 52.3 days
   d. Diameter = 12 ft-0 in.; height = 24 ft-9 in.
3. a. Flow rate = 0.842 kg/(h)(m²)
   b. Liters per kg = 12,800
   c. Diameter = 1.64 m
5. Dry resin = 766 kg; moist resin = 1320 kg

Chapter 8

1. a. Efficiency = 50 percent; number of membranes = 202; power = 134,000 watts; power costs = $0.80/1000 gal
   b. Brine flow = 7450 gal/day and product water flow = 98,300 gal/day
3. a. Area = 2780 m²
   b. Space = 1.10 m³

Chapter 9

1. Nondegradable COD = 92 mg/ℓ, pseudo–first order reaction, rate constant = 0.2244 ℓ/(gm MLSS) (h) or 0.2619 ℓ/(gm MLVSS) (h)

3. a. Reaction time for a plug-flow reactor = 5.33 hr
   b. Reaction time for a dispersed plug-flow reactor = 7.60 hr
   c. Reaction time for a completely mixed reactor = 25.0 hr

5. a. Aeration time = 15.3 hr
   b. Tank volume = 159,000 gal
   c. Sludge production = 3350 lb/day

7. a. Aeration time = 6.38 hr
   b. Tank volume = 166,000 gal
   c. Sludge production = 3290 lb/day

9. a. Aeration time = 6.68 hr; tank volume = 371,000 gal
   b. Sludge production = 859 lb/day
   c. Flow = 10,200 gal/day
   d. Oxygen demand = 1950 lb/day = 81.4 lb/hr. or 26.3 mg/ℓ-h
   e. First quarter = 28.5 lb/hr; second quarter = 21. 2 lb/hr; third quarter = 16.3 lb/hr; fourth quarter = 15.5 lb/hr. First quarter = 36.8 mg/ℓ-h; second quarter = 27.3 mg/ℓ-h; third quarter = 21.0 mg/ℓ-h; fourth quarter = 20.0 mg/ℓ-h
   f. First quarter = 454 scfm; second quarter = 336 scfm; third quarter = 259 scfm; fourth quarter = 246 scfm
   g. Air/gal = 2.09 ft^3; air/lb BOD$_5$ = 1720 ft^3

11. Y = 0.48 mg MLSS/mg COD or 0.42 mg MLVSS/mg COD; k_e = 0.181 day^{-1}; Y' = 0.40 mg O$_2$/mg COD; k_e' = 0.26 mg O$_2$/mg MLVSS-day

13. a. $C_{12}H_{22}O_{11} + 12\ O_2 \rightarrow 12\ CO_2 + 11\ H_2O$
    b. $5\ C_{12}H_{22}O_{11} + 12\ NH_4^+ \rightarrow 12\ C_5H_7O_2N + 31\ H_2O + 12\ H^+$
    c. $C_{12}H_{22}O_{11} + 5.42\ O_2 + 1.315\ NH_4^+ \rightarrow 1.315\ C_5H_7O_2N + 3.40\ H_2O + 1.315\ H^+$
    d. TOD = 1.12 lb O$_2$/lb sucrose
    e. BOD/TOD = 0.46 lb/lb

Chapter 10

1. First quarter = 18.8 scfm or a total of 659 scfm; second quarter = 17.3 scfm or a total of 485 scfm; third quarter = 15.4 scfm or a total of 367 scfm; fourth quarter = 14.5 scfm or a total of 348 scfm

3. Constant = 0.00103673

5. Transfer rate = 2.21 lb/(hp)(hr)

7. a. Free air flow = 1280 cfm
   b. Motor hp = 47.0

Chapter 11

1. Diameter = 110 ft; depth = 6 ft-5 in. Since the depth may range from 5 to 8 ft, there may be other diameters and depths for this problem.
3. BOD_5 = 11.7 mg/ℓ
5. Diameter = 80 ft; depth = 6.34 ft. Since the depth is from 5 to 8 ft, there may be other diameters and depths for this problem.
7. a. Volume = 0.4386 ac-ft per filter
   b. Diameter = 65 ft; depth = 5 ft-9 in. Since the depth may vary from 5 to 6 ft, there may be other diameters and depths for this problem.
9. Sludge = 0.294 lb SS/lb BOD_5 removed
11. a. Diameter = 60 ft
    b. Depth = 18.3 ft

Chapter 12

1. a. Temperature = 55.41°F or 13.0°C
   b. Area = 75.3 acres
   c. Time = 40.3 days
   d. Loading = 28.8 lb BOD_5/ac-day
   e. Width = 906 ft; length = 1812 ft
3. a. Area = 31.0 acres
   b. Width = 581 ft; length = 1162 ft; depth = 3.5 ft.
   c. Time = 17.67 days
5. Area = 558,000 m^2; depth = 1m

Chapter 13

1. a. Time = 23 days
   b. Diameter = 80 ft; wall height = 25 ft-3 in.
   c. Fuel value = 21,100,000 Btu/day
3. a. Time = 23 days
   b. Diameter = 27.5 m; depth = 6.85 m
   c. Fuel value = 53,000 watts
5. Volume = 15,900 ft³
7. a. Volume = 120,000 ft³
   b. Volume = 6.0 ft³/capita

Chapter 14

1. a. Oxygen = 2620 lb/day
   b. Air required = 1730 scfm or 41.8 scfm per 1000 ft^3
   c. Solids = 0.0493 lb vss/(ft^3)(day)
   d. Q_D = 5,440 gal/day; Q_E = 29,600 gal/day
   e. $Q_{R(min)}$ = 73,900 gal/day; $Q_{R(max)}$ = 87,500 gal/day
   f. Area = 4,300 ft^2; diameter = 75 ft for standard size
3. a. Depth = 5.77 m; width = 11.54 m; length = 69.24 m
   b. Q_D = 47.8 m^3/day, 65,000 ppm
      Q_S = 230 m^3/day, 30,000 ppm
      Q_E = 183 m^3/day
   c. $Q_{R(min)}$ = 411 m^3/day; $Q_{R(max)}$ = 519 m^3/day
   d. Air = 15,000 m^3/h or 326 m^3(h)(1000 m^3)
   e. Time = 16.6 days
   f. Time = 66.0 days

Chapter 15

1. a. 99,700 gal/day
   b. Diameter = 75 ft for standard size
3. Minimum solids = 22.6 percent
5. a. Cake = 75,200 kg/day
   b. Filtrate = 124,800 ℓ/day; solids = 9620 mg/ℓ
   c. Diameter = 3.53 m; length = 7.06 m
7. Filter yield = 4.49 lb/(hr)(ft^2)
9. Specific resistance = 3.10 × 10^7 sec^2/gm

Chapter 16

1. Head loss = 0.05 ft
3. Acid = 950 lb/day
5. a. Area = 92.6 ft^2
   b. Air pressure = 2.39 atmos. or 35.1 psia.

Index

Kinetics (continued)
  chemical, 268–273
  of trickling filters, 371–377

L

Lag phase, in bacterial growth, 259–261
Lagoons, sludge, 466, 484, 492. See also Aerated
  lagoons
Lamella separator, 118
Laminar flow, in mixing, 38
Laminar range, in settling, 73, 74
Land application, of wastewaters, 517, 518
Langmuir isotherm for adsorption, 189
Lawrence, A. W., 273, 310
Lewis and Whitman two-film theory, 349, 350
Lime, 28, 61–64
Lime recovery, 493
Lime recalcining, 489, 493
Lime slakers, 30, 31
Lime sludges, 488, 489
Lime-soda softening, 3, 4, 56–60
Lime-soda softening plant, 3, 4
Log death phase, in bacterial growth, 259–261
Longitudinal mixing, in reactors, 270, 285
Low-rate anaerobic digesters, 409–414, 418–421
Low-rate trickling filters, 368, 374–376

M

Mass transfer coefficient
  in dialysis, 236
  in oxygen transfer, 350
  in reverse osmosis, 244
Maximum stationary phase, in bacterial growth,
  259, 261
McCarty, P. L., 273, 310, 407, 408
Mean cell residence time
  in activated sludge, 280, 281, 305–310
  in aerated lagoons, 395, 396
  in anaerobic digestion, 419
Mechanical aeration
  brush-type, 356, 357
  pump-type, 353
  turbine, 354–356
  turbine and sparger, 349, 356, 357
Media. See Filtration media
Membrane processes, 235–250
  dialysis, 235–237
  electrodialysis, 237–242
  reverse osmosis, 242–248
Mesophilic anaerobic digestion, 409–421
Mesophils, 266, 411
Methane-producing bacteria, 405–407
Michaelis-Menton equation, 274–276, 310
Microorganisms, 256–268
  cell physiology of, 256, 257
  growth phases of, 259–264
  respiration of, 264, 265
Miscellaneous unit operations and processes
  in wastewater treatment, 499, 501–518
  in water treatment, 499–501
Mixed liquor suspended solids (MLSS), 251
Mixed-media filters, 131, 153, 154, 161, 162
Mixing
  in baffle basins, 31, 40, 43
  longitudinal, 270, 285
  mechanical, 33–43
  paddle, 33–39
  pneumatic, 39–43, 48
  propeller, 33–39
  turbine, 33–43
Mixing requirements
  in activated sludge, 335, 359, 360
  in aerated lagoons, 397
  in aerobic digestion, 359, 360, 444, 447
Mixing systems, in anaerobic digestion, 415, 416
MLSS, 251
MLVSS, 280

Modified aeration activated sludge, 294
Moisture-weight relationships for sludges, 422
Monod equation, 305–310
Moving-bed countercurrent columns, 193–195,
  207, 208, 220
Multimedia filters, 152–154, 161, 162
Multiple hearth incinerators, 485–487

N

National Research Council (NRC) equations
  for activated sludge, 311, 312
  for trickling filters, 377–379
Nematodes, 256
Neutralization, of wastewaters, 511, 512
Newton's law, 72
Nitrification
  by activated sludge, 182–184, 281
  by trickling filters, 368
Nitrification-denitrification, for ammonia removal,
  182–184
Noncarbonate hardness, 57–60
NRC equations. See National Research Council
  equations
Nutrients, for bacterial growth, 264, 265

O

Ohm's law, 240
Oil separation, in wastewaters, 517
Open filters, 132, 139, 154–156
Operating dissolved oxygen concentration
  in activated sludge, 266, 281, 335
  in aerobic digesters, 444, 447
Operating problems
  in activated sludge, 333–337
  in aerobic digesters, 448
  in anaerobic digesters, 417, 418
  in filtration, 151, 152
  in trickling filters, 382
Operation, of anaerobic digesters, 417, 418
Organic wastewater sludge, 459–488
Orthokinetic coagulation, 23
Osmosis, 242. See also Reverse osmosis
Osmotic pressure, 242, 243
Outlet hydraulics, for settling basins, 119–123
Overaeration, in activated sludge, 335
Overflow rates, 81–86
  in wastewater treatment, 109, 112, 114
  in water treatment, 108, 109
Oxidation ditch, 297–302
Oxidation ponds. See Stabilization ponds
Oxygen coefficient (Y')
  for activated sludge, 321–323, 325–333
  for aerated lagoons, 397
Oxygen content of air versus temperature and
  elevation, 547
Oxygen requirements
  in activated sludge, 321–324
  in aerated lagoons, 397
  in aerobic digestion, 442, 444, 447, 451, 454
Oxygen solubility table, 546
Oxygen solubility versus tank depth, 353
Oxygen transfer, 349–365
  compressor requirements in, 360, 361
  description of, 349–350
  by diffused compressed air, 350–353
  Lewis and Whitman two-film theory of, 349, 350
  by mechanical means, 353–356
Ozone, 532–534

P

Packing
  for ammonia stripping towers, 171–173
  for trickling filters, 367, 373, 380
Paddle impellers, 34, 36–39
Pence, R. F., 198
Perikinetic coagulation, 23
Peripheral feed circular clarifiers, 105, 107

CROSS COUNTRY

AN EXCLUSIVE GUIDE TO

WISCONSIN
UPPER MICHIGAN-NO. ILLINOIS
SKI TOURING AREAS

By John & Midge
Schweitzer

CROSS COUNTRY SKEE

FOURTH EDITION, 1984

ISBN 0-9613771-0-0

Cover Design by Gordon Neitzke
Cover Graphics by Graphic Industries, Inc., Appleton, WI 54911
Cover photo Wisconsin Division of Tourism
Cover photo skier Jiri Gabera, former Olympic Ski Coach, Appleton, WI

Published by
Cross Country Skee
638 S. Mayflower Drive
Appleton, WI 54915
Phone 414-733-7042

INTRODUCTION

In the span of seven years we are entering 1984-85 with our 4th Edition of Cross Country Skee. Back in 1977, when we became ski enthusiasts, it was then we thought how great it would be if there were a directory published for cross country skiers.

After months of interviews, trips, phone calls and much help from our eleven children and their spouses and much hard work, Volume 1 hit the stands with 120 trails and 150 pages and an overwhelming reception.

As people became addicted to cross country skiing and more trails were carved out of beautiful tracts of land, we had no choice but to update in 1979 and '80.

While attending a ski conference in Rhinelander and being told repeatedly that our guide was a "Bible" for skiers we gave it another shot in 1981 with many new trails in Wisconsin. It was then that we included Upper Michigan and Northern Illinois.

After our 1982 edition we enjoyed and deserved a break from editing. So — here we are again in 1984-85 with another revision. Back in November '83 we began again with our noses to the grindstone breathing, worrying, and living Cross Country Skee Guide. With the many new trails we feel our work is not in vain.

We hope you the skier will find our "Bible" both informative and pleasurable. Hopefully you will enjoy the beauty of skiing in the glorious peacefulness God has given us as His gift.

Would like to close with an ancient Indian prayer given to us by our grandchild Amy Meltz.

> Beauty is before me
> And beauty behind me
> Above and below me
> Hovers the beauty.
> I am surrounded by it
> I am immersed in it
> In my youth I am aware of it
> And in my old age I shall walk quietly,
> The Beautiful Trail.
>
> Navajo Indians

Peace,
Midge & John Schweitzer

⊛ Kimberly-Clark

Benefits From Cross Country Skiing

1 **Heart** Aerobic exercise increases the heart's stroke volume (the amount of blood pumped with each contraction of the heart). Exercise will help maintain the condition of the blood vessels which bring blood to the heart muscle.

2 **Muscles** Builds strength and endurance in upper body muscles of back, shoulders, arms. Also develops buttocks, quadricep and hamstring muscles in upper leg, calf muscles in lower leg.

3 **Joints** Movement of a joint is necessary for its cartilage to be supplied with important nutrients. Skiing maintains the condition of the knees, ankles, shoulders, hips and elbows.

4 **Fat** Besides taking off excess pounds, exercise can decrease your percent body *fat* by burning calories and maintaining muscle mass.

5 **Coordination** The diagonal stride technique used in cross country skiing requires timing and rhythm and a refinement of the *cooperation* between muscles and nerves to increase coordination and skill.

6 **Blood** Aerobic exercise increases the level of HDL (good cholesterol) in the blood. Because aerobic exercise increases the level of hemoglobin in the blood, more oxygen can be transported throughout the body.

7 **Metabolism** Skiing is an activity of long duration, allowing more calories to be expended during each workout.
Increased metabolism in the muscles during and after exercise will burn up extra calories.

8 **Mind** Outdoor exercise is especially good for relieving physical tension and mental stress.

The beneficial effects from aerobic workouts are only found if workouts are done at least three times per week on nonconsecutive days.

These workouts should be at least 30 minutes in length while your heart rate is at least 70% of its maximum speed as determined on the bicycle or treadmill exercise test.

Health Management Program

BASIC TECHNIQUES OF CROSS COUNTRY SKIING

Our book is designed to help you with the "where" of skiing. Your continuing enjoyment depends, too, on the "HOW". Your progression from the first tentative shuffles, to confidently skiing a hilly trail in the woods, depends on good equipment, good instructor, and practice. There is no substitute for lessons from a certified instructor, but you will get a good start by following these suggestions.

AT HOME, BEFORE SKIING

Practice putting your shoes into the bindings, observing right and left marks if any.

Learn correct **Pole Grip**. Insert your whole hand, thumb included, up from underneath the strap. Adjust strap length so that with your gloves on, the straps meet the pole handle just under the web between your thumb and forefinger. The strap should lie flat against your palm and around the back of your hand. Try pushing the pole into the ground. Notice that, with the correct grip, you can push hard and yet your fingers can be relaxed. Many skiers have never been taught how to put on their poles so their arms and shoulders suffer. Get this part right and you are on your way!

FIRST TIME ON SNOW

A level area with ski tracks is best. Put on your skis but leave your poles off for now. You'll learn the fundamental cross country stride **(the Diagonal Stride)** and the step turn. With bent knees and a forward lean, shuffle along. Let your arms

swing freely foreward and backward in time to your leg movements. Just as in walking, running, and skating as one leg swings forward, its opposite arm swings forward. As you practice you'll achieve a forward glide with each stride. Lean forward, let all your weight ride on the forward, gliding ski. Experiment with faster and slower stride rates. There is a tempo that produces a nice momentum, as you spring from one ski to the other, driving one leg forward then the other. You are skiing!

DIAGONAL STRIDE

To turn to the right, lift the top of your right ski and swing it to the right. Now put your weight on the right ski and bring the left ski along side it. You are doing a step turn. To turn around, just repeat the turn several times. Try it in both directions. You can turn while striding ahead, by taking one or more step turns in the desired direction.

HERRINGBONE

Now that you've got extra forward push from the poles you can go up hills! A slight hill can often be climbed using the **Diagonal Stride**. Just shorten and speed up your stride. A steeper grade (or tired skier) calls for the **Herringbone**. Spread the tips of the skis as far apart as you can without crossing the tails. Roll your ankles inward, to dig the inside edges of your skis into the snow. Keeping your poles pointed backwards as usual, just take one step at a time up the hill.

The same skis will get you down the hill. Take a couple diagonal strides over the top of the hill, then crouch and lean forward. Tuck your poles under your armpits, keeping your hands out in front and let gravity do the work. A knees bent, forward leaning, arms ahead crouch is a very stable position for skiing down hills. Keeping one ski a couple of inches ahead of the other adds even more stability.

Snow-plow

To slow down and stop, learn the **Snowplow**. While coasting in the knees bent, crouched position, slide the tails of the skis outward as far as possible. If the tips cross a bit it's OK. The more you press the inside edges of the skis into the snow, the greater the breaking effect. You can also control your breaking rate by varying the width of the V your skis are making.

Try these uphill and downhill methods on open slopes. Soon you'll be ready to turn the skis while skiing downhill. If you're not going too fast, try the step turn, taking small, quick steps. When you want to slow down and turn at the same time, use the **Snowplow Turn**. Slide the skis with the snowplow position. To turn right shift your weight onto the inside edge of the left ski. Try this turn in both directions, with more and less weight shift. You can control the direction and the speed of your skis with practice.

Getting up from a fall is easy. Remember to position your skis across the hill, not pointed up or down the hill. Put the skis parallel, down slope from your body. Put your body above your feet and stand up. Brush snow off yourself, and recheck your grip on the poles.

These basic skills will enable you to take on the challenge of skiing a trail cut through the woods. You'll usually find ski tracks there instead of untracked open areas. Professionally groomed, tracked trails will help keep your skis parallel when you diagonal stride. On downhill runs, the tracks will help guide you through the turns if you concentrate on your body position. If you want to slow down while skiing in tracks, use the snowplow. You'll first have to lift a ski out of the track, then put it into snowplow position. Often, keeping one ski in the track, and snowplowing with the inside edge of the other ski gives you just enough braking. On a steep downhill turn everybody snowplows with both skis, so there is no track for the skis to follow. Your snowplow turn and edging skills pay off handsomely here.

Spend plenty of time practicing your diagonal stride and step turn without poles, even after the first lesson. Your confidence and balance will improve more quickly, and a graceful fore and aft arm swing comes more easily if you practice without poles. But yes, there is a time and purpose for poles. They are propellers, not outriggers, and the time to put them on is after you can diagonal stride and step turn without them. Remember the correct grip: hands up from under.

Don't wrap your fingers tightly around the handle. Stride along, arms swinging, as before. As your arm swings and pushes back, let your hand open up so you're holding the pole between your thumb and first finger only. As you swing your arm forward keep the pole tip aimed backward. Don't let the tip swing out ahead of you.

When you want to take a breather step off to the side of the trail. Pull your hat down so you don't cool off too fast. Now is a good time to people watch. Learn from the more accomplished skier. Look at the forward leaning position, the full swing and push of the arms; the way each stride and turn maintains momentum. Look at the clothing: stretchy, non-binding, thin layers, gloves rather than mittens, always a hat.

Reflect on what you have learned from practicing these fundamentals. Those long skinny boards and sticks don't feel so foreign anymore. You can balance, glide and steer. With the help of your poles you can ski on the flats and up the hills. You can enjoy downhill runs that used to scare you. You may even have turned winter into a season to be met and enjoyed!

Contributed by
Mike Shannon
Janus Ski & Specialty Sports

How to use the Book

Our guide has been arranged by counties. By using the Wisconsin County Map on inside front cover, you can determine which area you would like to ski in. Most times this has a bearing on snow conditions. The trails are arranged alphabetically by counties (Bayfield, Brown, etc.) and the trails within each county are also alphabetical. Our thinking on this is if you have selected a place to ski and find that the trail is not to your liking or is in poor condition, you can use the guide to find other trails in the near vicinity.

We also have an index in the back that lists **all** trails alphabetically. If you have heard the name of a trail, look it up in the index to see what it looks like and where it is located.

For your convenience in looking for accommodations, we have also arranged the lodges, resorts and motels by counties with phone numbers and addresses.

The trails are depicted in blue on pages opposite the general information about each trail. The small state map in upper corner shows the county in blue ink, the enlarged road map at bottom designates how to find the trail once you are in the vicinity. The snowflake in white shows the trailhead. The description gives you pertinent information as to length of trails, loops and difficulty.

Information on individual trails change quite rapidly so all information is as up to date as possible.

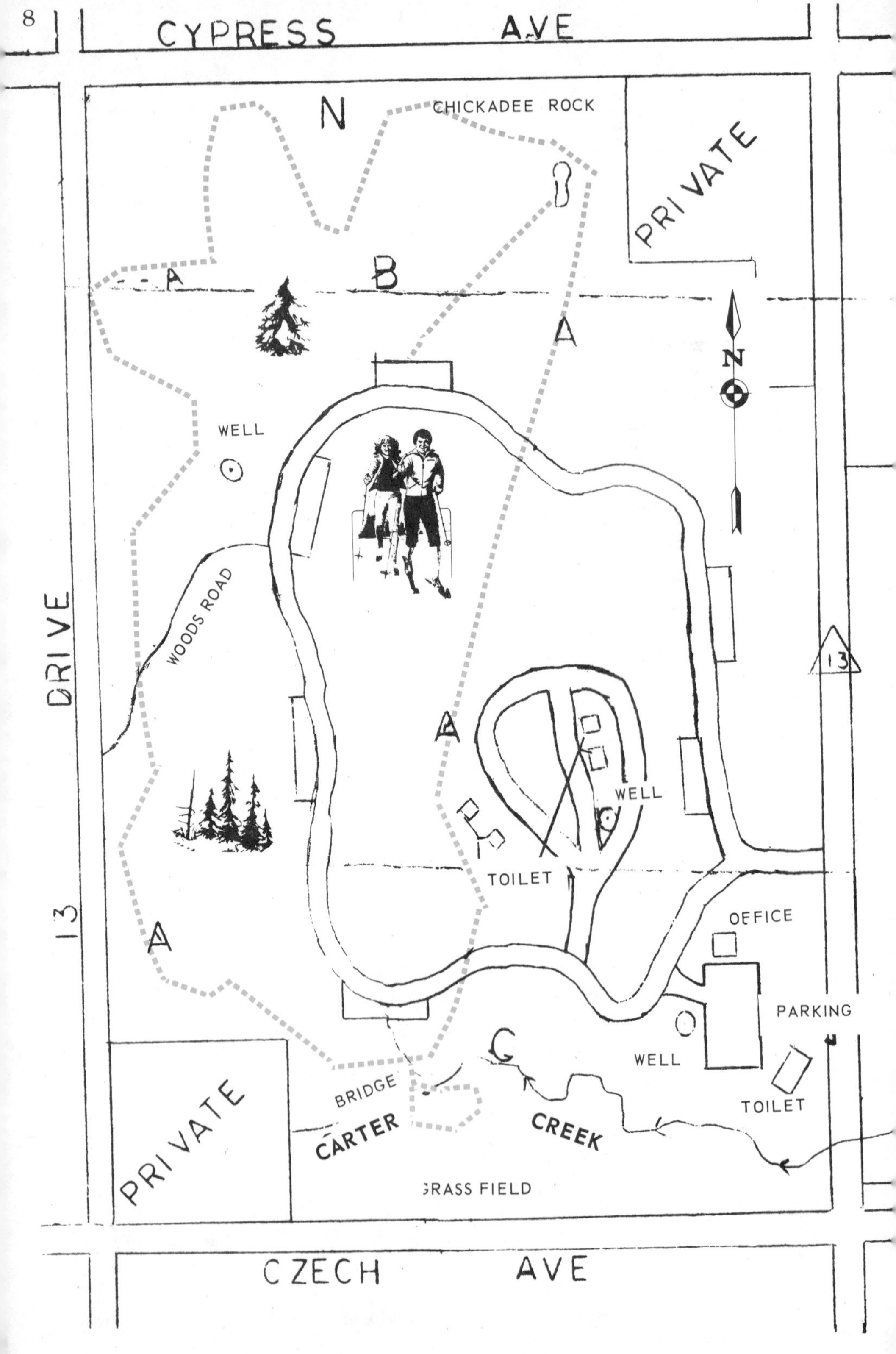

8

CYPRESS AVE

CHICKADEE ROCK

PRIVATE

N

A   B

A

N

WELL

WOODS ROAD

DRIVE

13

13

WELL

A

TOILET

A

OFFICE

PARKING

WELL

C

BRIDGE

CARTER   CREEK

PRIVATE

TOILET

GRASS FIELD

CZECH AVE

Snowbird County Ski Trail

Snowbird cross-country ski trail was developed in the summer of 1979, at that time it was only a one way trail, in need of a lot of work. Upgrading the trail started in 1980 and has continued into 1981. The trail is now 12 feet wide and 3 miles long. It has 19 nature signs for you to see as it winds through poplar, oak, and pine groves. It follows along and crosses Carter Creek by way of a foot bridge. The bridge takes you to the open field on the south end of the park, where you can get a good look at Roche a Cri Mound and its beauty. The trail also goes by Chickadee Rock and Roche a Cri Mound, both of which were formed in the glacier age.

The trail itself has gradual curves, small hills and some straightaways.

I hope you will come and enjoy what we have tried to provide.

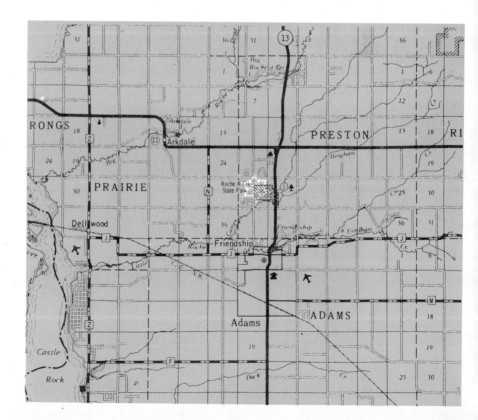

10

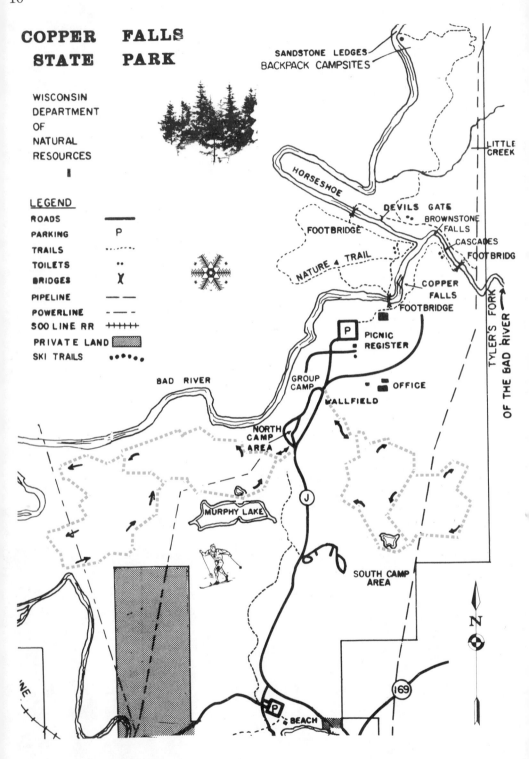

COPPER FALLS
STATE PARK

WISCONSIN
DEPARTMENT
OF
NATURAL
RESOURCES

LEGEND

ROADS
PARKING   P
TRAILS   ....
TOILETS   ..
BRIDGES   X
PIPELINE   —— ——
POWERLINE   – – –
SOO LINE RR   +++++
PRIVATE LAND   �
SKI TRAILS   •••••

SANDSTONE LEDGES
BACKPACK CAMPSITES

LITTLE
CREEK

HORSESHOE

DEVILS GATE
BROWNSTONE
FALLS
FOOTBRIDGE
CASCADES
FOOTBRIDG

NATURE 4 TRAIL

COPPER
FALLS
FOOTBRIDGE

TYLER'S FORK
OF THE BAD RIVER

P   PICNIC
    REGISTER

GROUP
CAMP   OFFICE
BAD RIVER

BALLFIELD

NORTH
CAMP
AREA

MURPHY LAKE

J

SOUTH CAMP
AREA

N

169

P
BEACH

Copper Falls State Park

Ashland County

Located 3 miles norhteast of Mellen, Copper Falls State Park offers four loops on two trails. The trails are marked, groomed and tracked.

Total trail lengths are about 10km (6.2 miles) in gentle rolling wooded areas.

Skiing hours are from 6 A.M. to 6 P.M.

Parking area is located near the office.

If more information is needed in regard to this scenic trail please phone 715-274-5123.

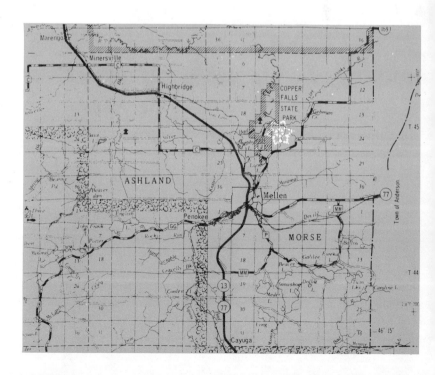

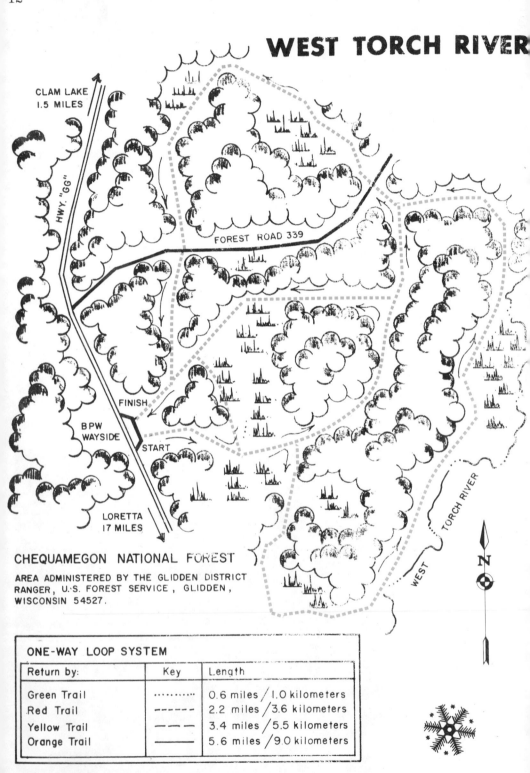

12

WEST TORCH RIVER

CLAM LAKE
1.5 MILES

HWY. "GG"

FOREST ROAD 339

WEST TORCH RIVER

FINISH

BPW
WAYSIDE

START

LORETTA
17 MILES

CHEQUAMEGON NATIONAL FOREST

AREA ADMINISTERED BY THE GLIDDEN DISTRICT
RANGER, U.S. FOREST SERVICE, GLIDDEN,
WISCONSIN 54527.

N

ONE-WAY LOOP SYSTEM

Return by:	Key	Length
Green Trail	....	0.6 miles / 1.0 kilometers
Red Trail	------	2.2 miles / 3.6 kilometers
Yellow Trail	— — —	3.4 miles / 5.5 kilometers
Orange Trail	——	5.6 miles / 9.0 kilometers

West Torch River

This one way trail system is composed of 4 loops varying from 0.6 miles to 8.9km (5.6 miles). Part of the orange trail runs along the West Torch River and the rest is thru heavily wooded areas. The trails are marked by tree blazes in various colors. Plowed parking is available at the trailhead. The trailhead is a natural forest wayside & nature trail in the summer. The business and Professional Womens Club sponsored a coop forest plantation in this area.

The location is four miles south of Clam Lake on Cty. Hwy. GG.

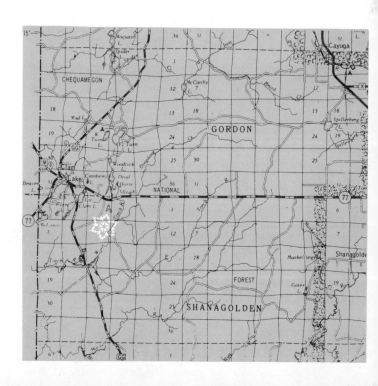

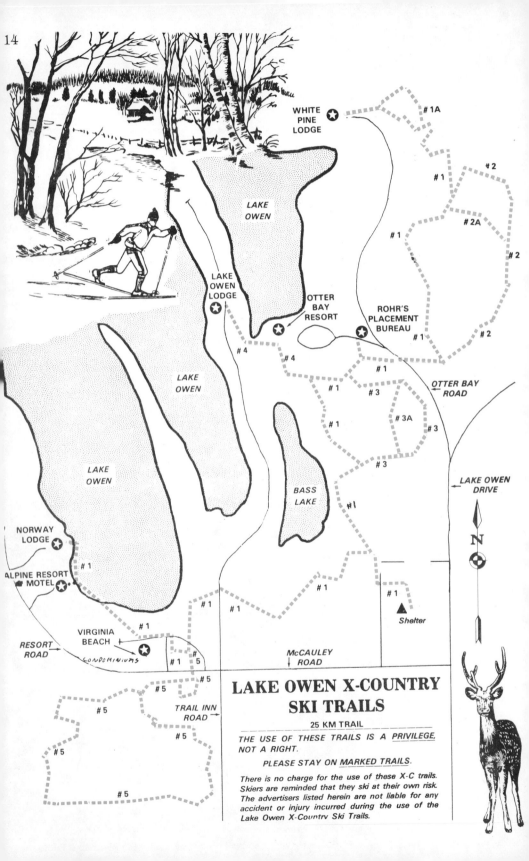

14

WHITE PINE LODGE

#1A

#1

#2

#2A

#2

LAKE OWEN

#1

#2

LAKE OWEN LODGE

OTTER BAY RESORT

ROHR'S PLACEMENT BUREAU

#1

#2

#4

#4

#1

OTTER BAY ROAD

LAKE OWEN

#1

#3

#1

#3A

#3

#3

LAKE OWEN DRIVE

BASS LAKE

#1

N

NORWAY LODGE

#1

ALPINE RESORT MOTEL

#1

#1

#1

#1

#1

Shelter

VIRGINIA BEACH

CONDOMINIUMS

#1

#5

McCAULEY ROAD

RESORT ROAD

#1

#5

#5

TRAIL INN ROAD

LAKE OWEN X-COUNTRY SKI TRAILS

#5

#5

25 KM TRAIL

#5

THE USE OF THESE TRAILS IS A *PRIVILEGE*, NOT A RIGHT.

PLEASE STAY ON *MARKED TRAILS*.

#5

There is no charge for the use of these X-C trails. Skiers are reminded that they ski at their own risk. The advertisers listed herein are not liable for any accident or injury incurred during the use of the Lake Owen X-Country Ski Trails.

Lake Owen Trails

The Lake Owen Trails were developed in 1974 and cover approximately 40km (25 miles). Trails run from resort to resort with varying lengths of loops in each area. Guests may use the trails at no charge and rental equipment is available. There are access points from White Pine Lodge, Otter Bay Resort, Lake Owen Lodge, Norway Lodge, Alpine Motel, and Virginia Beach Condominiums & Rohr's Placement Bureau. Travel east of County M out of Cable to Trail Inn Road, then north to resorts.

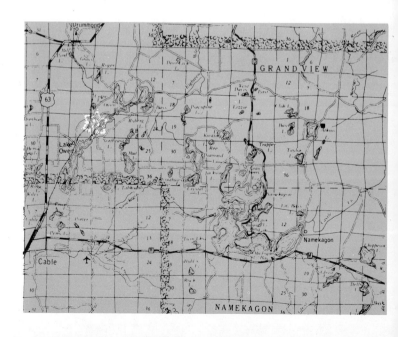

16

CONNECTION TRAIL

LOOP

WEST FORK

EAST FORK

CUT OFF

COREYS

PINES

GRAND TOUR

PULP ROAD

T-BAR

CHALET

RIDGE RUN

BASE TRAIL

SUGAR BUSH ROAD

N

CROSS COUNTRY TRAILS

Mt. Ashwabay

Bayfield County

Mt. Ashwabay is located 3 miles south of Bayfield on State Hwy. 13 and has 30km of cross-country trails.

This is a complete downhill and Nordic skiing center that is reminiscent of the great ski areas of New England because of its beautiful trails flanked by groves of maple, oak and pine. The area offers a perfect vantage point for a tremendous panoramic vista of the Apostle Island Chain and great stretches of Lake Superior shore line. This area produces the kind of weather that triggers continuous heavy snowfall throughout the season.

Services and facilities include a ski school, chalet, equipment rentals and ski shops. Cross-country guide service offers a unique opportunity to discover the winter splendor of the Bayfield Peninsula with a variety of woodland trails.

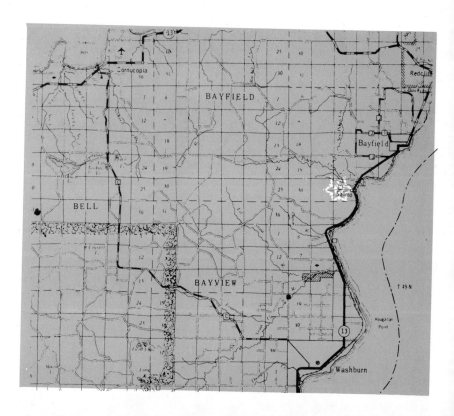

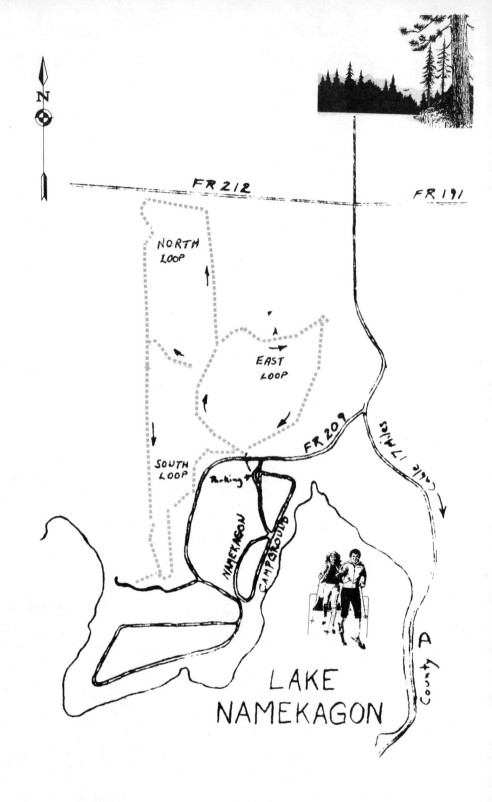

N

FR 212   FR 191

NORTH
LOOP

EAST
LOOP

FR 209

Calif 17 Miles

SOUTH
LOOP

Parking

NAMEKAGON

CAMPGROUND

County A

LAKE
NAMEKAGON

Namekagon Ski Trail

Bayfield County

Located north of Namekagon Lake, this trail is operated by the U.S. Forest Service. Suitable for the beginner and intermediate skier, it winds through flat to gently rolling hardwood forest. This well-marked, groomed trail is approximately 7km (4.3 miles) in length with a choice of three different loops: Loop A — 1.75 km, Loop B — 2.3 km, and Loop C — 3.45 km.

Parking is available at the entrance to Namekagon Campground. The trail also connects to Mogasheen Resort at the southwest corner of the trail system.

Located 17 miles NE of Cable. Take County M east from Cable to County D. Take D north to FR 209 and FR 209 to trail.

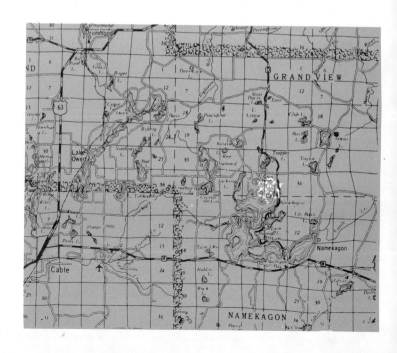

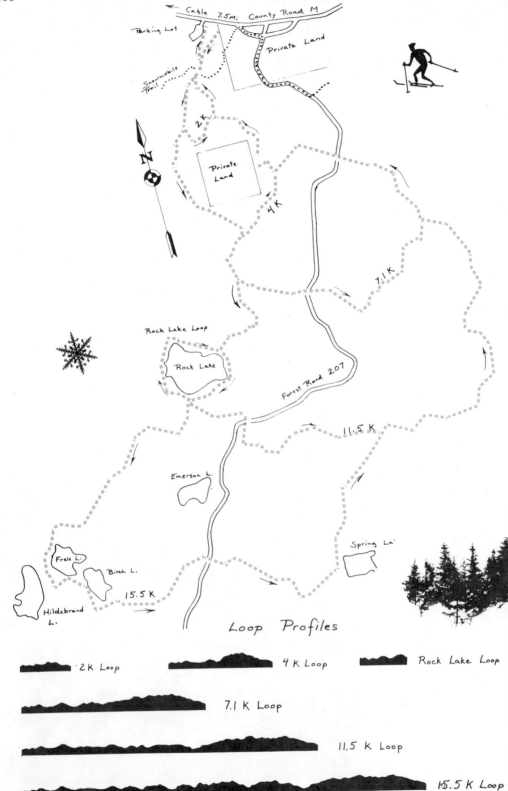

Cable 7.5 mi; County Road M

Parking Lot

Private Land

Snowmobile Trail

N

2 K

Private Land

4 K

7.1 K

Rock Lake Loop

Rock Lake

Forest Road 207

11.5 K

Emerson L.

Spring La'

Frels L.

Birch L.

15.5 K

Hildebrand L.

Loop Profiles

2 K Loop

4 K Loop

Rock Lake Loop

7.1 K Loop

11.5 K Loop

15.5 K Loop

Rock Lake Trail

Bayfield County

This 42km (26 miles) trail system operated by the U.S. Forest Service is located south of Lake Namekagon on Highway "M" near Lakewoods Resort. The trail originates at a plowed parking lot on Highway "M" and consists of a series of stacked loops.

The marked, groomed and tracked trails traverse gently rolling to hilly wooded terrain and are suitable for intermediate and expert skiers.

A toilet is available at the parking lot. No fee is charged for use of this trail. The trail serves as a hiking trail in the summer.

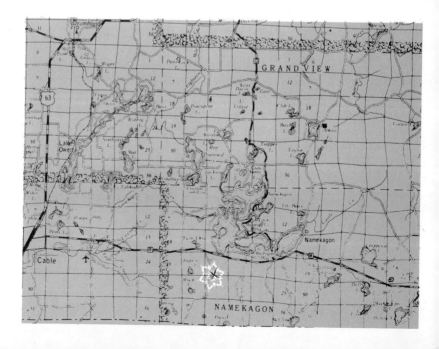

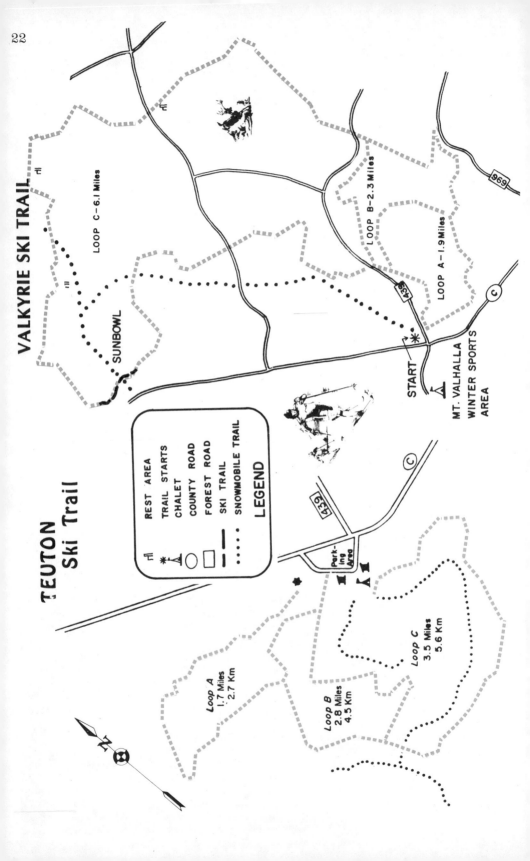

VALKYRIE SKI TRAIL

LOOP C – 6.1 Miles

LOOP B – 2.3 Miles

LOOP A – 1.9 Miles

SUNBOWL

969

435

START

MT. VALHALLA
WINTER SPORTS
AREA

TEUTON
Ski Trail

LEGEND

REST AREA

TRAIL STARTS

CHALET

COUNTY ROAD

FOREST ROAD

SKI TRAIL

SNOWMOBILE TRAIL

435

C

Park-
ing
Area

Loop A
1.7 Miles
2.7 Km

Loop B
2.8 Miles
4.5 Km

Loop C
3.5 Miles
5.6 Km

N

Valkyrie and Teuton Trail

Bayfield County

The Valkyrie and Teuton Cross-Country Ski Trails are located at the Mt. Valhalla Winter Sports Area. Mt. Valhalla is 8.5 miles northwest of Washburn on County Highway C. The trails are in the Chequamegon National Forest and are administered by the U.S. Forest Service. There are no fees for use of the trails.

The Valkyrie Trail consists of 3 loops with 10 miles of skiing. The Teuton Trail consists of 3 loops with 8 miles of skiing. While skiing, be on the lookout for snowmobile crossings in several places.

These Cross-Country Ski Trails are ideal for the novice or intermediate skier. The trails are groomed and tracked from December 1 through March 15. Vegetation types along the trails include red pine, jack pine, oak, aspen, and birch.

A Chalet, with electricity, is located at the parking area. Water and pit toilets are also available.

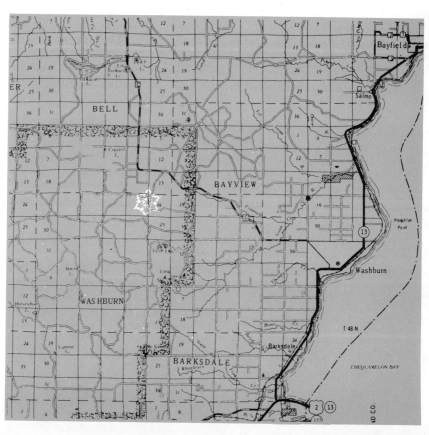

TELEMARK CROSS COUNTRY SKI TRAILS

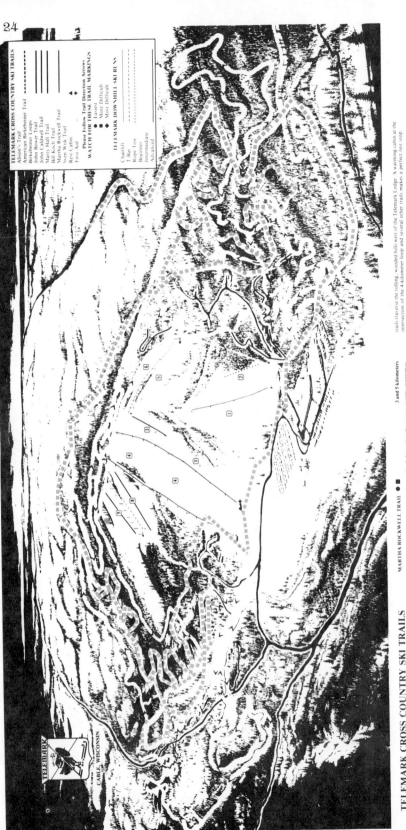

CABLE, WISCONSIN

TELEMARK

Legend:

TELEMARK CROSS COUNTRY SKI TRAILS
Alison's Trail
American Birkebeiner Trail
Birkebeiner Trail
John Bower Trail
John Caldwell Trail
Bill Koch Trail
Marty Hall Trail
Martha Rockwell Trail
Sven Wiik Trail
Rest Cabin
First Aid
Please Follow Trail Direction Arrows
WATCH FOR THESE TRAIL MARKINGS
◆ More Difficult
◆ Most Difficult
TELEMARK DOWNHILL SKI RUNS
Chairlift
T-Bar
Rope Tow
Beginner
Intermediate
Advanced

BILL KOCH TRAIL ■ ◆ 18 kilometers

This new intermediate/expert trail was designed by Olympic medalist Bill Koch. It follows the rolling contour of a spectacular glacial esker east of Telemark, winds its way around Farm Lake and through heavily wooded hills to the top of Mount Telemark, culminating in a series of expert-only downhills before winding its way back to the Lodge. A variety of cutoffs connect to other Telemark trails.

ALISON'S TRAIL ■ 3 and 5 kilometers

Formerly the Bill Koch Trail, Alison's Trail is now named in honor of Alison Owen, the American skier who won the first-ever F.I.S. World Cup race held at Telemark in 1978. The trail traverses a glacial ridge west of the Telemark Lodge and is designed for intermediate skiers. A 3-kilometer loop cuts across Telemark's rolling golf course and is an excellent trail for beginners or intermediate skiers.

MARTHA ROCKWELL TRAIL ● ■ 3 and 5 kilometers

This gentle trail system has given thousands of skiers their first cross country experience. The trail traverses the rolling Telemark golf course with the 5-kilometer loop extending across a wooded hill. The Martha Rockwell Trail is named in honor of the first American women to break into international competition.

JOHN BOWER TRAIL ■ 8 kilometers

The John Bower Trail honors the former U.S. Ski Team nordic director who helped boost the U.S. Ski Team to its best position ever in the world in the late 1970's. The trail is characterized by long, sweeping downgrades as it winds its way from the rest cabin intersection, west of the Telemark Lodge, around Mount Telemark, before connecting with the Martha Rockwell Trail.

BIRKEBEINER LOOPS ● ■ 1, 2, 4, 5, 7.5 and 10 kilometers

Although the American Birkebeiner race has outgrown this trail, the Birkebeiner loops continue to be Telemark's most popular ski trails with cutoffs for various levels of skiers. The trails traverse the rolling, wooded hills west of the Telemark Lodge. A warming cabin at the intersection of the 4-kilometer loop and several other trails makes a perfect rest stop.

AMERICAN BIRKEBEINER TRAIL ■ 55 kilometers

Much of the 30-foot-wide American Birkebeiner Trail is skiable throughout the season for skiers interested in training for America's largest cross country ski race. Skiers can connect to the trail on the 10-kilometer Birkebeiner loop and ski to the outskirts of Hayward. A pickup service can be arranged with guided tours offered through the ski school.

WORLD CUP TRAILS ◆

Sven Wiik Trail — 5 kilometers
John Caldwell Trail — 10 kilometers
Marty Hall Trail — 15 kilometers

Telemark's World Cup Trails are all named for former U.S. Olympic coaches. They are very challenging with ever-changing terrain, exhilarating downhills and thrilling uphills. The trails head west of the Telemark Lodge into wooded hills near the Birkebeiner loops, with the Marty Hall Trail extending all the way around Mount Telemark.

Telemark Nordic Center

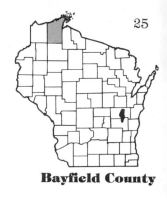

Bayfield County

The Telemark cross-country ski trail complex, part of the 2,000-acre Telemark Recreational Community three miles east of Cable, Wis., in the northwestern part of the state, was laid out by Sven Wiik and Marty Hall, former U.S. Olympic Nordic ski coaches.

Cross country trails usually have sufficient snow cover for touring by Dec. 1, and, protected by dense forests, can be skied until late March.

Telemark's 100km (62.1 miles) of cross-country trails, ranging from one to 15 kilometers in length, have been carved through acres of picturesque woodlands, around bogs and over moraines and eskers.Trails are available for all levels of skiers.

Telemark's cross-country trails are patrolled and are open from 9:00 A.M. to 4:30 P.M.

Rest cabins are equipped with a stove, and are located at various points on the trail.

Cross country passes, good all day on all trails, cost $7.00 per day on weekends.

Telemark provides complete X-C equipment rental services for both children and adults.

Both private and group classes are available, taught by certified X-C ski instructors.

Changing rooms with lockers and restrooms are available in the lodge for the convenience of day skiers.

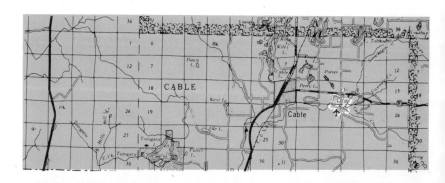

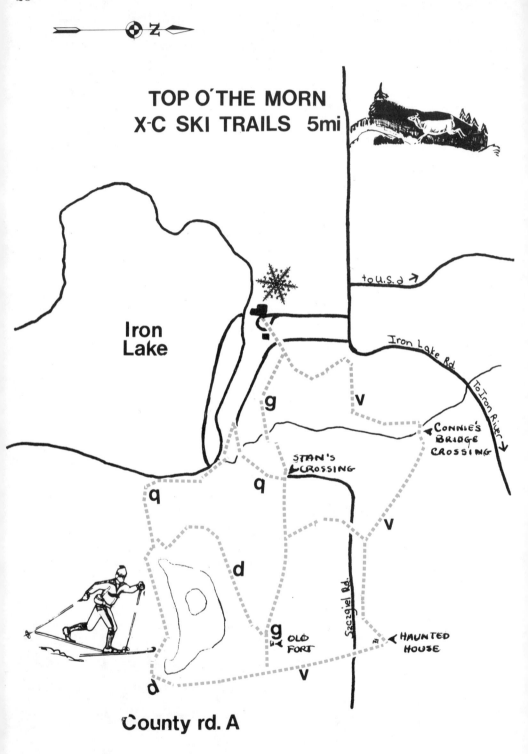

26

TOP O' THE MORN
X-C SKI TRAILS 5mi

Iron
Lake

to U.S.2 →

Iron Lake Rd

To Iron River

g   V

CONNIE'S
BRIDGE
CROSSING

STAN'S
CROSSING

q   q

V

d

Sacajwel Rd.

g OLD
   FORT

HAUNTED
HOUSE

d   V

County rd. A

Top O' The Morn Resort

Bayfield County

Oak, birch, maple, balsam, poplar, white pine engulf you as you ski along our 8km (5 miles) groomed and tracked intermediate type trails.

Crossing the Iron River on three different bridges you find yourself in beauty unlike any trails in northern Wisconsin. All trails are located on private land and no snowmobiles are allowed on the trails. The trails start and end at the Lodge where refreshments, toilets, hot heated showers and lodging are available. With the above average snowfall in our area, our trails are open from Dec. 1 thru March 15th.

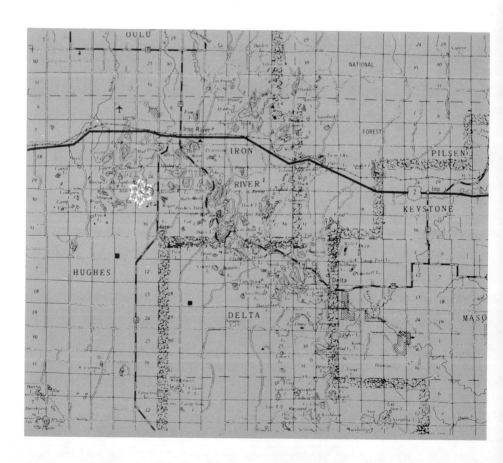

BAIRD CREEK PARKWAY
CROSS COUNTRY SKI TRAIL
GREEN BAY PARK & RECREATION DEPT.

▶ START
● FINISH

＊ CROSS COUNTRY SKI TRAIL — ONE WAY IN DIRECTION ARROWS INDICATE — APPROXIMATELY 1.3 MILES

No. 3 PARKING

TRIANGLE WINTER SPORTS AREA

No. 1 PARK.

No. 2 PARK.

BAIRD CREEK

R.R. TRACKS / PARK BOUNDARY

＊ NOTE PARKING LOTS 1 & 2 ARE OPEN MONDAY–FRIDAY 6:30 P.M.–10:00 P.M. & SATURDAYS & SUNDAYS 1:00 P.M.– 5:00 P.M. & 6:30 P.M. – 10:00 P.M.
PARKING LOT 3 WILL BE OPEN 7 DAYS A WEEK 6:00 A.M.– 10:00 P.M.

N

Baird Creek Parkway
Cross Country Ski Trail

Brown County

This ski trail is located in Baird Creek Parkway, a 165 acre city owned area, the majority of which is natural. The trail meanders approximately 2km (1.3 miles) through woods, open fields and along a picturesque stream. Future development will bing the total length of this trail to approximately 2.5 miles. The trail is groomed and tracked regularly and is best negotiated by the intermediate level skier.

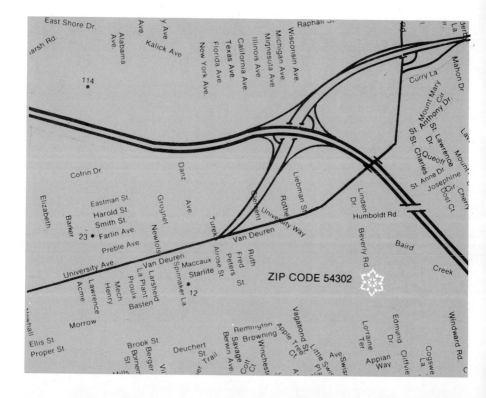

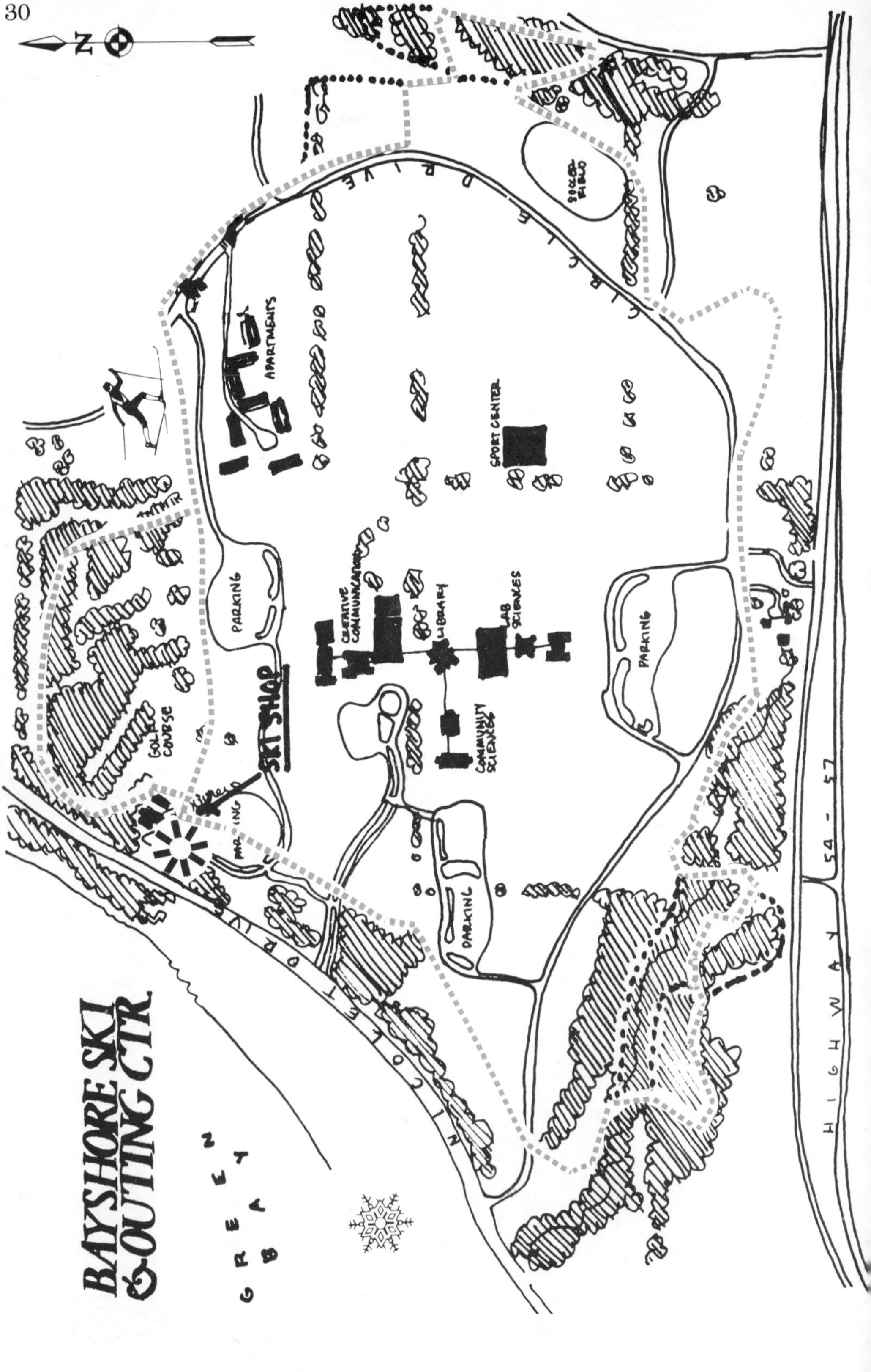
BAYSHORE SKI & OUTING CTR
N
GREEN BAY
NICOLET DRIVE
CIRCLE DRIVE
HIGHWAY 54 - 57
SKI SHOP
GOLF COURSE
PARKING
PARKING
PARKING
PARKING
APARTMENTS
SPORT CENTER
CREATIVE COMMUNICATION
LIBRARY
LAB SCIENCES
COMMUNITY SCIENCES
SOCCER FIELD

Bayshore Winter Center

Brown County

The Bayshore Outing Center has designed and developed 13km (8 miles) of winter wonderland skiing. These trails are open to the public and are on and around the University of Wisconsin Green Bay campus. Trails are well groomed and marked but at times do blow over due to the wind.

For more information call 414-465-2400. Rental equipment is available but reservations should be made 3 days in advance.

Hours of operation, Monday thru Friday 1:00 to 8:00 P.M. Saturday & Sunday 9 A.M. to 8:00 P.M.

HYPOTHERMIA

As in any sport, cross country skiing has its hazards.

Hypothermia is the least recognized, yet probably the most dangerous. Simply stated Hypothermia is the rapid and progressive mental and physical collapse resulting from lowering the inner temperature of the human body. It is caused by exposure to cold, aggravated by wet, wind and exhaustion. Hypothermia can result in death if untreated.

Symptoms: Fits of shivering, vague, slurred speech, memory lapses, fumbling hands, lurching walk, drowsiness and exhaustion, apparent unconcern about physical discomfort.

These symptoms are usually noticed by others before the victim is aware of them.

Treatment: Get the victim out of the wind and wet, into dry clothes and restore body temperature with warm drinks, quick energy food, body contact and a warm sleeping bag.

Proper use of clothing will greatly diminish the risk of succumbing to this phenomonon. It is not only wise, but imperative to follow these simple rules: avoid heavy bulky clothing. Rather dress lightly with layers of loose-fitting garments to form air pockets for insulation. For example: a thermal T shirt covered by a jogging suit jacket over which is pulled on unlined wind breaker. Trousers should be water resistant. (Blue jeans and corduroys absorb moisture and hasten heat loss through the legs.)

Always wear a hat no matter how warm the day. (You can lose up to one-half the body's total heat reduction without one.) Warm gloves or mittens are a must—with a dry pair in reserve. Two pairs of socks are essential, the under pair being absorbent wool. Leave plenty of room to wiggle your toes inside your boots.

Begin your tour at a leisure pace, picking up speed as your muscles loosen. If you feel yourself getting warm, stop to rest. You may need to remove a layer of clothing.

Nerver ski alone; and never tackle a trail that is beyond your endurance or is unfamiliar to you. If you begin to tire, turn back, unless you are closer to the end of the trail.

Travel with a back pack containing such lightweight items as an insulated blanket, dry gloves and socks, a first aid kit, high energy munchy foods, a tin of fresh water, a small pair of pliers, a jackknife, a box of matches and a spare ski tip.

These are basic common sense rules. Follow them for a season of exhilerating fun on the trails!

Bay Beach
Wildlife Sanctuary

Brown County

This 3.2km (2 miles) trail offers the cross country skier an opportunity to enjoy a few hours in a beautiful natural setting. The trail is a combination of wooded and flatland that is located near the heart of Green Bay.

Skiers are asked to minimize the recreational impact on the wildlife that is already coping with the rigors of winter. Leave your dogs at home. Follow the routes and stay off the lagoons.

Trail maps, heat, toilets are available at the Nature Center. Guided tours can also be arranged.

Take Highway 57 east out of downtown Green Bay to County Trunk A which swings past the sanctuary.

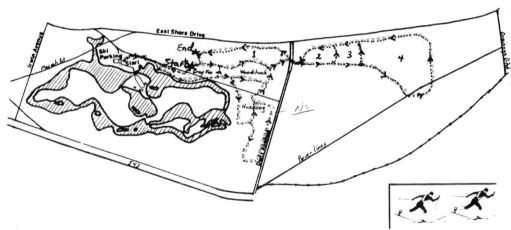

SKI TRAILS:

GRAY FOX - .75 mi. [1.2 km.]

HUSSONG - 1.25 mi. [2 km.]

WOODCHUCK - .5 mi. [.8 km.]

NEW LOOPS:

1 - .75 km.   3 - .5 km.

2 - .5 km.   4 - 2 km.

34

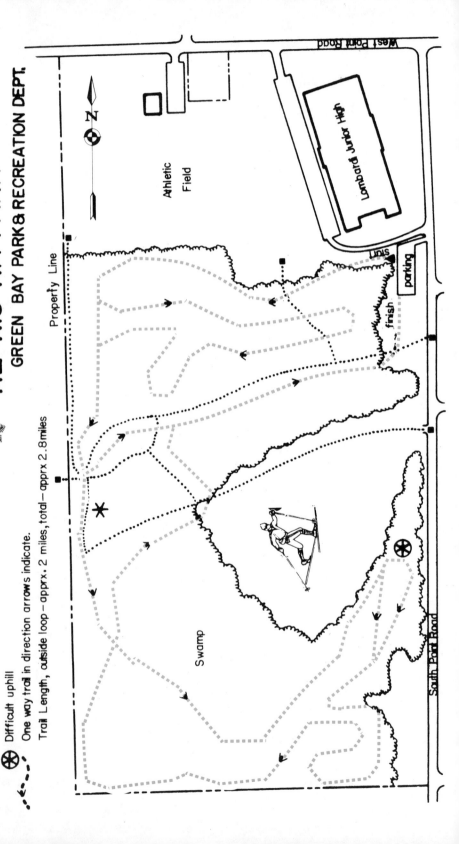

HE-NIS-RA PARK

GREEN BAY PARK & RECREATION DEPT.

Cross country ski trail (marked with blue & white arrows)

Trails not recommended for skier use

Difficult downhill

Difficult uphill

One way trail in direction arrows indicate.

Trail Length, outside loop—apprx. 2 miles, total—apprx 2.8miles

N

Property Line

Athletic Field

Lombard Junior High

West Point Road

start

finish

parking

South Point Road

Swamp

HE-NIS-RA Park
Ski Trail

Brown County

This trail is located in He-Nis-Ra Park adjacent to Lombardi Junior High School. The park contains over 44 acres of natural wooded area. The 4.6km (2.8 miles) trail passes through oak uplands, as well as through a unique wetland area. Passage through the wetlands is provided for by an upraised wood chip trail. The trail is generally useable by the intermediate level skier, except for several difficult spots as marked on the map. These spots require advanced skill levels. The trail is groomed and tracked regularly.

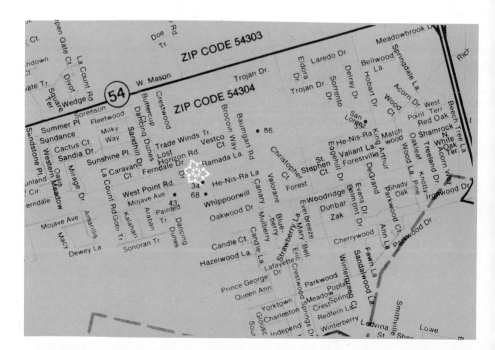

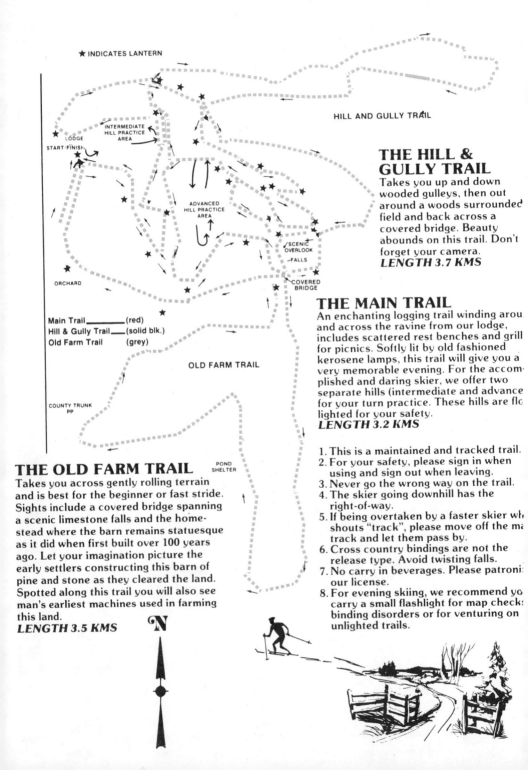

★ INDICATES LANTERN

HILL AND GULLY TRAIL

LODGE
START/FINISH

INTERMEDIATE
HILL PRACTICE
AREA

ADVANCED
HILL PRACTICE
AREA

SCENIC
OVERLOOK

FALLS

COVERED
BRIDGE

ORCHARD

Main Trail_____(red)
Hill & Gully Trail___(solid blk.)
Old Farm Trail ___(grey)

OLD FARM TRAIL

COUNTY TRUNK
PP

POND
SHELTER

THE HILL & GULLY TRAIL

Takes you up and down wooded gulleys, then out around a woods surrounded field and back across a covered bridge. Beauty abounds on this trail. Don't forget your camera.
LENGTH 3.7 KMS

THE MAIN TRAIL

An enchanting logging trail winding arou and across the ravine from our lodge, includes scattered rest benches and grill for picnics. Softly lit by old fashioned kerosene lamps, this trail will give you a very memorable evening. For the accom- plished and daring skier, we offer two separate hills (intermediate and advance for your turn practice. These hills are flo lighted for your safety.
LENGTH 3.2 KMS

1. This is a maintained and tracked trail.
2. For your safety, please sign in when using and sign out when leaving.
3. Never go the wrong way on the trail.
4. The skier going downhill has the right-of-way.
5. If being overtaken by a faster skier wh shouts "track", please move off the ma track and let them pass by.
6. Cross country bindings are not the release type. Avoid twisting falls.
7. No carry in beverages. Please patroni our license.
8. For evening skiing, we recommend yo carry a small flashlight for map check binding disorders or for venturing on unlighted trails.

THE OLD FARM TRAIL

Takes you across gently rolling terrain and is best for the beginner or fast stride. Sights include a covered bridge spanning a scenic limestone falls and the home- stead where the barn remains statuesque as it did when first built over 100 years ago. Let your imagination picture the early settlers constructing this barn of pine and stone as they cleared the land. Spotted along this trail you will also see man's earliest machines used in farming this land.
LENGTH 3.5 KMS

N

Hilly Haven

Brown County

The three trails total approximately 10km with each being about 3.5. they are maintained and tracked. Two practice hills are floodlighted for your safety. The main trail is softly lit by kerosene lamps for a memorable night skiing experience.

Ski rentals are available as well as warming spirits and sandwiches in the cozy lounge with fireplace.

Daily 10 AM to Dark, Saturday 9-10 PM, Sunday 9-To Dark. Located 6.5 miles south of Depere on C.H. PP off State Highway 32. Groups are welcomed by appointment.

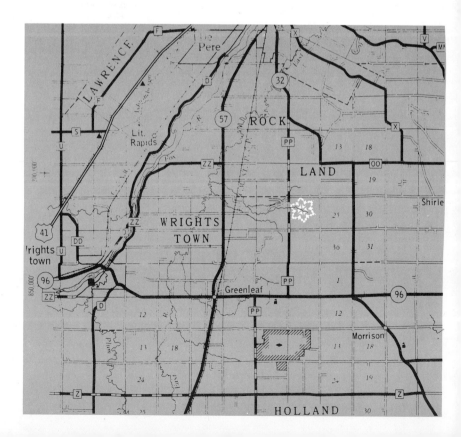

**BROWN COUNTY
REFORESTATION CAMP**

CROSS-COUNTRY SKI TRAILS

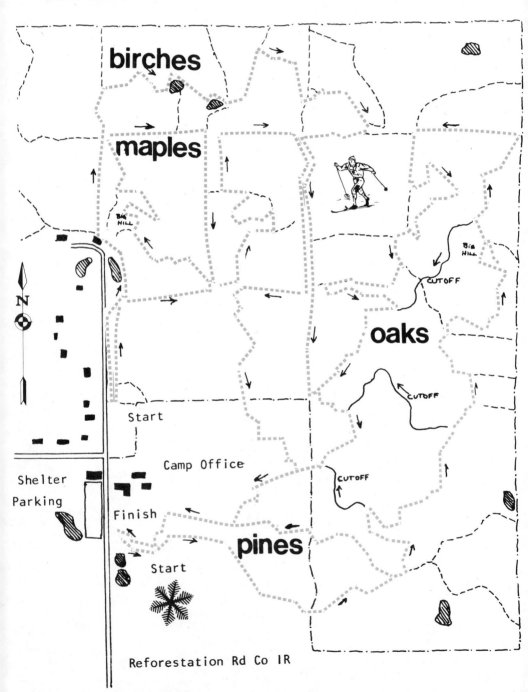

birches

maples

oaks

pines

N

Big Hill

Big Hill

CUTOFF

CUTOFF

CUTOFF

Start

Camp Office

Shelter

Parking

Finish

Start

Reforestation Rd Co IR

Reforestation Camp Trails

Brown County

The camp offers 21.7km (13.5 miles) of double-tracked, well groomed trails, which consists of four loops.

Skiing is offered from daylight until dark, seven days a week.

Take U.S. 41 north from Green Bay. Exit on Sunset Beach Rd. (County B). Go west to Reforestation Rd. (County 1R). Go north to Camp — 1 mile.

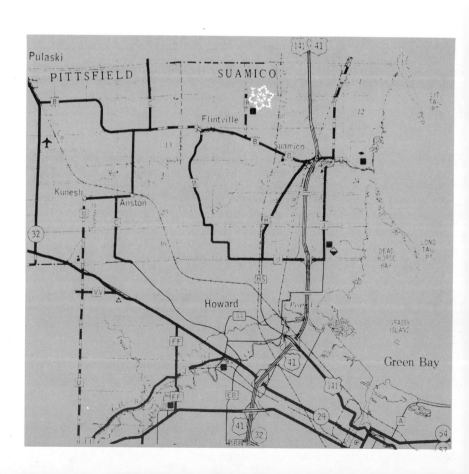

40

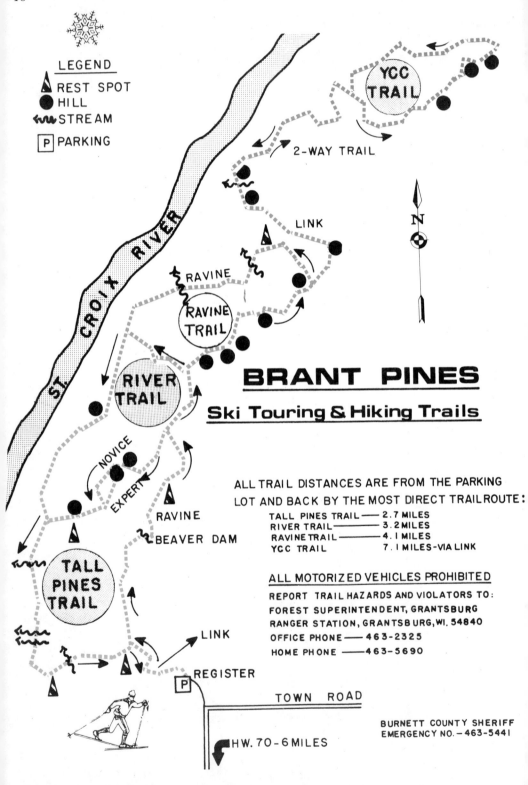

LEGEND

- ▲ REST SPOT
- ● HILL
- ∿ STREAM
- P PARKING

YCC TRAIL

2-WAY TRAIL

LINK

N

RAVINE

RAVINE TRAIL

RIVER TRAIL

NOVICE

EXPERT

RAVINE

BEAVER DAM

TALL PINES TRAIL

LINK

REGISTER

BRANT PINES

Ski Touring & Hiking Trails

ALL TRAIL DISTANCES ARE FROM THE PARKING
LOT AND BACK BY THE MOST DIRECT TRAIL ROUTE:

TALL PINES TRAIL	2.7 MILES
RIVER TRAIL	3.2 MILES
RAVINE TRAIL	4.1 MILES
YCC TRAIL	7.1 MILES - VIA LINK

ALL MOTORIZED VEHICLES PROHIBITED

REPORT TRAIL HAZARDS AND VIOLATORS TO:
FOREST SUPERINTENDENT, GRANTSBURG
RANGER STATION, GRANTSBURG, WI. 54840
OFFICE PHONE —— 463-2325
HOME PHONE —— 463-5690

ST. CROIX RIVER

TOWN ROAD

HW. 70 - 6 MILES

BURNETT COUNTY SHERIFF
EMERGENCY NO. — 463-5441

Brant Pines Trails

The Brant Pines Ski Touring and Hiking Trail system is located approximately 1½ miles west of Grantsburg on Highway 70, then north six miles to the trail head. A registration station is located on the trail just off the parking area and trail users are requested to sign the register.

These trails were designed and constructed specificially with cross-country skiers in mind. However, the trails are also used by backpackers and hikers in the summer months. All the trails run through woodlands and are gently to moderately rolling. There are four loops totaling 9.6km (5.5 miles) which are one way trails and a link of 1 mile which is two way. All trails are well marked, follow the white arrows. The trail system parallels the St. Croix National Scenic River and the trail has many scenic views.

Food and lodging is available in Grantsburg, which is located six miles south of the lodge.

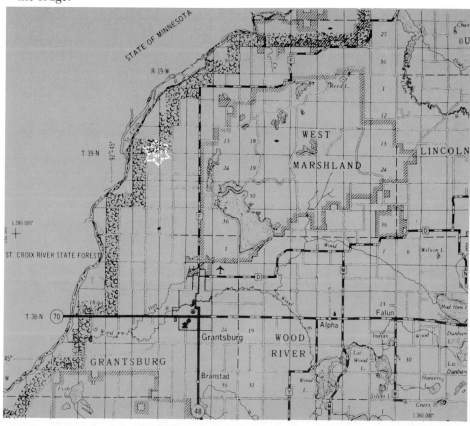

Nordic Ski Trail

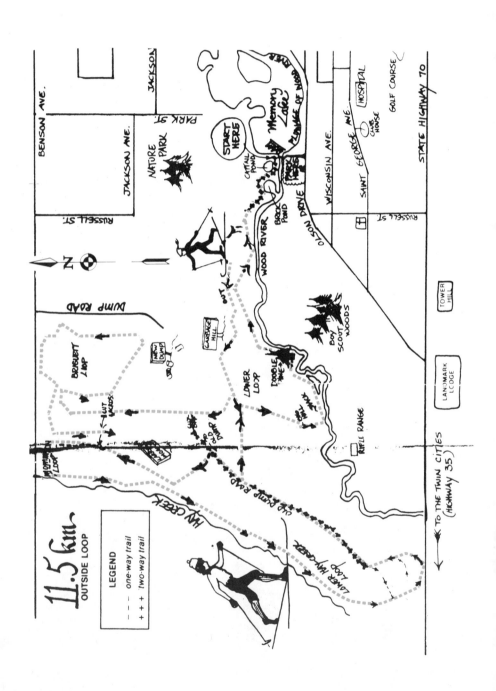

Grantsburg
Nordic Ski Trail

Burnett County

Called the "Big G" by many locals this readily accessible trail offers a great variety for the skier. Breath taking scenic winter setting. Long or short loops. Good speed on the larger hills and gentler "chicken runs" around them. Especially fun are the race track type banked corners on the bigger hills and the manmade series of moguls. Good layout and good grooming make this trail the most popular in the area.

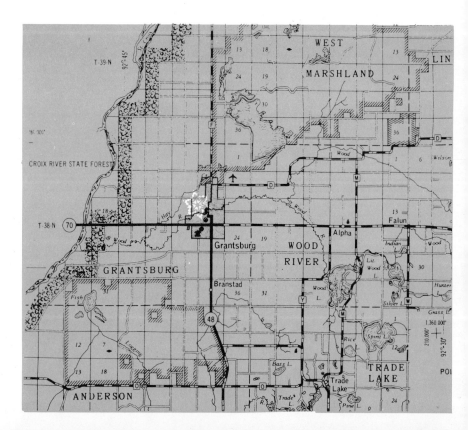

Grettum Ski Trail

Burnett County

This short figure eight loops gently through rolling terrain and its absence of large steep hills is perfect for those preferring less physical trails. Grettum is the local favorite for moonlight skiing. Look for the bear claw marks on the tall pines on the back loop of the trail.

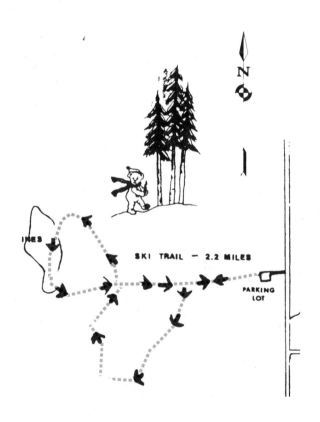

Grettum Ski Trail

Sandrock Trail

Burnett County

This is ideal for a short bushwhack trail of 2.5 mile. Because of the two moderate hills we recommend you being on the west side of the loop.

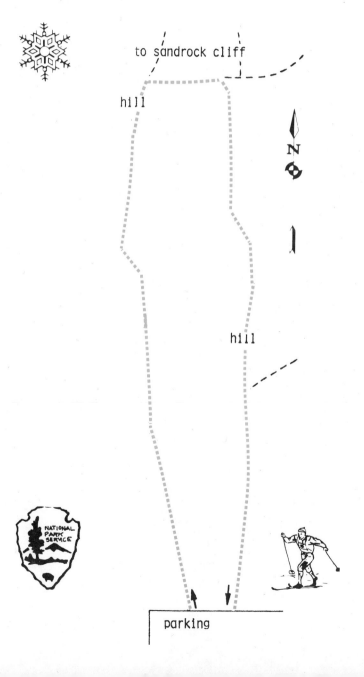

Loon Creek Ski Trails

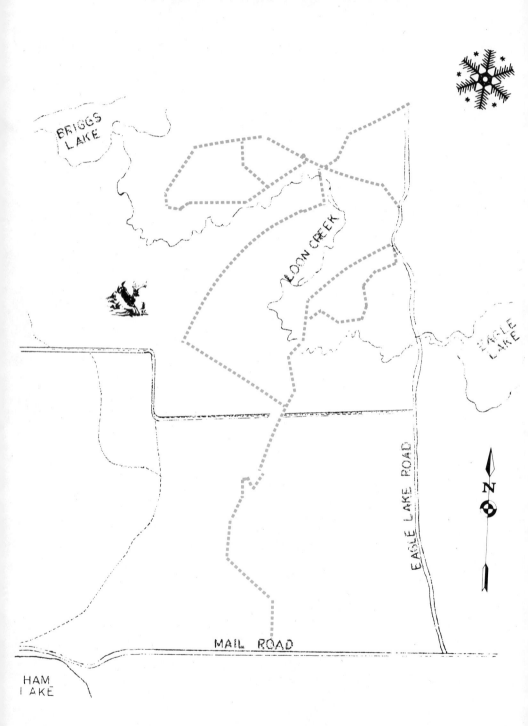

77 Series Ski Trails

These two trails are maintained by the county. They are being improved this summer by the Wisconsin Conservation Crew.

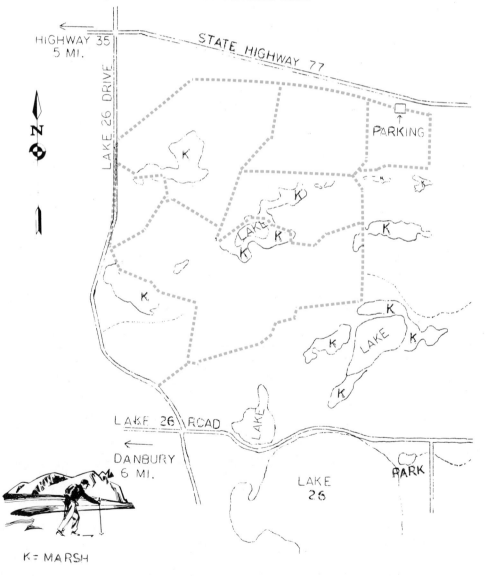

K = MARSH

Webster Ski Trail

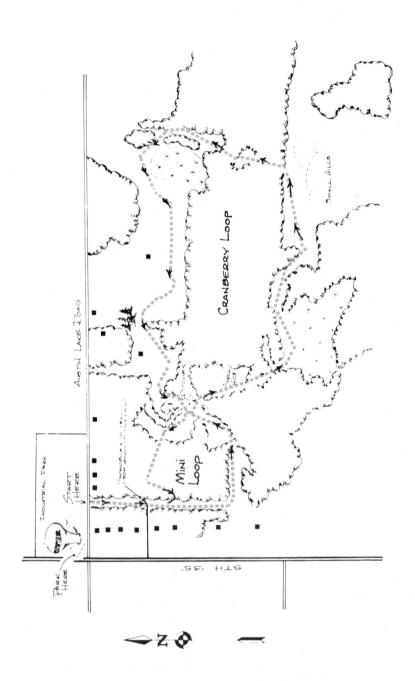

Webster Ski Trail

Burnett County

The Webster Jaycee's maintained this trail near the Industrial Park. Parking on Austin Lake Road and State 35.

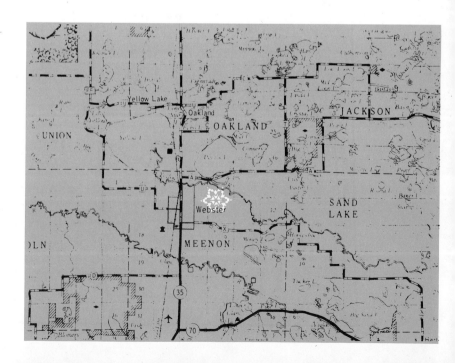

50

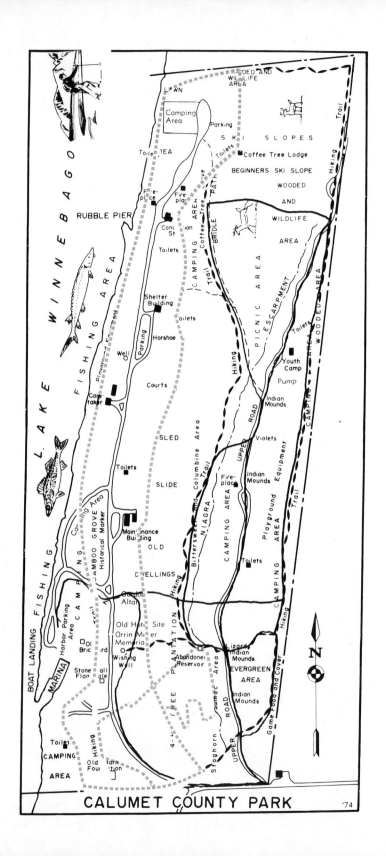

CALUMET COUNTY PARK '74

Calumet County Park

The Calumet Trail travels approximately 4.8km (3 miles) through wooded, gently rolling terrain that is geared to the novice and intermediate skier. A ½ mile trail was added '81, which enhances the historic past of the park.

Refreshments, shelter and toilets are available at the Coffee Tree Lodge. Parking is ample and ski equipment can be rented. The park is open from 6 A.M. to 11:00 P.M. and is located 17 miles southeast of Appleton on Hwy. 55, turning west on County E to Lake Winnebago.

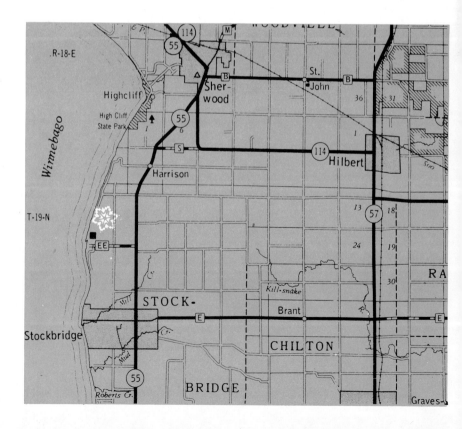

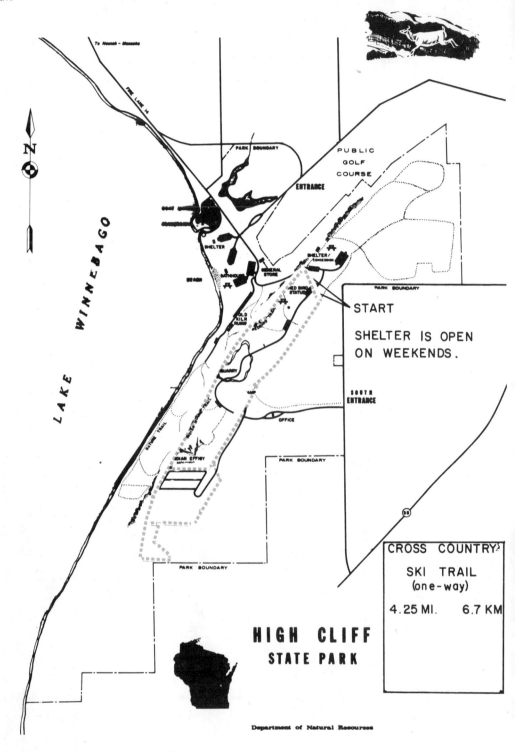

N

To Neenah - Menasha

FIRE LANE H

PARK BOUNDARY

PUBLIC
GOLF
COURSE

ENTRANCE

GOLF SHELTER

SHELTER

BATHHOUSE

GENERAL
STORE

BEACH

SHELTER/
CONCESSION

RED BIRD
STATION

START

SHELTER IS OPEN
ON WEEKENDS.

OLD
KILN
RUINS

QUARRY

CAMP

SOUTH
ENTRANCE

OFFICE

LAKE WINNEBAGO

HIKING TRAIL

INDIAN EFFIGY

PARK BOUNDARY

PARK BOUNDARY

PARK BOUNDARY

CROSS COUNTRY
SKI TRAIL
(one-way)

4.25 MI.   6.7 KM

HIGH CLIFF
STATE PARK

Department of Natural Resources

High Cliff State Park Trail

The 3 loops wind through 80 primarily wooded acreage.

It's mostly flat to gently rolling terrain of 6.7km (4.25 miles) takes the skier down into the quarry, east of the ridge and thru an open field.

A shelter is open on week-ends and is heated with a wood fireplace and is used as a warming house.

Access to the trail is available near the Red Bird Statue. Trails are marked and a park sticker is required. Park office is open from 7:45 A.M. to 4:30 P.M. daily and week-ends.

54

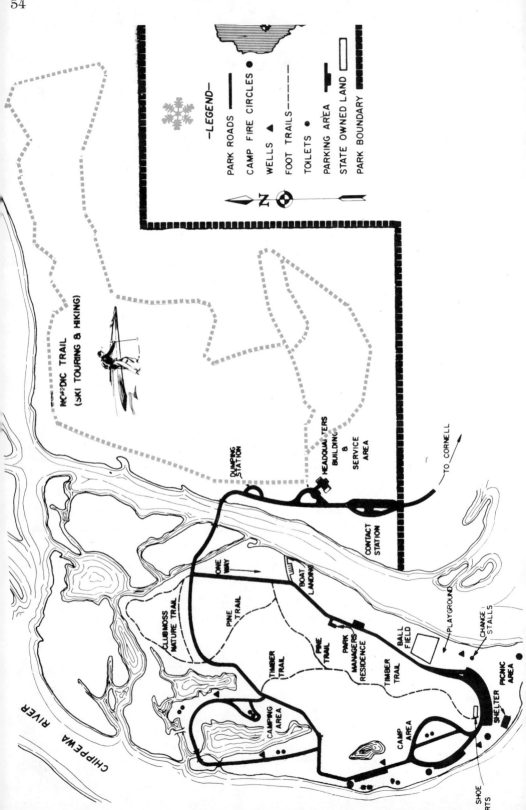

—LEGEND—

PARK ROADS ▬▬▬▬
CAMP FIRE CIRCLES ●
WELLS ▲
FOOT TRAILS ▬ ▬ ▬
TOILETS ●
PARKING AREA ▬▬
STATE OWNED LAND ▭
PARK BOUNDARY ▪▪▪▪

N

NORDIC TRAIL
(SKI TOURING & HIKING)

CHIPPEWA RIVER

PUMPING STATION

HEADQUARTERS
BUILDING &
SERVICE AREA

CONTACT STATION

TO CORNELL

ONE WAY

CLUBMOSS NATURE TRAIL

PINE TRAIL

PINE TRAIL

BOAT LANDING

PARK MANAGERS RESIDENCE

TIMBER TRAIL

TIMBER TRAIL

BALL FIELD

PLAYGROUND

CHANGE STALLS

PICNIC AREA

SHELTER

CAMPING AREA

CAMP AREA

SHOE RTS

Brunet Island State Park

Chippewa County

Adjacent to the park headquarters is the head of a marked 4.8km (3 miles) trail which is located on the "mainland" area of the park. The wooded and somewhat hilly loop includes orientation signs along the trail.

A plowed parking lot is available at the headquarters building. Park office is not open in winter. Location of park is 1 mile north of Cornell. A State Park admission sticker is not required from Nov. 1 to April 15.

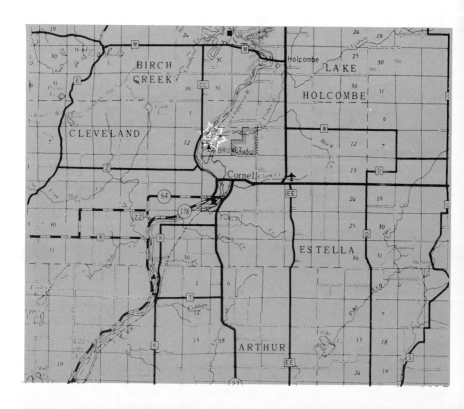

56

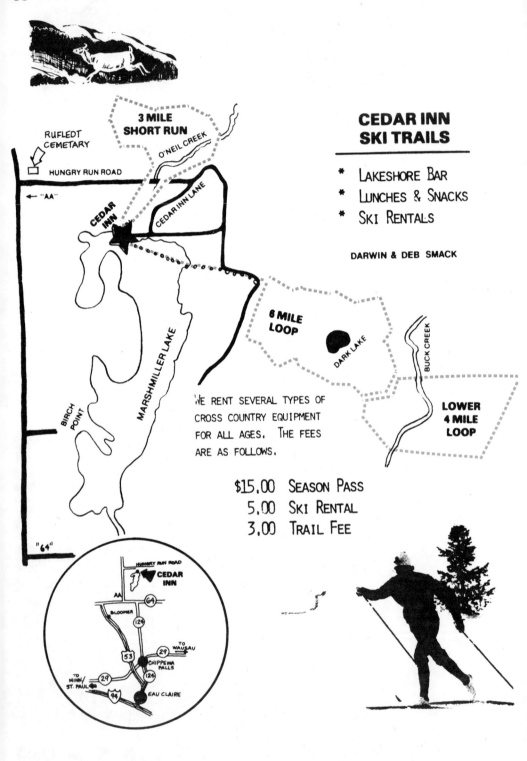

3 MILE SHORT RUN

RUFLEDT CEMETARY

O'NEIL CREEK

HUNGRY RUN ROAD

← "AA"

CEDAR INN

CEDAR INN LANE

CEDAR INN SKI TRAILS

* LAKESHORE BAR
* LUNCHES & SNACKS
* SKI RENTALS

DARWIN & DEB SMACK

6 MILE LOOP

DARK LAKE

BUCK CREEK

LOWER 4 MILE LOOP

MARSHMILLER LAKE

BIRCH POINT

WE RENT SEVERAL TYPES OF CROSS COUNTRY EQUIPMENT FOR ALL AGES. THE FEES ARE AS FOLLOWS.

$15.00 SEASON PASS
5.00 SKI RENTAL
3.00 TRAIL FEE

"64"

HUNGRY RUN ROAD

CEDAR INN

AA

64

BLOOMER

124

53

29

TO WAUSAU

CHIPPEWA FALLS

TO MINN/ ST. PAUL

29

124

94

EAU CLAIRE

Cedar Inn Ski Trails

Chippewa County

Cedar Inn Resort has approximately 14 miles of mapped, marked, and groomed trails. There are trails for skiers of all abilities.

The trails are divided into three loops (see map). Novice skiers enjoy the shortest run which is about 3 miles. For the more adventurous beginner, and intermediate to expert skiers, the 6 mile loop is the trail to ski. The lower 4 mile trail, because of its many hills, is recommended only for those considering themselves expert cross country skiers.

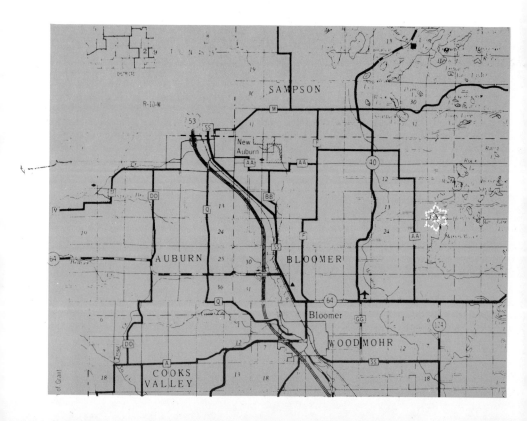

58

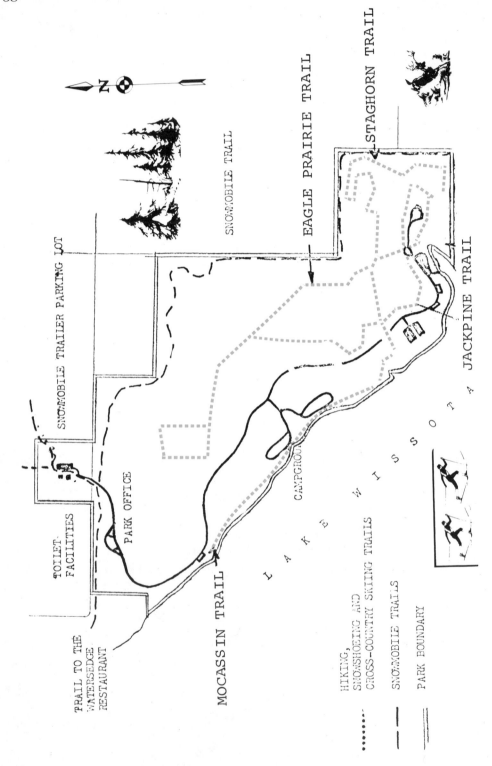

N

SNOMOBILE TRAILER PARKING LOT

SNOMOBILE TRAIL

EAGLE PRAIRIE TRAIL

STAGHORN TRAIL

JACKPINE TRAIL

L A K E   W I S S O T A

CAMPGROUND

PARK OFFICE

TOILET FACILITIES

TRAIL TO THE WATERSEDGE RESTAURANT

MOCASSIN TRAIL

HIKING,
SNOWSHOEING AND
CROSS-COUNTRY SKIING TRAILS

SNOMOBILE TRAILS

PARK BOUNDARY

Lake Wissota State Park Trails

Chippewa County

Lake Wissota offers the skier 4 trails of various lengths with a total of about 17.7km (11 miles). It is located 6 miles northeast of Chippewa Falls.

Beaver Meadow Nature Trail is lowland forest and marsh suited for the beginner to intermediate skier and is 1 mile in length.

Moccasin Trail is a 1.7 mile trail that offers beautiful lake vistas and is suited for all levels of skiers.

Staghorn Trail, a proven favorite of skiers, this 2 mile trail winds through mixed conifers and hardwoods. The gentle rolling terrain makes it suitable for beginners and yet is challenging enough for the intermediate.

Eagle Prairie Trail, taking in the Park's extensive open prairie areas as well as a jaunt around a pine plantation, this 4 mile trail is suited to the hardiest skier looking for a longer trail.

Jackpine Trail, one mile loop through wooded area adds length and diversity to the Staghorn Trail and is a good warmup trail. All trails will be color coded for the new season.

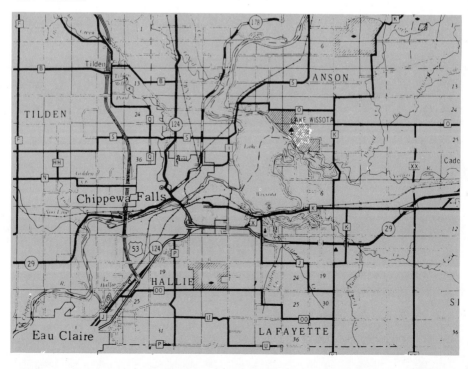

LEVIS MOUND
BRUCE MOUND
X-COUNTRY
SKI TRAILS

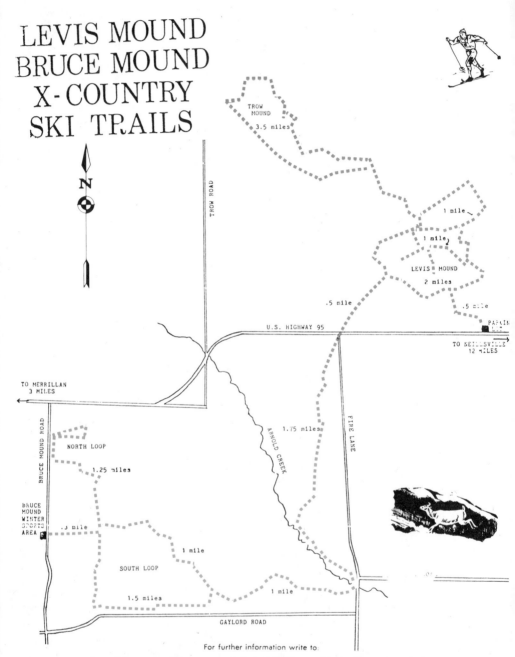

N

TROW
MOUND
3.5 miles

1 mile

1 mile

LEVIS MOUND
2 miles

.5 mile

.5 mile

TROW ROAD

U.S. HIGHWAY 95

PARKING
LOT

TO NEILLSVILLE
12 MILES

TO MERRILLAN
3 MILES

BRUCE MOUND ROAD

NORTH LOOP

1.25 miles

ARNOLD CREEK

1.75 miles

FIRE LANE

BRUCE
MOUND
WINTER
SPORTS
AREA

.3 mile

1 mile

SOUTH LOOP

1 mile

1.5 miles

GAYLORD ROAD

For further information write to:

CLARK COUNTY FORESTRY and PARKS DEPT.

DON KIRN, Administrator

Neillsville, Wisconsin

Phone 743-2490 (office) or 743-2296 (Bruce Mound during operating hours)

Levis Mound

The mound trails offer a total of 22.3km (14.5 mles) of pleasurable skiing.

It is geared to all type sof skiers for the trails are for beginners, intermediates and experts.

These gentle, rolling trails are well marked.

Plowed parking lots are available, also rentals and food.

The trails are located 3 miles east of Merridan or 15 miles southwest of Neillsville on Hwy. 95.

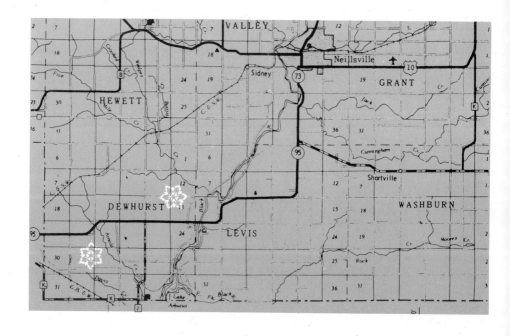

62

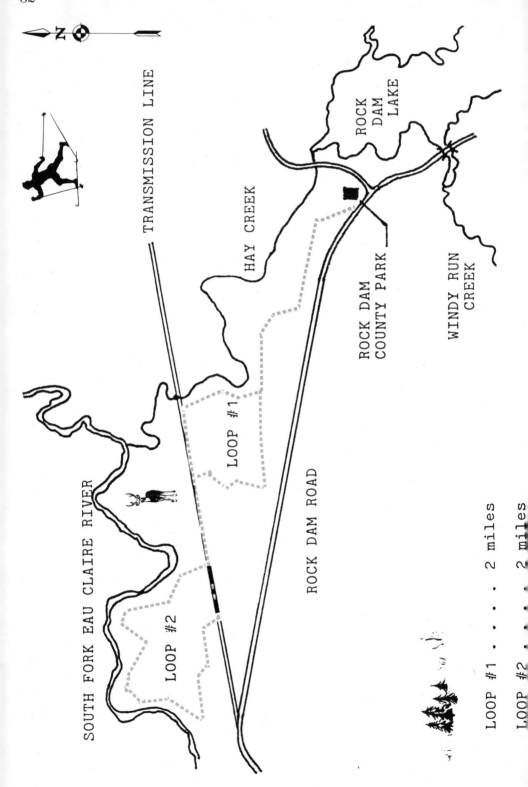

N

TRANSMISSION LINE

HAY CREEK

ROCK DAM LAKE

ROCK DAM COUNTY PARK

WINDY RUN CREEK

LOOP #1

ROCK DAM ROAD

SOUTH FORK EAU CLAIRE RIVER

LOOP #2

LOOP #1 . . . . . 2 miles

LOOP #2 . . . . . 2 miles

Rock Dam Trail

Clark County

Rock Dam is located 15 miles west of Greenwood or 14 miles south of Thorp at Rock Dam off County Highway M.

6.4km (4 miles) of marked trails are available. Trails are gentle and rolling.

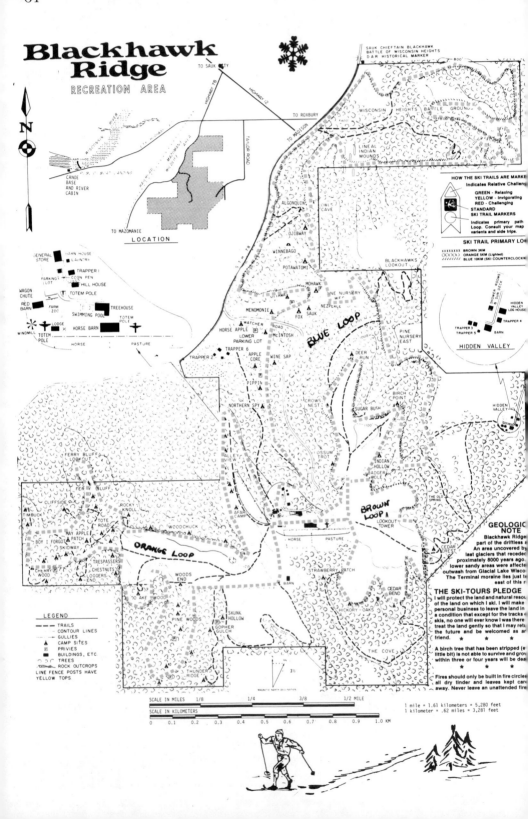

Blackhawk Ridge

RECREATION AREA

Blackhawk Ridge

Private membership only!

Dane County

Blackhawk Ridge is the first area in Wisconsin, exclusively devoted to cross-country skiing in the winter.

Blackhawk Ridge has over 55km (35.2 miles) of mapped trails; 25km (15.5 miles) are mapped, marked and groomed, divided into three loops, starting and ending at the trial head where a warming-house with a fireplace and snack bar is located. Double track is set on over 12 km. when appropriate.

The trails meander over unglaciated river bluffs overlooking the Wisconsin River Valley, thru forest of oak, pine, hickory and birch and over meadows. A lovely 5km (3.2 miles) trail, considered handle-able by beginners and intermediates and enjoyed by experts, is lighted for night-time skiing.

Rental equipment in all sizes from very little to very large is available. Group lessons are offered on a scheduled basis and private lessons by appointment.

Blackhawk Ridge also has the abilty to create man-made snow for many of the trails assuring good skiing even when Mother Nature forgets.

A limited number of overnite accommodations ranging from the roughing it rustic Trapper Cabins to the civilized Hill House and Barn House. In addition a "Bath House" with a sauna, therapeutic pool, showers and sun lamps. A great way to relax, unwind and loosen muscles tired from skiing.

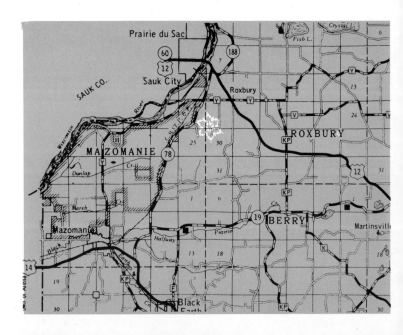

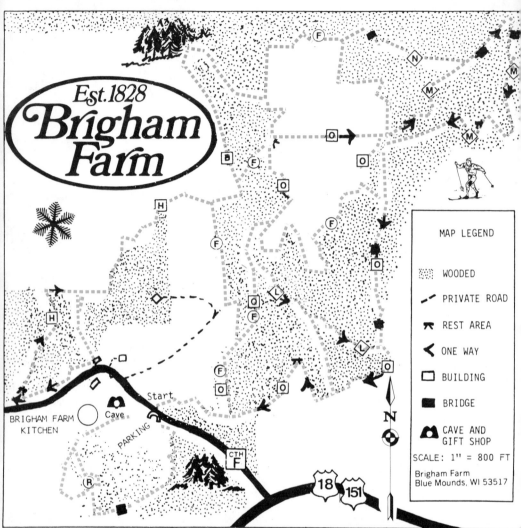

Est. 1828
Brigham Farm

MAP LEGEND

WOODED
PRIVATE ROAD
REST AREA
ONE WAY
BUILDING
BRIDGE
CAVE AND GIFT SHOP

SCALE: 1" = 800 FT

Brigham Farm
Blue Mounds, WI 53517

BRIGHAM FARM KITCHEN

Cave

Start

PARKING

CROSS COUNTRY SKI TRAIL MAP

EASIEST

(F) FARM TRAIL 2.6 miles (4.2 km.)

(R) RUNESTONE TRAIL .6 miles (1 km.)

MORE DIFFICULT

[H] HOG WOODS 1.5 miles (2.4 km)

[O] ORCHARD 2.3 miles (3.7 km)

[B] BRIGHAM .3 miles (.5 km)

ADVANCED

(N) NORTHLAND .28 miles (.45 km

(L) LOGGING .23 miles (.37 km

(M) MILLSTONE 1.2 miles (1.9 km

FOR RESERVATIONS AND TRAIL CONDITIONS CALL

(608) 437-3038

Brigham Farm

Dane County

Brigham Farm, located on the easternmost of the two Blue Mounds, was founded in 1828 by Ebenezer Brigham, the first pioneer to settle in Dane County. Today this 700 acre farm is the site of the Cave of the Mounds, the Brigham Hearth restaurant, Brigham Farm Gifts and Brigham Farm Cross-Country Ski Trails.

16.1km (10 miles) of marked and groomed cross-country trails wind through deep woods, along hillside fields, old orchards, logging roads, meadows and deep ravines. The Blue Mounds unglaciated area provides a wide variety of terrain for skiers of all abilities. A daily fee is charged. Rental skis are available. Open 9 a.m. to 4:30 p.m. daily except Monday. Located 20 miles west of Madison, just off U.S. Hwys. 18-151.

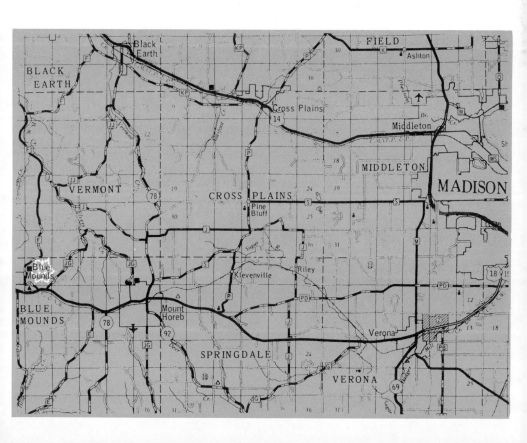

68

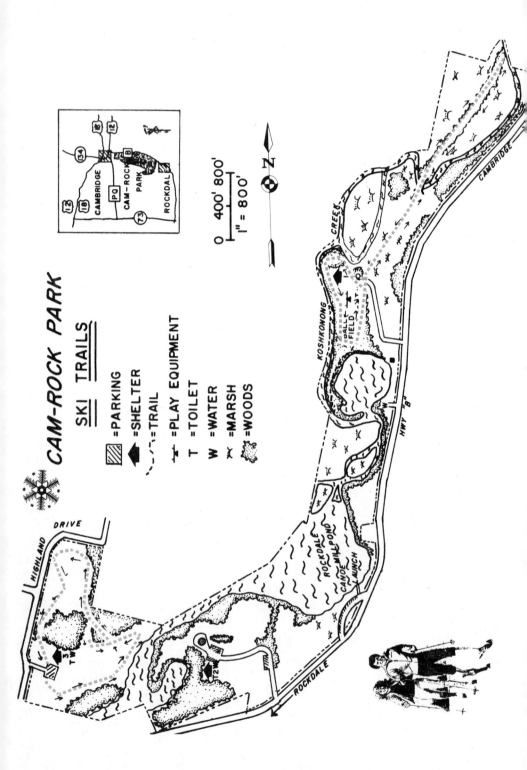

CAM-ROCK PARK

SKI TRAILS

▨ = PARKING

◢ = SHELTER

〰 = TRAIL

⊤ = PLAY EQUIPMENT

T = TOILET

W = WATER

☀ = MARSH

🌳 = WOODS

N

0 400' 800'

1" = 800'

CAMBRIDGE

ROCKDALE

CAMBRIDGE

ROCKDALE PARK

PQ

134

26 19

12 18

73

HIGHLAND DRIVE

ROCKDALE MILLPOND

CANOE MILLPOND LAUNCH

ROCKDALE

KOSHKONONG CREEK

BALL FIELD

HWY "B"

W

Cam-Rock Park

This 300-acre park site was purchased through the efforts of the Cambridge Foundation and the Dane County Park Commission. The park borders over two miles of the Koshkonong Creek from the Village of Cambridge south to the historical Village of Rockdale. At the present time facilities include three shelter houses, picnic areas, play equipment, a softball field, bike and nature trail, canoe launch, group camp area and cross country ski trails.

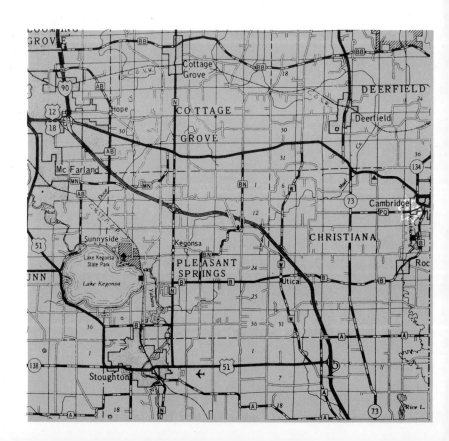

70

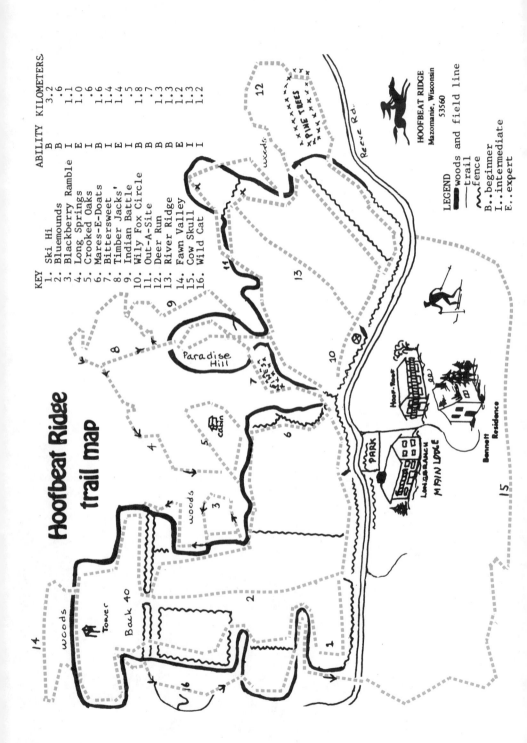

Hoofbeat Ridge
trail map

KEY		ABILITY	KILOMETERS
1.	Ski Hi	B	3.2
2.	Bluemounds	B	.6
3.	Blackberry Ramble	I	1.1
4.	Long Springs	E	1.0
5.	Crooked Oaks	I	.6
6.	Mares-E-Doats	B	1.6
7.	Bittersweet	I	1.4
8.	Timber Jacks'	E	1.4
9.	Indian Battle	I	1.5
10.	Wily Fox Circle	B	1.8
11.	Out-A-Site	I	.7
12.	Deer Run	B	1.3
13.	River Ridge	B	1.3
14.	Fawn Valley	E	1.2
15.	Cow Skull	I	1.3
16.	Wild Cat	I	1.2

HOOFBEAT RIDGE
Mazomanie, Wisconsin
53560

LEGEND

▬▬▬ woods and field line
▬▬▬ trail
∿∿∿ fence
B..beginner
I..intermediate
E..expert

Hoofbeat Ridge

Dane County

Hoofbeat is located just off Hwy. 14 about 27 miles west of Madison. 40km (25 miles) of trails on 250 acres of rolling hills and woodland is the setting for Hoofbeat. Trails range from smooth and easy for the beginner to winding and steep for the pro. Gliding over the soft snow in the moonlight is a delightful experience! Hoofbeat boasts the highest elevation in Dane County so it's naturally illuminated by the stars and moon.

Rental equipment and instructions are available. Daily trail use is charged.

There is hospitality, hot cocoa, hot cider and snacks at the end of your ski tour in the Longbranch Lounge. For information about the Family Ski Day, weekend packages for groups up to 75 people, or snow conditions call (608) 767-2593.

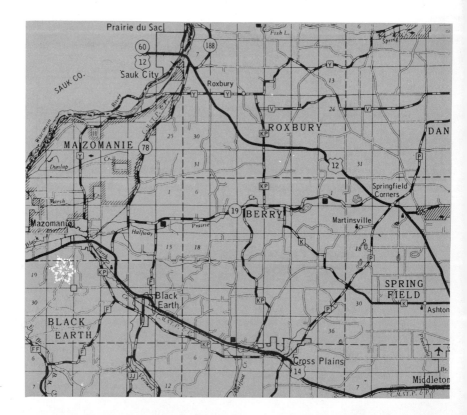

72

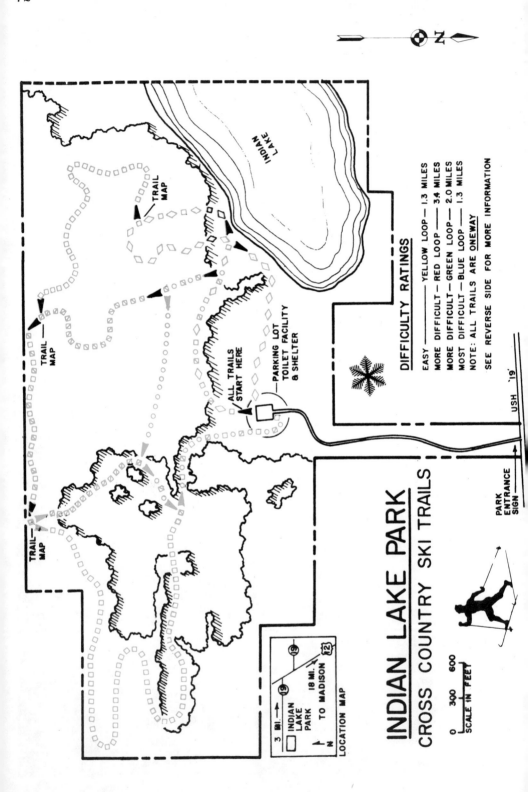

INDIAN LAKE
LAKE

TRAIL MAP

TRAIL MAP

TRAIL — MAP

TRAIL — MAP

ALL TRAILS
START HERE

PARKING LOT
TOILET FACILITY
& SHELTER

N

INDIAN LAKE PARK

CROSS COUNTRY SKI TRAILS

USH '19'

PARK
ENTRANCE
SIGN

LOCATION MAP

3 MI.
INDIAN
LAKE
PARK
19
19
12
18 MI.
TO MADISON
N

0   300   600
SCALE IN FEET

Indian Lake Park

Dane County

Indian Lake is the largest of the County Parks containing 442 acres. The size, outstanding natural resources and historic significance of this site offers a vast amount of diversity and recreational potential. Presently the park is undeveloped but does contain a historic chapel built in 1857 on a hilltop which commands a beautiful view of Indian Lake and the surrounding valley. In addition to the chapel, the park provides nature and cross country ski trails. Group camping is available upon request. It is located on State Highway 19 approximately four miles West of U.S. Highway 12.

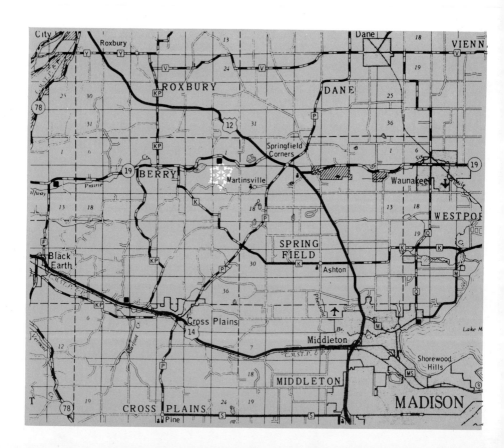

74

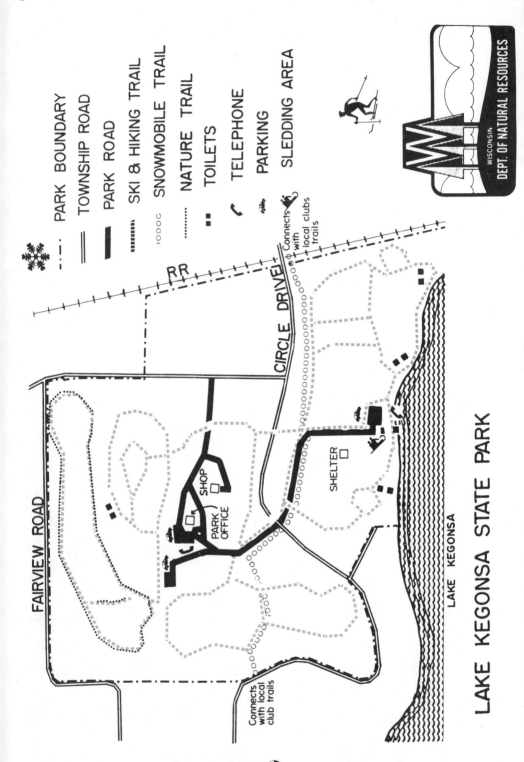

LAKE KEGONSA STATE PARK

PARK BOUNDARY
TOWNSHIP ROAD
PARK ROAD
SKI & HIKING TRAIL
SNOWMOBILE TRAIL
NATURE TRAIL
TOILETS
TELEPHONE
PARKING
SLEDDING AREA

WISCONSIN
DEPT. OF NATURAL RESOURCES

FAIRVIEW ROAD
RR
CIRCLE DRIVE
SHOP
PARK OFFICE
SHELTER
LAKE KEGONSA

Connects with local club trails
Connects with local clubs trails

Lake Kegonsa State Park

Dane County

Lake Kegonsa State Park offers about seven miles of trails to cross country skiers. The trail system transverses prairie, oakwood and marsh areas within this 342 acre park.

Parking areas are provided for park users and a state park admission sticker is required on all vehicles.

The park is located about three miles south of I-90 north of Stoughton on the northeast shore of Lake Kegonsa.

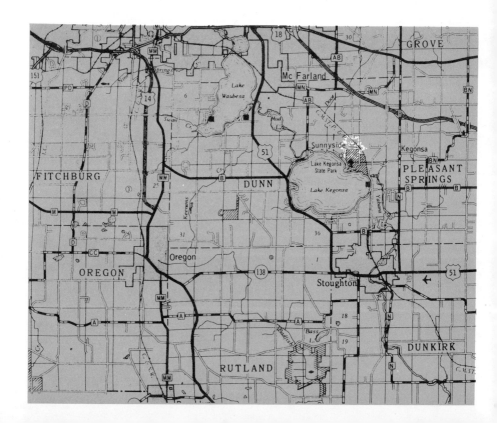

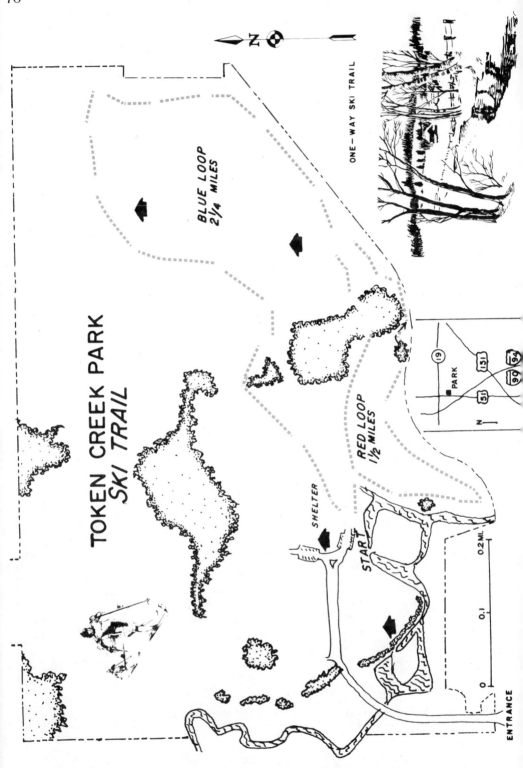

TOKEN CREEK PARK
SKI TRAIL

BLUE LOOP
2¼ MILES

RED LOOP
1½ MILES

SHELTER

START

ENTRANCE

0   0.1   0.2 MI.

ONE—WAY SKI TRAIL

N

PARK
19
51
90
94
51
90
94
N

Token Creek Park

Dane County

This 387-acre park located northeast of the City of Madison has been under construction in recent years. Although not fully completed the park offers a variety of recreational facilities including four shelter houses. Winter recreation facilities include a two and one-half mile cross country ski trail. The park entrance is on U.S. Highway 51, one quarter mile North of the Interstate Highway 90-94 Interchange.

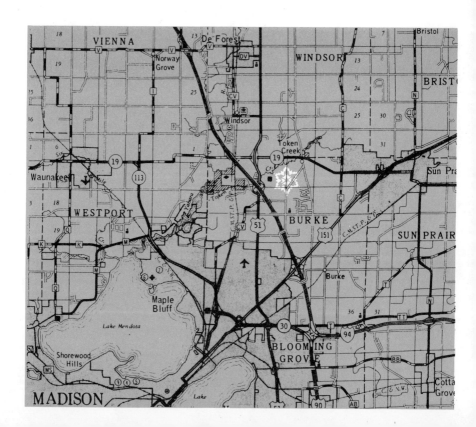

78

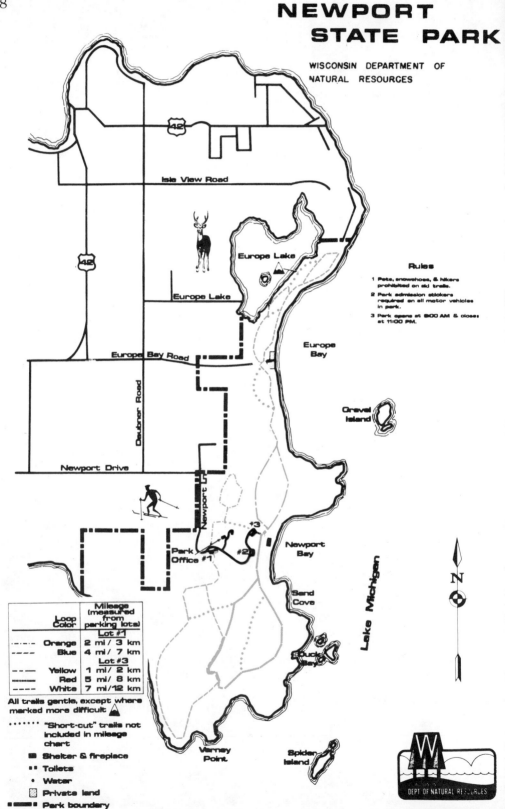

NEWPORT STATE PARK

WISCONSIN DEPARTMENT OF NATURAL RESOURCES

Isle View Road

Europe Lake

Europe Lake

Europe Bay Road

Deubner Road

Newport Drive

Newport Ln

Europe Bay

Gravel Island

Park Office #1

#3

#2

Newport Bay

Sand Cove

Duck Bay

Varney Point

Spider Island

Lake Michigan

N

Rules

1 Pets, snowshoes, & hikers prohibited on ski trails.

2 Park admission stickers required on all motor vehicles in park.

3 Park opens at 8:00 AM & closes at 11:00 PM.

Loop Color	Mileage (measured from parking lots)
	Lot #1
Orange	2 mi / 3 km
Blue	4 mi / 7 km
	Lot #3
Yellow	1 mi / 2 km
Red	5 mi / 8 km
White	7 mi / 12 km

All trails gentle, except where marked more difficult ▲

•••••• "Short-cut" trails not included in mileage chart

▣ Shelter & fireplace

∷ Toilets

• Water

▦ Private land

▰▰▰ Park boundary

DEPT OF NATURAL RESOURCES

Newport State Park

Door County

Newport State Park is comprised of 2000 acres of primarily wooded terrain. It is a very beautiful and secluded wilderness with forests of sugar maple, beech, aspen, birch, and pines. There are approximately 8 miles of Lake Michigan shoreline with numerous bays, long stretches of sand dunes, and a white cedar swamp. The ski trails, which are former logging roads, consist of 12 different trails and loops ranging from 0.5 miles to 3.5 miles with a total distance of 20 miles. They are gentle, relaxing trails with virtually no hils, ideal for the novice skier.

Thirteen wilderness campsites are available for back-packers. Grills and firewood are available and there are numerous toilets along the trails. No fee is currently required and snow conditions can be obtained by calling the Park Office at (414) 854-2500 on weekdays or Peninsula Park Office on weekends (414) 868-3258.

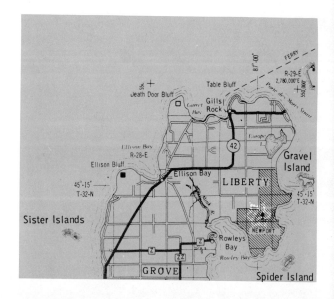

WELCOME TO
PENINSULA STATE PARK

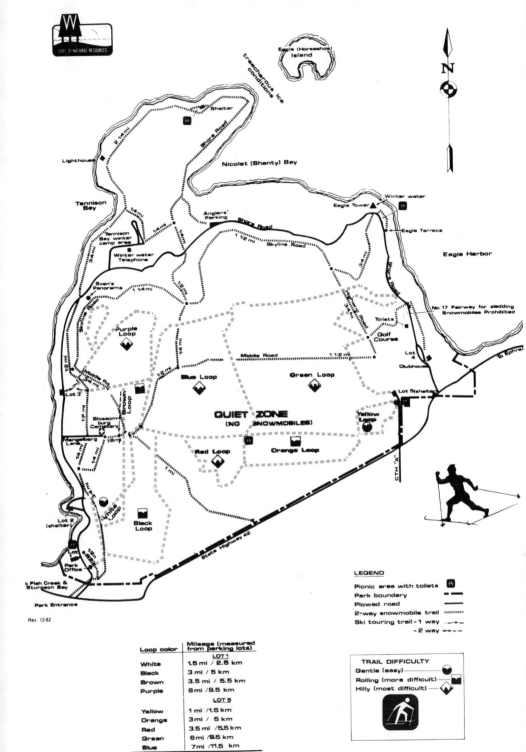

DEPT OF NATURAL RESOURCES

N

Eagle (Horseshoe) Island

treacherous ice conditions

Shelter

Shore Road

2 1/4 mi

Lighthouse

Nicolet (Shanty) Bay

Winter water

Eagle Tower

Tennison Bay

Anglers' Parking

Shore Road

Eagle Terrace

1/4 mi

1/4 mi

1/4 mi

Skyline Road

1 1/2 mi

Eagle Harbor

Tennison Bay winter camp area

3/4 mi

Winter water Telephone

Sven's Panorama

1 1/4 mi

Purple Loop

No 17 Fairway for sledding
Snowmobiles Prohibited

Toilets

Golf Course

Lot 4

Middle Road

1 1/2 mi

1/4 mi

Clubhouse

1/2 mi

Middle Rd

Lot 3

Blue Loop

Green Loop

Lot 5 (shelter)

Brown Loop

1/2 mi

Blossom-burg Cemetery

1 1/4 mi

QUIET ZONE
(NO SNOWMOBILES)

Yellow Loop

Mengelberg Lane

1/4 mi

1/4 mi

Red Loop

Orange Loop

1 mi

CTH "N"

White Loop

Black Loop

Lot 2 (shelter)

Lot 1

Park Office

State Highway 42

Fish Creek & Sturgeon Bay

to Ephraim

Park Entrance

Rev. 12-82

LEGEND

Picnic area with toilets	🚻
Park boundary	— ▪ —
Plowed road	————
2-way snowmobile trail	··········
Ski touring trail - 1 way	←—·→
- 2 way	←—·—→

Loop color	Mileage (measured from parking lots)
LOT 1	
White	1.5 mi / 2.5 km
Black	3 mi / 5 km
Brown	3.5 mi / 5.5 km
Purple	6 mi / 9.5 km
LOT 5	
Yellow	1 mi /1.5 km
Orange	3 mi / 5 km
Red	3.5 mi /5.5 km
Green	6 mi /9.5 km
Blue	7 mi /11.5 km

TRAIL DIFFICULTY

Gentle (easy) — ⬤

Rolling (more difficult) — ◼

Hilly (most difficult) — ◆

Peninsula State Park

Door County

Peninsula State Park encompasses 3800 acres of thickly wooded and, for the most part, gently rolling terrain. It is one of Wisconsin's most beautiful state parks with its picturesque Green Bay shoreline and numerous bluffs and bays. The trail consists of 9 different loops with varying degrees of difficulty, offering a total of 27.4km (17 miles) of rolling, and hilly trails.

A Wisconsin DNR sticker is required of all motor vehicles entering the Park. Dogs, snowshoes, and hikers are prohibited on the ski trails. The Park opens at 6:00 A.M. and closes at 11:00 P.M. daily. Winter camping is available at the Tennison Bay Campground located in the Park. Fires are permitted in the camping and picnic areas. Phone the Park Office (414) 868-3258 for current snow conditions. Two warming shelters are available on weekends.

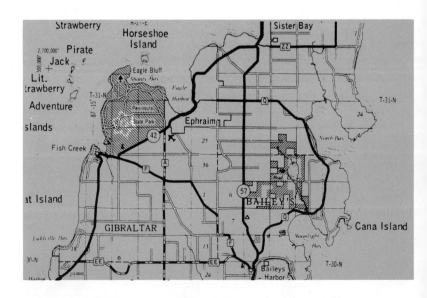

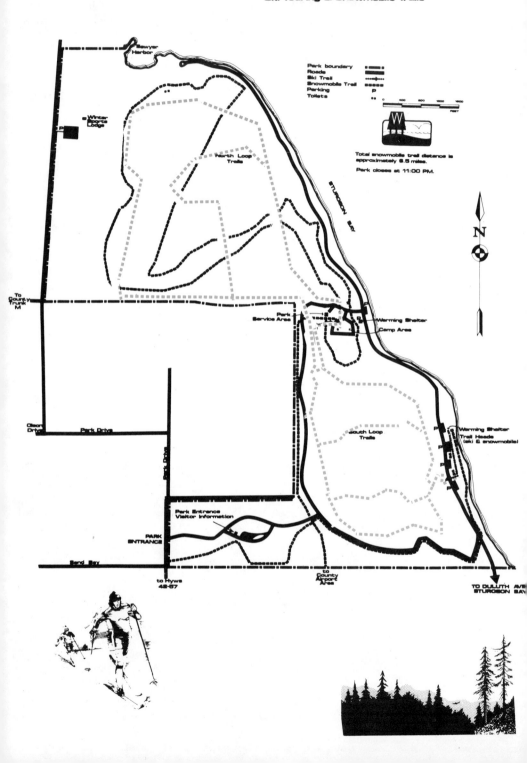

Welcome to
POTAWATOMI STATE PARK
Ski Touring & Snowmobile Trails

Potawatami State Park

Door County

Potawatami State Park is located 4 miles west of Sturgeon Bay off County Highway C.

A total of 10 miles of touring is offered. All trails are considered intermediate. Two warming shelters are available with firewood provided. Plowed parking is also provided.

A Wisconsin park sticker is required of all motor vehicles entering the park. Dogs, snowshoes and hikers are prohibited on ski trails. Winter camping is available in the electrical area of the campground. Water is available.

Phone No. 414-743-5123.

84

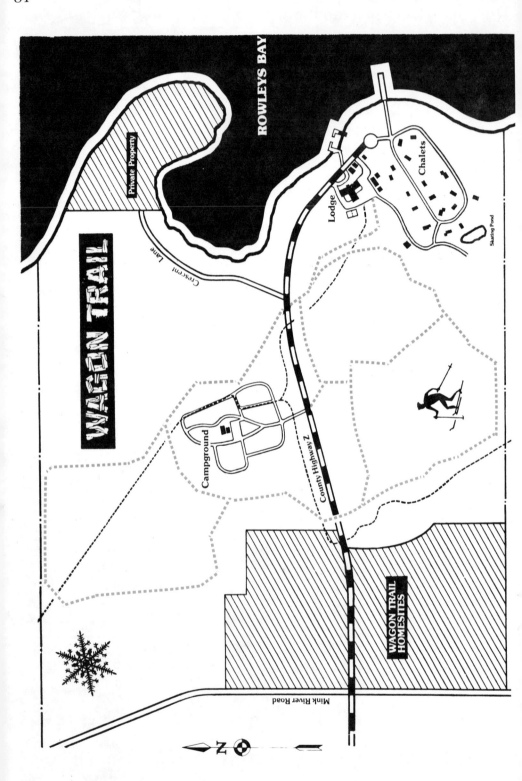

WAGON TRAIL

ROWLEYS BAY

Private Property

Crescent Lane

Lodge

Chalets

Skating Pond

Campground

County Highway Z

WAGON TRAIL HOMESITES

Mink River Road

N

Wagon Trail

Door County

Wagon Trail Resort and touring center is located five miles northeast of Sister Bay at the end of County Z. The resort complex overlooks Rowleys Bay on the Lake Michigan side of the Door Peninsula.

The trails are almost entirely wooded with virtually no hills, offering ideal skiing for the novice skier. There are approximately 8km (5 miles) of trails with loops ranging from .5 miles to 5.6km (3.5 miles). All trails are marked and maintained. Ice skating is also available on the premises. The complex features a 41 unit lodge with heated swimming pool, whirlpool, sauna, exercise gym, recreation room, styling salon and a gift-pro shop. Other accommodations include 21 year-round house-keeping chalets. The facility also provides dining in the lodge, offering good home cooking and baking.

Wagon Trail also provides complete cross-country and ice skating rental equipment for all ages, as well as ski instruction.

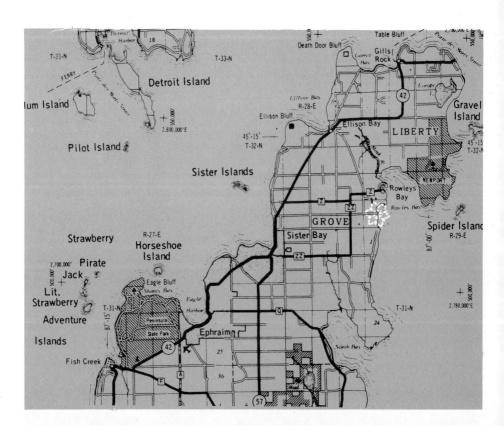

Brule River State Forest

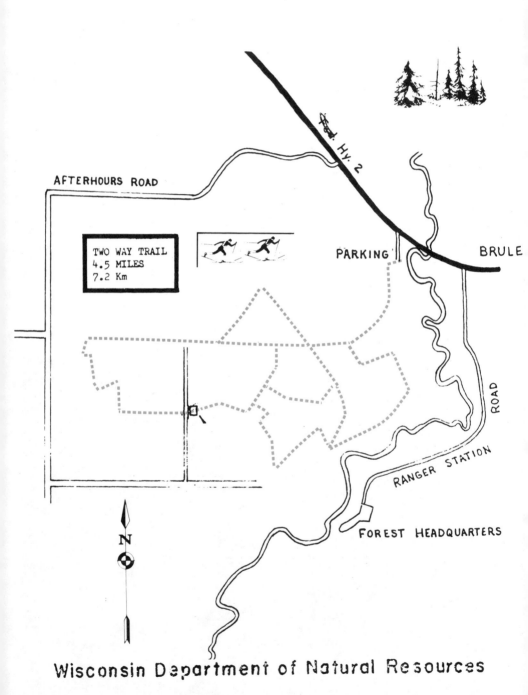

Hy. 2

AFTERHOURS ROAD

TWO WAY TRAIL
4.5 MILES
7.2 Km

PARKING

BRULE

RANGER STATION

ROAD

FOREST HEADQUARTERS

N

Wisconsin Department of Natural Resources

Brule River State Forest Trail

Douglas County

The loop consists of 7.2km (4.5 miles) of wooded two-way marked trails. Additional trails are being planned.

It is located 1 mile west of Brule on U.S. Highway 2.

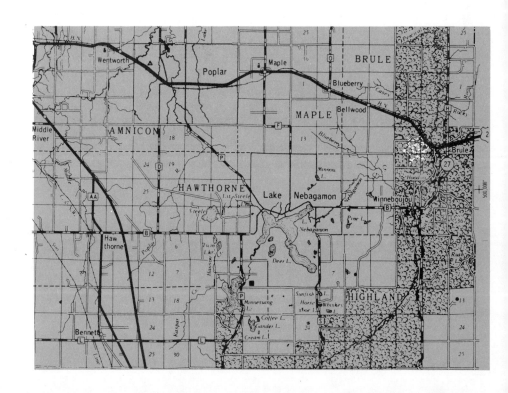

FOREST POINT RESORT

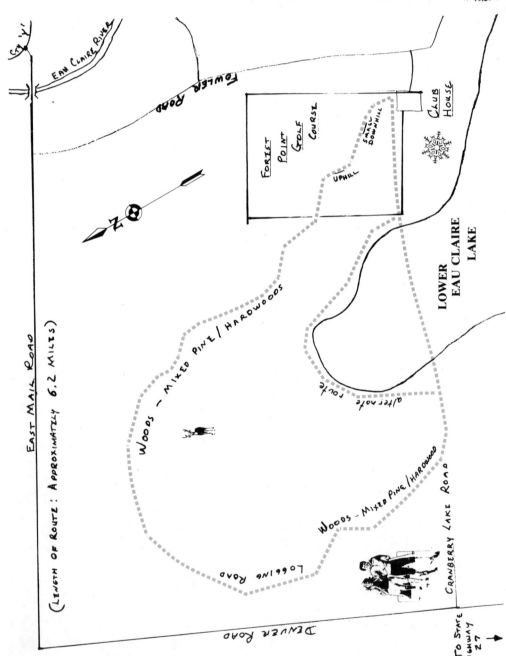

EAU CLAIRE RIVER

Fowler Road

FOREST POINT GOLF COURSE

SMALL DOWNHILL

UPHILL

CLUB HOUSE

N

EAST MAIN ROAD

(LENGTH OF ROUTE: APPROXIMATELY 6.2 MILES)

WOODS – MIXED PINE/HARDWOODS

LOWER EAU CLAIRE LAKE

alter nate route

WOODS – MIXED PINE/HARDWOOD

LOGGING ROAD

CRANBERRY LAKE ROAD

DENVER ROAD

TO STATE HIGHWAY 27

Forest Point Resort

The trail which begins on the golf course is 9.9km (6.2 miles) with some hilly slopes. It then leads off into old logging trails, which have a varied terrain. The return route back crosses a part of the lake, or the lake can be skirted on an alternate route. The trail, which is groomed and tracked, is surrounded by natural forests.

Cottages are available with fireplace and fully equipped kitchen.

Location — Traveling north on U.S. 53 turn right and east at Gordon onto County Highway Y. Follow their signs. If you come north·on Highway 27, go 17 miles north from Hayward, turn left and west onto Denver Rd. Follow signs.

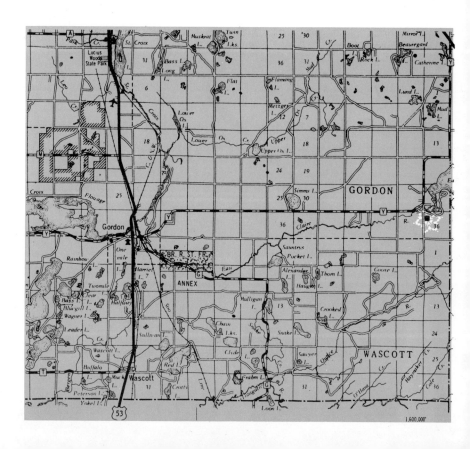

PATTISON STATE PARK
WISCONSIN DEPARTMENT OF NATURAL RESOURCES

CROSS COUNTRY SKI TRAIL

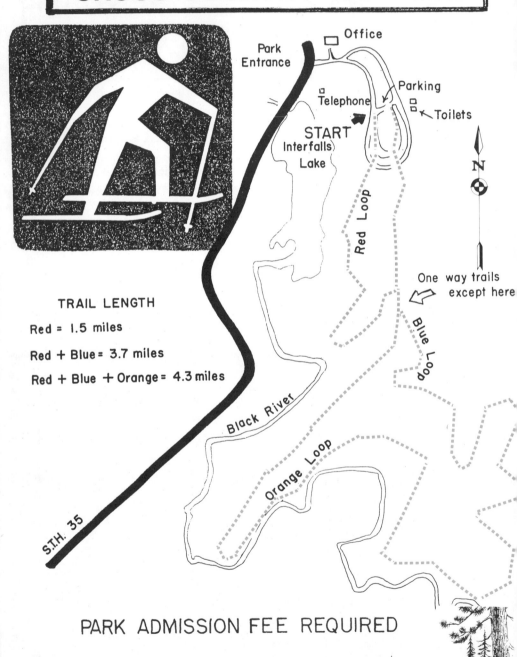

Office

Park
Entrance

Parking

Telephone

Toilets

START
Interfalls
Lake

Red Loop

N

One way trails
except here

Blue Loop

TRAIL LENGTH

Red = 1.5 miles

Red + Blue = 3.7 miles

Red + Blue + Orange = 4.3 miles

Black River

Orange Loop

S.T.H. 35

PARK ADMISSION FEE REQUIRED

Pattison State Park

Douglas County

Pattison Park Trail spreads over approximately 300 acres of land and has 3 one-way loops. The Red Trail is 2.4km (1.5 miles), Blue Trail 6.4km (4 miles), and the Orange Trail consists of 8.0km (5 miles). Follow the colored blazes on the trees for the loop you wish to ski. These trails are both rolling and wooded. This scenic park is known to have the highest waterfall in the State of Wisconsin.

Plowed parking and toilet facilities are available. It is located 12 miles south of Superior on State Highway 35. More information may be obtained by calling 715-399-8073.

State Park admission fees are required.

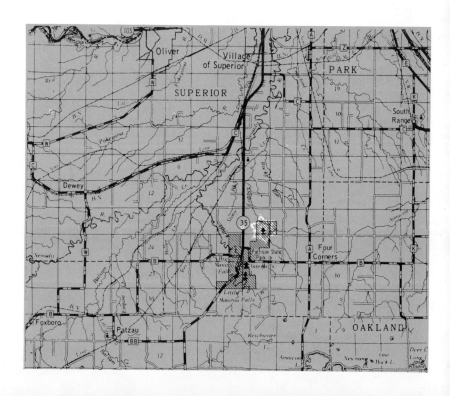

SUPERIOR CITY
FOREST

SKI TRAILS

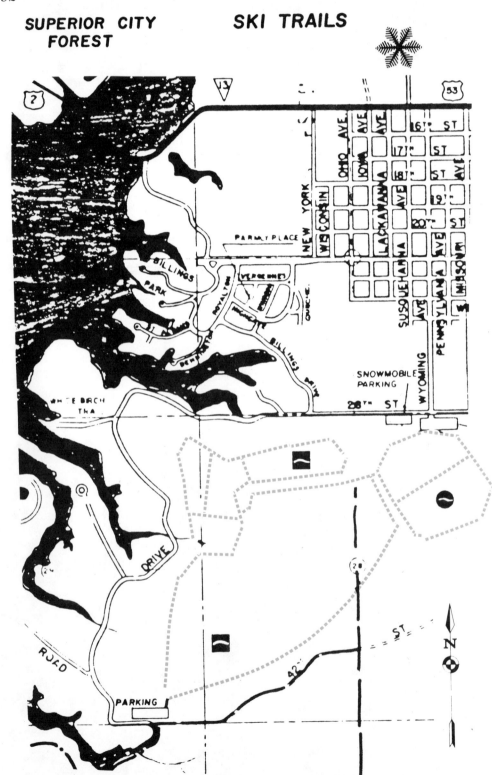

Superior City
Forest Trails

Douglas County

Trails are in the city of Superior ½ mile off Tower Avenue on N. 28th Street. 24km (15 miles) of marked groomed and tracked trails. Rolling, wooded with 2 plowed parking areas.

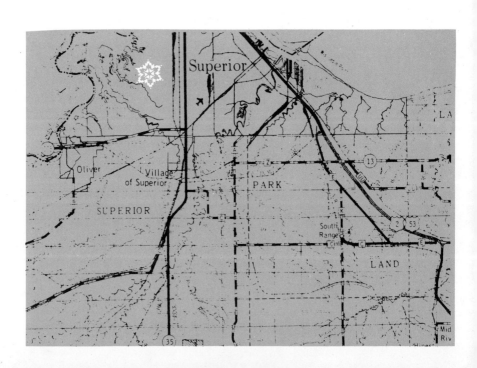

94

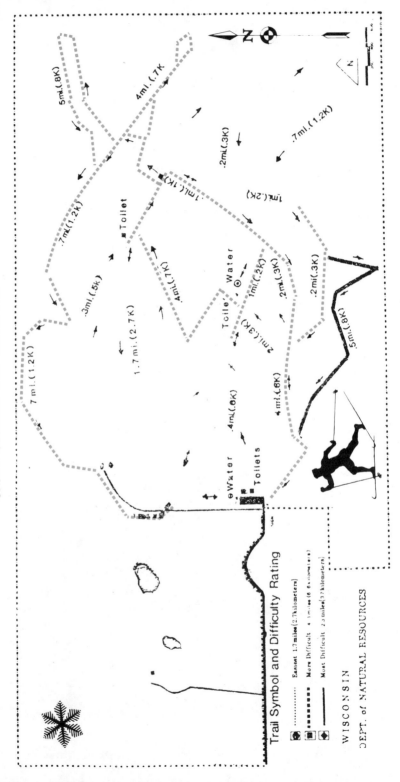

HOFFMAN HILLS STATE RECREATION AREA

CROSS-COUNTRY SKI TRAIL MAP

Trail Symbol and Difficulty Rating

········· Easiest 1.7 miles (2.7 kilometers)
▪▪▪▪▪▪▪▪ More Difficult 4.1 miles (6.6 kilometers)
━━━━━━ Most Difficult 2.3 miles (3.7 kilometers)

WISCONSIN
DEPT. of NATURAL RESOURCES

Hoffman Hills
State Recreation Area
Ski Trails

Dunn County

This 280 acres forms the nucleus of Hoffman Hills Recreation Area today, consisting of 400 acres with future expansion probable. It includes 8 miles of cross-country ski trails that wind atop hardwood ridges, climb up and down slopes covered with aspen and birch, and sneak through pine plantations.

Hoffman Hills offers skiing to all levels of skiers. The area is maintained by the state park system of Wisconsin and trails are groomed and tracked on a fairly regular basis.

A parking/picnic area marks the entrance to Hoffman Hills. To get there, take County Highway E west from State Highway 40 or E north from State Highway 29 in Dunn County. Then go north on Cedar Valley Road for 1¼ miles.

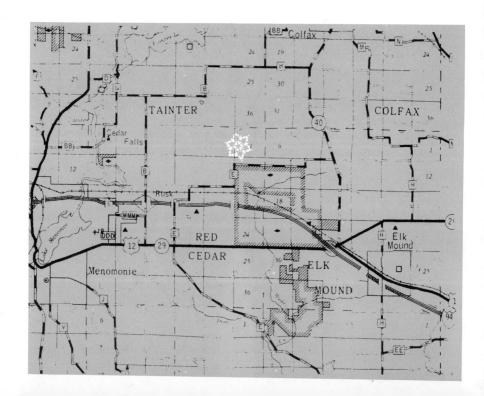

RED CEDAR
State Park Trail

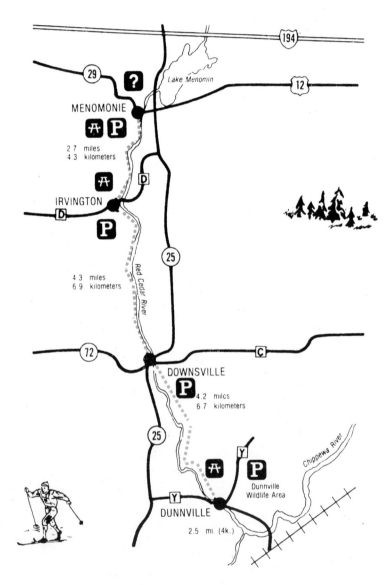

Red Cedar Trail

Dunn County

The Red Cedar Trail boarders the Red Cedar River throughout most of its run from Menomonie to the Chippewa River in Southern Dunn County. The Trail has been surfaced with limestone screenings and compacted to form a firm surface for bicycle riding and hiking.

In winter the Trail is an extremely popular spot for cross-country skiing because of its winter scenery. An abandoned railroad with little grade, the trail is especially suited to novice skiers.

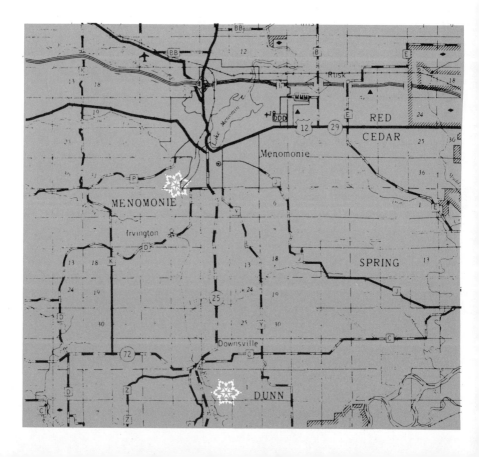

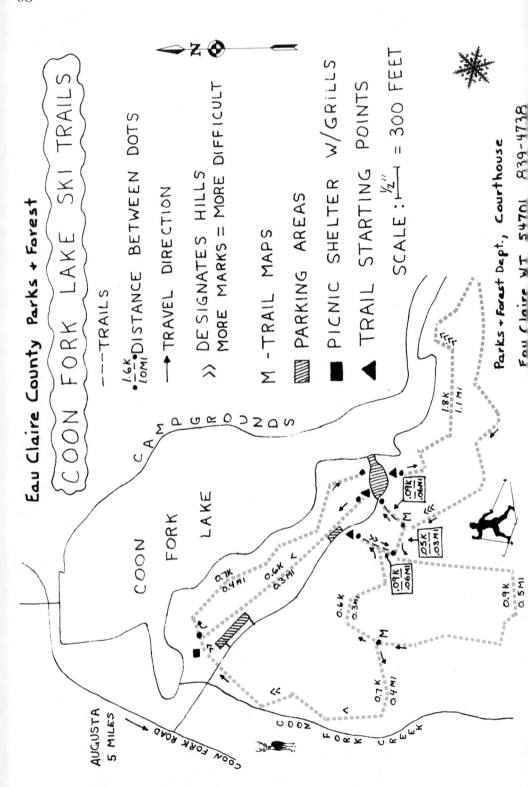

Eau Claire County Parks + Forest

COON FORK LAKE SKI TRAILS

---TRAILS

1.6K•——•10MI DISTANCE BETWEEN DOTS

➤ TRAVEL DIRECTION

≫ DESIGNATES HILLS
MORE MARKS = MORE DIFFICULT

M -TRAIL MAPS

▨ PARKING AREAS

■ PICNIC SHELTER W/GRILLS

▲ TRAIL STARTING POINTS

SCALE: ½" |——| = 300 FEET

Parks + Forest Dept., Courthouse

Eau Claire WI 54701 839-4738

CAMPGROUNDS

COON FORK LAKE

AUGUSTA
5 MILES

COON FORK ROAD

COON FORK CREEK

1.8K
1.1MI

.07K
.06MI

M

.05K
.03MI

.09K
.06MI

0.7K
0.4MI

0.6K
0.3MI

0.9K
0.5MI

0.6K
0.3MI

M

0.7K
0.4MI

Coon Fork Lake Ski Trail

Eau Claire County

Coon Fork Lake Ski Trail is designed for the beginner and intermediate skier and offers 5.1km (3.3 miles) of trail. The trails wind through Coon Fork Lake County Park, a pine forest, and offer a scenic view of Coon Fork Lake.

Facilities include parking areas, picnic shelter with grills, a "primitive" campground that is open year round, and pit toilets.

Coon Fork Lake is located 30 miles east of Eau Claire or 5 miles east of Augusta. Take Highway 12 east through Augusta approximately one mile and watch for Coon Fork Lake Park directional signs.

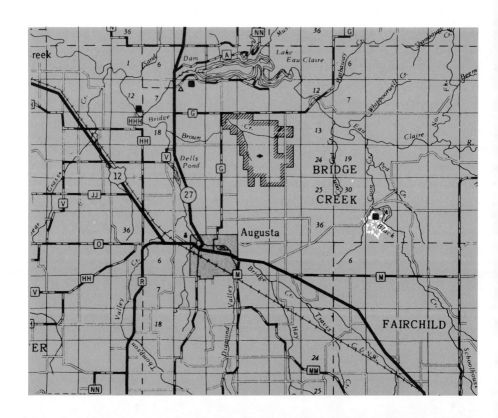

CROSS COUNTRY SKI TRAILS ❄

CARSON PARK
EAU CLAIRE, WISCONSIN

LEGEND

★ PARKING - TRAIL HEAD
1 WINTER DUCK REFUGE
2 FIRE PIT
3 PARK RANGER-MAINT. BLDGS
4 PAUL BUNYAN CAMP
5 CHIPPEWA VALLEY MUSEUM
6 WILD AREA

TRAIL LENGTH: APROX. 2 MILES

NOTE: SOME SECONDARY PARK ROA ARE CLOSED FOR THE WIN

SKING ON HALF MOON LA DONE AT YOUR OWN RISK

WELLS AREA
EAU CLAIRE, WISCONSIN

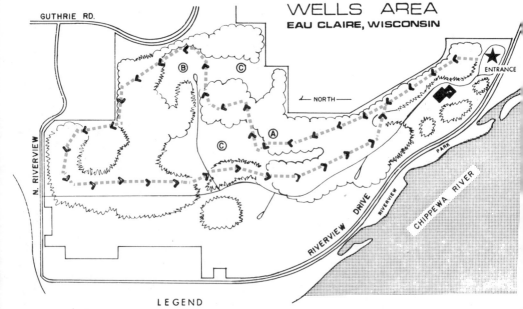

LEGEND

 PARKING TRAIL HEAD   C. PRAIRIE
A. DOWN HILL SLIDE   DECIDUOUS
B. RANGE BUILDINGS   CONIFEROUS

TRAIL LENGTH APROX. 3.5 MILES

Carson Park

Eau Claire County

The trail, which is marked and tracked, is approximately 3.2km (2 miles) in length.

Located in the city of Eau Claire, it offers a lovely wooded trail and is open 9:00 A.M. to 11:00 P.M.

Wells Area

Located in the city of Eau Claire, Wells Area offers 5.8km (3.5 miles) of wooded, well marked and tracked trail.

The park is open for skiing from 9:00 A.M. to 11:00 P.M.

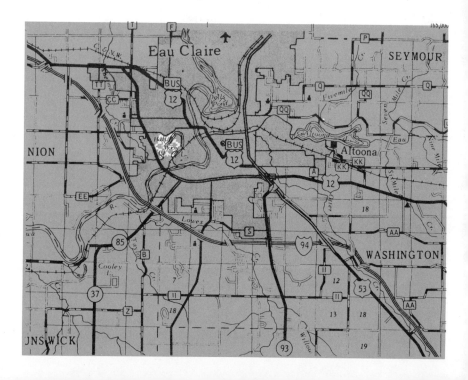

Eau Claire County Parks & Forest Department

LOWES CREEK COUNTY PARK SKI TRAIL

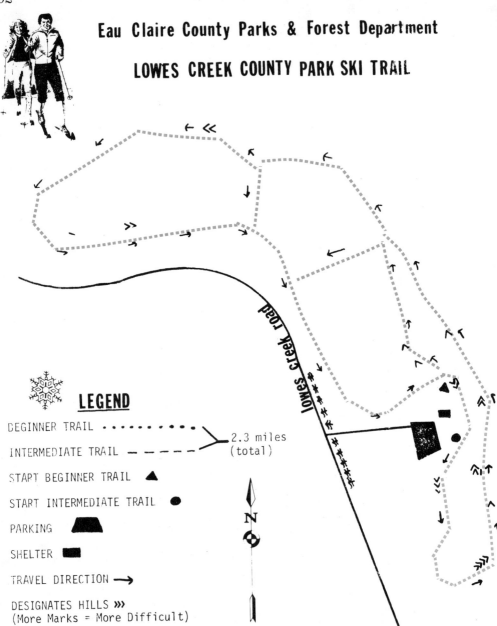

lowes creek road

LEGEND

BEGINNER TRAIL · · · · · · ·

INTERMEDIATE TRAIL — — — —

2.3 miles (total)

START BEGINNER TRAIL ▲

START INTERMEDIATE TRAIL ●

PARKING ▰

SHELTER ▬

TRAVEL DIRECTION →

DESIGNATES HILLS »
(More Marks = More Difficult)

N

ALL TRAILS ARE ONE WAY FOR SKIING, FOLLOW COLORED MARKERS

REPORT ALL TRAIL HAZARDS AND VIOLATORS TO:

Eau Claire County Parks & Forest Department
721 Oxford Avenue - Courthouse
Eau Claire, WI 54701   Tel. (715) 839-4738

Lowes Creek

County Park Ski Trail

Eau Claire County

Lowes Creek County Park ski trail in Eau Claire County offers 3.7km (2.3 miles) of trail for the beginner and intermediate skiier. The trail winds through a natural hardwood and pine forest along the banks of Lowes Creek.

Facilities include an information center and parking area.

Lowes Creek is located 15 minutes south of the City of Eau Claire and is accessible by driving south on State Street from the City to the junction of CTH F and Lowes Creek Road. Make a left turn on Lowes Creek Road and follow it to the park. Watch for park entrance signs.

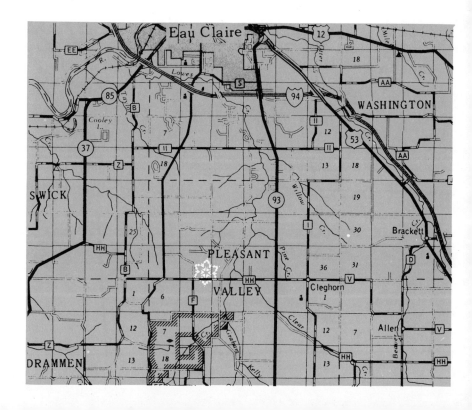

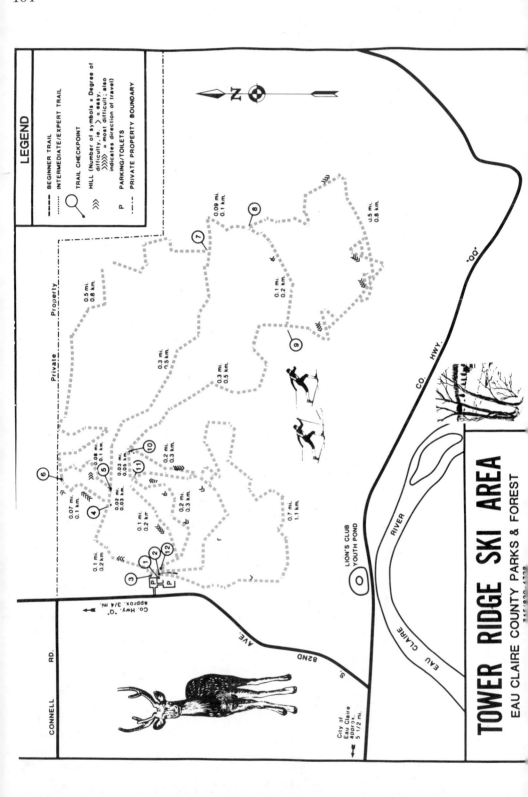

TOWER RIDGE SKI AREA
EAU CLAIRE COUNTY PARKS & FOREST

LEGEND

- – – – BEGINNER TRAIL
- ••••• INTERMEDIATE/EXPERT TRAIL
- TRAIL CHECKPOINT
- HILL (Number of symbols = Degree of difficulty, i.e. ≫ = easy, ≫≫≫ = most difficult; also indicates direction of travel)
- P PARKING/TOILETS
- –••– PRIVATE PROPERTY BOUNDARY

Private Property

0.5 mi.
0.8 km.

0.09 mi.
0.1 km.

0.5 mi.
0.8 km.

0.1 mi.
0.2 km.

0.3 mi.
0.5 km.

0.3 mi.
0.5 km.

0.08 mi.
0.1 km.

0.03 mi.
0.05 km.

0.2 mi.
0.3 km.

0.02 mi.
0.03 km.

0.2 mi.
0.3 km.

0.07 mi.
0.1 km.

0.1 mi.
0.2 km.

0.1 mi.
0.2 km.

0.7 mi.
1.1 km.

CONNELL RD.

Co. Hwy. 'Q'
approx. 3/4 mi.

82ND AVE.

S.

LION'S CLUB
YOUTH POND

City of
Eau Claire
approx.
5 1/2 mi.

CO. HWY.

'QQ'

EAU CLAIRE RIVER

Tower Ridge Ski Area

Eau Claire County's Tower Ridge Ski Area offers 18km (11.3 miles) of scenic trails for the beginner, intermediate and expert skier. The trails offer a peaceful, yet challenging, tour of the western end of the County forest. The trails wind through hardwood and pine forests and also a meadow.

Facilities include an information center, pit toilets, and a gravelled parking lot.

Tower Ridge is located 15 minutes east of Eau Claire. Take Highway 53 (Hastings Way) north to the Birch Street exit, make a right turn (east) on Birch Street to Highway "Q". Follow "Q" to South 82nd Avenue and turn right (south) on South 82nd Avenue. The parking area is three-quarters (¾) of a mile south of the "Q"-turn off.

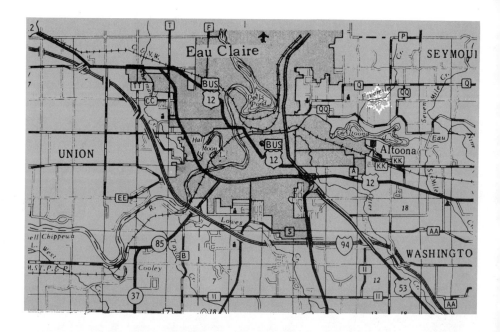

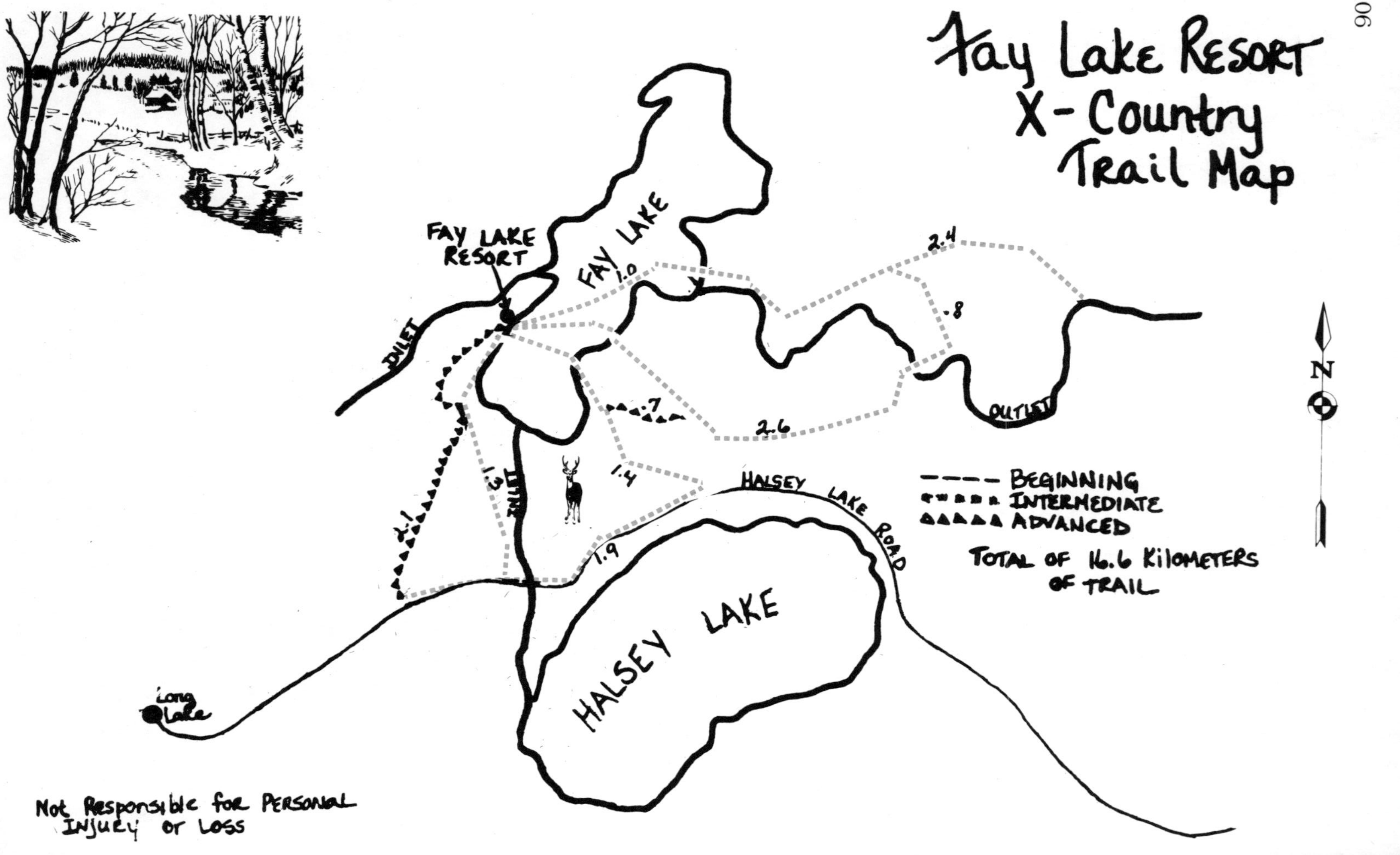

Fay Lake Resort
X-Country
Trail Map

N

Fay Lake Resort X-Country Trail Map

——— BEGINNING
••••• INTERMEDIATE
▲▲▲▲ ADVANCED

TOTAL OF 16.6 KILOMETERS
OF TRAIL

OUTLET

2.4

.8

2.6

HALSEY LAKE ROAD

FAY LAKE

1.0

.7

1.1

.9

FAY LAKE RESORT

INLET
1.3

INLET

1.7

HALSEY LAKE

Long Lake

NOT RESPONSIBLE FOR PERSONAL
INJURY OR LOSS

Fay Lake Resort

Located 2 miles north of Long Lake on 139. Resort offers accomodations with whirlpool, sauna, skating and food and beverage bar. Total of 16.6km of trails for beginners — intermediate and advanced skiers. Trails are well groomed.

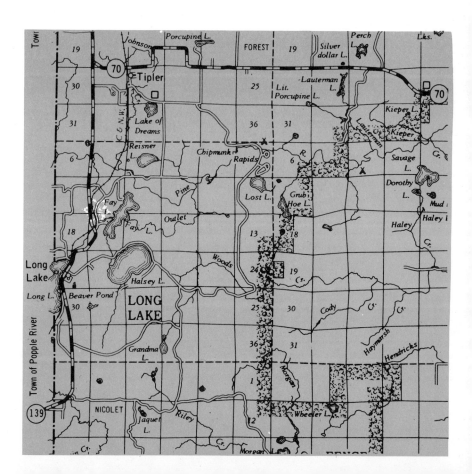

FLORENCE COUNTY CROSS COUNTRY SKI TRAILS

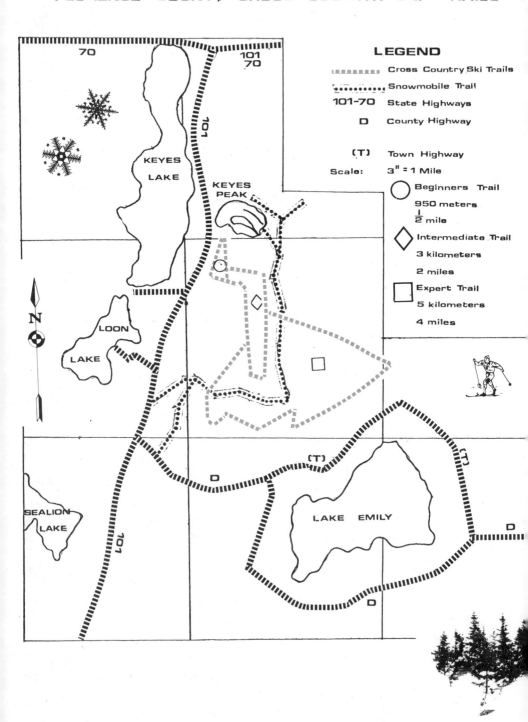

LEGEND

▪▪▪▪▪▪▪▪	Cross Country Ski Trails
•••••••••	Snowmobile Trail
101-70	State Highways
D	County Highway
(T)	Town Highway

Scale: 3" = 1 Mile

○ Beginners Trail
950 meters
½ mile

◇ Intermediate Trail
3 kilometers
2 miles

□ Expert Trail
5 kilometers
4 miles

70

101
70

101

KEYES
LAKE

KEYES
PEAK

N

LOON
LAKE

SEALION
LAKE

101

D

(T)

(T)

D

LAKE EMILY

D

Keyes Peak Trail

There are 3 groomed and marked trails offered at Keyes Peak for the beginner, intermediate and expert.

A total of 10km (6.4 miles) of gentle, hilly and wooded loops.

The trail is open 24 hours daily, with a ski chalet available on weekends.

Keyes Peak Trail is located 1 mile south of Lakewood on State Highway 32.

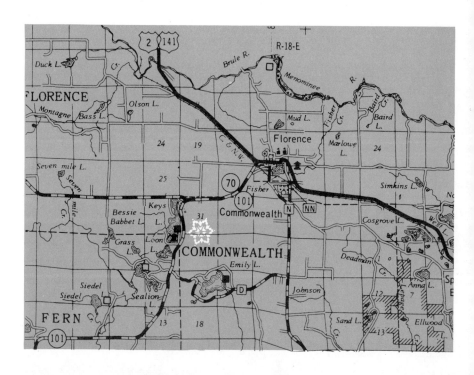

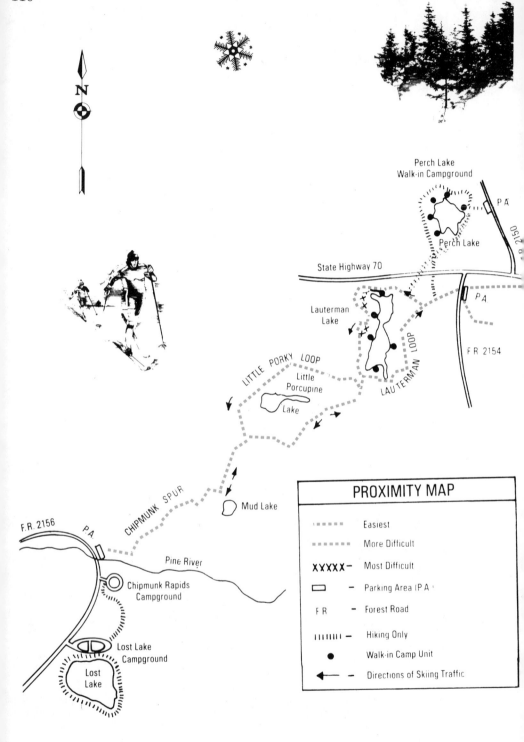

Perch Lake
Walk-in Campground

P A

F.R. 2150

Perch Lake

State Highway 70

P A

Lauterman
Lake

F R 2154

LITTLE PORKY LOOP

LAUTERMAN LOOP

Little
Porcupine
Lake

CHIPMUNK SPUR

Mud Lake

F.R. 2156

P.A.

Pine River

Chipmunk Rapids
Campground

Lost Lake
Campground

Lost
Lake

PROXIMITY MAP

▪ ▪ ▪ ▪ ▪	Easiest
▬ ▬ ▬ ▬	More Difficult
XXXXX —	Most Difficult
▭ —	Parking Area (P A)
F R —	Forest Road
‖‖‖‖‖ —	Hiking Only
●	Walk-in Camp Unit
◀ —	Directions of Skiing Traffic

Lauterman Trail

Florence County

This trail was constructed during 1978-1979 by members of the Youth Conservation Corps. and the Young Adult C.C.

Although portions of the 12.9km (8 miles) trail provide easy skiing as a whole, the trail requires intermediate skill. A trail section's degree of difficulty is indicated on the map. Arrows point the direction of skiing traffic. The trail is marked and tracked.

A 1.9km (1.2 mile) beginners loop was added in 1983.

Lauterman is located 13 miles west of Florence on State Highway 70 to Forest Road 2154.

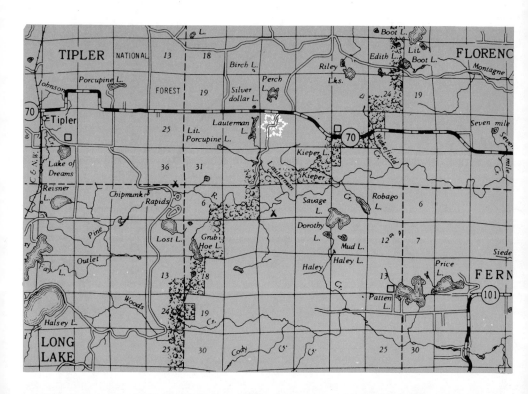

112

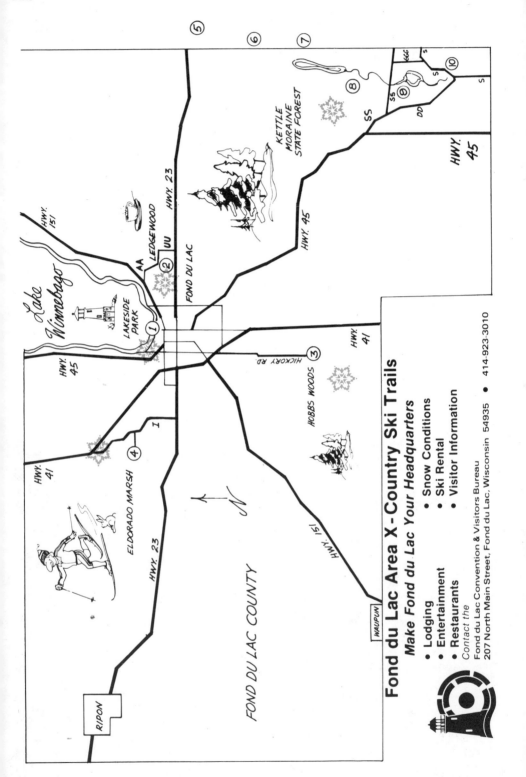

Fond du Lac Area X-Country Ski Trails

Make Fond du Lac Your Headquarters

- **Lodging**
- **Entertainment**
- **Restaurants**
- **Snow Conditions**
- **Ski Rental**
- **Visitor Information**

Contact the
Fond du Lac Convention & Visitors Bureau
207 North Main Street, Fond du Lac, Wisconsin 54935 • 414-923-3010

Fond du Lac Area X-Country Ski Trails

Fond du Lac County

Time a trip to Fond du Lac during our snow season and find yourself at the hub of a whole world of cross-country skiing! In town, around town, at the edge of town and just a few minutes beyond, there's ski terrain to challenge you whatever your level of accomplishment.

1. Lakeside Park — Fond du Lac's summertime showplace becomes wintertime's ski-wonderful straightaway for the beginner or intermediate, but don't be surprised to see experts working out on speed and endurance, all the while enjoying the snow-frosted beauty of 137.5 acres of park with winding canals, stone bridges and the magic that wintertime works on a Victorian bandstand and sentinel lighthouse. Lakeside Park embraces the southernmost reaches of Lake Winnebago, the largest inland lake in the state, 30 miles long and 18 miles at its widest...that's a lot of miles to ski on!

2. Ledgewood Country Club — Four and one-half miles of marked trails thread through a setting of hardwood and fir forest that flourishes on the Niagara Escarpment east of Fond du Lac. Here is varied terrain to ski with your own gear or, if you wish, rent skis, poles and boots at the pro shop. Instruction is available and afterward, there's apres ski atmosphere complete with food and drink and roaring hearthside.

3. Hobbs Woods Nature Study Area — Head for Hobbs Woods south of the city via Hickory Street Road. It's Fond du Lac's favorite summertime nature study spot, transformed to a favorite ski tour site with the first substantial snowfall. Families find it first-rate fun...easy trails, one to two miles in length, suited to young skiers just getting acquainted with their equipment.

4. Eldorado Marsh Wildlife Area — Here's territory for the stout-hearted! Begin your tour at the Fond du Lac River just off Seymour Street on the southwest side of the city. Ski the river for the next six to eight miles and you enter its headwaters, the 5,968 acres Eldorado Marsh. **Caution**...These unmarked trails test the mettle of the expert, but the rewards are well worth it. Underbrush and wetlands foster habitat for wildlife unequalled so near a city. Alternative access to the marsh may be gained by traveling west on WIS 23 to County Trunk I. The marsh lies to the west of County Trunk I north of 23 for about ten miles.

Kettle Moraine State Forest — Not too different from the way departing glaciers left it, so take time to call at Forest Headquarters and equip yourself with maps. Here are over 50 miles of trails that wind through miles of hardwood and evergreen forest, broad open meadows and around kettles and kames. Forest Headquarters, reached via U.S. 45 south from Fond du Lac, to County Trunk F on County Trunk G, is open from 8 a.m. to 4:30 p.m. weekdays and from 10 a.m. to 5 p.m. weekends. All the trails listed here are easily accessible from Kettle Moraine Scenic Drive with the exception of Tamarack Nature Trail which may be entered through the Forest Headquarters or Mauthe Lake Recreation Area.

See next 6 pages for these trails.

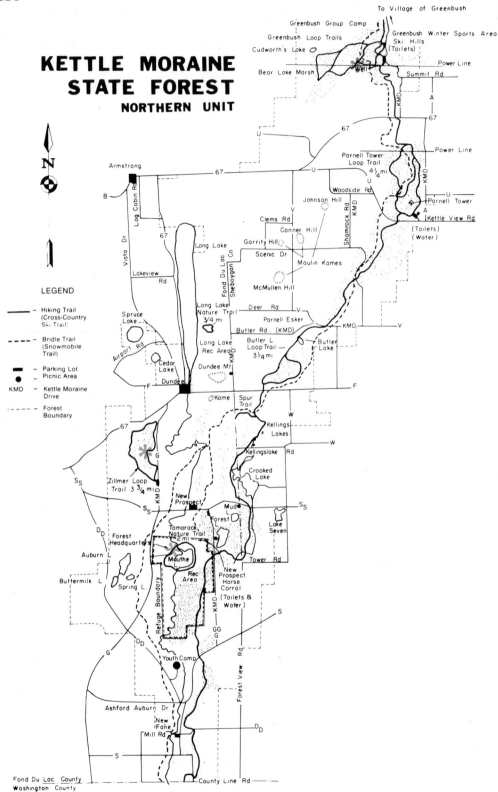

KETTLE MORAINE STATE FOREST
NORTHERN UNIT

N

LEGEND

— Hiking Trail (Cross-Country Ski Trail)

--- Bridle Trail (Snowmobile Trail)

▬ Parking Lot
● Picnic Area

KMD — Kettle Moraine Drive

--- Forest Boundary

To Village of Greenbush
Greenbush Group Camp
Greenbush Loop Trails
Greenbush Winter Sports Area Ski Hills (Toilets)
Cudworth's Lake
Bear Lake Marsh
Well
Power Line
Summit Rd
Power Line

Armstrong
Log Cabin Rd
Vista Dr
Lakeview Rd
Spruce Lake
Airport Rd
Cedar Lake
Dundee
Kame
67
B
U
U
67
Long Lake
Fond Du Lac Co
Sheboygan Co
Long Lake Nature Trail 3/4 mi
Long Lake Rec Area
Dundee Mt
F
F

Parnell Tower Loop Trail 4 1/4 mi
Woodside Rd
Johnson Hill
Clems Rd
Conner Hill
Garrity Hill
Scenic Dr
Moulin Kames
McMullen Hill
Deer Rd
Parnell Esker
Butler Rd (KMD)
Shamrock Rd
KMD
V
KMD
V
Parnell Tower
Kettle View Rd
(Toilets)
(Water)
U
A

Butler L Loop Trail 3 1/4 mi
Butler Lake
Spur Trail
Kellings Lakes
Kellingslake Rd
Crooked Lake
W
W

Zillmer Loop Trail 3 3/4 mi
KMD
New Prospect
Mud Forest
Lake Seven
S S
S S
D D
Forest Headquarters
Auburn L
Buttermilk L
Spring L
Tamarack Nature Trail 2 mi
Mauthe
Rec Area
New Prospect Horse Corral
(Toilets & Water)
Tower Rd
G
S

Refuge Boundary
GG
D D
G
Youth Camp
Forest View Rd
S

Ashford Auburn Dr
New Fane
Mill Rd
D D
S
County Line Rd

Fond Du Lac County
Washington County

Kettle Moraine State Forest
Northern Unit

Fond du Lac County

The Northern Unit of the Kettle Moraine was formed 12,000 to 15,000 years ago when Wisconsin was in the grip of the Ice Age. Monstrous, icy glaciers, miles thick at their points of origin, had moved out of Canada to cover much of Wisconsin. As the Great Wisconsin Glacier invaded the present day location of the Kettle Moraine Forest, two tongue-shaped lobes of ice, the Green Bay Lobe on the west and the Lake Michigan Lobe on the east, met along a line 120 miles in length! The resulting pressures, friction and buckling, as the two massive ice lobes "collided," caused the ice to melt and deposit tremendous loads of rocks, gravel and sand between the lobes. Thus, a long sinuous ridge was formed — the famous Interlobate Moraine.

Following is a list of the cross country trails in this area and a brief description of each trail. Maps of each are on the following pages.

GREENBUSH TRAIL — The Greenbush Trail is at the northern end of the Park, 2½ miles south of the town of Greenbush and 1.5 miles north of Hwy. 67. It begins and ends at the picnic area. Trails consist of one long loop of five miles with 3 cutoffs to make additional trails of 3.6 miles, 2 miles and .07 miles. Each trail is marked by a distinct color. The three shorter trails and much of the longer trail follow old logging roads with a beautiful view of Bear Lake. Trails are rated intermediate.

RAY ZILLMER TRAIL —

Green Loop — Beginners 2.9km (1.8 mile), Red — Intermediate is 4.8km (3.0 miles), Yellow — Advanced trail 8.6km (5.4 miles). Parking lot. Trails are one way marked and groomed.

The parking lot, located off FF, will accommodate 125 cars.

TAMARACK NATURE TRAIL — This trail is for the beginner, mostly flat and scenic. Ski thru three pine forests, over a walking bridge that spans a river. A nice relaxing trail of 2 miles that completely circles Mauthe Lake.

This trail is not groomed.

GREENBUSH TRAIL

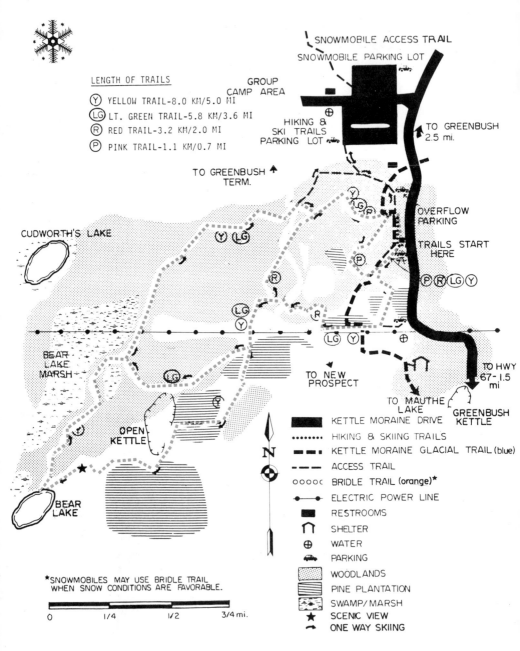

LENGTH OF TRAILS

- (Y) YELLOW TRAIL-8.0 KM/5.0 MI
- (LG) LT. GREEN TRAIL-5.8 KM/3.6 MI
- (R) RED TRAIL-3.2 KM/2.0 MI
- (P) PINK TRAIL-1.1 KM/0.7 MI

GROUP CAMP AREA

SNOWMOBILE ACCESS TRAIL
SNOWMOBILE PARKING LOT

TO GREENBUSH 2.5 mi.

HIKING & SKI TRAILS PARKING LOT

TO GREENBUSH TERM.

OVERFLOW PARKING

TRAILS START HERE

CUDWORTH'S LAKE

BEAR LAKE MARSH

TO NEW PROSPECT

TO MAUTHE LAKE

TO HWY 67-1.5 mi

GREENBUSH KETTLE

OPEN KETTLE

BEAR LAKE

N

KETTLE MORAINE DRIVE
HIKING & SKIING TRAILS
KETTLE MORAINE GLACIAL TRAIL (blue)
ACCESS TRAIL
BRIDLE TRAIL (orange)*
ELECTRIC POWER LINE
RESTROOMS
SHELTER
WATER
PARKING
WOODLANDS
PINE PLANTATION
SWAMP/MARSH
SCENIC VIEW
ONE WAY SKIING

*SNOWMOBILES MAY USE BRIDLE TRAIL WHEN SNOW CONDITIONS ARE FAVORABLE.

0   1/4   1/2   3/4 mi.

Zillmer Skiing Trails

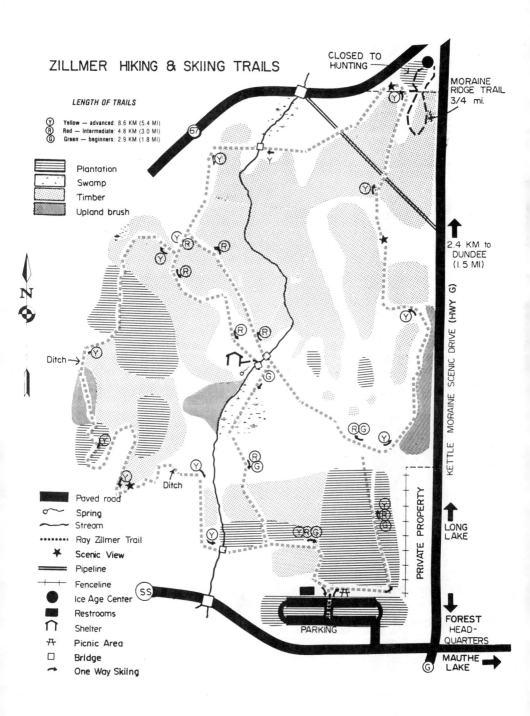

ZILLMER HIKING & SKIING TRAILS

LENGTH OF TRAILS

Ⓨ Yellow — advanced: 8.6 KM (5.4 MI)
Ⓡ Red — intermediate: 4.8 KM (3.0 MI)
Ⓖ Green — beginners: 2.9 KM (1.8 MI)

Plantation
Swamp
Timber
Upland brush

CLOSED TO HUNTING

MORAINE RIDGE TRAIL 3/4 mi.

2.4 KM to DUNDEE (1.5 MI)

KETTLE MORAINE SCENIC DRIVE (HWY G)

PRIVATE PROPERTY

LONG LAKE

FOREST HEAD-QUARTERS

Ditch →

← Ditch

Paved road
Spring
Stream
Ray Zillmer Trail
Scenic View
Pipeline
Fenceline
Ice Age Center
Restrooms
Shelter
Picnic Area
Bridge
One Way Skiing

PARKING

MAUTHE LAKE

ED'S LAKE

❄ SKIING & HIKING TRAIL

NICOLET NATIONAL FOREST

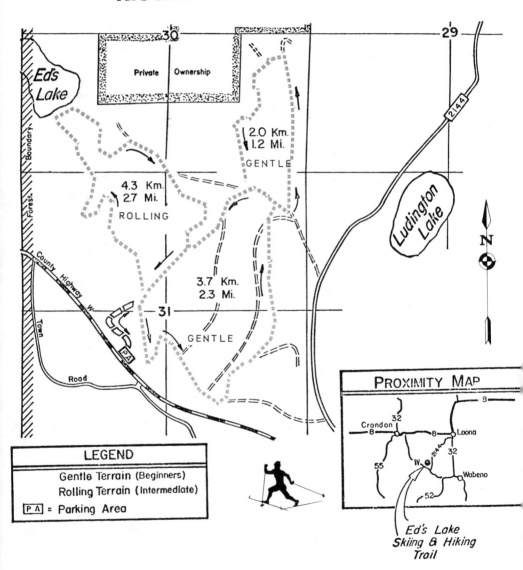

30

Private   Ownership

29

2144

Ed's
Lake

Boundary

Forest

2.0 Km.
1.2 Mi.

GENTLE

4.3 Km.
2.7 Mi.

ROLLING

Ludington
Lake

N

County   Highway   W

3.7 Km.
2.3 Mi.

31

Town

P A

GENTLE

Road

LEGEND

Gentle Terrain (Beginners)
Rolling Terrain (Intermediate)
P A = Parking Area

PROXIMITY MAP

8

Crandon
8

32

8

Loona

2144

32

W

55

Wabeno

52

Ed's Lake
Skiing & Hiking
Trail

Ed's Lake

Forest County

Ed's Lake has 3 trails totaling 10km (6 miles) of gentle, rolling terrain through wooded lands in the Nicolet National Forest. Trails are one way and are for beginners and intermediate skiers.

Take County Highway W south out of Crandon about 10 miles to plowed parking lot. Toilets are available at trail head.

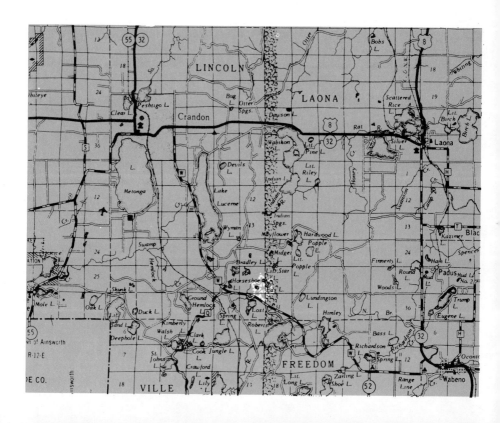

120

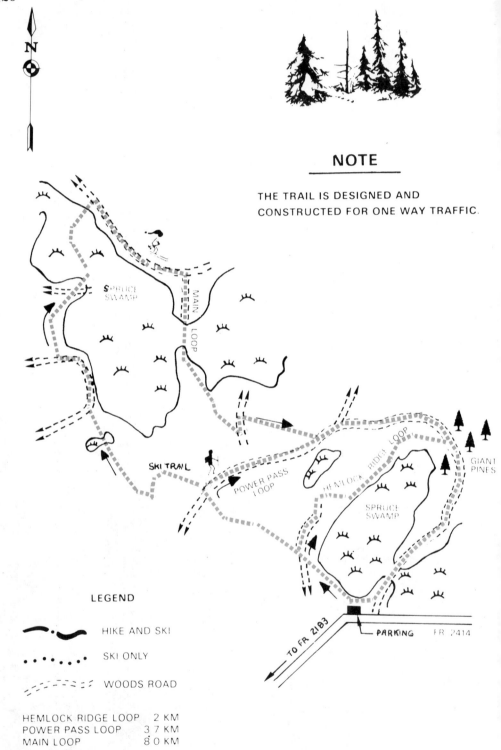

NOTE

THE TRAIL IS DESIGNED AND
CONSTRUCTED FOR ONE WAY TRAFFIC.

N

SPRUCE SWAMP

MAIN LOOP

SKI TRAIL

POWER PASS LOOP

HEMLOCK RIDGE LOOP

GIANT PINES

SPRUCE SWAMP

TO FR 2183

PARKING

FR 2414

LEGEND

HIKE AND SKI

SKI ONLY

WOODS ROAD

HEMLOCK RIDGE LOOP   2 KM
POWER PASS LOOP   3 7 KM
MAIN LOOP   8 0 KM

Giant Pine Trail

Located in the Nicolet National Forest, the Giant Pine Trail offers three loops for the beginner and intermediate skier. Hemlock Ridge loop is 2km (1.1 miles), Power Pass loop 3.7km (2.2 miles) and Main loop 8.0km (12.9 miles). The trails are designed for one way traffic. It has very scenic wooded and flat land. Parking lot is limited to about 10 cars. No shelter or toilets are available at this time.

Travel south on State Highway 32 from Three Lakes, Wisconsin for about 5 miles. You must cross Big Lake Bridge and continue south past Forest Road 2178 and over the Virgin Lake Bridge to Forest Road 2183. Turn left onto Forest Road 2183. Stay on Forest Road 2183 past Shelp and Scott Lakes to Forest Road 2414. Turn left onto Forest Road 2414 and go north for about 1.6 miles to the Trail Parking Lot.

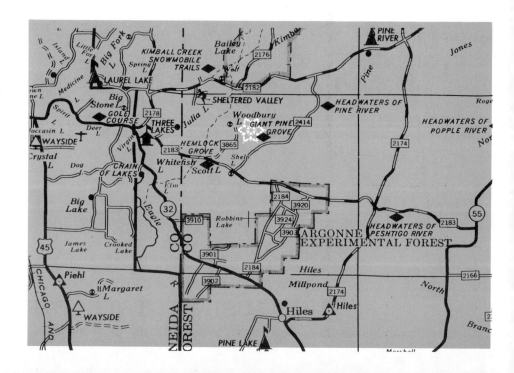

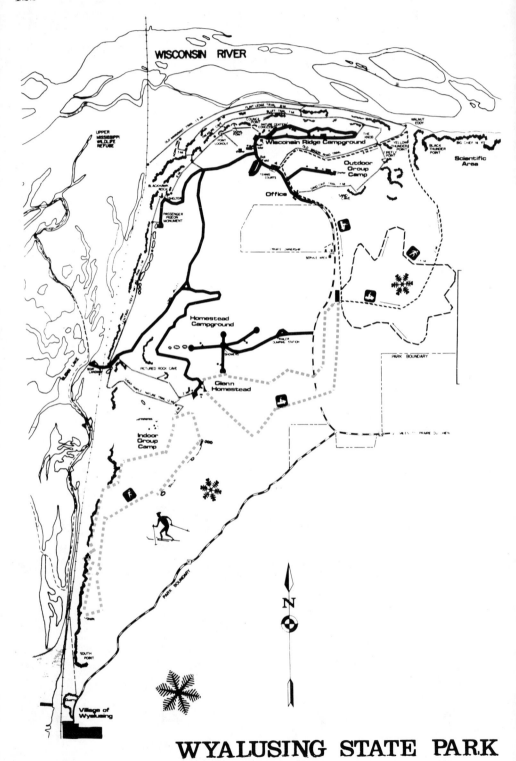

WYALUSING STATE PARK

Wyalusing State Park

The length of the Mississippi Ridge Trail is 4.8km (3 miles). It begins near the group camp and cuts through lovely wooded ridges with occasional scenic vistas to enjoy. The trail then follows an unplowed roadway back to the trail head. Park stickers are required for all vehicles.

All snowmobile trails have been closed and developed into hiking trails. Some are excellent for skiing trails and eventually could be developed.

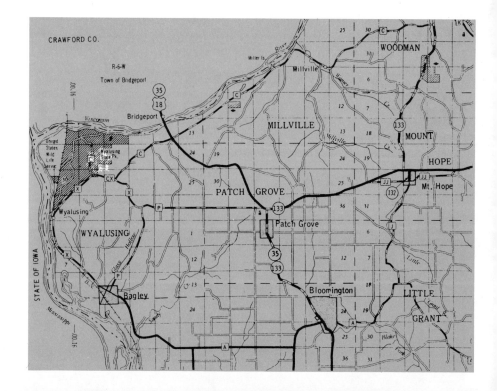

greenlake
Conference Center
American Baptist Assembly
Green Lake, WI 54941

Green Lake Ski Trails

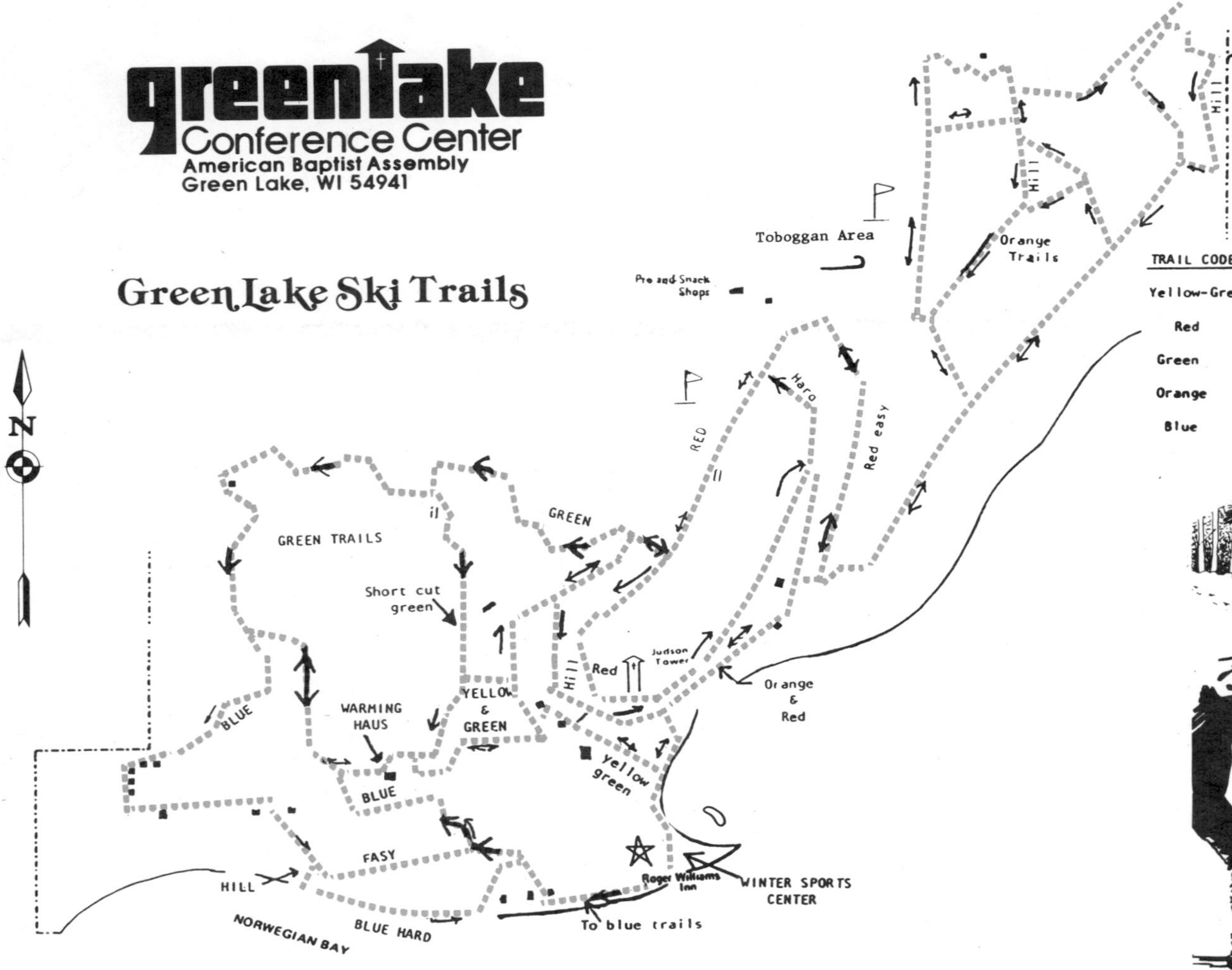

TRAIL CODE	MILES	KM	
Yellow-Green	2.4	4.0	Beginner
Red	2.5	4.4	Intermediate
Green	4.0	6.6	Intermediate
Orange	4.8	8.0	Intermediate
Blue	3.6	6.0	Advanced

WILDERNESS TRAIL
EXPERTS ONLY!

HILL

Orange Trails

HILL

Toboggan Area

Red easy

Hard

Pro and Snack Shops

RED

Red

Judson Tower

Orange & Red

WINTER SPORTS CENTER

Roger Williams Inn

To blue trails

yellow green

GREEN

HILL

il

Short cut green

GREEN TRAILS

YELLOW & GREEN

WARMING HAUS

BLUE

EASY

BLUE HARD

HILL

NORWEGIAN BAY

BLUE

N

Green Lake

Green Lake County

Located three miles west of Green Lake on Hwy. 23 this Ski Center offers well marked ski trails for the beginner as well as the expert and in-between skiers.

For the beginners there are two trails featuring open and wooded areas with stands of pine, oak, maple and cherry with a length of 7 miles. Another trail for the Intermediate skier features stands of pine and oak on terrain that includes some advanced slopes and is about 2¾ miles. The Expert trail includes part of the other trails plus challenging steep sections along the Green Lake shore totaling about 5 miles in length.

Trail use fee is charged which includes ski lodge, warming shelter, swimming pool and snack shop. Skis can be rented by reservation and lessons are available by appointment.

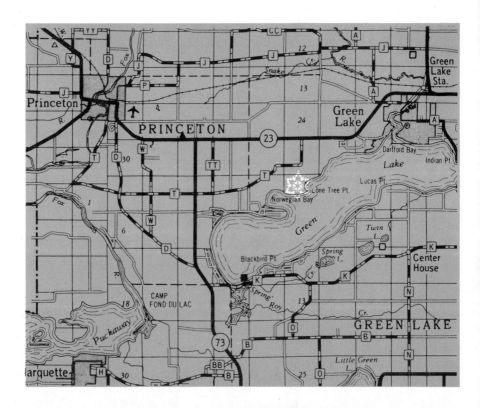

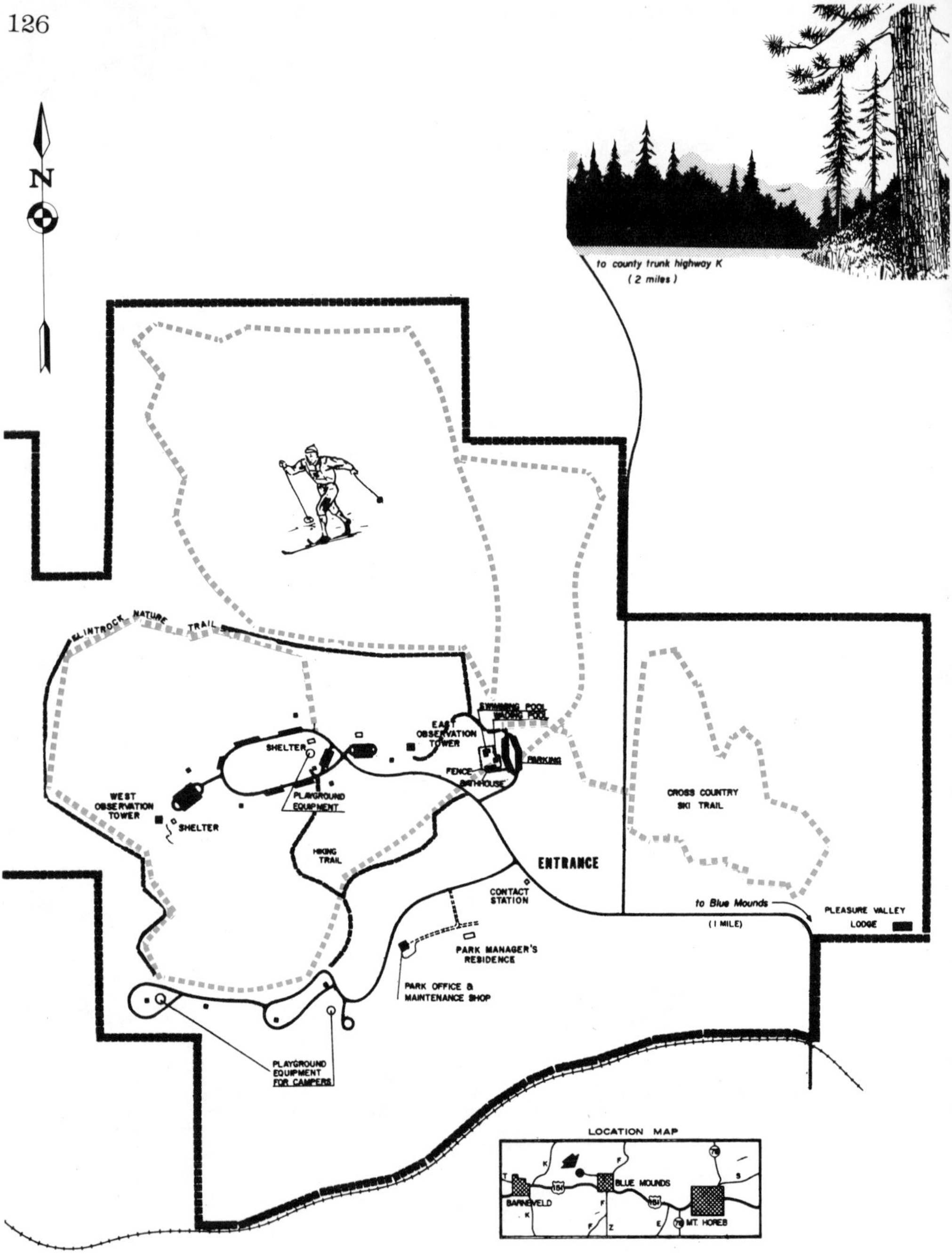

126

to county trunk highway K
(2 miles)

N

FLINTROCK NATURE TRAIL

SWIMMING POOL
WADING POOL

EAST
OBSERVATION
TOWER

SHELTER

PARKING

FENCE
BATHHOUSE

WEST
OBSERVATION
TOWER

PLAYGROUND
EQUIPMENT

SHELTER

CROSS COUNTRY
SKI TRAIL

HIKING
TRAIL

ENTRANCE

CONTACT
STATION

to Blue Mounds

(1 MILE)

PLEASURE VALLEY
LODGE

PARK MANAGER'S
RESIDENCE

PARK OFFICE &
MAINTENANCE SHOP

PLAYGROUND
EQUIPMENT
FOR CAMPERS

LOCATION MAP

BLUE MOUNDS

BARNEVELD

MT. HOREB

Blue Mound State Park

Iowa County

Blue Mound Park is located about 25 miles southeast of Madison on 1100 acres of state owned property. It is the highest point in southern Wisconsin with an elevation of 1716 feet above sea level. It has an abundance of peace and tranquility that will please the skier. Total length of trail is about 11.2km (8 miles) and varies from flat to intermediate slope. Trails are marked and groomed.

Maps are posted at trail heads and junctions. Parking lots are plowed on top of mound on east level. Restrooms and observation towers are also located in this area. Water is available near the maintenance shop. A vehicle sticker is required.

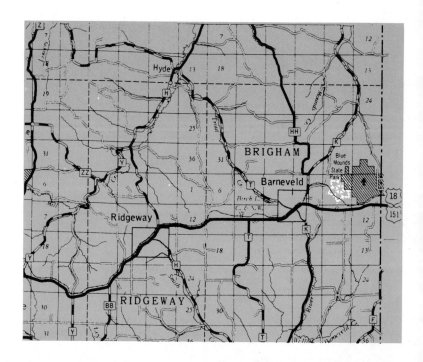

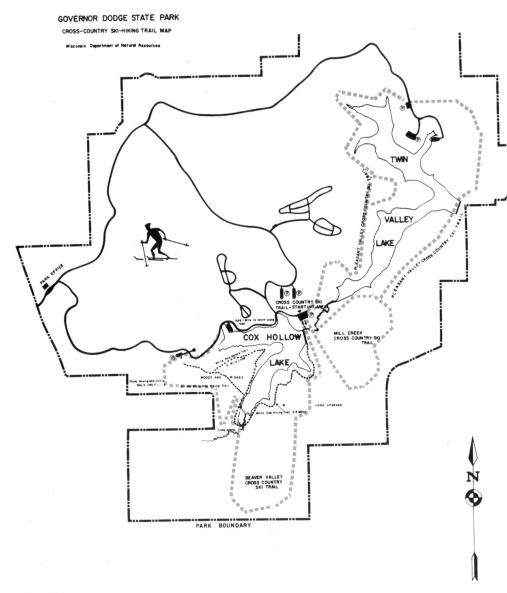

GOVERNOR DODGE STATE PARK

CROSS-COUNTRY SKI-HIKING TRAIL MAP

Wisconsin Department of Natural Resources

PARK OFFICE

CROSS COUNTRY SKI
TRAIL-STARTING AREA

Add 1 Mile to return along
road

COX HOLLOW

Steep downgrade, Curve
WALK ONLY

WOODS AND RIDGES

SKI and White Oak Hiking Trail

White Oak Hiking Trail
2 Mile Loop

LAKE

Long upgrade

LONG UPGRADE

White Oak Hiking Trail 3.5 Miles

TWIN

VALLEY

LAKE

MILL CREEK
CROSS COUNTRY SKI
TRAIL

BEAVER VALLEY
CROSS COUNTRY
SKI TRAIL

PARK BOUNDARY

N

ROADS ━━━
PARKING Ⓟ
CROSS COUNTRY SKI TRAIL — ← — ←
HIKING & NATURE TRAILS ●●●●●

SKI TRAILS ━━━
FOLLOW BLUE BLAZERS

Governor Dodge State Park

Iowa County

The varying terrain of Governor Dodge State Park offers opportunities for all ages of skiers.

The shortest trail for the intermediate is Mill Creek which is 4km (2.5 miles). This trail takes the skier thru natural stands of hardwoods and across gently rolling hills.

An additional trail has been added recently which is defined as more difficult. Pleasant Valley Trail is 12km (7.5 miles).

The Beaver Valley Trail consists of 6.4km (4 miles) of hilly, wooded touring which is geared for the advanced skier, it is most difficult.

All the trails are marked, groomed and tracked.

The Park is open from 6:00 A.M. to 11:00 P.M.

Spacious parking is available and pit toilets, picnic tables and grills are provided at the trailhead. A vehicle admission sticker is required. Park entrance is 3 miles north of Dodgeville on Hwy. 23 on right-hand side of road.

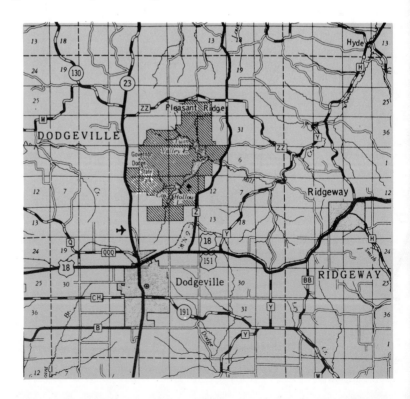

500 Acres of Tillable + Wooded Terrain
- 5 open Fields plus miles of trails
Open Field
(Top of Ridge)
Open Fields
(Marked Trails)
Open Fields
Spring Valley Trails
Office
Spring Valley Rd.
Hwy 130
N
E
Mill Rd.

Spring Valley

Iowa County

Located in beautiful Iowa County, Spring Valley is situated on a 500 acre tract of open fields and wooded terrain. It consists of 3 loops — approximately 4.8km (3 miles), 8km (5 miles) and 12.9km (8 miles) — which are marked and well-kept.

Shelter, toilets, parking, and refreshments are available on a year-round basis along with winter campsites. Trail fee is charged and ski rental is available.

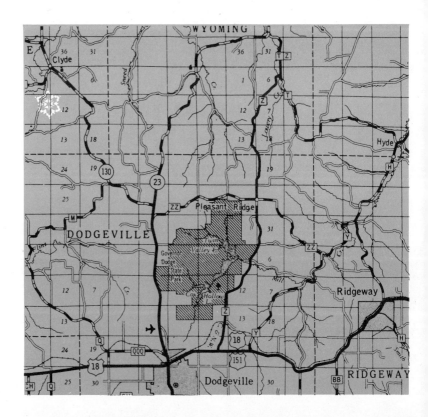

WINTERGREEN TOURING CENTER TRAIL MAP

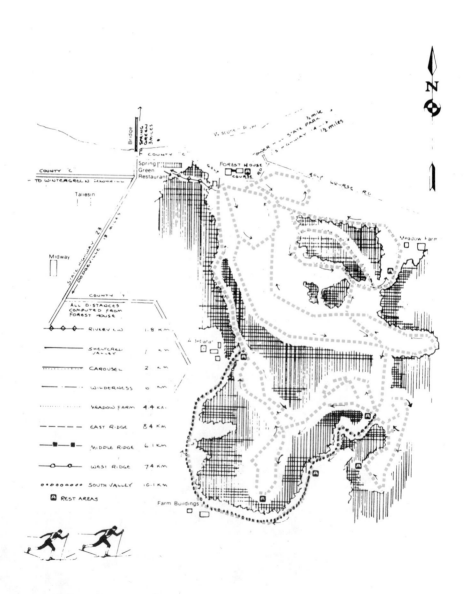

Wintergreen Touring Center

Iowa County

Set in the Wisconsin River Valley bordering Taliesin, the home and school of Frank Lloyd Wright, the Wintergreen Touring Center has 11 kilometers of trails for beginners and intermediates as well as 17 kilometers of wilderness trails for experts, and all wilderness trails are 14 feet wide:

Packed and double-tracked, the trails wind through the scenic wooded area surrounding the Springs Golf Course and run adjacent to the Spring Green Restaurant, the only one in the world designed by Frank Lloyd Wright. Wintergreen alpine area is one mile from the touring center.

The ski shop, located in a charming turn of the century house, has a fireplace, rentals, lessons, a retail shop and snack bar.

Housing is available in a wide range of accommodations, most with fireplaces, from interesting remodeled houses, unusual small cottages, 'A' frame vacation homes, European style rooms with shared baths, hostel like dorms, to motels all from right on the trails to within 6 miles of the tour center.

Open weekends only, open all holiday periods and any weekday by appointment.

Location — 38 miles west of Madison, WI. From the village of Spring Green, go south 3 miles on State Highway 23 over the Wisconsin River, then turn left on County Trunk 'C', in .2 mile turn right onto Golf Course Road. Center is .2 miles farther on the right.

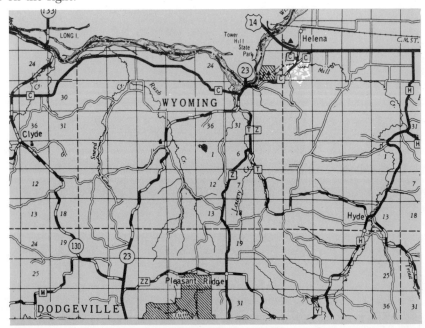

Cedar Trail

Located 12 miles west of Hurley, WI off U.S. Highway 2. Trailhead and parking at the Frontier Bar or at the Harbor Lights Bar on Lake Superior. Terrain: novice-intermediate. This trail skirts the edges of deep cedar lined ravines that open into Lake Superior. Deer are commonly seen and the views of Lake Superior in the winter are stunning! The trail is track-set.

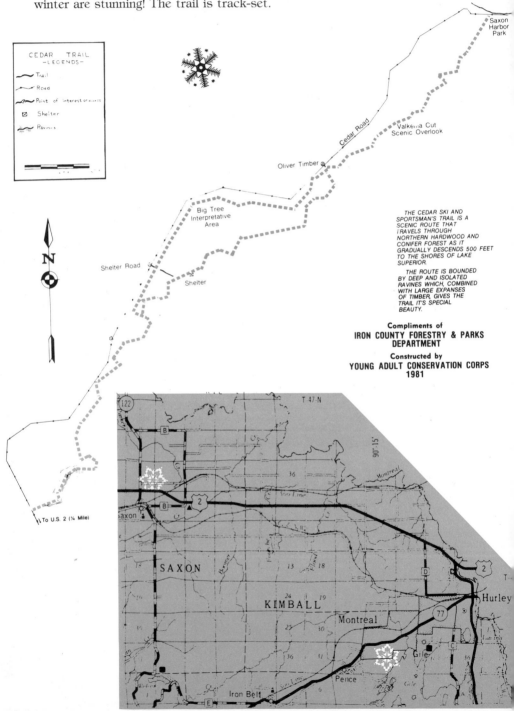

CEDAR TRAIL
—LEGENDS—

- Trail
- Road
- Point of interest or exits
- ⊠ Shelter
- Ravines

Saxon Harbor Park

Valkenia Cut Scenic Overlook

Cedar Road

Oliver Timber

Big Tree Interpretative Area

Shelter Road

Shelter

N

THE CEDAR SKI AND SPORTSMAN'S TRAIL IS A SCENIC ROUTE THAT TRAVELS THROUGH NORTHERN HARDWOOD AND CONIFER FOREST AS IT GRADUALLY DESCENDS 500 FEET TO THE SHORES OF LAKE SUPERIOR.

THE ROUTE IS BOUNDED BY DEEP AND ISOLATED RAVINES WHICH, COMBINED WITH LARGE EXPANSES OF TIMBER, GIVES THE TRAIL IT'S SPECIAL BEAUTY.

Compliments of
IRON COUNTY FORESTRY & PARKS DEPARTMENT

Constructed by
YOUNG ADULT CONSERVATION CORPS 1981

To U.S. 2 (¼ Mile)

Saxon

SAXON

KIMBALL

Montreal

Hurley

Gile

Pence

Iron Belt

Montreal Jaycees Trail

Iron County

Located in the small, old mining village of Montreal, WI along Highway 77. The trail passes historic mining areas of Montreal, WI, once called "White City" because the standard color of the mining company homes in the town was white. Terrain: beginner-intermediate. A very pleasant trail with good opportunities to meet other local cross country enthusiasts. It is track-set.

TERRAIN : PLEASANTLY HILLY

LENGTH : 15 KM   9 MI

PARKING : ⊠

SPONSOR : MONTREAL JAYCEE'S

GROOMING : PENOKEE RANGERS

136

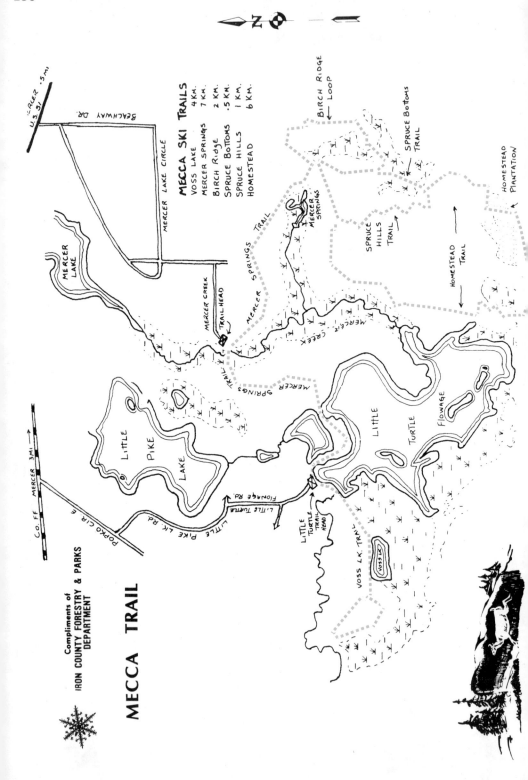

MECCA TRAIL

Compliments of
IRON COUNTY FORESTRY & PARKS
DEPARTMENT

MECCA SKI TRAILS

VOSS LAKE	4 KM.
MERCER SPRINGS	7 KM.
BIRCH RIDGE	2 KM.
SPRUCE BOTTOMS	.5 KM.
SPRUCE HILLS	1 KM.
HOMESTEAD	6 KM.

Mecca Ski Trail

Iron County

Mecca I is a 16km (10 mile trail) that winds through the breathtaking beauty of majestic snow-capped pines, over natural beaver dams, and the Little Turtle Flowage, a wildlife sanctuary. It is a one-way trail with gentle, rolling terrain and a few hills. You may ski the entire trail or only parts of it, entering at either Trailhead or alternate points.

Located 2 miles northwest of Mercer, WI off Highway 51 on Popko Circle and Little Pike Lake Road. (directions are signed from Highway 51). Total length of all trail loops — 17.5km. Rating — Novice. Trail is track set.

The annual MECCA SKI TOURING RACE (USSA-CD SANCTIONED) is held every January and is open to racers and citizen tourists. Further information and entry blanks can be obtained from the Chamber of Commerce, Box 368, Mercer, WI 54547.

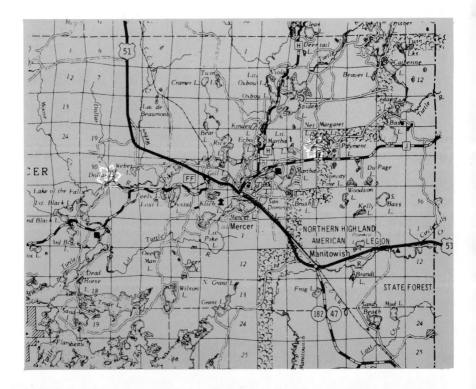

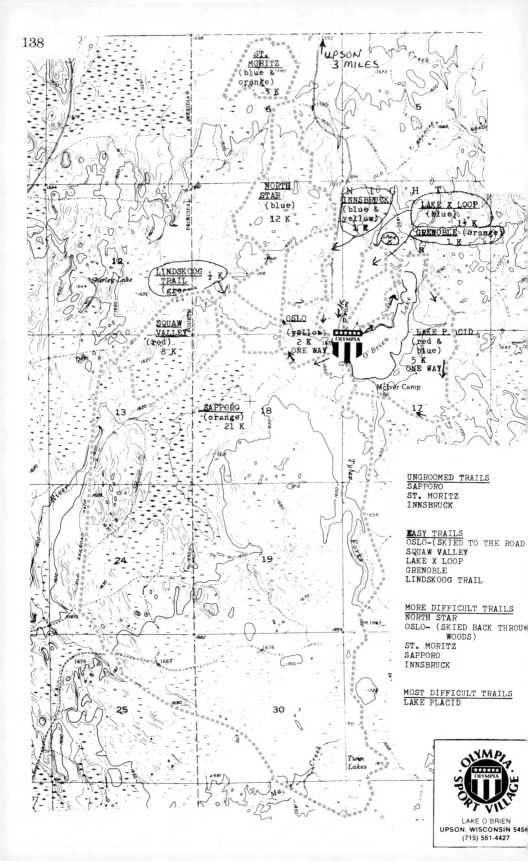

138

ST. MORITZ (blue & orange) 3 K

NORTH STAR (blue) 12 K

INNSBRUCK (blue & yellow) 1 K

LAKE X LOOP (blue) 1½ K

GRENOBLE (orange) 1 K

LINDSKOOG TRAIL (green)

SQUAW VALLEY (red) 8 K

OSLO (yellow) 2 K ONE WAY

OLYMPIA

LAKE PLACID (red & blue) 5 K ONE WAY

SAPPORO (orange) 21 K

UNGROOMED TRAILS
SAPPORO
ST. MORITZ
INNSBRUCK

EASY TRAILS
OSLO-(SKIED TO THE ROAD
SQUAW VALLEY
LAKE X LOOP
GRENOBLE
LINDSKOOG TRAIL

MORE DIFFICULT TRAILS
NORTH STAR
OSLO- (SKIED BACK THROUGH WOODS)
ST. MORITZ
SAPPORO
INNSBRUCK

MOST DIFFICULT TRAILS
LAKE PLACID

OLYMPIA SPORT VILLAGE
OLYMPIA
LAKE O'BRIEN
UPSON, WISCONSIN 5454
(715) 561-4427

Olympia Sport Village II

Olympia Sport Village is located in the wilderness of Northern Wisconsin, 5 miles south of Upson on Lake O'Brien.

The trails at Olympia consist of a series of loops of varying length totaling about 54km (34 miles) of groomed and marked trails. The more advanced trail winds around the lake, while the other trails explore the woods. Olympia provides a lodge for shelter which has a bar and restroom facilities.

Lodging with meals available to groups.

Since the mid-60's Olympia Sport Village has been the occasional site of USSA Central Division Cross-Country Ski Camps, National Biathlon and Olympic tryouts, ski touring and racing.

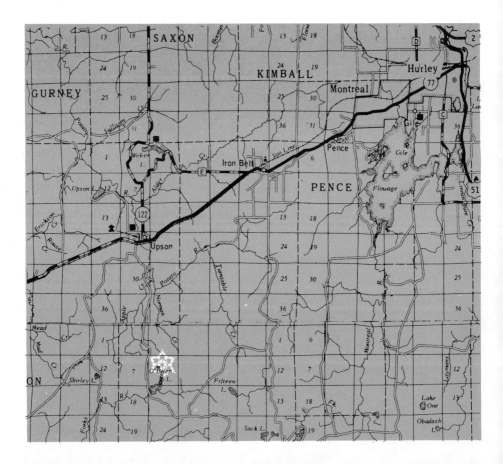

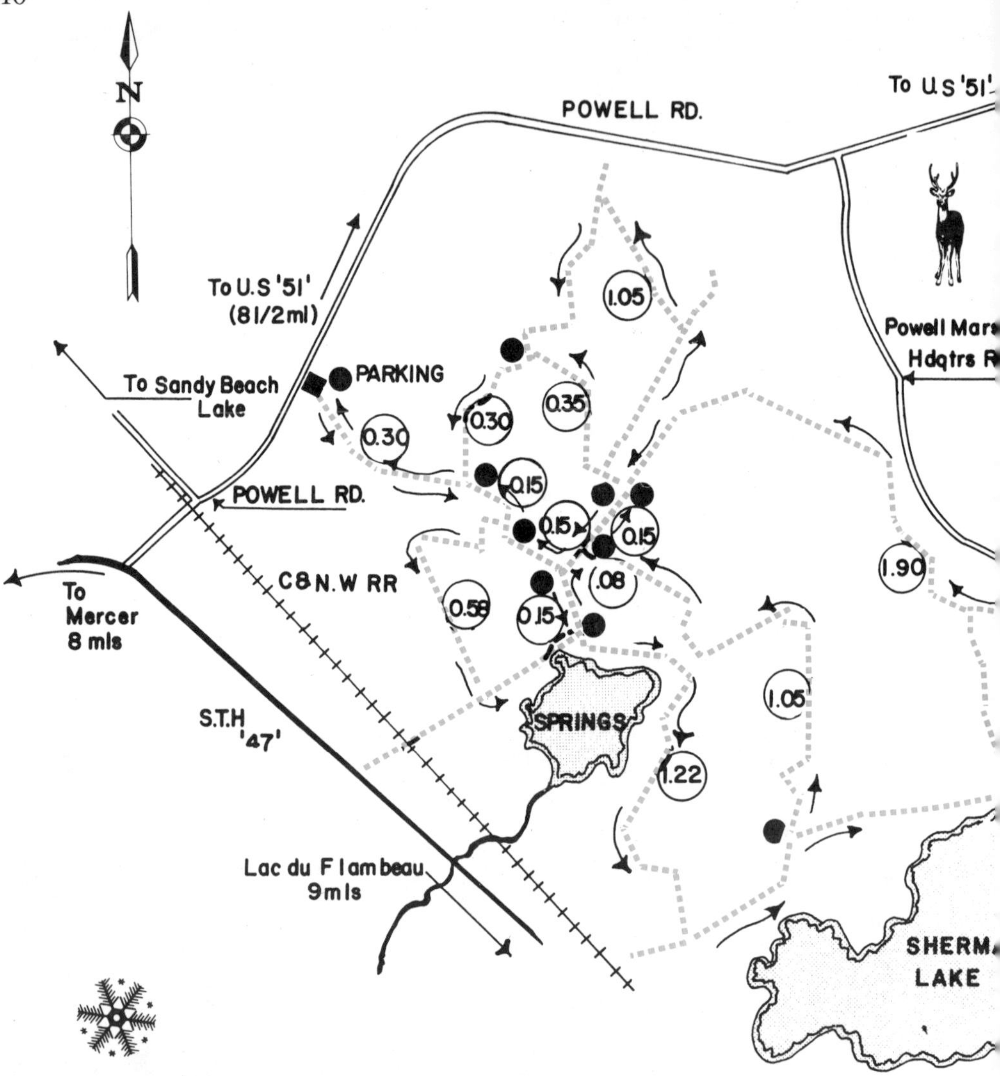

N

To U.S '51'

POWELL RD.

To U.S '51'
(8 1/2 ml)

PARKING

To Sandy Beach
Lake

1.05

0.30

0.35

0.30

0.15

POWELL RD.

0.15

0.15

Powell Mars
Hdqtrs R

To Mercer
8 mls

C 8 N.W RR

.08

0.58

0.15

1.90

S.T.H
'47'

SPRINGS

1.05

Lac du Flambeau
9 mls

1.22

SHERM
LAKE

POWELL MARSH SKI TRAIL

NORTHERN HIGHLAND—AMERICAN LEGION STATE FOREST
Wisconsin Department of Natural Resources

LEGEND

- - - - - TRAIL
DIRECTION of TRAVEL
● DISTANCE MARKER
approx scale —

0.25 miles
400 meters

EMERGENCY NO.— MINOCQUA AMBULANCE–356-6200

Powell Marsh Trail

Approximately half of the trail winds through an area that was logged 5-10 years ago. The remainder of the 11.3km (7 miles) trail travels through a variety of older timbers. Skiers will enjoy the scenic view of Sherman Lake, and some remote springs.

The trail is located 8½ miles southeast of Mercer — take Hwy. 47 to Powell Rd.

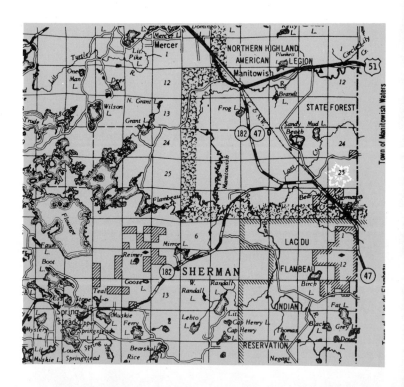

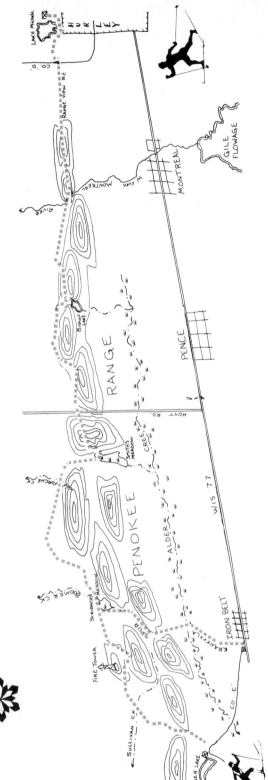

ULLER TRAIL

TERRAIN : MODERATE, REMOTE

LENGTH : 30 KM, 19 MI = ULLER
5 KM, 3 MI = KRANKKALA SPUR

GROOMING : PENOKEE RANGERS

Uller Trail

Uller traverses the heart of the Penokee Range, from Weber Lake Park, west of Iron Belt, to Highway 51 on the NW side of Hurley. It is 16 miles of some of the most remote high country in Wisconsin.

Uller is no trail to take lightly. Its terrain might be considered tame, but add 16 miles, changeable weather, and the fact that parts of the trail are three to four miles from the nearest house, and the challenge is there.

The trail is generously marked with signs and blazes. It will be groomed on a regular basis, but with the weekly snow falls in this country there will be times when the trail has no tracks.

You are never more than 3 miles from a road while skiing Uller. If you get in trouble and feel you must "bail out", ski either north or south. You will encounter a road within 3 miles and a house or small town shortly after that. The Hoyt Road, approximately half way on the trail is a good route out. North on the Hoyt Road are a couple of farm houses. Three miles south on the Hoyt Road is State Highway 77. A mile east on 77 is the Town of Pence, a mile west on 77 is the Town of Iron Belt.

Uller can be one of the most impressive skiing experiences one can have in the mid-west. The Iron County Forestry and Parks Committee wishes you the best of times on this public trail.

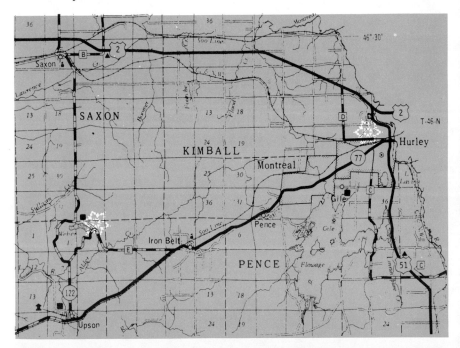

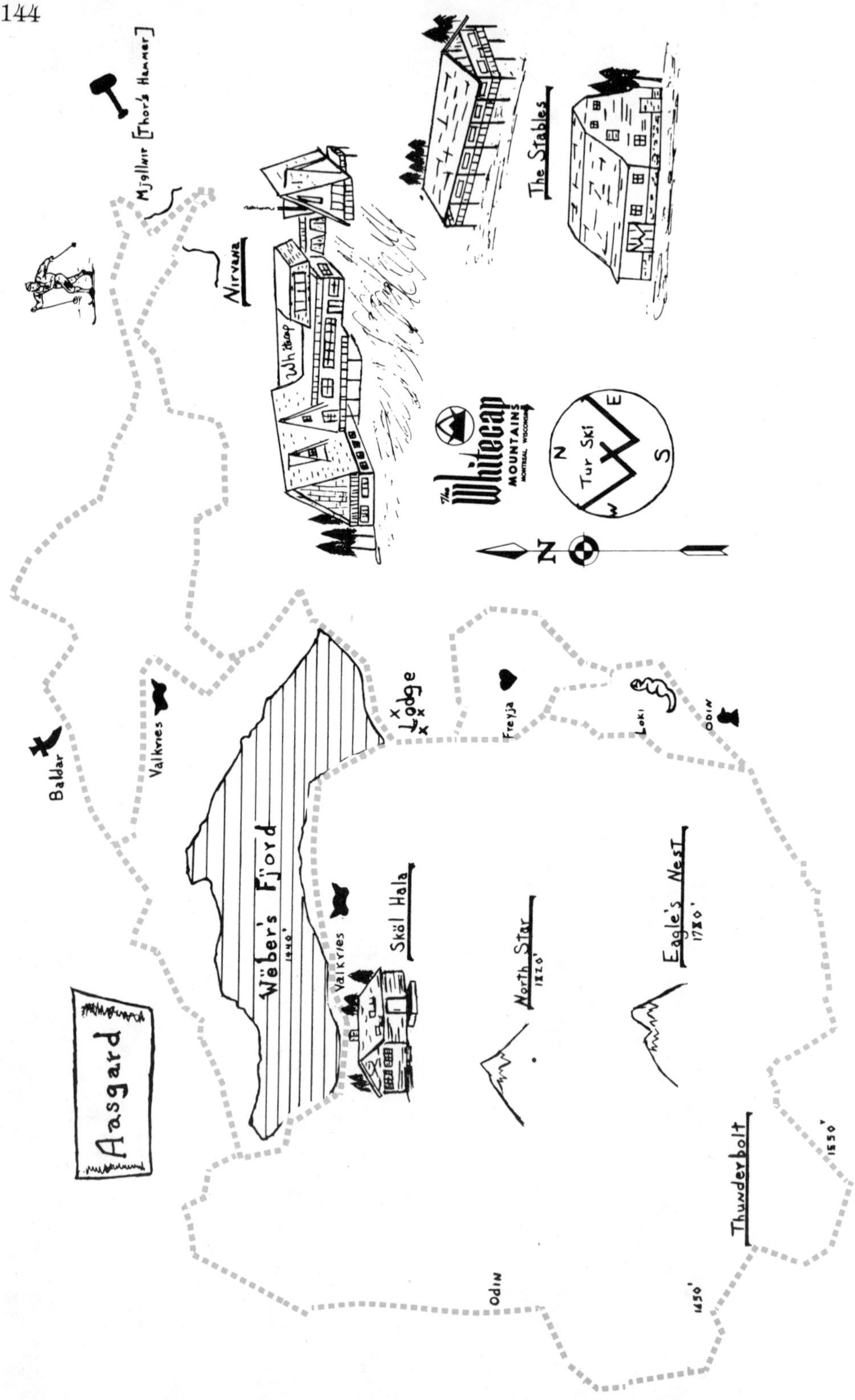
Aasgard
Mjollnir [Thor's Hammer]
Nirvana
Whitecap
The Stables
The Whitecap Mountains
Montreal Wisconsin
Tur Ski
N
E
S
W
N
Baldar
Valkries
Wëber's Fjord
1440'
Lodge
Valkries
Sköl Hala
Freyja
Loki
Odin
North Star
1820'
Eagle's Nest
1780'
Thunderbolt
1550'
Odin
1450'

Whitecap Mountain Trails

The cross-country ski trails at Whitecap Mountains are nestled deep in the forested hills surrounding beautiful Weber Lake, just 12 miles from Hurley, Wisconsin. They're a world apart from the hectic pace and crush of modern living, with 50km (31 miles) of tracked and groomed trails through spectacular scenery and breathtaking solitude. Whitecap offers the very best of a cross-country skiing tradition that goes back nearly a hundred years in this part of the country.

Whitecap offers cross-country ski rentals, a ski shop where top quality equipment may be purchased, cross-country ski instruction by certified instructors, a special chalet for cross-country skiers, a lodging complex, and other features to make the area's most complete and comprehensive facility for the Nordic skier. A trail fee is charged.

There are evening tours on moonlit trails, midnight tours for swashbucklers and romantics, and lunch hour jaunts to the Wine Hut.

Weekend packages and ski weeks are available.

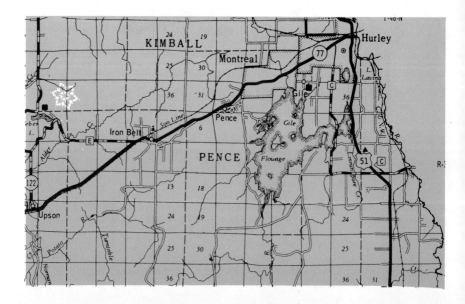

Best Western

Arrowhead Lodge Inc.

I-94 & Hwy. 54
BLACK RIVER FALLS, WISCONSIN 54615

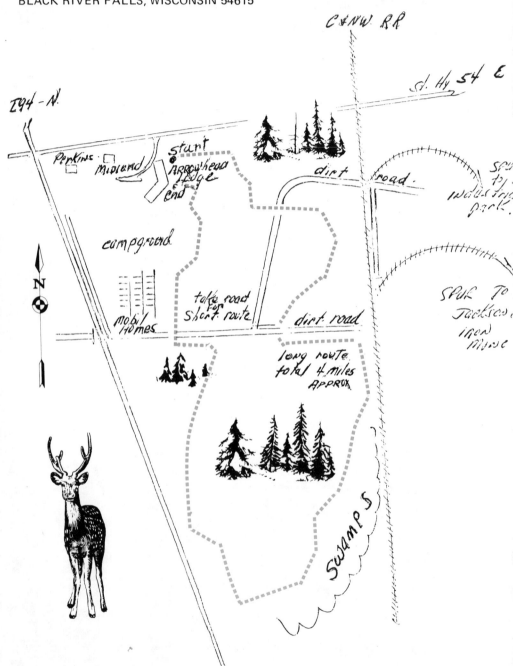

C&NW. RR

St. Hy 54 E

I94 - N.

Perkins
Midland
stunt
Arrowhead
Lodge
end

dirt road.

SPUR
To
Industry
park

campground

N

take road
for
Short route

dirt road

Mobil
Homes

long route
total 4 miles
APPROX

SPUR TO
Jackson
Iron
Mine

Swamp S

Arrowhead Lodge

The lodge has 6.4km (4 miles) of marked, groomed trails from the back door. They are gentle & wooded. There is a cut off about ⅓ of the way for a shorter route. The Black River Forest Trails at Millston are only 12 miles from the lodge and they have 38km (24 miles) of some of the best trails in the state.

Rentals, food & lodging are available with deluxe accomodations, ski chalet, indoor pool, sauna, whirlpool.

Location is 1 mile east of Black River Falls off I-94 at State Hwy. 54.

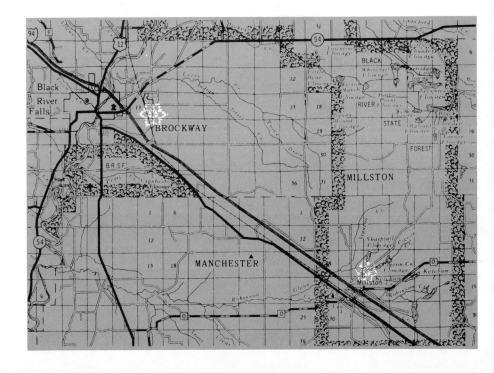

148

BLACK RIVER STATE FOREST SKI TRAILS

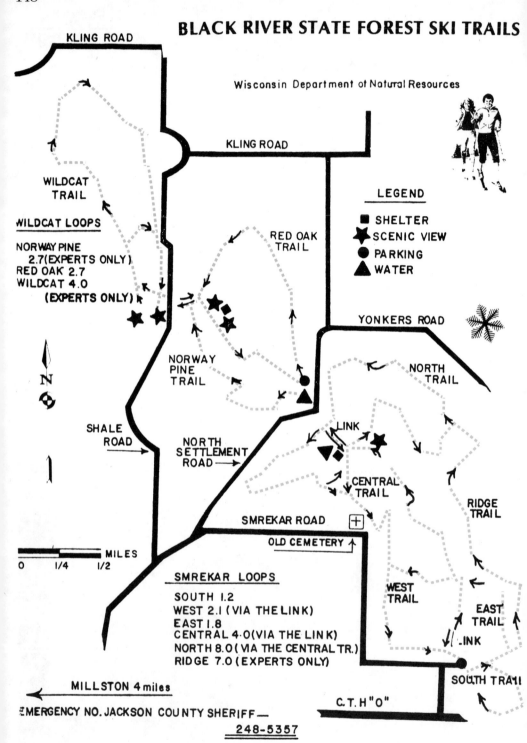

Wisconsin Department of Natural Resources

KLING ROAD

KLING ROAD

WILDCAT
TRAIL

WILDCAT LOOPS

NORWAY PINE
  2.7 (EXPERTS ONLY)
RED OAK 2.7
WILDCAT 4.0
  (EXPERTS ONLY)

RED OAK
TRAIL

LEGEND
■ SHELTER
★ SCENIC VIEW
● PARKING
▲ WATER

YONKERS ROAD

NORTH
TRAIL

N

SHALE
ROAD

NORWAY
PINE
TRAIL

NORTH
SETTLEMENT
ROAD

LINK

CENTRAL
TRAIL

RIDGE
TRAIL

MILES
0   1/4   1/2

SMREKAR ROAD

OLD CEMETERY

WEST
TRAIL

EAST
TRAIL

SMREKAR LOOPS

SOUTH 1.2
WEST 2.1 (VIA THE LINK)
EAST 1.8
CENTRAL 4.0 (VIA THE LINK)
NORTH 8.0 (VIA THE CENTRAL TR.)
RIDGE 7.0 (EXPERTS ONLY)

LINK

SOUTH TRAIL

MILLSTON 4 miles

C.T.H "O"

EMERGENCY NO. JACKSON COUNTY SHERIFF—
248-5357

Black River State Forest

Jackson County

Many people consider these to be the finest ski-touring trails in Wisconsin. The cross-country ski trails are located approximately five miles northeast of Millston. Two parking lots are provided: Wildcat lot, approximately 4 miles northeast of Millston on North Settlement Road and Smrekar lot, approximately 4 miles east of Millston on County Highway "O", then North about one mile on Smrekar Road. Two rest areas, as shown on the map, have adirondack shelters, small fireplaces, and picnic tables.

The trails all run through woodlands and rolling terrain. The ridge, Norway pine, and wildcat trails are quite steep in places and recommended for experts. There are non loops totaling 24 miles in length. The trails are all designed and marked for one way travel with exception of the links.

Scenic overlooks can be found on several trails, (see map). Overnite camping is recommended at these sites. Park stickers are required.

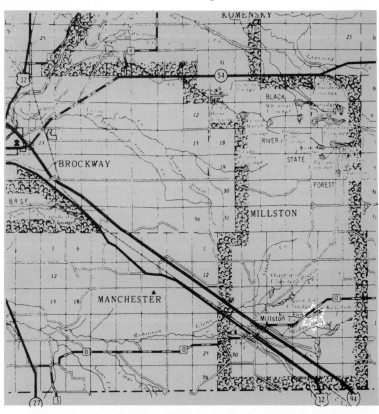

TYRANENA GOLF COURSE

800 S. MAIN STREET
LAKE MILLS, WIS. 53551

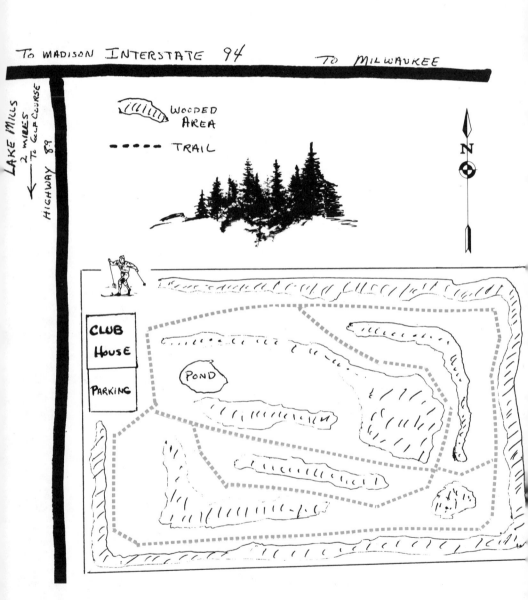

To MADISON INTERSTATE 94   TO MILWAUKEE

LAKE MILLS
2 MILES
TO GOLF COURSE
HIGHWAY 89

WOODED AREA

— — — — TRAIL

N

CLUB HOUSE

PARKING

POND

Tyranena Golf Course

The trail consists of 6.6km (4.2 miles) located on 80 acres of rolling terrain geared for the basic beginner and intermediate skier.

It is open daily (except Monday) from 8:00 a.m. to 5:00 p.m.

Trails are marked, groomed and tracked with rentals and instructions available.

The course is located on 800 S. Main St., Lake Mills, 2 miles South off I-94 on STH 89.

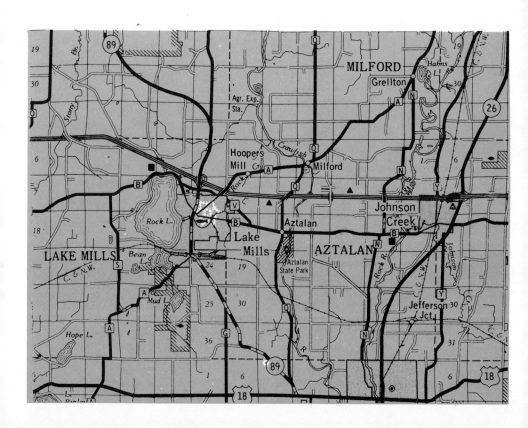

152

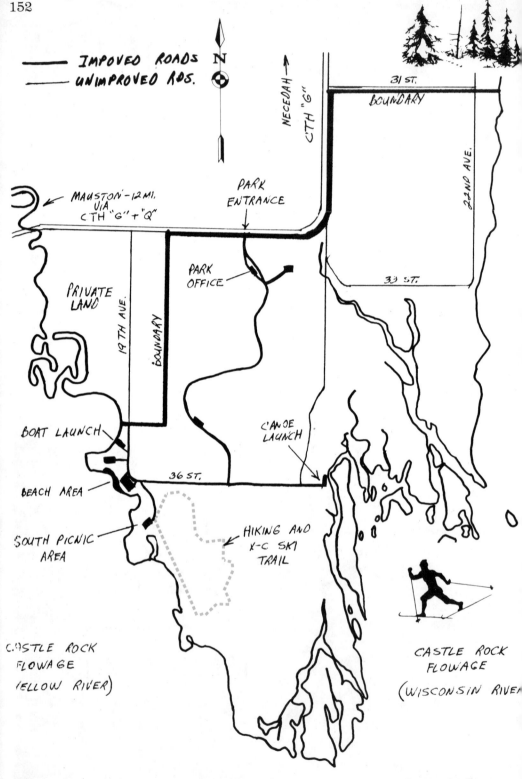

IMPOVED ROADS
UNIMPROVED RDS.

N

NECEDAH →
CTH "G"

31 ST.
BOUNDARY

22ND AVE.

PARK ENTRANCE

MAUSTON - 12 MI.
VIA CTH "G" + "Q"

PARK OFFICE

33 ST.

PRIVATE LAND

19TH AVE.

BOUNDARY

BOAT LAUNCH

CANOE LAUNCH

BEACH AREA

36 ST.

SOUTH PICNIC AREA

HIKING AND X-C SKI TRAIL

CASTLE ROCK FLOWAGE (YELLOW RIVER)

CASTLE ROCK FLOWAGE (WISCONSIN RIVER)

Buckhorn State Park

Juneau County

Buckhorn Park is located on the Castle Rock Flowage of the Wisconsin River in south central Wisconsin. It is approximately 12 miles northeast of Mauston and I-90-94 via CTH Q and G.

The park is one of the newest state parks and trail development will continue for the next few years. The present trail is a 3.2km (2 mile) loop beginning at the south picnic area and leads through the oak and pine forests on the southern end of the peninsula. The trail maps are posted at the trail head and junctions. The terrain is gently rolling and the trail offers views of the flowage where bald eagles or rough-legged hawks may be observed soaring above the park.

The vehicle sticker will be required in the near future, and no camping is allowed at the present time. Toilets and water are available at the trail head parking lot.

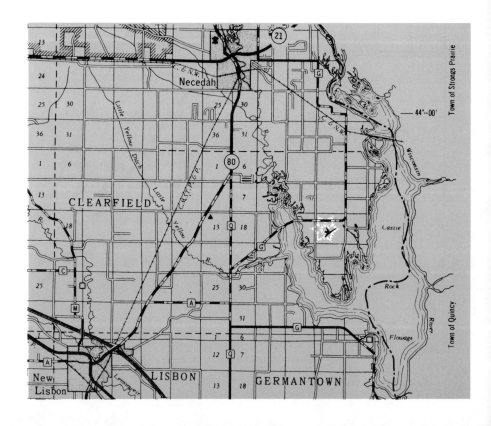

OAK RIDGE TRAIL
Juneau County Forest

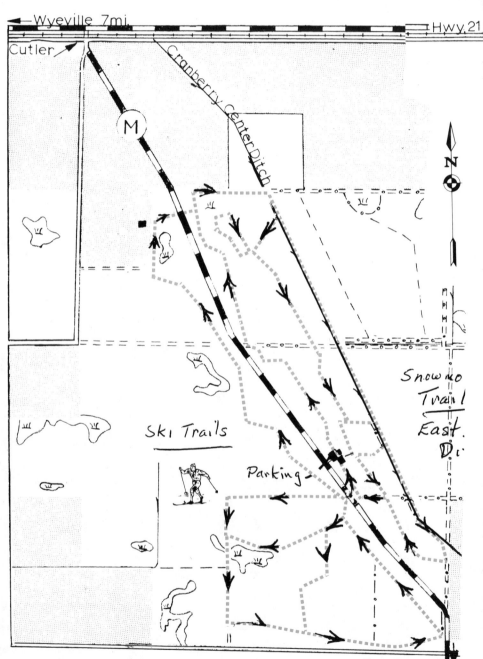

← Wyeville 7 mi.

Hwy. 21

Cutler

Cranberry Center Ditch

M

N

Ski Trails

Parking

Snowmo Trail East Dr.

New Lisbon 8 mi.

Oakridge Trail

Juneau County

Oakridge offers approximately 4.8km (3 miles) of gentle, wooded marked trail. A small cabin is located near the parking area, which can be used for shelter.

It is located 8 miles north of New Lisbon on CTH M.

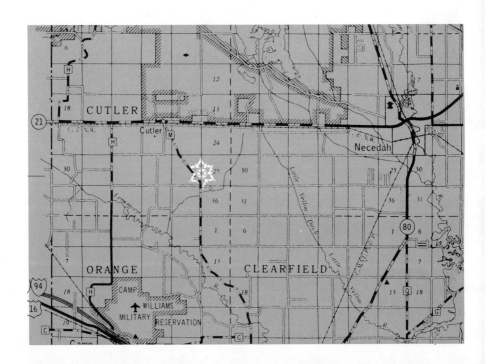

156

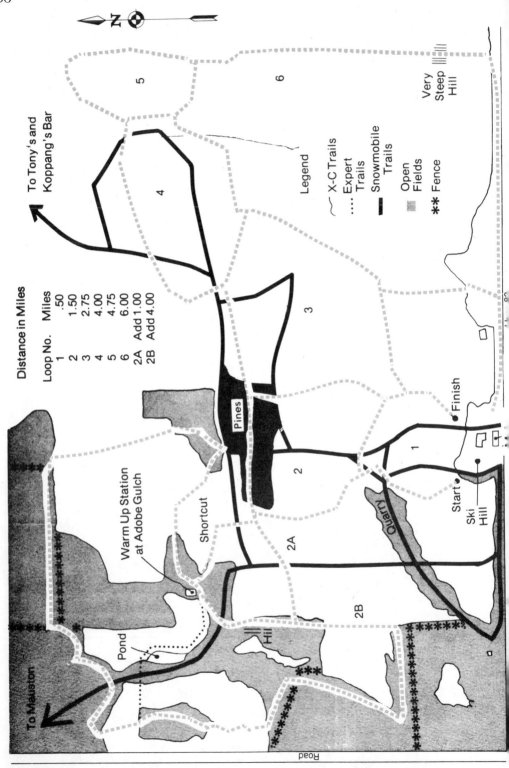

Woodside Ranch

Juneau County

Woodside Ranch offers a varied terrain for the skier. The groomed trails consist of 6 loops with totals 16km (10 miles) of skiing.

You can glide through a stand of pines or totally enjoy the expert trails.

Snowmobiles have separate trails but cross over the ski trail at different points.

Parking, water, shelter and toilets are available with cocktail lounge, snack bar and fireplace log cabins.

Skiing hours are from 9:00 a.m. to 10:00 p.m., 7 days a week.

A trail charge is requested from unregistered guests, rentals are available.

Woodside is located 5 miles east of Mauston on State Hwy. 82.

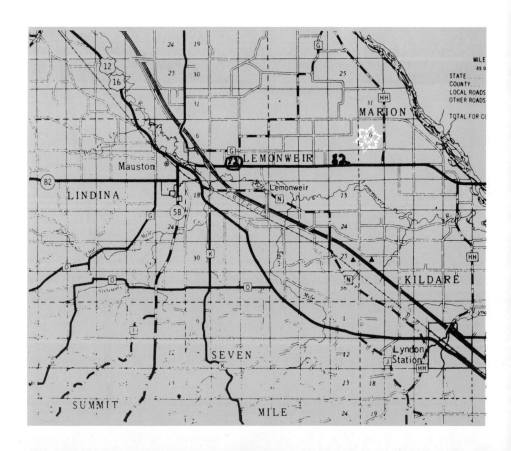

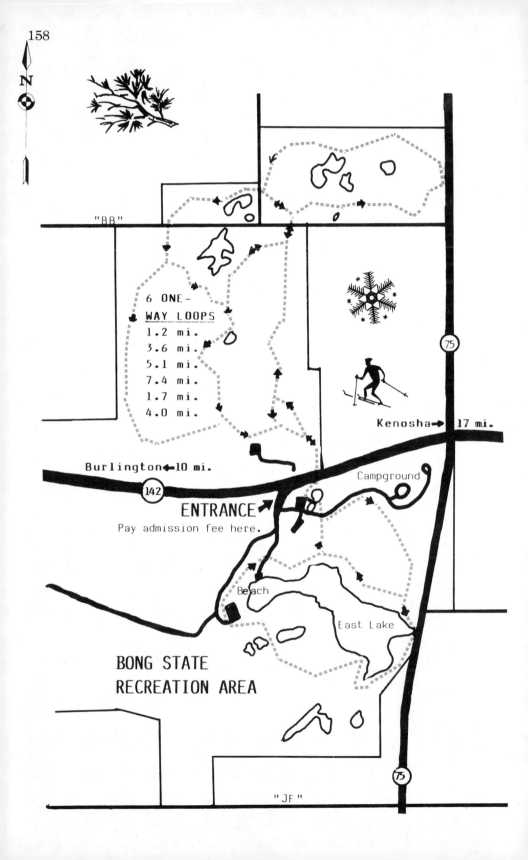

N

"BB"

6 ONE-
WAY LOOPS
1.2 mi.
3.6 mi.
5.1 mi.
7.4 mi.
1.7 mi.
4.0 mi.

Kenosha→ 17 mi.

Burlington←10 mi.

142

ENTRANCE→

Pay admission fee here.

Campground

75

Beach

East Lake

BONG STATE
RECREATION AREA

"JF"

75

Bong State Recreation Area

Located in northwestern Kenosha County, Bong is only 15 minutes west of I-94 on Highway 142. The entrance is located on Highway 142, one mile west of Highway 75 on the south side of the highway. Bong Recreation Area offers six loops with lengths varying from 1.2 to 7.4 miles; total trail mileage is 13.3. Trails are marked, tracked and groomed. All trails are rated "novice" with a few short, steep hills. We recommend that you dress properly for windy conditions and bring your own refreshments. Toilets are available. A State Park Vehicle Admission sticker is required to use Bong and may be purchased upon arrival. If you have further questions on Bong's facilities please call 414-878-4416.

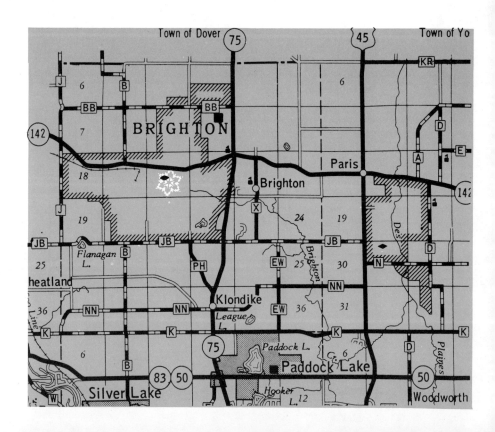

160

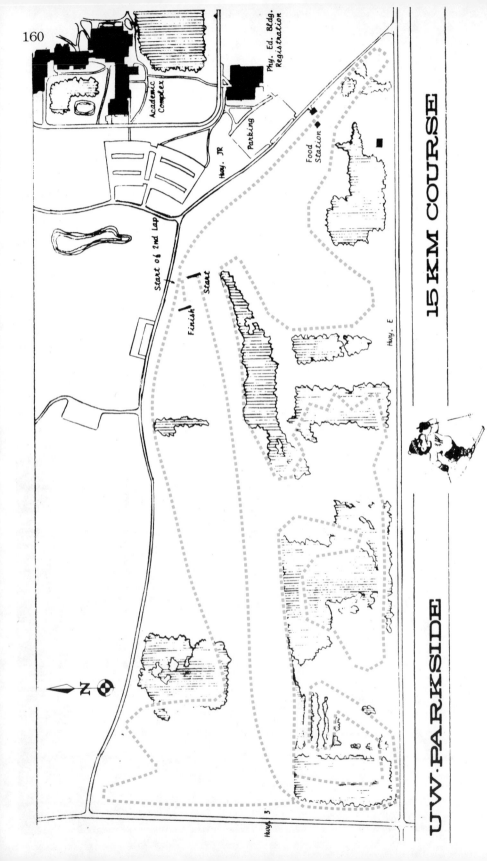

UW·PARKSIDE

15 KM COURSE

U. W. Parkside – Kenosha

Kenosha County

The UW-Parkside has developed the only National cross country course in the U.S. It is completely self contained within the boundaries of the University property and is open to recreational skiers.

This trail has a marked 16km (10 miles) course as well as spacious areas suitable for recreational skiing off the trails. There is no trail fee. Whenever possible, all trails are groomed and tracked. New trails are under development.

In addition, immediately north of the Parkside Course, extensive bridal paths are available for skiing in Petrifying Springs Park.

The course is located at the Junction of Wis. Hwy. 31 (Green Bay Road) and Kenosha County E.

The Parkside 15km cross country ski race is held annually on the first weekend in February. Call (414) 553-2446 for details.

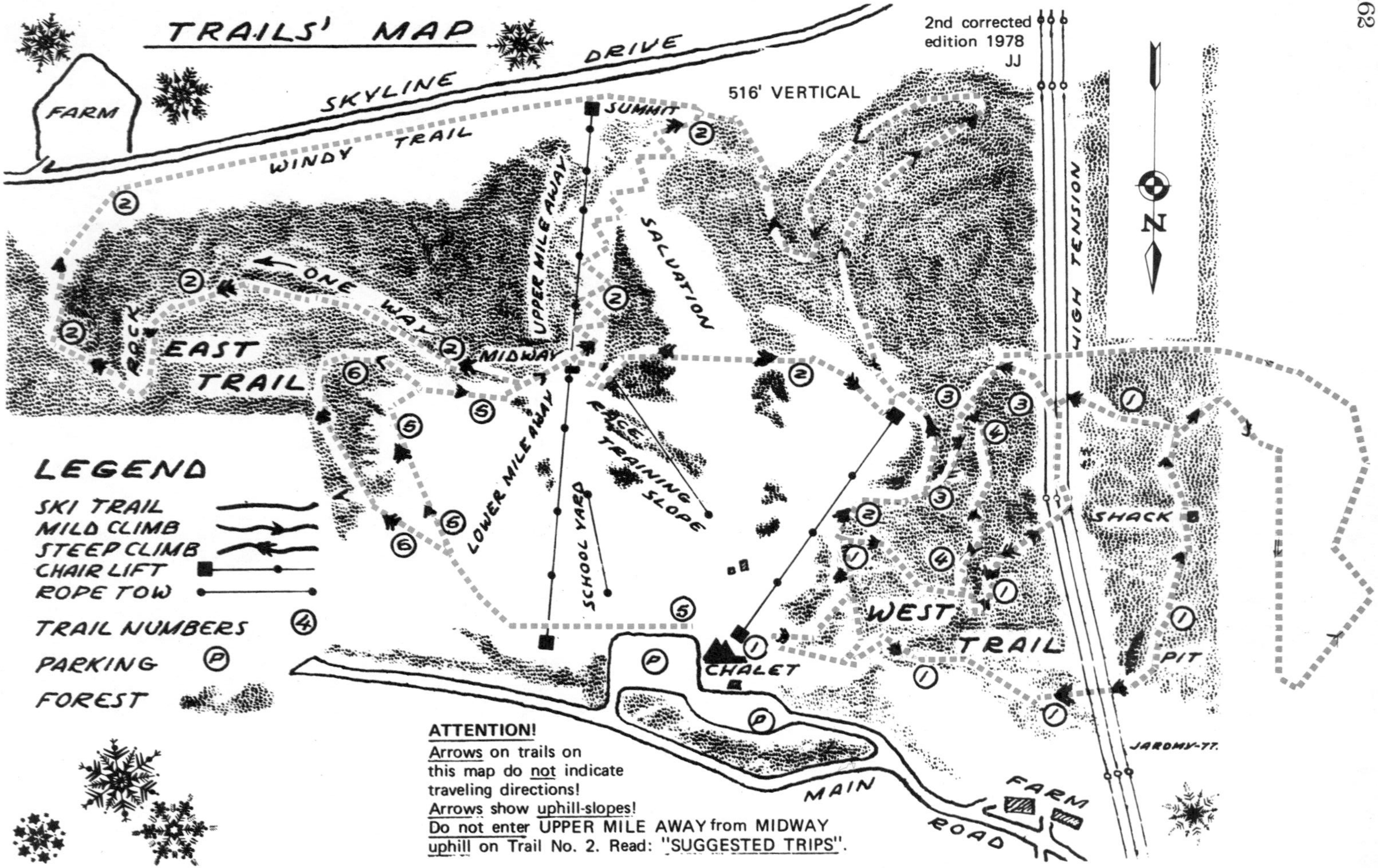

TRAILS' MAP

2nd corrected
edition 1978
JJ

N

516' VERTICAL

SKYLINE DRIVE

WINDY TRAIL

FARM

SUMMIT

SALVATION

UPPER MILE AWAY

MIDWAY

ONE WAY

ROCK

EAST TRAIL

LOWER MILE AWAY

FACE

TRAINING SLOPE

SCHOOL YARD

CHALET

WEST TRAIL

HIGH TENSION

SHACK

PIT

FARM ROAD

MAIN

LEGEND

SKI TRAIL	
MILD CLIMB	
STEEP CLIMB	
CHAIR LIFT	
ROPE TOW	
TRAIL NUMBERS	
PARKING	
FOREST	

ATTENTION!
Arrows on trails on
this map do not indicate
traveling directions!
Arrows show uphill-slopes!
Do not enter UPPER MILE AWAY from MIDWAY
uphill on Trail No. 2. Read: "SUGGESTED TRIPS".

Mt. La Crosse

La Crosse County

Located in the scenic Mississippi River Valley, Mt. La Crosse offers 3.2km (2 miles) of groomed trails.

There are trails not shown on the map, which are on private property and which provide very beautiful views over the Coulee Region and the Mississippi Valley. Such trips range from 4 to 5 miles and our guides will lead you safely through your tour.

There's a comfortable atmosphere in the A-frame Chalet, with a ski shop that reflects the flair of the area.

Mt. La Crosse is located 2 miles south of La Crosse on Hwy. 35.

Ample parking is available at the Chalet. Rentals, lessons and guide service are at your disposal. Trail fee is charged.

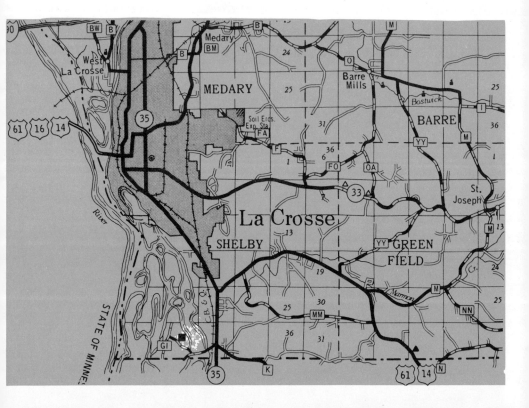

164

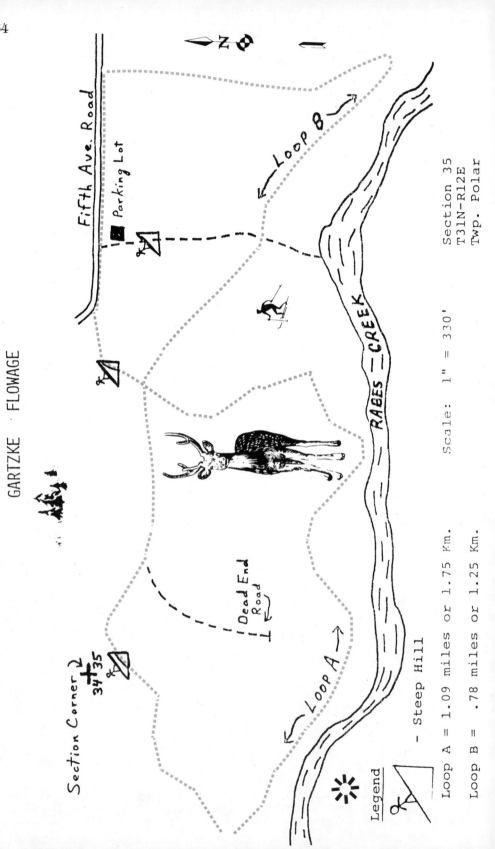

GARTZKE · FLOWAGE

Fifth Ave. Road

Parking Lot

Loop B

Loop A

Dead End Road

Section Corner
34 35

RABES CREEK

Legend

※ - Steep Hill

Loop A = 1.09 miles or 1.75 Km.

Loop B = .78 miles or 1.25 Km.

Scale: 1" = 330'

Section 35
T31N-R12E
Twp. Polar

Gartzke Flowage Ski Trail

This cross country ski trail in the Town of Polar is on county land. The trail is located 9 miles east of Antigo off Fifth Avenue Road just west of the junction of Rabe Creek and Drew Creek. This trail is a relaxing short jaunt for anyone who has a few minutes for a little exercise.

There are two "wee" hills with one at the beginning of Loop A and again midway through Loop A. A steep hill heading from the parking lot to Rabe Creek is not recommended for beginners.

These ski trails lead you through a norway pine plantation to a northern hardwood stand then into a swamp environment along Rabe Creek.

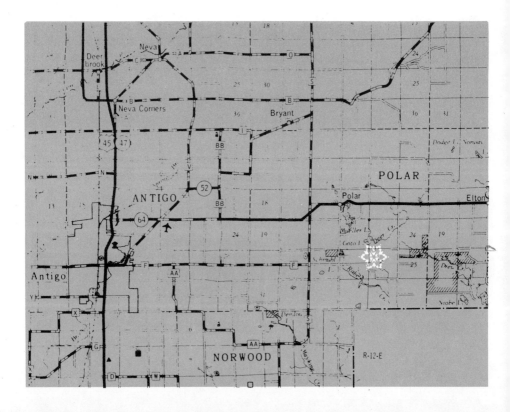

166

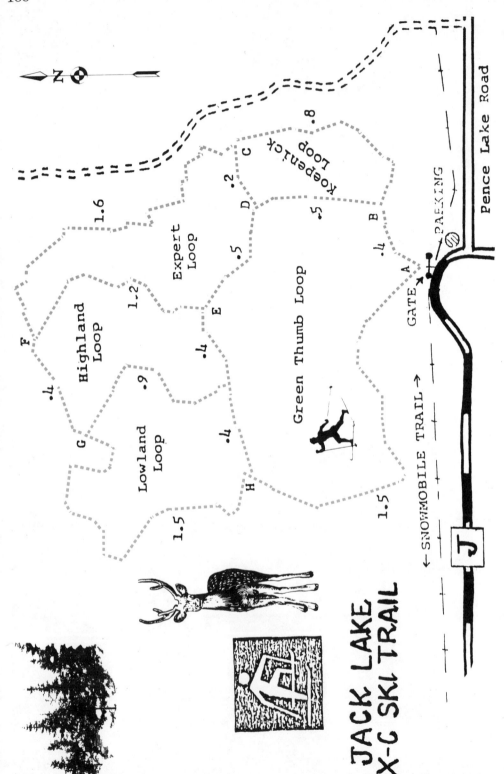

JACK LAKE
X-C SKI TRAIL

N

Pence Lake Road

PARKING

GATE A

← SNOWMOBILE TRAIL →

J

Koepenick Loop
.8
C
.2
D
.5
B
.4

Green Thumb Loop
.5
E
.4
H
1.5

Expert Loop
1.6
1.2
.5
.4

Highland Loop
F
.4
G
.9

Lowland Loop
1.5
.4

Jack Lake Trail

The Jack Lake Ski Trail is located on the Langlade County Forest approximately 12 miles north of Antigo on Hwy. "45" then east on County Trunk "J" for 2 miles to the junction of Pence Lake Road and the road to Veteran's Memorial Park.

Green Thumb Loop (The Basic Trail) — Starting at point "A" at the parking lot, this trail makes a loop returning to the parking lot. This is a good loop for beginners with only a few gently hills. You have the option of taking any of the other four loops which branch off from the Green Thumb Loop.

The Expert Loop — Caution: The segment of the Expert Loop between points C and F has steep hills and difficult curves. DO NOT ski this segment unless you are truly an Expert Skier.

One Way Ski Trail — The mileage for each segment is indicated on the map.

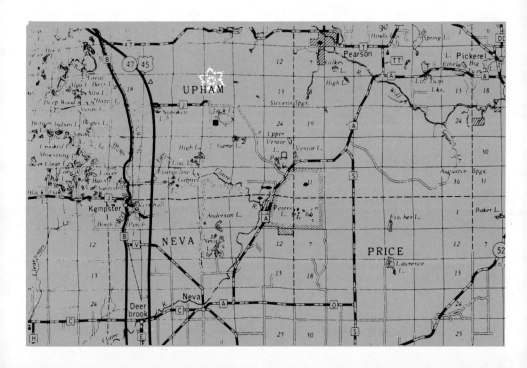

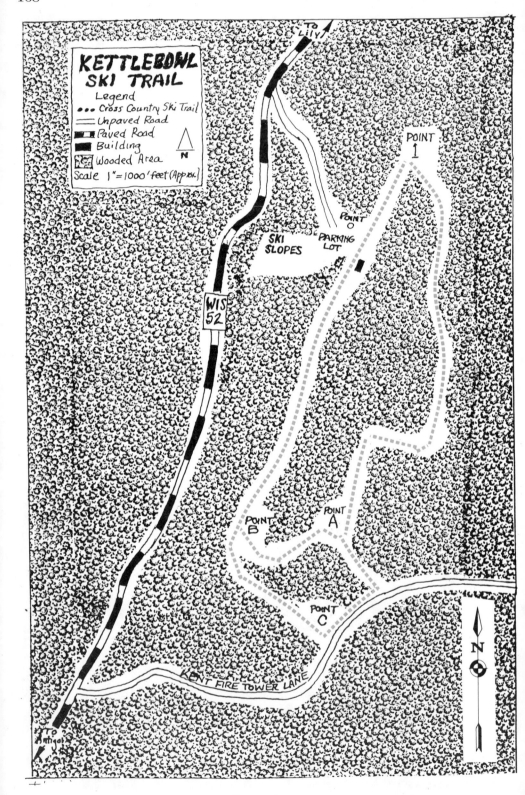

KETTLEBOWL
SKI TRAIL

Legend
••• Cross Country Ski Trail
═══ Unpaved Road
▬▬ Paved Road
■■ Building
▓▓ Wooded Area
Scale 1"=1000' feet (Approx.)

N

TO LILY

POINT

POINT O

SKI SLOPES

PARKING LOT

WIS 52

POINT B

POINT A

POINT C

KENT FIRE TOWER LANE

TO Antigo

N

Kettlebowl Ski Trail

This trail is located at Kettlebowl Ski Area. The winter sports area is located on Wisconsin Highway 52 about 15 miles northeast of Antigo. The trailhead is identified by the map case with the cross country ski symbol in the northeast corner of the parking lot. Just beyond the parking lot is Point 1. At this point follow the trail on the right.

The first part of the trail is a gradual climb with one steep ridge. At Point "A" skiers desiring the longer loop follow the blue and white tape to the Kent Fire Tower Road and turn right for a quarter mile to Point "C". At this point the ski trail joins the Ice Age Trail marked in pink paint for the return trip to the parking lot.

Skiers following the shorter loop will join the Ice Age Trail marked in pink paint at Point "B".

The Ice Age stretch of the loop is an exciting downhill return to the parking lot. Some experience and skill is required. The Kettlebowl loop is located on land owned by Langlade County, with exception of a short stretch on privately owned land. The ski area is open during daylight hours on most winter weekends.

Short Loop   1.75 miles   3.8 kilometers
Long Loop   2.5 miles   5.0 kilometers

ELCH X-C SKI CLUB

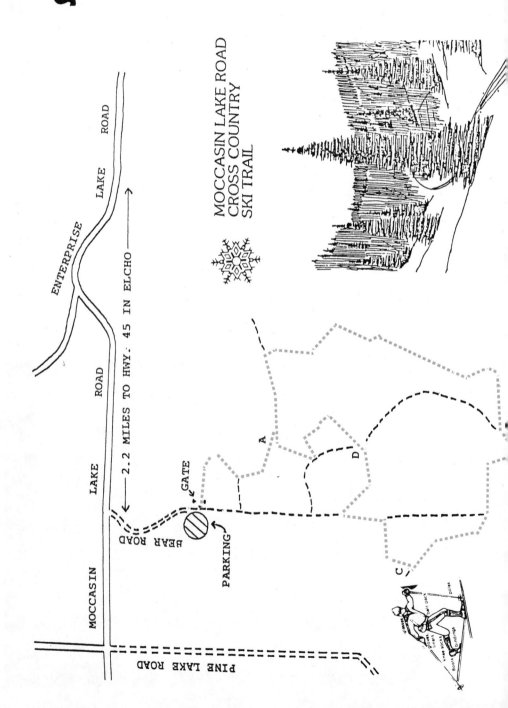

MOCCASIN LAKE ROAD
CROSS COUNTRY
SKI TRAIL

Moccasin Lake Road Ski Trails

Langlade County

This trail was developed by the Langlade County Forestry Department and the Elcho CC Ski Club. The trail is groomed with a track setter. Maintenance is paid entirely by Ski Club. Any donation would be appreciated.

The 5.0 mile double tracked trail is rated as an intermediate class. It weaves its course through rolling hardwoods. A portion of the trail offers a view of a vast black spruce swamp.

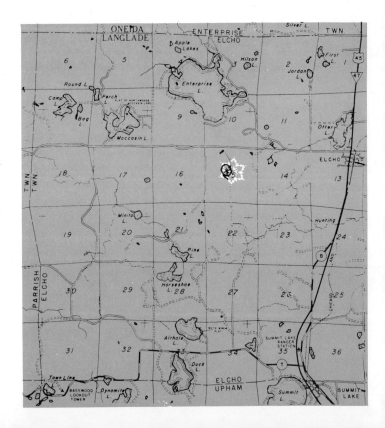

POST LAKE CROSS COUNTRY SKI TRAIL

A PUBLIC SKI TRAIL CREATED FOR WINTER USE

by

Langlade County Forestry Department
P.O. Box 460
Antigo, WI 54409

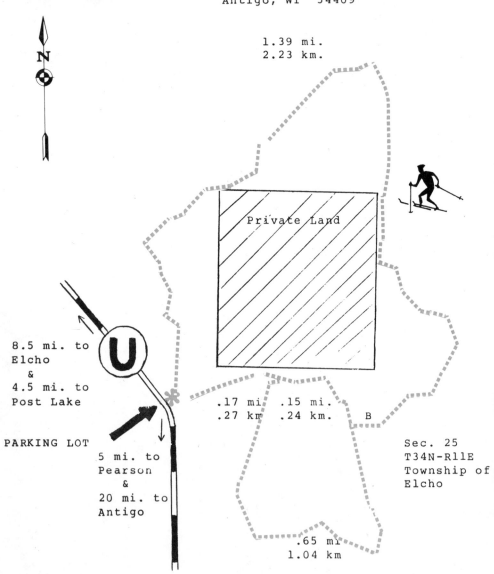

N

1.39 mi.
2.23 km.

Private Land

8.5 mi. to
Elcho
&
4.5 mi. to
Post Lake

PARKING LOT

5 mi. to
Pearson
&
20 mi. to
Antigo

.17 mi.   .15 mi.
.27 km   .24 km.   B

Sec. 25
T34N-R11E
Township of
Elcho

.65 mi.
1.04 km

Outer Loop - 2.21 miles
             3.55 kilometers

Inner Loop - 1.71 miles
             2.75 kilometers

Post Lake Trail

Langlade County

Two loops comprise this easy, family trail. The scenery of the woodlands and the Wolf River make this a trail to be remembered.

The main loop is 1.71 miles and a smaller loop to the south is 1.14 miles. Trail starts and returns to the parking lot. By leaving the parking lot on the north trail (left) you will encounter the only downhill spot. This hill is shortly after you leave the parking lot.

Comments on the trail can be sent to the Forestry Dept., P.O. Box 460, Antigo, WI 54409

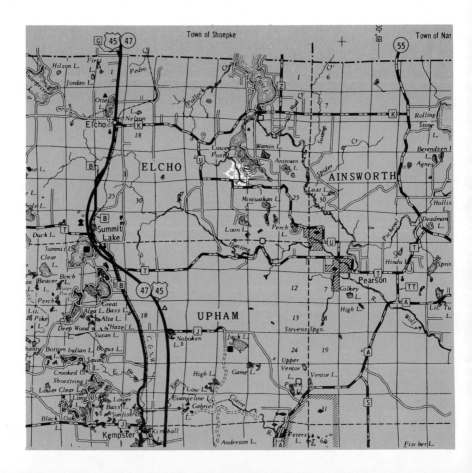

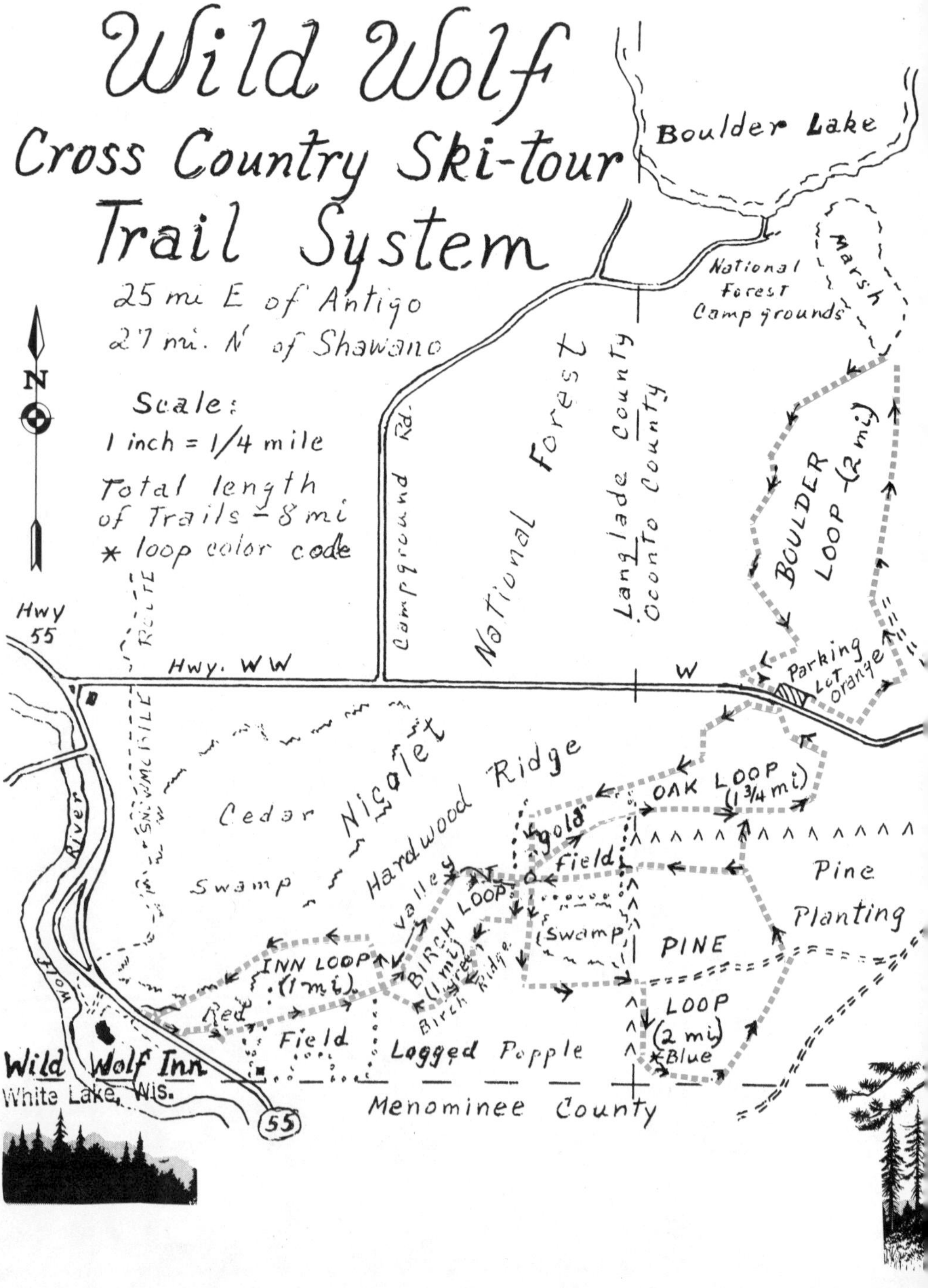

Wild Wolf
Cross Country Ski-tour
Trail System

25 mi. E of Antigo
27 mi. N of Shawano

Scale:
1 inch = 1/4 mile
Total length
of Trails = 8 mi
* loop color code

N

Hwy 55

Hwy. WW

Boulder Lake

National Forest Campgrounds

Marsh

National Forest

Langlade County
Oconto County

BOULDER LOOP (2 mi)

W

Parking Lot Orange

Campground Rd.

SNOWMOBILE ROUTE

River

Wolf River

Cedar Swamp

Nicolet

Hardwood Ridge

Gold Field

OAK LOOP (1¾ mi)

Pine Planting

Valley

BIRCH LOOP

Birch Ridge

Swamp

PINE

INN LOOP (1 mi)

Red Field

Birch Ridge (¼ mi)

LOOP (2 mi) Blue

Logged Popple

Wild Wolf Inn
White Lake, Wis.

55

Menominee County

Wild Wolf Trails

Langlade County

The Nicolet National Forest is the setting for these trails and they wind thru Pines, Birch and Hardwood Stands. Parts of the trails are flat and others have some good hills that will please the intermediate. The Oak loop has a good long downhill run from where the orange parking lot is on County W. Regular parking is at the Wild Wolf Inn on Hwy. 55. The ski trails are across the road from the Inn and snowmobile trails are well separated on the opposite side. Trails are well marked and are one way. The Boulder Loop is on the other side of County W. Total length of trails is 12.8km (8 miles).

The Inn has a good motel complex with complete meals, bar and dance floor. Rentals are available at the office.

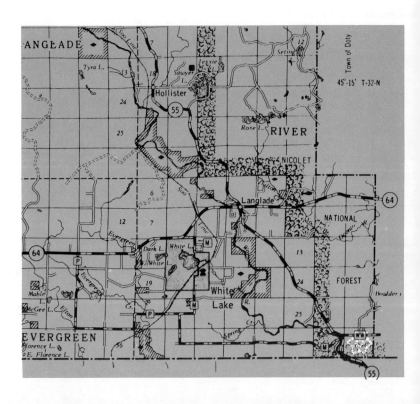

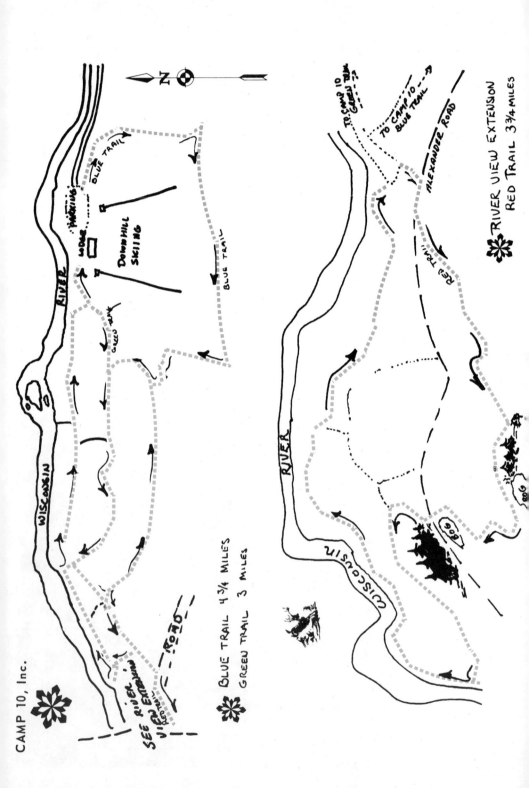

CAMP 10, Inc.

N

RIVER

WISCONSIN

BLUE TRAIL

PARKING

LODGE

DownHill SkiING

GREEN TRAIL

BLUE TRAIL

SEE RIVER VIEW EXTENSION

ROAD

BLUE TRAIL 4¾ MILES
GREEN TRAIL 3 MILES

TO CAMP 10
GREEN TRAIL

TO CAMP 10
BLUE TRAIL

ALEXANDER ROAD

RED TRAIL

RIVER

WISCONSIN

BOG

BOG

RIVER VIEW EXTENSION
RED TRAIL 3¾ MILES

Camp 10

The unique five loop 18.5km (11½ mile) layout provides an excellent selection of terrain of gentle-hilly, wooded, marked, groomed and tracked trails. The forest trails will enhance your tour with beautiful scenery and the treat of seeing an occasional deer, beaver or eagle along the way.

Lodge and rental shop are available. A small trail fee is required.

Camp 10 also offers downhill skiing.

It is located 10 miles S.W. of Rhinelander on CTH A.

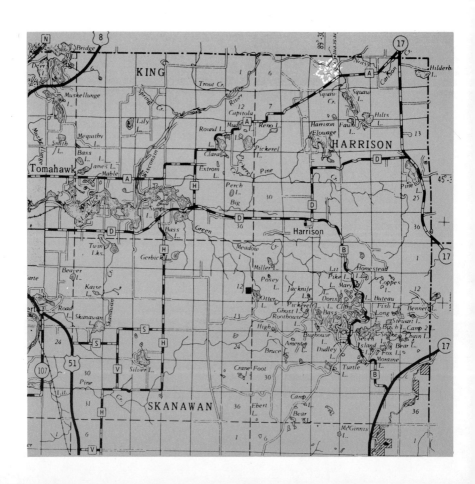

178

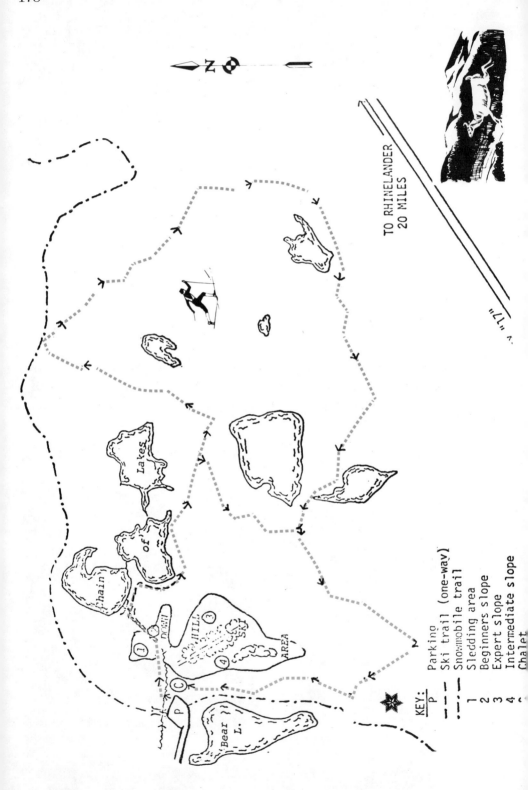

Harrison Hills Ski Trail

Lincoln County

The Harrison Hills ski trail consists of two loops which begin at the parking area. The trail traverses lakes and beautiful rolling hills. The extremely scenic trail is designed for an intermediate skier but a beginner should be able to ski the shorter loop. The west loop is approximately 2 miles long and the perimeter of both loops together is approximately 5 miles. The trails are marked and groomed.

The Harrison Hills Winter Recreation Area can be reached from: Merrill — Highway 17 north to County Trunk B, north on B to Chain O Lakes Road. Tomahawk — County Trunk D east to County Trunk B, south on B to Chain O Lakes Road. Rhinelander — Highway 17 south to County Trunk B, north on B to Chain O Lakes Road.

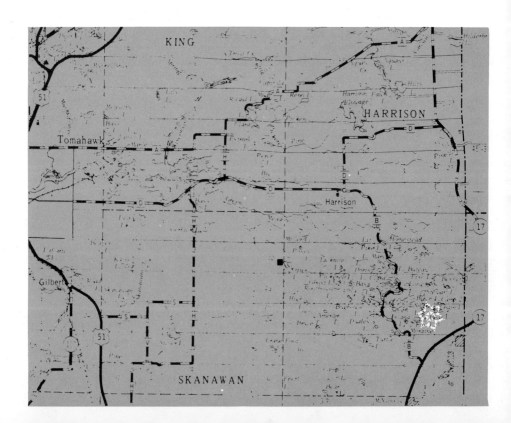

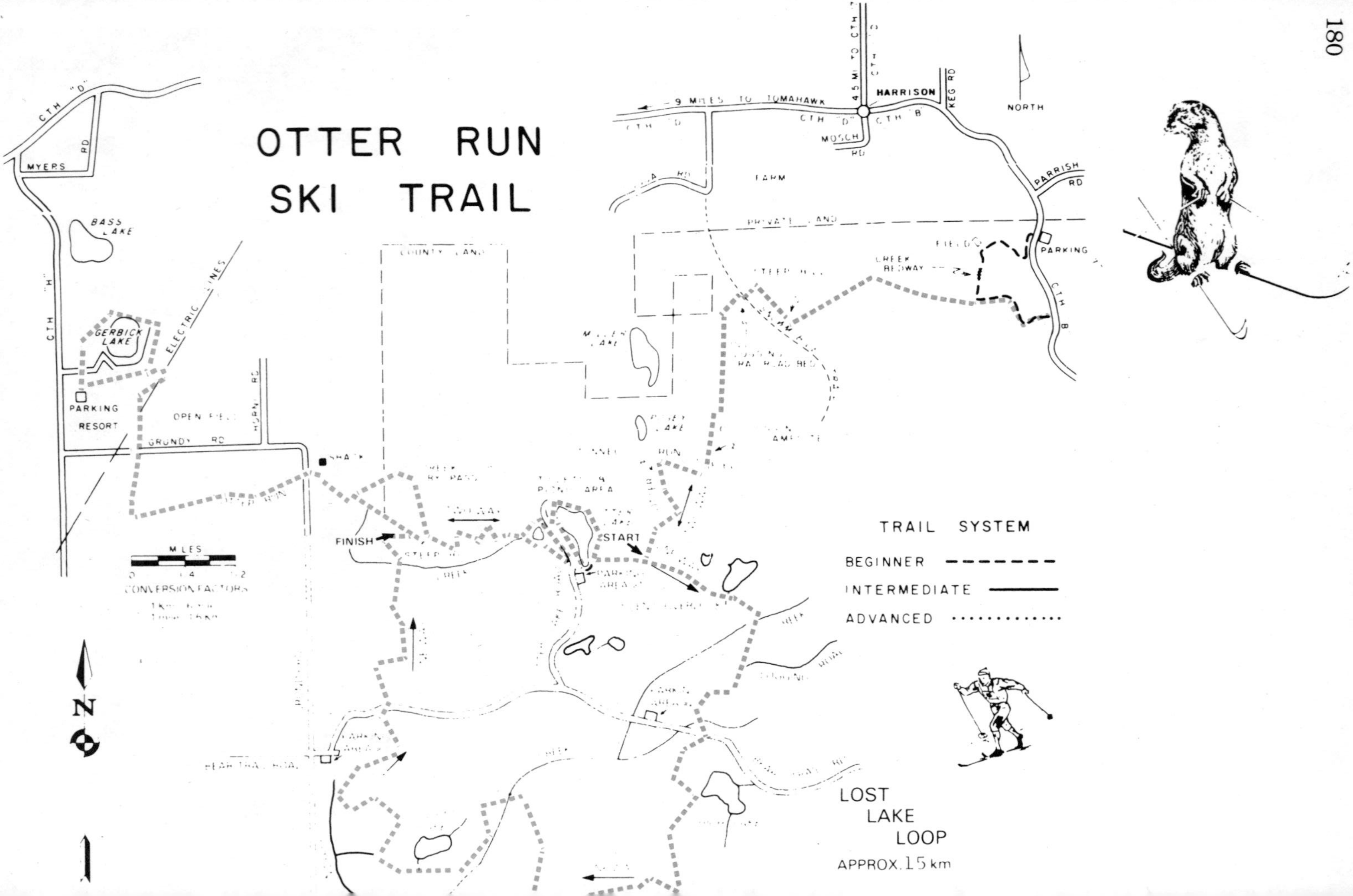

OTTER RUN
SKI TRAIL

TRAIL SYSTEM

BEGINNER ▬ ▬ ▬
INTERMEDIATE ▬▬▬▬
ADVANCED ·········

LOST
LAKE
LOOP
APPROX. 1.5 km

NORTH

HARRISON

9 MILES TO TOMAHAWK

45 MI. TO CTH

CTH B

KEG RD

PARRISH RD

PARKING

START

FINISH

MILES

CONSERVATION. TRHS.

BASS LAKE

GERBICK LAKE

PARKING

RESORT

ELECTRIC LINES

OPEN FIELDS

MYERS RD

CTH H

CTH O

N

Otter Run Trail

The area's unique ecologic features make the Otter one of the most awe-inspiring experiences in cross-country skiing.

The trail may be skied from 3 points.

1. Gerbick Lake (Berkinsee Resort) — This end connects to the recreation area via 12km (7.5 miles) of point.to point intermediate trail. The two-way trail is marked for night skiing and is tracked and groomed.

2. Otter Lake Recreation Area to Hwy. B or, 3. Hwy. B to Otter Lake Recreation Area. This is a two-way advanced 12km (7.5 miles) trail for the experienced skier, which is marked with reflectorized signs. A 1.6km (1 mile) beginners loop is located at the Hwy. B trailhead. Water, toilets and picnic area is offered at Otter Lake.

The Otter Run ski trail can be reached from three directions.

Tomahawk — Cty. Hwy. D east to Harrison — Cty. Hwy. B to marked parking lot (2 miles).

Merrill — Hwy. 51 north to Cty. Hwy. S, east on S to Cty. Hwy. H (north on H to Berkinsee Resort) or continue east on Town Hall Rd. 1½ miles to Grundy Rd. — follow directional signs to Otter Lake Recreation Area.

Rhinelander — Hwy. 17 to Cty. Hwy. D to Harrison, turn left on Cty. Hwy. B to marked parking lot (2 miles).

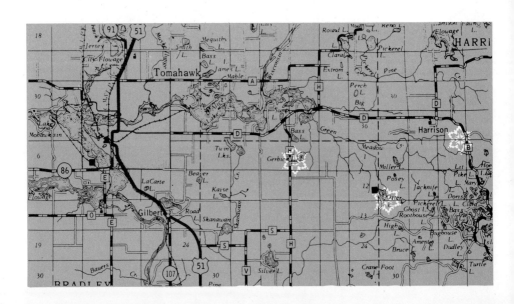

UNDERDOWN SKI TRAIL

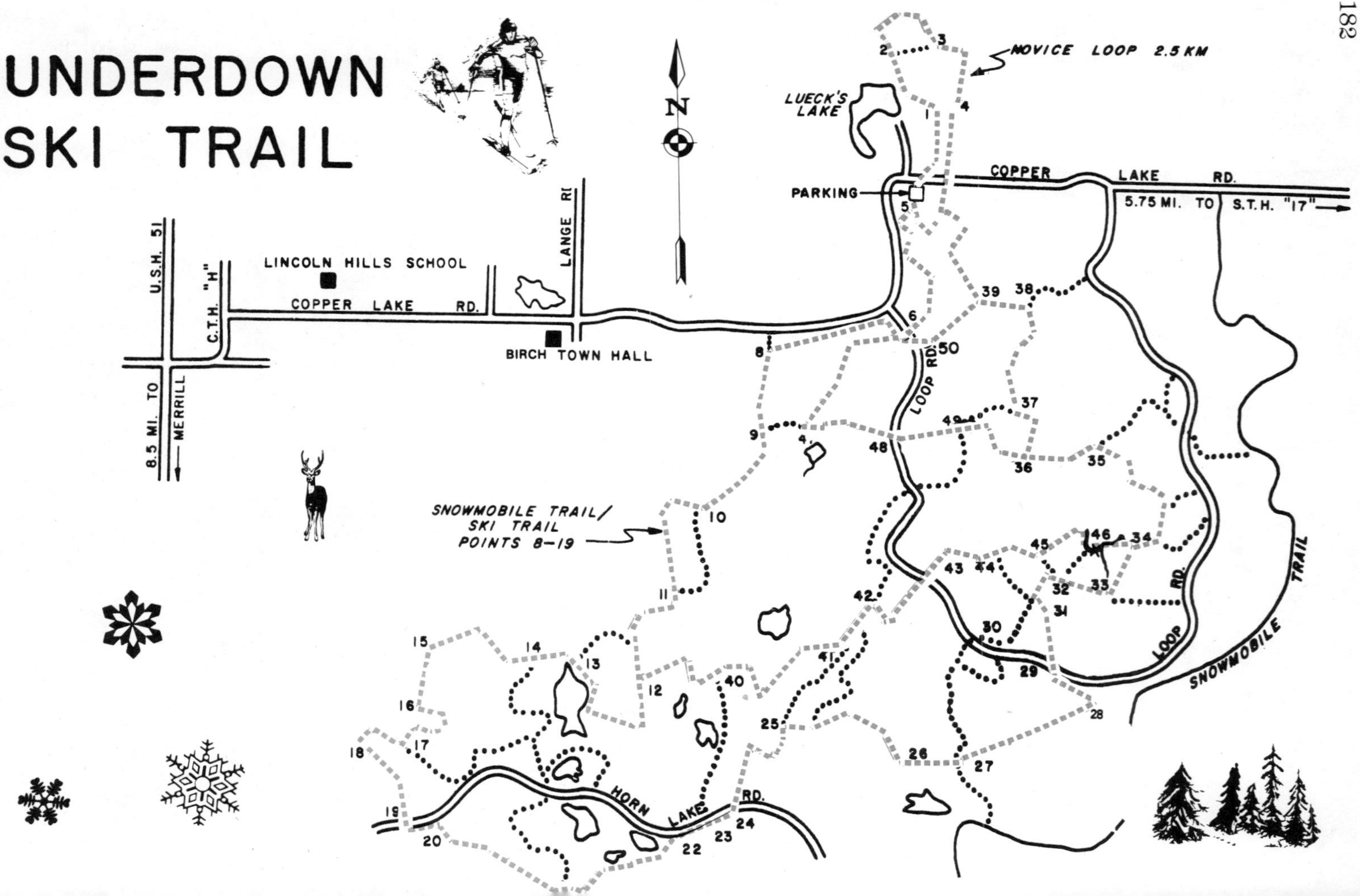

NOVICE LOOP 2.5 KM
LUECK'S LAKE
PARKING
COPPER LAKE RD.
5.75 MI. TO S.T.H. "17"
N
LANGE RD
LINCOLN HILLS SCHOOL
COPPER LAKE RD.
BIRCH TOWN HALL
U.S.H. 51
C.T.H. "H"
8.5 MI. TO MERRILL
LOOP RD.
SNOWMOBILE TRAIL/
SKI TRAIL
POINTS 8–19
HORN LAKE RD.
SNOWMOBILE TRAIL
1
2
3
4
5
6
8
9
10
11
12
13
14
15
16
17
18
19
20
22
23
24
25
26
27
28
29
30
31
32
33
34
35
36
37
38
39
40
41
42
43
44
45
46
48
50

Underdown Trail

Lincoln County

The Underdown ski area consists of 33km of marked trails, which traverse some 4,600 acres of Lincoln Forest.

The trail network consist of 3 loops. Located north of the parking lot is a 2.5km (1.6 miles) beginners loop. The blue trail is a 20km (12.4 mile) loop which circles the perimeter of the ski area. The blue trail is connected, at numbered intersections to the red trails, a 5.5km (3.4 mile) loop which is also connected at numbered intersections to the green trail, a 4.5km (2.8 mile) loop.

The loop system is well suited to the experienced novice and advanced skier.

Yellow blaze marks on trees designate the ice age hiking trail which does not make a loop.

Underdown is located 8 miles north of Merrill on U.S. 51, then 1 mile east and north on CTH H to Copper Lake Rd., 3.5 miles east and north.

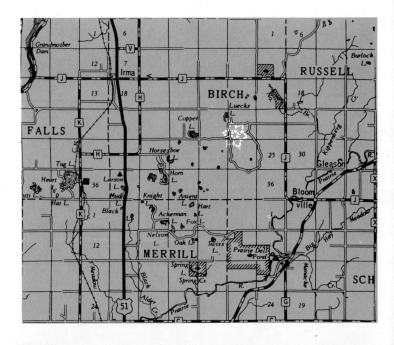

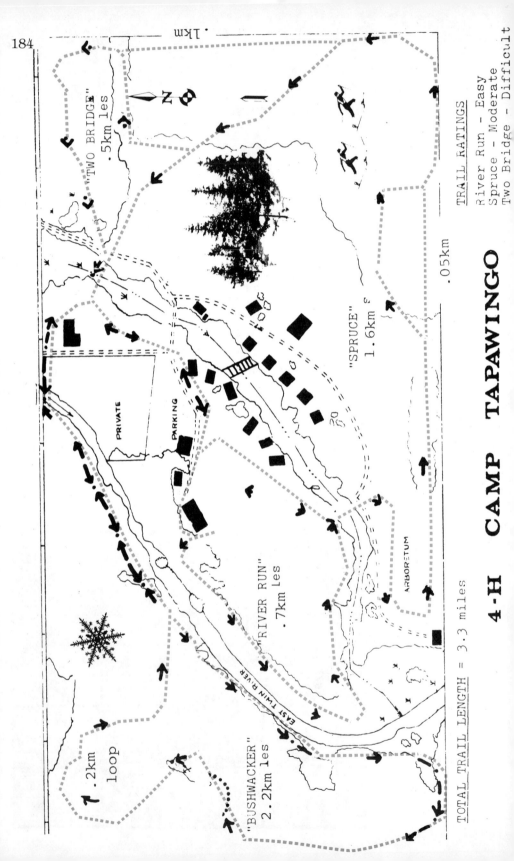

184

TRAIL RATINGS

River Run - Easy
Spruce - Moderate
Two Bridge - Difficult

"TWO BRIDGE"
.5km les

N

"SPRUCE"
1.6km s

PRIVATE

PARKING

ARBORETUM

"RIVER RUN"
.7km les

EAST TWIN RIVER

.2km loop

"BUSHWACKER"
2.2km les

.05km

4-H CAMP TAPAWINGO

.1km

TOTAL TRAIL LENGTH = 3.3 miles

Camp TaPaWingo

Manitowoc County

Camp TaPaWingo is located three and one half miles north of Mishicot on highway 163, one half mile west on TaPaWingo Road, signs at entrance.

Four adjoining loops, from gentle to hilly, pass through a variety of "natural" areas — mixed forest, restored prairie, conifer plantation, cedar swamp, and managed wildlife habitat.

Trails, lessons, and rentals are available weekdays, weekends by reservation. Parking, toilets, shelter, and lodging for groups are also available.

For information or brochure contact: Director; Camp TaPaWingo; Route 1; Mishicot, WI 54228; (414) 755-2785.

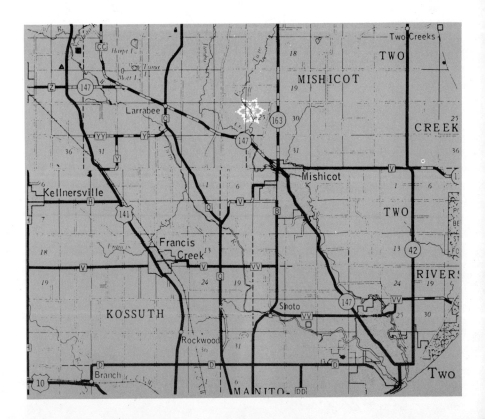

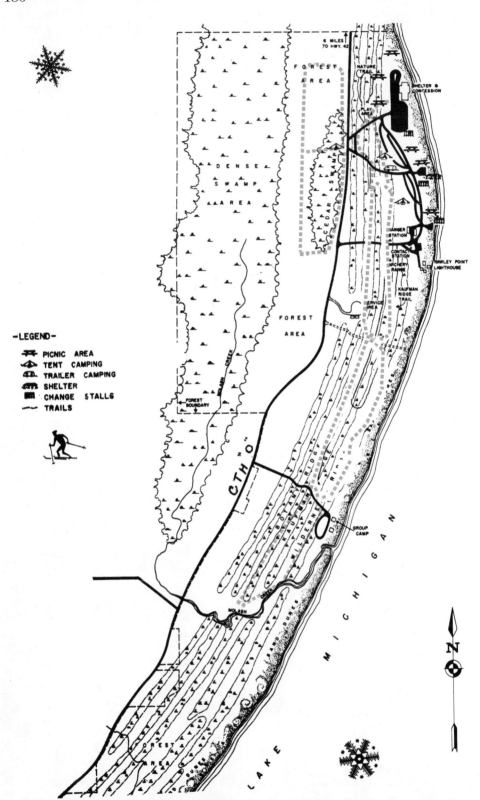

Point Beach
State Forest

Manitowoc County

All trails begin near concession building at north end of campground.

There is also an enclosed shelter with a fireplace at the concession building. It is open on weekends, or others days by appointment. Shelter is an open pavillion w/roof. Charcoal grills are available in pavillion area. Rest rooms are available at start of trail and at the group camp near the south end of trail.

South end of trail turns toward Lake Michigan for return of starting point. This is a good stopping area for lunch, refreshments, & rest. At this point you are on the banks of Molash Creek. This creek is skiable and skiers have cut some trails down & back along the creek itself.

The return trip on the shoreline winds in and out of the forest. The shore itself has small sand dunes which makes for an interesting site.

There is also 3 miles of trail west of CTH "O". This trail is wooded and terrain varies from flat to quite hilly.

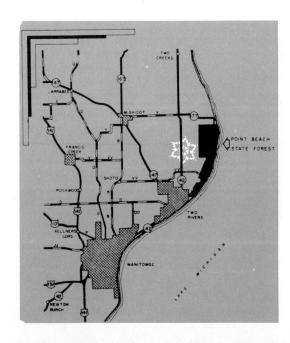

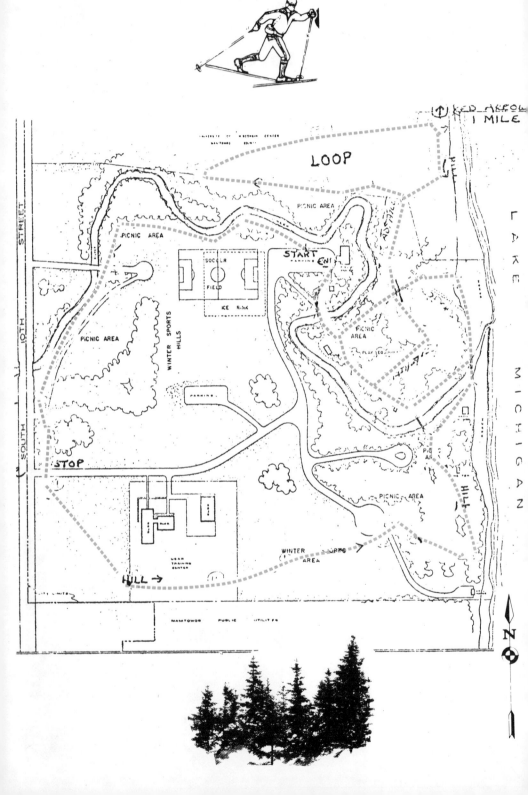

Silver Creek Park

Manitowoc County

The outside loop is 3½ miles of wooded generally flat with a few hills. Plowed parking and restrooms. The Henry Schuette Park is under development and should also be ready by 1984.

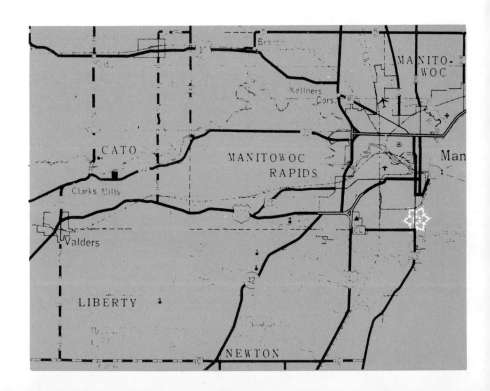

NATURE CENTER

- BIRD-BANDING STATION
- WILDLIFE RESEARCH AREA
- WILD FLOWER SANCTUARY

Division of
NATURAL AREAS PRESERVATION, INC.
P. O. Box 740, Appleton, Wisconsin 54911

Goodwin Rd

------- Black Cherry Trail

············ Trillium Trail

BLACK CHERRY TRAIL

TRILLIUM TRAIL

N

Woodland Dr

Northwestern R.R.

Memorial Dr

Hwy 42

Lake Michigan

Woodland Dunes Nature Center

Manitowoc County

A unique ski tour awaits the skier in the nature center located between Manitowoc and Two Rivers. Two one way trails are contained in 160 acres of snow-laden land which winds in and out of forests. Colored owls mark the 1¾ mile and the 2½ mile trails.

Woodland Dunes is unique in being a Tension Zone that has both southern and northern birds and animals. The Center is open 24 hours a day, 7 days a week, with ample parking available.

A trail fee is not required, but either a donation or membership would be greatly appreciated.

As a member you enjoy over twenty-five events held throughout the year admission free. Members receive a quarterly "Dunesletter" packed with information. If you are interested in a membership please write: Woodland Dunes, P. O. Box 763, manitowoc, WI 54220.

These trails make an interesting tour in the summer on foot and then a comparison ski tour in the winter.

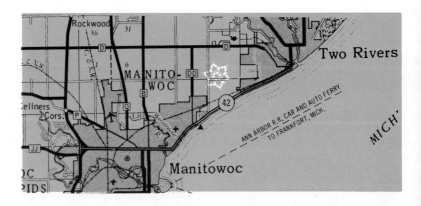

Big Eau Pleine Park
Ski Trail

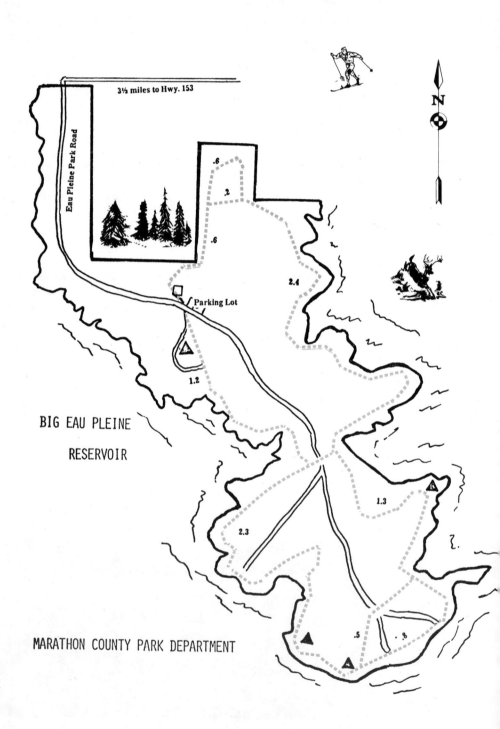

3½ miles to Hwy. 153

Eau Pleine Park Road

N

.6

.2

.6

2.4

Parking Lot

1.2

BIG EAU PLEINE

RESERVOIR

1.3

2.3

.5

.8

MARATHON COUNTY PARK DEPARTMENT

Big Eau Pleine Park Trail

Marathon County

The Big Eau Pleine trail follows the shoreline of the Big Eau Pleine Reservoir for 15 kilometers within the boundary of the Big Eau Pleine County Park. Wildlife abounds on this wilderness trail. It is not well marked or groomed.

Directions: Take Highway 153 West from Mosinee for about 8 miles. Then turn left or South on Eau Pleine Park Road. Follow signs to park.

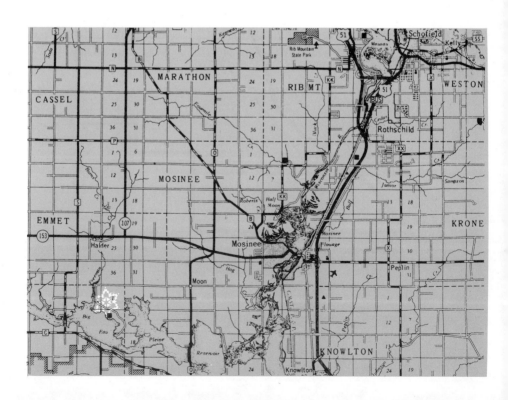

Brokaw Heights
X-C Ski Trail

DISTANCES

TOTAL DISTANCE	9.1 mi/15.1 km
LOOP A	1.8 mi/ 3.0 km
TRAIL SEGMENT B	.6 mi/ 1.0 km
LOOP C	1.48 mi/ 2.4 km
LOOP D	4.04 mi/ 6.5 km
LOOP E	.9 mi/ 1.4 km
WAXING LOOP	.3 mi/ .5 km

LEGEND

SKI TRAIL ———

WAUSAU PAPERS LAND ━━━

WISCONSIN RIVER

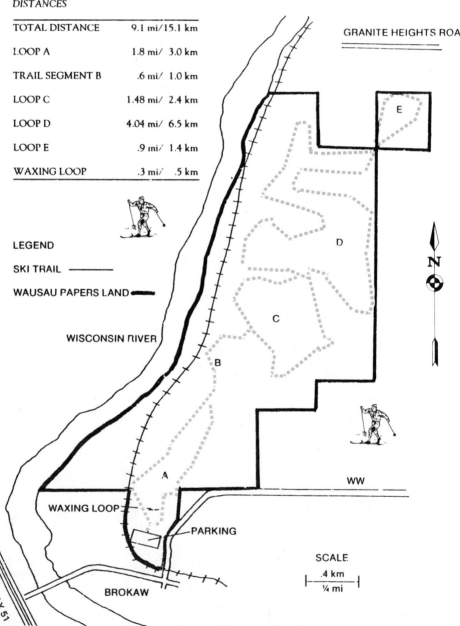

GRANITE HEIGHTS ROAD

E

D

C

B

A

N

WW

WAXING LOOP

PARKING

HIGHWAY 51

BROKAW

SCALE

.4 km
¼ mi

Brokaw Heights Trail

Brokaw Heights is a challenging trail for advanced skiers. The trail winds through a scenic hardwood forest owned by Wausau Paper Mills Company. The total length of the trail is about 15 kilometers, with cut-off loops available. It is regularly groomed and well maintained. Donations will be appreciated. No warming hut or toilets are available at the trail site, and be advised that there are no services available in Brokaw.

Directions: Take Highway 51 North from Wausau for about 5 miles to WW. Follow WW East through the Village of Brokaw. The trail head is on the left hand side of WW after crossing the railroad track.

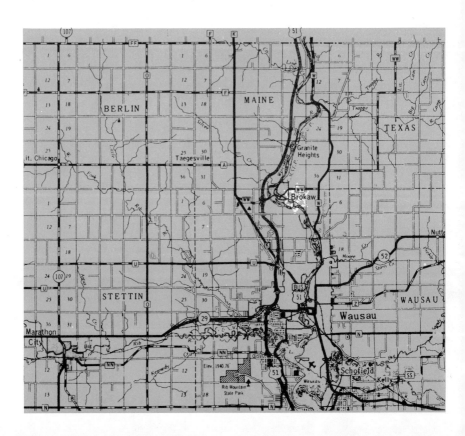

Kronenwetter
Ski Touring Trail

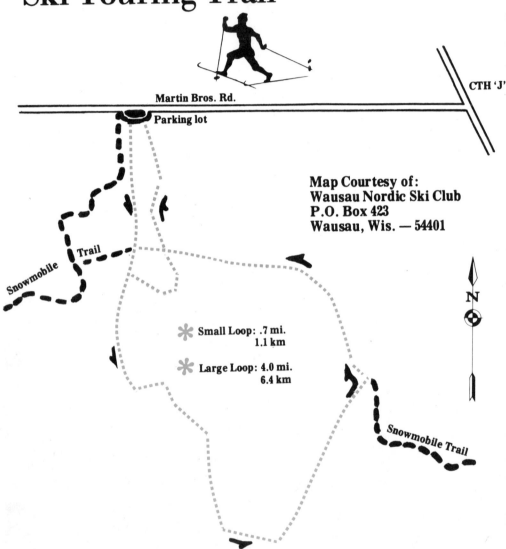

Martin Bros. Rd.

CTH 'J'

Parking lot

Map Courtesy of:
Wausau Nordic Ski Club
P.O. Box 423
Wausau, Wis. — 54401

Snowmobile   Trail

Snowmobile

✻ Small Loop: .7 mi.
              1.1 km

✻ Large Loop: 4.0 mi.
              6.4 km

N

Snowmobile Trail

Kronenwetter

Marathon County

The Kronenwetter Ski Touring Trail is a 6.4 kilometer wilderness trail for beginners that goes through a hardwood forest. The trail may not be tracked.

Directions: From Wausau, take Highway 29 East until J. Turn South on J, and proceed 4 miles to Martin Brothers Road. Turn right or West on Martin Brothers Road. Trail head is about 1 mile down this road.

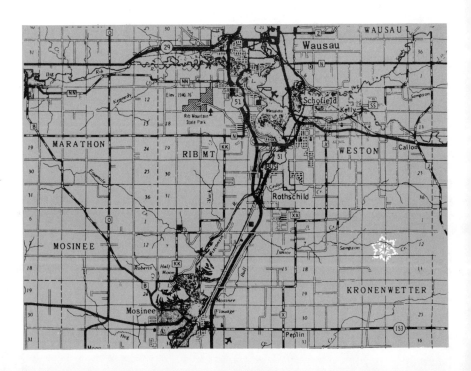

198

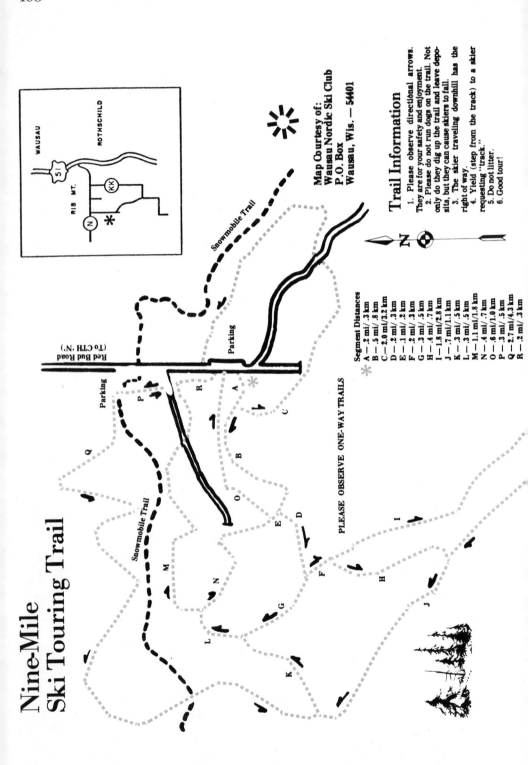

Nine-Mile Ski Touring Trail

Snowmobile Trail

Snowmobile Trail

Red Bud Road (To CTH 'N')

Parking

Parking

Parking

PLEASE OBSERVE ONE-WAY TRAILS

P
Q
R
O
A
B
C
D
E
F
G
H
I
J
K
L
M
N

Segment Distances

A — .2 mi/.3 km
B — .5 mi/.8 km
C — 2.0 mi/3.2 km
D — .2 mi/.3 km
E — .1 mi/.2 km
F — .2 mi/.3 km
G — .3 mi/.5 km
H — .4 mi/.7 km
I — 1.8 mi/2.8 km
J — .7 mi/1.1 km
K — .3 mi/.5 km
L — .3 mi/.5 km
M — 1.1 mi/1.8 km
N — .4 mi/.7 km
O — .6 mi/1.0 km
P — .3 mi/.5 km
Q — 2.7 mi/4.3 km
R — .2 mi/.3 km

WAUSAU
ROTHSCHILD
51
RIB MT.
KK
N

N

Map Courtesy of:
Wausau Nordic Ski Club
P.O. Box
Wausau, Wis. — 54401

Trail Information

1. Please observe directional arrows. They are for your safety and enjoyment.
2. Please do not run dogs on the trail. Not only do they dig up the trail and leave deposits, but they can cause skiers to fall.
3. The skier traveling downhill has the right of way.
4. Yield (step from the track) to a skier requesting "track."
5. Do not litter.
6. Good tour!

Nine Mile Cross Country Ski Trail

Marathon County

The Nine Mile Trail offers a network of trails totaling about 20 kilometers, with loop distances varying from 2 to 12 kilometers. Skiers of all abilities may choose from a variety of loops that meander through a conifer and hardwood forest owned by the Marathon County Forestry Department. The trail is well maintained and groomed regularly. Donations will be appreciated. Toilet facilities are available. Some changes and improvements in the trail and facilities are anticipated. Plans are not yet finalized at the time of printing.

Directions: Take Highway N exit west off of Highway 51. Proceed West on N approximately 4 miles, and then South on Red Bud Road for 1½ miles to the parking lot.

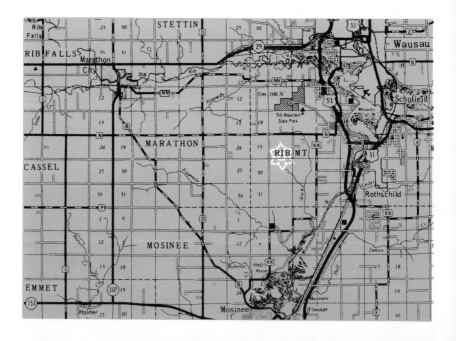

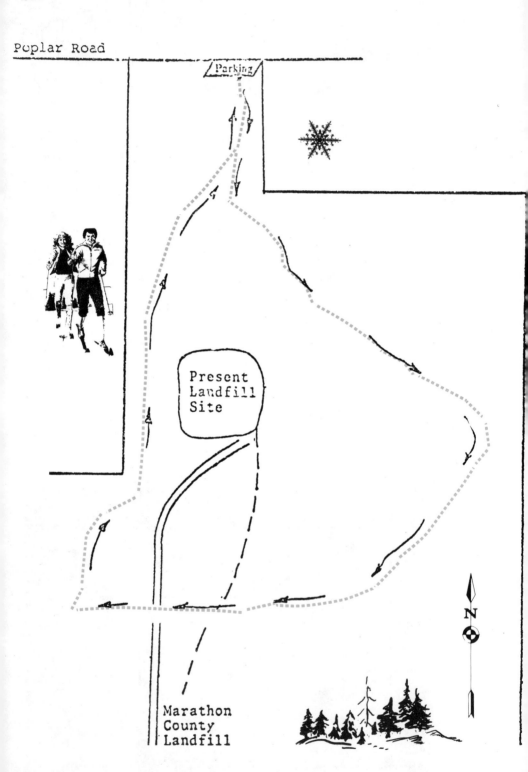

Poplar Road

Parking

Present
Landfill
Site

Marathon
County
Landfill

N

Ringle Ski Touring Trail

Marathon County

The Ringle Ski Touring Trail is well suited to beginning skiers. It is a 6.4 kilometer trail set on gently rolling and varied terrain owned by the Solid Waste Board. The trail is groomed regularly.

Directions: Take Highway 29 East from Wausau to Ringle. Turn left or North on Q. and proceed for 1 mile to Poplar Road. The trail head is 1 mile East on Poplar Road.

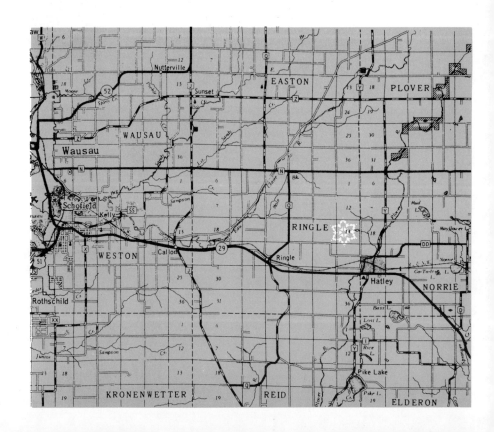

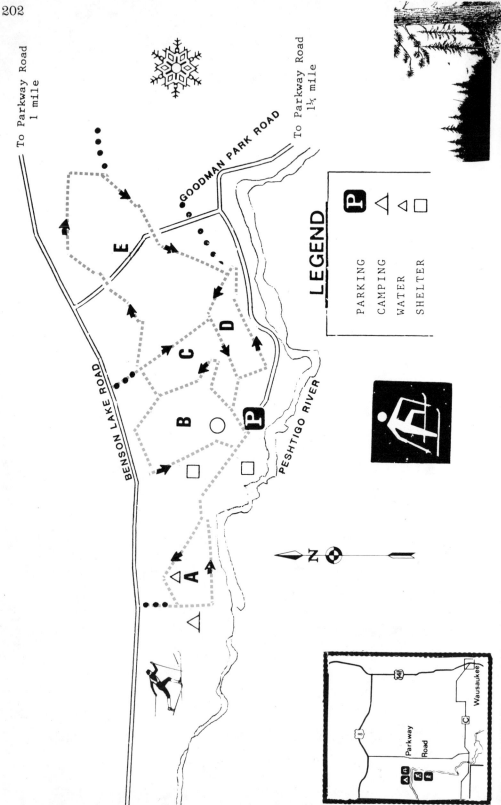

To Parkway Road
1 mile

To Parkway Road
1¼ mile

GOODMAN PARK ROAD

BENSON LAKE ROAD

PESHTIGO RIVER

E

C

D

B

A

P

LEGEND

P   PARKING

△   CAMPING

▵   WATER

□   SHELTER

N

Parkway
Road

Wausaukee

Goodman Park Trail

Marinette County

The trailhead is at Goodman Park where a plowed parking lot is available with toilets and shelter.

The five loops comprise 3½ miles of trail.

Trails are marked with signs and blue flagging.

The trail runs to McClintock Park which is 8km (5 miles) and returns to Goodman Park. The setting on these trails is very rustic and natural. Trails are mostly within a wildlife refuge.

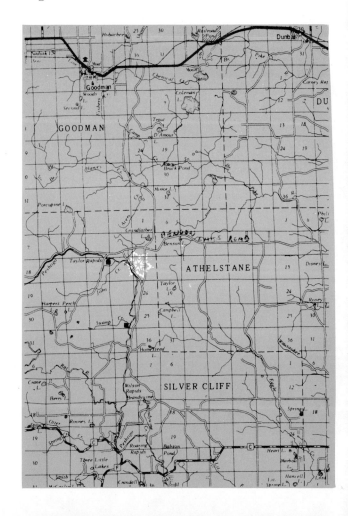

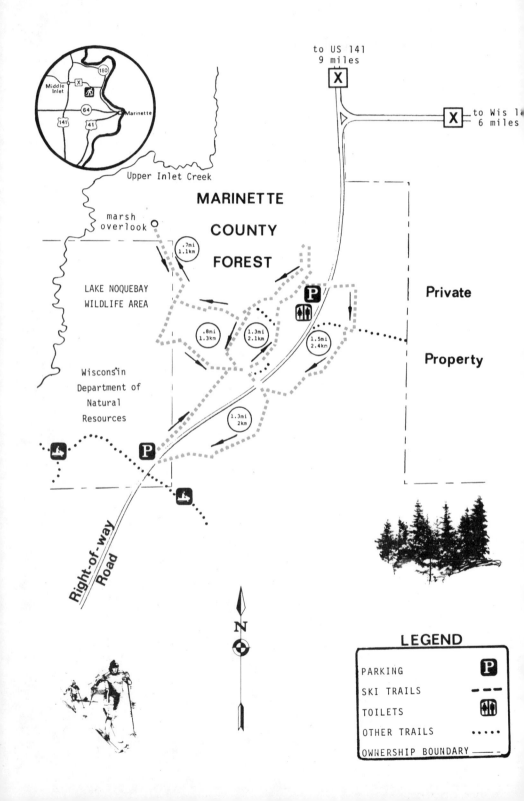

to US 141
9 miles

to Wis 1
6 miles

Upper Inlet Creek

MARINETTE
COUNTY
FOREST

marsh
overlook

.7mi
1.1km

LAKE NOQUEBAY
WILDLIFE AREA

.8mi
1.3km

1.3mi
2.1km

1.5mi
2.4km

Private

Property

Wisconsin
Department of
Natural
Resources

1.3mi
2km

Right-of-way
Road

N

LEGEND

PARKING	P
SKI TRAILS	- - -
TOILETS	🚻
OTHER TRAILS	·····
OWNERSHIP BOUNDARY	———

Middle
Inlet

Marinette

Lake Nocquebay

Marinette County

Development of the trail in the Lake Nocquebay area was a joint project with the M&M Ski Club under the leadership of Everett Waugus. The trail is in the Marinette County Forest where several scenic ridges are traversed.

There is a total of 7.2km (4.5 miles) of different trails traveling over light ridges and down through wooded swamps.

Some downhill runs are steep and should be walked by other than expert skiers.

Parking, trail guides and toilets are available at the trail origin and signs guide the

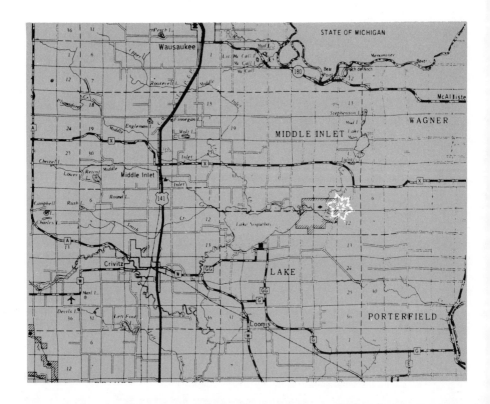

206

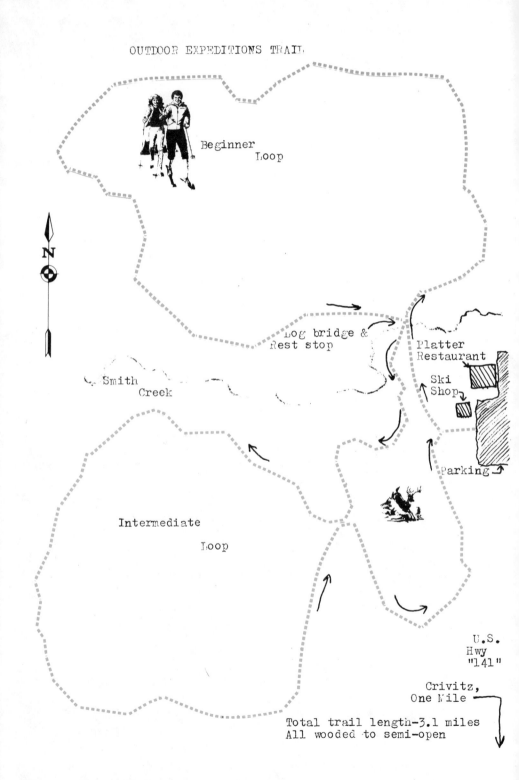

OUTDOOR EXPEDITIONS TRAIL

Beginner
Loop

N

Log bridge &
Rest stop

Platter
Restaurant

Smith
Creek

Ski
Shop

Parking

Intermediate

Loop

U.S.
Hwy
"141"

Crivitz,
One Mile

Total trail length-3.1 miles
All wooded to semi-open

Outdoor Expeditions

Marinette County

Outdoor Expeditions machine groomed-tracked trail begins at the XC ski shop and the Platter Restaurant just north of Crivitz on Hwy. 141. The well-marked, 4.8km (3 mile) double loop, one way, trail is built on the legendary "Issac Stephenson" logging trail; evidence of Wisconsin frontier history all but forgotten. A pleasant log bridge crosses Smith Creek as the trail winds up, down, around, and through pleasant wooded scenery. There are separate loops designed respectively for beginners and intermediate skiers. All types of Wisconsin wildlife can be seen on these trails and snowmobiling is strictly forbidden in the area. Ample parking, rental equipment, instruction, as well as food, beverages, and toilet facilities are available at the trailhead. Arrangements can be made in advance for lodging, rental, and meals by calling Outdoor Expeditions Inc.

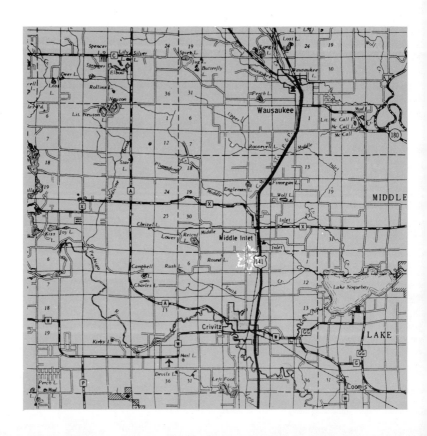

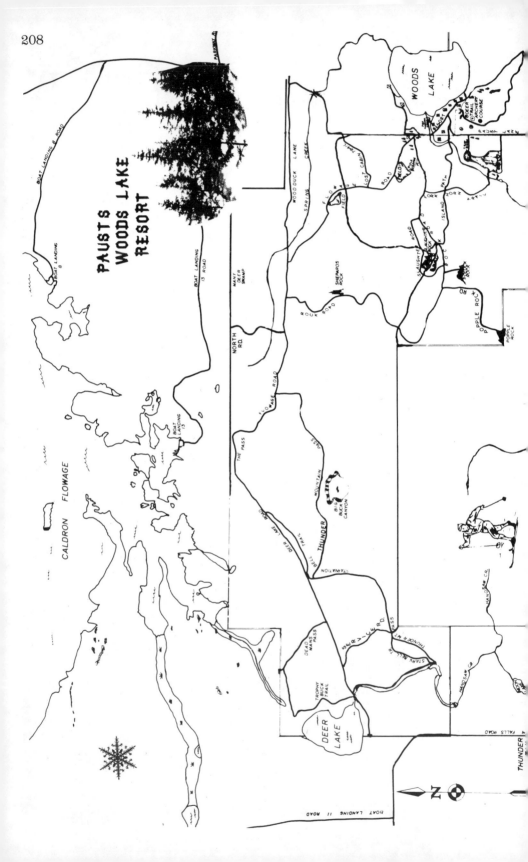

Paust's Woods Lake Resort

Paust's Woods Lake Resort is located 16 miles northwest of Crivitz. The Resort has 3000 acres, of private land. There are 30 miles of marked, groomed, tracked ski trails. The trails are for the beginner, intermediate and experienced skier. The trails are a combination of gradual slopes, open and wood terrain to challenging step hill and turns.

The Resort's trails are open to the public with a trail fee. Ski rental is not available but can be rented in Crivitz. The Resort offers cottage and motel units with housekeeping and modified American plan.

The annual USSA sanctioned Thunder Mt. Classic citizen's ski race is held the end of January on the Resort's property and is open to the public. The Thunder Mt. Classic citizen's ski race is part of the Great Lakes challenging race series.

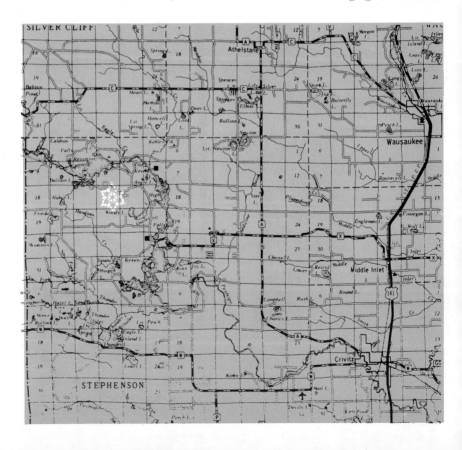

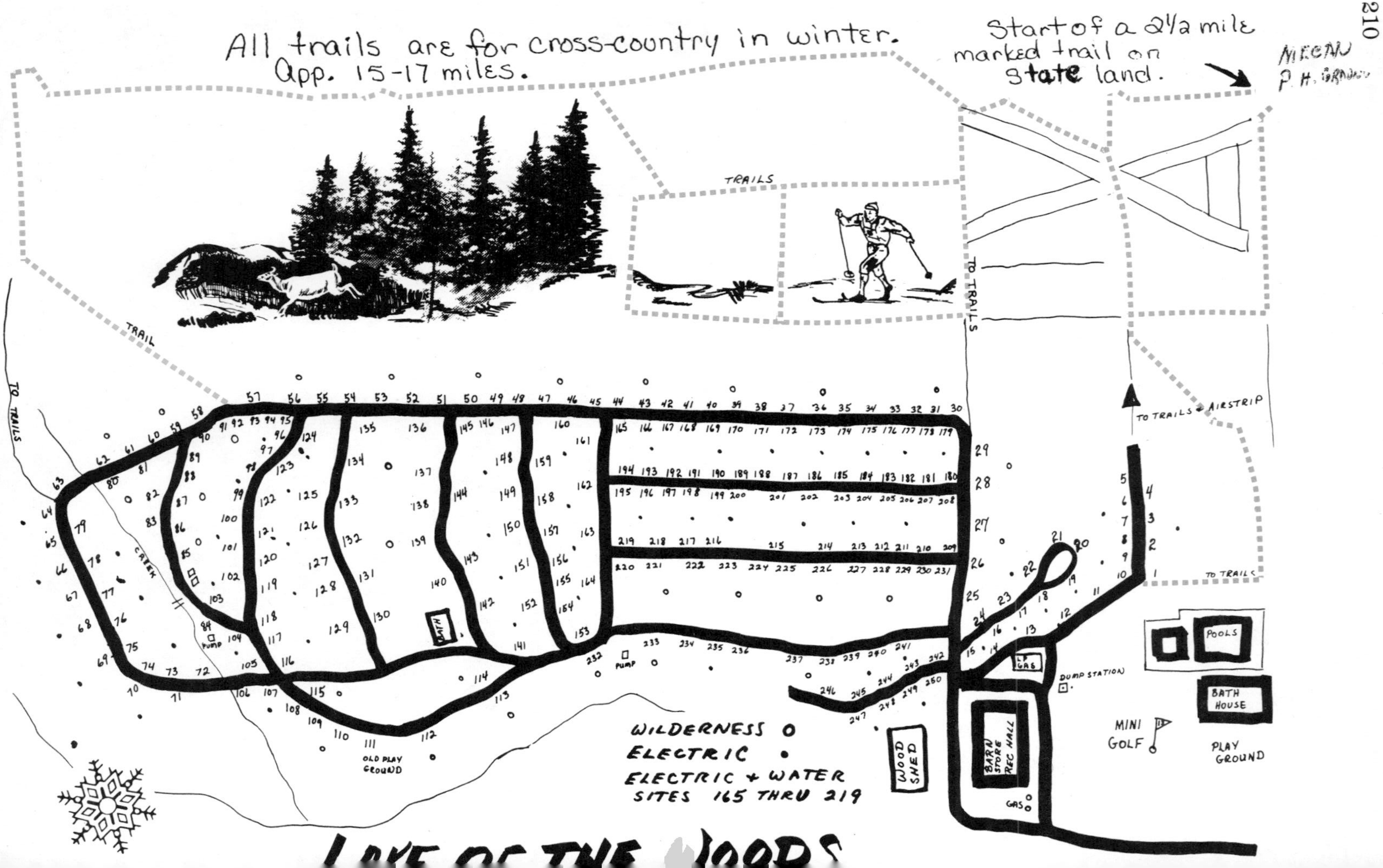

MEGAN
P.H. señorz

Start of a 2¼ mile
marked trail on
State land.

All trails are for cross-country in winter.
App. 15-17 miles.

TRAILS

TO TRAILS

TRAIL

TO TRAILS

TO TRAILS AIRSTRIP

TO TRAILS

POOLS

BATH
HOUSE

PLAY
GROUND

MINI
GOLF

DUMP STATION

GAS

GAS

BARN
STORE
REC HALL

WOOD
SHED

WILDERNESS ○
ELECTRIC ●
ELECTRIC + WATER
SITES 165 THRU 219

LAKE OF THE WOODS

OLD PLAY
GROUND

PUMP

PUMP

CREEK

LAVT

Lake of the Woods & Mecan Trail

Marquette County

Lake of the Woods is a year 'round campground that has 15 to 17 miles of trails for skiing cross-country. The trail fee is $2 per car with a plowed parking lot, shelter, toilets and refreshments available.

This set of trails connects with the State Mecan Cross-Country Trail on the south end of the campground. This state trail offers an additional 2.5 miles of terrain that is both level and hilly.

Lake of the Woods can be reached by taking State 22 south 8 miles out of Wautoma to JJ, then west on JJ one mile and follow signs.

BROWN DEER PARK

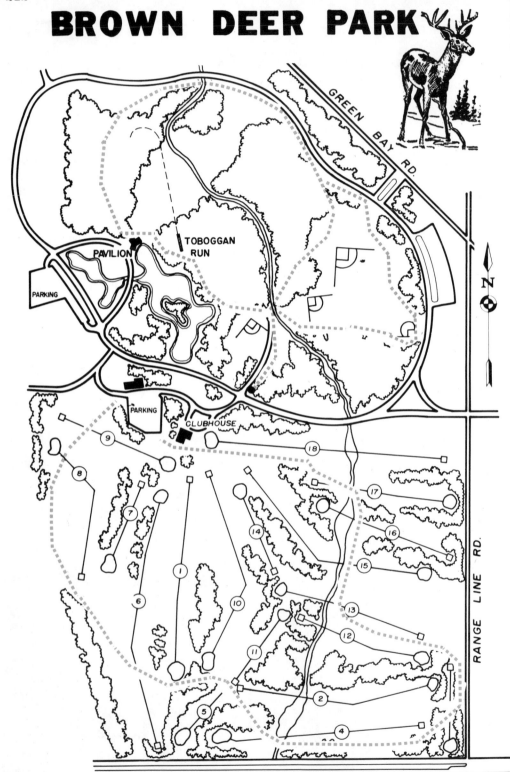

GREEN BAY RD.

N

PAVILION

TOBOGGAN
RUN

PARKING

PARKING

CLUBHOUSE

RANGE LINE RD.

9

8

7

6

1

10

5

14

11

18

17

16

15

13

12

2

4

GOOD HOPE ROAD

Brown Deer Park

Milwaukee County

Located at 7835 N. Green Bay Road in Milwaukee, Brown Deer offers 3 loops of marked, groomed trail.

West Short Loop — 2.2km (1.4 miles). East Short Loop — 1.65km (1 mile). Long Loop — 2.7km (1.7 miles).

Trails are gentle and wooded. Rentals and instructions are available. Shelter and toilets.

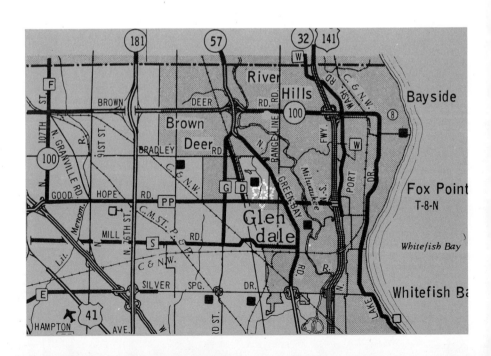

CURRIE 🌲 PARK

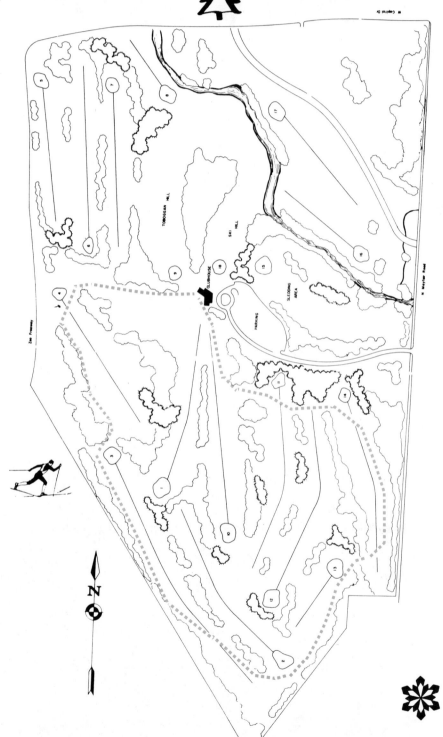

Currie Park

Milwaukee County

The Park has 2.1km (1.3 miles) of gentle, open trail. It is marked, groomed and tracked with rentals and instructions available. Ample parking, shelter and toilets are available.

Currie Park is located at 3535 N. Mayfair Rd.

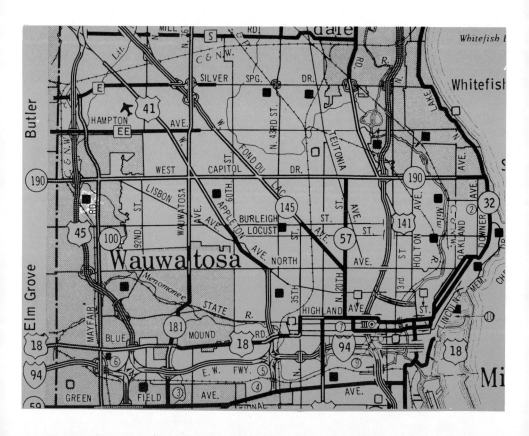

CLUBHOUSE

BRADLEY RD.

Dretzka Park

One of Milwaukee's County Park systems touring trails, Dretzka Park is located on Milwaukee's north side at 120th St. (State Highway 145) and Bradley Rd. The Park offers 2 loops — Short Loop 1km (¾ mile), Long Loop 4km (2.5 miles) of gentle, wooded, marked, groomed trail. Shelter, refreshments and toilets are available.

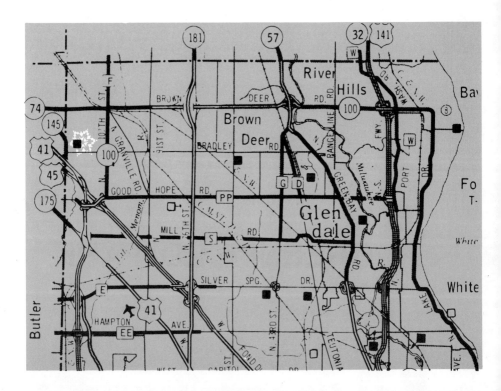

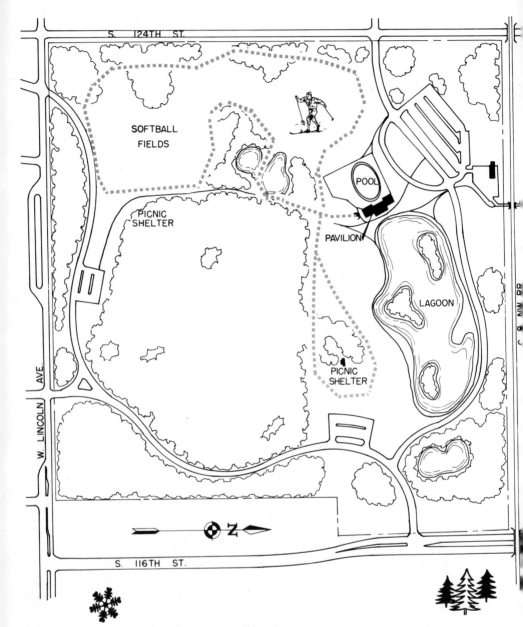

GREENFIELD PARK

MILW. CNTY. DEPARTMENT OF PARKS, RECREATION, AND CULTURE

APPROXIMATE TRAIL LENGTH -- SHORT LOOP, .4 mi.
LONG LOOP, 1.0 mi.

Greenfield Park

Milwaukee County

Located at W. Lincoln Ave., South 124th St. Greenfield Park consists of 2 loops, Short Loop — ¼ mile, Long Looop — 1.6km (1 mile).

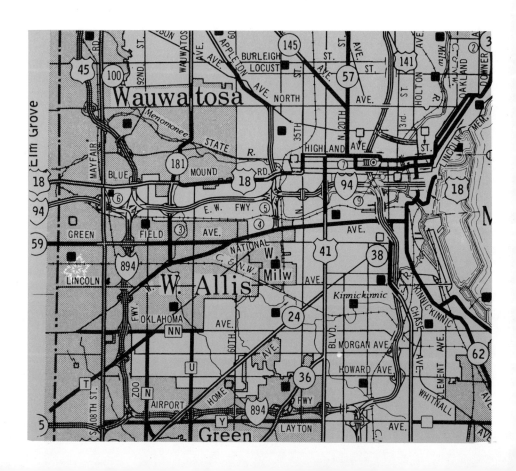

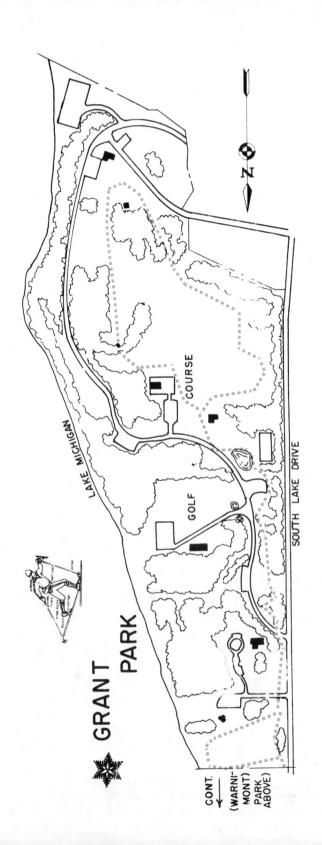

GRANT PARK

LAKE MICHIGAN

GOLF COURSE

SOUTH LAKE DRIVE

N

CONT.
(WARNI-
MONT)
PARK
ABOVE)

Grant Park

Milwaukee County

The golf course loop is 1.05 mile. From clubhouse to north end of park and back is 3.06 miles. Park is located at 100 Hawthorne Avenue on the shores of Lake Michigan.

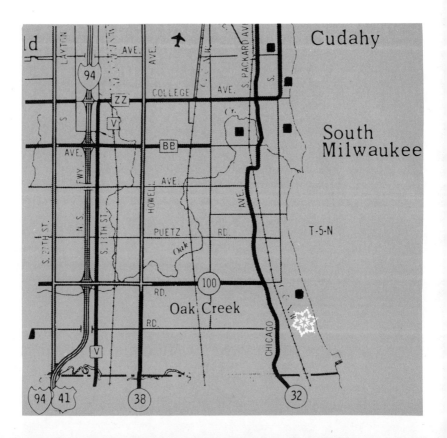

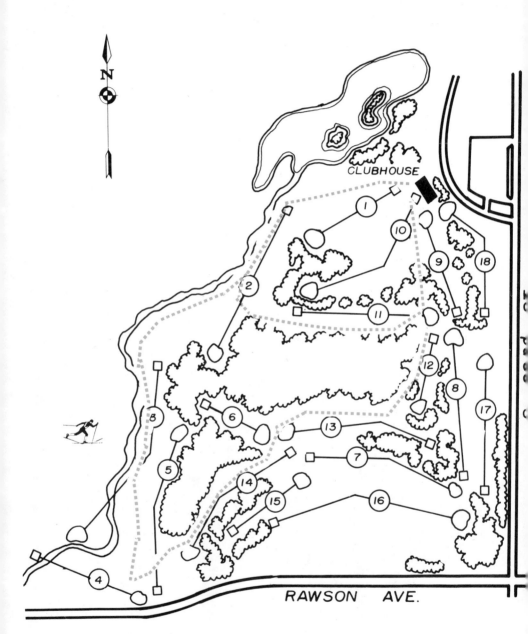

CLUBHOUSE

RAWSON AVE.

Whitnall Park

Milwaukee County

Established in late 1976, Whitnall Trail is located on the golf course, it is groomed to wind through rolling hills offering the skier both a short loop of ¾ miles and a long loop 2.7km (1.7 miles).

The trail is geared for the beginner and intermediate-expert skier. Rentals, instructions, food, shelter and toilets are offered for your comfort and convenience.

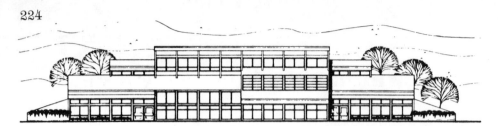

EDUCATIONAL CENTER SCHEDULED FOR COMPLETION IN DEC. 1984

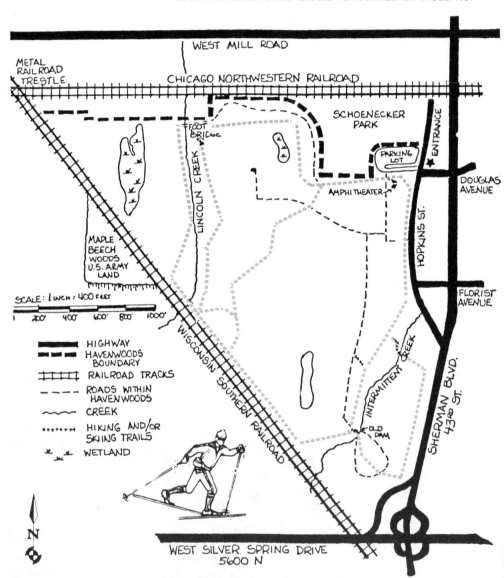

WEST MILL ROAD

METAL RAILROAD TRESTLE

CHICAGO NORTHWESTERN RAILROAD

SCHOENECKER PARK

ENTRANCE

FOOT BRIDGE

PARKING LOT

DOUGLAS AVENUE

LINCOLN CREEK

AMPHITHEATER

HOPKINS ST.

MAPLE BEECH WOODS U.S. ARMY LAND

FLORIST AVENUE

SCALE: 1 INCH = 400 FEET

200' 400' 600' 800' 1000'

HIGHWAY
HAVENWOODS BOUNDARY
RAILROAD TRACKS
ROADS WITHIN HAVENWOODS
CREEK
HIKING AND/OR SKIING TRAILS
WETLAND

WISCONSIN SOUTHERN RAILROAD

INTERMITTENT CREEK

SHERMAN BLVD. 43RD ST.

OLD DAM

N

WEST SILVER SPRING DRIVE
5600 N

HAVENWOODS
ENVIRONMENTAL AWARENESS CENTER

Havenwoods Ski Trails

Milwaukee County

This environmental awareness center is 237 acres of open space inside the city of Milwaukee. Since 1980 the Department of Natural Resources with the help of many volunteers has been working to return Havenwoods to a more natural state.

The three mile trail system is used for hiking and cross country skiing. It is not exceptional for a challenging cross country trail but does offer a basic opportunity to enjoy cross country skiing in conjunction with nature study.

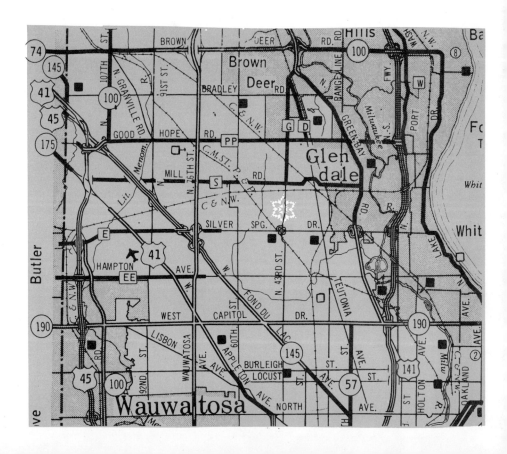

JONES SPRING AREA

Four Seasons, Nonmotorized Recreation

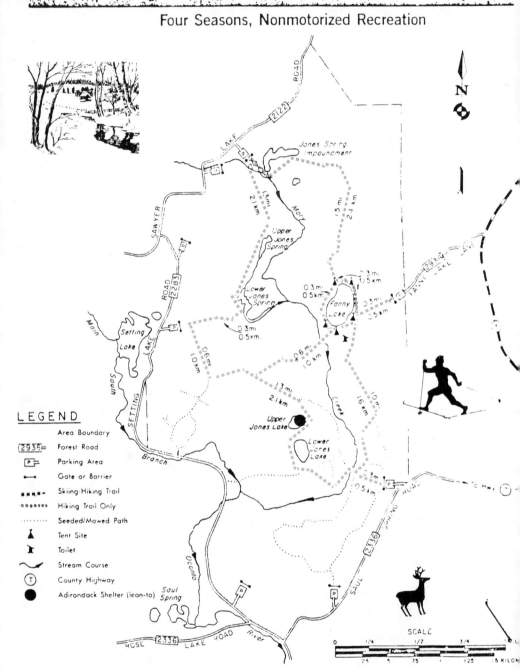

LEGEND

	Area Boundary
2935	Forest Road
P	Parking Area
•—•	Gate or Barrier
▪▪▪▪▪	Skiing-Hiking Trail
∘∘∘∘∘∘∘	Hiking Trail Only
··········	Seeded/Mowed Path
▲	Tent Site
⚑	Toilet
〜	Stream Course
Ⓣ	County Highway
●	Adirondack Shelter (lean-to)

Jones Springs Area

The Jones Spring Area is located in the southwestern portion of the Nicolet National Forest near Townsend and Lakewood, Wisconsin. The area contains 2,000 acres which have been set aside for nonmotorized recreation use. Jones Spring area contains seven miles of developed two-way skiing trails that are of intermediate degree of difficulty. The entire system is signed with blue diamond shaped markers. Four different parking areas provide access to the developed trail system.

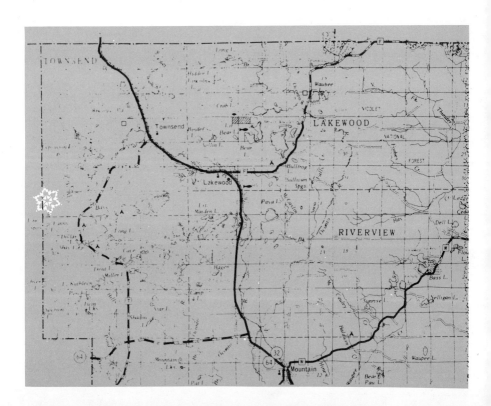

LAKEWOOD CROSS COUNTRY SKI & HIKING TRAILS

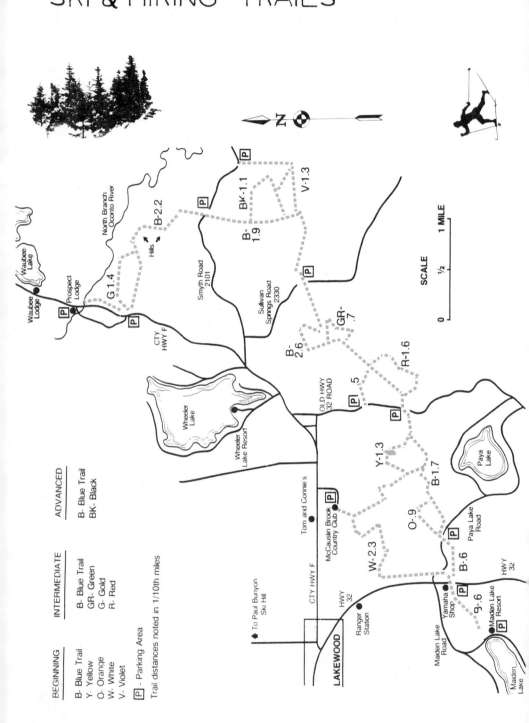

BEGINNING

B- Blue Trail
Y- Yellow
O- Orange
W- White
V- Violet

INTERMEDIATE

B- Blue Trail
GR- Green
G- Gold
R- Red

ADVANCED

B- Blue Trail
BK- Black

P - Parking Area

Trail distances noted in 1/10th miles

SCALE

0   ½   1 MILE

Lakewood Ski Trails

Oconto County

The Lakewood ski trails are set in a wooded area with rolling hills. The hills offering the greatest challenge are found in the northern section of the trail.

Geologically, the land is highly granitic and is part of the Glaciated Canadian Shield. A variety of deciduous trees are seen as is an occasional stand of conifers.

The main trail, known as the Blue Trail, is continuous between Maiden Lake Resort and Prospect Lodge. Several secondary trails or loops branch off from and return to the Blue Trail.

Trails are marked and color coded. They are signed with regard to degree of difficulty. There are approximately equal trail distances for beginning, intermediate, and expert skiers.

Trails are groomed in accord with snow conditions. Double tracks are set.

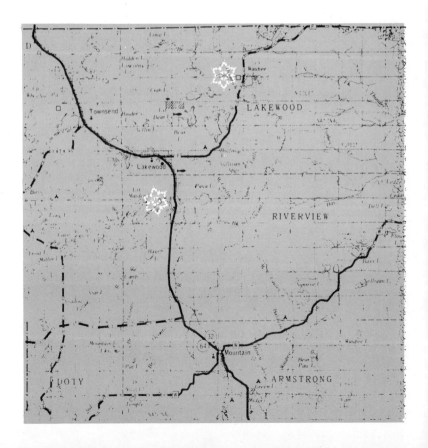

230

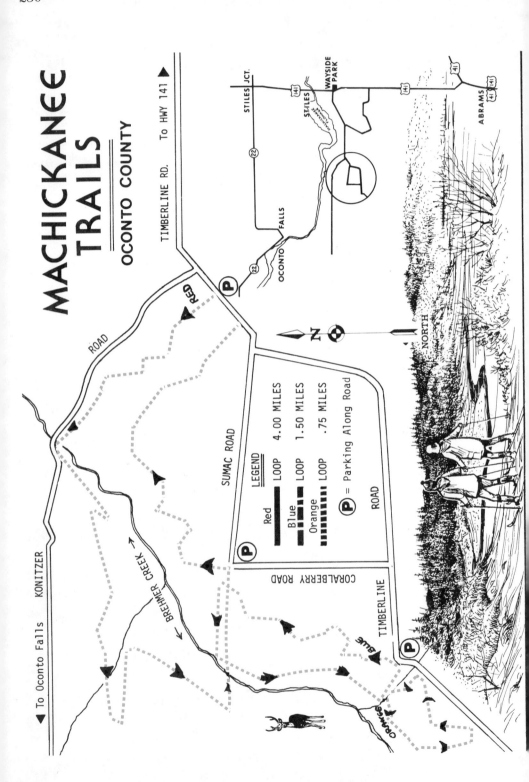

MACHICKANEE
TRAILS
OCONTO COUNTY

TIMBERLINE RD.   To HWY 141 ▶

STILES JCT.

WAYSIDE PARK

STILES

22

141

141

41

141

41 141

ABRAMS

NORTH

N

ROAD

RED

KONITZER

▶ To Oconto Falls

BREHMER CREEK

SUMAC ROAD

CORALBERRY ROAD

TIMBERLINE

BLUE

Orange

ROAD

OCONTO FALLS

22

22

P

P

P

P

LEGEND

Red   LOOP   4.00 MILES
Blue   LOOP   1.50 MILES
Orange   LOOP   .75 MILES

P = Parking Along Road

Machickanee Trails

Oconto County

Machickanee Trails is comprised of three loops totaling approximately 10km (6.2 miles). The trails set in the heart of the Machickanee County Forest travel over high ridges, logging roads, through wooded swamps, pine plantation and cross several branches of Brehmer Creek.

The Red trail is recommended for the intermediate skier.

Recommended for the beginner and intermediate skiers are the Blue & Orange trail.

Location — 4 miles north of Abrams on U.S. 141, west on Timberlane Rd. — 3 miles.

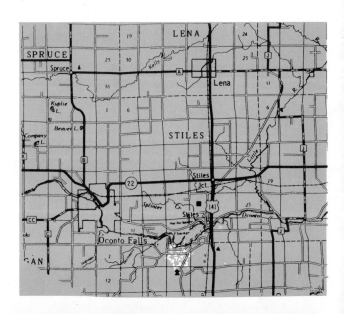

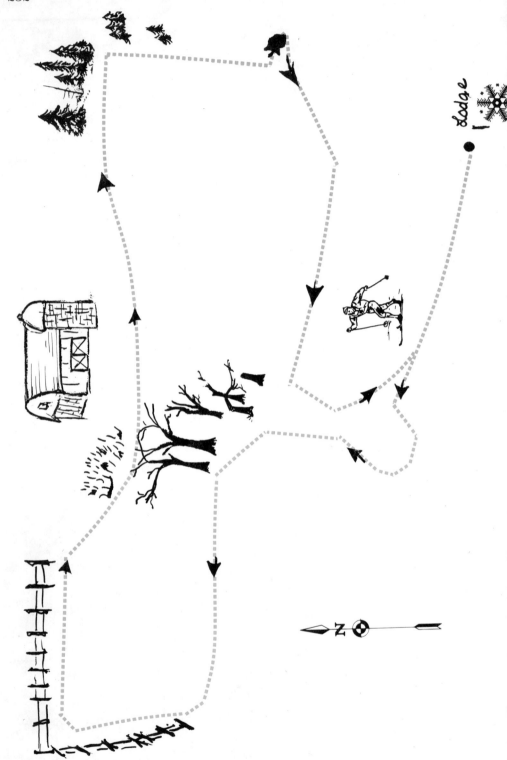

lodge

Mt. LeBett Ski Trail

Oconto County

Mt. LeBett is mainly a downhill ski area but they do have a 2.4km (1.5 mile) gentle hilly trail which is groomed and tracked. A large lodge with fireplace, restaurant, public seating, toilets, and pro shop are offered. Cross country ski instructions are available from registered instructor with special group rates. Skis can be rented at the pro shop. No trail fee is charged. Mt. LeBett can be reached by taking County B 6 miles west out of Coleman.

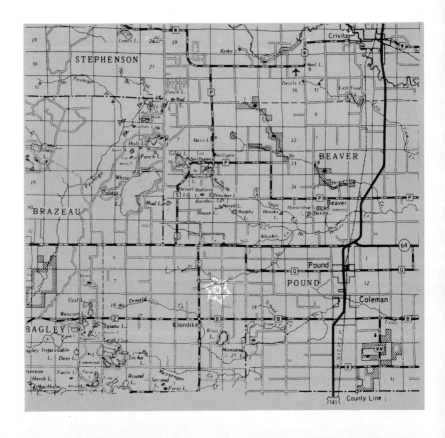

Cassian – Two Way
Ski Trail

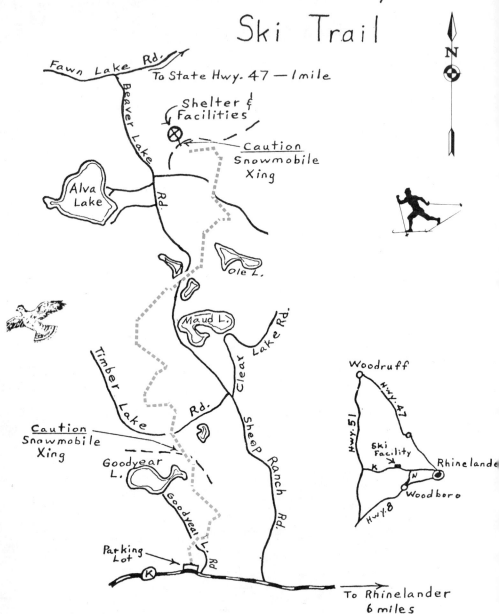

Fawn Lake Rd.

To State Hwy. 47 — 1 mile

N

Shelter & Facilities

Caution
Snowmobile Xing

Beaver Lake Rd.

Alva Lake

Ole L.

Maud L.

Clear Lake Rd.

Timber Lake Rd.

Caution
Snowmobile Xing

Goodyear L.

Sheep Ranch Rd.

Goodyear L. Rd.

Parking Lot

K

To Rhinelander
6 miles

Woodruff

Hwy. 51

Hwy. 47

Ski Facility

K

N

Rhinelander

Woodboro

Hwy. 8

Trail Length: 15 miles
(round trip)

Cassian Two Way Trail

Oneida County

Cassian Trail is unique for the skier who does not want to complete the 11.2km (7 mile) trail. The skier can either turn around at any spot or leave a car at one of the road crossings.

The trail is marked and groomed and is rolling and wooded.

Please use caution on the curves as this is a two way trail.

Cassian trail is located 6 miles west of Rhinelander on Hwy. K. Parking lot is on the north side of the road.

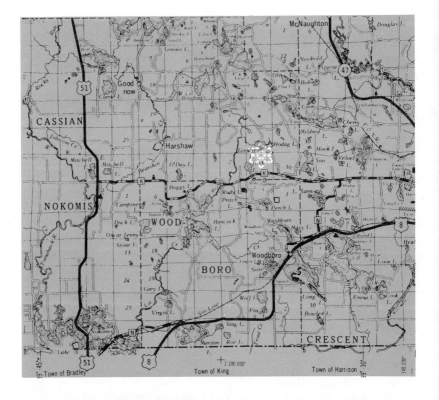

236

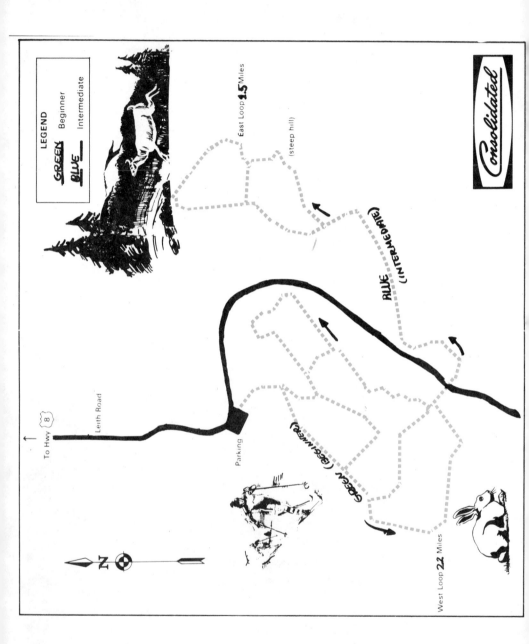

LEGEND

GREEN — Beginner

BLUE — Intermediate

Consolidated

East Loop **1.5** Miles

(steep hill)

BLUE (INTERMEDIATE)

To Hwy 8

Leith Road

Parking

GREEN (BEGINNER)

West Loop **2.2** Miles

N

Monico Cross-Country Ski Trail

The Monico cross-country ski trail system is 10.5 miles east of Rhinelander on Hwy. 8-47. The trail provides various options to the skier as to length and degree of difficulty, and for the most part follows old logging roads.

Variations in timber types include 25 year-old red pine plantations, mixed hardwoods, white pine and aspen. Some of this area has been logged in the past, and presents visual evidence of a forest's ability to renew itself.

The 2.2 mile western portion of the trail (green) is gently sloping and will challenge the beginning skier. The 1.5 eastern portion (blue) contains considerably more topographical variation, and in cross-country skiing terms qualifies as an intermediate trail.

Be alert for wildlife on this trail system. Deer, ruffed grouse and snowshoe hare are common, and the quiet skier on these trails will have an opportunity to observe a great variety of winter-active wildlife. Good tour!

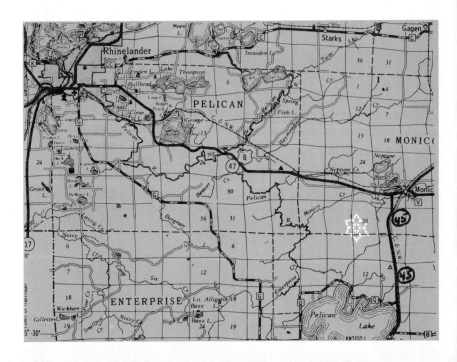

238

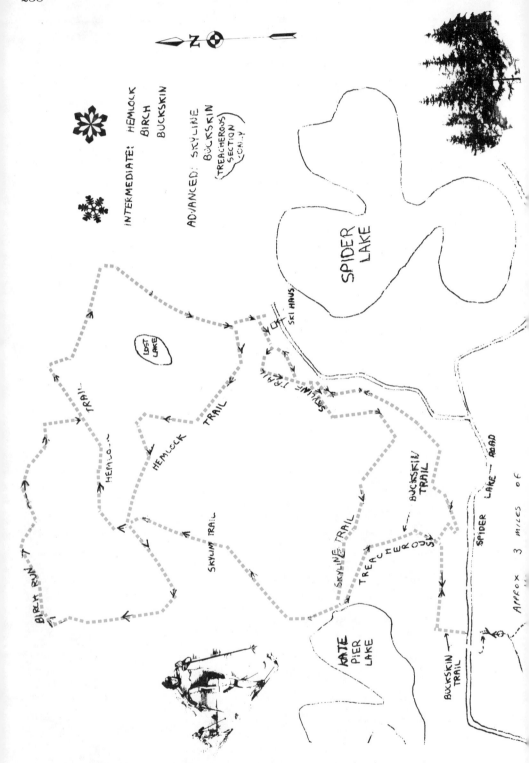

N

INTERMEDIATE: HEMLOCK
BIRCH
BUCKSKIN

ADVANCED: SKYLINE
BUCKSKIN
(TREACHEROUS SECTION ONLY)

SPIDER LAKE

SKI HAUS

LOST LAKE

BIRCH TRAIL

HEMLOCK

HEMLOCK TRAIL

SKYLINE TRAIL

SKYLINE TRAIL

BUCKSKIN TRAIL

TREACHEROUS

SPIDER LAKE ROAD

BIRCH RUN

KATE PIER LAKE

BUCKSKIN TRAIL

APPROX 3 MILES OF

Fort Wilderness Ski Trail

Oneida County

Fort Wilderness offers approximately 12 miles of trails which wind through stands of hemlock, birch, and aspen in a complete wilderness. There are four loops to the marked and groomed trail system. They consist of two beginner trails on gently rolling terrain, plus an intermediate, more hilly trail, and a more challenging expert trail. Because their system loops to an expanse of 50,000 of wilderness State Forests, there is an unlimited possibility of exploring. Snow camping and trail packing are available to remote forests and lakes. Fort Wilderness has easy access to other public trail systems.

Fort Wilderness offers a heated ski center and lodge, parking, refreshments, ski rentals, instructions, and bathroom facilities. It is open Tuesday through Sunday from 9AM to 5PM. There is a charge for trail use and alcoholic beverages are prohibited. Directions: Follow Highway 8 to one mile west of Rhinelander to Highway 47 north. Follow 10 miles north to McNaughton. Turn at "Fredrich's" and follow Fort Wilderness signs about 5 miles to camp.

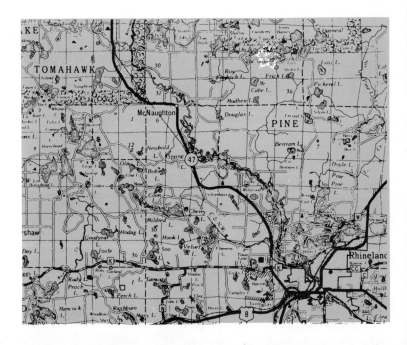

240

Holiday Acres Trails

Oneida County

This picturesque trail system in Wisconsin's Northwoods originates at Holiday Acres Resort. It is comprised of three different loops, totalling 17.7km (11 miles). For the novice skier there is the Goat Farm Trail: 3¾ miles of gently rolling terrain with some easy curves and slopes. The intermediate trail — Plantation Trail — consists of 4 miles of challenging hilly terrain. The expert skier will enjoy the Roller Coaster Trail which spans 4 miles and also has a 1 mile shortcut leading back to the intermediate trail. It has fast hills and sharp curves and will challenge even the seasoned skier. All trails are one-way except where indicated.

Holiday Acres provides parking as well as complete facilities for meals, cocktails, and overnight accomodations. Ski rental is available. These well-marked and groomed trails are open to the public and free of charge, but all skiers must register at Holiday Acres.

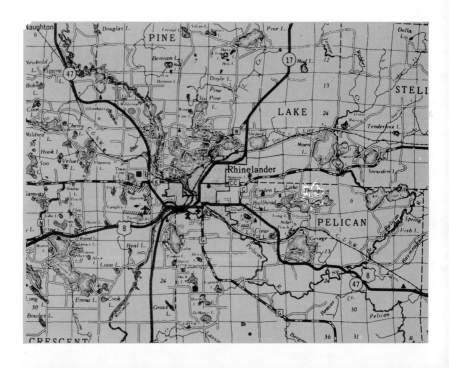

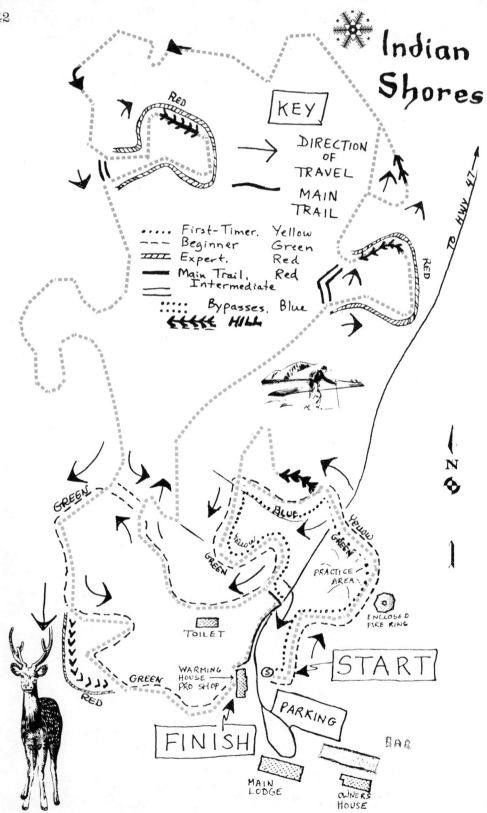

Indian Shores

KEY

→ DIRECTION OF TRAVEL

━━ MAIN TRAIL

····· First-Timer. Yellow
- - - Beginner. Green
//// Expert. Red
━━ Main Trail. Red
Intermediate
:::: Bypasses. Blue
≪≪≪ HILL

RED

RED

TO HWY 47

N

GREEN

BLUE

YELLOW

GREEN

YELLOW

GREEN

PRACTICE AREA

ENCLOSED FIRE RING

START

TOILET

WARMING HOUSE PRO SHOP

GREEN

RED

FINISH

PARKING

BAR

MAIN LODGE

OWNERS HOUSE

Indian Shores Trails

Oneida County

Indian Shores is located on 100 acres of wooded hilly land on the shores of Lake Tomahawk. The trails are of varying lengths with a total of about 5km (8 miles). There are sections that are for the expert and some for the beginner. All expert sections have cutoffs and all trails are marked with colored tapes.

Toilet facilities, ample parking with warming house, bar and restaurant. Lodging for groups is available and you are within a few miles of many fine DNR Trails: Escanaba Lake, Madeline Lake, Schlecht Lake, and McNaughton Lake.

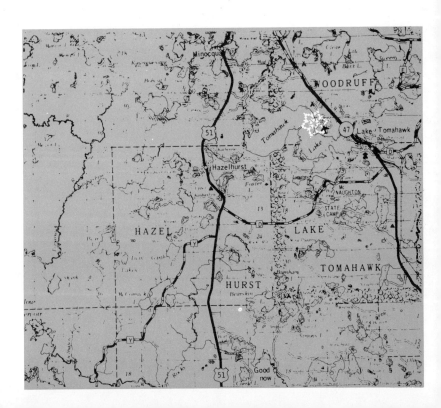

244

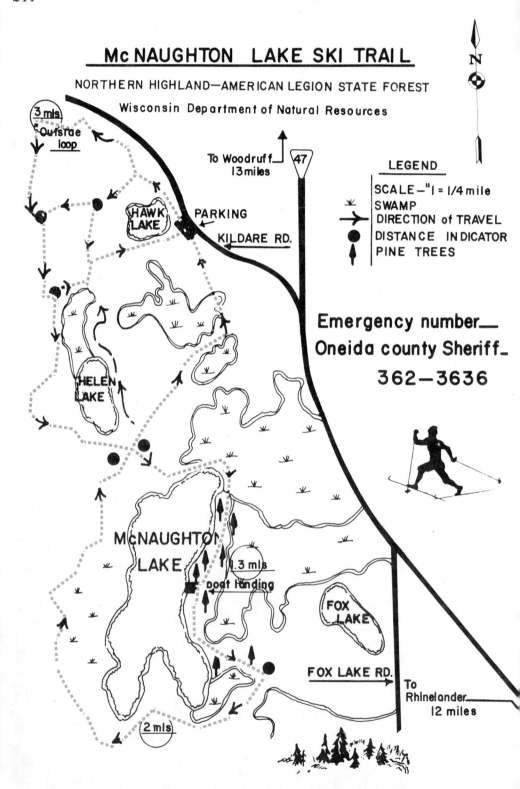

McNaughton Lake Ski Trail

Oneida County

Located in Northern Highland-American Legion State Forest this trail is a one way system consisting of 6 loops with a total of 11.3km (7 miles). McNaughton is accessible to rank beginners except for one downhill and steep uphill climb as noted, 1, on map. No real danger, just inconvenience. The northern 5 loops are the prettiest with a beautiful dense red pine grove North East of Helen Lake. Good variation in woods, mostly birch, aspen, with scattered pines. Trail has high usage as many as 750 skiers a day. Trail deteriorates rapidly but since it's so flat this doesn't produce the same degree of treacherous skiing as a more hilly trail.

Medium sized parking lot with well plowed roads — no shelter or toilets. Location is 12 miles North of Rhinelander on Hwy. 47. Turn left on Kildare Road to parking lot. No trail fees. Skies can be rented in Rhinelander and Minocqua.

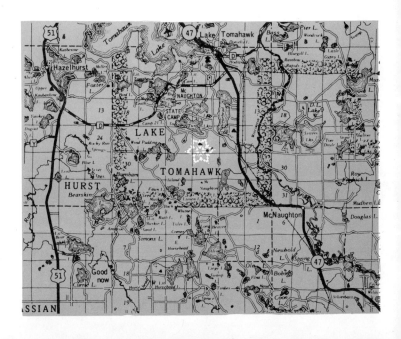

MINOCQUA WINTER PARK AND NORDIC CENTER

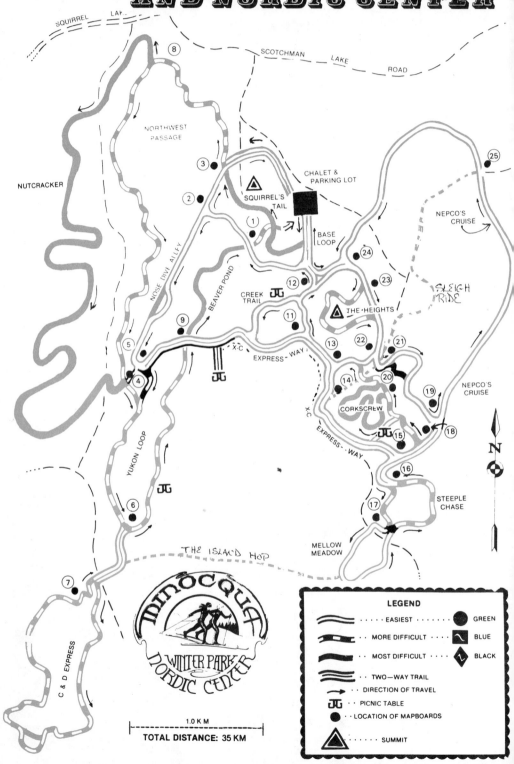

Minocqua Winter Park and Nordic Center

Oneida County

The center is located in the Township of Minocqua. It features one of the best groomed & tracked trail systems in the Midwest.

The center can be reached from U.S. 51 west on 70 for 6½ miles, south on Squirrel Lake Road for 4½ miles, west on Scotchman Lake Road for ½ mile, south on Ski Hill Road for ½ mile to the chalet. There are clear signs from Hwy. 70.

The system has over 40km of beautifully groomed trails with heated chalet, snack bar, tubing, telemarking and a 2.5km lighted trail on the base loop. Open 9-5 every day with ski rentals, retail shop and PSIA Certified Ski School.

See AD Section at end of book for Accomodations

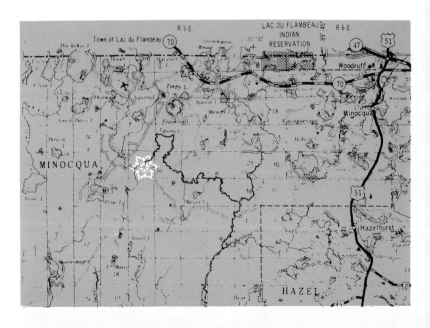

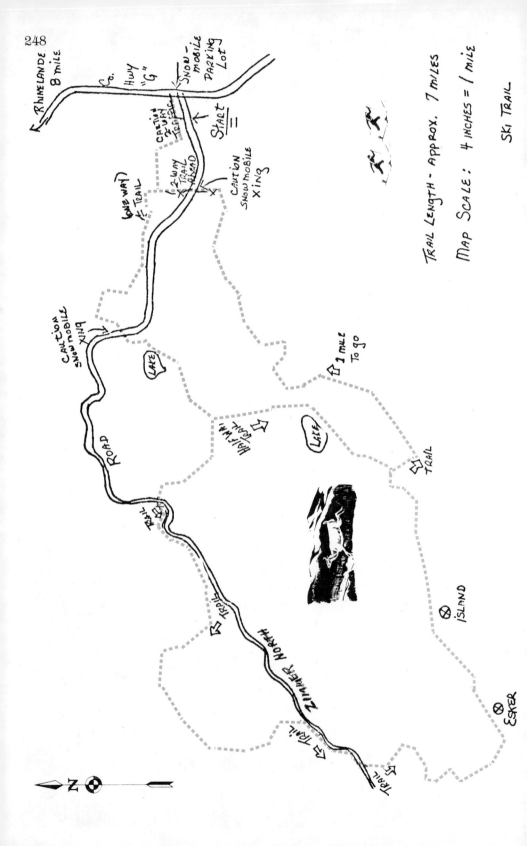

248

Enterprise Trail Unit I

Oneida County

Enterprise Trail is comprised of two trails totaling 11.3km (7 miles) of rolling, wooded, scenic, marked and groomed trail.

It is located 8 miles southeast of Rhinelander on County Hwy. G. Parking is available in the snowmobile lot on the west side of the road.

Open 24 hours.

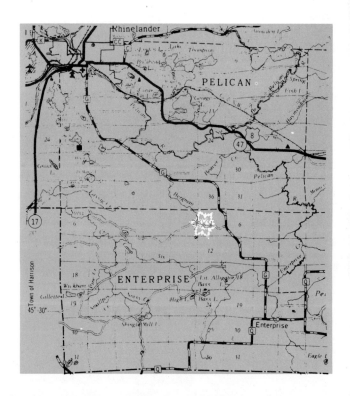

250

RAVEN SKI TRAIL

NORTHERN HIGHLAND—AMERICAN LEGION STATE FOREST
Wisconsin Department of Natural Resources

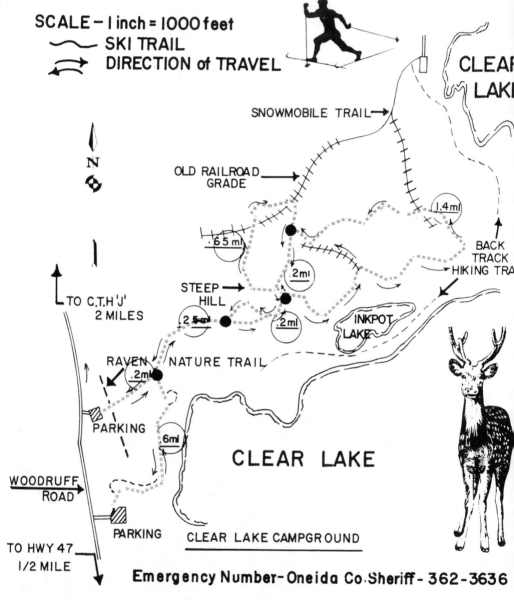

SCALE – 1 inch = 1000 feet
SKI TRAIL
DIRECTION of TRAVEL

SNOWMOBILE TRAIL →

N

OLD RAILROAD GRADE →

CLEAR LAKE

1.4 ml

BACK TRACK
HIKING TRAIL

.65 ml

.2 ml

STEEP HILL →

.25 ml

.2 ml

INKPOT LAKE

TO C.T.H 'J'
2 MILES

RAVEN NATURE TRAIL

.2 ml

PARKING

6 ml

CLEAR LAKE

WOODRUFF ROAD →

PARKING

CLEAR LAKE CAMPGROUND

TO HWY 47
1/2 MILE

Emergency Number – Oneida Co. Sheriff – 362-3636

Raven Nature Trails

Oneida County

The majority of the 6.4km (4 mile) trail winds through an area that was logged 5-10 years ago. Both Inkpot and Clear Lakes can be seen from the trail.

There are several steep hills, although the majority of the trail is relatively easy. Good trail for novice skiers.

Location — Hwy. 47 to Woodruff Rd. at Raven Nature Trail.

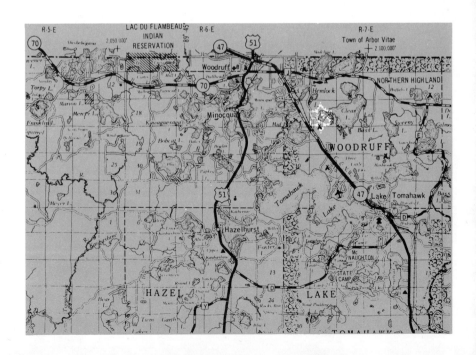

SCHLECHT LAKE SKI TRAIL

NORTHERN HIGHLAND—AMERICAN LEGION STATE FOREST
Wisconsin Department of Natural Resources

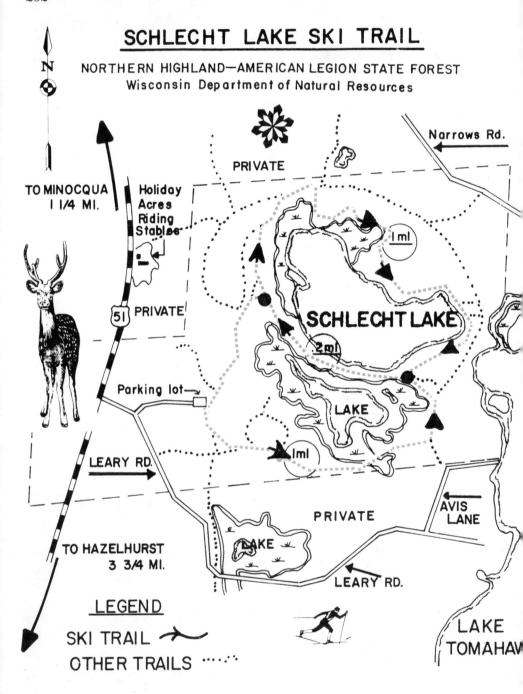

N

Narrows Rd.

PRIVATE

TO MINOCQUA
1 1/4 MI.

Holiday
Acres
Riding
Stables

51 PRIVATE

PRIVATE

1 ml

SCHLECHT LAKE

2 ml

LAKE

Parking lot

LEARY RD.

1 ml

PRIVATE

AVIS
LANE

LAKE

TO HAZELHURST
3 3/4 MI.

LEARY RD.

LAKE
TOMAHAW

LEGEND

SKI TRAIL

OTHER TRAILS

Scale 8" = 1 Mile
Emergency Number—Oneida County Sheriff – 362–3636

Schlect Lake Ski Trail

Oneida County

The 0.6 mile portion is the most difficult section of this trail. No special danger exists except that several sections if run in reverse direction are SUPER EXPERT. The main feature that is readily evident is the technique challenge of this trail. Length is not an enduring factor as the trail consists of 3.5km (2.2 miles) of trail.

Schlect Lake is designed to be a "close in" (close to town) trail for a very challenging "quickie" tour. Good trail to practice on though it is an EXPERT trail.

Width of trail is 5-6 feet with about 70% cut for the ski trails and final 30% is old logging roads. Scenery is best around Schlect Lake itself. Wildlife is relatively scarce. No charge for trail use, trails are not groomed, no toilets or shelter.

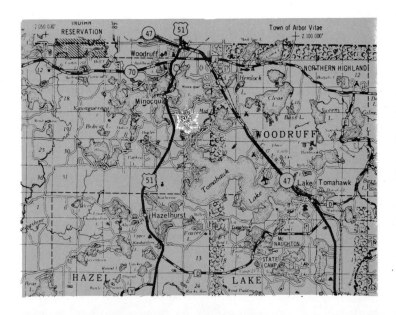

Woodboro Ski Touring Trail

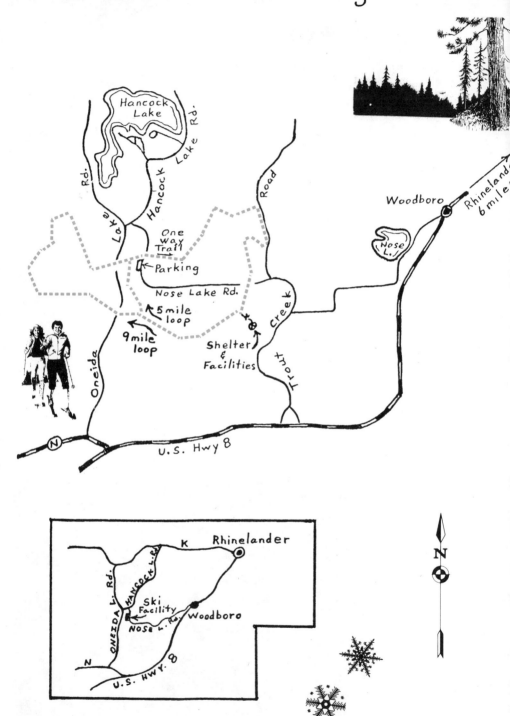

Woodboro Ski Touring Trail

Woodboro Trail is for the intermediate skier. Both trails cover the same first 6.4km (4 mile) with rolling and wooded terrain. The combination is a total of 14.4km (9 miles).

The Forestry Dept. grooms and tracks both the trails regularly.

Shelter and facilities are available.

The trail is 9 miles west of Rhinelander on CTH K, then 3.5 miles south of Hancock Lake Road, then 3 miles east on Nose Lake Road.

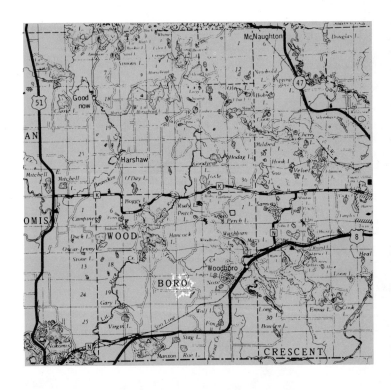

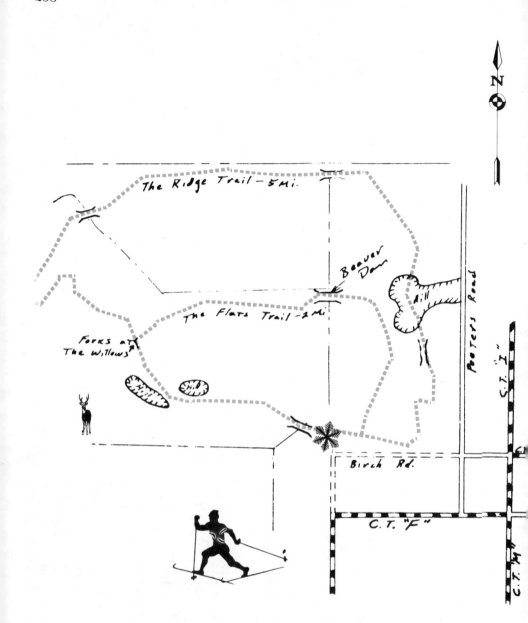

Deer Creek Wildlife Ski Area

Outagamie County

Aspen, birch, maple balsam, red pine, and tag alder swales line the path of these trails located on 1400 acres of forest along the boundary of Outagamaie and Waupaca counties. The Ridge Trail is 8.8km (5.5 miles) in length and the Flats Trail about 3.2km (2 miles). It was designed as a rustic skiing and hiking trail in 1974, aimed at preserving its' natural beauty. No snowmobiles can use the trail as there is room only enough for a skier or hiker in some spots. The parking area is somewhat difficult and there are no toilets or shelter available.

This area is being logged and bulldozers are creating havoc with the original trails. Parts of the trail are enjoyable.

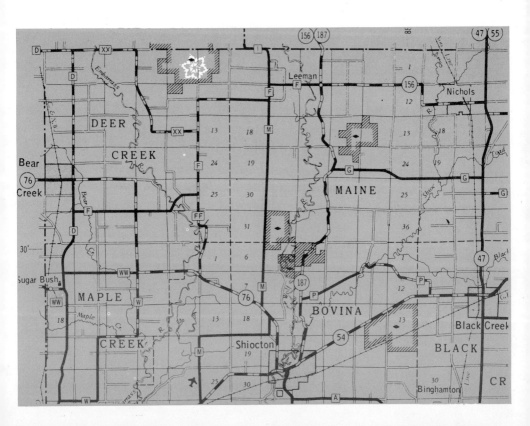

258

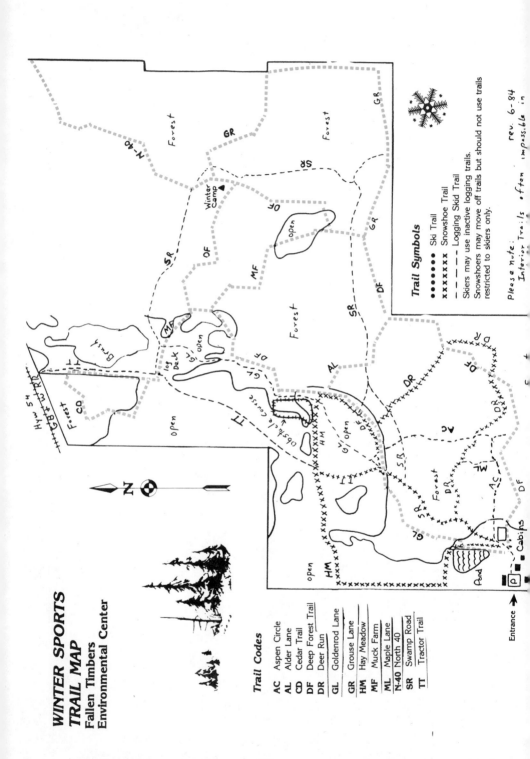

WINTER SPORTS TRAIL MAP
Fallen Timbers
Environmental Center

Trail Codes

AC Aspen Circle
AL Alder Lane
CD Cedar Trail
DF Deep Forest Trail
DR Deer Run
GL Goldenrod Lane
GR Grouse Lane
HM Hay Meadow
MF Muck Farm
ML Maple Lane
N-40 North 40
SR Swamp Road
TT Tractor Trail

Trail Symbols

●●●●●●● Ski Trail
✗✗✗✗✗✗✗ Snowshoe Trail
– – – – Logging Skid Trail

Skiers may use inactive logging trails.
Snowshoers may move off trails but should not use trails restricted to skiers only.

Please note:
Interior trails often impossible in
rev. 6-84

Fallen Timbers Environmental Center

The Center is a School Forest and Outdoor Camp operated by the School district.

Ski trails at Fallen Timbers are nearly level, well marked, and rated for the beginner. The trails are suited for an easy family style traverse through a mixed wooded country. There are approximately 8.0km (5 miles) of trail, plus 4km (2.5 miles) of logging trails available when logging operations end.

A minimal charge is requested of the skier.

Location — Entrance located ⅝ miles South of Wis. Hwy. 54 on County Hwy. P.P. Town of Black Creek, Wi.

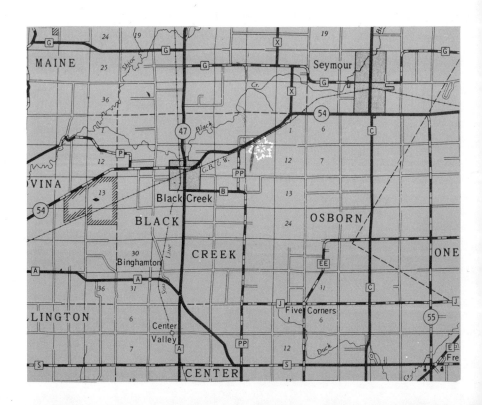

260

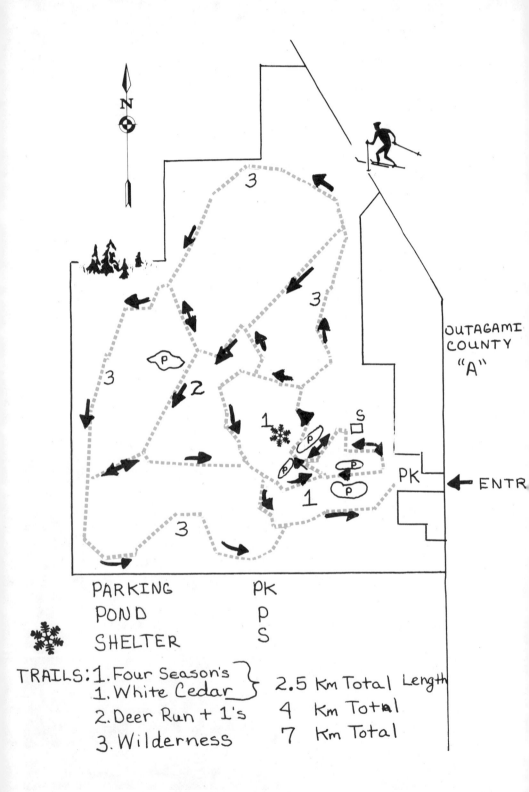

N

3

3

OUTAGAMI
COUNTY
"A"

P

3

2

3

1

S

P

P

P

PK

ENTR

P

1

3

PARKING   PK

POND   P

SHELTER   S

TRAILS: 1. Four Season's ⎫
        1. White Cedar   ⎬ 2.5 Km Total Length
        2. Deer Run + 1's   4 Km Total
        3. Wilderness   7 Km Total

Gordon Bubolz Nature Preserve

Outagamie County

"The Gordon Bubolz Nature Preserve" is located on County Highway "A", 1.7 miles north of County OO. The 657 acre Preserve contains 6 miles of One way, tracked trails through a lowland coniferous forest. The gently rolling to flat terrain is ideal for the beginner, for families, or the skier who enjoys a leisurely pace through a beautiful natural setting.

The trail system is open daylight hours daily and a minimal fee is charged. (1984 — $1.50 adults, $1.00 students and senior citizens). The Nature Awareness Center is open regular business hours and 12:30 — 4:30 p.m. on Saturdays and Sundays. The center features natural history displays, a hot drink concession and ski rentals for ages 6 — 14. Also available at the center are 70 pairs of snowshoes for all ages.

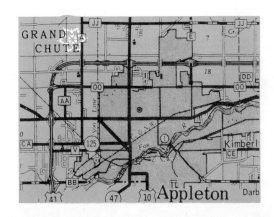

262

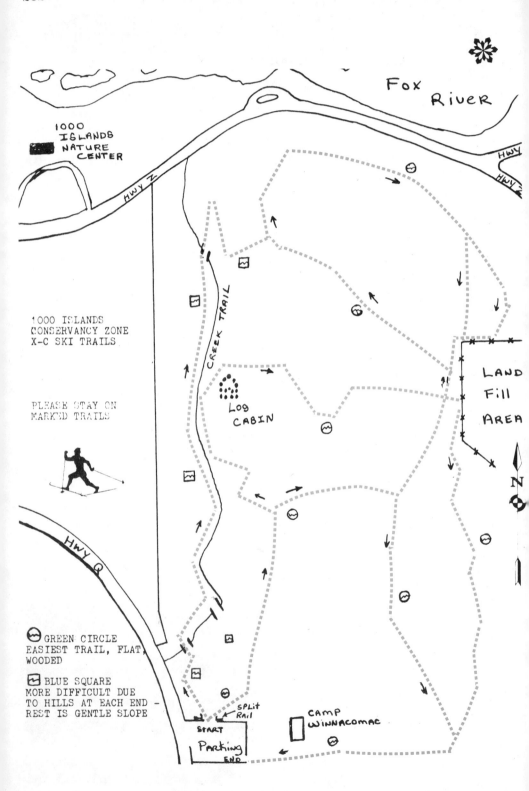

1000 ISLANDS
CONSERVANCY ZONE
X-C SKI TRAILS

PLEASE STAY ON
MARKED TRAILS

GREEN CIRCLE
EASIEST TRAIL, FLAT,
WOODED

BLUE SQUARE
MORE DIFFICULT DUE
TO HILLS AT EACH END –
REST IS GENTLE SLOPE

Thousand Islands Environmental Center and Conservancy Zone

Outagamie County

The past of the Thousand Islands is steeped in massive glaciers, several Indian tribes and countless generations of scenic beauty.

This trail offers a total of 7.2km (4.5 miles) skiing for the beginner, intermediate skier.

The center offers a fireplace and can provide programs and a meeting place for interested groups. It is open 7 days a week 9:00 A.M. to 5:00 P.M.

More information may be obtained by calling 766-4733.

Directions: Past the Girl Scout Camp off County Trunk Q or turn off on Co. Trunk Z, lowside of zone.

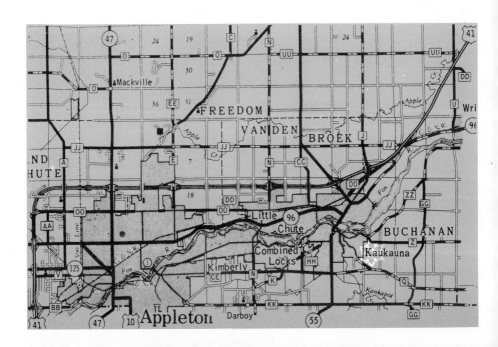

264

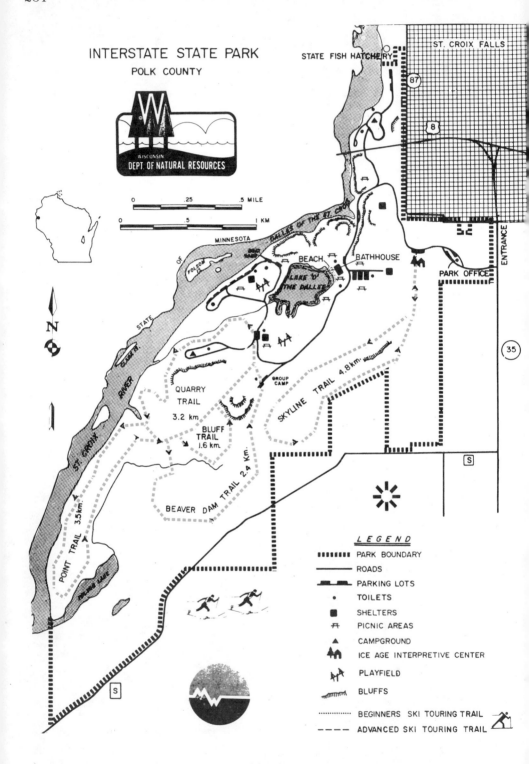

INTERSTATE STATE PARK
POLK COUNTY

WISCONSIN
DEPT OF NATURAL RESOURCES

ST. CROIX FALLS

STATE FISH HATCHERY

87

8

35

S

MINNESOTA

FOLSOM IS.

BEACH

BATHHOUSE

PARK OFFICE

ENTRANCE

LAKE O' THE DALLES

GROUP CAMP

QUARRY TRAIL
3.2 km

SKYLINE TRAIL 4.8 km

BLUFF TRAIL
1.6 km

BEAVER DAM TRAIL 2.4 Km.

POINT TRAIL 3.5 km.

ST. CROIX RIVER

DALLES OF THE ST. CROIX

STATE OF MINNESOTA

N

0   .25   .5 MILE
0   .5   1 KM

S

S

LEGEND

▪▪▪▪▪▪▪▪ PARK BOUNDARY
——————— ROADS
▄▄▄▄▄▄ PARKING LOTS
• TOILETS
■ SHELTERS
🀫 PICNIC AREAS
▲ CAMPGROUND
🎄 ICE AGE INTERPRETIVE CENTER
🌲🌲 PLAYFIELD
⋃⋃⋃ BLUFFS
············ BEGINNERS SKI TOURING TRAIL
– – – – ADVANCED SKI TOURING TRAIL

Interstate State Park

Polk County

The ski trails at Interstate Park traverse a 1300 acre area rich in glacial history. Five loops of rolling, wooded terrain offer an abundance of wildlife as well as vistas of the St. Croix River gorge, which was blasted out of solid rock by the melting floodwaters of the last glacier.

Skiers can choose from a total of 15.5km (9.33 miles) of marked, groomed and tracked trails.

The Ice Age Interpretive Center is open daily throughout the year from 9 am to 5 pm. The center offers a fireplace, a 20-minute film about glaciers, and an exhibit room filled with displays describing Wisconsin during the great Ice Age.

Programs are scheduled throughout the winter, and interested groups can arrange special programs with the park naturalist. More information can be obtained by calling 715-483-3747.

Interstate Park is located south of St. Croix Falls on Hwy 35. An admission sticker is required for each vehicle entering the park.

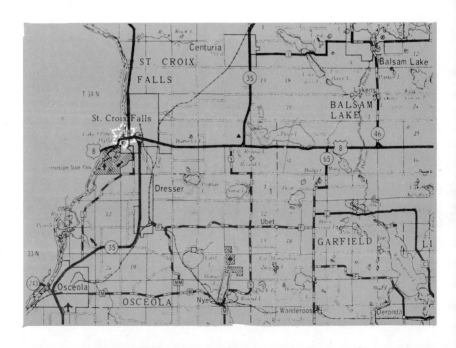

266

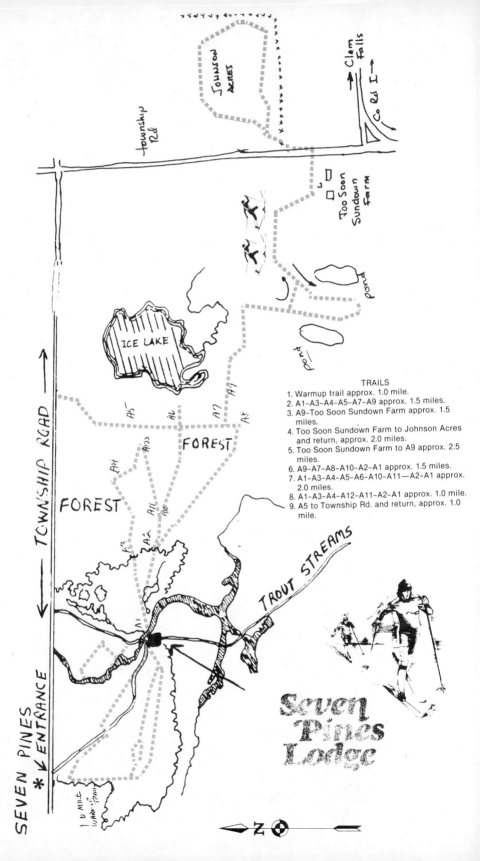

TRAILS

1. Warmup trail approx. 1.0 mile.
2. A1–A3–A4–A5–A7–A9 approx. 1.5 miles.
3. A9–Too Soon Sundown Farm approx. 1.5 miles.
4. Too Soon Sundown Farm to Johnson Acres and return, approx. 2.0 miles.
5. Too Soon Sundown Farm to A9 approx. 2.5 miles.
6. A9–A7–A8–A10–A2–A1 approx. 1.5 miles.
7. A1–A3–A4–A5–A6–A10–A11—A2–A1 approx. 2.0 miles.
8. A1–A3–A4–A12–A11–A2–A1 approx. 1.0 mile.
9. A5 to Township Rd. and return, approx. 1.0 mile.

Seven Pines Lodge

Polk County

Seven Pines Lodge & Retreat, once the private summer home of C. E. Lewis of Minneapolis, Minn., provides a unique experience for the cross-country skier and nature lover. The Lodge in its rustic elegance offers a family atmosphere with tasty country cuisine. After a day of clear air and miles of skiing, the warm fireplace offers a center for gathering to share experiences, enjoying a hot beverage and singing a tune or two. After your dinner is served, the old grandfather clock will remind you of the comfortable and cozy bed awaiting you while the stream running by lulls you into a refreshing sleep. Arise to a great new day!

We encourage groups so the Lodge can be yours for your stay. 14 miles of trails. At Lewis turn right (east) at the Shell Station for 4 blocks, turn right (south) one mile. Take left fork at Jones Game Farm to Seven Pines entrance.

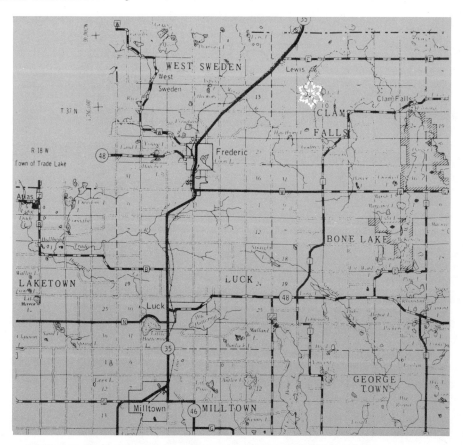

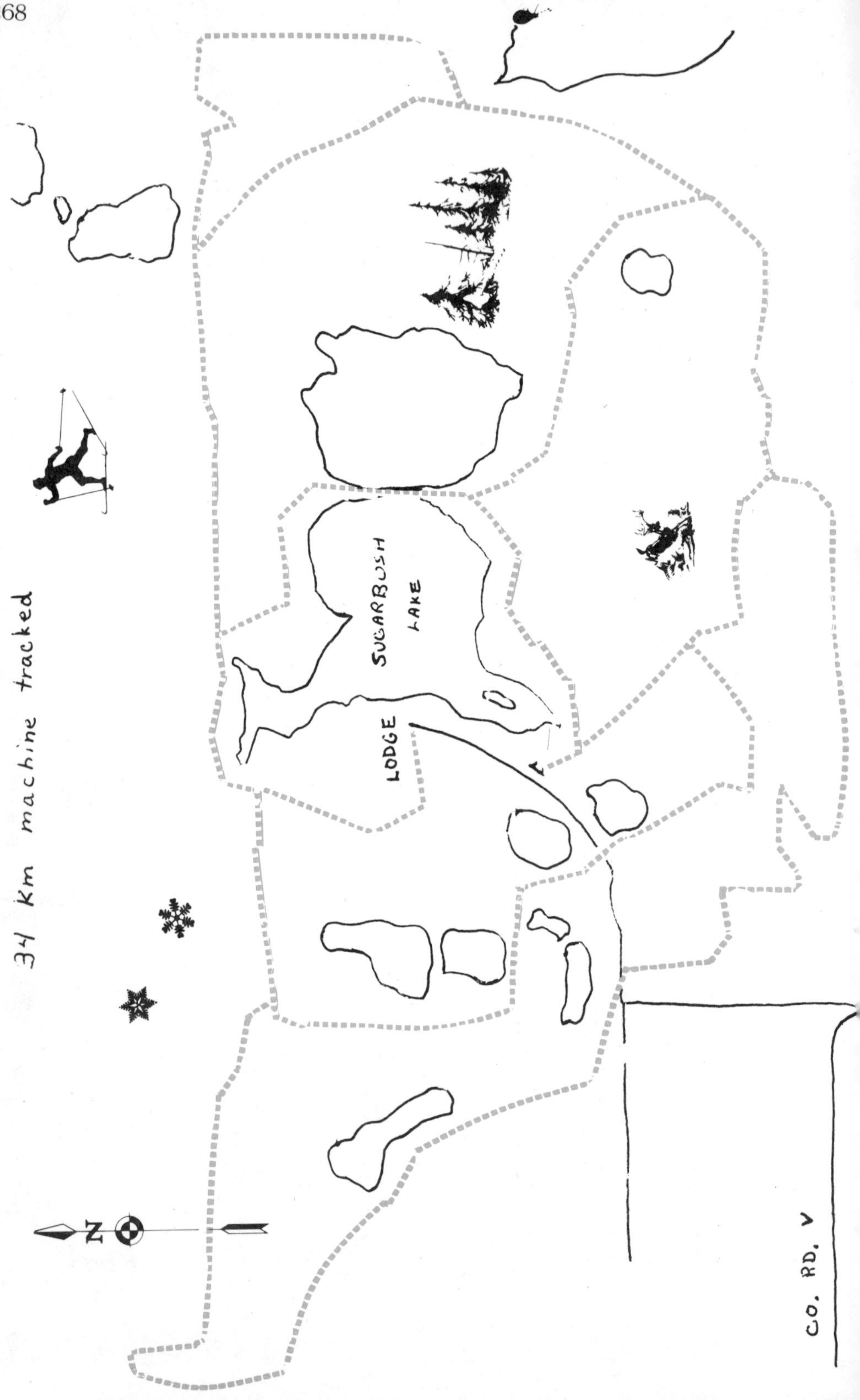

268

34 km machine tracked

SUGARBUSH LAKE

LODGE

CO. RD. V

N

Timberlake Lodge Trail

Timberlake Lodge is a recreational facility located on 950 acres of rolling, glacial terrain. It includes natural trails winding through hardwood forests studded with lakes and abounding with wildlife.

More than twenty miles of trails are marked and machine tracked, and they carry a complete line of rental equipment and accessories. Wilderness camping accessible only by ski trail.

The lodge with fireplace and snack bar overlooks picturesque Sugarbush Lake. Group accommodations with bunkhouse lodging or family room lodging now available. Kitchen facilities available with group and family rates also. Full accommodations are located within 10 miles at Turtle Lake.

Timberlake Lodge is open for camping year-around. Skiing starts with the snow!

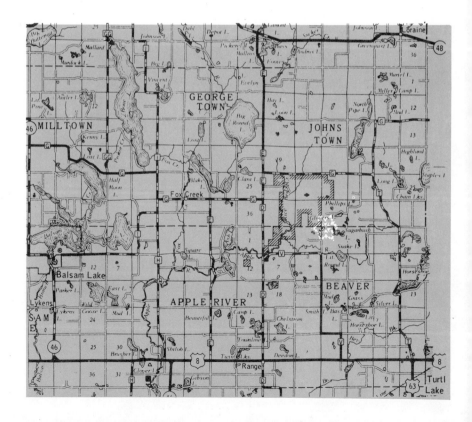

JORDAN PARK SKI TRAIL

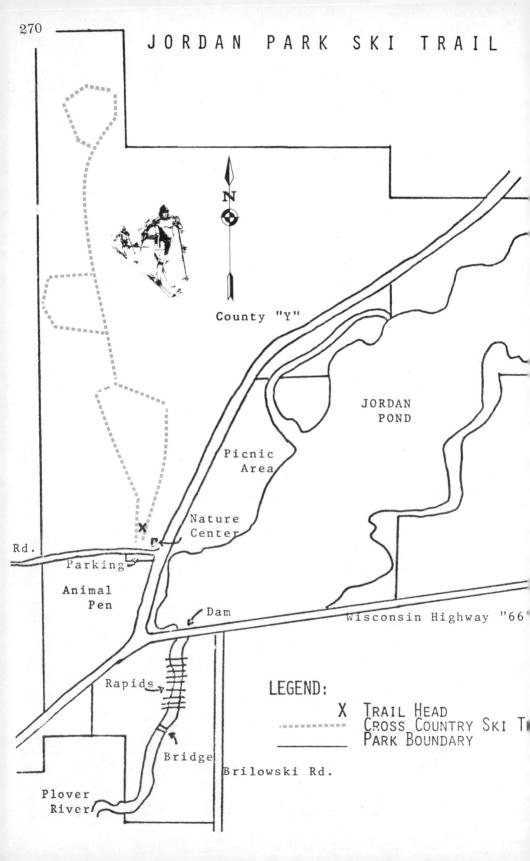

N

County "Y"

JORDAN POND

Picnic Area

Nature Center

Rd.

Parking

Animal Pen

Dam

Wisconsin Highway "66"

Rapids

LEGEND:

X TRAIL HEAD
CROSS COUNTRY SKI T
PARK BOUNDARY

Bridge

Brilowski Rd.

Plover River

Jordan Park

Jordan Park has a cross-country ski trail with loops of varying lengths for the beginning skier. A total of 3.8km (2.4 miles) of trail is available, part of which includes a self-guiding nature trail. Parking and winter camping are available. Jordan Park is located about 5 miles northeast of Stevens Point on Highway 66. For more information call 715-346-1433.

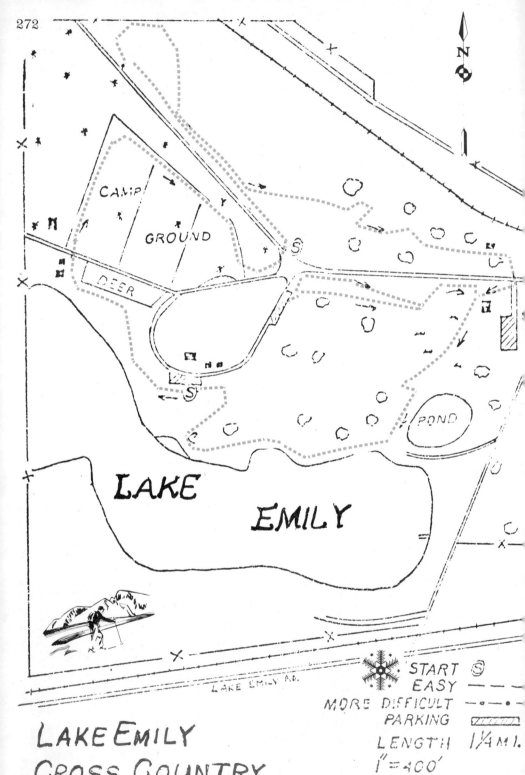

N

CAMP

GROUND

DEER

S

S

POND

LAKE

EMILY

LAKE EMILY RD.

START Ⓢ
EASY ———
MORE DIFFICULT ---•---•
PARKING ▨▨▨
LENGTH 1¼ MI.
1"=400'

LAKE EMILY
CROSS COUNTRY
SKI TRAIL

Lake Emily Ski Trail

Portage County

The trail winds through the hilly terrain on the south side of Lake Emily. The trail is 2km (1.25 miles) in length. Winter camping with plowed parking is available. The park is located just off Hwy. 10 halfway between Waupaca and Stevens Point. For more information, call 715-346-1433 or 715-824-3175.

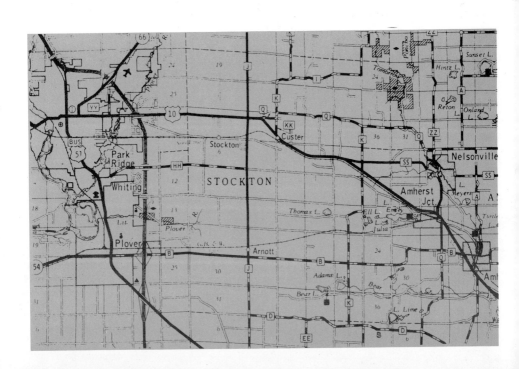

274

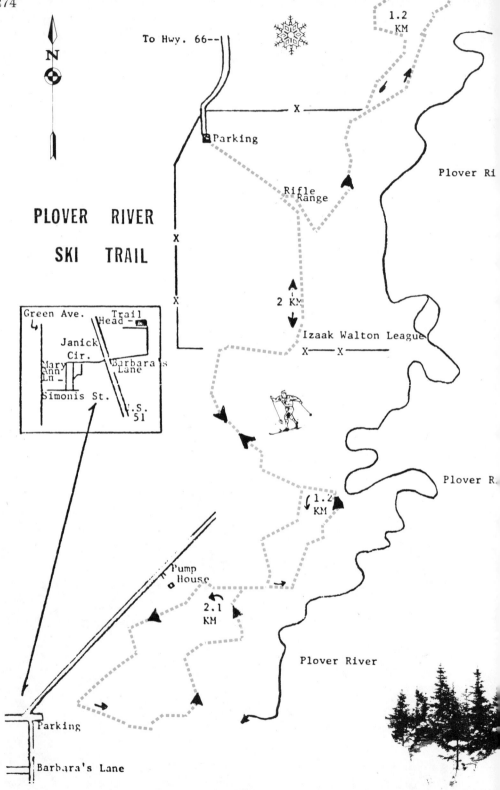

PLOVER RIVER
SKI TRAIL

N

To Hwy. 66--

Parking

1.2 KM

Rifle Range

Plover Ri

X

X

2 KM

Izaak Walton League

X — X

Green Ave.
Trail Head --
Janick Cir.
Mary Ann Ln.
Barbara's Lane
Simonis St.
U.S. 51

Plover R.

1.2 KM

Pump House

2.1 KM

Plover River

Parking

Barbara's Lane

Plover River Ski Trail

The Plover River Ski Trail has three one-way loops connected by two-way trails. A total of over 4 wooded miles of trail for beginners and intermediate skiers is available. Entry to the trail can be gained from Stevens Point (see map inset) or the Bill Cook Chapter — Izaak Walton League property off Highway 66.

276

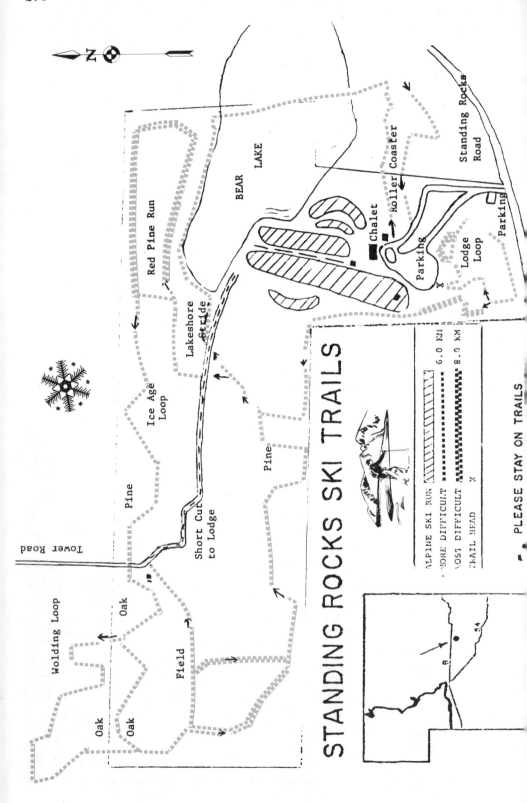

STANDING ROCKS SKI TRAILS

PLEASE STAY ON TRAILS

Standing Rocks Park

Portage County

Standing Rocks Park has a cross-country ski trail with loops of varying lengths for the intermediate and expert skier. A total of 8km (5 miles) of trail is available. The one way trails wind thru wooded and open areas. This is also a downhill ski area with 5 runs and three rope tows. A large rustic chalet provides a place to rest and keep warm. Hot food and beverages and downhill and cross-country ski rentals are available during open hours which are from 9:00 A.M. to 4:30 P.M. on weekends and during the Christmas holidays. The ski trail is open through the week. The park is located about 10 miles southeast of Stevens Point. For further information call 715-346-1430 or 715-824-3949.

278

WOLF LAKE SKI TRAIL

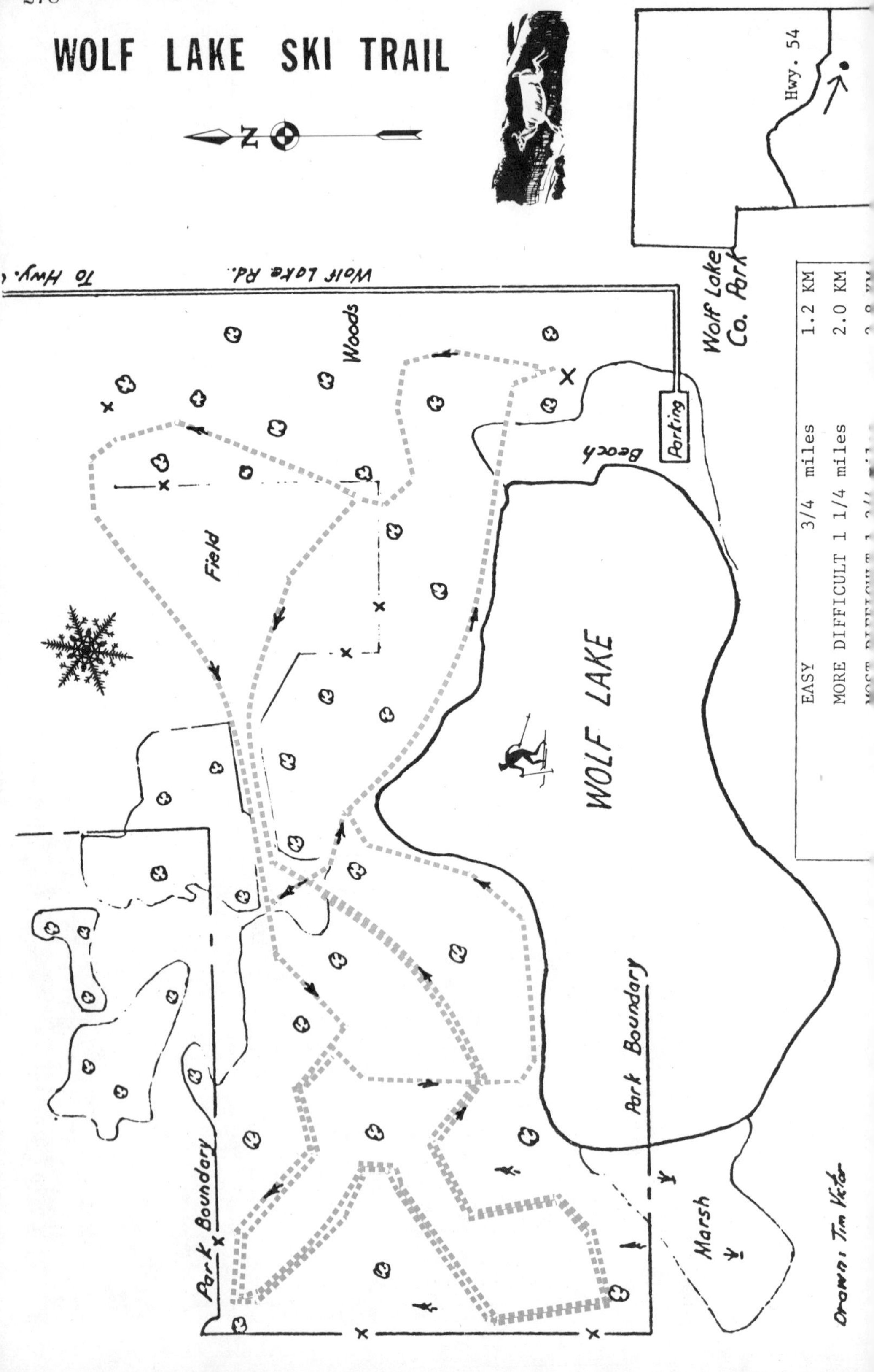

N

Wolf Lake Rd. To Hwy.

To Hwy.

Woods

Field

Park Boundary

Park Boundary

WOLF LAKE

Beach

Parking

Wolf Lake Co. Park

Marsh

Hwy. 54

Drawn: Tim Veto

EASY	3/4 miles	1.2 KM
MORE DIFFICULT	1 1/4 miles	2.0 KM
MOST DIFFICULT	1 3/4 miles	2.8 KM

Wolf Lake Park

Portage County

Wolf Lake Park has a ski trail with three loops — one each for the beginner, intermediate and advanced skier. A total of 3.2km (2 miles) of trail winds through woods and fields overlooking Wolf Lake. A parking area and toilets are located near the trail head. Wolf Lake is located about fifteen miles southeast of the village of Plover or about three miles from Wisconsin Highway 54. For more information call 715-346-1433

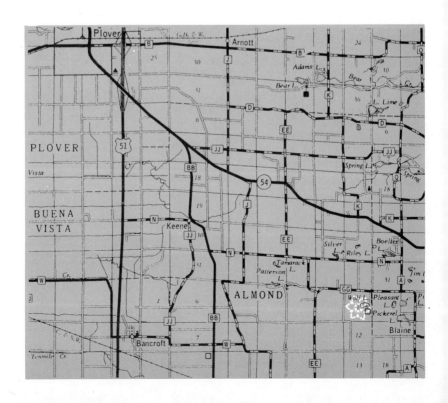

BOYD'S NORDIC SKI TRAILS

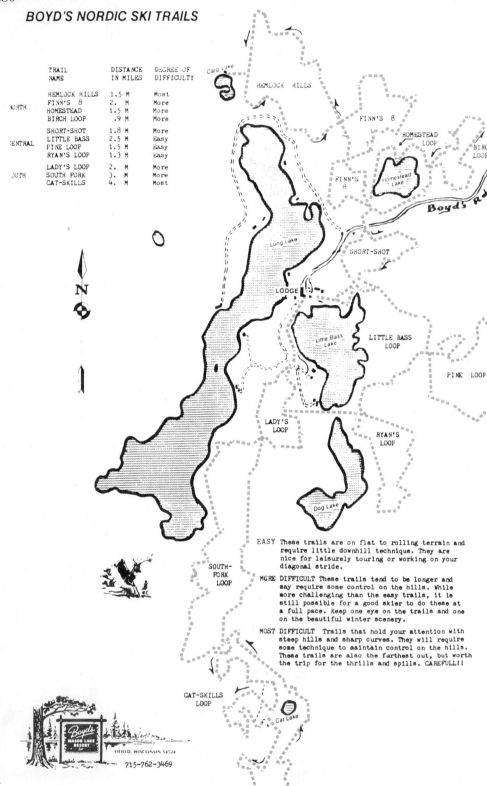

TRAIL NAME	DISTANCE IN MILES	DEGREE OF DIFFICULTY
NORTH		
HEMLOCK HILLS	1.5 M	Most
FINN'S 8	2. M	More
HOMESTEAD	1.5 M	More
BIRCH LOOP	.9 M	More
CENTRAL		
SHORT-SHOT	1.8 M	More
LITTLE BASS	2.5 M	Easy
PINE LOOP	1.5 M	Easy
RYAN'S LOOP	1.3 M	Easy
SOUTH		
LADY'S LOOP	2. M	More
SOUTH FORK	3. M	More
CAT-SKILLS	4. M	Most

Carp Lake

HEMLOCK HILLS

FINN'S 8

HOMESTEAD LOOP

BIRCH LOOP

FINN'S 8

Homestead Lake

Boyd's Rd.

N

Long Lake

SHORT-SHOT

LODGE

Little Bass Lake

LITTLE BASS LOOP

PINE LOOP

LADY'S LOOP

RYAN'S LOOP

Dog Lake

SOUTH-FORK LOOP

CAT-SKILLS LOOP

Cat Lake

EASY These trails are on flat to rolling terrain and require little downhill technique. They are nice for leisurely touring or working on your diagonal stride.

MORE DIFFICULT These trails tend to be longer and may require some control on the hills. While more challenging than the easy trails, it is still possible for a good skier to do these at a full pace. Keep one eye on the trails and one on the beautiful winter scenery.

MOST DIFFICULT Trails that hold your attention with steep hills and sharp curves. They will require some technique to maintain control on the hills. These trails are also the farthest out, but worth the trip for the thrills and spills. CAREFULL!!

Boyd's
MASON LAKE
RESORT

FIFIELD, WISCONSIN 54524

715-762-3469

Boyd's Mason Lake Resort

Price County

Located 12 miles west of Fifield on Highway 70, this quiet and secluded site is ideal for ski groups. The resort offers over 50 kilometers of groomed and tracked trails for your exclusive enjoyment as our guest. Only one group is hosted at a time, ensuring private and uninterrupted use of all trails and facilities.

The well designed trail system, with its numerous well marked loops, allows groups to plan tours of varying difficulty and length. The trails meander over 6,000 acres of private land, and are engulfed in the beauty of winter as they circle lakes, fringe fields, and wind amongst the tall timber. The terrain is varied enough to accomondate the beginner, or challenge the experienced skier. PSIA certified instruction, guide services, skills critiques, video analysis, and of course trail maps are all available to our guest.

Accomodations are in the main lodge which includes a lobby with fireplace, dining room, and daily maid service. Just outside is our new wood fired sauna to ease away the fatigue of a days skiing. All your meals are included and are prepared fresh to satisfy the hearty appetite. Groups of 20 to 40 are comfortably accomodated for stays of 2 nights or more. For more information call Boyd's Mason Lake Resort 715-762-3469.

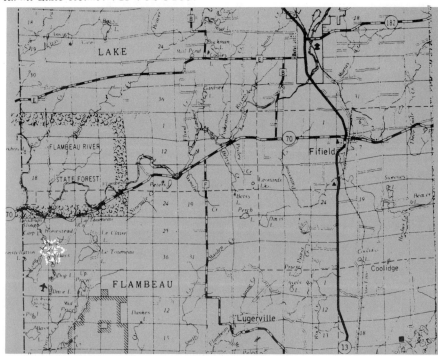

Camp 5 Ski Trail

PRICE COUNTY FOREST
Two-way Trails

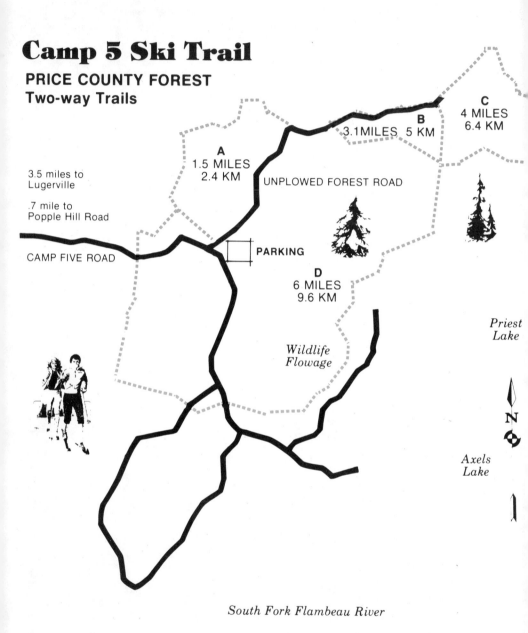

A
1.5 MILES
2.4 KM

B
3.1 MILES 5 KM

C
4 MILES
6.4 KM

UNPLOWED FOREST ROAD

3.5 miles to
Lugerville

.7 mile to
Popple Hill Road

CAMP FIVE ROAD

PARKING

D
6 MILES
9.6 KM

*Wildlife
Flowage*

*Priest
Lake*

N

*Axels
Lake*

South Fork Flambeau River

Camp 5 Cross-Country Ski Trail

Price County

Camp 5 Cross-Country Ski Trail is located on Price County Forest Crop land in the heart of the 4,078 acre Flambeau Unit. The trail consists of four loops varying in length from 2.1km (1.3 miles) to 7km (4.2 miles). The loops all offer a variety of relief that can be mastered by the amateur and enjoyed by the expert.

Many tree species line the trails from towering pines to the beautiful white birch. From the trail between loops B and C can be seen a thousand acre black spruce bog. White tail deer, ruffed grouse, and squirrels are common in the area.

The trail is presently being expanded to keep up with future demand. More information can be obtained by contacting Price County Forestry Dept., Normal Building, Philips, WI 54555, (715) 339-2255.

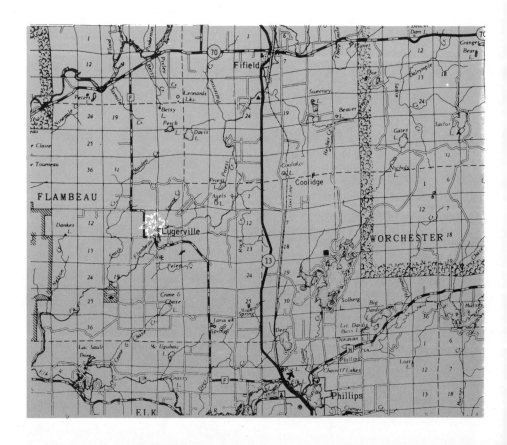

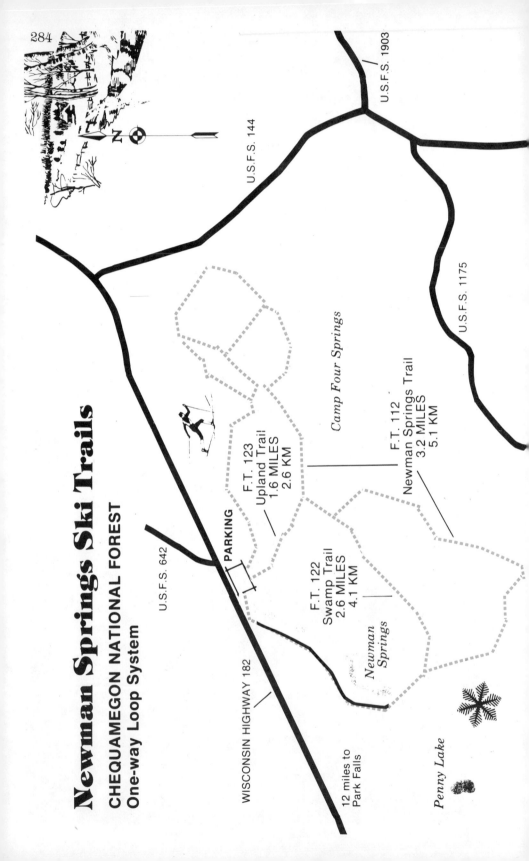

Newman Springs
Ski Trail

The Newman Springs Ski Trail is for your winter skiing enjoyment. The trail is located 12 miles east of Park Falls on Hwy. 182. A large parking area and toilet facilities are at the trailhead. The trail offers the skier a variety of routes through glaciated terrain ranging from upland ridges to lowland marshes with scenic vistas of several spring ponds. The trail passes through areas of lowland conifers, marshes, red pine plantations, northern hardwoods, and white birch. The diversity of the trail provides a pleasant and unique ski tour for the beginner and experienced skier. As you ski the trail, enjoy the winter beauty of your National Forest.

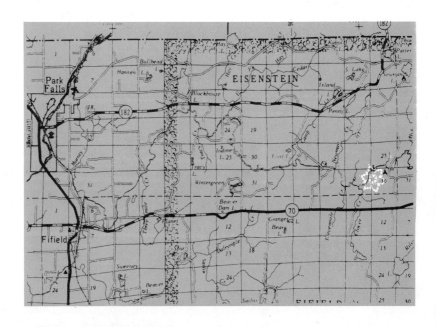

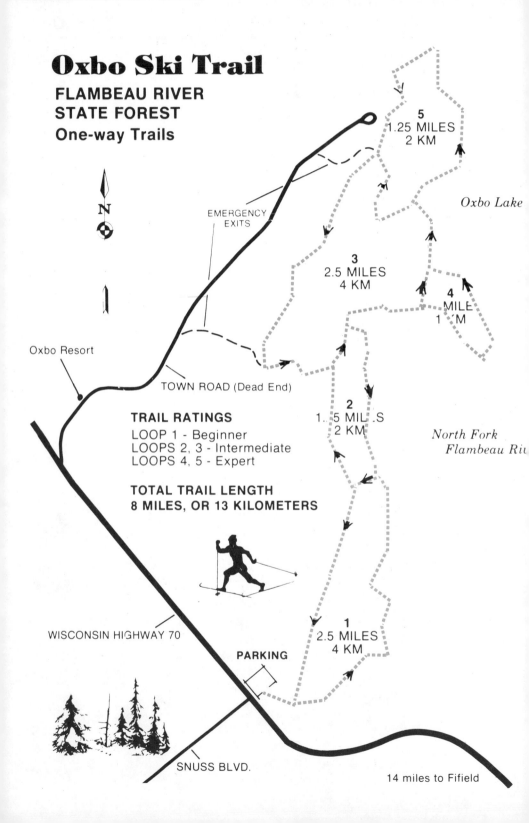

Oxbo Ski Trail

**FLAMBEAU RIVER
STATE FOREST
One-way Trails**

N

EMERGENCY
EXITS

Oxbo Resort

TOWN ROAD (Dead End)

TRAIL RATINGS
LOOP 1 - Beginner
LOOPS 2, 3 - Intermediate
LOOPS 4, 5 - Expert

**TOTAL TRAIL LENGTH
8 MILES, OR 13 KILOMETERS**

WISCONSIN HIGHWAY 70

PARKING

SNUSS BLVD.

5
1.25 MILES
2 KM

Oxbo Lake

3
2.5 MILES
4 KM

4
MILE
1 KM

2
1.25 MILES
2 KM

*North Fork
Flambeau Riv*

1
2.5 MILES
4 KM

14 miles to Fifield

Oxbo Ski Trail

Price County

Located in the Flambeau River State Forest the Oxbo has 8 miles of scenic trail with five loops. These one-way trails are as follows: Loop 1 Beginner, Loop 2 & 3 Intermediate, Loop 4 & 5 for the Expert.

To find these trails take Hwy. 70 west out of Fifield about 14 miles to Snuss Blvd. Parking lot is on the right side of the road.

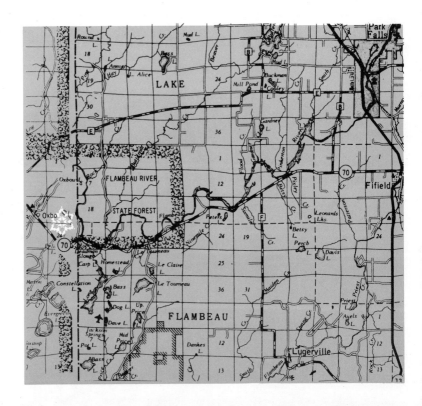

South Fork Flambeau River

10 Miles to
Park Falls

ROCK CREEK ROAD

8 Miles to
Phillips

WISCONSIN
HIGHWAY 13

2.5 MILES
4KM

Rock Creek

EAST ROCK CREEK ROAD

EXPERT HILL

1.5 MILES
2.4 KM

PARKING

N

2 Miles to
Lugerville

1.5 MILES
2.4 KM

SOUTH
ROCK CREEK
ROAD

Rock Creek
Ski Trails

PRICE COUNTY FOREST
Two-way Trails

7 Miles to
Phillips

Rock Creek

Price County

These trails for beginners and intermediate skiers have a rolling topography with heavy forested areas. It can be skied in five separate loops if desired. It is maintained by the Price County Forestry Department. Take Hway 13, 8 miles north of Phillips to Rock Creek Road.

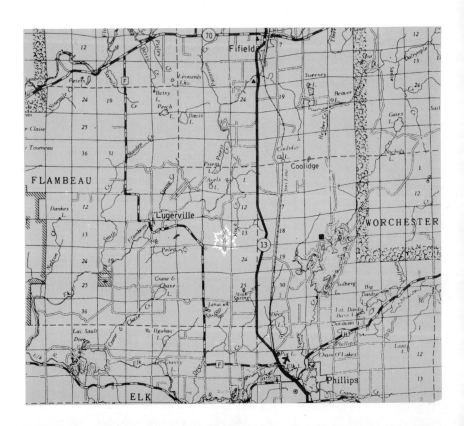

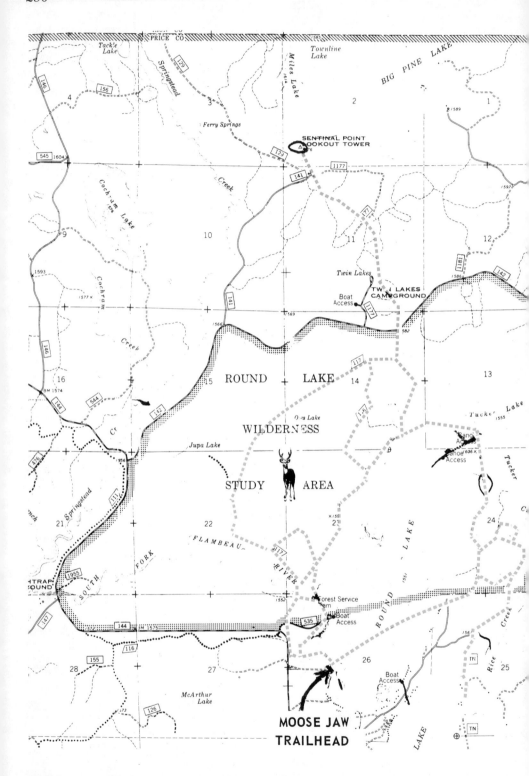

Round Lake Trail

Price County

The trail begins and ends at Moose Jaw Lodge on Round Lake. Trails go thru the Round Lake Wilderness study area and are generally level or gently hilly. The trails are geared for weekend skiers or people who wish to get off the beaten path and explore the unlimited interesting sights.

For a day of gentle skiing, beautiful sights and peaceful feelings or for the adventurous who wish to explore, Round Lake trails are ideal.

After skiing, relax in the Moose Jaw Lodge where food and cocktails are served.

Trail lengths — loops of: Beginners 1.6km (1 mile), 3.2km (2 miles); Intermediate 9.6km (6 miles) 12.8km (8 miles); Advanced 16km (10 miles).

Round Lake is located east of Fifield on 70 to Forest Road 144, turn north to lodge.

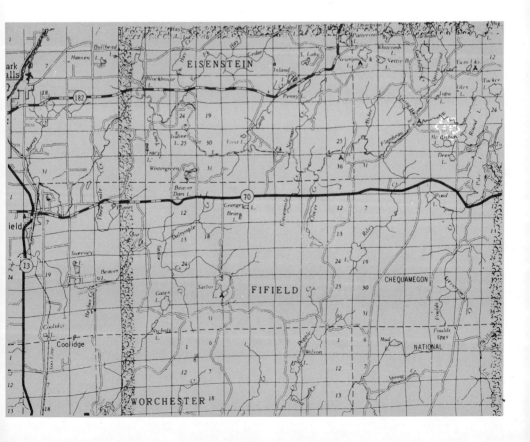

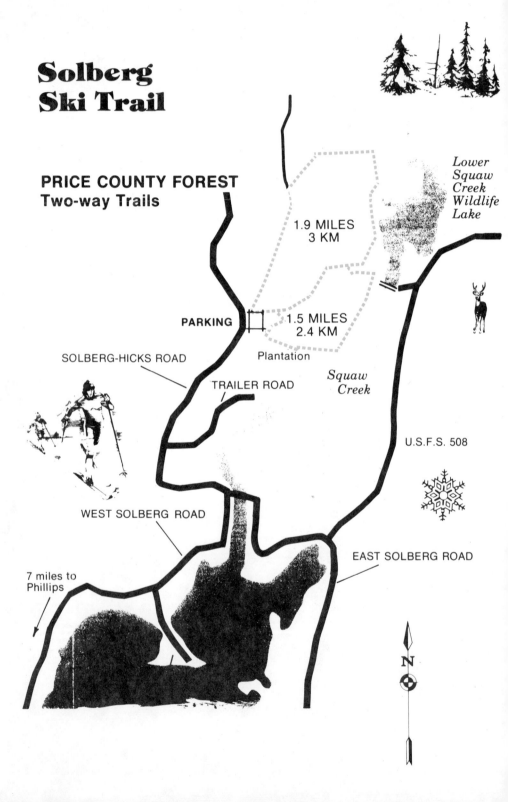

Solberg Ski Trail

PRICE COUNTY FOREST
Two-way Trails

1.9 MILES
3 KM

1.5 MILES
2.4 KM

Lower Squaw Creek Wildlife Lake

PARKING

Plantation

SOLBERG-HICKS ROAD

TRAILER ROAD

Squaw Creek

U.S.F.S. 508

WEST SOLBERG ROAD

EAST SOLBERG ROAD

7 miles to Phillips

N

Solberg Trail

Located 9 miles northwest of Phillips, the Solberg Trail consists of 2 loops with a total of 5km (3.1 miles) of gentle, wooded marked trail. It is recommended for the beginner skier.

294

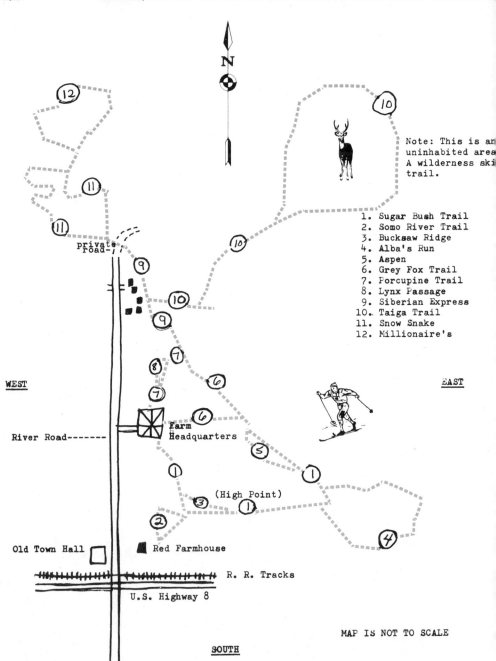

Note: This is an
uninhabited area
A wilderness ski
trail.

1. Sugar Bush Trail
2. Somo River Trail
3. Bucksaw Ridge
4. Alba's Run
5. Aspen
6. Grey Fox Trail
7. Porcupine Trail
8. Lynx Passage
9. Siberian Express
10. Taiga Trail
11. Snow Snake
12. Millionaire's

private road

WEST

EAST

River Road-------

Farm
Headquarters

(High Point)

Old Town Hall

Red Farmhouse

R. R. Tracks

U.S. Highway 8

MAP IS NOT TO SCALE

SOUTH

Palmquist's "The Farm"

Price County

Palmquist's "The Farm" offers groomed and tracked trails, food and lodging, equipment rental, and the atmosphere of a large old northwoods farm ranch.

Explore remote trails that wind through dense forests, and over birch-covered ridges. Terrain varies from gently rolling to moderately challenging. Ski right from your cabin door or try one of the off-the-farm trails. Other activities include a Finnish sauna, sing-a-longs, and light melodrama. Lodging consists of pleasant cabins and inn rooms in the main farmhouse.

Palmquist's "The Farm" is located 20 miles west of Tomahawk of Highway 8, then ½ mile north on River Road.

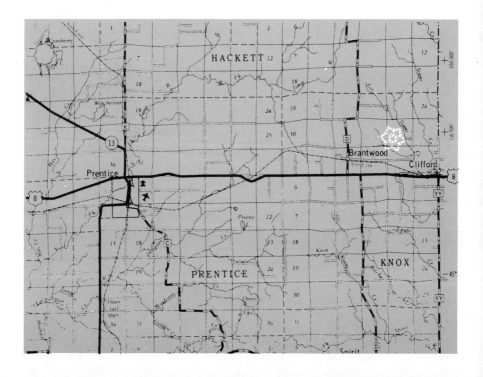

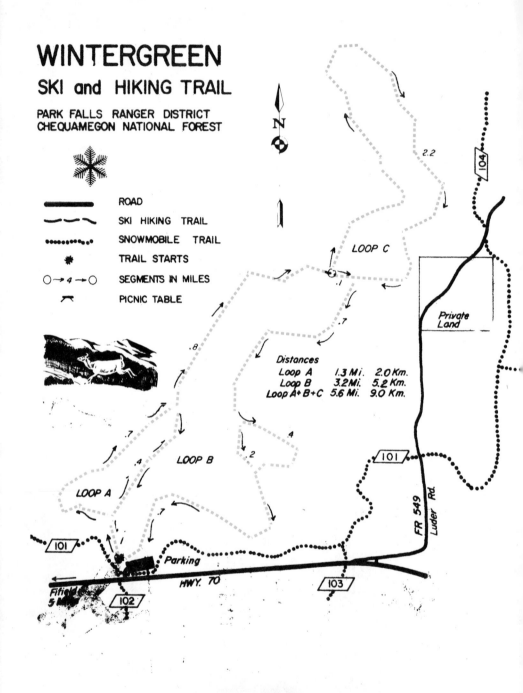

WINTERGREEN
SKI and HIKING TRAIL

PARK FALLS RANGER DISTRICT
CHEQUAMEGON NATIONAL FOREST

————	ROAD
∼∼∼∼	SKI HIKING TRAIL
••••••••••	SNOWMOBILE TRAIL
✳	TRAIL STARTS
○→4→○	SEGMENTS IN MILES
⊼	PICNIC TABLE

N

LOOP C

2.2

104

Private Land

.1

.7

.8

Distances
Loop A   1.3 Mi.   2.0 Km.
Loop B   3.2 Mi.   5.2 Km.
Loop A+B+C 5.6 Mi.   9.0 Km.

.4

.7

.4

.2

LOOP B

LOOP A

.7

101

FR 549

Luder Rd.

101

Parking

HWY. 70

Fifield
5 Mi.

102

103

Wintergreen
Ski & Hiking Trail

Price County

The Wintergreen Cross Country Skiing and Hiking Trail is for your year round enjoyment. The trail is approximately 5 miles East of Fifield, Wisconsin, North of State Highway 70. A large parking area and toilet facilities are available at the trailhead. The trail offers both the hiker and skier a variety of routes through rolling terrain, ranging from upland ridges to lowland marshes. Scenic vistas overlooking several ponds and bogs are present. The trails pass through a mixture of forest types from lowland conifers and red pine plantations, to northern hardwoods and young aspen stands (ideal for spotting wildlife). The trail loop system provides a variety of opportunities for both beginner and experienced skiers and hikers. This trail network has been constructed for your enjoyment of your National Forest.

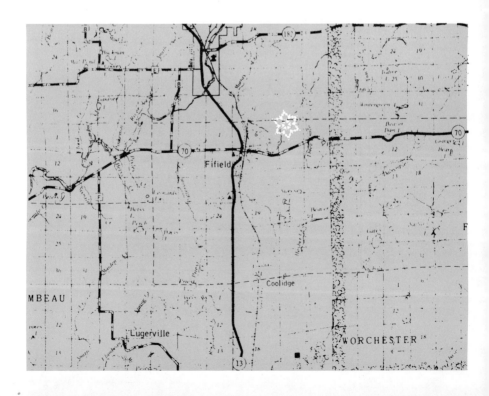

Arboretum Ski Trails

One mile west of Janesville on W. Memorial Drive (County A), turn right at Austin Road, ½ mile to parking lot.

1½ miles of trail in 2 loops. Trails gentle/rolling through oak and hickory woods. Plowed Parking Lot. Marked & Tracked Trails. Maps. Toilet.

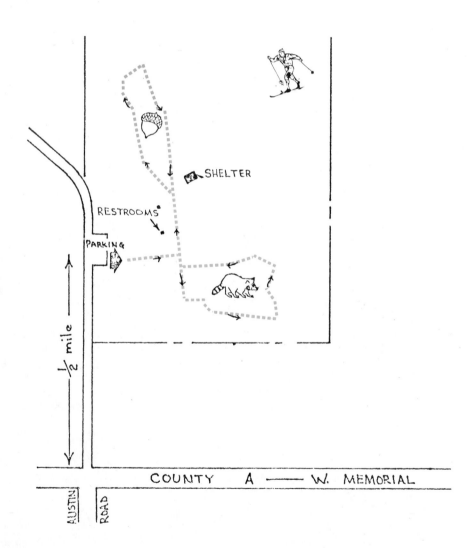

Lustig Park Trails

Rock County

Trail begins in parking lot on Riverside Drive in front of rest room facilities.

.8 miles of trail in 1 loop. Gentle rolling terrain leads through park area. This is a very nice beginners trail.

Plowed Parking Lot. Marked Trail. Maps.

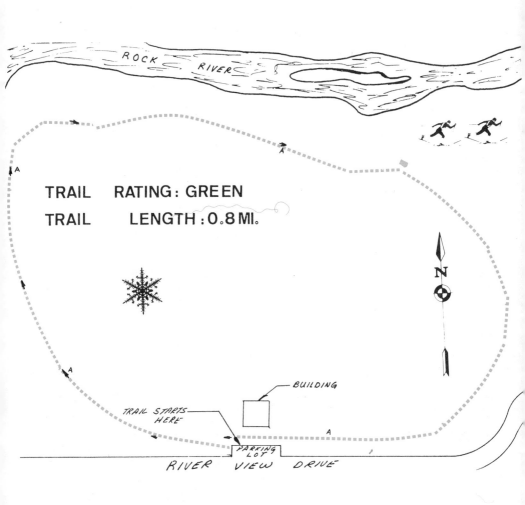

ROCK RIVER

TRAIL RATING: GREEN

TRAIL LENGTH: 0.8 MI.

N

BUILDING

TRAIL STARTS HERE

PARKING LOT

RIVER VIEW DRIVE

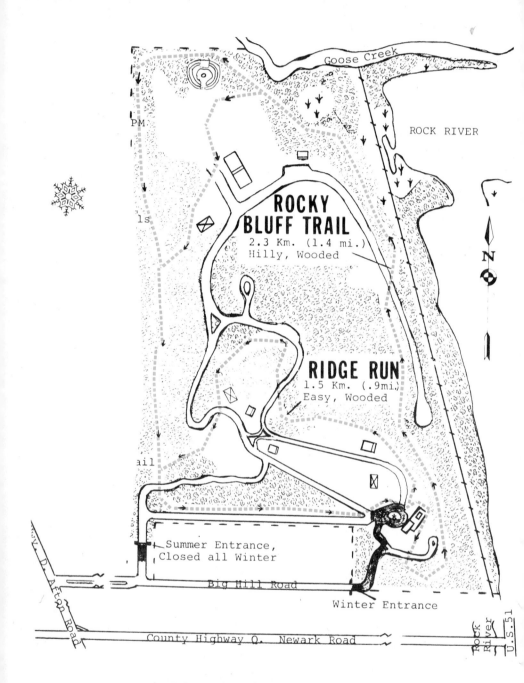

ROCKY BLUFF TRAIL
2.3 Km. (1.4 mi.)
Hilly, Wooded

RIDGE RUN
1.5 Km. (.9mi.)
Easy, Wooded

Goose Creek

ROCK RIVER

N

PM

ls

ail

Summer Entrance,
Closed all Winter

Big Hill Road

Winter Entrance

County Highway Q. Newark Road

Afton Road

Rock River

U.S. 51

Big Hill Park Trails

Beginners can enjoy the gently sloped trail through the woods while more experienced skiers will be challenged by the woody hilly trail traversing the park.

Beloit Trails

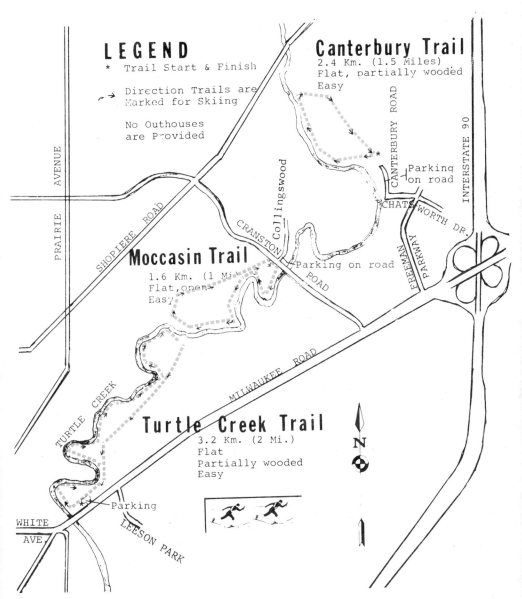

LEGEND
* Trail Start & Finish

↷→ Direction Trails are Marked for Skiing

No Outhouses are Provided

Canterbury Trail
2.4 Km. (1.5 Miles)
Flat, partially wooded
Easy

Moccasin Trail
1.6 Km. (1 Mi.)
Flat, open
Easy

Turtle Creek Trail
3.2 Km. (2 Mi.)
Flat
Partially wooded
Easy

Turtle Creek Floodplain

Three trails totaling 4½ mile for beginners that wind along Turtle Creek through open and partially wooded countryside.

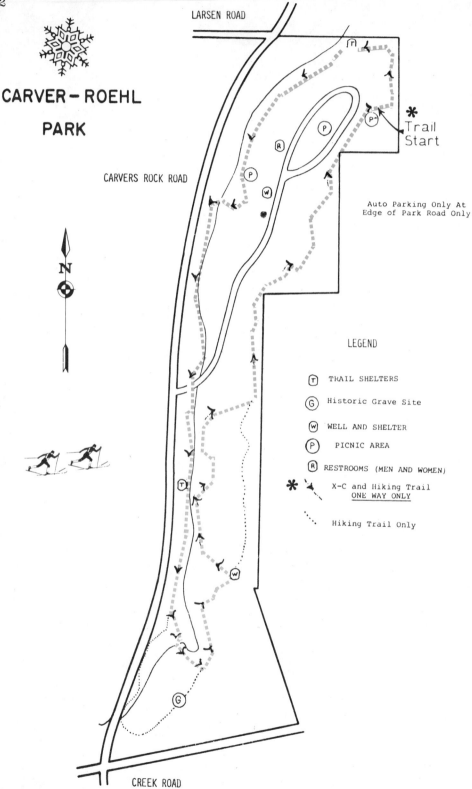

CARVER – ROEHL
PARK

LARSEN ROAD

CARVERS ROCK ROAD

N

Trail
Start

Auto Parking Only At
Edge of Park Road Only

LEGEND

Ⓣ TRAIL SHELTERS

Ⓖ Historic Grave Site

Ⓦ WELL AND SHELTER

Ⓟ PICNIC AREA

Ⓡ RESTROOMS (MEN AND WOMEN)

✳ X-C and Hiking Trail
ONE WAY ONLY

.... Hiking Trail Only

CREEK ROAD

Carver-Roehl Park

This trail is approximately 4.4km (2.7 mles) in length. It is for intermediate skiers with some portions a little bit challenging for the beginner. This park contains areas of extreme topography, and anyone using this trail should stay on the trail for their own safety. Because of this extreme change in topography and wide diversity of plant life it is by far one of our most picturesque parks, whatever the season.

Parking is provided at the north end of the park loop road, with the trail starting immediately to the east. Toilets, picnic tables and grills are provided at the trailhead. This trail is set up to be skied in one direction only, counter-clockwise. It is designated by red marker posts, and is open to public use every day from dawn to dusk.

Besides being an extremely scenic trail, it also offers an opportunity for the advanced skier to traverse it a number of times to build endurance.

Carver-Roehl County Park is located 14½ miles east of Janesville. Take Highway 11-14 going east to the intersection of Carver-Rock Road; turn right (south) and proceed the final 4½ miles to the park entrance, which is on the left side of the road (east); stay on the internal park road until you get to the end of its loop.

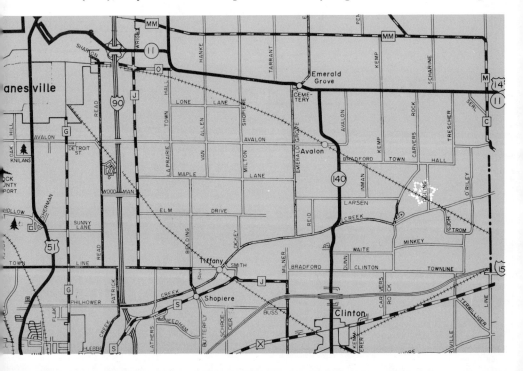

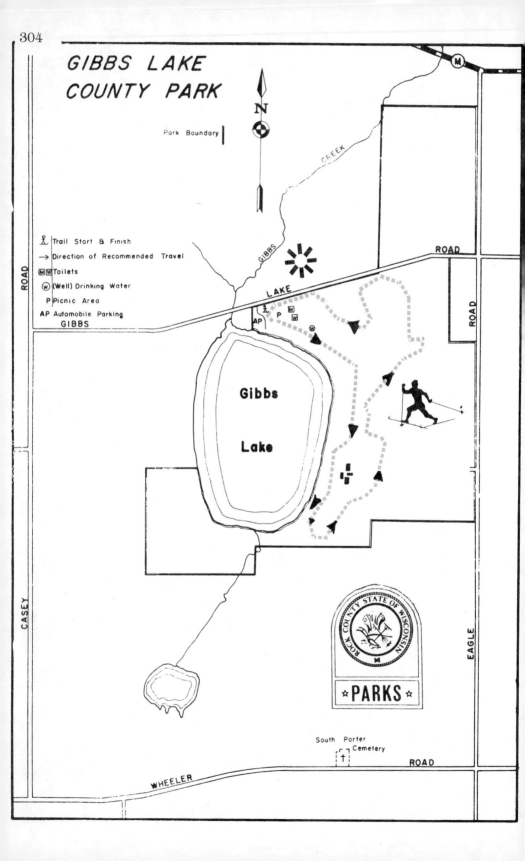

304

GIBBS LAKE
COUNTY PARK

Park Boundary

N

Trail Start & Finish
→ Direction of Recommended Travel
MW Toilets
W (Well) Drinking Water
P Picnic Area
AP Automobile Parking

GIBBS

CREEK

GIBBS

LAKE

ROAD

ROAD

ROAD

Gibbs

Lake

ROAD

CASEY

EAGLE

ROCK COUNTY STATE OF WISCONSIN

☆ PARKS ☆

South Porter
Cemetery

ROAD

WHEELER

Gibbs Lake County Park

Rock County

This trail is 2.97km in length; however, by utilizing the short-cut, the beginner can ski ony 1.53km. By skiing the short-cut circuit, and entire trail using the "high leg" and re-skiing the entire trail using the "low leg", the total distance is 7.4km (4.5 miles).

Parking is provided at the park entrance and trail starting area. Toilets, picnic tables and grills are provided at the trailhead.

This trail has been set up to be skied in a counter-clockwise fashion. It is defined with orange marker posts, and is open to public use every day from dawn to dusk. It has been designed for the novice and intermediate skier.

Gibbs Lake County Park is located 12 miles from the center of Janesville. Take Hwy. 14 west 7 miles and turn right (north) on to Eagle Road; travel 3 miles to Gibbs Lake Road, turn left (west) and travel ½ mile to the park entrance.

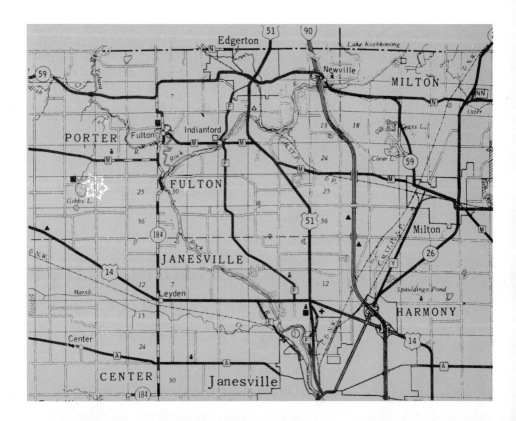

Greenbelt Trail

Trail Rating: Green
Trail Length: 5.1 mi.

✳ Tunnel – Low Clearance

⌒ Bridge

⊗ Street Crossing

Greenbelt Trail

Trail head begins on the corner of Ruger Avenue and Greendale. Parking is along the street.

5.1 miles of trails in 6 loops. Gentle/rolling terrain leads through residential area. Low clearance necessary for tunnel passes.

Marked and Tracked trails.

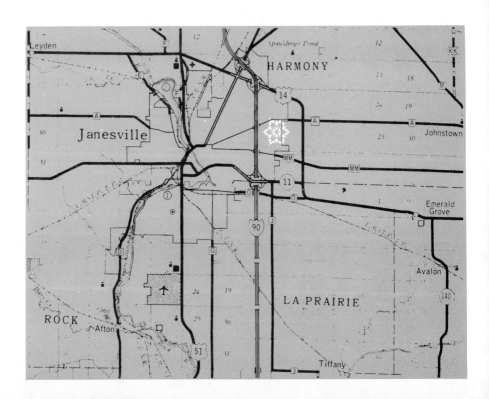

308

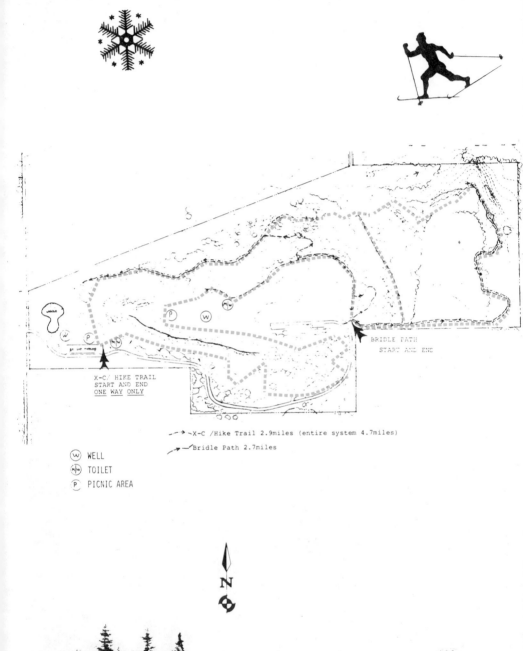

X-C/ HIKE TRAIL
START AND END
<u>ONE WAY ONLY</u>

BRIDLE PATH
START AND END

-·→ X-C /Hike Trail 2.9miles (entire system 4.7miles)

→ Bridle Path 2.7miles

(W) WELL
(M) TOILET
(P) PICNIC AREA

N

PARKS

MAGNOLIA BLUFF

COUNTY PARK | REV. 9-28-81
| T.G.K.

Magnolia Bluff
County Park

Rock County

Magnolia Bluff County Park is located approximately 18 miles from central Janesville. The park is just south of Evansville, and located ½ mile south of the intersection of State Highway 59 and Croad Road, on Croad Road.

This 120-acre park features beautiful rock outcroppings and vistas. Because this park has steep areas and rock outcroppings, extreme caution must be used by all park visitors.

The X-C skiing and hiking trail is 2.9 miles (4.67km) and has a short-cut (intermediate) length of 2.1 miles or (3.38km). By skiing the entire outside loop (expert) and then reskiing, using the intermediated short-cut, the entire distance is 4.7 miles (7.56km).

This X-C ski and hiking trail was constructed by the YACC (Young Adult Conservation Corps) in 1981 with funding for materials and supplies provided by Rock County.

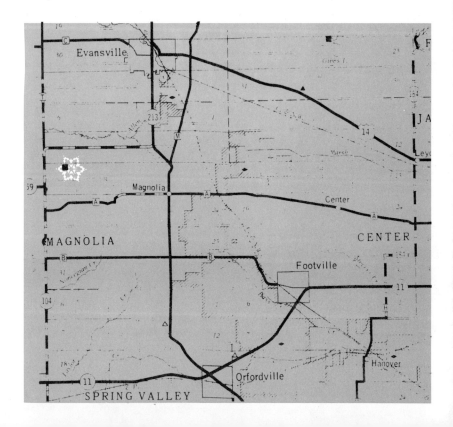

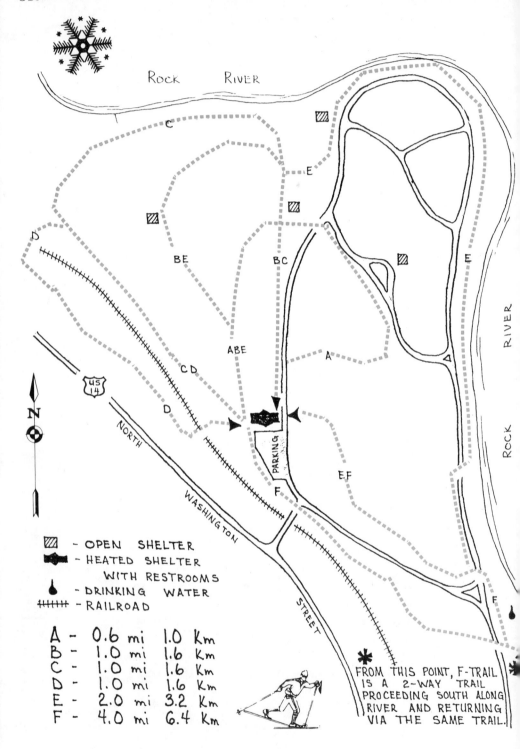

ROCK RIVER

C

E

D

BE   BC

ABE

A

CD

D

US
14

N

NORTH

PARKING

E

EF

F

WASHINGTON

STREET

RIVER

ROCK RIVER

F

- OPEN SHELTER
- HEATED SHELTER
  WITH RESTROOMS
- DRINKING WATER
- RAILROAD

A - 0.6 mi   1.0 Km
B - 1.0 mi   1.6 Km
C - 1.0 mi   1.6 Km
D - 1.0 mi   1.6 Km
E - 2.0 mi   3.2 Km
F - 4.0 mi   6.4 Km

✳ FROM THIS POINT, F-TRAIL
IS A 2-WAY TRAIL
PROCEEDING SOUTH ALONG
RIVER AND RETURNING
VIA THE SAME TRAIL.

Riverside Golf Course

Rock County

Six marked trails wind around the course in rolling, wooded terrain with excellent vistas of the Rock River.

Trails are open from dawn to dusk, nighttime skiing is not allowed. Please stay on trails to avoid greens and steep topography.

Plowed parking, shelter and restrooms are available.

Riverside is located on the northeast side of the city of Janesville, on Business Highway 14.

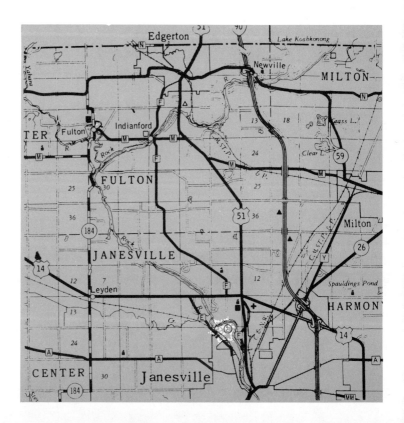

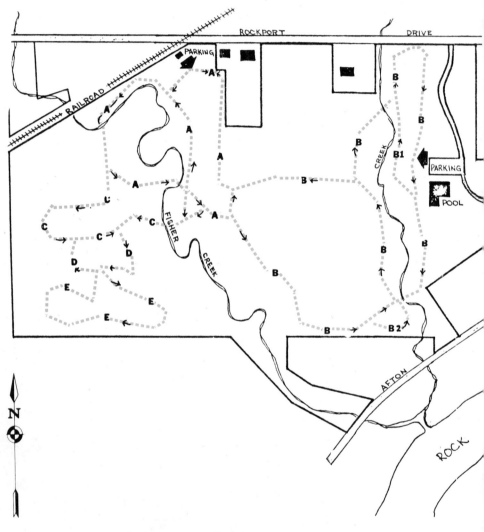

Trail	Length		Rating
A	1.5 mi -	2.4 km	Blue
B	1.7	2.7	Black
C	.5	.8	Blue
D	.3	.5	Green
E	.6	1.0	Black

Rockport Park

Rockport has two trail heads. One is located by the swimming pool ¼ mile west of Crosby Avenue on Rockport Road. The second is ¾ mile west of Crosby Avenue on Rockport Road.

6 miles of trails in 5 loops. Rolling/hilly terrain makes these trails more challenging. Trails wind through secluded meadows and wooded areas.

There is a Plowed Parking Lot. All Marked & Tracked Trails. Toilet.

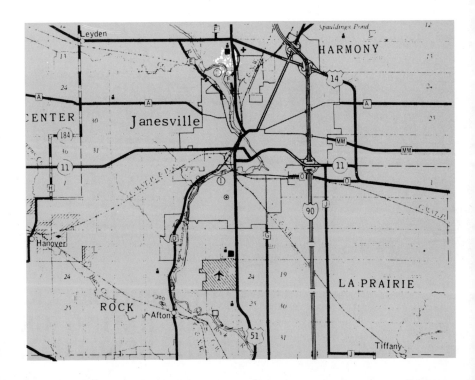

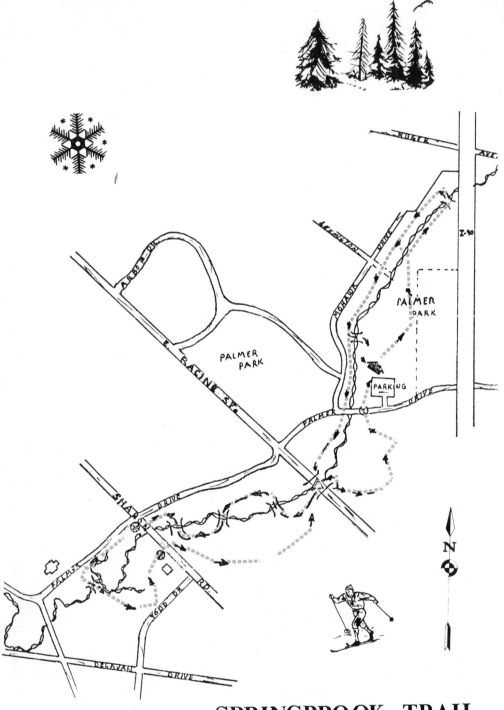

SPRINGBROOK TRAIL
TRAIL RATING - GREEN
TRAIL LENGTH - 3.5 mi.

Springbrook Trail

Trail head begins on parking lot at Palmer Park on Palmer Drive; 500 ft. east of Mohawk Drive.

3.6 miles of trail in 1 loop. Gentle rolling terrain leads through park and golf courses and also through undeveloped park land. Low clearance necessary for tunnel passes.

Plowed Parking Lot. Marked and Tracked Trails.

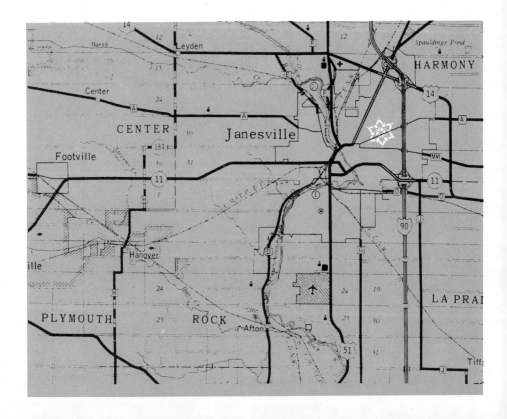

SAFETY SUGGESTIONS

1. Never ski alone—parties of three recommended.

2. Take along extra clothing and emergency equipment including *map, compass, spare ski tip, extra cork wax,* (always!) *a scraper.*

3. Notify friend of route and expected time of return. Leave note on dashboard of car & phone no.

4. Plan a pace of no more than 1^1/2 m.p.h. for your tour, especially with a larger group.

5. Plan to return 2 hours before sun set.

6. Be flexible in your planning—change goals in face of heavy weather or injury etc.

7. Observe trail directions.

8. Keep trail etiquette.

9. *Beware of hypothermia;* killer of the unprepared.

10. Check your equipment before you leave.

Enjoy the quiet, beautiful
world of winter!

Game Unlimited

St. Croix County

Game Unlimited's cross-country trails wind through 1000 acres of primordial hardwood forests. The groomed trails take you over 25.6km (16 miles) of rolling hills and woods in absolute quiet and tranquility as no snowmobiles or motor vehicles are allowed.

A lodge is available with shelter, refreshments, and kitchen facilities. Overnight accommodations by reservation only. Fees are charged for the use of this trail. Rental equipment is also available, and a trail map is posted in the lodge.

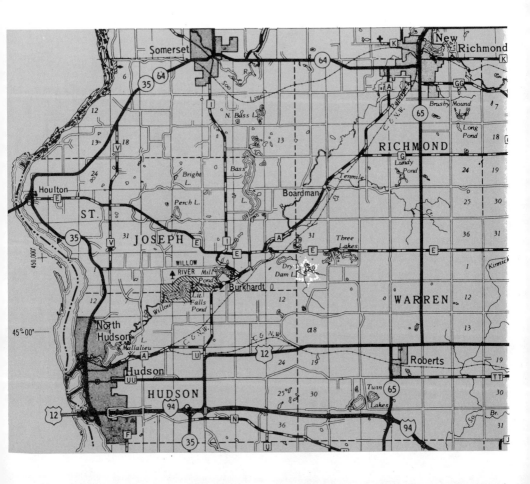

318

Snowcrest CROSS COUNTRY AREA

8 MILES OF TRAILS · 3 LIGHTED TRAILS · COMPLETE AND SEPARATE FACILITIES

- Certified cross-country ski instructors
- Country home chalet with 2 fireplaces ½ mile from Snowcrest Alpine Area
- Complete rental equipment for adults and children
- Track setter used on all trails
- Food service

ALPINE & CROSS-COUNTRY MAPS & STATISTICS

N

VESTMARKA 3.0 km

NORDMARKA 2.0 km

LYSLOYPA 2.5 km

CROSS-COUNTRY CHALET

LATRANS 1.0 km

Snowcrest

St. Croix County

Snowcrest is located on 250 acres of wooded and hilly terrain that has trails for beginners, intermediate and expert cross-country skiers. There are 6 trails with three of them lighted for night use. Total length of trails is 17.6km (11 miles). All trails are well marked, one-way and well groomed including track setting.

Rental skis are available per day and lessons per hour.

A trail fee is charged. Hours are from 9 A.M. to 10 P.M. daily. Except for lodging they have the usual complete facilities for serving meals and other refreshments. They have complete and separate facilities for downhill and cross-country skiers.

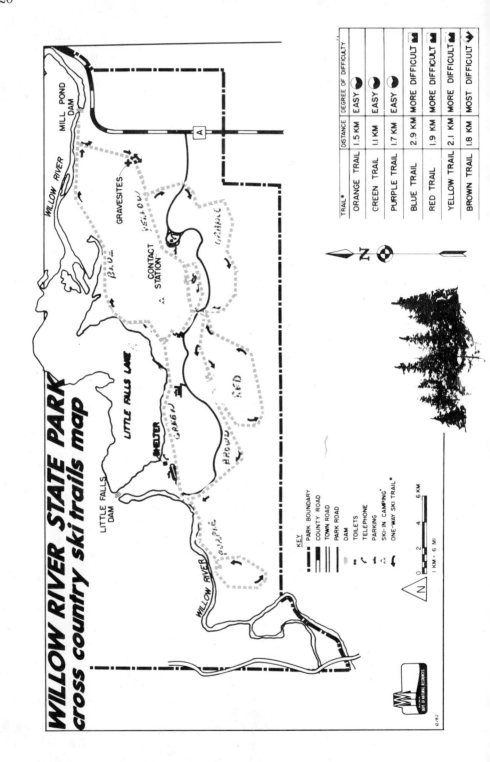

WILLOW RIVER STATE PARK
cross country ski trails map

TRAIL	DISTANCE	DEGREE OF DIFFICULTY
ORANGE TRAIL	1.5 KM	EASY
GREEN TRAIL	1.1 KM	EASY
PURPLE TRAIL	1.7 KM	EASY
BLUE TRAIL	2.9 KM	MORE DIFFICULT
RED TRAIL	1.9 KM	MORE DIFFICULT
YELLOW TRAIL	2.1 KM	MORE DIFFICULT
BROWN TRAIL	1.8 KM	MOST DIFFICULT

KEY

- PARK BOUNDARY
- COUNTY ROAD
- TOWN ROAD
- PARK ROAD
- DAM
- TOILETS
- TELEPHONE
- PARKING
- SKI-IN CAMPING*
- ONE-WAY SKI TRAIL*

1 KM = .6 MI

0 2 4 6 KM

WILLOW RIVER

MILL POND DAM

GRAVESITES

CONTACT STATION

LITTLE FALLS LAKE

LITTLE FALLS DAM

SHELTER

12/82

Willow River State Park

St. Croix County

This recently developed state park is nestled along the Willow River on 2700 acres. It is located northeast of Hudson on County A.

Skiing is permitted anywhere conditions are suitable. The hiking trail enables you to see magnificent river scenery, high overlooks and breathtaking panoramas. There is a total of 13.5km (8.5 miles) of marked, groomed and tracked trails. A secluded winter campground developed in a wooded area is available.

Vehicle admission sticker is required.

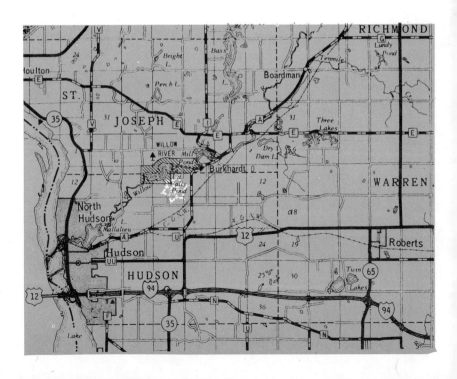

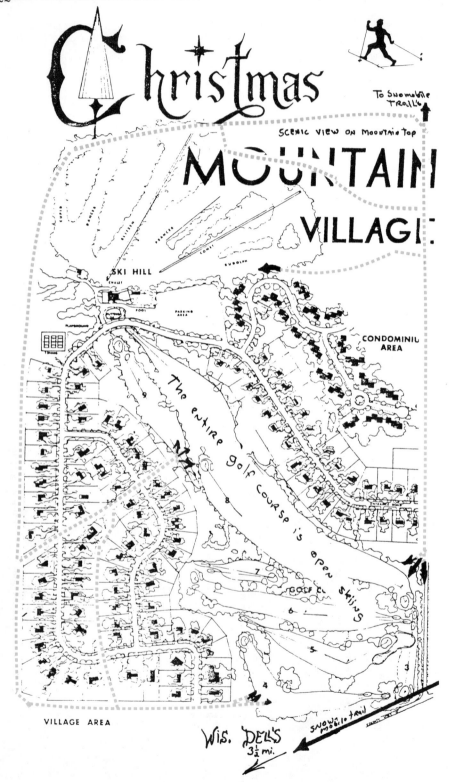

Christmas

To Snowmobile Trails

SCENIC VIEW ON MOUNTAIN TOP

MOUNTAIN

VILLAGE

SKI HILL

CHALET

POOL

PARKING AREA

PUTTING

PLAYGROUND

Tennis

CONDOMINIUM AREA

The entire golf course is open skiing

GOLF CO.

VILLAGE AREA

Wis. DELLS
3½ mi.

SNOWMOBILE TRAIL

Christmas Mountain

Sauk County

Located in the famous Wisconsin Dells area that was carved by a glacier over 1 million years ago, these trails are part of a large 200 acre recreation area. The trails are rolling and wooded with a length of about 11.2km (7 miles). They have a large downhill facility with good parking, shelter, warming area, toilets, and refreshments.

The trails wind around the golf course (which is open skiing) and up to the top of the downhill ski area which gives you a very scenic view. Rental equipment is available and there is no trail fee.

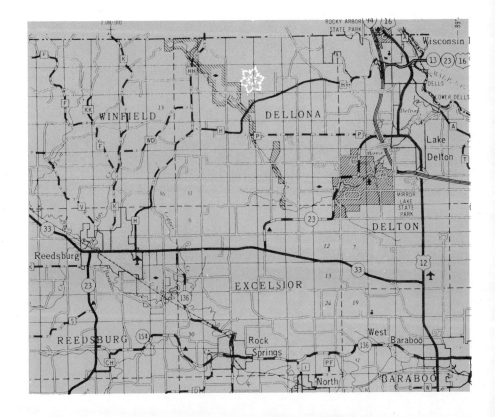

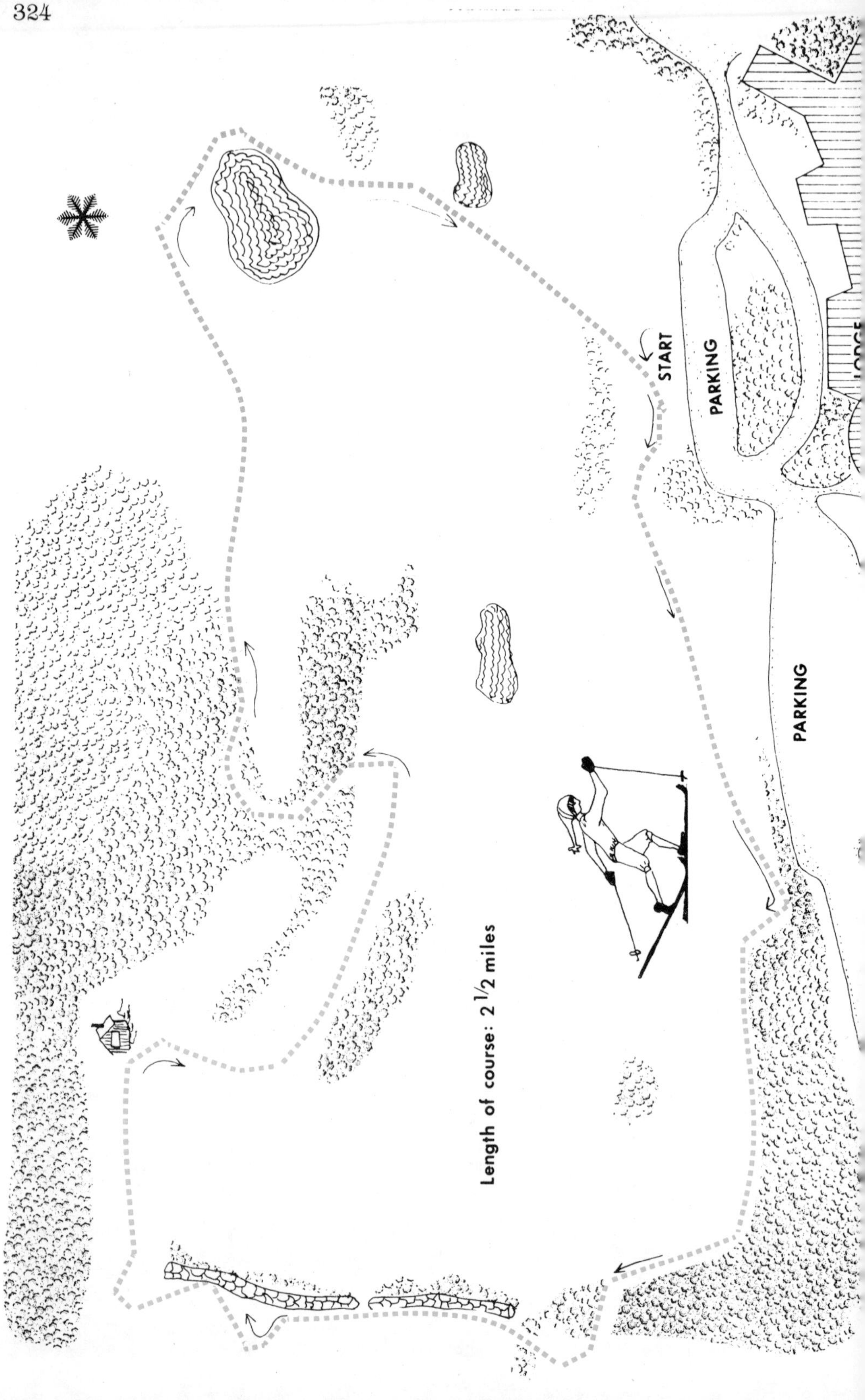

Length of course: 2 1/2 miles

START

PARKING

PARKING

LODGE

Devils Head

Sauk County

This trail is for the beginner and intermediate skier. It winds around the edge of the woods circling the golf course with 4km (2.5 miles) of rolling terrain.

Directions: Take Hwy. 113 10 miles out of Baraboo to County Trunk "D.L", also known as Parfay's Glen Rd. Go straight 2.5 miles east on 'D.L" to Devils Head.

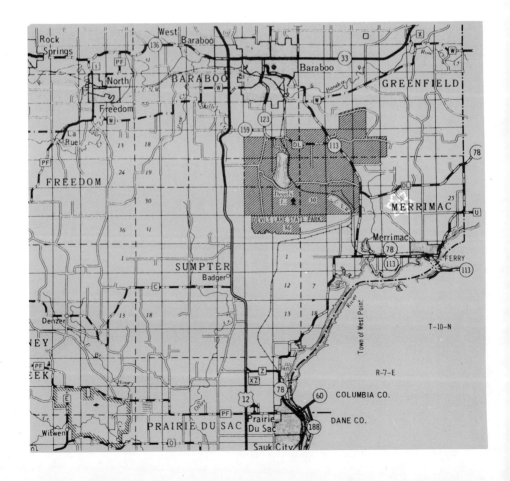

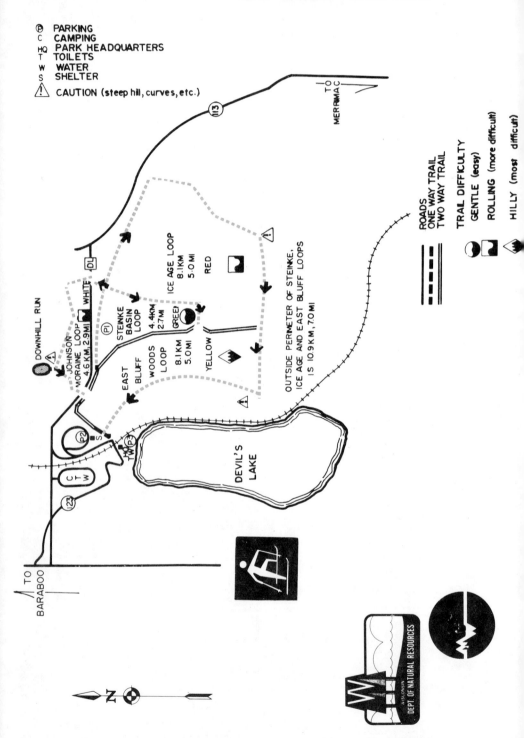

DEVIL'S LAKE STATE PARK
SKI TRAIL COLOR-CODED

P PARKING
C CAMPING
HQ PARK HEADQUARTERS
T TOILETS
W WATER
S SHELTER
⚠ CAUTION (steep hill, curves, etc.)

ROADS
ONE WAY TRAIL
TWO WAY TRAIL

TRAIL DIFFICULTY
GENTLE (easy)
ROLLING (more difficult)
HILLY (most difficult)

TO MERRIMAC

113

DL

DOWNHILL RUN

JOHNSON MORAINE LOOP
4.6KM 2.9MI

WHITE

STEINKE BASIN LOOP
4.4KM 2.7MI

GREEN

ICE AGE LOOP
8.1KM 5.0MI

RED

EAST BLUFF

WOODS LOOP
8.1KM 5.0MI

YELLOW

OUTSIDE PERIMETER OF STEINKE,
ICE AGE AND EAST BLUFF LOOPS
IS 10.9KM, 7.0MI

P1

P2

S

T W P3

DEVIL'S LAKE

C T W

123

TO BARABOO

N

WISCONSIN
DEPT. OF NATURAL RESOURCES

Devils Lake State Park

Sauk County

Located 5 miles southeast of Baraboo on County Highway DL. Devils Lake offers a total of 25.2km (15.6 miles) of one-way trails contained within 4 loops. 1)Steinke Basin Loop is 4.3km (2.7 miles) of gentle mixed trail which is easy to tour. 2) The scenic Ice Age Loop, which winds through 8km (5 miles) of wooded terrain is considerably more difficult and rolling. 3) East Bluff Woods Loop, also 8km (5 miles) is for the more experienced skier with its steep and curving trails. 4) Johnson Moraine Loop 4.6km (2.9 miles) is ungroomed and rolling wooded terrain.

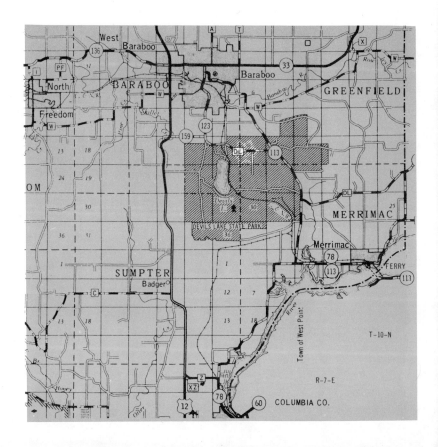

328

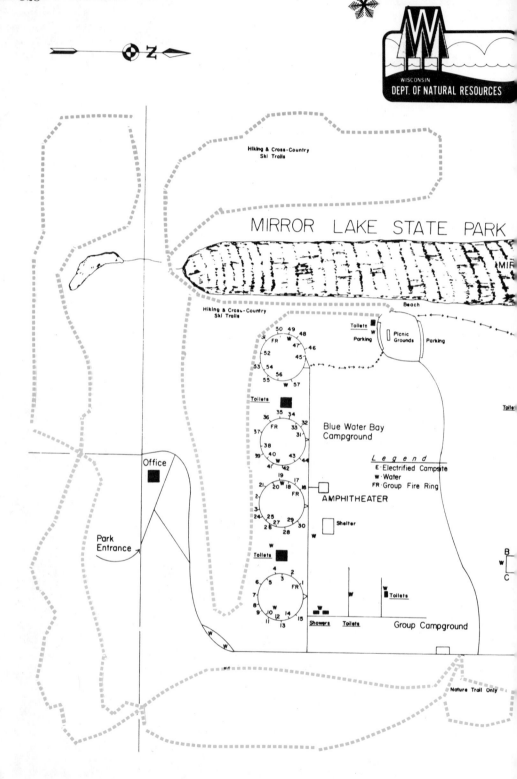

Mirror Lake State Park

Sauk County

Mirror Lake Trail offers 19.3km (12 miles) of cross-country skiing. The trail is very scenic as it circles around a portion of the lake. Future developments are being planned for additional trails. Parking, water, and toilet facilities are available at the beginning of the trail. Water is available at the hand pump by the shelter building. A vehicle admission sticker is required but no trail fee is charged. Mirror Lake Park is located 1.5 miles south of Lake Delton. Take Highway 12 exit off I90-84 and travel south to Fern Dell Road.

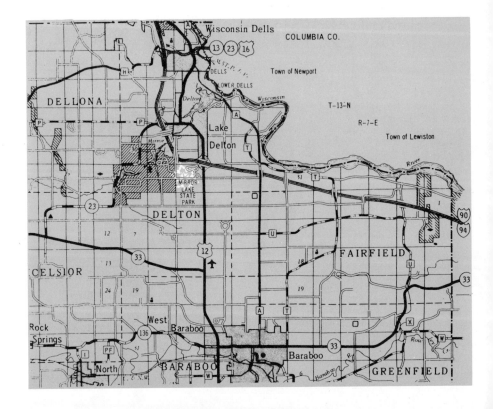

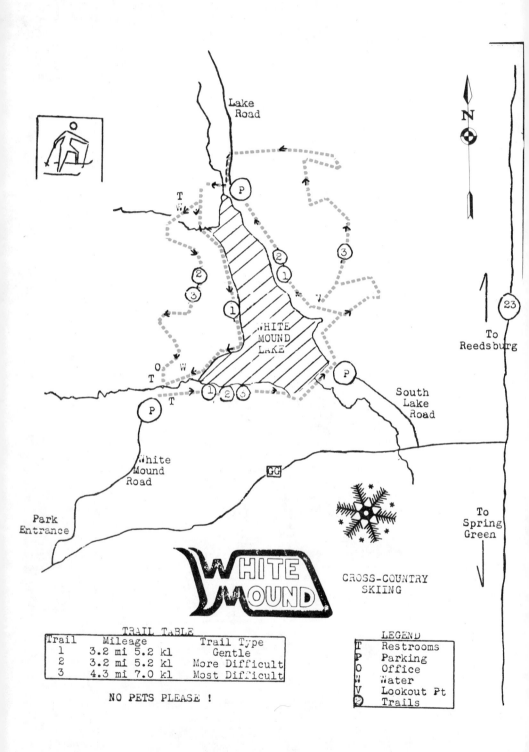

Lake Road

N

WHITE MOUND LAKE

South Lake Road

To Reedsburg

23

White Mound Road

GG

To Spring Green

Park Entrance

WHITE MOUND

CROSS-COUNTRY SKIING

Trail	Mileage		Trail Type
1	3.2 mi	5.2 kl	Gentle
2	3.2 mi	5.2 kl	More Difficult
3	4.3 mi	7.0 kl	Most Difficult

TRAIL TABLE

NO PETS PLEASE !

LEGEND	
T	Restrooms
P	Parking
O	Office
W	Water
V	Lookout Pt
⊘	Trails

White Mound

Sauk County

White Mound is located off of State Highway 23, 18 miles north of Spring Green. Take G.G. west to park entrance on White Mound Road. Parking is available at the Trail Head as well as South Lake Road and North Lake Road. All three trails begin at parking lots for a total of 10.7 miles. Trail one is gentle with trails 2 and 3 becoming more difficult. All trails are groomed and tracked. There is a scenic lookout point on trail 3. No pets allowed.

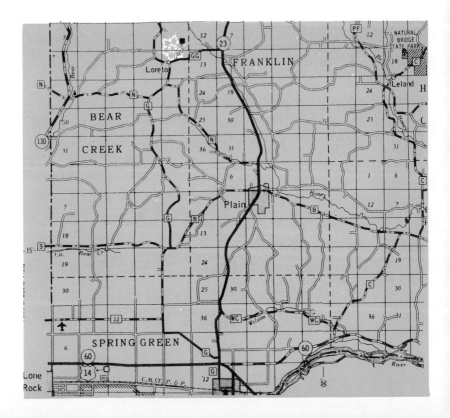

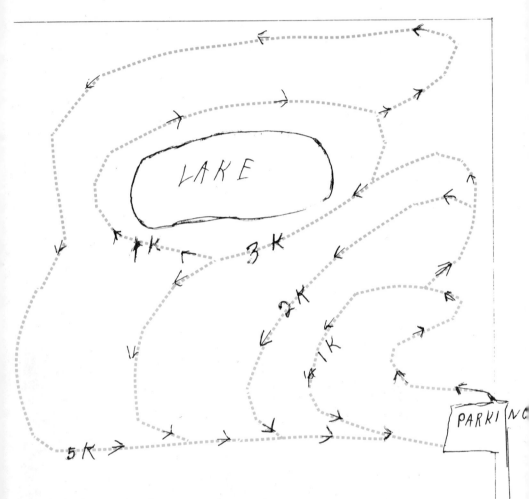

LAKE

1 K

3 K

2 K

1 K

5 K

PARKING

FARNSWORTH ROAD

Enchanted FOREST

CROSS-COUNTRY SKI TRAILS

Enchanted Forest
Ski Trails

Sawyer County

The Trails are located 12 miles east of Hayward off of State Highway 77 on 200 acres of private land. The trails are connected one-way loops of 1, 2, 3, 4 and 5km. Trails are well marked and groomed with alternate by-pass on steeper hills. Parking lot is kept open and firewood is furnished for outdoor fires.

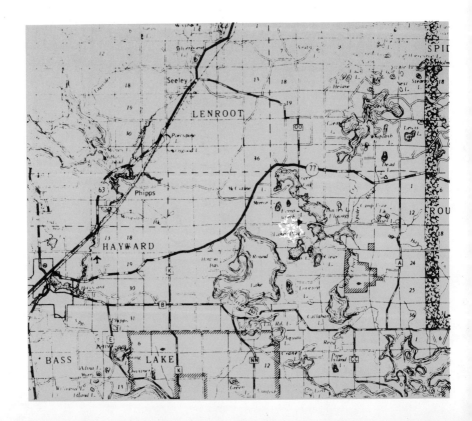

334

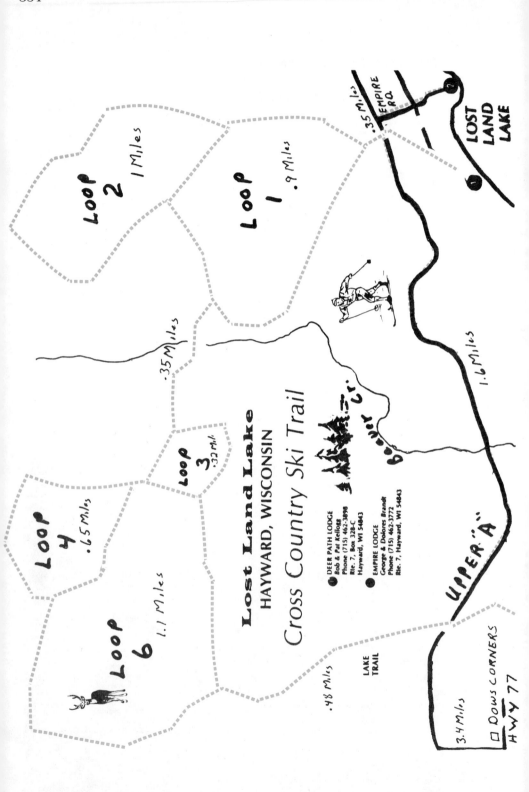

Lost Land Lake
HAYWARD, WISCONSIN

Cross Country Ski Trail

LOOP 2
1 Miles

LOOP 1
.9 Miles

LOOP 3
.32 Mi.

LOOP 4
.65 Miles

LOOP 6
1.1 Miles

.35 Miles

.35 Miles

.48 Miles

LAKE
TRAIL

Beaver Cr.

UPPER "A"

1.6 Miles

EMPIRE RD.

LOST LAND LAKE

2

1

3.4 Miles

DOWS CORNERS

HWY 77

● DEER PATH LODGE
Bob & Pat Kellogg
Phone (715) 462-3898
Rte. 7, Box 328-C
Hayward, WI 54843

● EMPIRE LODGE
George & Dolores Brandt
Phone (715) 462-3772
Rte. 7, Hayward, WI 54843

Lost Land Lake Trails

Sawyer County

The Lost Lake Trail is a beginner to intermediate trail with varied scenery ranging from hardwood forests to cedar swamps. The trail is groomed and tracked. Refreshments, food and lodging at trail access. If conditions permit the trail will extend to the south side of the lake adding 8km and another loop is being planned which will add 6km.

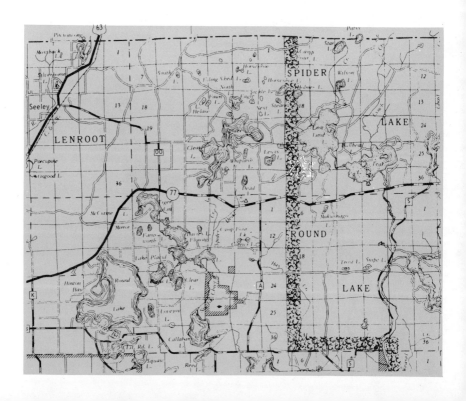

SAWYER COUNTY BIRKEBEINER

Cross Country Ski & Hiking Trail

HAYWARD, WISCONSIN U.S.A.

THE 4-SEASON VACATIONLAND

COMPLIMENTS OF THE HAYWARD CHAMBER OF COMMERCE HAYWARD, WI 54843

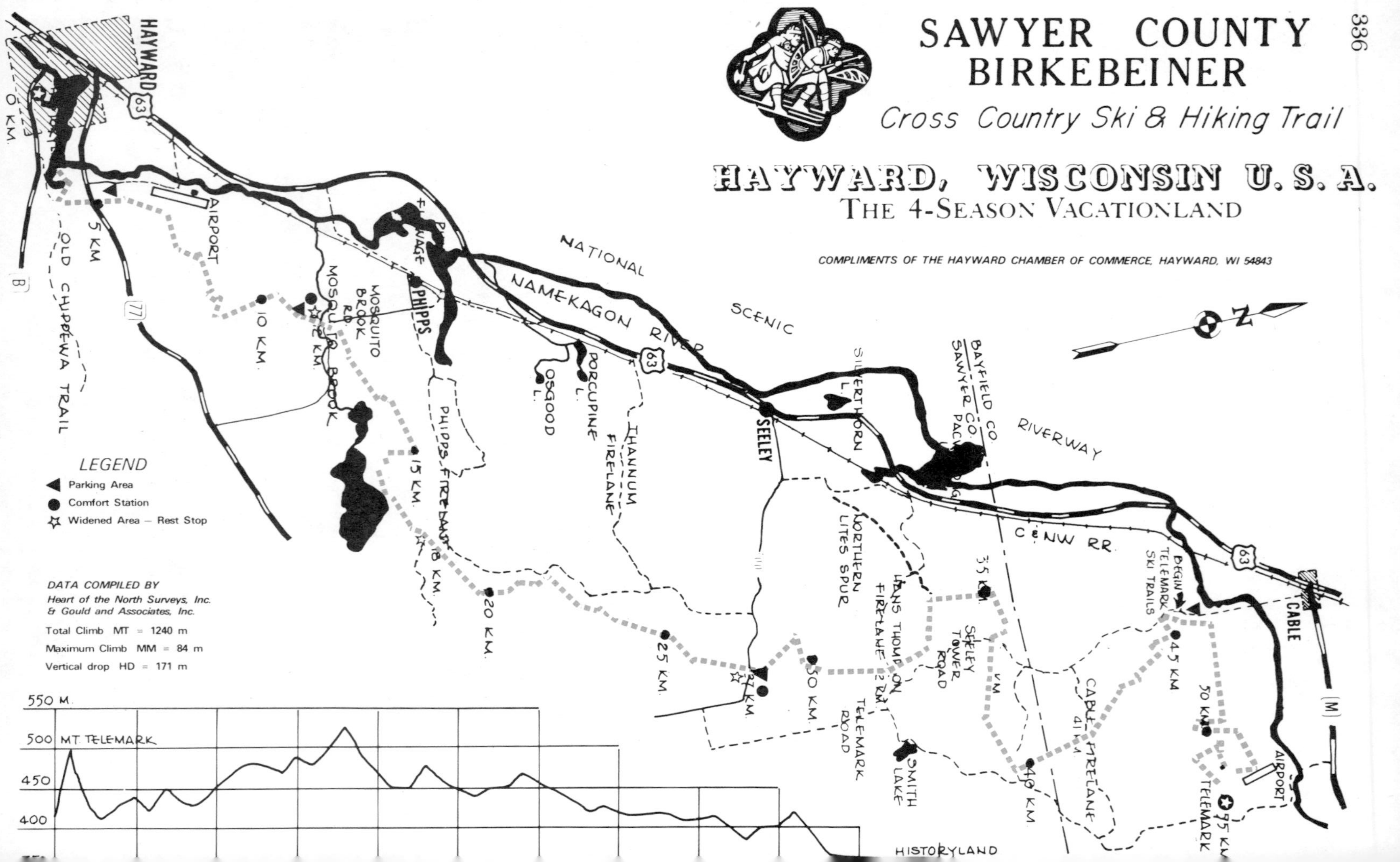

N

CABLE

63

AIRPORT

75 KM
50 KM
TELEMARK

45 KM

BEGIN
TELEMARK
SKI TRAILS

CABLE FIRELANE
41 KM

40 KM

C.&N.W. R.R.

35 KM

SEELEY
TOWER
ROAD

VM

SMITH
LAKE

HANS THOMPSON
FIRELANE 2 KM

TELEMARK
ROAD

30 KM

27 KM

25 KM

HISTORYLAND

RIVERWAY

BAYFIELD CO.
SAWYER CO.
PACWAWONG

SILVERTHORN
L.

NORTHERN
LITES SPUR

SEELEY

63

IHANNUM
FIRELANE

PORCUPINE
L.

OSGOOD
L.

20 KM

18 KM

PHIPPS FIRELANE

15 KM

SCENIC

NATIONAL

NAMEKAGON RIVER

PIKE
FLOWAGE

PHIPPS

MOSQUITO
BROOK
RD.

MOSQUITO BROOK

12 KM

10 KM

AIRPORT

HAYWARD

63

77

5 KM

OLD CHIPPEWA TRAIL

0 KM

B

LEGEND

▼ Parking Area
● Comfort Station
☆ Widened Area – Rest Stop

DATA COMPILED BY
Heart of the North Surveys, Inc.
& Gould and Associates, Inc.

Total Climb MT = 1240 m
Maximum Climb MM = 84 m
Vertical drop HD = 171 m

550 M

500 MT. TELEMARK

450

400

Sawyer County Birkebeiner

Sawyer County

The Sawyer County Birkebeiner Cross-Country Skiing and Hiking Trail is maintained 12 months a year for the sole purpose of X-C skiing during the winter and for hiking thru the summer. The trail runs for 48km (30 miles) from Hayward to Telemark or vice-versa. It can be skied either way with several access points along the way. The trail is described as rolling in the northern half and gentle in the southern half. The best access to the trail is at Seeley which is the half way mark. There is no trail fee charged at this point. Trail access at the north end is a Telemark where there is a trail fee but they do have good parking, rest rooms and other facilities. The trail is not groomed except for race time.

Each year this trail is the site of the AMERICAN—BIRKEBEINER CROSS COUNTRY SKI RACE starting at Telemark and ending at Historyland in Hayward.

In the beginning the race was patterned after the Birkebeiner race annually run in Norway between the villages of Rena and Lillehammer in the province of Osterdal.

The Norwegian Birkebeiner commemorates the historic rescuer of the Norwegian child, Price Haakon Haakonsson during the Norwegian Civil War of 1206. Prince Haakon's rescuers were of a Viking tribe, known in Norwegian as Birkebeiners because of their birch bark leggings and thus came the name of today's race.

What was a dramatic rescue in 1206 has become a challenge to the many skiers who have pitted themselves against the similar 55 kilometer course between Telemark and Hayward, Wisconsin. American-Birkebeiner winners of the touring and classified competitors classes receive a commemorative set of birch leggings made by the area's Ojibwa Indians and further increase the tie by travelling to Norway to join thousands of other skiers from many nations in the original trek from Rena to Lillehammer.

The race is open to both tourers and competitors 18 years of age or over. It is sanctioned by both the U.S.S.A. and the Federation Internationale de Ski. Skiers are divided into age and class categories in accordance with these sanctions.

Exclusively numbered American-Birkebeiner medals are presented at the finish line to all those who complete the marathon for the first time. Commemorative pins are presented to contestants who have completed the race in previous years. Clubs with the most amount of registered members entering and finishing the race receive cash prizes. The winners of the touring and competition classes are sent to Norway to join thousands of others in the original historical trek.

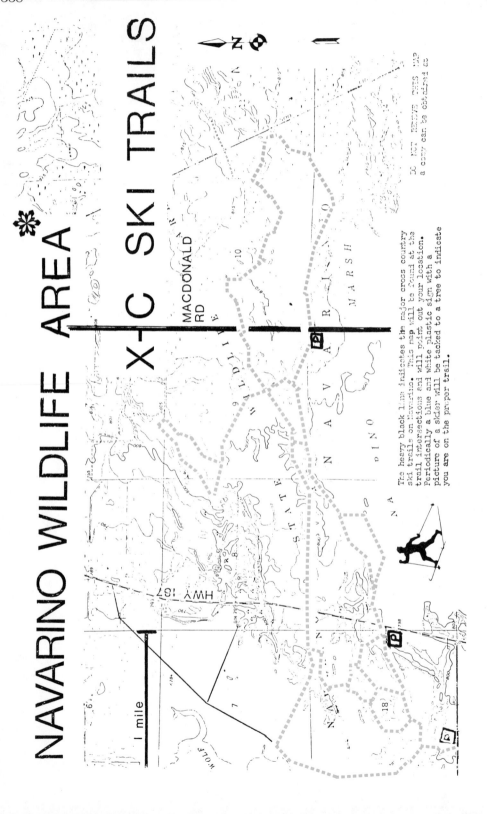

NAVARINO WILDLIFE AREA ❁ X-C SKI TRAILS

The heavy black line indicates the major cross country ski trails on Navarino. This map will be found at the trail intersections and will point out your location. Periodically a blue and white plastic sign with a picture of a skier will be tacked to a tree to indicate you are on the proper trail.

DO NOT REMOVE THIS MAP. A copy can be obtained at...

MACDONALD RD

HWY 187

1 mile

Navarino Wildlife

Shawano County

These DNR cross-country trails are located in the Navarino State Wildlife area near Shawano. There are three access points with parking areas. One is located on the southwest corner on Hwy. 156, another on Hwy. 187, and another on MacDonald Road.

The trails are very scenic but not very difficult. A few hills but mostly wooded flat land. During the week the trails have little use, but weekends are quite heavy. There is a total of approximately 12 miles with varying cutoffs and loops to choose. They are marked with standard blue and white plastic signs and there are trail maps at the intersections. No shelter, toilets, or water.

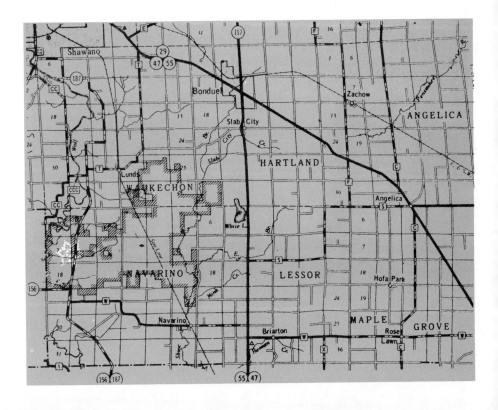

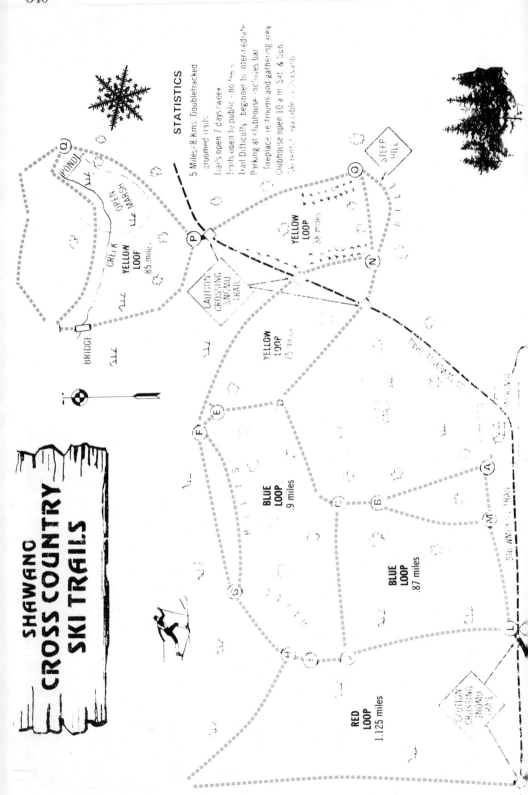

Shawano Ski Trails

Shawano County

Trails are double tracked and groomed. Five loops with total of 8km (5 miles). Trails are open 7 days with no trail fees. Rated beginner to intermediate. Parking at clubhouse includes bar, fireplace, restrooms and gathering area. Clubhouse open 10 am Saturday and Sunday.

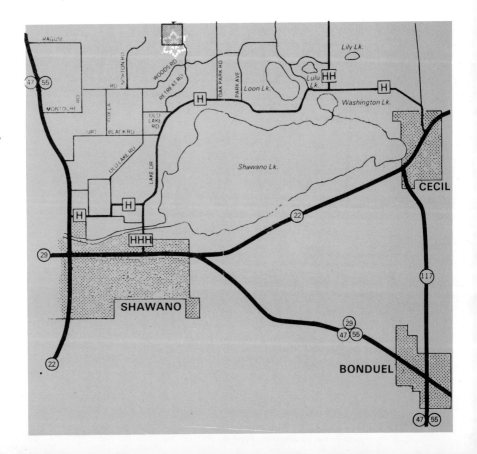

JOHN MICHAEL KOHLER/
TERRE ANDRE STATE PARK

WISCONSIN DEPARTMENT OF
NATURAL RESOURCES

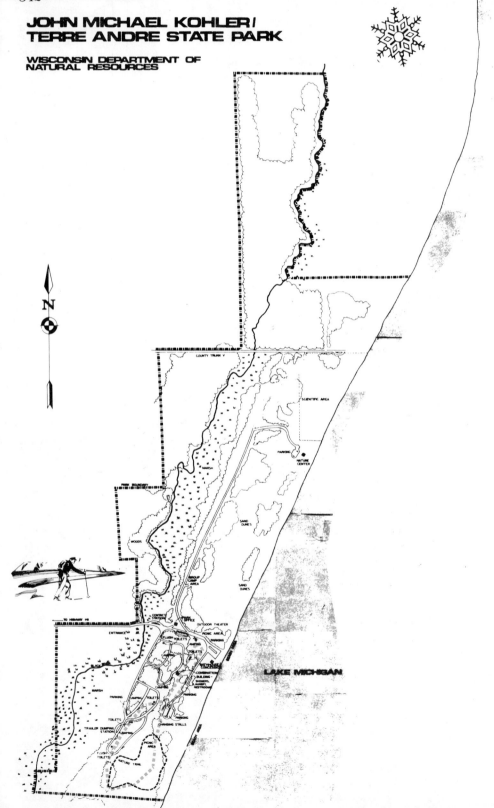

N

COUNTY TRUNK V

SCIENTIFIC AREA

PARKING

NATURE
CENTER

PARK BOUNDARY

MARSH

SAND
DUNES

WOODS

GROUP
CAMP
AREA

SAND
DUNES

TO HIGHWAY H8

CONTROL
STATION

OFFICE

OUTDOOR THEATER

PICNIC AREA

ENTRANCE

FLUSH TOILETS

PARKING

TOILETS

COMBINATION
BUILDING
SHOWER
NURSERY
RESTROOM

LAKE MICHIGAN

MARSH

PARKING

CAMPING

TOILETS

PARKING

TOILETS

PARKING

TRAILER DUMPING
STATION

CAMPING

CHANGING STALLS

PICNIC
AREA

FLUSH
TOILETS

NATURE TRAIL

Kohler·Andrae State Parks

This 760 acre state park has skiing on the summer hiking trail which is about 2.4km (1½ miles) in length. When snow is available you can ski anywhere in the park, although the staff encourages you not to ski in the fragile unstabilized dunes area where some of the parks rare and endangered plants are found. There are no snowmobiles to contend with. There are 28 winter campsites available but no showers or flush toilets. Pit toilets and electrical hook-ups are provided.

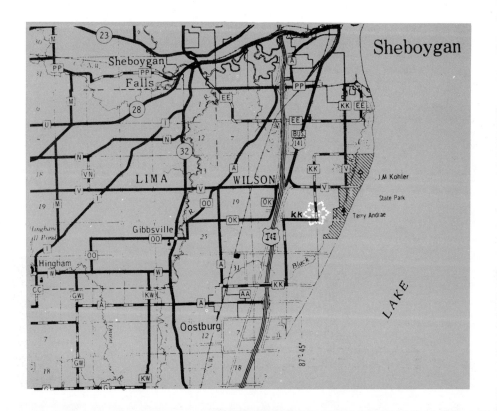

TRAIL	LENGTH	
316	0.75 mi	1.21km
318	2.50 mi	4.03km
318A	0.70 mi	1.13km
321	8.00 mi	12.90km
361	0.40 mi	0.65km
361A	0.40 mi	0.65km

Sitz-Mark Trail

Sitz-Mark has three loops. The major loop No. 321 is 12.9km (8 miles). Loop No. 318 and 318A combine 5km (3 miles). The third loop is rather mini consisting of 1.2km (¾ mile).

The total distance on the 3 loops is 19.3km (12 miles) of marked, rolling, wooded trail.

Maps are available at location.

Sitz-Mark is located 4 miles north of State Highway 64 on Forest Road 119 between Medford and Gilman.

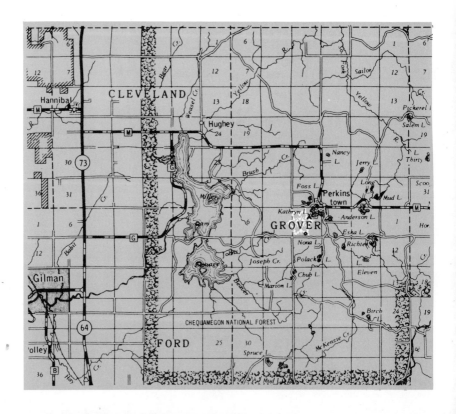

346

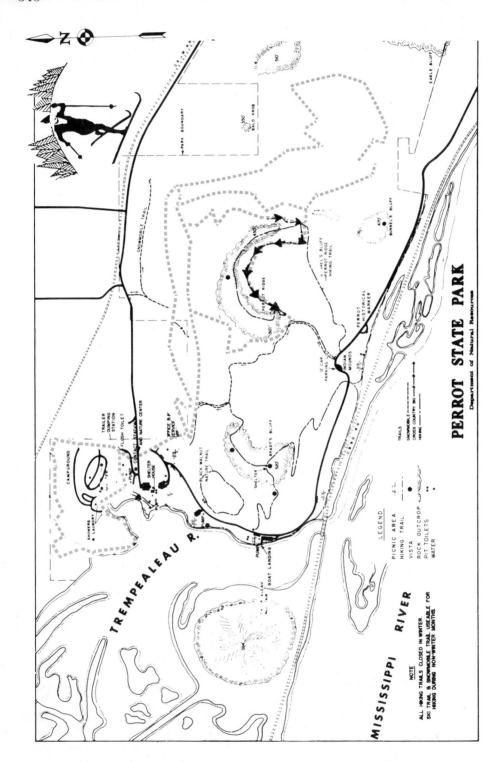

PERROT STATE PARK
Department of Natural Resources

LEGEND

PICNIC AREA	
HIKING TRAIL	
VISTA	
ROCK OUTCROP	
PIT TOILETS	
WATER	

TRAILS
SNOWMOBILE
CROSS COUNTRY SKI
HIKING

MISSISSIPPI RIVER

TREMPEALEAU R.

NOTE
ALL HIKING TRAILS CLOSED IN WINTER
SKI TRAIL & SNOWMOBILE TRAIL USEABLE FOR
HIKING DURING NON-WATER MONTHS

Perrot State Park

Perrot Park is 2½ miles north of Trempleau on 1393 acres bordering the Mississippi River. The 16.1km (10 miles) trail goes through mostly wooded areas between ridges that overlook the Trempleau and Mississippi Rivers. The other much longer loop takes you around Reeds Peak and Perro Ridge. Trails are well marked and there is a plowed parking lot as well as a shelter house with a fireplace. A State Park vehicle admission sticker is required. Trails are groomed and tracked.

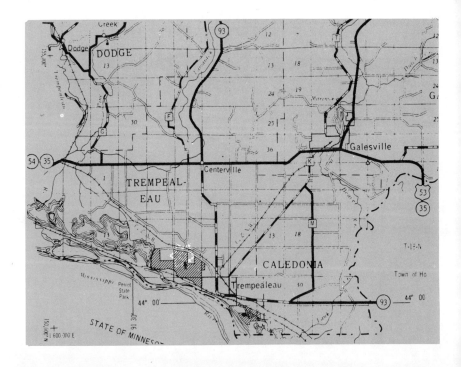

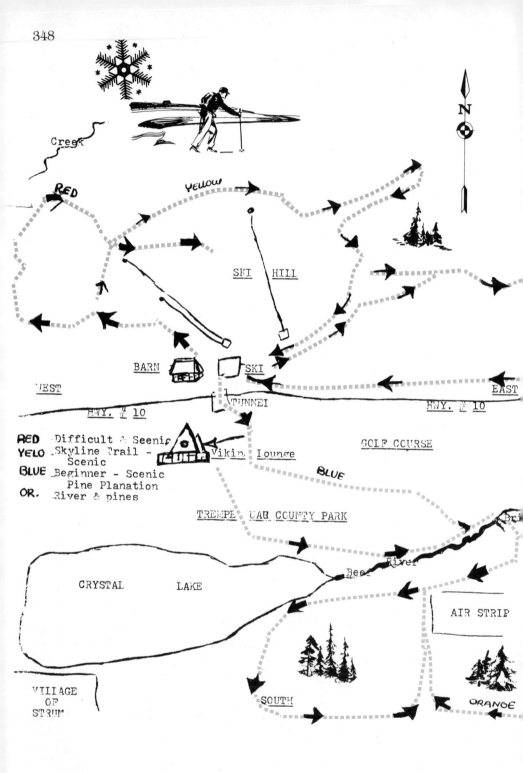

Creek

RED

YELLOW

SKI   HILL

BARN

WEST

HWY. # 10

SKI

TUNNEL

EAST

HWY. # 10

RED Difficult & Scenic
YELO Skyline Trail -
Scenic
BLUE Beginner - Scenic
Pine Planation
OR. River & pines

Viking Lounge

GOLF COURSE

BLUE

TREMPEALEAU COUNTY PARK

CRYSTAL   LAKE

River

Bear

AIR STRIP

VILLAGE
OF
STRUM

SOUTH

ORANGE

Viking Skyline

Trempleau County

Viking Skyline has 14.5km (9 miles) of cross-country trails that are color coded and well marked.

The blue beginner trails are either 2, 4, or 5 miles long. These trails start at the ski hill, cross the golf course, pass thru the county park, go over the Beef River and then wind thru a beautiful pine plantation and then back along the river bank.

The scenic yellow trail is either 1 or 3 miles long and starts at the top of the ski hill. From this point you can look down at the village of Strum, Crystal Lake and the numerous pine plantations in the area. This trail has a lot of variety for the beginner and intermediate skier as well.

The red trail is ¾ mile long and is for the expert and thrill-seeking skier.

A trail fee is charged. Lounge serving food and beverages is open every weekend during the snow season.

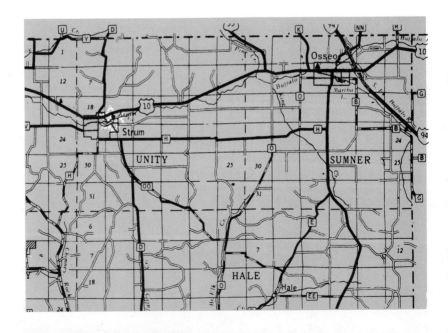

WILDCAT MOUNTAIN STATE PARK

RIDGETOP CROSS-COUNTRY SKI TOURING TRAIL

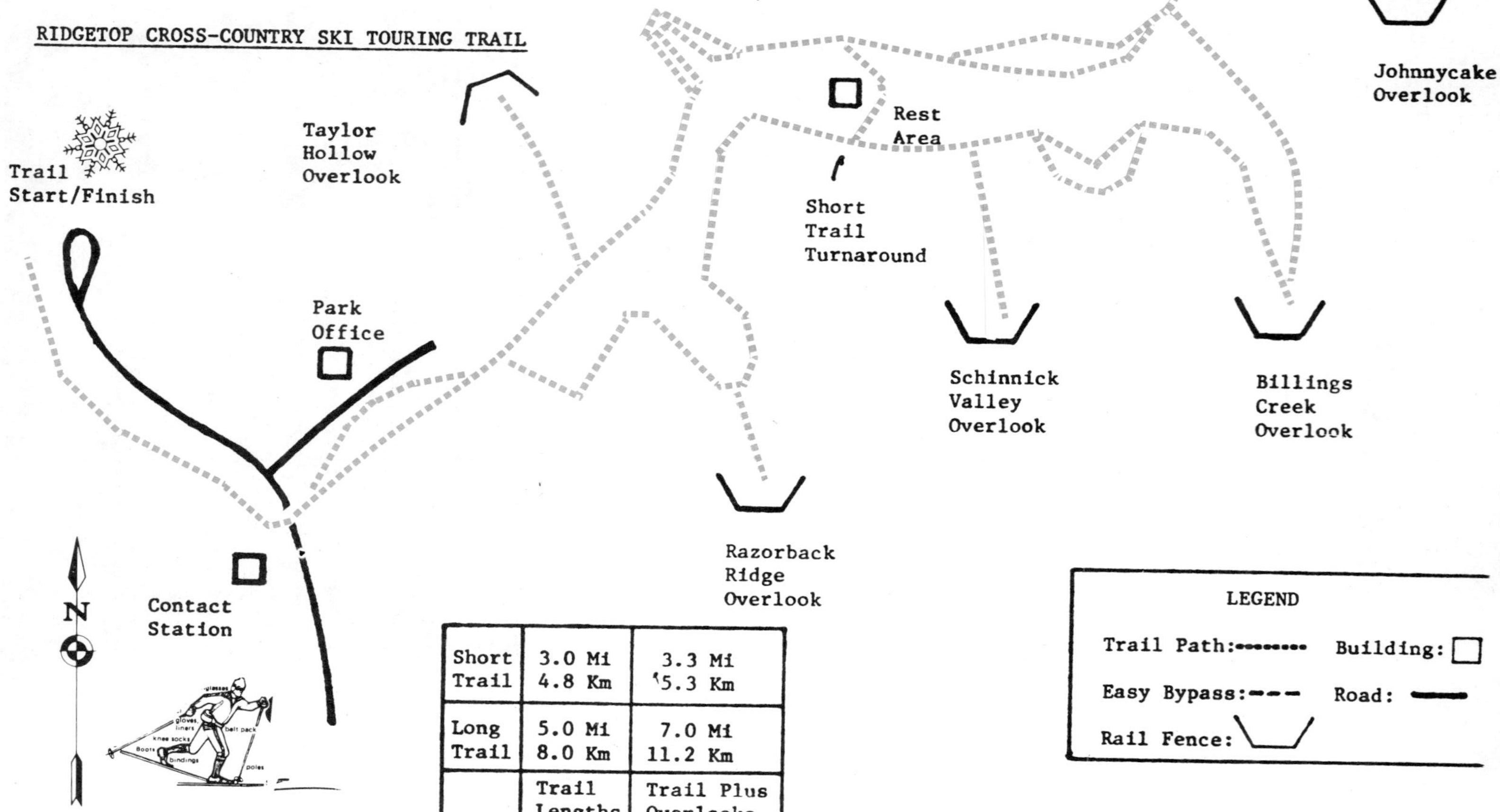

Johnnycake
Overlook

Billings
Creek Overlook

Schinnick
Valley
Overlook

Rest
Area

Short
Trail
Turnaround

Razorback
Ridge
Overlook

Taylor
Hollow
Overlook

Park
Office

Contact
Station

Trail
Start/Finish

LEGEND

Trail Path: ••••••• Building: ☐

Easy Bypass: --- Road: ▬▬

Rail Fence: ﹀

	Trail Lengths	Trail Plus Overlooks
Short Trail	3.0 Mi 4.8 Km	3.3 Mi 5.3 Km
Long Trail	5.0 Mi 8.0 Km	7.0 Mi 11.2 Km

N

Wildcat Mountain State Park

Vernon County

Wildcat Mountain State Park is located 2½ miles southeast of the village of Ontario on U.S. Highway 33.

A short loop and long loop, plus overlooks give a total of 11.3km (7 miles) of touring along the ridge tops. The groomed, well marked trails are popular with beginning and intermediate skiers. Plowed parking and toilet facilities are available.

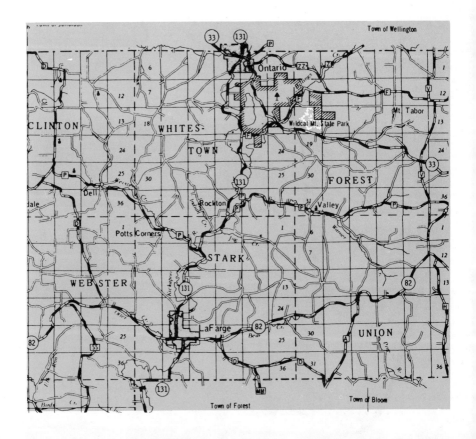

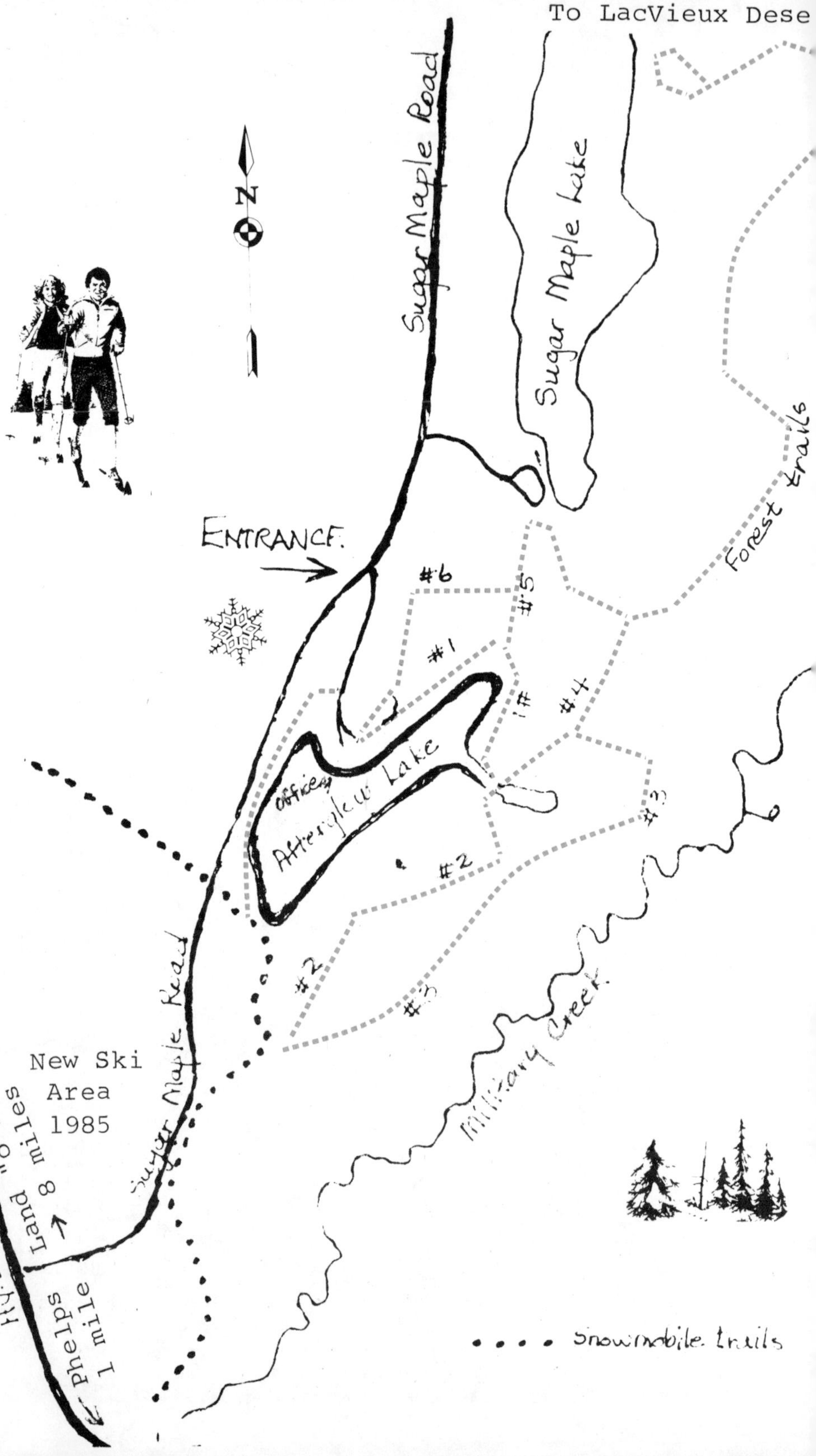

To LacVieux Dese

Sugar Maple Road

Sugar Maple Lake

Forest trails

N

ENTRANCE

#6
#5
#1
#4
#1
#3

Office
Afterglow Lake

#2

#2

#3
#3

Military Creek

New Ski
Area
1985

Hwy "E"
Land "o" Lakes
8 miles

Sugar Maple Road

Phelps
1 mile

•••• Snowmobile trails

Afterglow Resort Trails

6 private trails for beginner and intermediate skiers on 240 acres surrounded by Nicolet Nat'l Forest, into which trails extend for miles. Adventurers can bushwhack in adjacent Nicolet and Ottawa Nat'l Forests.

Located in the best snowbelt in Wisconsin, coupled with new grooming and tracking equipment, Afterglow has dependable skiing conditions from early December until late March. Skiers staying elsewhere are welcome to drive in and try the trails; donations toward grooming will be accepted. Check in at the office for a trail map before using the trails. For variety, enjoy their lighted ice rink, tobogganing and tubing hill, deluxe rec room plus new whirlpool and sauna. Numerous fine X-C areas nearby all add to the fun. Afterglow offers lodging in cozy year round cottages with fireplaces. Rates and further information can be had by calling or writing. (See directory at the end of this book).

Eighteen miles north of Eagle River on Hwy. 17. Take Hwy. E north out of Phelps for one mile, turn right on Sugar Maple Rd. to resort. (Approx. one mile).

To Sylvania Ski Area

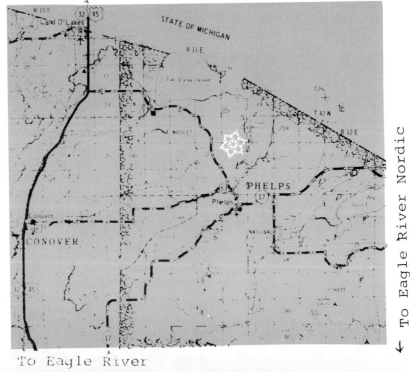

To Eagle River

Anvil Trail System

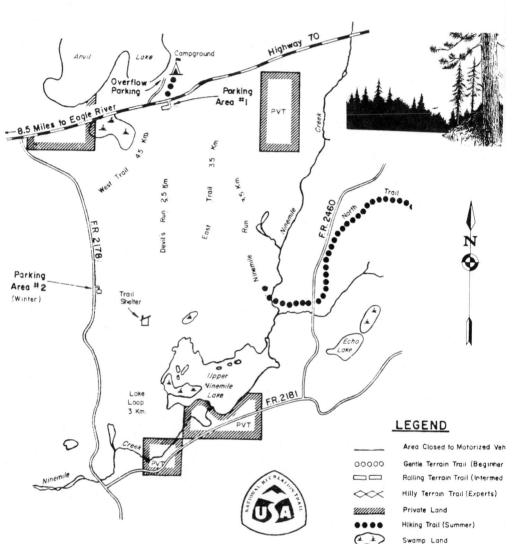

Anvil Lake

Campground

Highway 70

Overflow Parking

Parking Area #1

PVT

8.5 Miles to Eagle River

Creek

West Trail

Devil's Run 2.5 Km

East Trail 3.5 Km

Ninemile Run 4.5 Km

4.5 Km

Ninemile

Creek

F.R. 2460

North Trail

F.R. 2178

Parking Area #2 (Winter)

Trail Shelter

Echo Lake

Upper Ninemile Lake

Lake Loop 3 Km.

F.R. 2181

PVT

Creek

PVT

Ninemile

N

NATIONAL RECREATION TRAIL

USA

LEGEND

——	Area Closed to Motorized Veh
ooooo	Gentle Terrain Trail (Beginner
▭▭	Rolling Terrain Trail (Intermed
◇◇◇	Hilly Terrain Trail (Experts)
▨▨▨	Private Land
●●●●	Hiking Trail (Summer)
⟷	Swamp Land

ALL TRAILS CLOSED TO MOTORIZED VEHICLES

Anvil Trail System

Vilas County

Excellent and exciting trails for the novice, intermediate or expert skiers.

Located on the Eagle River district of the Nicolet National Forest, the Anvil Trail was originally constructed by the CCC in the mid-30's. New Construction by the U.S.D.A. Forest Service has expanded the trail system to 12 miles of beautiful skiing of which approximately 10 niles are groomed.

New signing has been designed along the trail. Green circles for beginners gentle slopes of 10% or less. Blue squares — intermediate moderate slope of 10-30%. Black diamonds — expert steep slopes exceeding 30%.

A cozy log cabin shelter with fireplace & firewood is located on the trail.

Take State Hwy. 70 east before entering Eagle River for 8½ miles.

Clearly marked on right side of roads.

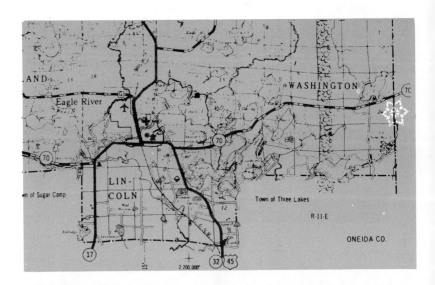

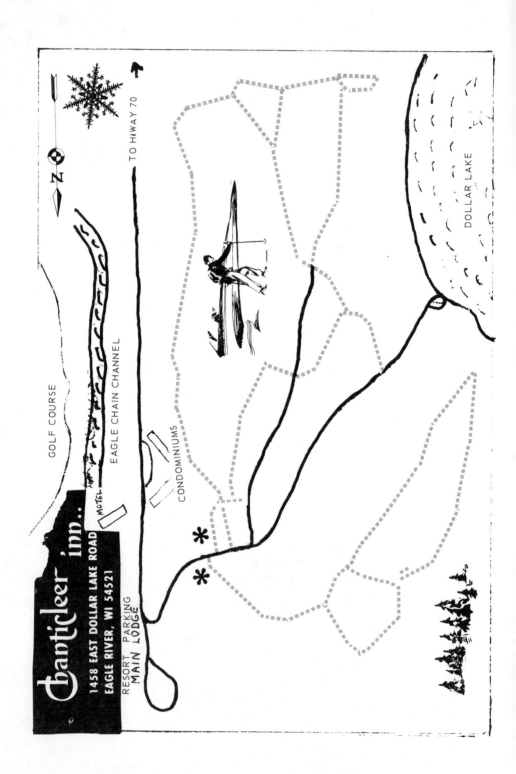

Chanticleer inn...

1458 EAST DOLLAR LAKE ROAD
EAGLE RIVER, WI 54521

RESORT PARKING
MAIN LODGE

N

TO HIWAY 70

GOLF COURSE

EAGLE CHAIN CHANNEL

MOTEL

CONDOMINIUMS

DOLLAR LAKE

Chanticleer Trails

Vilas County

Located about 3 miles east of Eagle River off Hwy. 70. Watch for signs and entry road on north side of Hwy. 70 approximately 6.4km (4 miles) of gently rolling trails. Lodging, dining and cocktails at economy rates during the week. Located very near Anvil Trail system for additional skiing opportunities. Also Eagle River Nordic Ski Center is a short 12 mile drive away. this si the ultimate X-C Ski Center.

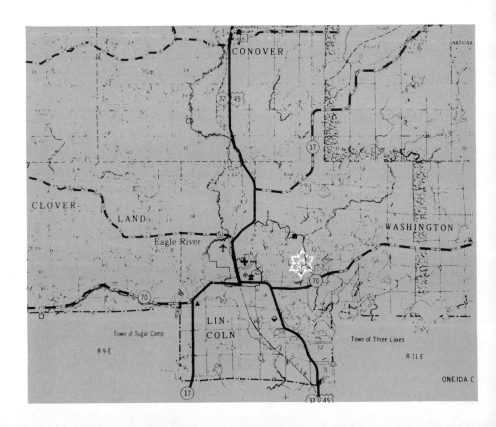

358

EAGLE RIVER NORDIC TRAIL SYSTEM

ALL TRAILS
CLOSED
TO MOTORIZED
VEHICLES.

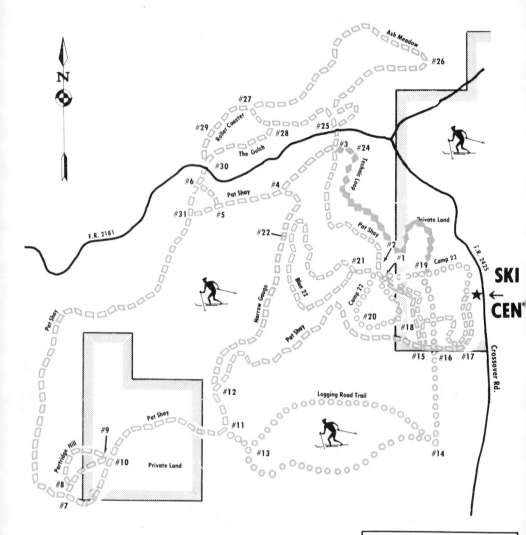

N

Ash Meadow
#26

#27
#29   Roller Coaster
#28   #25
The Gulch   #3   #24
#30   Technic Loop
#6   Pat Shay   #4
#31   #5   Pat Shay   Private Land
F.R. 2181   #2
#22   #1   #19   Camp 22
#21   F.R. 2425
Narrow Gauge   Camp 22   SKI
Blue 22   #20   CEN
Pat Shay   #18
#15   #16   #17
Pat Shay   Crossover Rd.
#12
Pat Shay   Logging Road Trail
#9   #11
Partridge Hill   #10   Private Land   #13   #14
#8
#7

LEGEND

ooooo	Gentle Terrain Trail (beginners)
▭▭▭▭▭	Rolling Terrain Trail (intermediate)
◆◆◆◆	Hilly Terrain Trail (experts)

Eagle River Nordic

Eagle River Nordic Ski Center has developed and maintains 55km (34 miles) of trails in the Nicolet National Forest. The trails were developed in 1977 under a co-operative agreement with the Forest Service and have been constantly expanded and improved. They are superbly groomed and double tracked and vary from a new gently rolling 3km (1.9 mile) beginners trail to a challenging 7.5km (3 mile) and 10km (6.4 mile) racing and training track. There is a new expert trail with super hills to challenge the best.

The NEW Warming Hut is the first earth sheltered commercial building in the state and provides trailhead rentals, instruction, guided tours, parking and hot meals. The south facing glass walled building is the home of the Ski Research Group, a blend of science and skiing which has pioneered in ski testing and wax research.

The NEW Warming Hut is reached by traveling 8 miles on Highway 70 East of Eagle River to Military Road (FR 2178) and then following the signs to Eagle River Nordic.

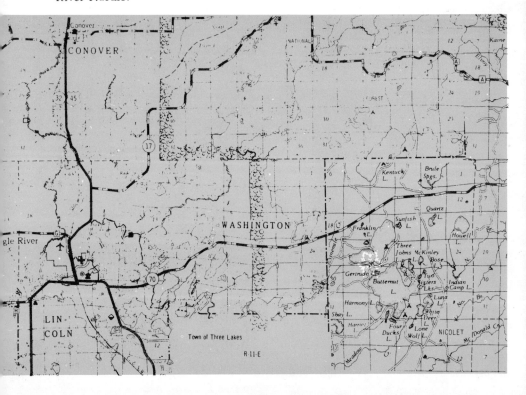

360

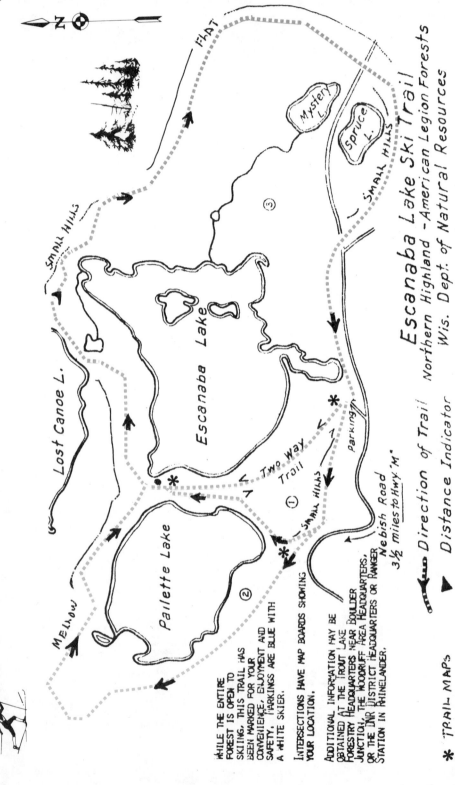

Escanaba Lake Ski Trail
Northern Highland - American Legion Forests
Wis. Dept. of Natural Resources

FLAT

Mystery L.

Spruce L.

Small Hills

Small Hills

③

Escanaba Lake

Lost Canoe L.

Two Way Trail

Parking

①

Small Hills

Nebish Road
3½ miles to Hwy "M"

Mellow

Pailette Lake

②

WHILE THE ENTIRE FOREST IS OPEN TO SKIING, THIS TRAIL HAS BEEN MARKED FOR YOUR CONVENIENCE, ENJOYMENT AND SAFETY. MARKINGS ARE BLUE WITH A WHITE SKIER.

INTERSECTIONS HAVE MAP BOARDS SHOWING YOUR LOCATION.

ADDITIONAL INFORMATION MAY BE OBTAINED AT THE TROUT LAKE FORESTRY HEADQUARTERS NEAR BOULDER JUNCTION, THE WOODRUFF AREA HEADQUARTERS, OR THE DNR DISTRICT HEADQUARTERS OR RANGER STATION IN RHINELANDER.

▬▬ = Direction of Trail

▶ = Distance Indicator

✳ TRAIL MAPS

Escanaba Lake Trail

Vilas County

One of the most popular trails whose reputation is state-wide will have up to 30-50 skiers on a good weekend day. It requires regular snowfall or the quality of the tracks deteriorates rapidly. This can lead to treacherous skiing on loop 2. The trail is very well designed for a challenging experience. There is no place on this trail for beginners unless they insist on crawling alot.

Loop 3 which is 12.1km (7.5 miles), is for the advanced intermediate. One stretch of this loop is a great spot for practicing double poling (slight downhill grade) and kick and glide uphill skiing (slight uphill grade). Loop 1 is for the intermediate with 3.2km (2 miles) of trail and loop 2 is for the expert with 7.2km (4.5 miles). The complete outer loop is total of 17.7km (11 miles).

Set in a wilderness area, the scenery is a rare privilege. Mostly a birch, aspen forest the trail passes through several pine groves and past marshes and lakes. Wildlife is abundant and especially interesting are the otter around Escanaba Lake in the spring.

Location is 7 miles south of Boulder Junction on County M to Nebish Road then East for 3½ miles of winding road to parking lot. Snow tires are nearly mandatory as there are steep hills on the roads and they are very treacherous. No shelters or toilets.

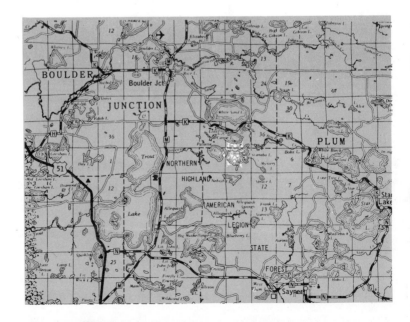

362

GATEWAY LODGE
& Recreation Center
Land O'Lakes, Wis.
(715) 547-3321

X-C SKI
TRAILS

GATEWAY
LODGE

Wisconsin River

1.7 kms.

Wisconsinaire
Motel

3 Fork to
Main Lodge

2 shortcut
to Chalet

U.S. 45

(RFD)

start
Ottawa
Trail

Ski
Area

4 start of Last "Lapp"
Expert Run

Bypass

Chalet

1 Shortcut
to Chalet

N

start
Deerpath Trail
(YELLOW)

Last "Lapp" Expert Run	1.3 Km.	Expert
Ottawa Trail (Red Markers)	9.4 Km.	Advance
Deerpath Trail (Yellow)	4.9 Km.	Beginne

$2.00 TRAIL FEE REQUESTED

Gateway Lodge

Vilas County

Gateway offers 27km (17.3 miles) of groomed and tracked trails for the beginner to the expert skier being described as gentle and rolling.

Rentals are available and lessons can be arranged.

The annual Gateway Cup X-C Citizens Race is usually held on the first week-end in March. Gateway offers a fun 10km. X-C race for all ages.

The Lodge and trail can be found 2 blocks East of Land O Lakes off Hwy. 45 on Hwy. B.

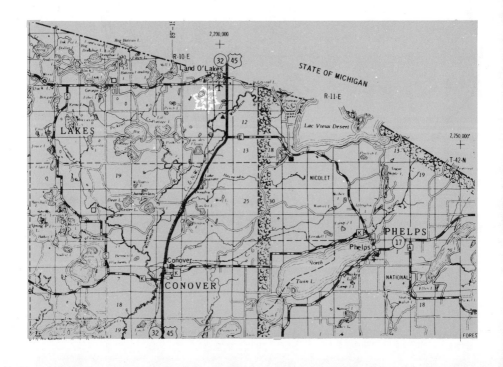

LAKE FOREST RECREATION AREA
Winter Sports and Trail Map

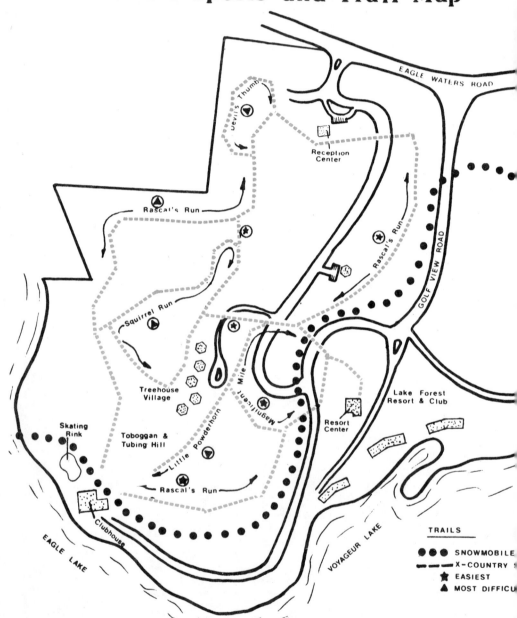

Devil's Thumb

Reception Center

EAGLE WATERS ROAD

Rascal's Run

Rascal's Run

Squirrel Run

GOLF VIEW ROAD

Treehouse Village

Magnificent Mile

Little Powderhorn

Rascal's Run

Lake Forest Resort & Club

Resort Center

Skating Rink

Toboggan & Tubing Hill

Clubhouse

EAGLE LAKE

VOYAGEUR LAKE

TRAILS

●●● SNOWMOBILE

— — X-COUNTRY

★ EASIEST

▲ MOST DIFFICU

Lake Forest

Recreation Area

Vilas County

Ski Lake Forest! The Lake Forest Recreation Area offers over five miles of groomed double-track trails through beautiful birch and pine forests and over snowy golf fairways. The varied terrain provides a nice mixture for everyone, whether a casual or avid skier; trails are well-marked for difficulty and directions. At the end of the trail, one can enjoy a warm drink in front of the fireplace at the Lake Forest Country Club. The trails and the Country Club are open to the public. Lodging available — 715-479-7483. 5 miles east of Eagle River on Highway 70; follow signs.

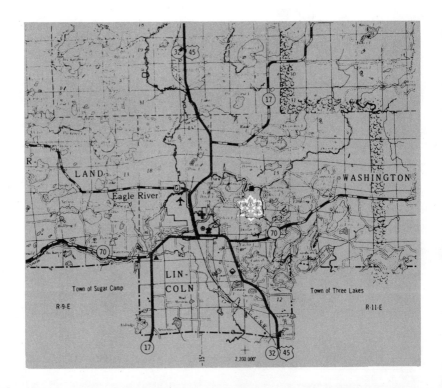

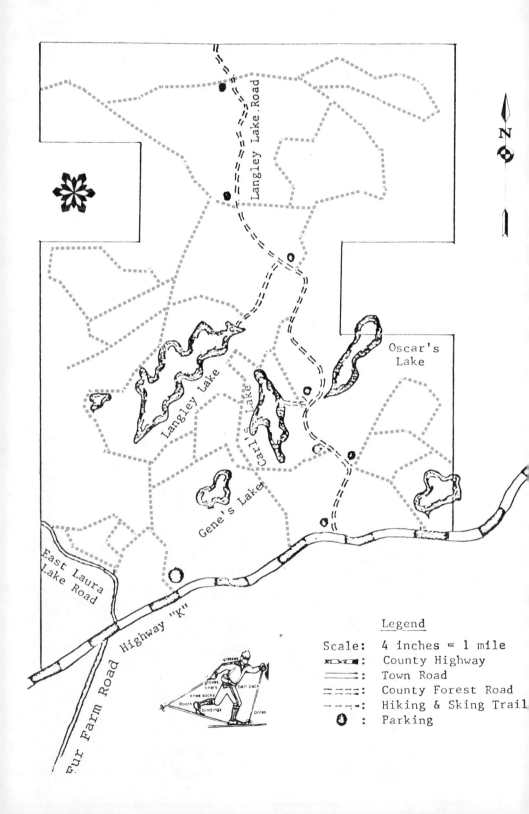

Langley Lake Road

Oscar's Lake

Langley Lake

Carl's Lake

Gene's Lake

East Laura Lake Road

Fur Farm Road Highway "K"

Legend

Scale: 4 inches = 1 mile

▱◻◻: County Highway

——: Town Road

═ ═ ═: County Forest Road

– – –: Hiking & Skiing Trail

🕭 : Parking

Langley Lake Trails

Vilas County

Located 7 miles west of Conover on Hwy. K. Marked trails in gentle, rolling wooded area. Trails are 14.2km (8.85 miles). Plowed parking, open anytime.

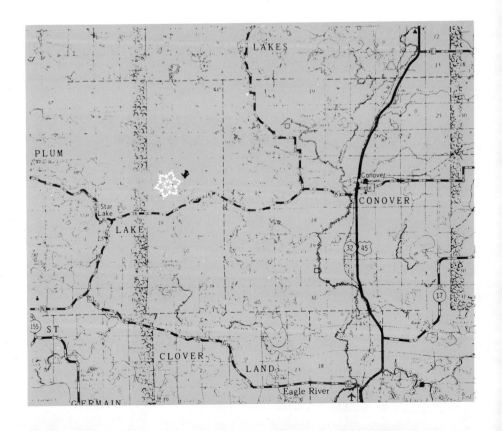

LUMBERJACK TRAIL
(SKI & BACKPACK)
a Two_way Trail

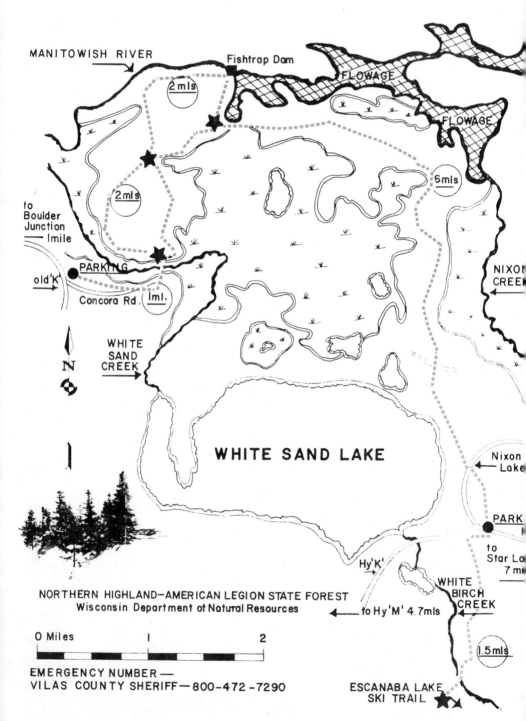

MANITOWISH RIVER

Fishtrap Dam

FLOWAGE

FLOWAGE

2 mls

6 mls

2 mls

to
Boulder
Junction
— 1mile

old 'K'

PARKING

Concora Rd.   1ml.

NIXON
CREEK

N

WHITE
SAND
CREEK

WHITE SAND LAKE

Nixon
Lake

PARK

to
Star La
7 m

Hy 'K'

WHITE
BIRCH
CREEK

to Hy 'M' 4.7mls

1.5 mls

NORTHERN HIGHLAND—AMERICAN LEGION STATE FOREST
Wisconsin Department of Natural Resources

O Miles   1   2

EMERGENCY NUMBER —
VILAS COUNTY SHERIFF — 800-472-7290

ESCANABA LAKE
SKI TRAIL

Lumberjack Trail

Vilas County

Two parking lots — one is 1 mile east of Boulder Junction — take "old K" to Concora Rd. — or 7 miles east of Boulder Junction — take Hwy. M to Hwy. K east — 5 miles Nixon Lake Road.

Wilderness trail winding through many different timber types. Approx. ½ of trail winds thru an area that was logged a few years ago. The rest of the trail is thru older timer. This trail connects with the Escanaba Trail. Backpacking is allowed along the Lumberjack.

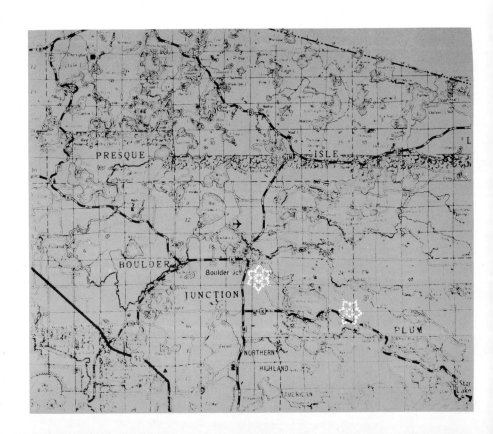

MADELINE LAKE SKI TRAIL

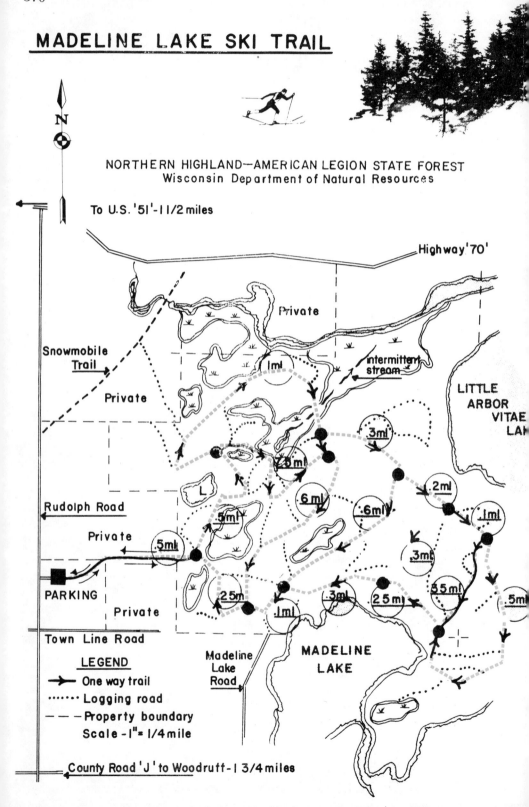

N

NORTHERN HIGHLAND—AMERICAN LEGION STATE FOREST
Wisconsin Department of Natural Resources

To U.S. '51'-11/2 miles

Highway '70'

Private

intermittent stream

Snowmobile Trail

Private

LITTLE ARBOR VITAE LAKE

1 ml

3 ml

7.0 ml

.2 ml

.1 ml

L

6 ml

.6 ml

Rudolph Road

.5 ml

3 ml

Private

.5 ml

.25 ml

.3 ml

.25 ml

.55 ml

.5 ml

PARKING

1 ml

Private

Town Line Road

Madeline Lake Road

MADELINE LAKE

LEGEND
→ One way trail
······ Logging road
— — Property boundary
Scale - 1" = 1/4 mile

County Road 'J' to Woodrutt-1 3/4 miles

Madeline Lake Trail

The Northern Highland — American Legion State Forest is the setting for this very popular ski trail in Vilas County. It can be reached by taking County J east to Rudolph Road, then north to the parking lot.

Total length of trails is approximately 15.2km (9.5 miles) with different loops of varying lengths. The 4 eastern loops are classed as intermediate. There are only 2 hills that present extreme challenges to beginners but they are wide and not dangerous. The 2 central trails are for beginners. Terrain is gentle and rolling with abundant wildlife. Some of the trail is in cutover areas but it still offers some unusual scenic winter views. The trail has the usual blue and white trailmarkers. Please observe directional arrows as this is a one-way system.

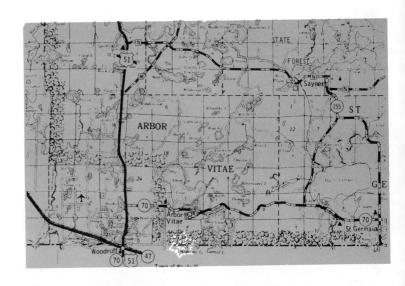

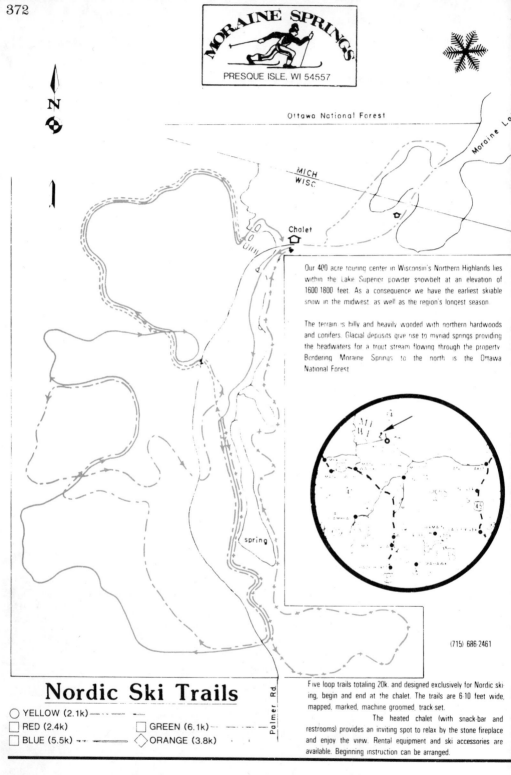

Moraine Springs

PRESQUE ISLE. WI 54557

N

Ottawa National Forest

MICH
WISC.

Moraine La

Chalet

Our 400 acre touring center in Wisconsin's Northern Highlands lies within the Lake Superior powder snowbelt at an elevation of 1600-1800 feet. As a consequence we have the earliest skiable snow in the midwest, as well as the region's longest season.

The terrain is hilly and heavily wooded with northern hardwoods and conifers. Glacial deposits give rise to myriad springs providing the headwaters for a trout stream flowing through the property. Bordering Moraine Springs to the north is the Ottawa National Forest.

spring

(715) 686-2461

Palmer Rd.

Nordic Ski Trails

○ YELLOW (2.1k)
□ RED (2.4k)
□ BLUE (5.5k)
□ GREEN (6.1k)
◇ ORANGE (3.8k)

Five loop trails totaling 20k. and designed exclusively for Nordic skiing, begin and end at the chalet. The trails are 6-10 feet wide, mapped, marked, machine groomed, track-set.

The heated chalet (with snack-bar and restrooms) provides an inviting spot to relax by the stone fireplace and enjoy the view. Rental equipment and ski accessories are available. Beginning instruction can be arranged.

Moraine Springs

Moraine Springs Nordic Center in Presque Isle, Wisconsin, offers some of the finest cross country skiing in the Midwest. Twenty kilometers of machine tracked trails wind through 400 acres heavily wooded with northern hardwoods and conifers. The gently rolling terrain offers something for all levels of skiing ability.

Located 20 miles southeast of Wakefield, Michigan, Moraine Springs reaps the benefits of the Superior Snowbelt with wno accumulations of over 150 inches annually. The touring center abuts the Ottawa National Forest which makes it an attractive home for many species of wildlife. An added attraction is a year-roun trout stream flowing through the property.

There are five loop trails totaling 20 kilometers and designed exclusively for Nordic skiing. All trails are mapped, marked, machine groomed and track-se There is a comfy chalet at trailhead where, after skiing, you can relax by the fireplace, warm up with a hot drink, a steaming bowl of homemade soup, or share a pizza with a friend. Rental equipment and ski accessories are available.

This is one of the most scenic trail systems in the state so be sure to take your camera.

Take Hwy. W west out of Presque Isle 4 miles to Old W (watch for our yellow sign), then right 1 mile to Palmer Road. Follow Palmer Road 2 miles to chalet parking. For current trail conditions or further information call (715) 686-2461 or write Moraine Springs, Presque Isle, WI 54557. Open 9:30 to dusk. Closed Wednesdays (except holidays).

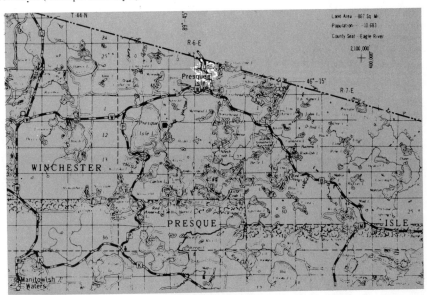

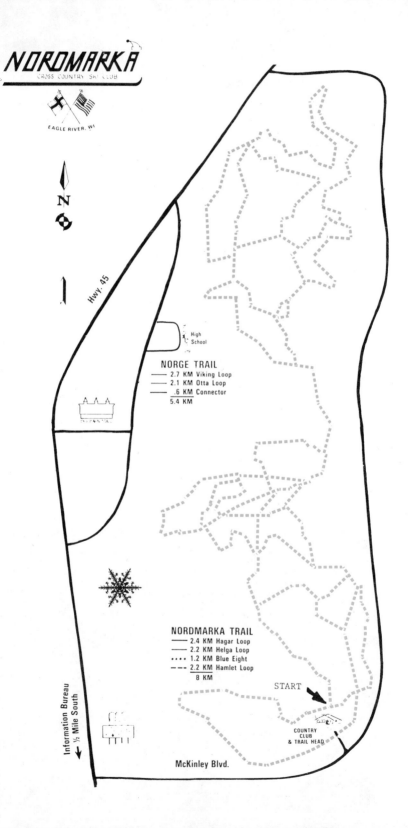

NORDMARKA

CROSS COUNTRY SKI CLUB

EAGLE RIVER, WI

N

Hwy. 45

High
School

NORGE TRAIL
——— 2.7 KM Viking Loop
——— 2.1 KM Otta Loop
——— .6 KM Connector
5.4 KM

NORDMARKA TRAIL
——— 2.4 KM Hagar Loop
——— 2.2 KM Helga Loop
•••• 1.2 KM Blue Eight
— — 2.2 KM Hamlet Loop
8 KM

START

COUNTRY
CLUB
& TRAIL HEAD

Information Bureau
½ Mile South

McKinley Blvd.

Nordmarka Ski Trails

Vilas County

The Nordmarka Cross Country Ski trails offer 13.4 kilometers of rolling, wooded terrain, ideal for beginner and intermediate skiers.

The trail head is at the Eagle River Country Club on the North edge of Eagle River. Take Hwy 45 North ½ mile and turn right on McKinley Boulevard for ¼ mile.

All trails are groomed and doubled tracked by the Nordmarka Cross Country Ski Club. Donations are appreciated.

The Eagle River Country Club is open daily for food and beverage.

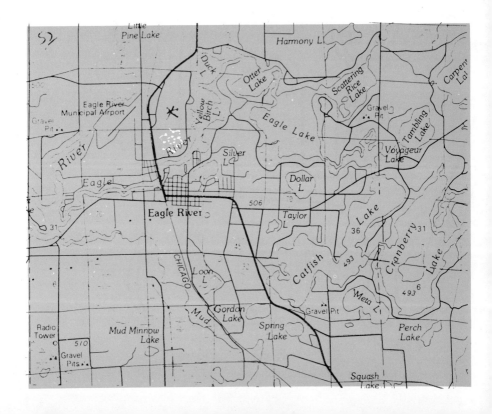

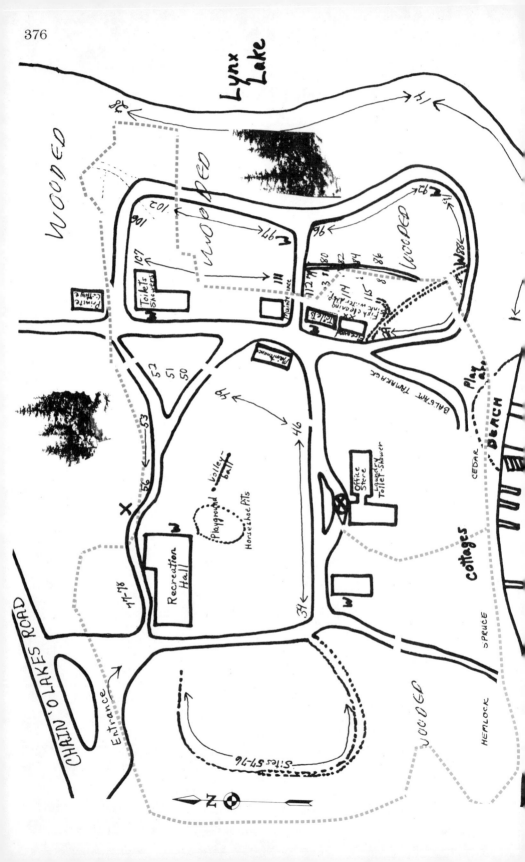

Pine Aire – Resort

Vilas County

Located two miles from Eagle River, Pine-Aire Resort offers a mile and a half of gently rolling ski trail on 30 acres of land.

Ideal for beginners, this marked trail is a combination of beautiful wooded areas and brisk exposed trails. Plowed parking lots, heated shelter with fireplace, winter campsites plus heated toilets and shower facility are all available. Pine-Aire also features some cozy housekeeping cottages with bath, and fireplaces.

There is no charge for trail use. Rental skis are available as well as instructions by appointment only. Additional 1 mile trail is available. Pine-Aire can be reached by taking Hwy. 45 two miles north of Eagle River and then turn right on Chain O' Lakes Road ⅛ mile.

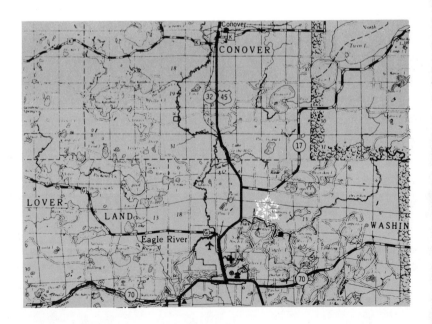

378

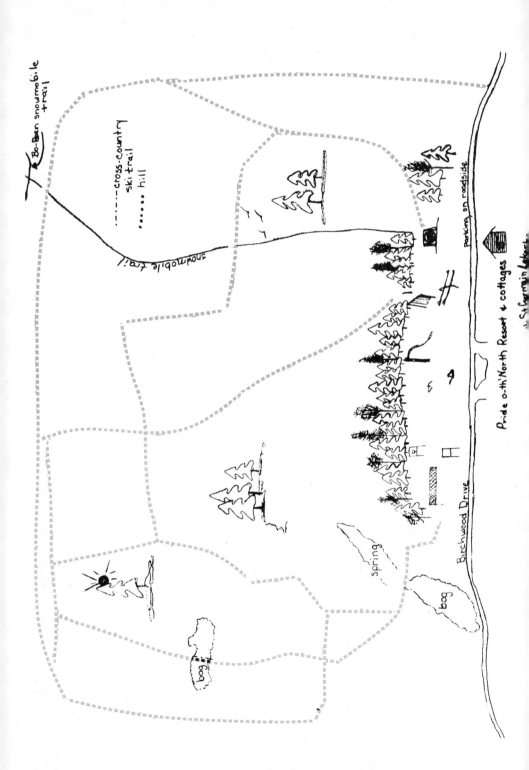

Bo-Ben snowmobile trail

------ cross-country ski trail

•••••• hill

snowmobile trail

spring

bog

bog

Birchwood Drive

Pride o-th'North Resort & cottages

Parking on roadside

St Germain Lakes

Pride O-Th' North Trails

Vilas County

These trails are located on 90 acres of wooded land and are good for beginners and intermediate skiers. The trailhead is at the playground and trails are marked with blue markers. Total of trail lengths is about 2km (2 miles) with additional mileage planned for the near future. The resort is open all year and skis, poles, and boots are furnished free to resort customers. No charge for use of the trails but parking is reserved for customers only. Other trail users must park along Birchwood Drive. Pride o-th' North can be reached by taking Highway 155 north of St. Germain to Birchwood Drive, then east to trailhead.

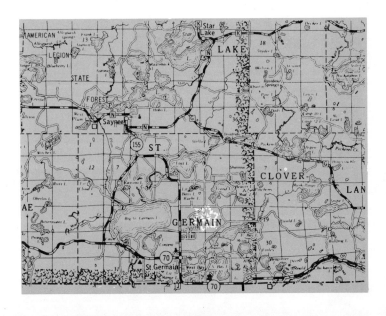

Ski Hill Trail

Located 5 miles north of Eagle River on Hwy. 45 to Ski Hill Road. There are 8km (5 miles) of gentle rolling trails in wooded area. Plowed parking. Open at all times.

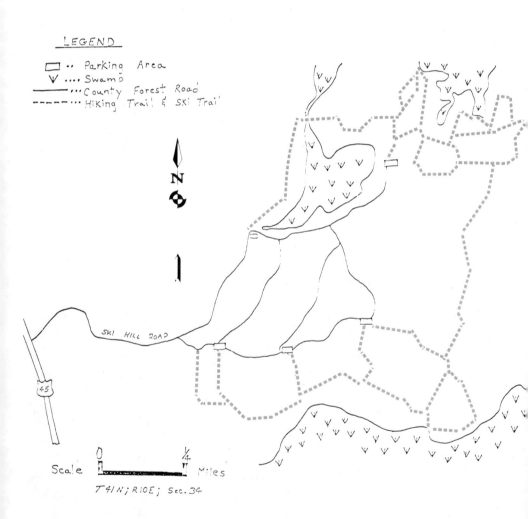

LEGEND
☐ .. Parking Area
V .... Swamp
———— ···County Forest Road
------ ··· Hiking Trail & Ski Trail

N

SKI HILL ROAD

45

Scale 0 ———— ¼ Miles

T41N; R10E; Sec. 34

Conover School Trail

Vilas County

Located 1 mile south of Conover on Hwy. 45. 10.4km (6.45 miles) of trails marked in gentle rolling, wooded area. Open anytime.

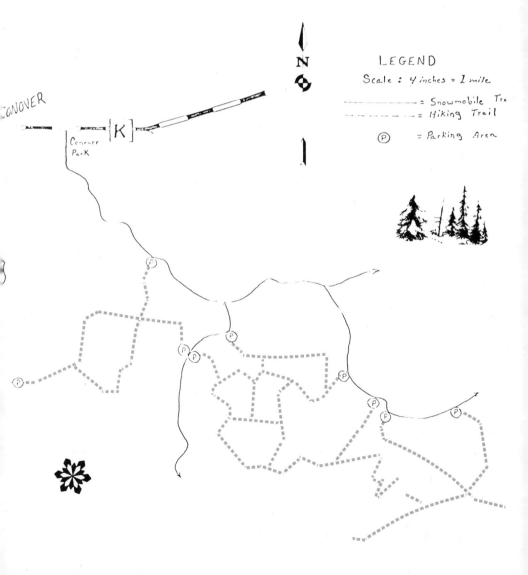

CONOVER

[K]

Conover
Park

N

LEGEND

Scale : 4 inches = 1 mile

——————— = Snowmobile Tra

—·—·—·—·— = Hiking Trail

Ⓟ = Parking Area

SHANNON LAKE SKI TRAIL

NORTHERN HIGHLAND—AMERICAN LEGION STATE FOREST

Wisconsin Department of Natural Resources

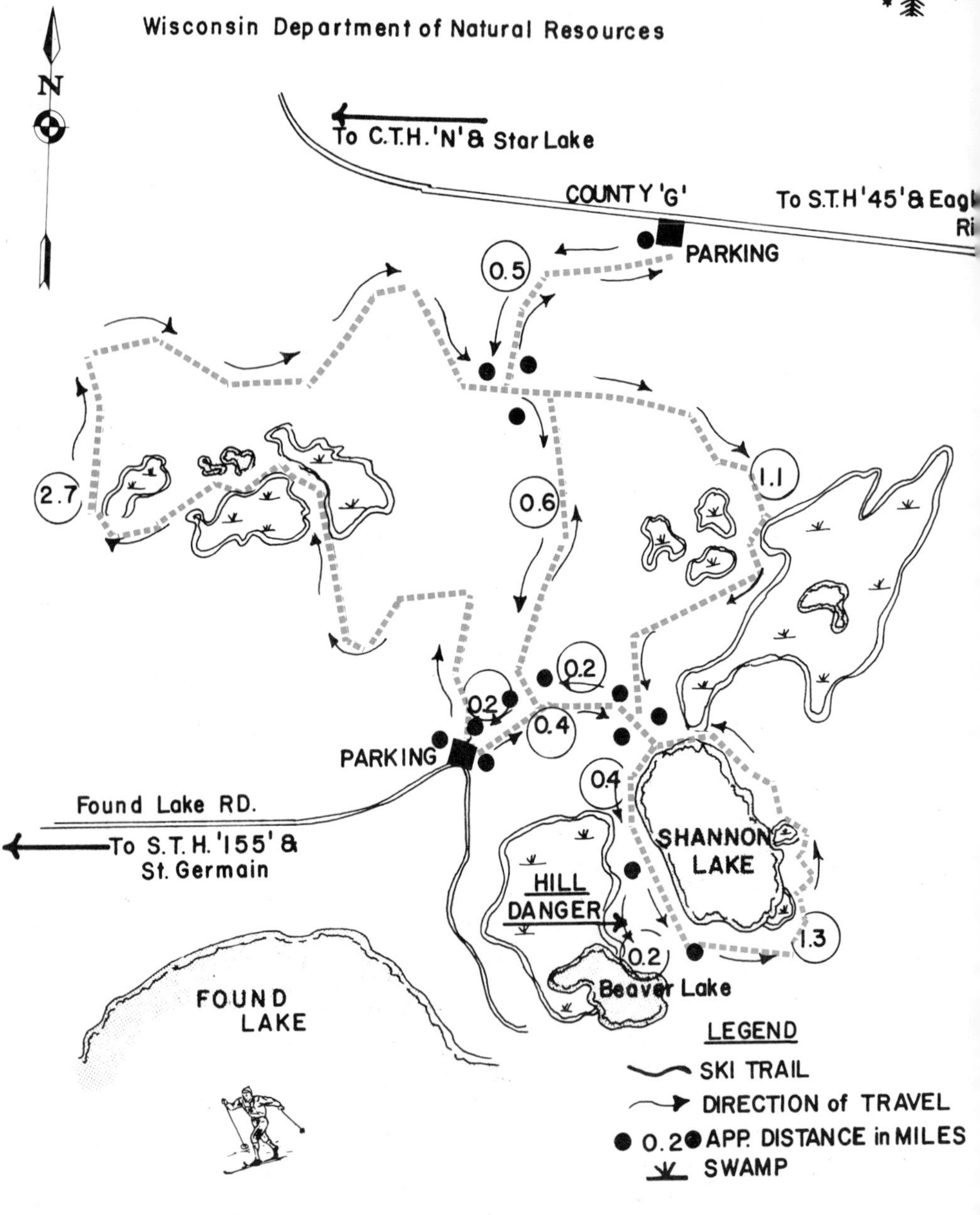

N

To C.T.H.'N'& Star Lake

COUNTY 'G'

To S.T.H '45'& Eagle Ri

PARKING

0.5

2.7

0.6

1.1

0.2

0.2

0.4

PARKING

0.4

Found Lake RD.

To S.T.H.'155'&
St. Germain

SHANNON
LAKE

HILL
DANGER

1.3

0.2

Beaver Lake

FOUND
LAKE

LEGEND

SKI TRAIL
DIRECTION of TRAVEL
0.2 APP. DISTANCE in MILES
SWAMP

Emergency Number_Vilas County Sheriff_800-472-7290

Shannon Lake Trail

Vilas County

Approximately ½ of the trail runs through an area that was just recently logged. The remainder of the trail meanders through a variety of Northern Hardwoods. Skiers will find the loop around Shannon Lake to be especially scenic. The 11.2km (7 mile) trail is recommended for all skiers.

It is located 2½ miles northeast of St. Germain off CTH G or off Found Lake Road.

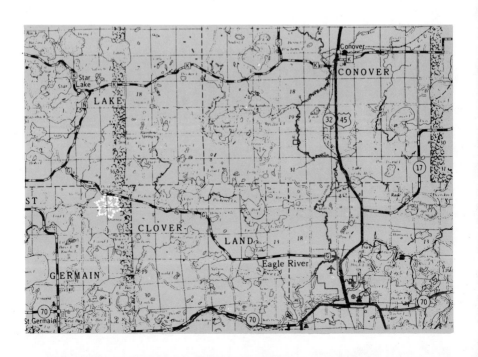

STATEHOUSE LAKE SKI TRAIL

NORTHERN HIGHLAND – AMERICAN LEGION STATE FOREST

Wisconsin Department of Natural Resources

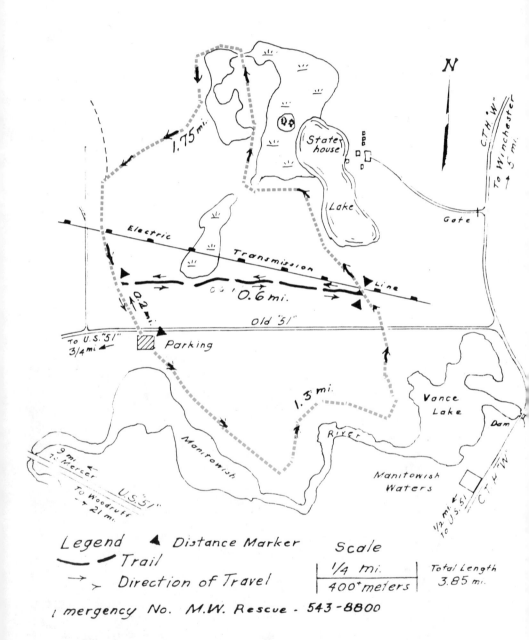

N

1.75 mi.

State house

Lake

Electric

Transmission Line

0.6 mi.

0.2 mi.

Old "51"

To U.S. "51"
3/4 mi.

Parking

Gate

C.T.H. "W"
To Winchester
5 mi.

1.5 mi.

Vance Lake

Dam

River

Manitowish

Manitowish Waters

9 mi.
To Mercer

To Woodruff
21 mi.

U.S. "51"

1/2 mi.
To U.S. "51"

C.T.H. "W"

Legend ◄ Distance Marker
━━ Trail
→ Direction of Travel

Scale
1/4 mi.
400± meters

Total Length
3.85 mi.

Emergency No. M.W. Rescue - 543-8800

Vilas County

Statehouse Lake Trail

Two miles west of Manitowish Waters off "Old 51".

Beautiful scenery along 1.5 miles of the Manitowish River. Very few steep hills. Good trail for novices.

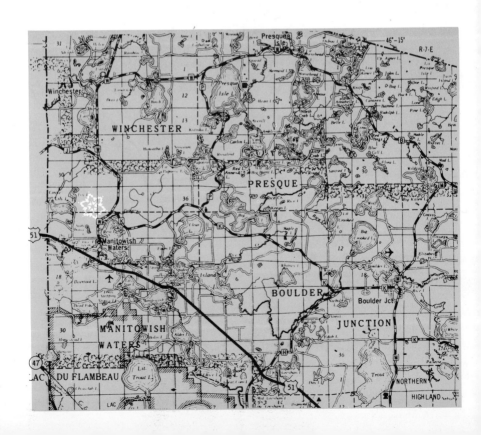

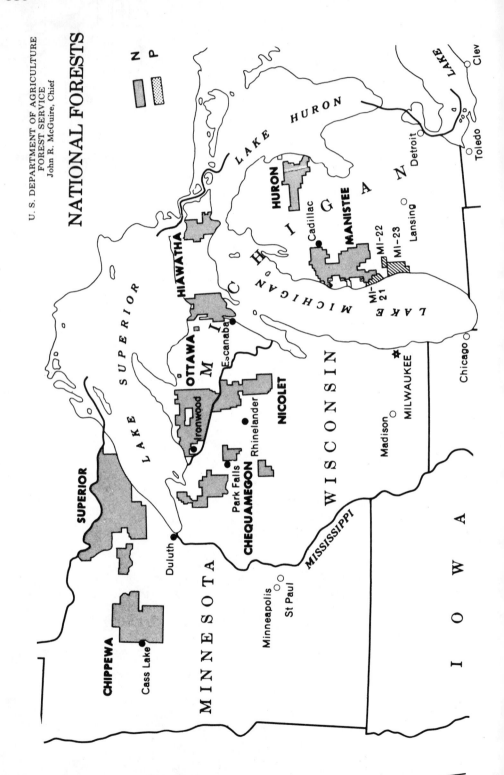

U. S. DEPARTMENT OF AGRICULTURE
FOREST SERVICE
John R. McGuire, Chief

NATIONAL FORESTS

Trees for Tomorrow Center

Vilas County

Trees for Tomorrow Resources Education Center is a non-profit organization open to the public membership. Founded in 1944 by the paper and power industries to reforest northern Wisconsin, it has since evolved to its present role as an innovator in resource education. Year around programs are offered in ecology to help our urbanized society understand the outdoor natural environment. In addition to ecology, lifetime outdoor skills are taught such as cross-country skiing, canoeing, nature photography, mushroom and wild edible identification, orienteering and wilderness survival to provide people with the skills necessary to explore the out-of-doors. Participants also gain a better understanding of natural resources and an awareness and appreciation of the social and economic aspects of the environment. Over 135,000 have attended environmental education programs at the Center since 1944.

Facilities, Accommodations and Meals — The Center is a forty acre complex located within the city limits of Eagle River. On-site facilities include four dormitories (linens and blankets provided) with separate rooms (2, 3 and 4 to a room) with a capacity of 104 people, dining hall, classroom, 170 year old Demonstration Forest and beautiful scenery. Nearby are over 60 miles of groomed public and private ski trails. Bus or van transportation to and from off-site trails is provided. Meals (all-you-can-eat) are served in the Center dining hall.

The Program — The Center offers a weekend package that includes two nights' lodging, six meals and ski instruction led by naturalists who will interpret the North Woods for participants. Optional evening programs include waxing clinics, winter clothing and safety, ski films, ecology and wine-tasting.

In addition, during the holidays and at other times during the winter, skiers can use "Trees" facilities and either ski on their own or join regularly scheduled guided interpretive tours of the forest. Ski tours vary in length between 5 and 15 miles depending on the individuals' physical condition.

For a free color brochure describing the facilities arid programs available write to: Ski Registrar, Trees for Tomorrow Resources Education Center, Box 609, Eagle River, WI 54521.

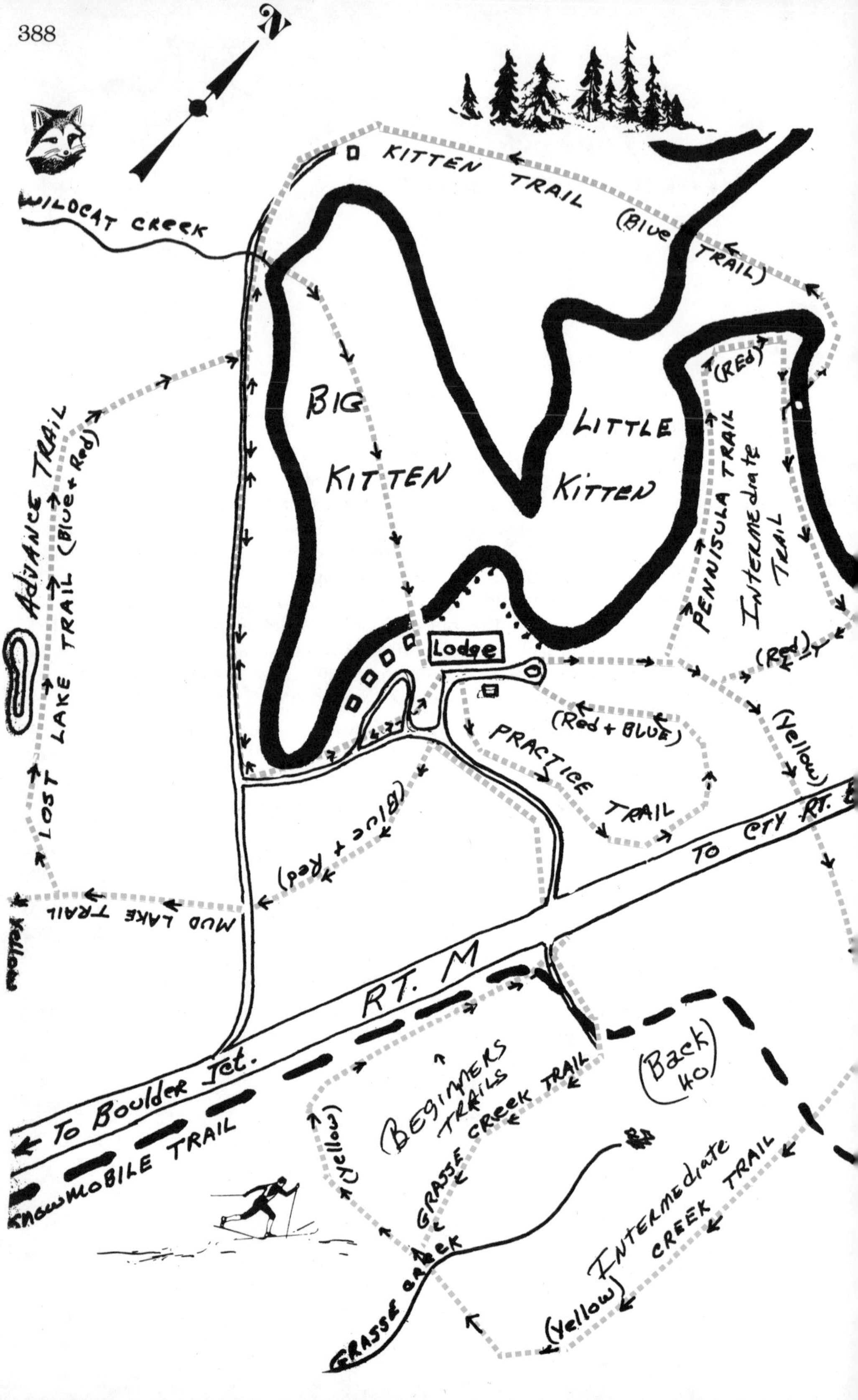

N

WILDCAT CREEK

KITTEN TRAIL (Blue TRAIL)

Advance TRAIL (Blue + Red)

LOST LAKE TRAIL

BIG KITTEN

LITTLE KITTEN

(Red)

PENNISULA TRAIL

Intermediate TRAIL

(Red)

Lodge

(Red + BLUE)

PRACTICE TRAIL

(yellow)

TO CITY RT. E

(Red + BLUE)

MUD LAKE TRAIL

yellow

RT. M

To Boulder Jct.

SNOWMOBILE TRAIL

(yellow)

BEGINNERS TRAILS

GRASSE CREEK TRAIL

(Back) 40

GRASSE CREEK

INTERMEDIATE CREEK TRAIL

(yellow)

Wildcat Lodge

The lodge offers 20 miles of trails from 1 kilometer to 10 kilometers on seven different trails for use primarily by their guests. Some trails are for beginners and some are for the more experienced. Most trails can be started right from the doors of the units. There are also many old logging roads and fire trails you can take off on, however these are not marked. Take Highway M north out of Boulder Junction. Lodge is on left before reaching County B. Rentals available in Boulder Junction.

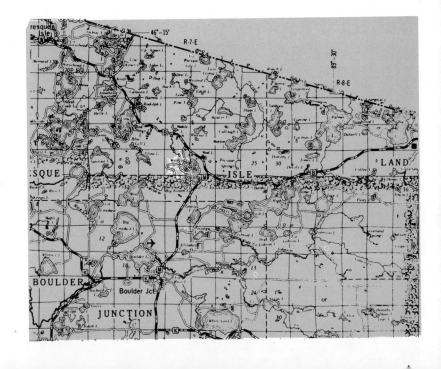

Winchester Cross-Country Ski Trail

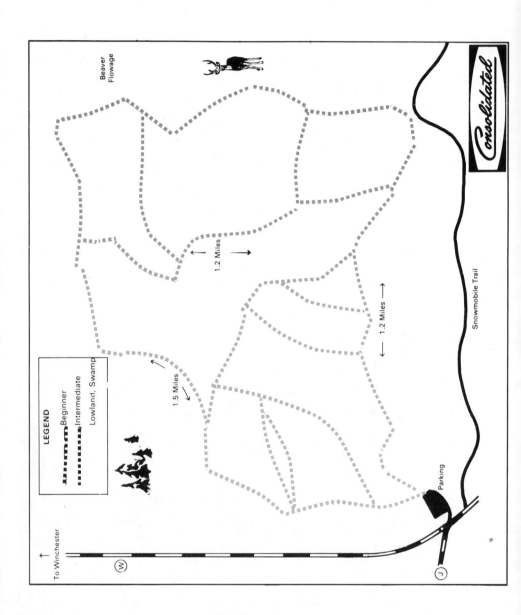

Winchester Trail

Vilas County

The Winchester cross-country ski trail is just south of Winchester on Highway W. The Consolidated Papers Inc. was approached by the North Lakeland Community Services with the idea of development of a trail system on Consolidated forestland. They were happy to cooperate and provided signs and maintenance of a parking lot at the trail head.

Trail Features. The outside trail circumventing the system is approximately 6.4km (4 miles) long, and courses through gently rolling forestland with a few challenging hills. Trail spurs within the system are also moderate in gradient, with the extreme western area containing the steeper slopes.

Several muskeg swamps and small lakes are scattered throughout the system, and the trails pass and even cross a number of beaver ponds. A birch and aspen forest dominates this area, intermixed with maple, pine and oak. On the trails you may encounter deer, ruffed grouse, snowshoe hares and occasionally a weasel, mink or porcupine.

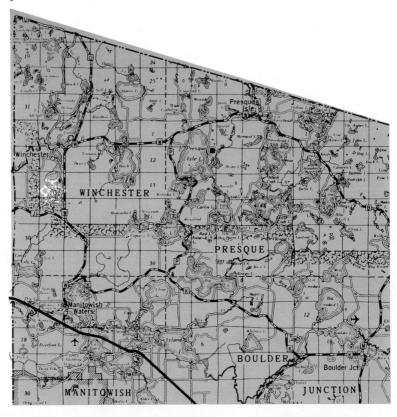

392

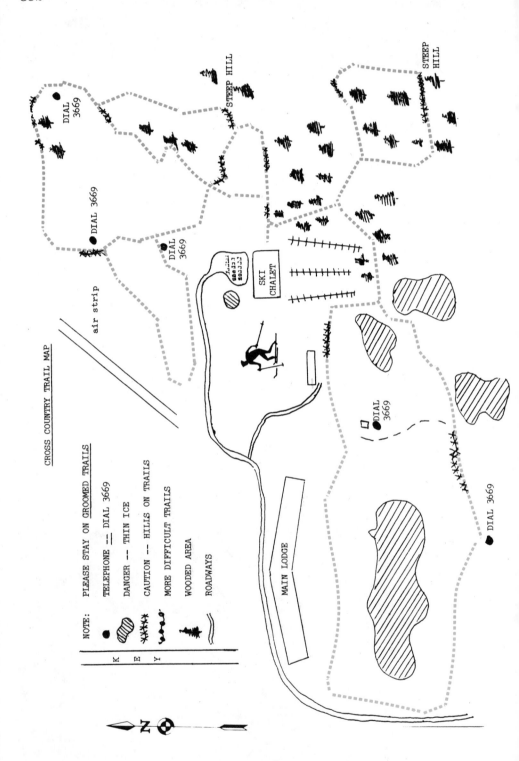

CROSS COUNTRY TRAIL MAP

NOTE: PLEASE STAY ON GROOMED TRAILS

KEY

TELEPHONE -- DIAL 3669

DANGER -- THIN ICE

CAUTION -- HILLS ON TRAILS

MORE DIFFICULT TRAILS

WOODED AREA

ROADWAYS

N

air strip

DIAL 3669

DIAL 3669

DIAL 3669

DIAL 3669

DIAL 3669

STEEP HILL

STEEP HILL

SKI CHALET

MAIN LODGE

Americana Ski Trails

Americana Ski Area offers 14 kilometers of groomed and marked cross country ski trails in addition to its 13 downhill runs. Cross country trails meander through the beautiful southern Wisconsin country side on the 1100 Americana acres.

Beginners, intermediates and advanced skiers can choose their trail according to their own ability. Beginner trails are relatively flat and short loops; advanced trails are longer and consist of steeper hills and woods.

Food and beverage may be found in the main ski lodge. Room accomodations are also available.

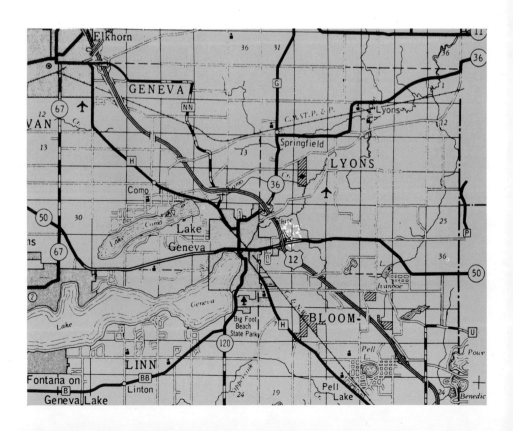

WASHBURN COUNTY FOREST
GULL LAKE SKI TRAIL

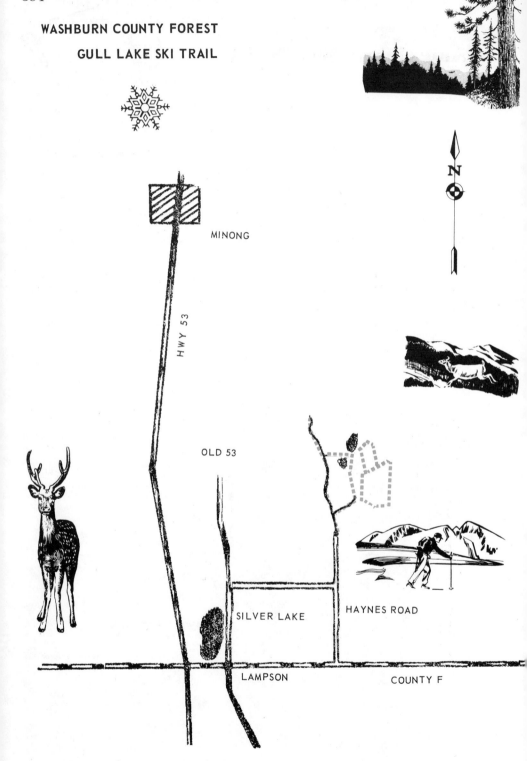

MINONG

HWY 53

OLD 53

SILVER LAKE

HAYNES ROAD

LAMPSON

COUNTY F

Gull Lake

Consisting of 3 loops, Gull Lake offers 8km (5 miles) of hilly and wooded trails, which are marked. These remote bushwack trails have several long glide slopes through a scenic ridgeland and ravine landscape.

It is located 6 miles north of Trego on State Highway 53 then 1.5 miles east on County Highway F and 2 miles on Haynes Road.

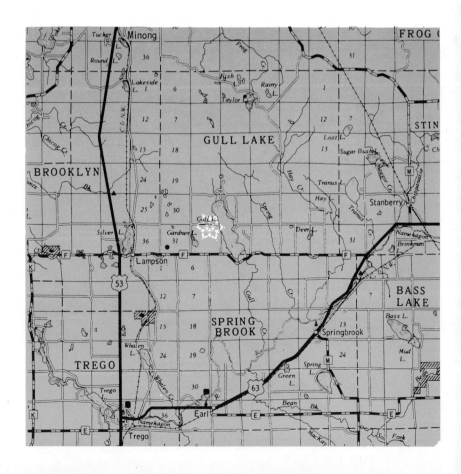

① Nordic Woods Ski Trail⸺

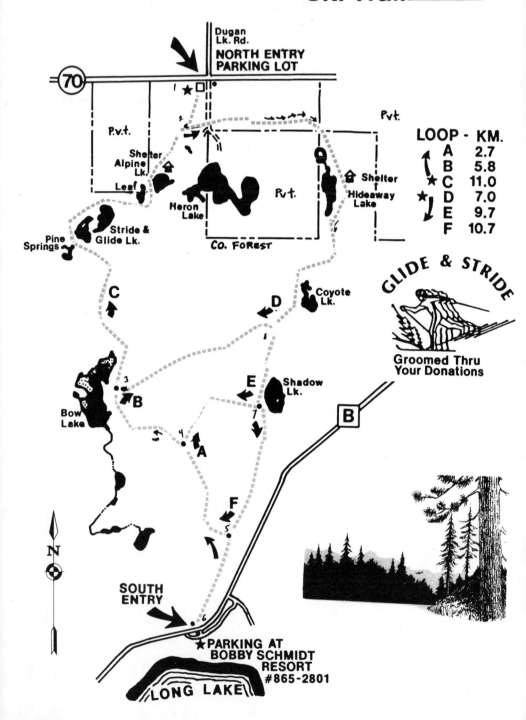

Dugan Lk. Rd.

NORTH ENTRY PARKING LOT

70

Pvt.

P.v.t.

Shelter
Alpine Lk.

Leaf

Heron Lake

Pvt.

Pvt.

Shelter
Hideaway Lake

Pine Springs

Stride & Glide Lk.

CO. FOREST

LOOP - KM.

	A	2.7
	B	5.8
★	C	11.0
★	D	7.0
	E	9.7
	F	10.7

C

D

Coyote Lk.

GLIDE & STRIDE

Groomed Thru Your Donations

B

Bow Lake

E

Shadow Lk.

A

B

F

N

SOUTH ENTRY

★ **PARKING AT BOBBY SCHMIDT RESORT**
#865-2801

LONG LAKE

Nordic Woods Ski Area

Washburn County

Considered by many to be the county's best trail, this general touring trail traverses through a wild and picturesque kettle moraine lake area of county forestland. Recently improved by the county in 1983 it is periodically groomed and tracked. Two adirondack shelters along the trail provide a spot for rest or picnic.

The trail can be reached by parking along Dugan Lake road just off Hwy. 70 about 5 miles west of Stone Lake. The trail can also be reached from the south by parking at Indian Hills Resort on County Trunk B on the northern end of Long Lake. The Lodge offers food, liquor bar and housing.

Oakridge Ski Trail

Located two miles southeast of Minong this intermediate trail runs through mostly second growth northern hardwoods. The trail is under development by Washburn County and is not regularly groomed.

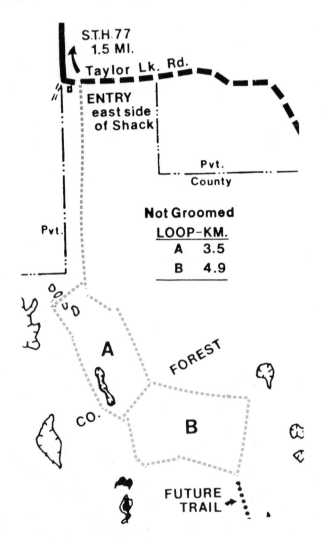

S.T.H. 77
1.5 MI.

Taylor Lk. Rd.

ENTRY
east side
of Shack

Pvt.
County

Pvt.

Not Groomed

LOOP-KM.

A   3.5
B   4.9

A

FOREST

CO.

B

FUTURE
TRAIL

Rodeo Pines Trail

Washburn County

This short 2.1 mile double loop trail in Spooner is on WDNR fish hatchery lands. Maintained, groomed and regularly tracked by the Glide and Stride Ski Club it is open to the public.

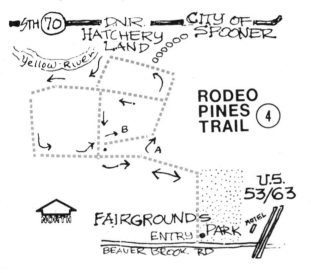

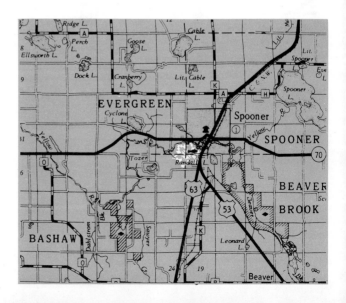

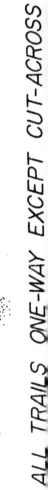

TREGO LAKE

SKI TOURING TRAIL

ST. CROIX NATIONAL SCENIC RIVERWAY

TRAIL LOOPS

LOCATION	MI.	KM.	DIFFICULTY
	1.2	1.9	Beginner
	1.3	2.1	Intermediate
	1.1	1.8	Advanced

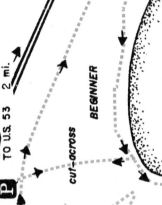

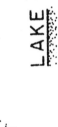

NORTH RIVER ROAD

TO U.S. 53 2 mi.

P

TO COUNTY "K" 3 mi.

BEGINNER

CUT–ACROSS

ADVANCED

INTERMEDIATE

LAKE

TREGO

powerline

N

ALL TRAILS ONE-WAY EXCEPT CUT-ACROSS

Trego Lake

The trail was designed by the National Park Service for beginning and intermediate skiers. The hills (except on the inner loop) are moderate and wipe-out areas are provided. The inner loop has steeper climbs and runs and should only be used by the more experienced skier. Trails are groomed and tracked and are all one-way except for the cut across.

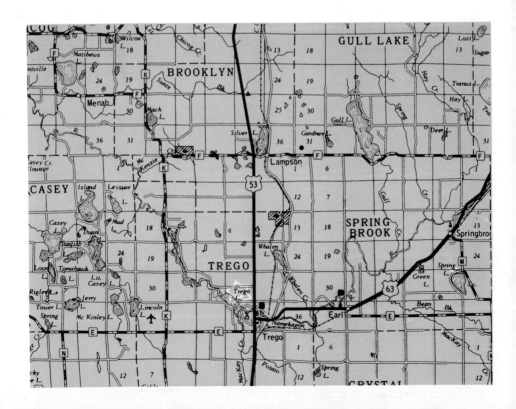

402

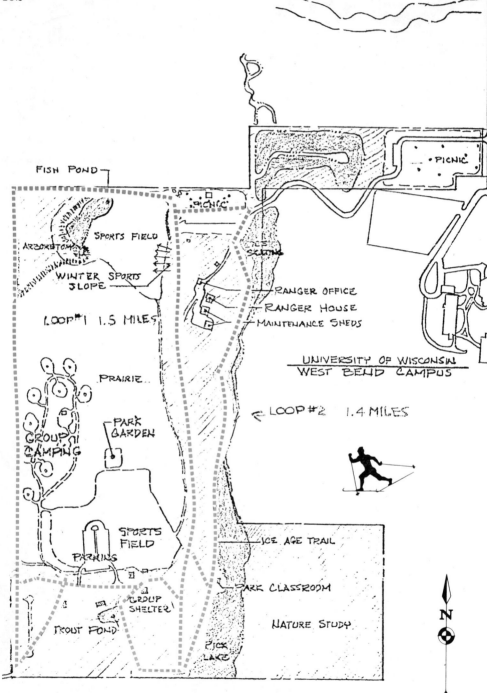

FISH POND

PICNIC

PICNIC

ICE SKATING

ARBORETUM

SPORTS FIELD

WINTER SPORTS SLOPE

RANGER OFFICE
RANGER HOUSE
MAINTENANCE SHEDS

LOOP #1 1.5 MILES

UNIVERSITY OF WISCONSIN
WEST BEND CAMPUS

PRAIRIE

LOOP #2 1.4 MILES

PARK GARDEN

GROUP CAMPING

SPORTS FIELD

ICE AGE TRAIL

PARKING

PARK CLASSROOM

GROUP SHELTER

NATURE STUDY

TROUT POND

PICK LAKE

N

UNIVERSITY DRIVE

RIDGE RUN COUNTY PARK

Ridge Run County Park

Washington County

This 100 acre County Park offers a natural setting of the Kettle Moraine with 3.2km (2 miles) of trails, a variety of wildlife, and three small lakes.

Loop one is about 1.5 miles of trail perhaps best designed to serve the local enthusiast. Although somewhat short in length, it bounds a winter sports area with toboggan run, group camping, and an afternoon of family fun.

Loop two, 1.4 miles, may provide a bit more scenery as it snakes its way along the upper ridges of Washington County. Locally known as the "ice age trail", this run provides the skier scenic topography and man's own handiwork.

Glacier Hills Trail

1664 Freiss Lake Road

Washington County

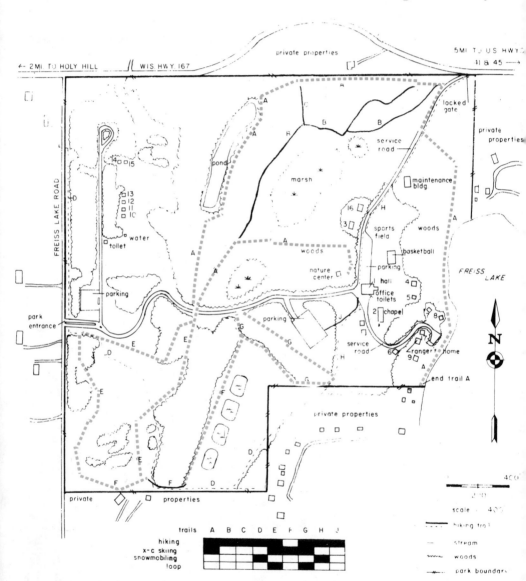

← 2 MI TO HOLY HILL   WIS. HWY 167

private properties

5 MI TO US HWY 41 & 45

FREISS LAKE

WASHINGTON COUNTY PARK SYSTEM

Homestead Hollow Trail

N120 W19809 Freistadt Rd.

Washington County

HOMESTEAD HOLLOW COUNTY PARK

N120 W19809 Freistadt Rd. ... Germantown ... acres

Dale Kluever - Resident Supervisor

FREISTADT RD.

park entrance

GOLDENDALE RD

MERKEL DR

toilet

parking

ranger's home
& park office

shelter HH1

shelter HH2

toboggan hill

North

400

scale 1-400

hiking trail

stream

woods

park boundary

WASHINGTON COUNTY PARK SYSTEM

WASHINGTON COUNTY LAND USE & PARK DEPARTMENT

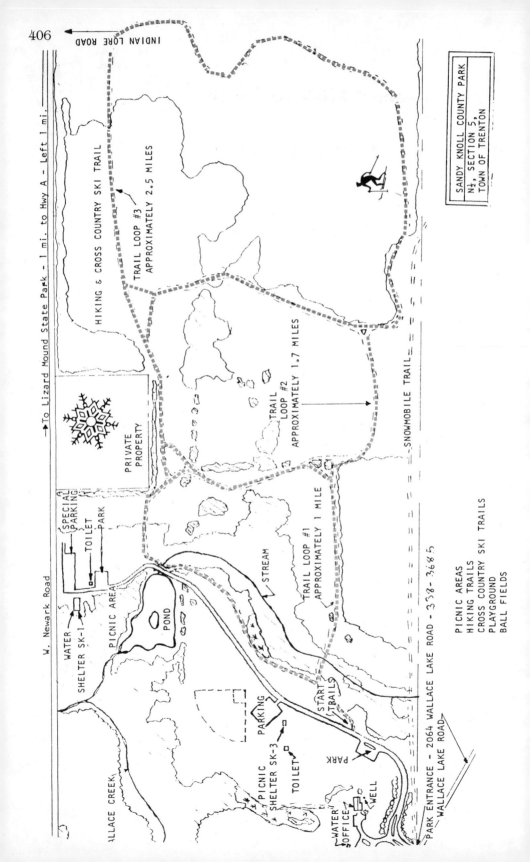

SANDY KNOLL COUNTY PARK
N½, SECTION 5,
TOWN OF TRENTON

W. Newark Road

←To Lizard Mound State Park - 1 mi. to Hwy A - Left 1 mi.

INDIAN LORE ROAD

HIKING & CROSS COUNTRY SKI TRAIL

TRAIL LOOP #3
APPROXIMATELY 2.5 MILES

PRIVATE
PROPERTY

SPECIAL
PARKING

TOILET

PARK

WATER

SHELTER SK-1

PICNIC AREA

POND

STREAM

TRAIL LOOP #2
APPROXIMATELY 1.7 MILES

SNOWMOBILE TRAIL

TRAIL LOOP #1
APPROXIMATELY 1 MILE

START
TRAILS

PARKING

PICNIC
SHELTER SK-3

TOILET

PARK

WELL

WATER
OFFICE

←PARK ENTRANCE - 2064 WALLACE LAKE ROAD - 338-3685
WALLACE LAKE ROAD→

WALLACE CREEK

PICNIC AREAS
HIKING TRAILS
CROSS COUNTRY SKI TRAILS
PLAYGROUND
BALL FIELDS

Sandy Knoll County Park

Washington County

Sandy Knoll has three trails that originate in the parking lot and wind thru wooded and clear areas. Loop 1 is approximately 1.6km (1 mile), loop 2 is approximately 2.6km (1.7 miles) and loop 3 is 4km (2.5 miles).

The park consists of 267 acres and was developed in 1976. Ample parking and toilet facilities are available at the trailhead. No charge for use of trails which are also hiking trails in the summer.

The park can be reached by taking Hiway 33 east out of West Bend to Trenton Road (just before the airport). Turn left for approximately 3 miles on Trenton Road to Wallace Lake Road then east a short distance to park entrance.

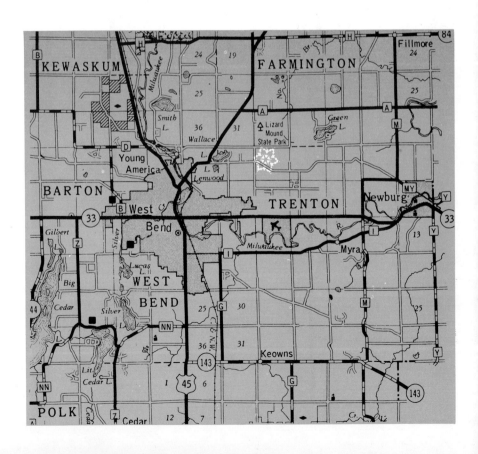

PIKE LAKE STATE PARK

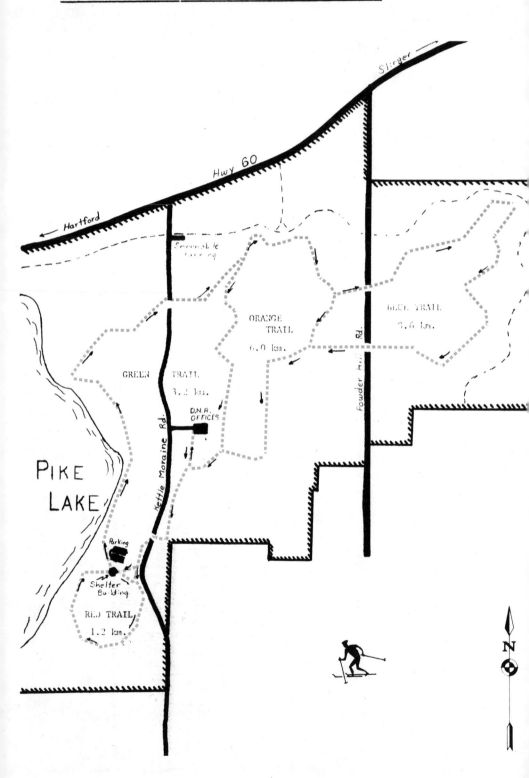

Pike Lake State Park

Located on 680 aces of hilly and wooded land, Pike Lake Trail is ideal for those who have a few hours for touring. Pine plantations, hardwood forests, and some open areas are scenes of 4 loops of marked, one-way trails which should be skied clockwise.

Due to the close proximity to Milwaukee, the trail has very heavy use on week-ends. Ample parking is provided and a vehicle admission sticker is required.

The State Park is located 3 miles west of Slinger off Highway 60 on Kettle Moraine Drive.

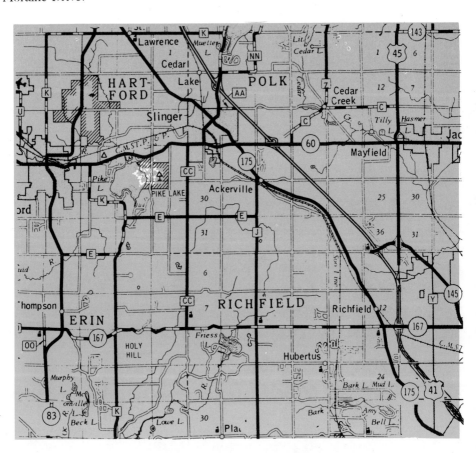

410

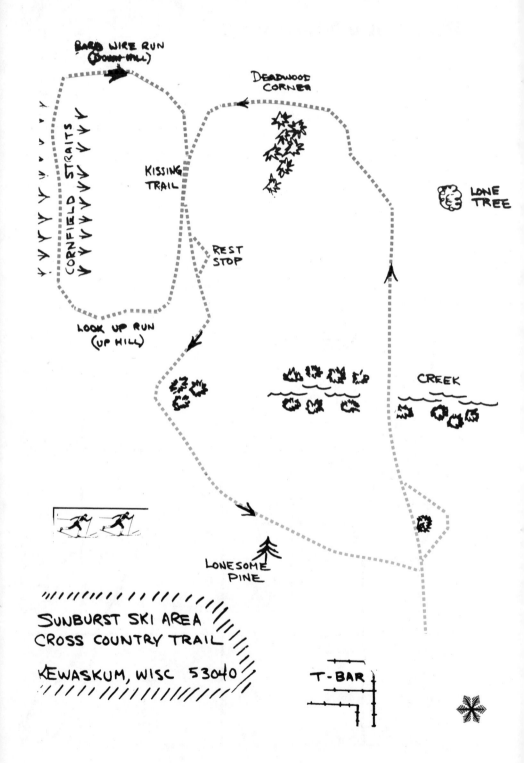

Sunburst

Sunburst is located just south of Kewaskum on Highway 45. This is primarily a downhill ski area, but they do have cross-country trails, natural snow permitting. Cross-country rental equipment is available as well as instructions. Food, beverages and toilets available in the Chalet.

This could be a good weekend for the beginner to spend one day with the instructor and practice on the Sunburst Trail. On the next day you could travel a few miles to the Kettle Moraine area for more advanced trails.

NASHOTAH PARK

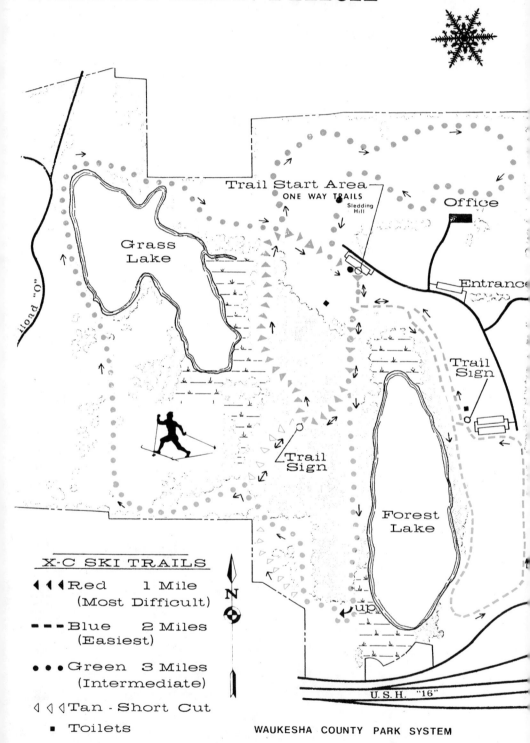

Trail Start Area
ONE WAY TRAILS
Sledding Hill

Office

Grass Lake

Entrance

Railroad "O"

Trail Sign

Trail Sign

Forest Lake

up

X-C SKI TRAILS

◀◀◀ Red   1 Mile
        (Most Difficult)

- - - Blue   2 Miles
        (Easiest)

●●● Green   3 Miles
        (Intermediate)

◁◁◁ Tan - Short Cut

■ Toilets

N

U.S.H. "16"

WAUKESHA COUNTY PARK SYSTEM

Nashota Park Trails

Waukesha County

Nashotah Park is located north of the Village of Nashotah on C.T.H. 'C' just off of U.S.H. '16'. The entrance is about ¼ mile from this intersection on the west side of C.T.H. 'C'. This 420 acre park was just opened in 1977 and has parking, toilet and picnic facilities. The X-C ski trails originate at the north parking lot and are color coded according to length and difficulty: Red — 1 mile (most difficult). Blue — 2 miles (easiest). Green — 3 miles (intermediate). These trails are also groomed and tracked as they wind around two lakes and thru woods, fields, farm lanes and pines. The topography is typical Kettle Moraine offering some challenging hills and beautiful views.

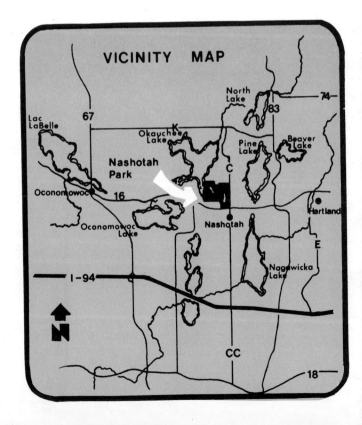

Menominee Park Trails

Menomonee Park is located within the villages of Menomonee Falls and Lannon 1½ mile north of State Highway 74 on CTH V. Two trails totaling 4 miles on 397 acres of rolling field, maple woods, cottail marsh and Quarry Lake.

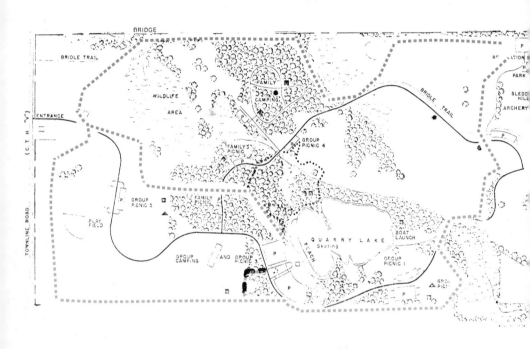

Trail Information
- FOUR MILES OF SKI TRAILS
- TWO – WAY TRAILS
- NO DOGS OR HORSES ALLOWED ON SKI TRAILS
- NO SLEDS ALLOWED ON SKI TRAILS

L E G E N D
P PARKING
🚻 TOILET
☐ BEACH HOUSE
▲ SHELTER BLDG.
◉ DRINKING WATER
–·– PROPOSED DEVELOPMENT
···· HIKING & NATURE TRAIL
▬▬ X-C SKI TRAILS

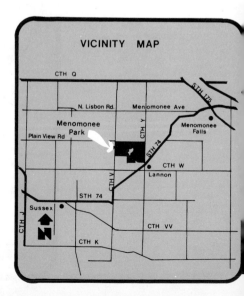

VICINITY MAP

Muskego Park Trails

This 160 acre park has 2 ski trails totaling about 2 miles in length. Located on State Highway 24 approximately ½ mile west of Racine Avenue.

Waukesha County

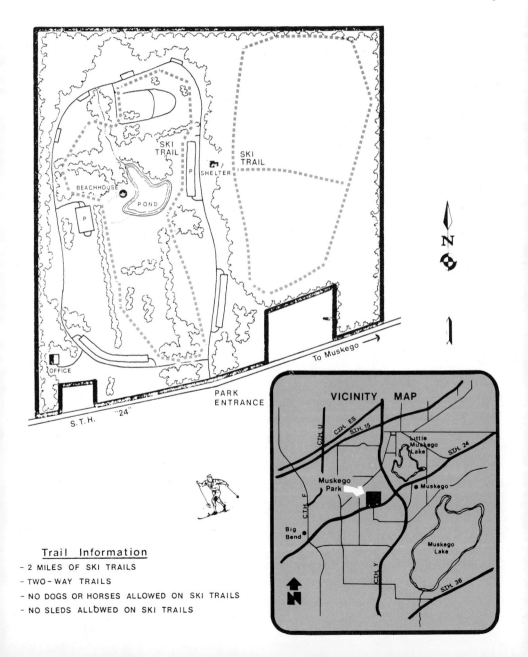

Trail Information

- 2 MILES OF SKI TRAILS
- TWO-WAY TRAILS
- NO DOGS OR HORSES ALLOWED ON SKI TRAILS
- NO SLEDS ALLOWED ON SKI TRAILS

MINOOKA
PARK

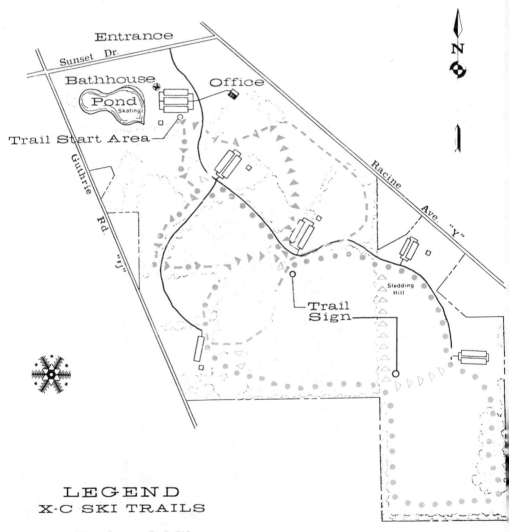

Entrance

Sunset Dr.

Bathhouse

Pond Skating

Office

Trail Start Area

Guthrie Rd. "J"

Racine Ave. "Y"

Sledding Hill

Trail Sign

LEGEND
X-C SKI TRAILS

◀◀◀ Red   1 Mile

▬▬▬ Blue   2 Miles

●●● Green   3 Miles

◁◁◁ Tan - Short Cut

▫ Toilets

WAUKESHA COUNTY PARK SYSTEM

Minooka Park Trails

Waukesha County

Monooka Park is located south-east of the City of Waukesha on Racine Avenue (C.T.H. 'Y'). To reach it, take I-94 to U.S.H. "18", then C.T.H. 'A' south to Racine Avenue. The park entrance is on Sunset Drive, west of Racine Avenue. The park is 300 acres in size, containing rolling terrain, some steeper hills and a combination of woods and open fields. Trails are marked with a color code system: Red — 1 mile, Blue — 2 miles, Green — 3 miles, and tracked with a track setter. The trails originate at the beach parking lot near the entrance where there is an ice skating warming house available to skiers.

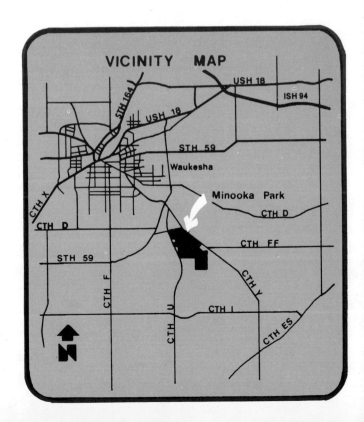

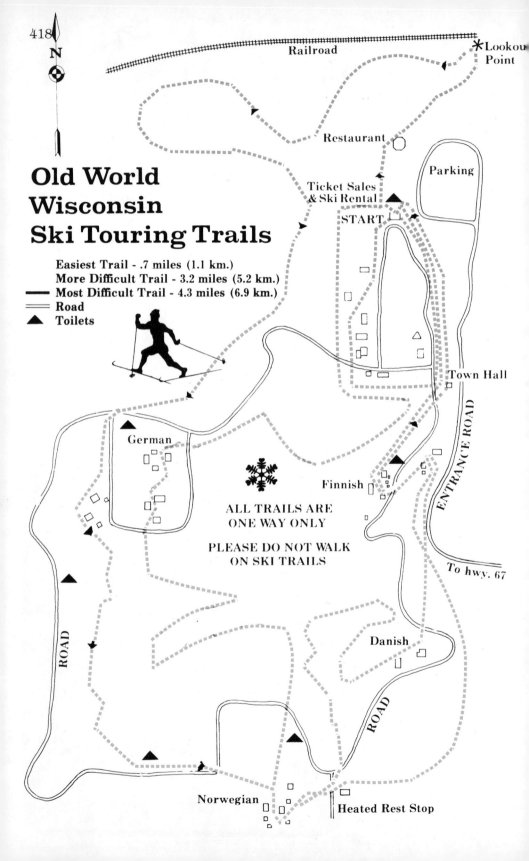

Old World Wisconsin

Waukesha County

Old World Wisconsin is open for cross-country skiing every Friday, Saturday, and Sunday from December 16 to March 4 (conditions permitting). Hours are from 9 AM to 4 PM (last tickets sold at 3 PM).

Ski rentals are available at the Ramsey Barn. Trails are well marked, and distances vary to suit your skill and endurance.

Location: Old World Wisconsin is 1½ miles south of Eagle in Waukesha County, Wisconsin, off state trunk highway 67.

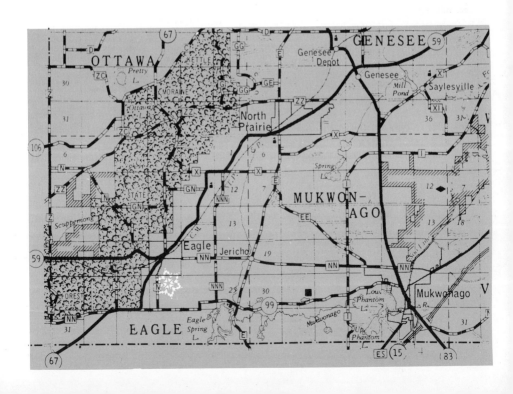

The Wisconsin Ice Age Trail

THE GREAT ICE AGE

The Great Ice Age began about a million years ago and reached its greatest expanse about 25,000 years ago when the last of a series of four major ice sheets bulged as far south as the Ohio river valley.

This was the Wisconsin glacier.

The massive ice sheets scraped and carved the earth, filling in lakes, rounding off hills, forcing rivers into new channels.

Potholes were gouged out, thousands of lakes, bogs, swamps created, particularly in Wisconsin and Minnesota.

Tremendous quantities of rocky debris were left as the sheet of ice melted, forming ridges, hills and mounds, known today as moraines, drumlins, kames and eskers.

The magnificent landscape that resulted is the heritage of the Great Ice Age. Nowhere is there more of it or a greater variety than in Wisconsin.

THE ICE AGE NATIONAL SCIENTIFIC RESERVE

Men who recognized the scientific and scenic values of this terrain urged that some be put into parks. A prime mover among them was Raymond T. Zillmer, Milwaukee, hiker and nature lover.

The first result was the state's Kettle Moraine Forest, established in 1926.

But Zillmer's dream was a national park. He worked tirelessly for it, enlisting the support of Wisconsin members of Congress and governors. To carry on the work, he incorporated the Ice Age Park and Trail Foundation in 1958, shortly before his death.

Two Congressional Acts transformed the dream into reality. One, signed by President Johnson in 1964, authorized a park. The other, signed by President Nixon in 197 provided funds.

A National Park Service study found feasible to include only scattered example of the vast glacial terrain, the accepted type of park was not practical.

So a new concept resulted: The Ice A National Scientific Reserve, a federal and state project. Nine important areas a included, totalling 32,500 acres. Much of the land already is public-owned; some still ha to be acquired.

THE ICE AGE TRAIL

The Reserve plan calls for a 600-mile tra along terminal moraines of the glacie linking the nine Reserve units. It is mappe on the other side.

In some places the trail must cross privat lands. And the trail-making will have to done by private citizens.

So the eventual existence of Wisconsin's I Age Trail, a national treasurer like th Appalachian Trail in the east, must depen on the co-operation of public-spirited citizens — landowners, sponsoring groups an trail makers.

A major trail segment already is in use the Northern Kettle Moraine Fores Another soon will be cut through th Southern Kettle Moraine Forest.

Citizen groups are preparing to route othe segments.

More such groups are needed to tackl remaining segments of the 600-mile trail.

Kettle Moraine State Forest
Southern Unit

Waukesha-Walworth County

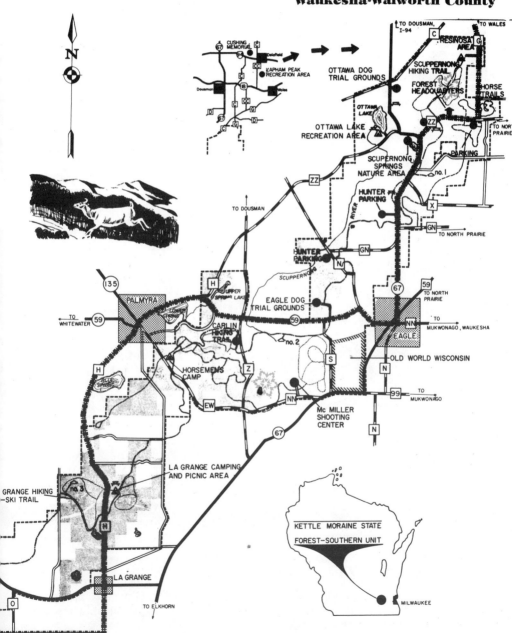

John Muir Trail

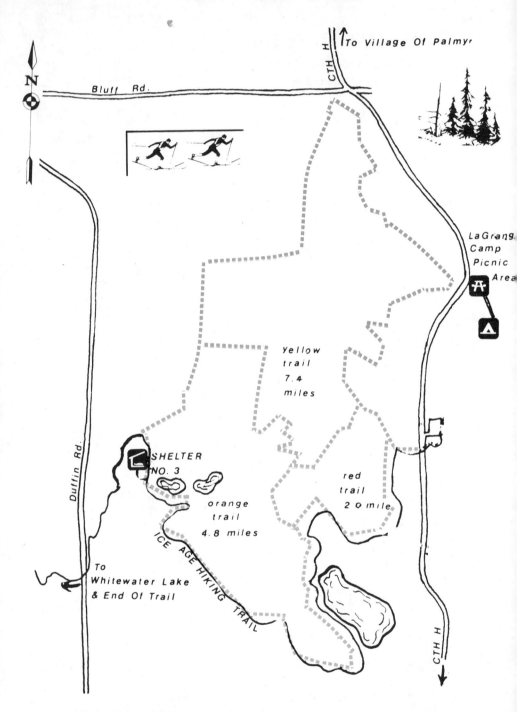

This trail is strictly for the skier who is interested in touring steep terrain. It is located in Walworth County on County H, west of the La Grange camping and picnic area. There are 3 loops and the longest is 11.8km (7.4 miles). These are ungroomed bushwacking trails.

NORDIC TRAIL

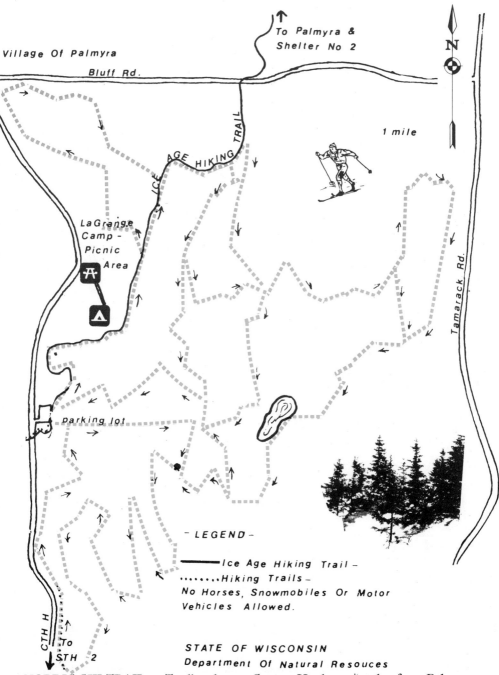

To Palmyra &
Shelter No 2

N

Village Of Palmyra

Bluff Rd.

1 mile

ICE AGE HIKING TRAIL

LaGrange
Camp -
Picnic
Area

Tamarack Rd.

parking lot

- LEGEND -

Ice Age Hiking Trail -

·······Hiking Trails -

No Horses, Snowmobiles Or Motor
Vehicles Allowed.

CTH H

To
STH 2

STATE OF WISCONSIN
Department Of Natural Resouces

NORDIC SKI TRAIL — Trailhead is on County H, about 4 miles from Palmyra. Located in Walworth County, this trail is the most popular of all the trails in the Southern Unit. The white portion is the only beginner trail in the forest. It is three miles in length and varies from gentle to gentle rolling. The red loop in the system is the shortest of the five loops, which encompasses 14.4km (9 miles) of total trail.

McMiller Ski Trail

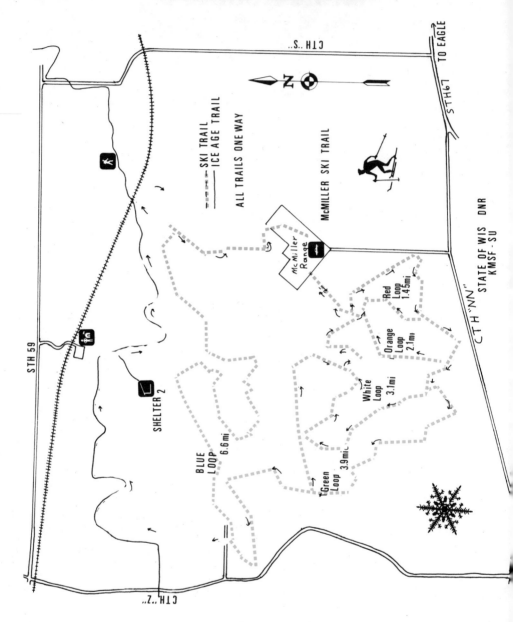

Located in Waukesha County, McMiller offers approximately 27.2km (17 miles) of groomed trails consisting of 5 loops. The Blue Loop is primarily a practice course for racers. It is located 4 miles southwest of Eagle on County Trunk NN.

Scuppernong Trail

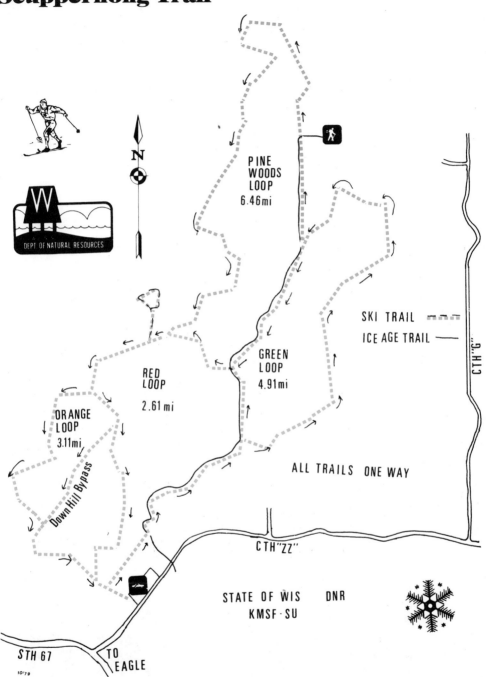

PINE WOODS LOOP 6.46mi

GREEN LOOP 4.91mi

RED LOOP 2.61 mi

ORANGE LOOP 3.11mi

Down Hill Bypass

SKI TRAIL

ICE AGE TRAIL

ALL TRAILS ONE WAY

CTH "G"

CTH "ZZ"

STATE OF WIS   DNR
KMSF·SU

STH 67   TO EAGLE

N

DEPT OF NATURAL RESOURCES

10/79

A very popular trail with 4 loops. Orange loop is 5.1km (3.2 miles), red loop is 4.6km (2.9 miles) and lime green loop is 8.3km (5.2 miles). Blue loop is 6.5 miles. There is an overlook in the northern part of the red trail. It has a variety of landscape for the intermediate to expert skier. Located in Waukesha County, the parking lot is on ZZ at Wayside.

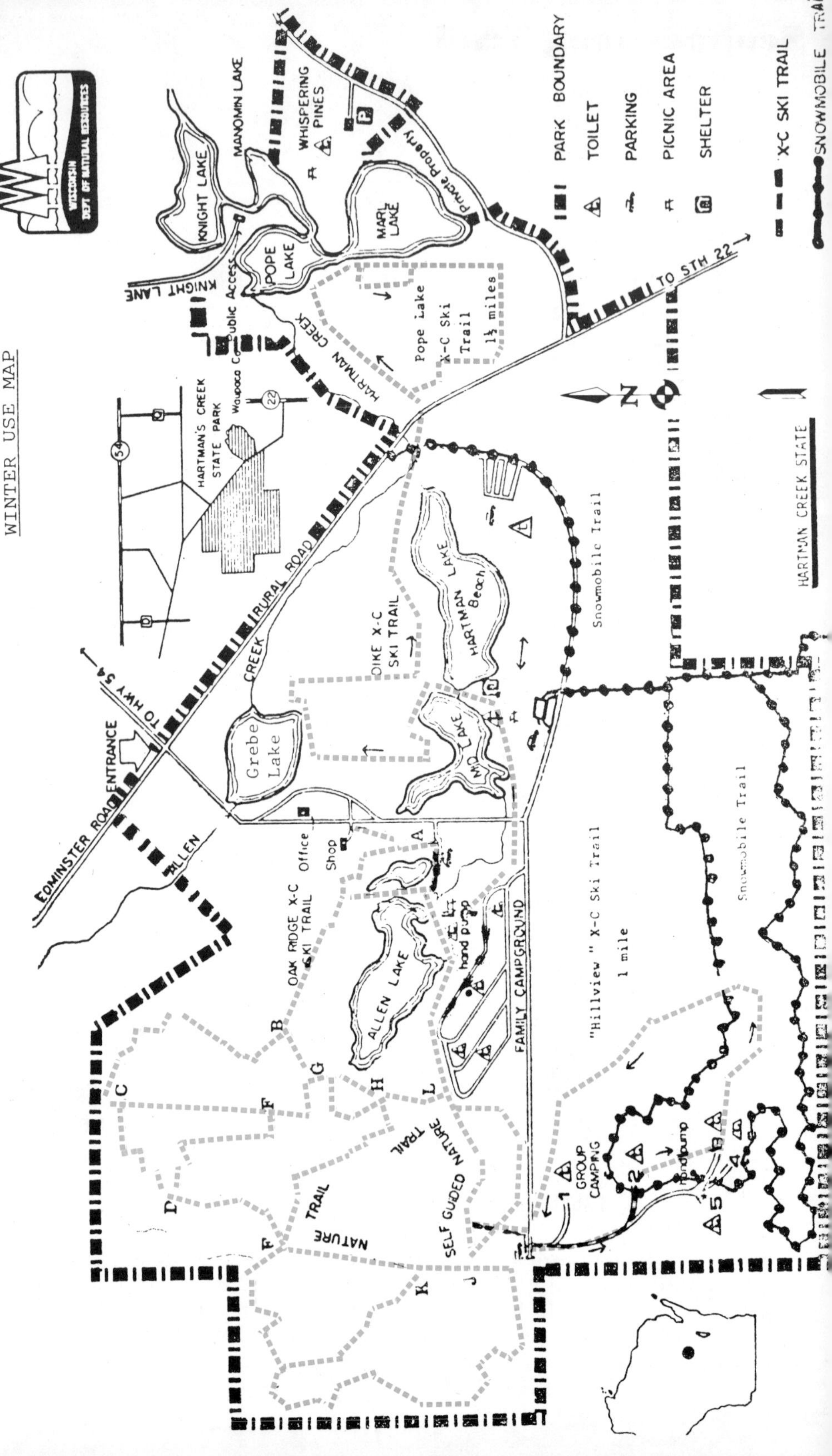

WINTER USE MAP

426

WISCONSIN DEPT OF NATURAL RESOURCES

| PARK BOUNDARY |
| TOILET |
| PARKING |
| PICNIC AREA |
| SHELTER |
| X-C SKI TRAIL |
| SNOWMOBILE TRAIL |

HARTMAN CREEK STATE

MANOMIN LAKE

WHISPERING PINES

KNIGHT LAKE

POPE LAKE

MAPLE LAKE

Private Property

KNIGHT LANE

Public Access

Waupaca County

HARTMAN'S CREEK STATE PARK

HARTMAN CREEK

Pope Lake X-C Ski Trail 1½ miles

TO STH 22

N

COMINSTER ROAD ENTRANCE

TO HWY 54

CREEK RURAL ROAD

Grebe Lake

ALLEN

DIKE X-C SKI TRAIL

HARTMAN LAKE

Hartman Beach

WHY LAKE

Office

Shop

OAK RIDGE X-C SKI TRAIL

ALLEN LAKE

hand pump

FAMILY CAMPGROUND

"Hillview" X-C Ski Trail 1 mile

Snowmobile Trail

Snowmobile Trail

NATURE TRAIL

SELF GUIDED NATURE TRAIL

GROUP CAMPING

hand pump

Hartman Creek State Park

Waupaca County

Hartman Creek is by far the best skiing area available in the Waupaca area. Very heavily used on weekends, these trails offer the beginner and intermediate skier a pleasant few hours of touring. The trails are approximately 15km (9.35 miles) in length. There are several exciting hills, but the remainder is very scenic with trails through a mature white pine plantation, natural hardwoods and open fields.

This 1200 acre park is located — five miles west of Waupaca on Highway 54.

Ample parking is available on a plowed lot with toilets nearby. Water is available as well as a telephone at the Park Office. A vehicle admission sticker is required, but no trail fees are charged.

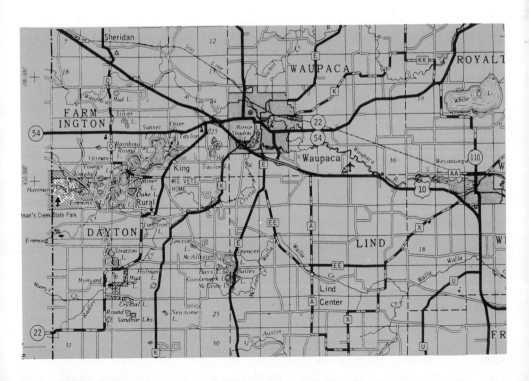

428

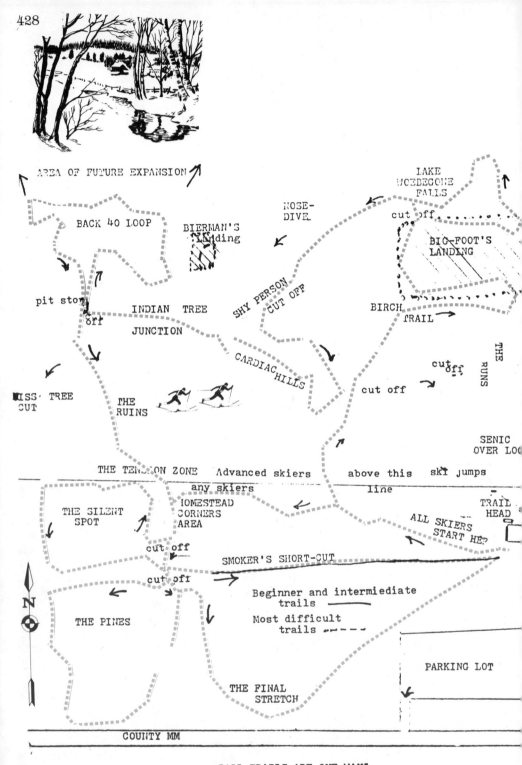

AREA OF FUTURE EXPANSION

LAKE WOEBEGONE FALLS

BACK 40 LOOP

BIERMAN'S Landing

NOSE-DIVE

cut off

BIG-FOOT'S LANDING

pit stop

cut off

INDIAN TREE JUNCTION

SHY PERSON CUT OFF

BIRCH TRAIL

THE RUNS

cut off

CARDIAC HILLS

cut off

MISS TREE CUT

THE RUINS

SENIC OVER LOOK

THE TENSION ZONE   Advanced skiers   above this   ski jumps
                    any skiers   line

THE SILENT SPOT

HOMESTEAD CORNERS AREA

TRAIL HEAD

ALL SKIERS START HERE

cut off

SMOKER'S SHORT-CUT

cut off

Beginner and intermiediate trails ———

Most difficult trails ------

N

THE PINES

PARKING LOT

THE FINAL STRETCH

COUNTY MM

ALL TRAILS ARE ONE WAY
OBEY ALL SIGNS AND MARKINGS

Iola Winter Sports Area

Waupaca County

The Iola Winter Sports Area, located five miles northwest of Iola on County MM, offers 12.5km of groomed and double tracked trails that are sure to please any skier from the novice to the training racer. The course is laid out over challenging glacial terrain and is all located in pine and hardwood forest land.

4km of novice trails prepare the skier for the 5.5km of intermediate trails that follow. For the most seasoned skiers there are 3km of thrilling cutoffs and challenging up hill climbs.

Situated on the trails are scenic areas and a shelter house. Overnight accommodations can be provided in our overflow warming house which can be rented by groups for that unique skiing weekend.

A plowed parking lot is located at the trailhead adjacent to the wood heated shelter house where one can find snacks and soda, clean indoor restrooms, ski rentals and PSIA certified lessons. For the 1984-85 season, the trail fees are $4.00 adult and $2.00 under 12. Family rates and season passes are available. The phone is 715-445-3411, the manager of the area is Roger Miller. Mail will reach him at P.O. Box 234, Iola, WI 54945.

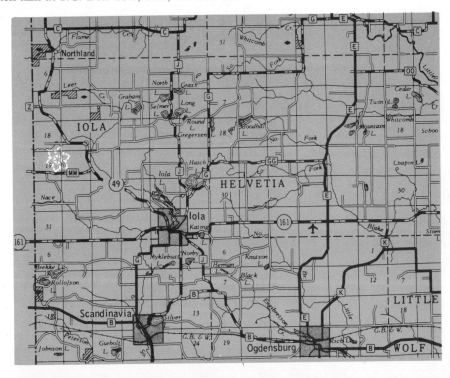

NORDIC MT. TRAIL

INTERMEDIATE WITH ADVANCED OPTIONS

7.2 KM.

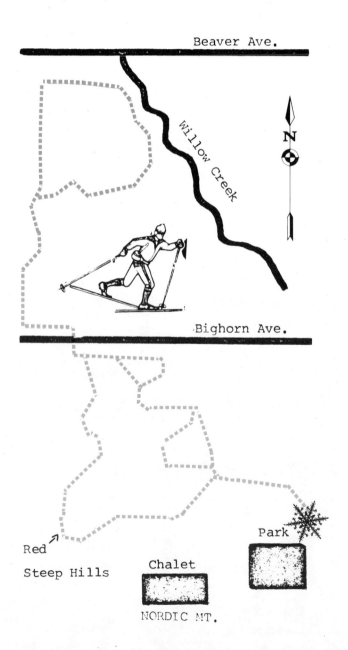

Beaver Ave.

Willow Creek

N

Bighorn Ave.

Red

Steep Hills

Park

Chalet

NORDIC MT.

Nordic Mountain Willow Creek Trails

These trails are approximately 7.2km (4½ miles) long and begin at the Nordic Mountain Ski Area on County Trunk W in Mount Morris. It follows Willow Creek making a loop and returns to Nordic Mountain. There is a connecting trail to the Kusel Lake Trails.

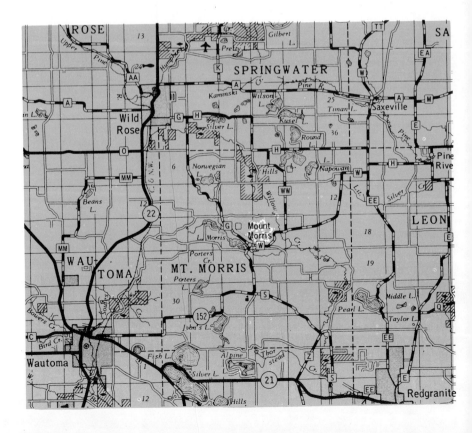

432

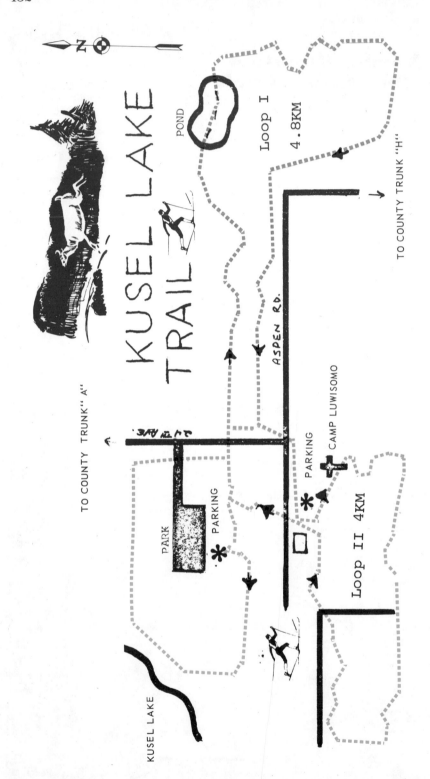

KUSEL LAKE TRAIL

POND

Loop I 4.8KM

TO COUNTY TRUNK "A"

TO COUNTY TRUNK "H"

ASPEN RD.

PARKING

PARKING

PARK

CAMP LUWISOMO

Loop II 4KM

KUSEL LAKE

Kusel Lake Trial

Waushara County

Trails can be reached by taking County A out of Wild Rose east to 24th Ave. to Kusel Lake Park. Parking is available at the park or at Camp Luwisomo on Aspen Rd. Trails are about 11.3km (7 miles). All are one-way with blue and white trail markers. Trails are intermediate with several good downhill runs. Toilet facilities are available at the park and at Camp Luwisomo.

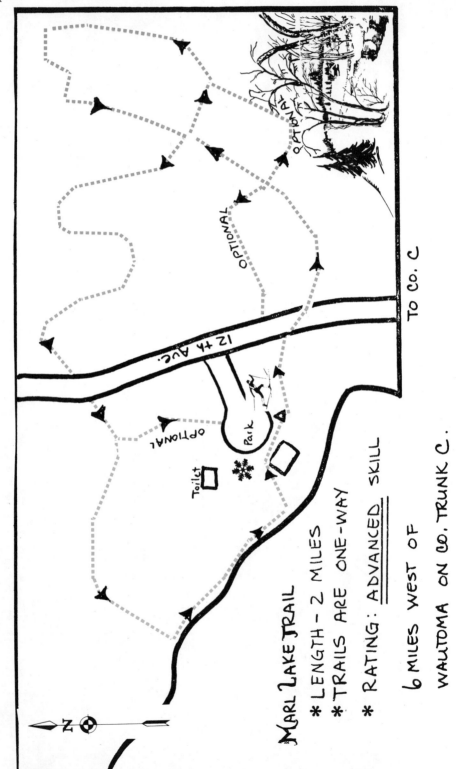

MARL LAKE TRAIL

* LENGTH – 2 MILES
* TRAILS ARE ONE-WAY
* RATING: ADVANCED SKILL

6 MILES WEST OF
WAUTOMA ON CO. TRUNK C.

Marl Lake County Park Trail

This trail is a very short trail but it is very steep in some places and is rated for the advanced or expert skier. The trail is a one-way trail with several options or cutoffs. Total length is 3.2km (2 miles).

The trailhead can be located by going 6 miles west of Wautoma on County C to 12th Ave.

436

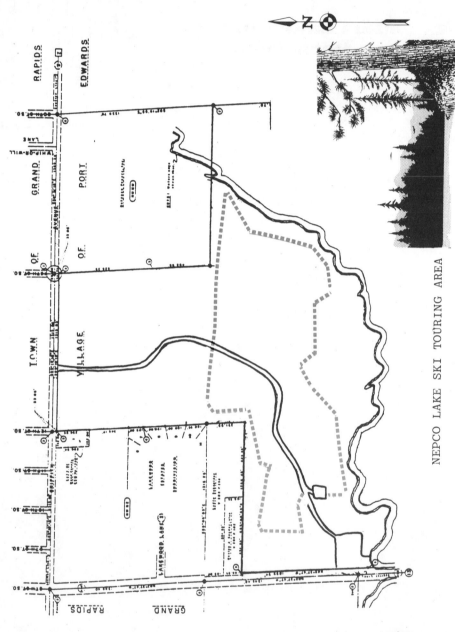

NEPCO LAKE SKI TOURING AREA

Nepco Lake Ski Touring Trail

Wood County

The Nepco Lake Trail was laid out and brushed in the summer and fall of 1980.

Although a short trail (2.7km (1.6 miles) the location of the trail is ideal, being just outside the city limits of Wisconsin Rapids.

The terrain is gently rolling. Parking is available within the Nepco County Park.

This area has a great variety of wildlife, along with scenic views of Nepco Lake.

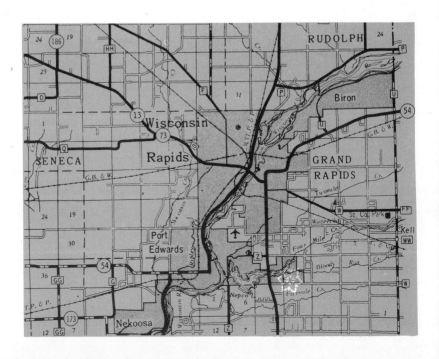

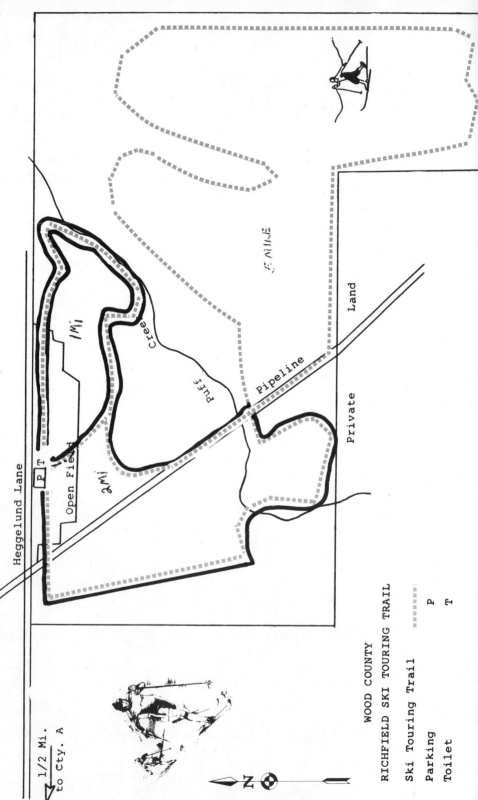

RICHFIELD SKI TOURING TRAIL

1/2 Mi.
to Cty. A

Heggelund Lane

Open Field

P   T

1 Mi.

2 Mi.

Pull

Creek

Pipeline

3 Mi.

Private   Land

N

WOOD COUNTY

RICHFIELD SKI TOURING TRAIL

Ski Touring Trail

Parking   P

Toilet   T

Richfield Ski Touring Trail

Wood County

The winter trail includes a ski touring trail, along with a parking lot.

The ski touring trail is approximately 8km (5 miles) in length. It is routed across open fields, through forest, and along and across Puff Creek to provide additional scenic value.

The trail was constructed in the fall of 1975, through the fall of 1976, with brushing, bulldozing and leveling of the trail now completed. The crossing on Puff Creek has also been installed. This trail will be marked and groomed during the winter season.

Location — 6 miles north of Pittsville on County Trunk Hwy. A, then ½ mile east on town road.

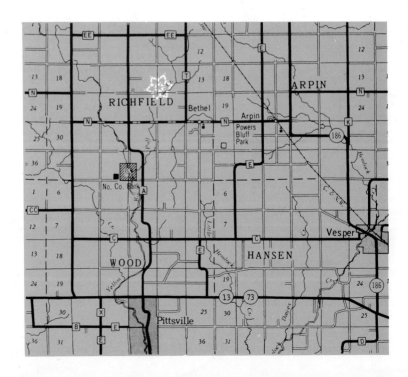

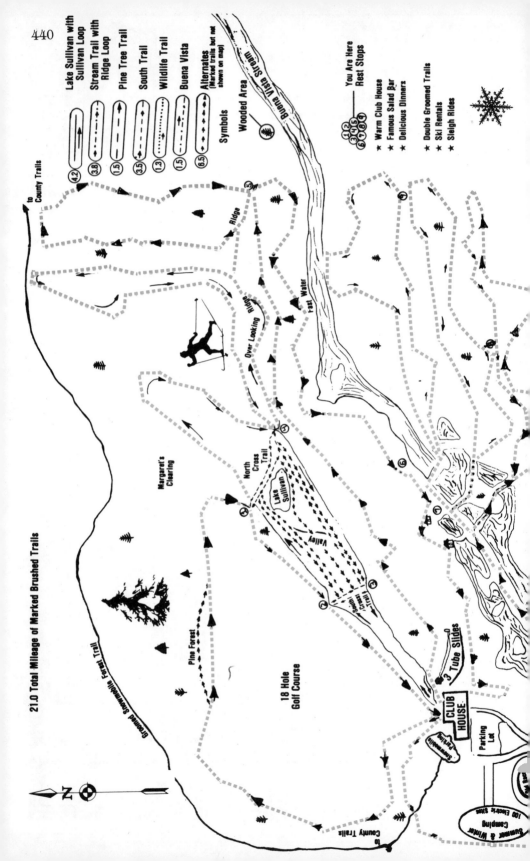

440

The Ridges

Wood County

'The Ridges offer 33.8km (21 miles) of groomed trails over hills, past a winding river, and through a pine forest. These trails were laid out with the right mixture of uphill, downhill, and straight going to give you a great day in the outdoors. Beginners have an easy-going 1½ mile trail with gentle slopes, a 3½ mile wooded and open trail for intermediate and a 6 mile trail with a choice of steep or gentle hills for the more advanced skiers. Each trail returns to clubhouse.

Trail fee is charged and ski equipment rentals are available at the pro shop. Refreshments, food, warming and toilets also available.

Located 1 mile south of Wis. Rapids on Hwy. 13, then ¾ mile east on Griffith Ave.

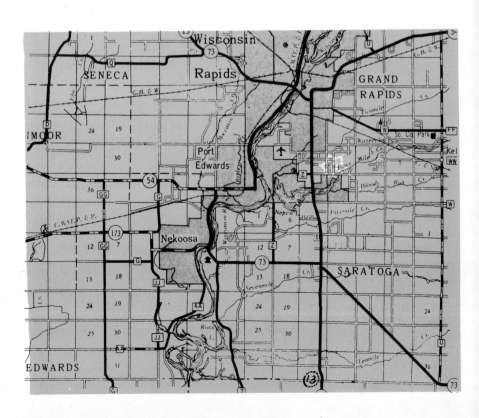

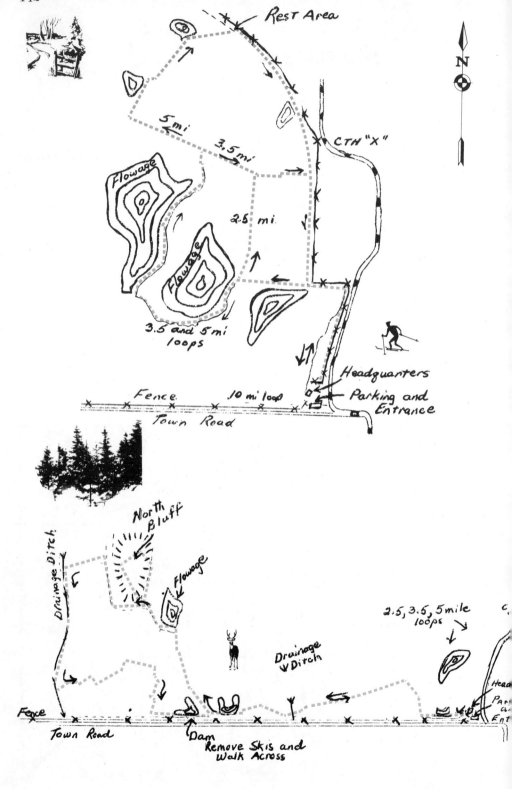

Rest Area

5 mi

3.5 mi

CTH "X"

N

Flowage

2.5 mi.

Flowage

3.5 and 5 mi loops

Headquarters

Parking and Entrance

Fence

10 mi loop

Town Road

North Bluff

Drainage Ditch

Flowage

2.5, 3.6, 5 mile loops

Drainage Ditch

Head

Par

En

Fence

Town Road

Dam
Remove Skis and Walk Across

Sandhill Ski Trail

Wood County

This area is a demonstration and research area for wildlife management run by the State of Wisconsin. Cross-country skiing is encouraged to demonstrate to the public the important work being done on such species as deer, waterfowl, and ruffed grouse. Sandhill covers over 9100 acres of forest and marshland with such timber species as aspen, white birch, red and black oak, and jack pine. The terrain is level to gently rolling except for the North Bluff in the southwest part of the area. There are four loops with lengths of 3.8km (2.5 miles), 5.2km (3.5 miles), 8km (5 miles), and 16km (10 miles). The 10 mile trail encompasses a portion of the 300 foot North Bluff.

Trails are well marked and each passes at least one man-made flowage. No snowmobiles are allowed in Sandhill. Please register before entering area.

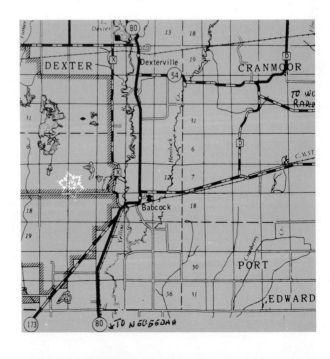

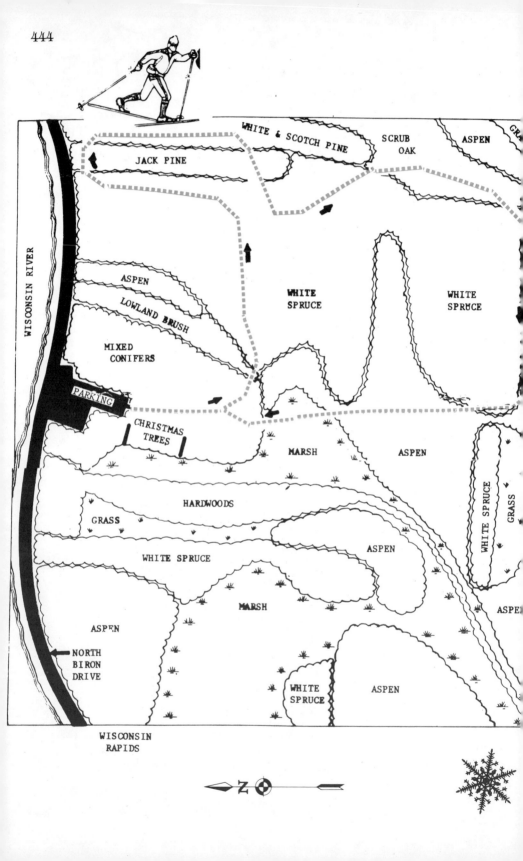

Biron Cross-Country Ski Trail

Wood County

The Biron cross-country ski trail is just east of Biron. The trail is used as an educational forest tour, and winds through 1.5 miles of picturesque woodland adjacent to the Biron Flowage of the Wisconsin River. This is the site of Consolidated's first tree nursery established in the early 1930's.

The Biron Trail offers the skier a tour through a variety of forest types and tree species. Numbered posts along the trail correspond to information in the Forest Tour II booklet, available at the trail entrance. Though some of these features may be hidden by snow, we encourage you to use the booklet to pinpoint your location on the trail and to increase your knowledge of forestry.

The trail is generally level and courses through conifer plantations, aspen regeneration areas and a variety of forest types common to Wisconsin. You may encounter deer, snowshoe hares, ruffled grouse, snowbirds and other inhabitants of the forest who remain active in winter.

We're sure you'll enjoy the Biron Trail, and we hope that your knowledge and appreciation of our natural resources is enhanced by your visit here. Good tour!

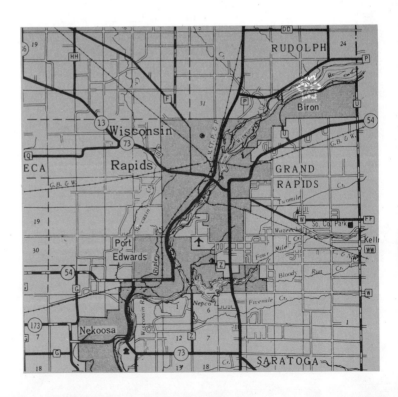

MICHIGAN'S UPPER PENINSULA

THE CLEAN AIR, CLEAR WATER COUNTRY

**USE, DON'T MISUSE
OUR ENVIRONMENT**

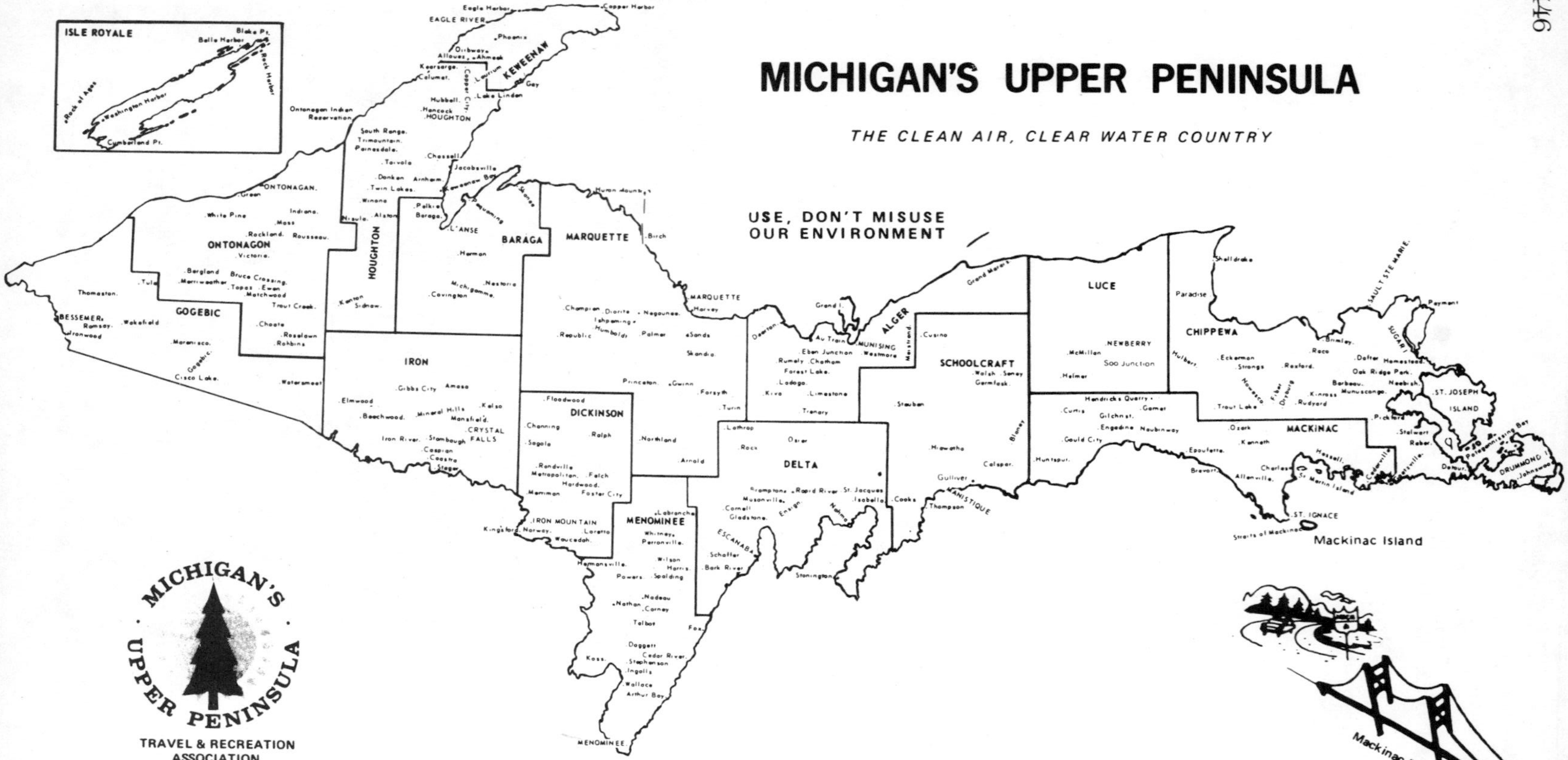

Mackinac Bridge

Mackinac Island

TRAVEL & RECREATION
ASSOCIATION
P.O. Box 400
Iron Mountain, Michigan 49801
PHONE 906/774-5480

MICHIGAN'S UPPER PENINSULA

Bruno's Run

Alger County

Nordic skiers can experience some of the wild character of Michigan's Upper Peninsula on this "Trail of the North" as they glide through white pine, hemlock and hardwood forests. Under the towering trees, winter white wraps all in solitude. Glistening lakes, swift-moving streams and glacial formations add interest to the landscape of ice-capped rocks and frozen shoreline. All around are reasons and opportunities to pause and reflect on the area's natural wonders.

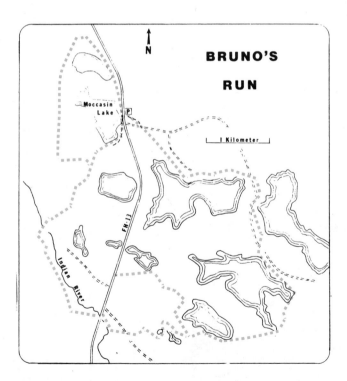

This is an ungroomed trail.

Grand Marais Trail

Trap Trail is 3 miles long and is the most difficult of the two trails. The difficulty is the short, relatively steep and narrow sections. There are only a few of these and can be avoided by skiing in a clockwise direction.

The Lost Loop Trail begins near the Grand Sable Center and takes off the Trap Trail. It is a short loop of 1½ miles.

GRAND MARAIS SKI/SNOWSHOE TRAILS

Pictured Rocks National Lakeshore

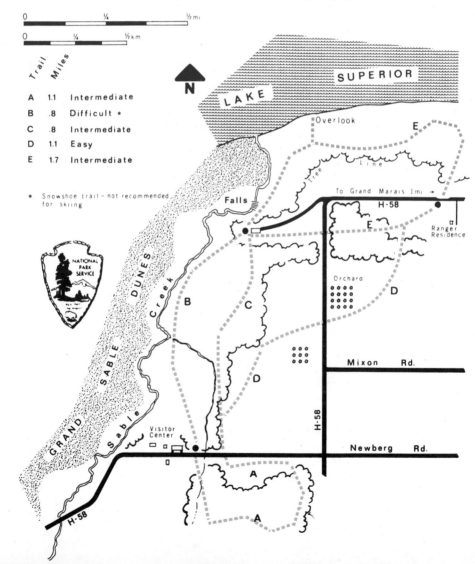

Munising Trail

Alger County

Skiers don't have to range far from Munising to take in winter's glory in the area's rolling landscape. Follow this trail through the area north of Munising to view scenic hardwood-covered slopes and remnants of the past at old abandoned farmsites. Whatever your skill, this trail offers loops to captivate your interest as a Nordic skier; beginners will find the trail's easiest loop to their liking. The groomed trail is kept in prime condition throughout the cross-country ski season.

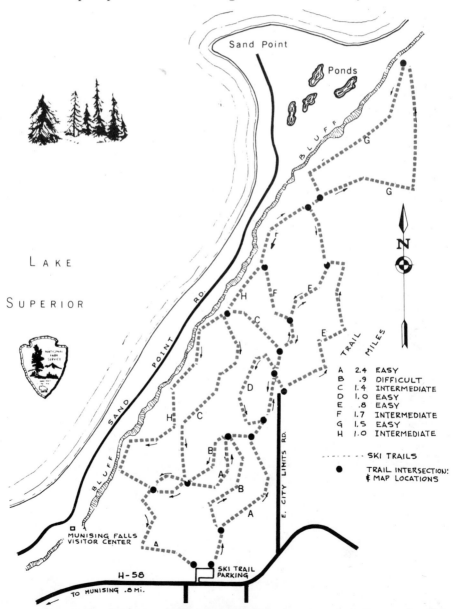

TRAIL MILES

A	2.4	EASY
B	.9	DIFFICULT
C	1.4	INTERMEDIATE
D	1.0	EASY
E	.8	EASY
F	1.7	INTERMEDIATE
G	1.5	EASY
H	1.0	INTERMEDIATE

- - - - - - SKI TRAILS

● TRAIL INTERSECTION & MAP LOCATIONS

Valley Spur Trail

As the last glacier receded northward, it paused over what is now Munising, leaving deposits that created the area's hills. Glacial streams cut ravines through the hills, and, over the years, a lush stand of hardwoods grew on this glacial end moraine.

Centuries of change have created the ideal setting for the Valley Spur ski trail, which has been constructed to best utilize the area's variety in elevations and picturesque beauty.

Easier sections of the trail travel through gentler terrain, while the more difficult segments take advantage of the hills and twisting ravines to provide adventure in Nordic skiing. Combined, they offer challenges to skiers of all abilities. Trail designers capitalized on Nature's "engineering" of the area's unique physical features to provide something for most every skier. Accept the challenges of this quality groomed cross-country ski area.

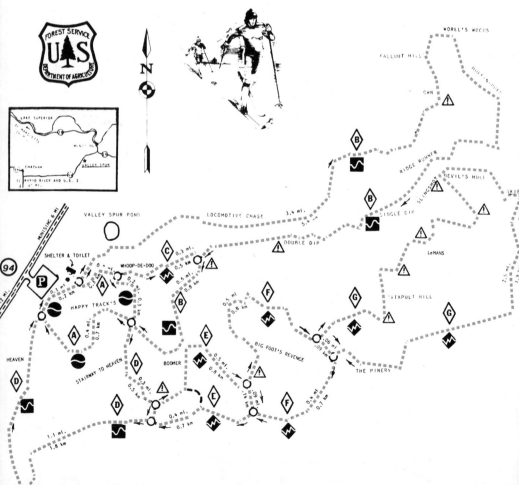

Natural Area Pathway

Chippewa County

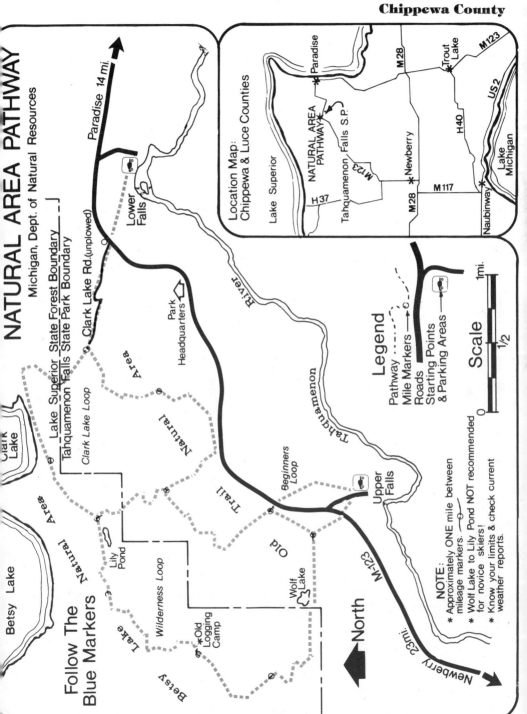

NATURAL AREA PATHWAY

Michigan, Dept. of Natural Resources

Paradise 14 mi.

Lower Falls

Clark Lake Rd.(unplowed)

Lake Superior State Forest Boundary
Tahquamenon Falls State Park Boundary

Park Headquarters

River

Clark Lake Loop

Natural

Clark Lake

Area

Betsy Lake

Follow The
Blue Markers

Natural Area

Betsy Lake

Trail

Old

Wilderness Loop

Lily Pond

Old Logging Camp

Beginners Loop

Wolf Lake

Tahquamenon

Upper Falls

North

M-123

Newberry 23 mi.

Legend

Pathway
Mile Markers
Roads
Starting Points
& Parking Areas

Scale

0   1/2   1 mi.

Location Map:
Chippewa & Luce Counties

Lake Superior

Paradise

M 28

Trout Lake

M123

NATURAL AREA PATHWAY

Tahquamenon Falls S.P.

M123

Newberry

H 37

M 28

H40

US 2

M 117

Naubinway

Lake Michigan

NOTE::
* Approximately ONE mile between mileage markers.
* Wolf Lake to Lily Pond NOT recommended for novice skiers!
* Know your limits & check current weather reports.

452

Paradise Pathway

Luce County

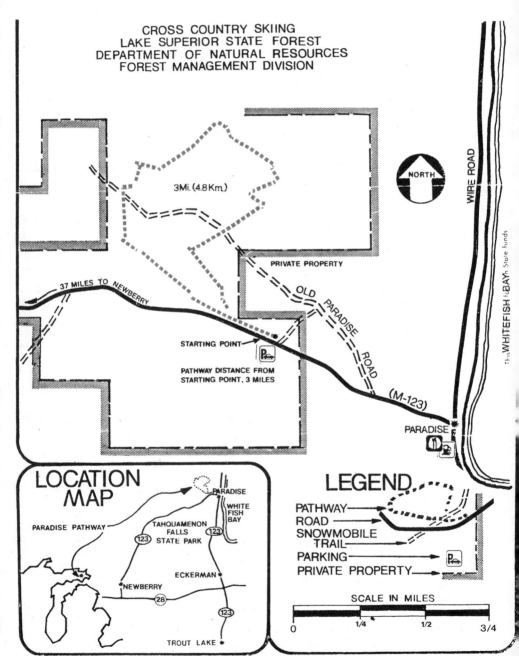

CROSS COUNTRY SKIING
LAKE SUPERIOR STATE FOREST
DEPARTMENT OF NATURAL RESOURCES
FOREST MANAGEMENT DIVISION

3 Mi. (4.8 Km.)

NORTH

WIRE ROAD

This WHITEFISH BAY th State funds

PRIVATE PROPERTY

OLD PARADISE ROAD

37 MILES TO NEWBERRY

STARTING POINT

PATHWAY DISTANCE FROM
STARTING POINT, 3 MILES

(M-123)

PARADISE

LOCATION MAP

PARADISE PATHWAY

PARADISE

WHITE FISH BAY

TAHQUAMENON FALLS STATE PARK

123

123

ECKERMAN

NEWBERRY

28

123

TROUT LAKE

LEGEND

PATHWAY

ROAD

SNOWMOBILE TRAIL

PARKING

PRIVATE PROPERTY

SCALE IN MILES

0   1/4   1/2   3/4

Pine Bowl Pathway

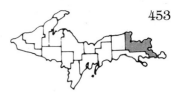

Chippewa County

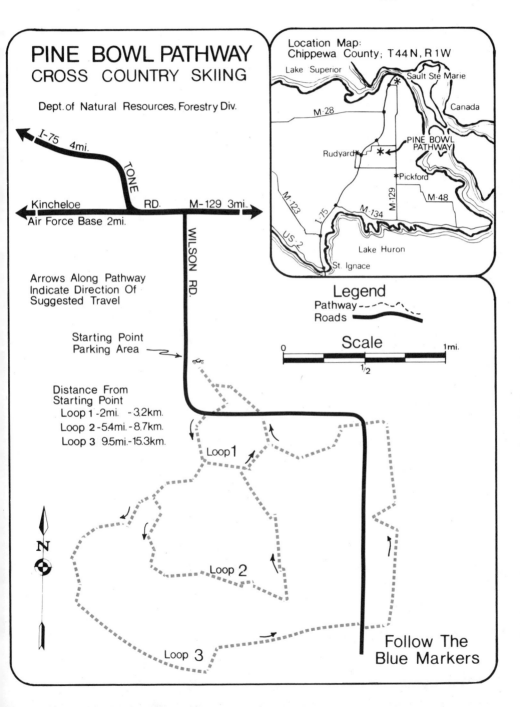

PINE BOWL PATHWAY
CROSS COUNTRY SKIING

Dept. of Natural Resources, Forestry Div.

I-75 4mi.

TONE

Kincheloe   RD.   M-129 3mi.
Air Force Base 2mi.

WILSON RD.

Arrows Along Pathway
Indicate Direction Of
Suggested Travel

Starting Point
Parking Area

Distance From
Starting Point
Loop 1 - 2mi. - 3.2km.
Loop 2 - 5.4mi. - 8.7km.
Loop 3 9.5mi. - 15.3km.

Loop 1

Loop 2

N

Loop 3

Location Map:
Chippewa County; T44N, R1W

Lake Superior

Sault Ste Marie

M-28

Canada

Rudyard

PINE BOWL
PATHWAY

Pickford

M-123

M-129

M-48

I-75

M-134

US-2

Lake Huron

St. Ignace

Legend
Pathway
Roads

Scale
0   1mi.
½

Follow The
Blue Markers

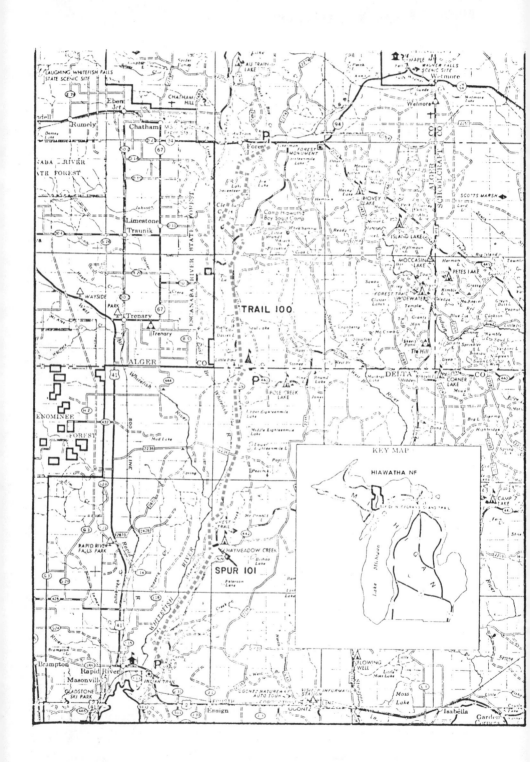

Bay de-Noc Grand Island Trail

Alger & Delta Counties

The trail closely approximates the old foot trail used by Indians and trappers to travel between Lakes Superior and Michigan. The trail parallels the Whitefish River for most of its length. Several vistas allow scenic views of the river valley and surrounding area. Trail is easy to moderate over many flat areas and occasional rolling hills, and is not presently groomed.

Trail beginning — 3 miles east of Rapid River (town of).

Trail ending — ten miles southwest of Munising at Ackerman Lake Public Access along M-94.

Access #1: Two miles east of Rapid River on U.S. 2. Turn left onto Cty. Rd. 509 and travel 1.5 miles north. Parking lot on west side of road.

Access #2: Two miles east of Rapid River on U.S. 2. Turn left onto Cty. Rd. 509 and travel 16 miles north. Parking lot on east side of road.

Access #3: Ten miles southwest of Munising on M-94. Parking lot on north side of road opposite Ackerman Lake.

Days River Pathway

Delta County

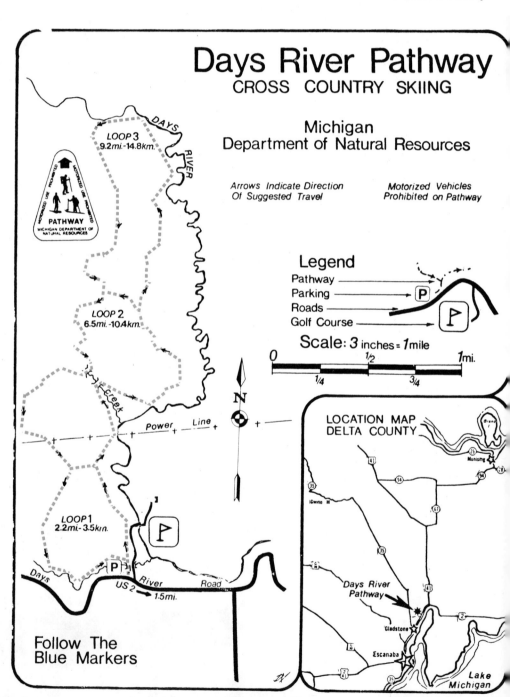

Rapid River

Delta County

The trail crosses an area of high sand ridges and wetlands. You glide by russet trunks of red pines on the ridges and rush downhill into the dark green groves of cedar, spruce and fir in the low areas. The trail is groomed but there will be uneven spots, twigs and branches falling from the trees, so be aware.

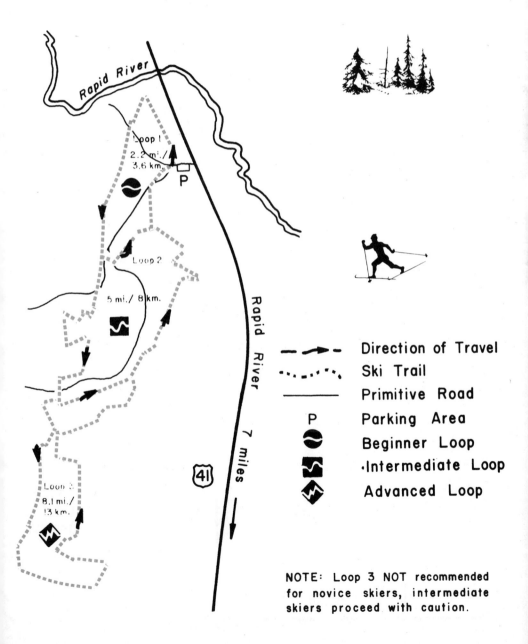

Rapid River

Loop 1
2.2 mi./
3.6 km.

P

Loop 2

5 mi./ 8 km.

Rapid River

7 miles

41

Loop 3
8.1 mi./
13 km.

Direction of Travel

•••• Ski Trail

——— Primitive Road

P Parking Area

Beginner Loop

Intermediate Loop

Advanced Loop

NOTE: Loop 3 NOT recommended for novice skiers, intermediate skiers proceed with caution.

Merriman East Pathway

Dickinson County

HIKING & CROSS COUNTRY SKIING
COPPER COUNTRY STATE FOREST- DEPARTMENT OF NATURAL RESOURCES
FOREST MANAGEMENT DIVISION

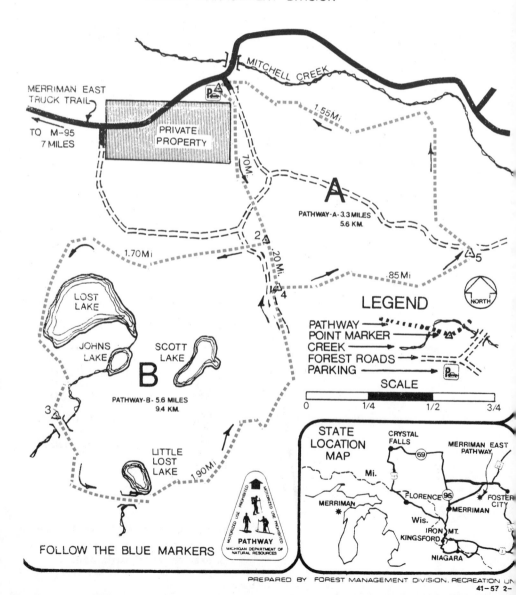

MITCHELL CREEK

MERRIMAN EAST
TRUCK TRAIL

TO M-95
7 MILES

PRIVATE
PROPERTY

1.55 Mi

.70 Mi.

A

PATHWAY-A-3.3 MILES
5.6 KM.

1.70 Mi

.85 Mi

2 20 Mi.

4

5

LOST
LAKE

JOHNS
LAKE

SCOTT
LAKE

B

PATHWAY-B- 5.6 MILES
9.4 KM.

3

LITTLE
LOST
LAKE

1.90 Mi

LEGEND

NORTH

PATHWAY
POINT MARKER
CREEK
FOREST ROADS
PARKING

SCALE

0   1/4   1/2   3/4

MOTORIZED USE PROHIBITED
PATHWAY
MICHIGAN DEPARTMENT OF
NATURAL RESOURCES

FOLLOW THE BLUE MARKERS

STATE
LOCATION
MAP

CRYSTAL
FALLS

69

Mi.

MERRIMAN EAST
PATHWAY

MERRIMAN

FLORENCE 95

FOSTER
CITY

MERRIMAN

Wis.

IRON MT.
KINGSFORD

NIAGARA

PREPARED BY FOREST MANAGEMENT DIVISION, RECREATION UN
41-57 2-

Johnson's Ski Trails

WAKEFIELD, MICHIGAN

Gogebic County

Tamarack Flats	3.2 KM	
Loop Trail	1.6 KM	
Meander Trail	4.0 KM	
Deer Yard Trail	4.5 KM	
Tom's Trail	1.6 KM	
Cliff's Ridge Trail	3.2 KM	
Triple Threat Trail	2.4 KM	

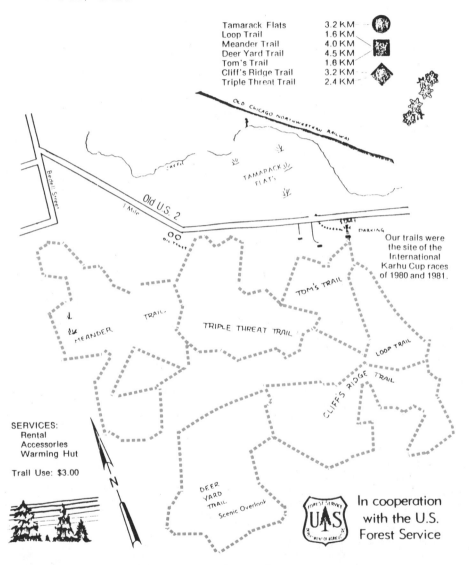

Our trails were the site of the International Karhu Cup races of 1980 and 1981.

SERVICES:
Rental
Accessories
Warming Hut

Trail Use: $3.00

In cooperation with the U.S. Forest Service

Located east of Wakefield on old U.S. 2 these trails were the site of the International Karhu Cup Races in 1980 — and 1981. 7 trails for a total of 20.5km for beginners — intermediate and expert skiers. Ski Rentals, Accessories and Warming Hut. Trail use fee is charged.

460

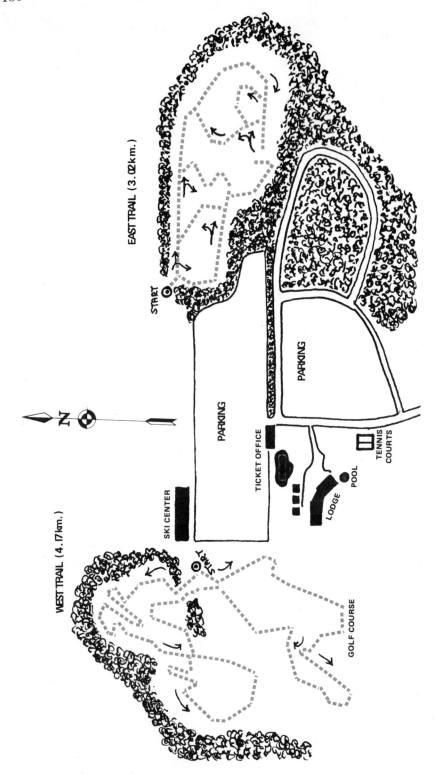

Indianhead Mountain Resort

Gogebic County

Meandering across the top of a 600-foot mountain, the Indianhead Mtn. Prima Vista Ski Touring Trail System has added a new dimension to winter recreation at the popular Upper Michigan resort.

The trails have been improved and expanded to more than seven kilometers, with gentle cruising over Indianhead's golf course, more challenging skiing through rolling terrain or demanding skiing through the woods on the advanced loop.

Rental equipment and instruction is available at the ski center. Complete lodging, dining, bar and rest facilities are found at either the ski center or the lodge. Indianhead has long been known for its downhill skiing so the expanded ski touring trail system makes the recreation picture complete. The trails were designed to not only provide a quality skiing experience but to take advantge of the view of the Black River Valley and Lake Superior to the north.

In the immediate area there is a total of 100km (62.1 miles) of trails.

Cross country skiers are asked to stop at the ticket office or ski school desk in the ski center for information.

Indianhead is located two miles west of Wakefield, Mich., via U.S. 2, to Indianhead Rd.

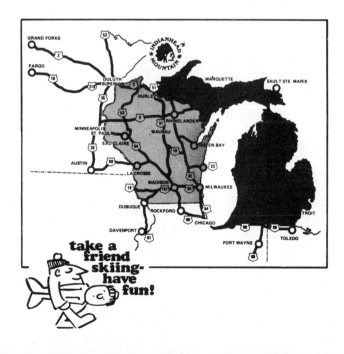

462

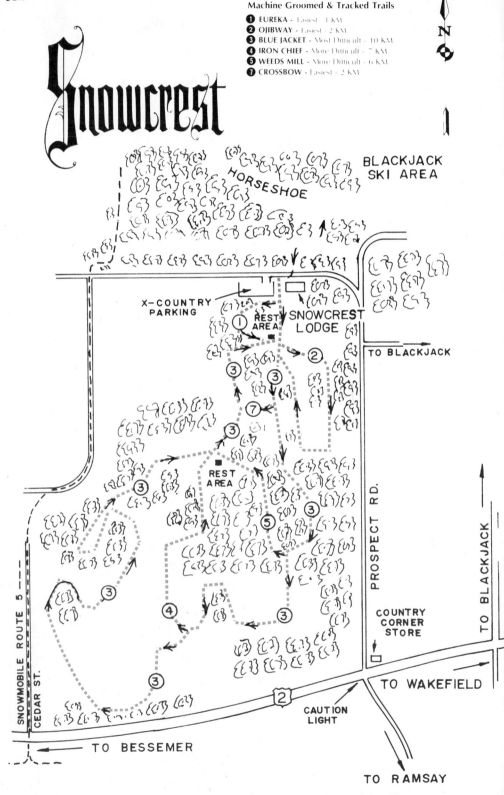

Snowcrest

Machine Groomed & Tracked Trails
1 EUREKA - Easiest - 1 KM
2 OJIBWAY - Easiest - 2 KM
3 BLUE JACKET - Most Difficult - 10 KM
4 IRON CHIEF - More Difficult - 7 KM
5 WEEDS MILL - More Difficult - 6 KM
7 CROSSBOW - Easiest - 2 KM

N

HORSESHOE

BLACKJACK
SKI AREA

X-COUNTRY
PARKING

REST
AREA

SNOWCREST
LODGE

TO BLACKJACK

REST
AREA

PROSPECT RD.

TO BLACKJACK

COUNTRY
CORNER
STORE

SNOWMOBILE ROUTE 5

CEDAR ST.

CAUTION
LIGHT

TO WAKEFIELD

TO BESSEMER

TO RAMSAY

Snowcrest (Blackjack)
X-C Ski Trails

Snowcrest ski trails consist of 29 kilometers of machine groomed and tracked trails. The trail system has five different loops with varying difficulty to meet the need of the skiers ability.

Rental equipment and x-c accessories are available at Snowcrest lodge with rest facilities and lodging on the trail site. Additional lodging, dining, bar and rest facilities are located at near-by Blackjack Mountain ski resort.

Snowcrest is located on Prospect Road, Bessemer, Michigan. Heading east on U.S. 2, turn left at the Country Corner Store.

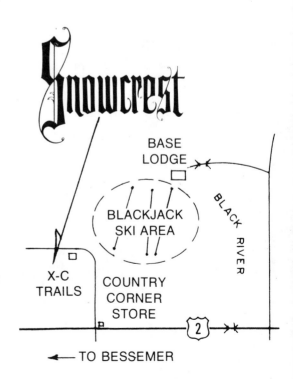

464

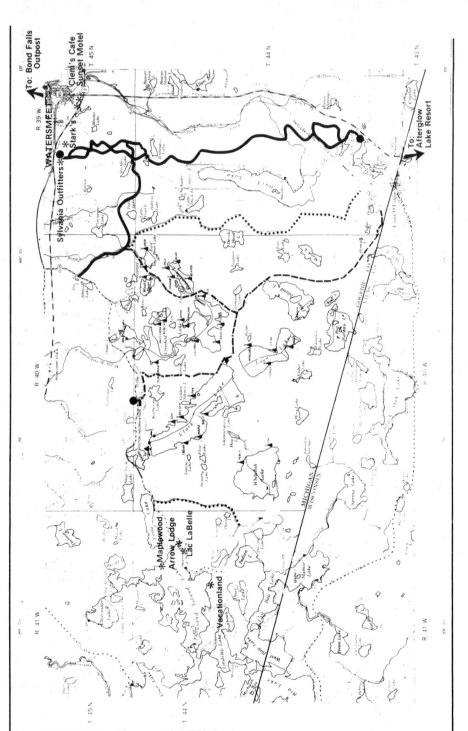

To: Bond Falls Outpost

Cieri's Cafe
Sunset Motel

Stark's

WATERSMEET

Sylvania Outfitters

To: Afterglow Lake Resort

Maplewood
Arrow Lodge
Lac LaBelle

Vacationland

Groomed Trails —————   Skied Trails — — —   Unplowed Forest Roads ············   Access ●   Sylvania Recreation Area

Sylvania

Gogebic County

The Sylvania Recreation area is a combination of lakes and old growth forests spread out over 21,000 acres. It is part of the Ottawa National Forest and the boundaries are shown on the map by the _____ lines. The forests consisting of maple, birch, hemlock and scattered pine, spruce and fir are still largely virgin timber. Wildlife is abundant.

The short dash & dot trails are groomed trails that originate at Sylvania Outfitters, the ski touring headquarters, which is located 1 mile west of the intersection of Hwy. 2 & 45 at Watersmeet. 20 miles of trail are groomed in cooperation with the Ottawa National Forest. These trails connect with an additional 20 miles of ungroomed trails in the Sylvania Wilderness area. These trails are shown as a long dash & dot.

Snow season is from December to March with snowfall measuring at least 100 inches. The Sylvania Outfitters handle equipment and supplies for the skier. Fundamental lessons are offered. Ski rentals are available. Trails can be transversed by the beginner but most of the trail development had the experienced skier in mind.

Trails connect or lead to lodging at PineAire Lodge or Sunset Motel.

KEY MAP

Wolverine Ski Trail

Gogebic County

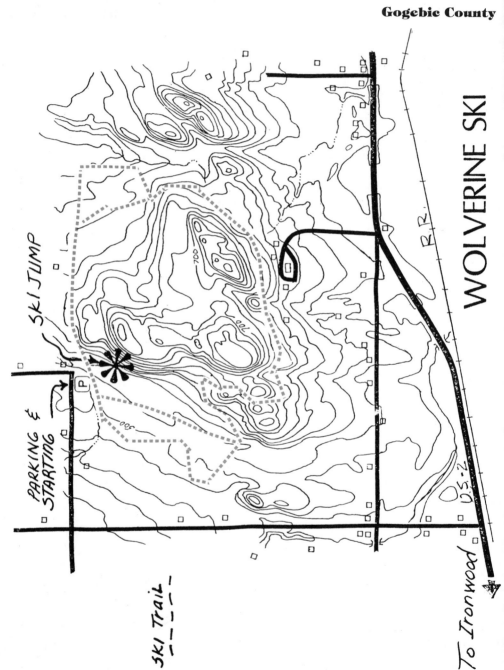

WOLVERINE SKI

SKI JUMP

PARKING & STARTING

P

SKI Trail - - -

To Ironwood

U.S. 2

R.R.

Old Grade Ski Trail

Located 7 miles west of Nisula on M38 to Forest Trail #203. Main trail is 3.9 miles with a cut off spur of 1.8 miles. Trail is around Courtney Lake and Six Mile Lake.

This trail is excellent for beginners but can be enjoyed by all skiers. The trailhead parking is at Lakeview Cafe. Part of trail is alongside a snowmobile trail so stay in designated lane.

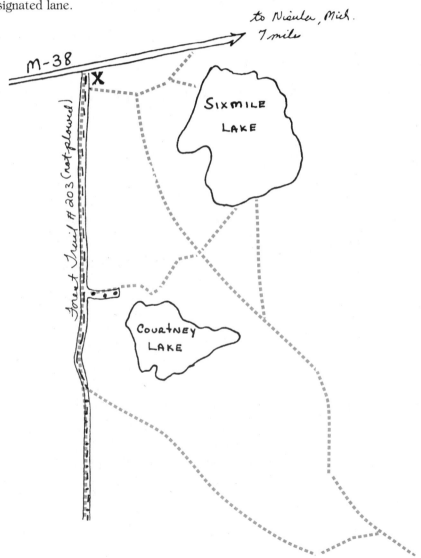

SAFE SKIING TIPS
from the
SUPERIOR NORDIC SKI PATROL

1. *Whenever you venture out into cold weather, it is important to do everything possible to protect the heat producing capacity of your body. This capacity depends on:*
    - good physical conditioning;
    - an ample supply of warm food, including plenty of carbohydrates to enhance heat production;
    - adequate intake of fluids to make up for perspiration and respiration of dry air;
    - ample rest and avoidance of general fatigue.

2. *Take heed of "HYPOTHERMIA WEATHER".*
    - Stay dry.
    - Beware of the wind.
    - Understand the cold. Ask yourself, "How cold is the moisture against my body?"
    - Use your clothes and the layering system *before* you get wet and *before* you start shivering.

3. *Watch for FROSTNIP and FROSTBITE on fellow skiers.*
    - FROSTNIP: The skin will appear yellowish-white and waxy, and is doughy to the touch.
    - FROSTBITE: The skin becomes hard, with the telltale sign of a solid, woody feeling.
    - In case of frostnip or frostbite, the person should be protected from further exposure until proper thawing is complete.

4. *Select ski trails that match skiing ability, and discourage beginners from skiing on advanced trails.*

5. *If skiing alone, inform someone of the exact skiing location and estimated time of return.*

6. *Fill in "SITZMARKS" with snow after a fall on the trail.*

7. *Observe all signs of safety and direction.*

Certified by the AMERICAN RED CROSS
Marquette County Chapter
213 N. Front St., Marquette 228-3659

Maasto Hiihto
Cross Country Ski Trail

Houghton County

Maasto Hiihto

Cross Country Ski Trail

Hancock, Michigan

Yooper Looper 1.5 km

Australia Loop 4.5 km

Out Trail

Gorge Loop 6.5 km

Quincy Loop 2.5 km

Railroad Ravine

Anaerobic Gulch

Swedetown Creek

Sisu Hill

2 way trail

Skiers Only One Way Trail

Parking and Access Points

Steep Hill

Trail cut off

M203

Houghton County Arena

Ingot St.

Poplar St.

N. Lincoln Dr. U.S. 41

From Calumet

Birch St.

S. Lincoln

Quincy St.

Portage Canal

Quincy St. U.S. 41

Hancock St.

Hospital

Maasto Hiihto - The City of Hancock's 15 kilometer cross country ski trail. The trail can be accessed at several points and there's plenty of plowed parking. The trail head, located at Houghton County Arena, takes skiers to the first loop - Quincy Loop - which offers a 2.5 km run with return to the arena on the trail's only two way section. By using loop cut-offs, skiers can enjoy a 6.5 km trail; for the more ambitious, the groomed course offers 10 miles of scenic enjoyment.

MTU Ski Trails

Houghton County

SKI TRAIL MAP
Michigan Technological University
HOUGHTON, MICHIGAN

Trails are located on Michigan Technological University Campus in Houghton. Two loops offer various degrees of experience with the upper loop for beginners and lower loop for the more advanced.

Parking lot is on access road off of Sharon Avenue.

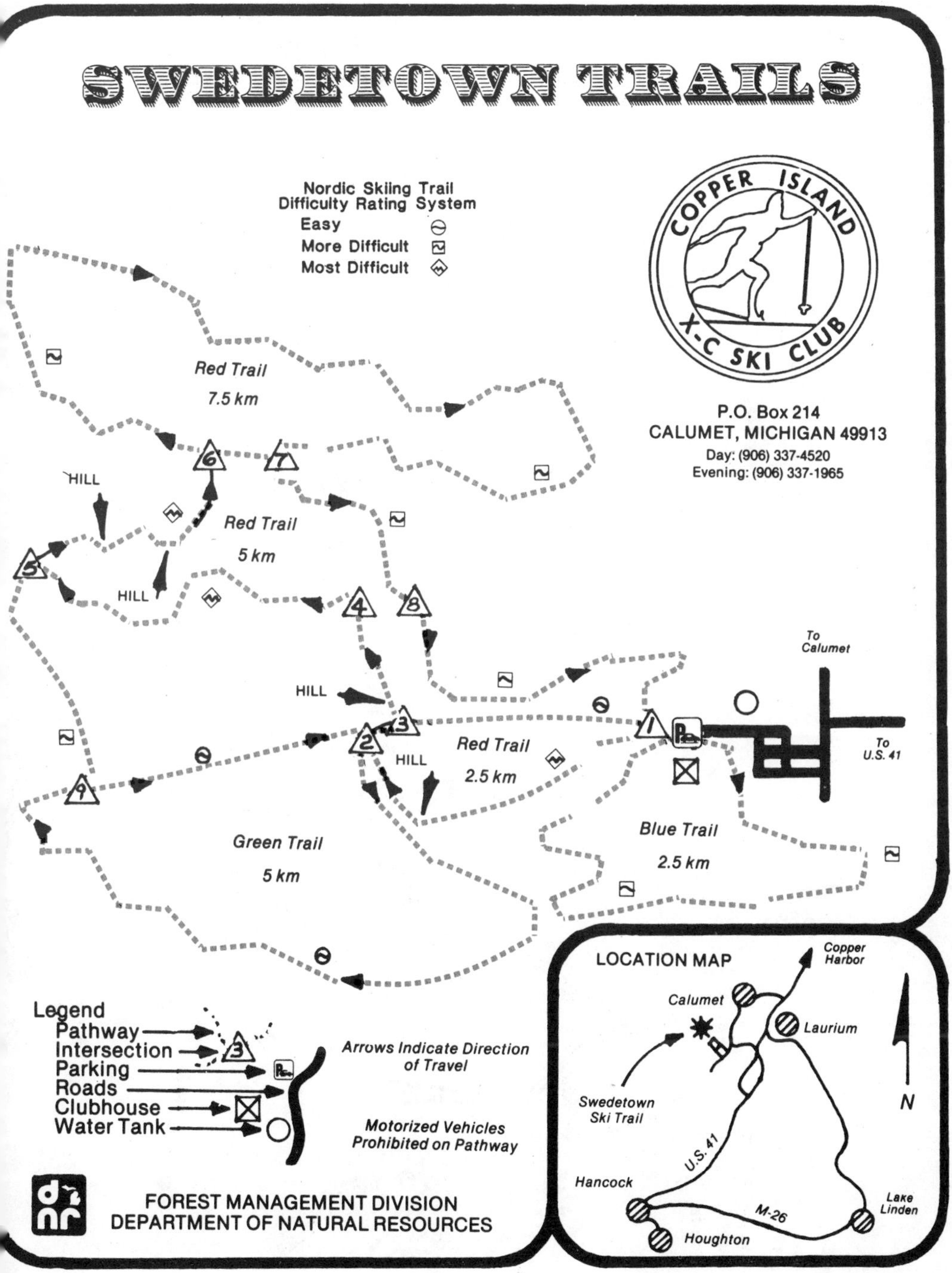

SWEDETOWN TRAILS

Nordic Skiing Trail
Difficulty Rating System
Easy ⊖
More Difficult ▣
Most Difficult ◈

COPPER ISLAND
X-C SKI CLUB

P.O. Box 214
CALUMET, MICHIGAN 49913
Day: (906) 337-4520
Evening: (906) 337-1965

Red Trail
7.5 km

HILL

Red Trail
5 km

HILL

HILL

Red Trail
2.5 km

To
Calumet

To
U.S. 41

Blue Trail
2.5 km

Green Trail
5 km

Legend
Pathway
Intersection ③
Parking
Roads
Clubhouse ☒
Water Tank ○

Arrows Indicate Direction
of Travel

Motorized Vehicles
Prohibited on Pathway

FOREST MANAGEMENT DIVISION
DEPARTMENT OF NATURAL RESOURCES

LOCATION MAP

Copper
Harbor

Calumet

Laurium

Swedetown
Ski Trail

U.S. 41

Hancock

M-26

Lake
Linden

Houghton

N

472

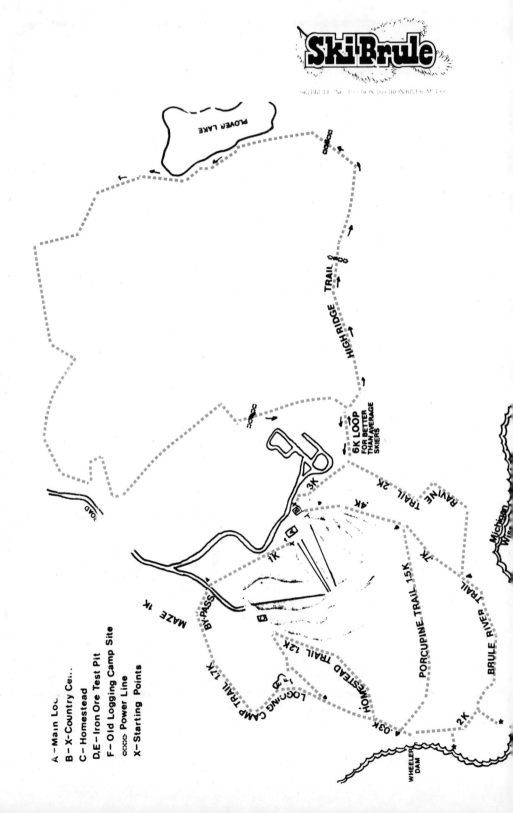

Brule Mountain

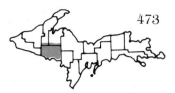

Brule Mountain offers the cross country skier miles of trails starting at the lodge. Trails are groomed and double tracked. Trails are 1½km, 3km, and 5km in length on terrain that varies from flat to hilly.

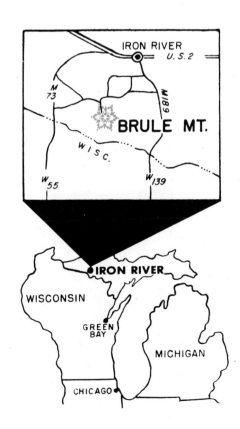

Ge-Che Ski Trail

Iron County

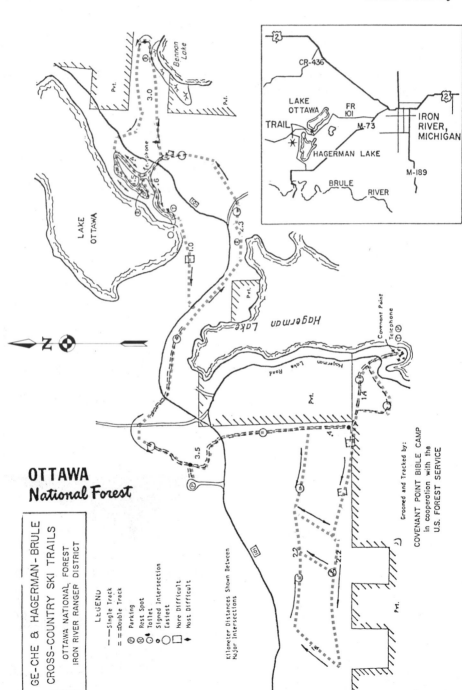

OTTAWA
National Forest

GE-CHE & HAGERMAN-BRULE
CROSS-COUNTRY SKI TRAILS
OTTAWA NATIONAL FOREST
IRON RIVER RANGER DISTRICT

LEGEND
— Single Track
= Double Track
Ⓟ Parking
⊗ Rest Spot
⊙ Toilet
○ Signed Intersection
● Easiest
□ More Difficult
◆ Most Difficult

Kilometer Distances Shown Between
Major Intersections

Groomed and Tracked by:
COVENANT POINT BIBLE CAMP
in cooperation with the
U.S. FOREST SERVICE

Lake Mary Plains

Iron County

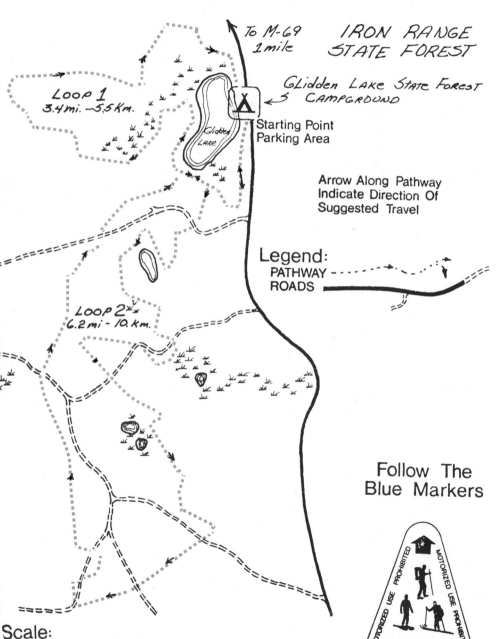

To M-69
1 mile

IRON RANGE
STATE FOREST

GLidden Lake State Forest
CAMPGROUND

LOOP 1
3.4 mi. ~5.5 Km.

Glidden
Lake

Starting Point
Parking Area

Arrow Along Pathway
Indicate Direction Of
Suggested Travel

Legend:
PATHWAY
ROADS

LOOP 2
6.2 mi - 10. Km.

Follow The
Blue Markers

MOTORIZED USE PROHIBITED
MOTORIZED USE PROHIBITED
PATHWAY
MICHIGAN DEPARTMENT OF
NATURAL RESOURCES

Scale:
0   1 Mile

1/2

Dept. of Natural Resources, Forestry Division

476

Copper Harbor
Pathway
Lake Fanny Hooe

Keweenaw County

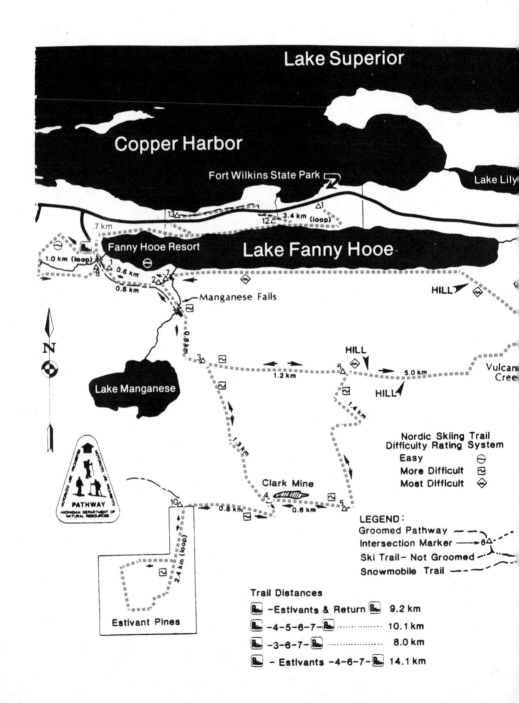

Lake Superior

Copper Harbor

Fort Wilkins State Park

Lake Lily

13 △   1   12 △   3.4 km (loop)   △1

.7 km

Fanny Hooe Resort   **Lake Fanny Hooe**

1.0 km (loop)   1   0.5 km   2 △ 7   HILL

0.5 km

Manganese Falls

0.8 km

HILL

Lake Manganese   3 △   1.2 km   6 △ HILL   5.0 km   Vulcan Cree

HILL

1.4 km

1.3 km

N

**Nordic Skiing Trail
Difficulty Rating System**
Easy ⊖
More Difficult ▣
Most Difficult ◇

PATHWAY
MICHIGAN DEPARTMENT OF NATURAL RESOURCES

Clark Mine

10 △   0.8 km   4 △   0.6 km   5 △

LEGEND:
Groomed Pathway ─ ─ ─
Intersection Marker ──→ 6△
Ski Trail– Not Groomed ─··─
Snowmobile Trail ─ ─ ─

2.4 km (loop)

Estivant Pines

Trail Distances

🥾 –Estivants & Return 🥾	9.2 km
🥾 –4–5–6–7– 🥾	10.1 km
🥾 –3–6–7– 🥾	8.0 km
🥾 – Estivants –4–6–7– 🥾	14.1 km

Canada Lakes Pathway

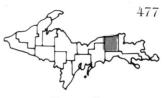

Luce County

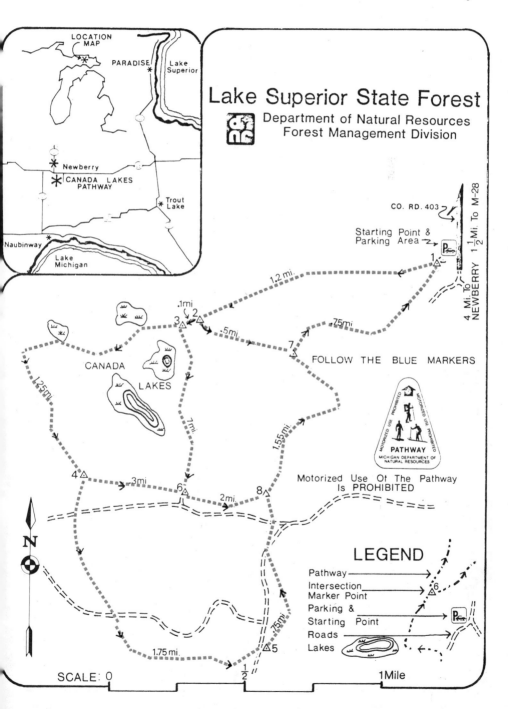

Lake Superior State Forest

Department of Natural Resources
Forest Management Division

LOCATION MAP

PARADISE

Lake Superior

Newberry

CANADA LAKES PATHWAY

Trout Lake

Naubinway

Lake Michigan

CO. RD. 403

Starting Point & Parking Area

4 Mi. To NEWBERRY   1½ Mi. To M-28

1.2 mi.

.1mi

.5 mi.

.75mi

CANADA LAKES

1.25 mi.

.7mi.

1.55 mi.

FOLLOW THE BLUE MARKERS

MOTORIZED USE PROHIBITED
MOTORIZED USE PROHIBITED

PATHWAY
MICHIGAN DEPARTMENT OF NATURAL RESOURCES

Motorized Use Of The Pathway Is PROHIBITED

4   3mi   6   2mi   8

N

LEGEND

Pathway
Intersection Marker Point
Parking & Starting Point
Roads
Lakes

.75mi

1.75 mi.   5   ½

SCALE: 0   1 Mile

Anderson Lake Pathway

Marquette County

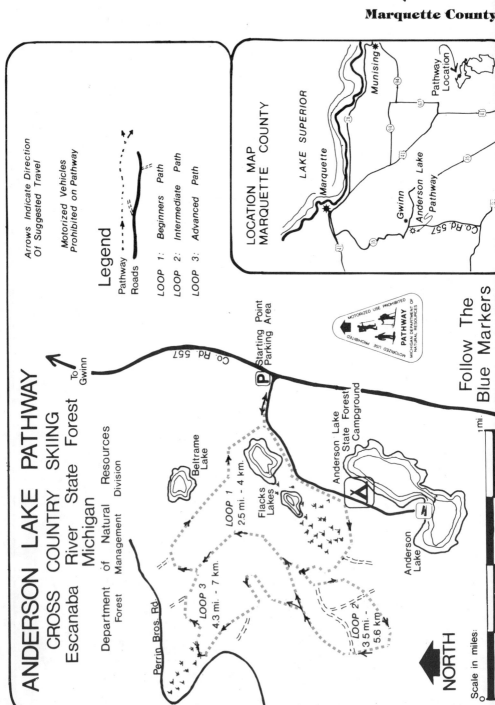

ANDERSON LAKE PATHWAY

CROSS COUNTRY SKIING

Escanaba River State Forest
Michigan

Department of Natural Resources
Forest Management Division

To Gwinn

Arrows Indicate Direction
Of Suggested Travel

Motorized Vehicles
Prohibited on Pathway

Legend

Pathway
Roads

LOOP 1: Beginners Path
LOOP 2: Intermediate Path
LOOP 3: Advanced Path

LOCATION MAP
MARQUETTE COUNTY

LAKE SUPERIOR

Munising

Marquette

Gwinn

Anderson Lake
Pathway

Co. Rd 557

Pathway
Location

Starting Point
Parking Area

Co. Rd. 557

MOTORIZED USE PROHIBITED
PATHWAY
MICHIGAN DEPARTMENT OF
NATURAL RESOURCES

Follow The
Blue Markers

Beltrame
Lake

Flacks
Lakes

LOOP 1
2.5 mi. - 4 km.

LOOP 3
4.3 mi. - 7 km.

LOOP 2
3.5 mi. -
5.6 km.

Perrin Bros. Rd.

Anderson Lake
State Forest
Campground

Anderson
Lake

NORTH

Scale in miles:
0 · · · 1 mi.

Cleveland Trails

Marquette County

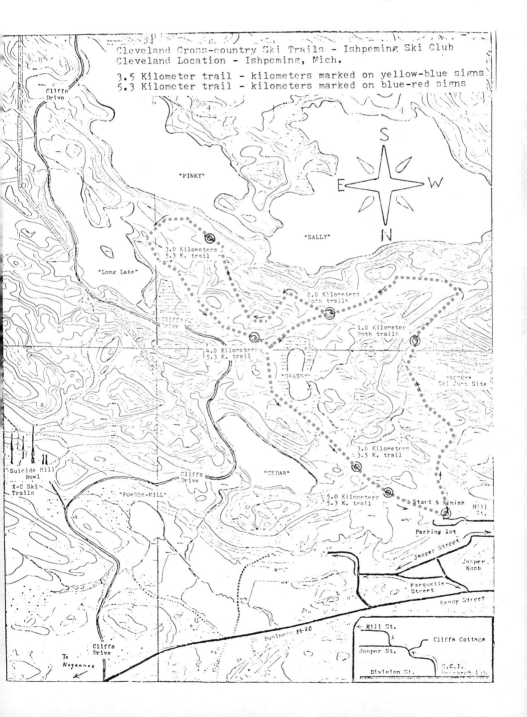

Cleveland Cross-country Ski Trails - Ishpeming Ski Club
Cleveland Location - Ishpeming, Mich.
3.5 Kilometer trail - kilometers marked on yellow-blue signs
5.3 Kilometer trail - kilometers marked on blue-red signs

Black River Pathway

Marquette County

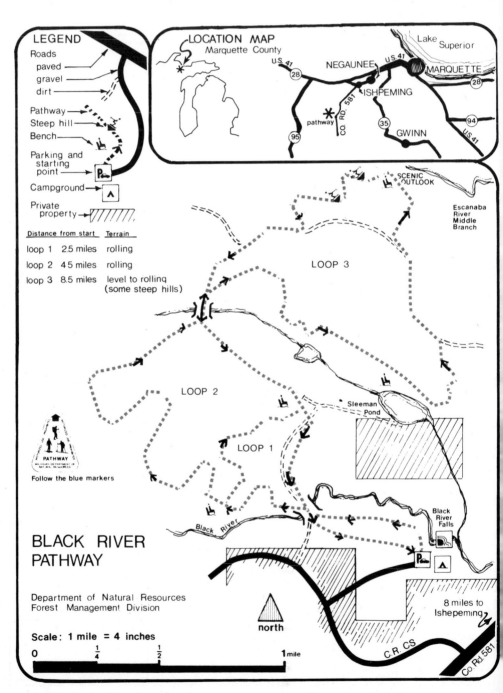

LEGEND

Roads
paved
gravel
dirt

Pathway
Steep hill
Bench

Parking and starting point

Campground

Private property

Distance from start		Terrain
loop 1	2.5 miles	rolling
loop 2	4.5 miles	rolling
loop 3	8.5 miles	level to rolling (some steep hills)

PATHWAY
MICHIGAN DEPARTMENT OF NATURAL RESOURCES

Follow the blue markers

LOCATION MAP
Marquette County

Lake Superior
U.S. 41
NEGAUNEE
U.S. 41
MARQUETTE
28
ISHPEMING
CO. RD. 581
pathway
35
GWINN
94
US 41
95

SCENIC OUTLOOK

Escanaba River Middle Branch

LOOP 3

LOOP 2

LOOP 1

Sleeman Pond

Black River Falls

Black River

BLACK RIVER PATHWAY

Department of Natural Resources
Forest Management Division

Scale: 1 mile = 4 inches

0   ¼   ½   1 mile

north

8 miles to Ishpeming

C.R. C.S.

Co. Rd. 581

Blueberry Ridge Pathway

Marquette County

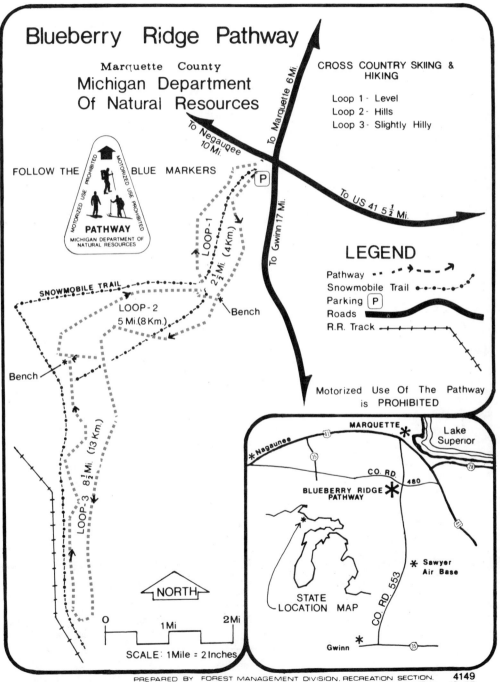

Harlow Lake Pathway

Marquette County

DNR

Escanaba River State Forest
Forest Management Division

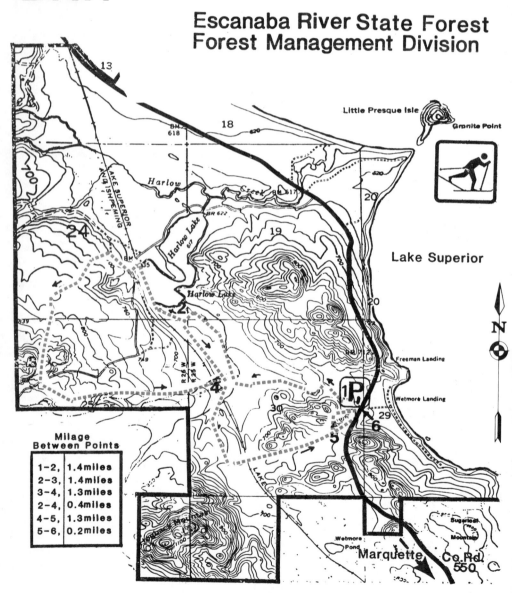

Lake Superior

Milage Between Points

1–2,	1.4 miles
2–3,	1.4 miles
3–4,	1.3 miles
2–4,	0.4 miles
4–5,	1.3 miles
5–6,	0.2 miles

Little Presque Isle

Granite Point

Freeman Landing

Wetmore Landing

Marquette Co. Rd. 550

Wetmore Pond

Superior Mountain

N

TRAIL ETIQUETTE

1. When skiing a trail, stay in the track.

2. Always yield the track to faster skiers.

3. Downhill—do not try to pass, do not go down until trail is clear.

4. Uphill—keep your eyes open, downhill skiers have right of way.

5. Do not walk in tracks; if needed, walk to side of trail.

6. Ski only in the marked directions on one way trails.

7. When falling on a hill, get off to side A.S.A.P., to allow others to pass.

8. If necessary to pass, yell "track". Passed skier should step off to right. Please avoid passing on downhills.

9. Fill in your sitzmarks!

10. Keep together with your ski party.

11. Do not litter, do not urinate near the trail, it ruins the beauty.

12. Do not trespass, respect private property.

13. When required, pay area usage fees gladly, somebody has to maintain the track you are enjoying.

14. Leave your dogs at home—or in the car. Paw prints can ruin a trail.

15. Be friendly in the woods. The classic Norwegian greeting always works; "good tour".

Marquette Cross-Country Ski Trail

The Marquette Cross-Country Ski Trail is located on forty acres on the west edge of the Park Cemetery.

The beginners loop is level and the intermediate loop is slightly hilly. The site is heavily wooded providing a very natural setting for skiers. There are two groomed tracks on the 1.25 miles of trails.

The trail is open daily and is lighted for night skiing.

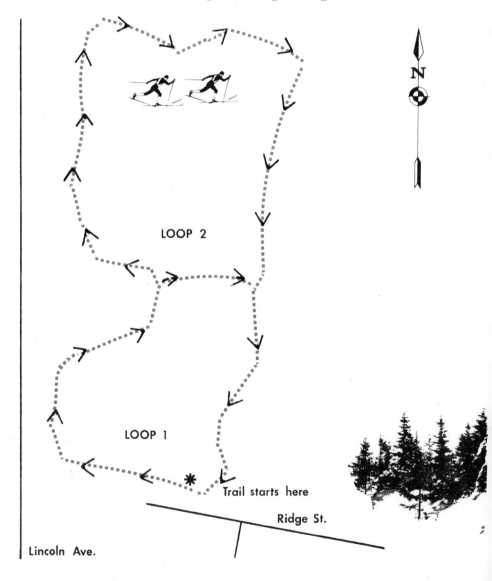

Presque Isle Nature Trail

Marquette County

Presque Isle is famous throughout the United States for its natural beauty. Skiing can be enjoyed on Peter White Drive or the natural trails, with a total of 3½ miles of groomed trails.

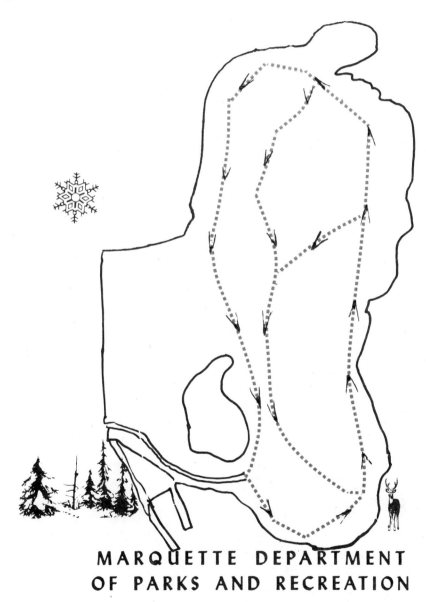

MARQUETTE DEPARTMENT OF PARKS AND RECREATION

Suicide Bowl

Marquette County

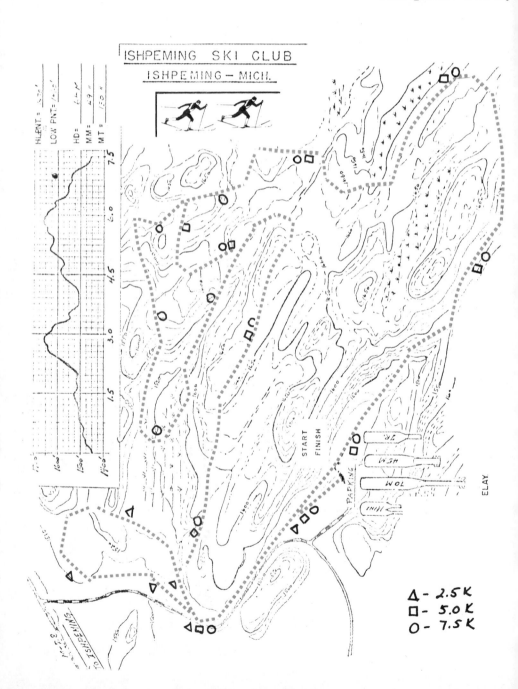

ISHPEMING SKI CLUB
ISHPEMING — MICH.

△ – 2.5 K
▢ – 5.0 K
O – 7.5 K

Cedar River Pathway

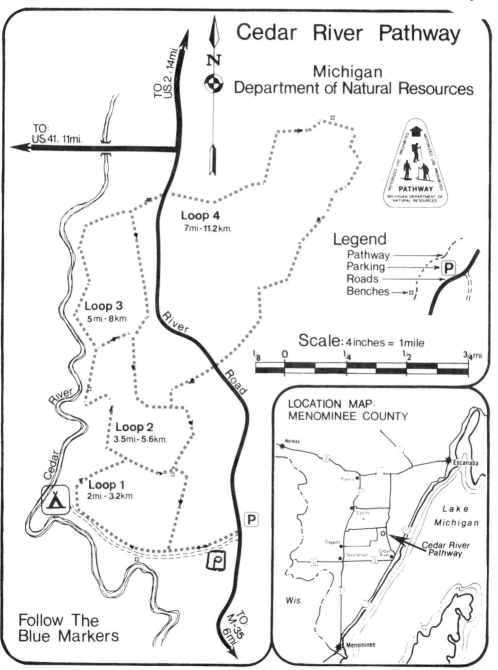

Cedar River Pathway

Michigan
Department of Natural Resources

TO: US.2, 14mi.

TO: US.41, 11mi.

N

Loop 4
7mi-11.2km.

Loop 3
5mi.-8km.

River Road

Loop 2
3.5mi.-5.6km.

Loop 1
2mi.-3.2km.

Cedar River

P

Follow The
Blue Markers

TO: M-35 6mi.

Legend
Pathway
Parking — P
Roads
Benches — ☐

PATHWAY
MICHIGAN DEPARTMENT OF
NATURAL RESOURCES

Scale: 4 inches = 1 mile

⅛ 0 ¼ ½ ¾mi.

LOCATION MAP:
MENOMINEE COUNTY

Norway
Escanaba
Powers
Carney
Daggett
Stephenson
Cedar River

Lake Michigan

Cedar River
Pathway

Wis.

Menominee

J. W. Wells State Park

Menominee County

This park offers seven miles of groomed trails within the park boundary and more trails on nearby state land.

Camping is available with water, pit toilets and electricity.

Two frontier cabins, rustic in design, are available to rent. The trail starts at the cabin door.

A motor vehicle permit is required.

For further information call the manager at J. W. Wells State Park at 1-906-863-9747.

Location is 25 miles north of Menominee on Hwy. M-35.

LEGEND

Pathway

Parking ─────→ P

Roads

Shelter ─────→

International Trail Markings:

easiest ── O
more difficult ── □

N
W E
S

M-35

G-12

CEDAR RIVER TRAIL 1.6 MI

EVERGREEN TRAIL 1.1 MI

TIMBER TRAIL 1.3 MI

2.25 MI.

GREEN BAY

J.W. WELLS STATE PARK

CROSS-COUNTRY SKI TRAILS

MOTOR VEHICLE PERMIT FEES:
ANNUAL ········· 10.00
DAILY ········· 2.00
SENIOR RESIDENT
ANNUAL ········· 1.00

Camping Available
water
Pit Toilets & Electricity
$4.00 per night

Trail Shelter

Bergland Trails

Ontonagon County

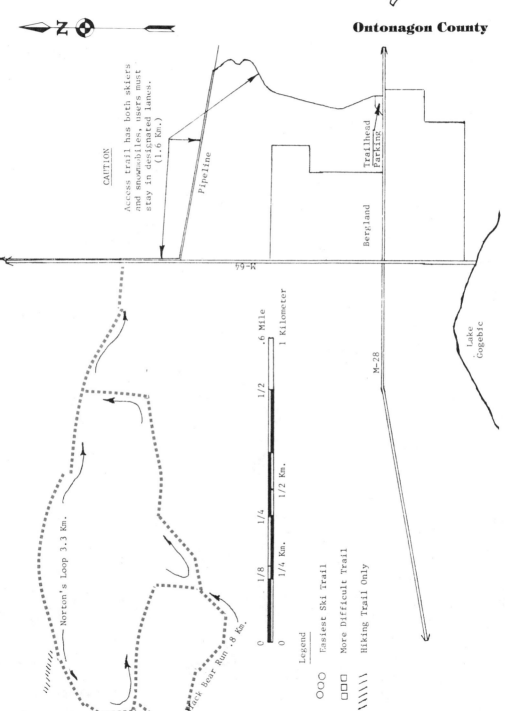

CAUTION

Access trail has both skiers and snowmobiles, users must stay in designated lanes. (1.6 Km.)

Pipeline

Trailhead Parking

Bergland

M-64

M-28

Lake Gogebic

Norton's Loop 3.3 Km.

Black Bear Run .8 Km.

| 0 | 1/8 | 1/4 | 1/2 | .6 Mile |
| 0 | 1/4 Km. | 1/2 Km. | | 1 Kilometer |

Legend

OOO   Easiest Ski Trail

□□□   More Difficult Trail

\\\\   Hiking Trail Only

490

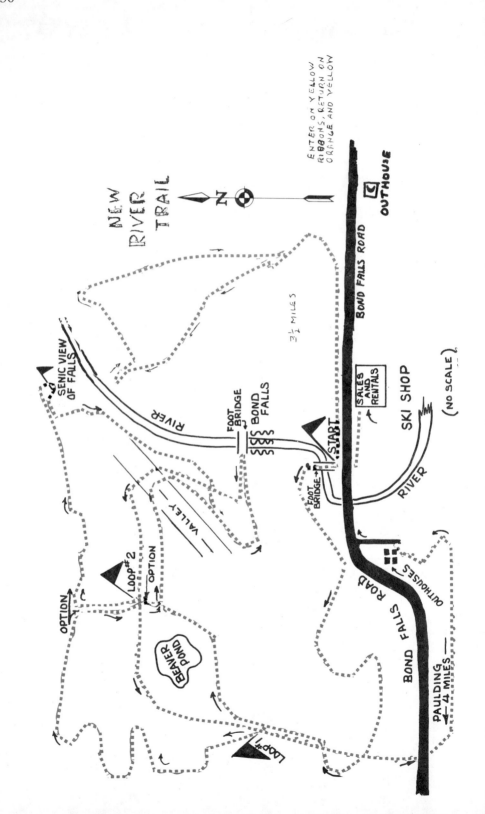

Bond Falls Ski Area

Ontonagon County

The land for this trail is provided thru the courtesy of the Upper Peninsula Power Co. Development and maintenance is furnished by   Ski Shop at head of trails.

Red flags take you to falls. Return by blue flags. Trail is 5.6km (3½ miles). Option loops add approximately 3 additional miles. Another trail of 5.6km (3.5 miles) is now open to the east of the river.

Ski shop is open Thursday, Friday, Saturday and Sunday from 10 A.M. to 5 P.M. Food and warmup area as well as ski sales and rentals are available.

Wind, Temperature, and Moisture

These weather factors can greatly affect the safety of a winter traveler. Each contributes to the loss of body heat. The "Wind Chill" chart illustrates the effect of wind and temperature on a dry, properly clothed person. If clothing is wet from perspiration or precipitation, the net effect of wind and temperature is much greater.

WIND SPEED	COOLING POWER OF WIND EXPRESSED AS "EQUIVALENT CHILL TEMPERATURE"											
MPH	TEMPERATURE (°F)											
CALM	40	30	20	10	5	0	−10	−20	−30	−40	−50	−60
	EQUIVALENT CHILL TEMPERATURE											
5	35	25	15	5	0	−5	−15	−25	−35	−45	−55	−70
10	30	15	5	−10	−15	−20	−35	−45	−60	−70	−80	−95
15	25	10	−5	−20	−25	−30	−45	−60	−70	−85	−100	−110
20	20	5	−10	−25	−30	−35	−50	−65	−80	−95	−110	−120
25	15	0	−15	−30	−35	−45	−60	−75	−90	−105	−120	−135
30	10	0	−20	−30	−40	−50	−65	−80	−95	−110	−125	−140
35	10	−5	−20	−35	−40	−50	−65	−80	−100	−115	−130	−145
40	10	−5	−20	−35	−45	−55	−70	−85	−100	−115	−130	−150
	LITTLE DANGER			INCREASING DANGER (Flesh may freeze within 1 min.)				GREAT DANGER (Flesh may freeze within 30 seconds)				

Frosbite or hypothermia are the common dangers due to wind, temperature, and moisture.

Porcupine Mountain

Ontonagon County

The park is composed of 58,000 acres of remote and wild areas with towering virgin pine and hemlock. The trail system consists of 24½ miles of well marked, groomed trails. The Deer Yard and River trails are geared for the novice skier. Union Springs Trail is for the advanced skier due to its long uphill climbs and exhilarating downhill runs.

A small fee is required for parking. Rentals and food services are available at the chalet. Location — 17 miles west of Ontonagon via M64 and M107.

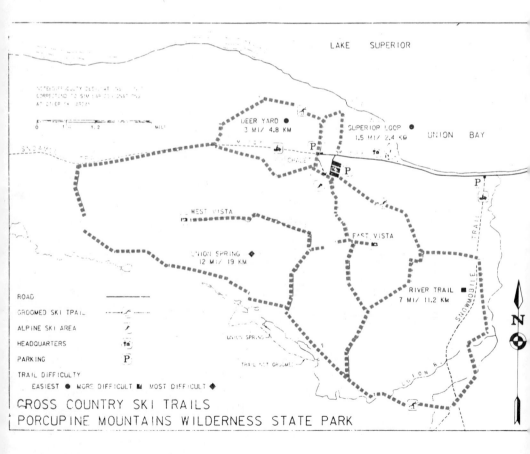

CROSS COUNTRY SKI TRAILS
PORCUPINE MOUNTAINS WILDERNESS STATE PARK

Ashford Lake Pathway

Schoolcraft County

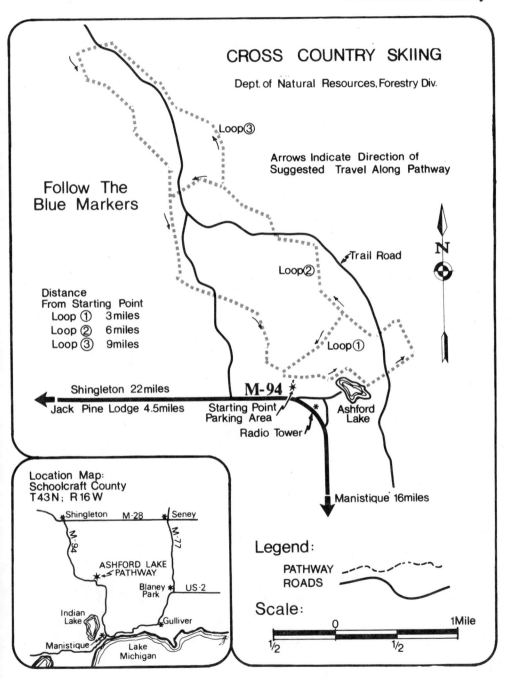

CROSS COUNTRY SKIING

Dept. of Natural Resources, Forestry Div.

Loop③

Arrows Indicate Direction of
Suggested Travel Along Pathway

Follow The
Blue Markers

Trail Road

N

Loop②

Distance
From Starting Point
Loop ① 3 miles
Loop ② 6 miles
Loop ③ 9 miles

Loop①

M-94

Shingleton 22 miles

Jack Pine Lodge 4.5 miles

Starting Point
Parking Area

Radio Tower

Ashford
Lake

Location Map:
Schoolcraft County
T 43 N ; R 16 W

Shingleton   M-28   Seney

M-94

M-77

ASHFORD LAKE
PATHWAY

Blaney   US-2
Park

Indian
Lake

Gulliver

Manistique

Lake
Michigan

Manistique 16 miles

Legend:

PATHWAY
ROADS

Scale:

0   1 Mile

½   ½

The Canadian Near North

Canada sounds far away, but it is much closer than you think. Sault Ste. Marie is a 6 hour drive from East-Central Wisconsin, making it a reasonable alternative to the 5 hours to Telemark.

The main attraction is the snow. Even in bad years in the lower Midwest, the edge of Superior north of Sault Ste. Marie will have 3 to 6 feet of cold, dry powder. The terrain is unlike anything in the states; massive granite hills, thick forest, and, of course, the frozen shore of Superior. Snow season is early December to early April.

The opportunities for wilderness skiing are endless, especially if you go north along highway 17 to Superior Provincial Park, about 2 hours drive north of the Sault. Ski any of the logging roads or unplowed park roads, or along the Superior shore. If you're more adventurous, simply head into the woods. (With a Topo map and compass, please.) With their normal snow, most brush is covered, and when you run your first hill, you'll find out why the locals call it "pin-ball" skiing.

For the less adventurous, there are marked trails in the park; you'll see the signs for them along highway 17. But if you want tracked, groomed trails, two of the best systems in North America are hidden up here.

Hiawatha Lodge, on the northeast side of Sault Ste. Marie, is the home of the Soo Finn Club. The Lodge is set up for day skiers, with dining room, fast food area, and lounge. They have 50km of trails, with a 2km loop lighted for night skiing. All trails are groomed and double-tracked.

Stokely Creek Lodge is one of those gems that you'd like to keep to yourself. Hidden in the wilderness north of the Sault, they have over 70km of groomed, double-tracked trails, ranging from gently rolling beginner-intermediate trails, to long and fast, expert loops. They have a small lodge for Stokeley Creek Club members only, and a day skiers cabin with instant soups and hot drinks. There is a small trail fee for non-club members. Stokely Creek is reached by taking the Buttermilk Road exit off highway 17, 18 miles north of Sault Ste. Marie. Watch for signs.

There are plenty of motels in Sault Ste. Marie, and customs for U.S. travelers is no problem. There is an information center just past the Canadian customs booths with plenty of maps and helpful people to assist newcomers.

Helpful numbers:

Algoma Kinniwabi Travel Association, Suite 203, 616 Queen St. E., Sault Ste. Marie, Ontario: 705-254-4293

Stokely Creek Club, Karalash Corners, Goulais River, Ontario, Canada POS-IEO 705-659-3421

Superior Provincial Park, Ministry of Natural Resources, Wawa District Office, Box 1160, Wawa, Ontario, Canada POS IKO

Kwagama Lake Lodge

in the Heart of the Algoma Central Country

Imagine -

A Train Trip that attracts tens of thousands of tourists each year . .

That this exciting Train Trip is just the prelude to your excursion, the means for reaching your destination . . . . .

That you can take this Train Trip from Sault Ste. Marie, Ontario on a Sunday Morning, Travel 118½ Miles North in plush comfort, be met by an Expert Winter Guide, Ski 9 miles through Canyonland to the Lodge, spend 5 days exploring this remote paradise while enjoying the Best of Meals and comfortable accommodations, then on Saturday morning, make the trek back to the Train for your return to the humdrum.

That a Small Island is the Central Point from which a variety of cross-country Ski Trails emanate, wandering through some of the Most Scenic Wilderness East of The Rocky Mountains . . . . .

Now Stop Imagining - - -

Here Are The Facts

KWAGAMA LAKE LODGE can be reached in the Winter only by the Algoma Central Railway. The Winter Cross-Country Package at this Lodge calls for boarding the ACR "SNOW TRAIN" at 8:30 a.m. on Sunday, then settling for a Scenic Four Hour Trip North. You will be met at Mile 118½, and then set off for the Lodge.

A variety of groomed trails, from the "very easy" to the "expert only" through a Ruggedly Beautiful Wilderness are yours for the next 5 days. Saturday morning Breakfast precedes your trip to the Railroad and return to the "SOO."

For further information write:—

Kwagama Lake Lodge

176 MANITOU DRIVE
SAULT STE. MARIE, ONTARIO, CANADA
P6B 5L1

Telephone (705) 253-3075

Mac and Grace MacEwan

COUNTIES IN ILLINOIS

COUNTIES IN ILLINOIS

Mississippi Palisades

Carroll County

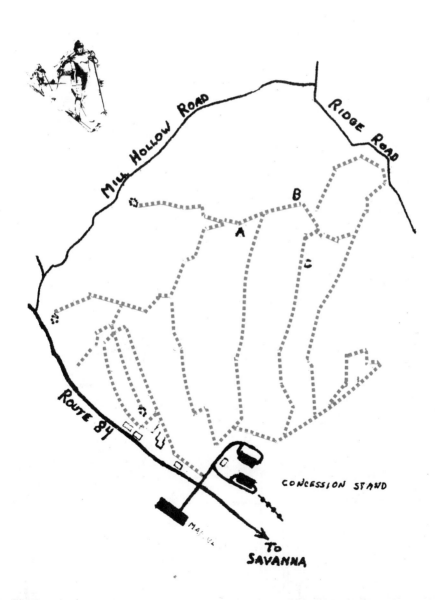

MILL HOLLOW ROAD

RIDGE ROAD

A

B

C

ROUTE 84

CONCESSION STAND

MARINA

TO SAVANNA

Palos & Sag Valley Division

Cook County

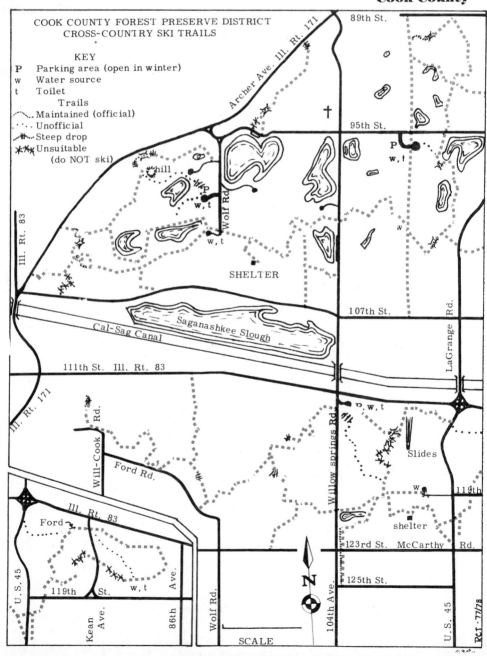

COOK COUNTY FOREST PRESERVE DISTRICT
CROSS-COUNTRY SKI TRAILS

KEY
P   Parking area (open in winter)
w   Water source
t   Toilet
        Trails
·····Maintained (official)
·····Unofficial
╫╫╫·Steep drop
✕✕✕Unsuitable
    (do NOT ski)

Fox Valley Country Club

Kane County

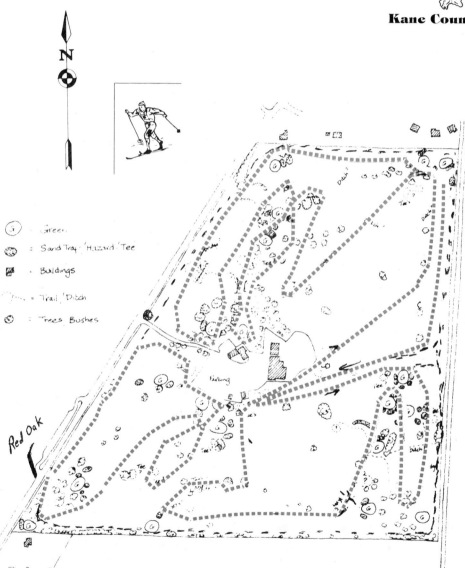

N

○ = Green

○ = Sand Trap / Hazard / Tee

▨ = Buildings

···· = Trail / Ditch

○ = Trees / Bushes

Red Oak

Chain O' Lakes State Park

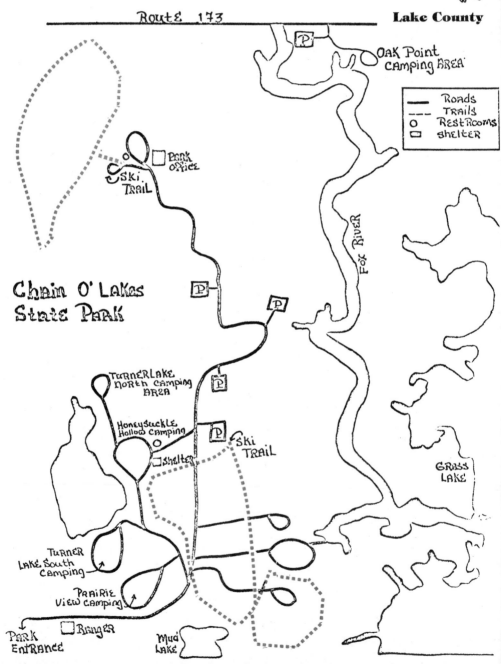

Route 173

Oak Point Camping Area

	Roads
	Trails
o	Restrooms
□	Shelter

Fox River

Ski Trail

□ Park Office

Chain O' Lakes State Park

Turner Lake North Camping Area

Honeysuckle Hollow Camping

□ Shelter

Ski Trail

Grass Lake

Turner Lake South Camping

Prairie View Camping

Park Entrance

□ Ranger

Mud Lake

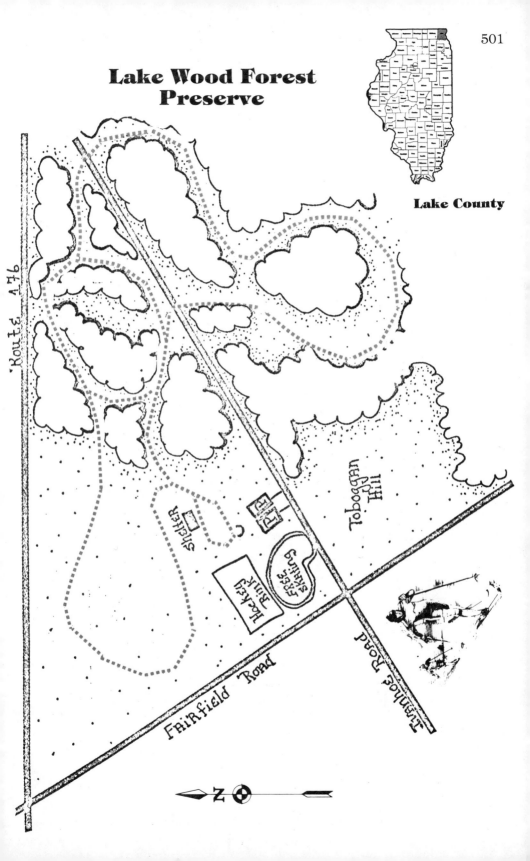

Lake Wood Forest Preserve

Lake County

Route 176

shelter

Hockey Rink

Ice Skating

Toboggan Hill

Fairfield Road

Trumbull Road

N

Ryerson Conservation Area

Lake County

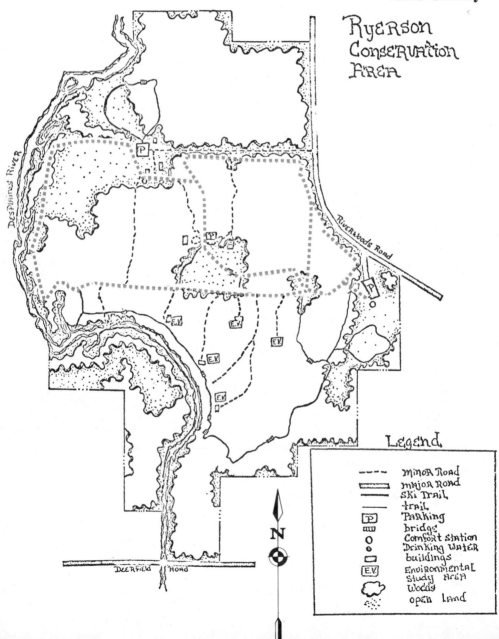

Harrison-Benwell

Location: 7055 McCullom
Lake Road,
Wonder Lake
Hours: 9:00 A.M. till Dark
Size: 80 acres

McHenry County

2 miles, EASIEST.
Winding trail through
forested area. Trail
begins across the
creek from parking
lot. Passes through
dense hawthorn and
dogwood, across flat
lowland with ponds,
then through more
mature forest of
oaks and black
cherries. Returns
to start via forest
edge overlooking
low wetlands.

SHORTCUT: Go left at
first fork for 1/2
mile circuit back to
start.

Hickory Grove

McHenry County

HIGHLANDS TRAIL: 1-1/2 miles, EASIEST. Trail starts downhill through a rolling, semi-open forested area, crosses a small creek and continues through flatter semi-open area with scenic oaks, winds through more rolling forest. Trail crosses a 300-foot wide marsh then loops back through rolling, open forest of oaks and hickories overlooking a natural prairie area. Trail forks: spur to right returns to parking lot.

Fox River Conservation Area -

Location: 500 Hickory Grove Road **Hours:** 8 A.M. — Sunset
Size: 220 Acres — Fox River Frontage

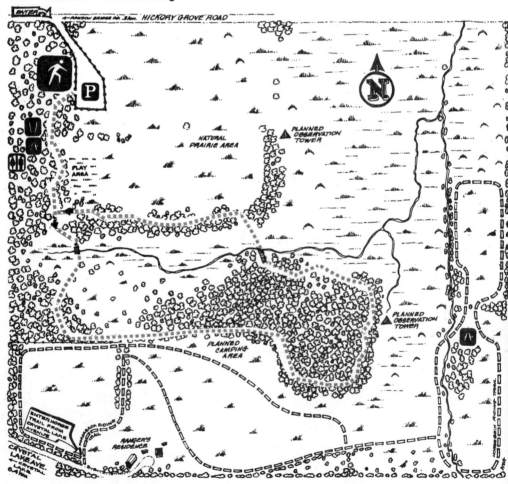

Marengo Ridge-Knude Woods

McHenry County

Kishwaukee Conservation Area

Location: 2411 S. Route 23, Marengo
Hours: 9:00 A.M. — Sunset
Size: 150 Acres

SHORT LOOP: 1-1/3 miles, DIFFICULT. Trail begins on west side of parking area "B"; gradual slopes through mature hardwood forest, then into rolling open land. Last 1/2 mile is through pine plantation with some challenging slopes and turns.

LONG LOOP: 2-2/3 miles, DIFFICULT. First section is same as short loop until northernmost point where trail leaves hardwood forest. Long loop forks left toward northern pine plantation. After passing through pines, begin ups and downs along forested glacial ridge. Watch out for small boulders along trail. Several creeks to cross on small footbridges. Upper loop completed at bridge, through pines again and then along second half of short loop.

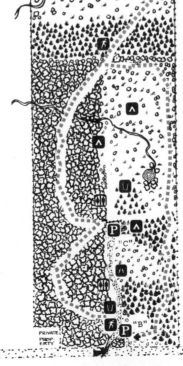

Morraine Hills

McHenry County

WARRIOR LAKE

Tamarow Lake

WILDERNESS LAKE

MORAINE HILLS

Nature Preserve

LAKE Defiance

Leather Leaf Bog Trail 3.5 mi.

LAKE Defiance TRAIL 4 mi.

River Road

Main Park Road

Lily Lake Road

Hickory Road

Fox River Trail 2.5 mi.

McHenry Dam

Wildlife Viewing Platform

Fox River

Fernview Lake

NATURE PRESERVE

	MAIN Road
	Minor Road
	trail
o	comfort station
o	Drinking Water
P	PARKing
	building

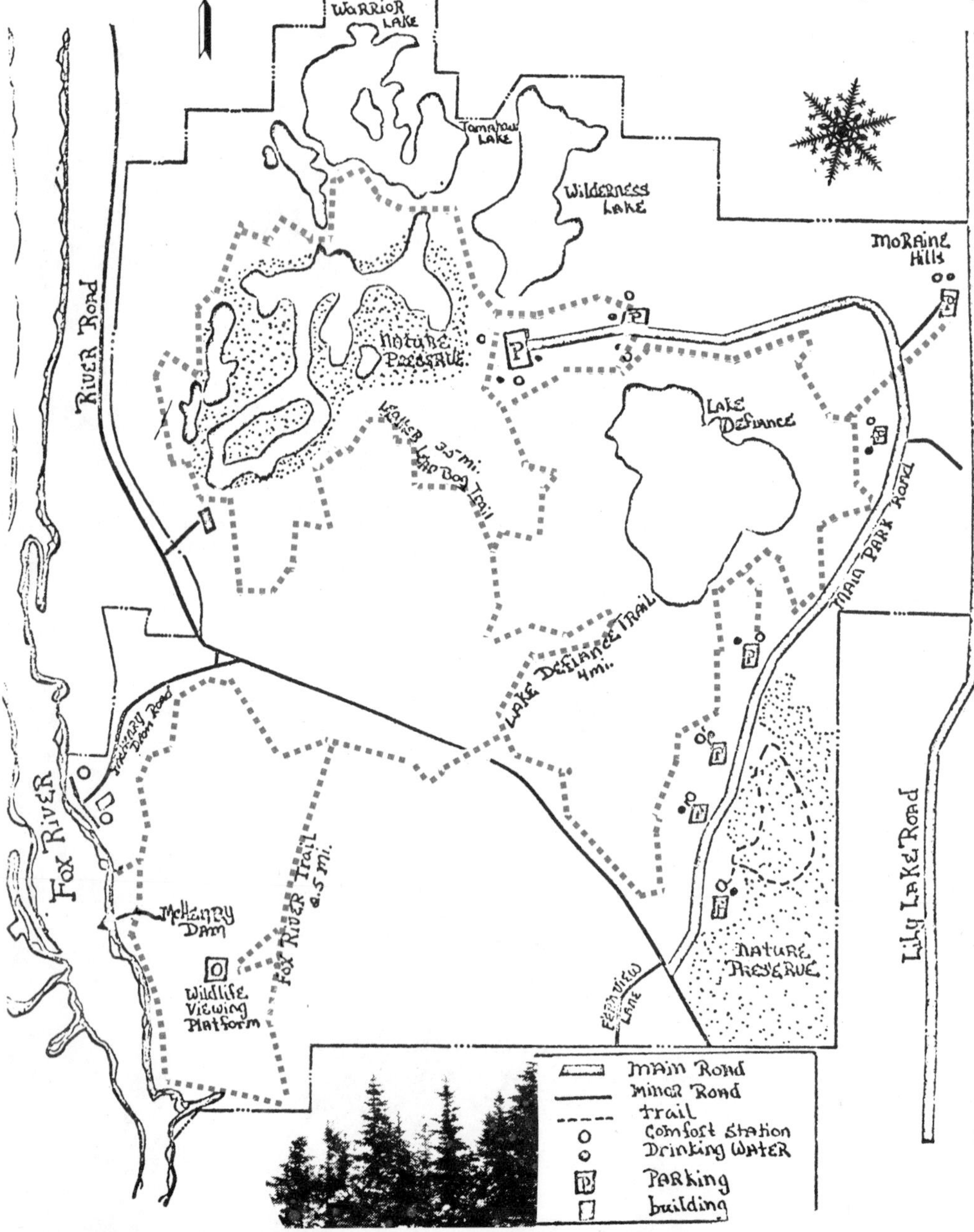

Nippersink
Glacial Park

McHenry County

COYOTE LOOP TRAIL: 3/4 mile, DIFFICULT. Mostly flat open terrain, but with several small hills at forest/field borders and rolling character at end of trail.

DEER PATH TRAIL: 2-1/3 miles, DIFFICULT. First quarter mile from parking lot coincides with Coyote Loop Trail, then forks north (right). Variety of forested, semi-open terrain. Typical rolling glacial topography, including an optional run over 100-foot high glacial kames.

BROKEN SKI EXTENSION: 1/4 mile, MOST DIFFICULT. Steep downslope through open hillside. FOR EXPERIENCED SKIERS ONLY.

Nippersink
Conservation Area

NIPPERSINK TRAIL: 5 miles, DIFFICULT. Entirely level trail through former pasture. From spur connecting with Deer Path Trail, Nippersink Trail extends 2 miles north to shelter with fire ring, and 1/2 mile south to Keystone Road and Barnard Mill Road junction. Return to spur along same route.

Location: 6512 Harts Road, Ringwood
Hours: 9:00 A.M. — Sunset

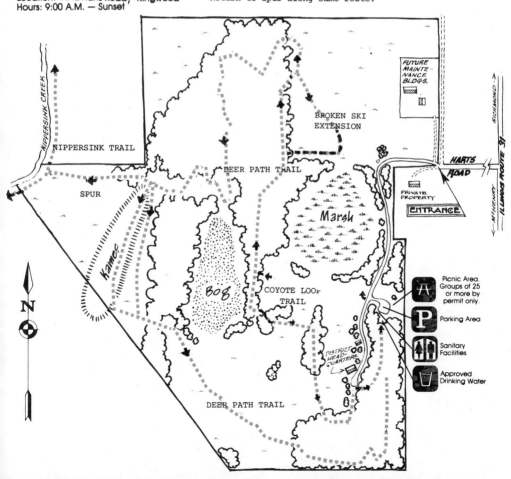

508

Rush Creek
Burrows Woods

McHenry County

Location: 20501 McGuire Road, Harvard, Illinois 60033.

Hours: 8:00 A.M. to Sunset

Size: 250 acres

SHORT LOOP: 1 mile, EASIEST. Trail starts on southeast side of creek, winds through fairly level forest, crosses first creek and reenters forest. Left at second creek to overlook open prairie area, then left again to rejoin long loop through slightly rolling forest and back to parking lot. Mostly level ground.

LONG LOOP: 2 miles, EASIEST. Same as short loop until second creek. Crosses second creek, skirts open prairie area, then passes through small pine plantation. More fairly level hardwood forest, past an old windmill and through wooded area with more rolling terrain. Past open prairie area again and across another bridge, then through forest to trail end.

RUSH CREEK CONSERVATION AREA

In July of 1974, the Trustees of the McHenry County Conservation District passed a resolution to purchase tracts of land in the area bounded by McGuire Road on the North, Lindwall Road on the East and U. S. Route 14 on the South and West, thereby creating the Rush Creek Conservation Area. The purpose of the acquisition was to preserve and protect flood plain and other watershed lands, unique geological, vegetative and historical sites, and to provide a site for recreational use and wildlife habitat.

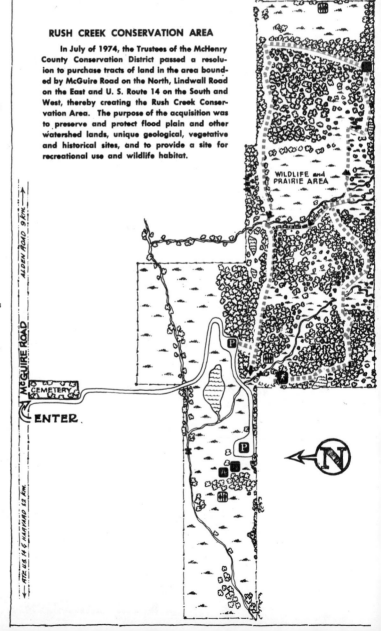

ADDITIONAL TRAILS IN WISCONSIN WITHOUT MAPS

OUNTY

ADAMS

SKYLINE SKI AREA — Primarily downhill. Area has 6 miles of gentle wooded trails. Ski rental, instruction, refreshments available at lodge. ¼ mile northwest of Friendship. 608-339-3421.

BARRON

HARDSCRABBLE — About 15 miles of trails in varying loops. This is primarily downhill area with all facilities. Located just outside of Rice Lake. No good trail map is available at the present time.

DOOR

WHITE FISH DUNES — Four different trails ranging in length from 1.8 — 4.2 miles — total of 10 miles. The trailhead has a heatroom for skiers. Located at Rt. 3 Sturgeon Bay. 414-823-2400.

DUNN

DEEPWOOD — Very secluded spot for the cross country skier, with modern chalet and bar. Deepwood offers 9 miles of groomed trails as well as a downhill ski area. It is located north of Wheeler on Juct. N.

JUNEAU

BUCKHORN LOOP — Gentle, wooded, water, shelter, toilets, plowed parking, map. 10 miles north of Mauston off County Q on County G, adjacent to Buckhorn State Park.

MARQUETTE

SKY LODGE CH. CAMP — 8 miles, marked, groomed, tracked, three loops. Open daily — Monday-Saturday, 12:30 to dusk-Sunday. Fee, rentals, chalet, lodging. 414-297-2566.

ONEIDA

MOCCASIN LAKE — Trailhead begins at Northland Marina in Three Lakes. Connecting trails lead to Northernaire. Trail sponsored and maintained by 3-Lakes Ski Club. For information call Northland Marina 715-546-2333.

SHELTERED VALLEY — Offers 8 KM (5 miles) of gentle hilly and wooded trail, consisting of two loops. The trail is located at STH 32, 4 miles south of Three Lakes and it is open daily.

OUTAGAMIE

PLAMANN PARK — Plamann consists of 3 miles of groomed, hilly and wooded trails, geared to the intermediate skier. Heated shelter is available. It is open for public use daily and Saturday and Sunday from 12:00 to 4:30. The Post Crescent Ski School will be held at Plamann in January.

POLK

TROLLHAUGEN — One-half mile east of Dresser on CTH "F". 4½ miles of rolling, mixed trail. Rentals, instructions and Chalet available. No charge to use trails. Primarily downhill ski area.

RACINE

RIVER NATURE CENTER — 4 miles, marked, rolling/wooded. 9:00-5:00 Tuesday-Saturday, 1:00-5:00 Sunday. Rentals, instruction, chalet. 1-414-639-0930. 3600 N. Green Bay Road, Racine.

WALWORTH

MT. FUJI SKI AREA — 2 miles northeast of Lake Geneva at Jct. of State Hwy. 36 and Krueger Road. Another downhill ski area that has added 3 miles of Cross Country trails. All facilities, including a lounge, are available.

"Ski Minnesota" is a Cross Country Skiers Guide by Elizabeth and Gary Noren that covers Minnesota and Western Wisconsin. It is available from Nodin Press — 519 N. Third Street, Minneapolis, MN 55401.

SKIERS MUST KNOW THE MEANING OF SIGNS

A standard trail marking system was initiated nationwide for Nordic ski areas. Watch for these signs. Like road signs when you are driving a car, these signs point out conditions, hazards and facilities to the skier.

TRAIL MARKINGS SIGNS—NORDIC

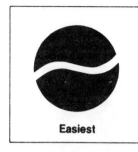

Easiest

GREEN CIRCLE

More Difficult

BLUE SQUARE

Most Difficult

BLACK DIAMOND

Caution

Closed

Emergency Telephone

Don't walk in front of me
I may not follow
Don't walk behind me
I may not lead
Walk beside me
And just be my friend

— Albert Camus

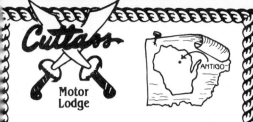

WALWORTH

THE QUIET HUT
X-Country Ski

- Sales
- Clothing
- Rentals
- Instructions

Just Minutes Away From X-Country Ski Trails
DOWNTOWN — WHITEWATER, WI 53190

414-473-2950

WAUPACA

RIVERSIDE MARINE & OUTDOOR SPORTS

X-C Sales & Rentals
Groomed Trails — Accessories

New London, Wis. 54961
414-982-4874

VILLAGE INN MOTEL
Located ½ Mi. West of Waupaca
Hwy. 22 & 54
Route 5, Box 452-A
WAUPACA, WI 54981
715-258-8526

WASHBURN

GREEN ACRES MOTEL

21 ULTRA MODERN UNITS RESTAURANT & LOUNGE NEARBY

- CABLE COLOR TV & HBO
- DIRECT DIAL PHONES
- NEW INDIVIDUAL CONTROL HEAT
- SOME WATER BEDS
- REASONABLE RATES
- CAR HEATER PLUG INS
  ON HWY 63 & 53, SPOONER, WI 54801
  715-635-2177

COUNTRY HOUSE MOTEL
ON U.S. 53 AND 63 SOUTH

RESTAURANT ON PREMISES

CLOSE TO CROSS COUNTRY
SKI TRAILS

OPEN YEAR ROUND

HOT WATER – HEAT – CAR PLUG INS
COLOR CABLE TV, HBO

P.O. BOX 367, SPOONER, WI 54801
715-635-8721

WOOD

THE RIDGES (Snowplace)

X-C Sales & Rentals
20 Miles of Trails — 3 lited
10 Miles of Double Groomed Trails
Winter Campground — Limited Ser.
Downhill Tubing — Uphill Tow
Delightful Food & Drinks
Access to State Snowmobile Trails
Nearby Lodging
Sno-Cat — Bob Sleigh Rides

2311 - U - Griffith Ave.
WISCONSIN RAPIDS, WI 54494
715-424-1111   715-424-1320

MICHIGAN

GOGEBIC

ARROW LODGE RESORT
Modern Heated Vacation Homes
Trail Connects with Sylvania Trails

THOUSAND ISLAND LAKE ROAD
BOX 66X, WATERSMEET, MI 49969

906-358-4390

SUNSET MOTEL
Near Sylvania and Bond Falls
XC SKI TRAILS
CORNER U.S. 2 & U.S. 45, WATERSMEET, MI 49969
906-358-4450

VACATIONLAND RESORT
Easy Access to Sylvania Ski Trails
Deluxe Fireplace Unit
Accommodates 15 People
Other One or Two Bedroom Units
with Fireplace Available

- COMPLETE KITCHENS
- EVERYTHING FURNISHED
  EXCEPT TOWELS
- SKI RATES AVAILABLE

BILL & JAN SMET, HOSTS
THOUSAND ISLAND LAKE ROAD
WATERSMEET, MI 49969
906-358-4380

MARQUETTE

A Touch of
Austria

150 CARP RIVER HILLS
MARQUETTE, MI 49855
906-226-7516

Sylvania
Outfitters
canoe trips & cross country ski center

- 20 MILES GROOMED TRAILS
- SKI SHOP
- REFRESHMENTS
- WARM UP ROOM
- FIREPLACE

LAKE FANNY HOOE RESORT AND CAMPGROUNDS
Copper Harbor, Michigan 49918

Deluxe Accommodations • Sauna
Kitchenettes • Club Room • Fireplace
906-289-4451

west u.s.-2
watersmeet, michigan 49969
for reservations phone 906-358-4766

INDEX

INDEX (Cont.)

INDEX (Cont.)

MICHIGAN TRAILS

CANADA TRAILS

ILLINOIS TRAILS